Polymer Science Library

Edited by A.D. Jenkins
University of Sussex,
The School of Molecular Sciences,
Falmer, Brighton BN1 9QJ, England

1. *K. Murakami* and *K. Ono*, Chemorheology of Polymers
2. *M. Bohdanecký* and *J. Kovář*, Viscosity of Polymer Solutions
3. *J. Wypych*, Polyvinyl Chloride Degradation
4. *J. Wypych*, Polyvinyl Chloride Stabilization
5. *P. Kratochvíl*, Classical Light Scattering from Polymer Solutions
6. *J. Bartoň* and *E. Borsig*, Complexes in Free-Radical Polymerization
7. *Yu.S. Lipatov*, Colloid Chemistry of Polymers
8. *F.J. Baltá-Calleja* and *C.G. Vonk*, X-Ray Scattering of Synthetic Polymers

Polymer Science Library 9

THERMODYNAMICS OF POLYMER SOLUTIONS

PHASE EQUILIBRIA AND CRITICAL PHENOMENA

Polymer Science Library 9

THERMODYNAMICS OF POLYMER SOLUTIONS

PHASE EQUILIBRIA AND CRITICAL PHENOMENA

Kenji Kamide

Fundamental Research Laboratory of Fibers and Fiber Forming Polymers, Asahi Chemical Industry Co., Ltd., 11-7, Hacchonawate, Takatsuki, Osaka 569, Japan

ELSEVIER

Amsterdam — Oxford — New York — Tokyo 1990

ELSEVIER SCIENCE PUBLISHERS B.V.
Sara Burgerhartstraat 25
P.O. Box 211, 1000 AE Amsterdam, The Netherlands

Distributors for the United States and Canada:

ELSEVIER SCIENCE PUBLISHING COMPANY INC.
655, Avenue of the Americas
New York, NY 10010, U.S.A.

ISBN 0-444-88184-0

This book is printed on acid-free paper.

Printed in The Netherlands

To my family

PREFACE

The embryotic stage of polymer science, in particular the physical chemistry of macromolecules, occurred in 1920s-1930s. We can never neglect the great contribution, at that time, of Staudinger and his collaborators at Freiburg who established the concept of "macromolecules". The sound basis of polymer physical chemistry can be considered to have actually been founded during the 1940s and 1950s, a period which included a fruitless, vacant period of war. Since those days, polymer scientists have rapidly increased in number and a gigantic realm of the science has been opened up.

Polymer solutions are the focus of this work. Some characteristic features of polymer solutions have been discovered by applying simple thermodynamic concepts of low molecular weight solutions to the polymer solutions, and as a result of this approach, the theoretical and experimental relationship between the chemical potential (deduced from the colligative properties), polymer concentration (low concentration) and polymer molecular weight were established, although only to a zeroth approximation. To these achievements must be added the valuable contributions by Guggenheim, Miller, Flory, Huggins and Fujishiro. In the 1940s, the thermodynamics of polymer solutions, in which the polymolecularity of the polymers was (at least qualitatively or semi-qualitatively) taken into account, was extended to two-liquid phase equilibria of the solution and then to molecular weight fractionation, based on the two-phase equilibria. However, we dare say that a more comprehensive study on the thermodynamics of phase equilibria and critical phenomena of polymer solutions, the main topics of this book, started only in the late 1960s, because this kind of study requires computer technology, which only at that time became readily available to polymer scientists. In particular, the establishment of a theory, which permits the accurate theoretical prediction of the spinodal, binodal, cloud point and critical point for multicomponent polymer solutions, is undoubtedly one of the most significant milestones achieved in polymer science over the last 20 years. In addition, these studies showed that the physical quantities such as the concentration dependence of polymer-solvent interaction parameter χ and the Flory enthalpy parameter at infinite dilution, which were evaluated by direct analysis on the phase equilibria, are in very good agreement with those estimated by other methods such as light scattering, osmometry and calorimetry. The experimental data obtained by analysing phase equilibria or critical phenomena of polymer solutions can be very consistently explained in the framework of a general thermodynamic theory of polymer solutions, a theory which was originally developed in order to extend one's understanding to the more concentrated range of polymer solutions.

The above mentioned physical quantities are not only very important from the standpoint of pure science but are also significant for industry: Underlying the basic principles of industrial manufacture of membranes by casting methods, fibers by wet spinning, and paints, are the thermodynamics of phase separation of polymer solutions. The establishment of this kind of science is of paramount importance for process control, development of new innovative processes and control of the polymer supermolecular structure which governs the performance of membranes, fibers, and paints to mention but a few applications.

Despite the importance of the thermodynamics of polymer solutions, until now, there has been no comprehensive book covering the theoretical fundamentals of phase equilibria and critical phenomena of multicomponent polymer solutions at an advanced level and which gives due attention to the application of polymer solution theory. This situation forces scientists who have a need for relevant information and technologists and experimentalists (almost all of them chemists and chemical engineers), to read the original scientific literature, an extremely difficult task for the inexperienced.

I had a keen interest in phase equilibria of polymer solutions when I was a university student in 1950s and my interest was strongly and continuously motivated to carry out the theoretical and experimental studies on phase equilibrium of polymer solutions. This has resulted in publication of more than 50 papers since the first published in 1968. I have felt some responsibility to bridge the significant gap between the primary fundamentals of my discipline and topics of technical significance involving the thermodynamics of phase equilibria; this book is the result.

In this book I have attempted to show how to rigorously solve the so-called modified Flory-Huggins theory under the conditions given by the Gibbs theory of phase equilibrium at a constant temperature and pressure. Consideration has been given to the fact that any real polymer is more or less a multicomponent system (sometimes involving thousands of different molecular sized species). I have not only derived the necessary equations without serious abbreviation in a systematic manner and but have also tried to present the results of computer experiments and the actual experiments as far as was possible.

Prof. Katsumasa Kaneko, ex-president of the University of Kanazawa (Japan), was the first great teacher to introduce me to the fascinating world of polymer science. In the initial stage of my career in research (1950s-1960s) I attempted, without a concrete perspective of the future, to use the "electronic computer for business" for phase equilibrium study. I would like to express my gratitude to my professor's lasting advice and encouragement without which this book could not have been realized. And also I would like to express my thanks to Prof. Toru Kawai, former professor at the Tokyo Institute of Technology, who guided me for many years as my collaborator.

In writing the book, two young coworkers Dr. S. Matsuda and Mr. H. Shirataki earnestly helped me. Without their talented contributions, the book could not have been completed. I am also very obliged to Prof. Manfred Gordon of Cambridge University and Prof. Anthony Johnson of Bradford University for their fruitful comments on the whole original manuscript. The author would like to express his thanks to Prof. Hidematsu Suzuki of Kyoto University and Dr. Takeshi Matsuura of National Research Council of Canada for kindly reading parts of the manuscript and making their useful comments.

Acknowledgement

The author would like to express his sincere gratitude to Ms. Makiko Matsumoto, Mr. Masahiko Sanada, Mr. Tsutomu Ogawa, Prof. Chozo Nakayama, Dr. Keiko Yamaguchi, Dr. Ken Sugamiya, Mr. Toshikazu Terakawa, Mr. Tasuku Hara, Mr. Nobuyuki Baba, Dr. Yukio Miyazaki, Prof. Toru Kwai, Prof. Ichiro Noda, Mr. Hidenobu Ishikawa, Mr. Tatsuyuki Abe, Prof. Motozo Kaneko, Dr. Toshiaki Dobashi, Dr. Masatoshi Saito, Mr. Keisuke Kowsaka, Prof. Hidematsu Suzuki, Mr. Hideki Iijima, Mr. Akira Kataoka, Dr. Hiroyuki Hanahata, Mr. Michitaka Iwata, Miss Kazuko Sogawa, Dr. Sei-ichi Manabe, Mrs. Yukari Kamata, Mrs. Sakae Nakamura, Mr. Hironobu Shirataki, and Dr. Shigenobu Matsuda for their helpful cooperation during this study. The author is grateful to thank Hüthig & Wepf Verlag for permission to reprint Figs 2.4, 2.11-2.15, 2.22,

2.23, 2.25-2.27, 2.34-2.50, 2.53 in "Die Makromolekulare Chemie", The Society of Polymer Science, Japan for permission to reprint Figs. 2.2, 2.3, 2.7-2.10, 2.16-2.18, 2.20, 2.24, 2.29, 2.30, 2.33, 2.51, 2.54-2.58, 2.78-2.83, 2.91-2.113, 3.1-3.57, 5.1-5.7, 6.36-6.38, 6.64-6.70, 6.95-6.97 published in "Polymer Journal" and Fig. 6.3 published in "Kobunshi Ronbunshu", American Chemical Society for permission to reprint the Fig. 6.7 published in ACS Symposium Series 269, and Elsevier Applied Science Publishers for permission to reprint published Figs 2.21, 2.28, 2.31, 2.32 in "British Polymer Journal".

CONTENTS

Chapter 1

INTRODUCTION

Polymer solutions, which are mixtures of high-molecular weight compounds (solutes) and low-molecular weight solvents, have been examined experimentally, since the early 1930s. These studies revealed some abnormalities when the results were compared with those for low-molecular weight solutions e.g. an unexpectedly small vapor depression, small boiling point elevation, small osmotic pressures and extremely high solution viscosities. These abnormalities became a strong motivating force in the search for a thermodynamic theory for polymer solutions. The fundamental quantity of the state for a solution is the chemical potential of the solvent $\Delta\mu_0$, from which other physical quantities related to entropy, volume and enthalpy can be derived mathematically. $\Delta\mu_0$ can be determined from vapor pressure P, vapor osmotic pressure (eqn (1.1)), membrane osmotic pressure π (eqn (1.1)), and by analyzing phase separation and critical solution phenomena (Fig. 1.1) (refs 1,2).

$$\Delta\mu_0 = \tilde{R}T \ln(P_0/P_0{}^0) = -\pi V_0 = \tilde{R}T \ln a_0 . \qquad (1.1)$$

Here $\tilde{R}$ is the gas constant, T, the Kelvin temperature, P_0, the vapor pressure of the solvent component in solution, $p_0{}^0$, the vapor pressure of the pure solvent, V_0, the molar volume of solvent, and a_0, the activity of the solvent in the solution. In polymer solutions, the vapor pressure deviates significantly from the Raoult's law ($n_0 = P_0/P_0{}^0$; n_0 = molar fraction of the solvent) and the degree of deviation is larger for polymers with larger molecular weight (ref. 3). The osmotic pressure can be expressed in virial expansion (refs 1,2);

$$\pi\ (= -\Delta\mu_0/V_0) = \tilde{R}Tc\,(1/M_n + A_2c + A_3c^2 + A_4c^3 + \cdots) \qquad (1.2)$$

where A_2, A_3, and A_4 are the second, third, and forth virial coefficients; c, the weight concentration (g/cm^3), and M_n, the number-average molecular weight. For polymer solutions, A_3 cannot be ignored even at relatively low concentrations of solute ($c \simeq$ several %) and A_2 should be considered over the whole range of

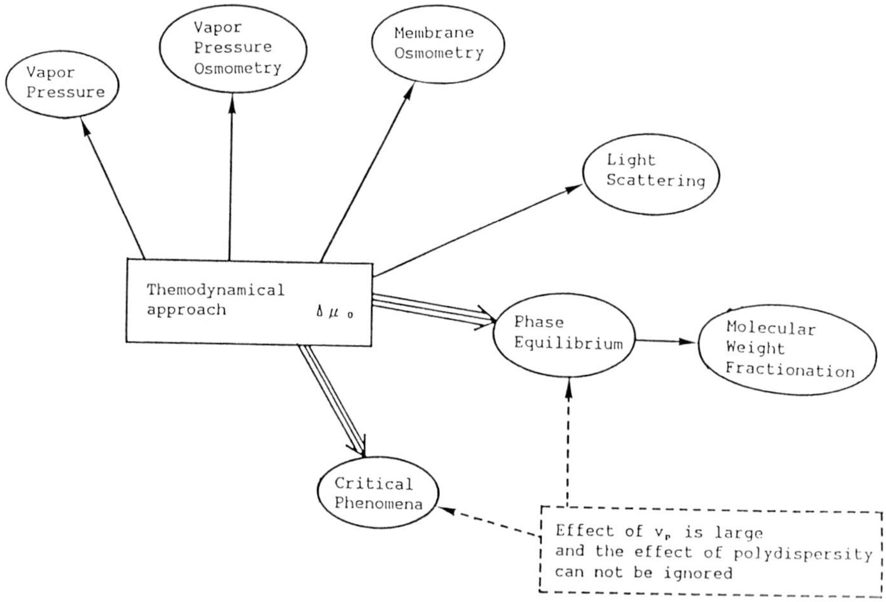

Fig. 1.1 Method of determining $\Delta\mu_0$.

concentration (ref. 4). In other words, the larger the molecular weight of the dissolved polymer, the larger the observed non-ideality. With the exception of some biopolymers, polymers are not strictly single components, but are a mixture of molecules of different molecular weights and (sometimes) chemical structures (polymolecurarity or polydispersity).

The Gibbs' free energy of mixing ΔG of a polydisperse polymer in a single solvent (quasi-binary) system can be expressed as (refs 1,2)

$$\Delta G = \Delta G^{id} + \Delta G^{E} \tag{1.3}$$

where ΔG^{id} and ΔG^{E} are the ideal and the excess ΔG. ΔG^{id} in eqn (1.3) can be divided into two parts (ref. 1):

$$\Delta G^{id} = \Delta H^{id} - T\Delta S^{id} \tag{1.4}$$

where

$$\Delta H^{id} = 0 \tag{1.5}$$

$$\Delta S^{id} = -\tilde{R}(N_0 \ln n_0 + \sum_{i=1}^{m} N_{xi} \ln n_{xi}). \tag{1.6}$$

Here, N_0 and N_{xi} are the mole numbers of the solvent and X_i-mer molecules, respectively; X_i, the degree of polymerization (strictly speaking, the molar volume ratio of polymer to solvent); n_0 and n_{xi}, mole fractions of solvent and X_i-mer; m, total number of different molecular weight components in the polymer sample. Eqn (1.6) can be rewritten in terms of the volume fractions of the components instead of mole fractions as follows (ref. 1).

$$\Delta S^{id} = -\tilde{R}(N_0 + \sum_{i=1}^{m} X_i N_{xi})(v_0 \ln v_0 + \sum_{i=1}^{m} \frac{v_{xi}}{X_i} \ln v_{xi}) \tag{1.7}$$

where volume fractions of solvent and X_i-mer, v_0 and v_{xi}, are given by:

$$v_0 = \frac{N_0}{N_0 + \sum X_i N_{xi}}, \tag{1.8}$$

$$v_{xi} = \frac{X_i N_{xi}}{N_0 + \sum X_i N_{xi}}. \tag{1.9}$$

Substitution of eqns (1.5) and (1.7) into eqn (1.4) yields

$$\Delta G^{id} = \tilde{R}T(N_0 + \sum_{i=1}^{m} X_i N_{xi})(v_0 \ln v_0 + \sum_{i=1}^{m} \frac{v_{xi}}{X_i} \ln v_{xi}). \tag{1.10}$$

The chemical potentials of the solvent and polymer (X_i-mer) for ideal solution, $\Delta\mu_0^{id}$ and $\Delta\mu_{xi}^{id}$, are derived from eqn (1.3) in the forms:

$$\Delta\mu_0^{id} = \frac{\partial \Delta G^{id}}{\partial N_0} = \tilde{R}T\{\ln(1 - v_p) + (1 - \frac{1}{X_n}) v_p\}, \tag{1.11}$$

$$\Delta\mu_{xi}^{id} = \frac{\partial \Delta G^{id}}{\partial N_{xi}} = \tilde{R}T\{\ln v_{xi} - (X_i - 1) + X_i(1 - \frac{1}{X_n}) v_p\}. \tag{1.12}$$

For non-ideal polymer solutions, the chemical potentials $\Delta\mu_0$ and $\Delta\mu_{xi}$ are given by the combination of eqns (1.3), (1.11) and (1.12);

$$\Delta\mu_0 = \frac{\partial \Delta G}{\partial N_0} = \tilde{R}T\{\ln(1 - v_p) + (1 - \frac{1}{X_n}) v_p\} + \Delta\mu_0^{E}, \tag{1.13}$$

$$\Delta\mu_{xi} = \frac{\partial \Delta G}{\partial N_{xi}} = \tilde{R}T\{\ln v_{xi} - (X_i - 1) + X_i(1 - \frac{1}{X_n})v_p\} + \Delta\mu_{xi}^E. \tag{1.14}$$

Here $\Delta\mu_0^E$ and $\Delta\mu_{xi}^E$ are the excess chemical potentials of the solvent and X_i-mer, respectively; X_n is the number-average of X_i for all species $(i = 1,\cdots,m)$; v_p is the overall polymer volume fraction $(\equiv \sum_{i=1}^{m} v_{xi})$. ΔG^E in eqn (1.3) is defined by

$$\Delta G^E \equiv \int_0^{N_0} \Delta\mu_0^E \, dN_0 \tag{1.15a}$$

$$\equiv \int_0^{N_{xi}} \Delta\mu_{xi}^E \, dN_{xi} \tag{1.15b}$$

with

$$\Delta\mu_0^E = \frac{\partial \Delta G^E}{\partial N_0} = \tilde{R}T\chi v_p^2, \tag{1.16}$$

$$\Delta\mu_{xi}^E = \frac{\partial \Delta G^E}{\partial N_{xi}} \tag{1.17a}$$

$$= \frac{\partial}{\partial N_{xi}} \int_0^{N_0} \left(\frac{\partial \Delta G^E}{\partial N_0}\right) dN_0 \tag{1.17b}$$

$$= \tilde{R}T\left\{\frac{\partial}{\partial N_{xi}} \int_0^{N_0} \chi v_p^2 \, dN_0\right\}, \tag{1.17c}$$

where χ is the polymer-solvent thermodynamic interaction parameter. Substituting eqn (1.16) and (1.17a) into (1.13) and (1.14), respectively, $\Delta\mu_0$ and $\Delta\mu_{xi}$ are obtained as follows:

$$\Delta\mu_0 = \tilde{R}T\{\ln(1 - v_p) + (1 - \frac{1}{X_n})v_p + \chi v_p^2\}, \tag{1.18}$$

$$\Delta\mu_{xi} = \tilde{R}T\{\ln v_{xi} - (X_i - 1) + X_i(1 - \frac{1}{X_n})v_p\} + \frac{\partial \Delta G^E}{\partial N_{xi}}. \tag{1.19}$$

This procedure for the derivation of $\Delta\mu_{xi}$ was first given by Kamide (1970), and, of course, eqn (1.18) and (1.19) satisfy the Gibbs-Duhem relations (see, eqn (2.14)) unconditionally. It's also possible to derive an expression for $\Delta\mu_{xi}^E$ independently in the form:

$$\Delta \mu_{X_i}{}^{E} = \tilde{R}T\{-\chi v_p(1-v_p) + \int_{v_p}^{1} \chi \, dv_p\} \tag{1.20}$$

(Kurata (1975) (ref. 2)). The nonideality of the polymer solution is described by the two second terms of righthand side of eqn (1.18). Note that eqn (1.18) is always valid independent of the theoretical models on which eqn (1.18) is derived, if χ is considered as a purely empirical parameter. We can describe χ accurately as functions of T, v_p, and X_i using various experimental data obtained for a given polymer solution, and accordingly, all thermodynamical phenomena can be explained very consistently. For example, using the value for χ, evaluated from the critical solution point, the experimental data obtained by vapor pressure and membrane osmotic pressure measurement should be explained reasonably.

Kurata (ref. 5) stated "It is now well known that the Flory entropy of mixing for chain polymer solution is straightforwardly derivable from the phase integral of the system without the aid of a special model such as the lattice model, if the correlation between segments arising from the chain connectivity is entirely ignored and this implies that the Flory entropy, although crude in itself, can be given a fundamental role in a more general polymer solution theory."

Fig. 1.2 Liquid-liquid phase separation of a polymer solution: Polymer, atactic polystyrene (a-PS) [the weight-average molecular weight $M_w = 13.1 \times 10^4$ and the number average molecular weight $M_n = 11.8 \times 10^4$]; solvent mixture, toluene-methanol; composition of the solution, a-PS / toluene / methanol = 7.3 / 63.8 / 28.9; temperature, 300K.

By lowering or rising the temperature of quasi-binary solution or by adding a nonsolvent to the solution (resulting in quasi-ternary solution), the polymer-rich (or -lean) phase particles separate from the mother solution (i.e., cloud point) and, after settling for a long time, the total solution separates into a two-liquid phase (Fig. 1.2). At constant temperature and constant pressure these two phases are in equilibrium with each other. We define hereafter the phase of smaller v_p as the polymer-lean phase and the phase of larger v_p as the polymer-rich phase.

When the two phases are in equilibrium under constant T and constant P, the well-known Gibbs' law applies, that is (refs 1,2),

$$\Delta\mu_{0(1)} = \Delta\mu_{0(2)}, \tag{1.21}$$

$$\Delta\mu_{X_i(1)} = \Delta\mu_{X_i(2)} \qquad (i = 1, \cdots, m). \tag{1.22}$$

Here, the suffixes (1) and (2) denote the polymer-lean and -rich phases, respectively: Gibbs discussed "the conditions of equilibrium for heterogeneous masses in contact when uninfluenced by gravity, electricity, distortion of the solid mass, or capillary tensions", showing that "the potential, defined by eqns (1.13) and (1.14), for each component substance must be constant throughout the whole mass" (ref. 6). From the vapor pressure and osmotic pressure measurements, $\Delta\mu_0$ can be calculated using eqn (1.1) and putting $\Delta\mu_0$ into eqn (1.18) allows M_n and χ to be evaluated. In phase equilibrium studies, an accurate knowledge of $\Delta\mu_{X_i}$ is necessary in addition to that of $\Delta\mu_0$. $\Delta\mu_{X_i}$ can be estimated from $\Delta\mu_0$ through use of the Gibbs-Duhem relation (see, eqn (2.14)), which, even for polymer solutions, should hold its validity at constant T and constant P (see, for example, eqn (1.19)). For a single component polymer / single solvent system numerous researchers, including Flory (ref. 1), performed theoretical studies on two-phase equilibrium mainly in 1950s. In these studies, the theory employed was based on a lattice model and the calculations were only made under very specific conditions. No comparisons were made with actual experimental data. It is interesting to note that Flory stated in his classical book "Principle of Polymer Chemistry" (1953) (ref. 1) that "we need not undertake the incomparably more involved calculation of σ (partition coefficient see, eqn (2.35)). Even if the latter step was carried out, the numerical value obtained for σ probably would be subject to a considerable error owing to imperfections of the theory." From this statement, it is clear that Flory did not consider his theory to be fully quantitative in nature. It seems important to point out that in 1944, Fujishiro (ref. 7) derived independently a theory similar to that of Flory (ref. 8) and Huggins (ref. 9).

In order to carry out the accurate simulation of two-phase equilibrium of actual polymer / solvent systems for all components (m is usually of order of $10^2 \sim 10^5$) eqns (1.21) and (1.22) should be solved concurrently. As early as 1968 Kamide and coworkers (refs 10,11) and Koningsveld et al. (refs 12-14) independently succeeded in carrying out accurate computer simulations of these equations. As will be shown in Fig. 2.78 (section 2.6), cloud point curves (CPC) of single component polymer / single solvent (i.e., strictly binary) system as well as multicomponent polymers / single solvent (i.e., quasi-binary) systems reveal larger asymmetry as X or X_n of the polymer increases. Thereafter, Kamide and his collaborators (refs 15-22) and Koningsveld et al. (refs 23-27) developed more advanced and more rigorous theories for the quasi-binary system than those first published in 1968 and they established the necessary computer simulation techniques in a more systematic way.

These quasi-ternary systems are important because they are employed widely in industry e.g. in the wet spinning of fibers, for the production of porous polymeric membranes and with polymer coatings. Therefore, in addition, the phase equilibrium of quasi-ternary solutions attracts our scientific interest in connection with selective adsorption of solvent onto the polymer molecule, and cosolvency. Theoretical studies on two-phase equilibrium of single component polymer / binary simple solvent systems (strictly ternary system), carried out by Flory (ref. 8), Scott (refs 28,29), Tompa (ref. 30), Nakagaki and Sunada (ref. 31), Krigbaum and Carpenter (ref. 32), Suh and Liou (ref. 33) are unfortunately very qualitative. Recently Kamide and Matsuda (refs 34-37) derived a rigorous theory of two-phase equilibrium of multicomponent polymers / binary simple solvent systems, and established the computer simulation algorithms for this theory.

Fig. 1.3 shows the external factors of prerequisites for the simulation studies and the phase equilibrium characteristics obtained by computer simulation of phase separation of quasi-binary and -ternary polymer solutions.

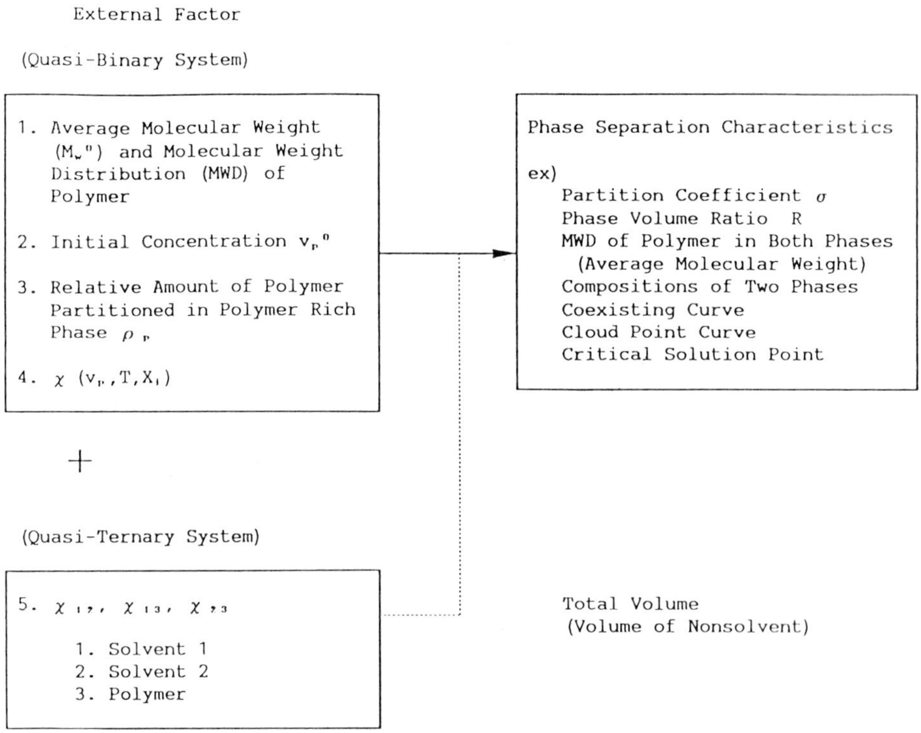

Fig. 1.3 External factors and phase equilibrium characteristics.

REFERENCES

1 P.J. Flory, Principle of Polymer Chemistry, Cornell Univ. Press, Ithaca, New York, 1953.

2 M. Kurata, Thermodynamics of Polymer Solutions, Harwood Academic Pubs. Chur, London, NY, 1982, Chapter 2.

3 See, for example, G. Gee and L.R.G. Treloar, Trans. Faraday Soc. 38 (1942) 147 and C.E. Bawn, R.F.J. Freeman and A.R. Kamaliddin, Trans. Faraday Soc., 46 (1950) 677.

4 See, for example, W.R. Krigbaum and P.J. Flory, J. Am. Chem. Soc., 65 (1943) 372.

5 M. Kurata, Ann. New York Acad. Sci., 89 (1961) 635.

6 J.W. Gibbs, The Scientific Papers of J. Willard Gibbs, Ph.D., LL. D., Vol. I, Thermodynamics, p65, Dover Pub., New York, 1961, unaltered republication, originally published by Longmans, Green, and Co., 1906.

7 R. Fujishiro, Nippon Kagaku Zasshi, 65 (1944) 519.

8 P.J. Flory, J. Chem. Phys., 12 (1944) 425.

9 M.L. Huggins, J. Am. Chem. Soc., 64 (1942) 1712.
10 K. Kamide, T. Ogawa, M. Sanada and M. Matumoto, Kobunshi Kagaku, 25 (1968) 440.
11 K. Kamide, T. Ogawa and M. Matumoto, Kobunshi Kagaku, 25 (1968) 788.
12 R. Koningsveld and A.J. Staverman, Kolloid-Z. Z. Polym., 218 (1967) 114.
13 R. Koningsveld and A.J. Staverman, J. Polym. Sci., A-2, 6 (1968) 305.
14 R. Koningsveld and A.J. Staverman, J. Polym. Sci., A-2, 6 (1968) 349.
15 K. Kamide, in L.H. Tung (Ed.), Fractionation of Synthetic Polymers, Marcel Dekker Inc., New York, 1977, Chapter 2.
16 K. Kamide, Y. Miyazaki and T. Abe, Polym. J., 9 (1977) 395.
17 I. Noda, H. Ishizawa, Y. Miyazaki and K. Kamide, Polym. J., 12 (1980) 87.
18 K. Kamide, K. Sugamiya, T. Kawai and Y. Miyazaki, Polym. J., 12 (1980) 67.
19 K. Kamide and Y. Miyazaki, Polym. J., 12 (1980) 205.
20 K. Kamide and Y. Miyazaki, Polym. J., 13 (1981) 325.
21 K. Kamide, Y. Miyazaki and T. Abe, Brit. Polym. J., 13 (1981) 168.
22 K.Kamide, T. Abe and Y. Miyazaki, Polym. J., 14 (1982) 355.
23 R. Koningsveld and A.J. Staverman, J. Polym. Sci., A-2, 6 (1968) 367.
24 R. Koningsveld and A.J. Staverman, J. Polym. Sci., A-2, 6 (1968) 383.
25 M. Gordon, H.A.G. Chermin and R. Koningsveld, Macromolecules, 2 (1969) 107.
26 R. Koningsveld, Adv. Polym. Sci., 7 (1970) 1.
27 R. Koningsveld, W.H. Stockmayer and J.W. Kennedy, L.A. Kleintjens, Macromolecules, 7 (1974) 731.
28 R.L. Scott, J. Chem. Phys., 13 (1945) 178.
29 R.L. Scott, J. Chem. Phys., 17 (1949) 268.
30 H. Tompa, Trans. Farady Soc., 45 (1949) 1142.
31 M. Nakagaki and H. Sunada, Yakugaku Zasshi, 83 (1963) 1147.
32 W.R. Krigbaum and D.K. Carpenter, J. Polym. Sci., 14 (1954) 241.
33 K.W. Suh and D.W. Liou, J. Polym. Sci., A-2, 6 (1968) 813.
34 K. Kamide, S. Matsuda and Y. Miyazaki, Polym. J., 16 (1984) 479.
35 K. Kamide and S. Matsuda, Polym. J., 16 (1984) 515.
36 K. Kamide and S. Matsuda, Polym. J., 16 (1984) 591.
37 S. Matsuda, Polym. J., 18 (1986) 993.

Chapter 2

QUASI-BINARY SOLUTIONS OF POLYDISPERSE POLYMERS AND A SINGLE SOLVENT

2.1 GIBBS' FREE ENERGY OF MIXING ΔG

First, the spinodal curve, the binodal curve, and the critical solution point of polymer solutions will be briefly explained for a single component polymer / single solvent (i.e., strict binary solution) system.

The mean molar Gibbs' free energy of mixing of the system, ΔG, is schematically plotted in Fig. 2.1 as a function of the solute (polymer) concentration (mole fraction) x_p and temperature T at constant pressure P. Here, the system is assumed to be one molar and in this case ΔG is Gibbs' free energy change per mole (i.e., molar Gibbs' free energy). In Fig. 2.1 the line connecting the points of inflection is the spinodal curve. In other words, this is the line at which the second differential of ΔG, $\partial^2 \Delta G/\partial x_p^2$ becomes zero. The line connecting the points of contact of the double tangent (this is the point at which $(\partial \Delta G/\partial x_p)_{(1)} = (\partial \Delta G/\partial x_p)_{(2)}$ holds) is the binodal curve (i.e., coexistence curve). The line connecting the points at which $\partial^3 \Delta G/\partial x_p^3 = 0$ holds in concave plane between two spinodal branches, is often described as the neutral equilibrium condition. The point at which the binodal curve coincides with the spinodal curve, is the critical point of the solution at a given pressure P. At the critical solution point, ΔG is a minimum and of course the neutral equilibrium condition is satisfied at the critical solution point. The critical point can be determined as the point at which the spinodal condition ($\partial^2 \Delta G/\partial x_p^2 = 0$) and the neutral equilibrium condition ($\partial^3 \Delta G/\partial x_p^3 = 0$) are concurrently satisfied. Projected curves on the x_p-T plane are also shown in Fig. 2.1. Thermodynamic equilibrium exists between two phases, whose compositions are given by two points in the binodal curve at given T (i.e., the two cross points of the binodal curve and T= constant line). When the solution of a given v_p is cooled (or warmed), the solution becomes turbid at a specific temperature i.e. the cloud point. The line connecting these points obtained for various v_p is the cloud point curve. Only in the case of a single component polymer / single solvent system does the binodal curve strictly coincide with the cloud point curve. The fact that this dose not apply for multicomponent polymer / single solvent systems, was first pointed out in 1968 (refs 1,2) by Kamide et al.

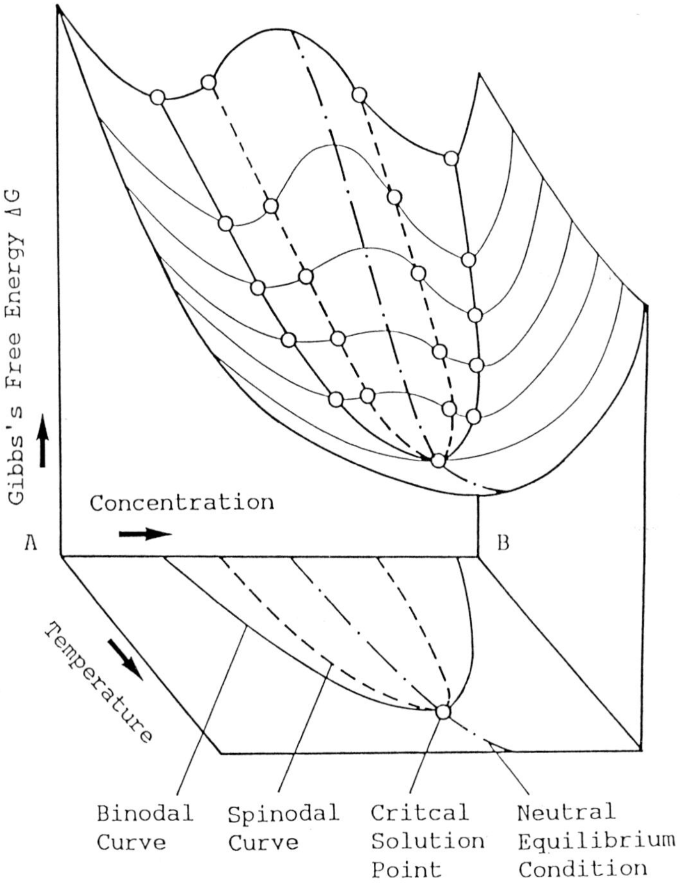

Fig. 2.1 Gibbs' free energy of mixing ΔG'.

2.2 CHEMICAL POTENTIALS $\Delta\mu_0$ AND $\Delta\mu_{Xi}$

The χ parameter in eqn (1.18) can be expressed phenomenologically as a power series of the concentration (refs 3-6),

$$\chi_i = \chi_{00}\{1+(k'/X_i)\}(1+\sum_{j=1}^{n} p_j v_p^{\,j}) \qquad (i=1,\cdots,m) \tag{2.1}$$

or simply (refs 7,8),

$$\chi_i = \chi_{00}\{1+(k'/X_i)\}(1+p_1 v_p) \qquad (i=1,\cdots,m) \tag{2.2}$$

where χ_{00} is a parameter independent of the polymer volume fraction v_p and the molar volume ratio of the i-th component to the solvent X_i (roughly equal to the degree of polymerization); k' and p_j are molecular weight- and the concentration-independent parameters. χ_{00} and k' are functions of temperature and probably of pressure; m is the total number of components. Kamide et al. confirmed experimentally that eqn (2.2) is accurate enough to represent the two-phase equilibrium of a multicomponent polymer / single solvent system and disclose the effects of the k' and p_1 parameters on the characteristics of coexisting curve in detail (refs 7,8). In other words, although a higher term containing v_p^2 (i.e., p_2) can be observed in the χ parameter by membrane osmometry and vapor pressure measurements, this term does not significantly contribute to the phase separation characteristics including coexistence curve, within experimental certainty. From eqn (2.1) χ can be written as (refs 3-6):

$$\chi = \chi_0(1+\sum_{j=1}^{n} p_j v_p^{\,j}) \tag{2.3}$$

where

$$\chi_0 = \chi_{00}\{1+(k'/X_n)\} \tag{2.4}$$

$$= (a+\frac{b}{T})\{1+\frac{k_0}{X_n}(1-\frac{\theta}{T})\}. \tag{2.4'}$$

Kamide et al. (ref. 8) showed from the phase separation experiment for atactic polystyrene in cyclohexane and methylcyclohexane that k' is empirically given as

$$k' = k_0 \{1 - (\theta / T)\} \qquad (2.5)$$

where k_0 is the parameter independent of T and X_i; θ, Flory theta temperature. The temperature dependence of χ_{00} can be expressed as:

$$\chi_{00} = a + (b/T). \qquad (2.6)$$

Here, a and b are the parameters independent of T, X_i and v_p (see, eqn (2.217)). The concentration and temperature dependence of the χ parameter are thus given as

$$\chi = (a' + \frac{b'}{T} + \frac{c'}{T^2})(1 + \sum_{j=1}^{n} p_j v_p{}^j) \qquad (2.7)$$

with

$$a' = a\{1 + (k_0/X_n)\}, \qquad (2.8a)$$

$$b' = b + (k_0/X_n)(b - a\theta), \qquad (2.8b)$$

$$c' = -(k_0 b\theta / X_n). \qquad (2.8c)$$

Eqn (2.7) is readily derived by comparing eqns (2.4)-(2.6). When $k_0 = 0$, a' reduces to a, b' to b and c' to zero.

Using eqns (1.18), (2.3) and (2.4), $\Delta\mu_0$ can be written as

$$\Delta\mu_0 = \tilde{R}T\{\ln(1 - v_p) + (1 - \frac{1}{X_n})v_p + \chi_{00}(1 + \frac{k'}{X_n})(1 + \sum_{j=1}^{n} p_j v_p{}^j)v_p{}^2\}. \qquad (2.9)$$

$\Delta\mu_0$ can also be rewritten by expanding the term $\ln(1 - v_p)$ using a Taylor series, and by expressing v_p (cm^3/cm^3) by c (g/cm^3) as:

$$\Delta\mu_0 = -\tilde{R}TcV_0\{\frac{1}{M_n} + \frac{\bar{v}^2}{V_0}(\frac{1}{2} - \chi_0)c + \sum_{j=1}^{n} \frac{\bar{v}^{j+2}}{V_0}(\frac{1}{j+2} - \chi_0 p_j)c^{j+1}\}. \qquad (2.10)$$

Here, $\bar{v}$ is the polymer specific volume.

Comparing eqn (2.10) with eqn (1.2), A_2, A_3 and $A_4, \cdots$ are given as follows (ref. 5),

$$A_2 = \frac{\bar{v}^2}{V_0}\left(\frac{1}{2} - \chi_0\right), \tag{2.11}$$

$$A_{j+2} = \frac{\bar{v}^{j+2}}{V_0}\left(\frac{1}{j+1} - \chi_0 p_j\right) \qquad (j=1,\cdots,n). \tag{2.12}$$

If we can assume that $A_2 = A_3 = A_4 = \cdots = 0$ at $T = \theta$, for the upper critical solution point (UCSP), we obtain (ref. 5)

$$\chi_0 = \frac{1}{2}, \tag{2.13a}$$

$$p_1 = \frac{2}{3}, \quad p_2 = \frac{2}{4}, \quad p_3 = \frac{2}{5}, \quad \cdots, \quad p_n = \frac{2}{n+2}. \tag{2.13b}$$

These are the theoretically predicted values for χ_0 and p_j $(j=1,\cdots,n)$. In a lower critical solution point (LCSP) region, p_1 is not influenced by free volume in contrast with χ_0 (refs 9,10) and $\chi_0 p_1 = 1/3$ is also satisfied (ref. 5). Kamide et al. confirmed experimentally that consideration of p_1 (accordingly A_3) in the χ parameter (see, eqn (2.2)) is accurate enough to represent the two-phase equilibrium of a multicomponent polymer / single solvent system (see, sections 2.3 and 2.4). However, p_2 is also necessary to describe the critical solution point (see, section 2.5), the cloud point curve (see, section 2.6) and entropy and enthalpy parameters (see, sections 2.7 and 2.8). It is possible to evaluate p_1 and p_2 (A_3 and A_4) by analyzing the critical solution point or cloud point curve data and an approximate value for p_1 (A_3) can be determined from the phase separation experiment.

$\Delta\mu_{X_i}$ is derived from eqn (2.9) when these potentials satisfy Gibbs-Duhem equation:

$$N_0 d(\Delta\mu_0) + \sum_{i=1}^{m} N_{X_i} d(\Delta\mu_{X_i}) = 0. \tag{2.14}$$

Combining eqns (1.16), (2.3) and (2.4), the excess chemical potential of solvent $\Delta\mu_0^E$ for a quasi-binary system consisting of multicomponent polymer in a single solvent becomes (refs 4,6),

$$\Delta\mu_0^E = \frac{\partial \Delta G^E}{\partial N_0} = \tilde{R}T\chi_0 (1+\sum_{j=1}^{n} p_j v_p^j) v_p^2 \tag{2.15}$$

$$= \tilde{R}T\chi_{00} (1+\frac{k'}{X_n}) (1+\sum_{j=1}^{n} p_j v_p^j) v_p^2 . \tag{2.16}$$

Integrating eqn (2.16) by N_0, the excess free energy of a quasi-binary system ΔG^E is obtained (refs 4,6):

$$\Delta G^E = \int_0^{N_0} (\frac{\partial \Delta G^E}{\partial N_0}) dN_0$$

$$= \tilde{R}T (N_0 + \sum_{i=1}^{m} X_i N_{Xi}) [\chi_{00} (1+\frac{k'}{X_n}) (1+\sum_{j=1}^{n} \frac{p_j}{j+1} \frac{1-v_p^{j+1}}{v_0})] v_0 v_p . \tag{2.17}$$

The Gibbs' free energy of ideal solution ΔG^{id} of quasi-binary system is expressed by eqn (2.18) (refs 11,12):

$$\Delta G^{id} = \tilde{R}TL [v_0 \ln v_0 + \sum_{i=1}^{m} \frac{v_{Xi}}{X_i} \ln v_{Xi}] \tag{2.18}$$

with

$$L = N_0 + \sum_{i=1}^{m} X_i N_{Xi} . \tag{2.19}$$

Substituting eqns (2.17) and (2.18) into eqn (1.3), ΔG is finally expressed by eqn (2.20) (refs 4,6):

$$\Delta G = \tilde{R}TL [v_0 \ln v_0 + \sum_{i=1}^{m} \frac{v_{Xi}}{X_i} \ln v_{Xi}$$

$$+ \chi_{00} (1+\frac{k'}{X_n}) (1+\sum_{j=1}^{n} \frac{p_j}{j+1} \frac{1-v_p^{j+1}}{v_0}) v_0 v_p] . \tag{2.20}$$

Differentiations of ΔG by N_{Xi} gives $\Delta\mu_{Xi}$ as follows (refs 4,6-8):

$$\Delta\mu_{Xi} = \Delta\mu_{Xi}^{id} + \Delta\mu_{Xi}^E$$

$$= \tilde{R}T [\ln v_{Xi} - (X_i - 1) + X_i (1-\frac{1}{X_n}) v_p$$

$$+X_i(1-v_p)^2\chi_{00}\left[\left(1+\frac{k'}{X_n}\right)\left\{1+\sum_{j=1}^{n}\frac{p_j}{j+1}\left\{\sum_{q=0}^{j}(q+1)v_p{}^q\right\}\right\}\right.$$

$$\left.+k'\left(\frac{1}{X_i}-\frac{1}{X_n}\right)\left\{\frac{1}{1-v_p}+\sum_{j=1}^{n}\frac{p_j}{j+1}\left(\sum_{q=0}^{j}\frac{v_p{}^q}{1-v_p}\right)\right\}\right]\right].\tag{2.21}$$

Here, we assume that (a) the polymer and solvent are volumetrically additive, and that (b) the densities of polymer and solvent are the same.

Eqn (2.14) can be converted into eqn (2.22):

$$N_0\left[\left(\frac{\partial\Delta\mu_0}{\partial v_p}\right)dv_p+\left(\frac{\partial\Delta\mu_0}{\partial(1/X_n)}\right)d(1/X_n)\right]$$

$$+\sum_{i=1}^{m}N_{Xi}\left[\left(\frac{\partial\Delta\mu_{Xi}}{\partial v_p}\right)dv_p+\left(\frac{\partial\Delta\mu_{Xi}}{\partial v_{Xi}}\right)dv_{Xi}+\left(\frac{\partial\Delta\mu_{Xi}}{\partial(1/X_n)}\right)d(1/X_n)\right]=0.\tag{2.22}$$

Substituting eqns (2.9) and (2.21) into eqn (2.22), we can readily confirm that eqn (2.14) is absolutely satisfied.

For quasi-binary systems, Koningsveld et al. (refs 13,14) introduced a thermodynamic interaction parameter g in the expression of ΔG:

$$\Delta G=\tilde{R}TL\left[v_0\ln v_0+\sum_{i=1}^{m}\frac{v_{Xi}}{X_i}\ln v_{Xi}+gv_0v_p\right].\tag{2.23}$$

g can be expanded as a series function of v_p in the form (refs 14-16),

$$g=\sum_{j=0}^{n}g_jv_p{}^j,\tag{2.24}$$

where g_j is the concentration dependent parameter. Here, they assumed that g_0 could be derived into temperature-independent and -dependent components given by (refs 14,17):

$$g_0=g_{00}+g_{01}/T.\tag{2.25}$$

Between χ (given by eqn (2.3)) and g, the following relation holds (refs 12,18):

$$g = \frac{1}{v_0}\int_{1-v_0}^{1} \chi \, d(1-v_0) \qquad (2.26)$$

$$= \chi_0 \left[1 + \sum_{j=1}^{n} \frac{p_j}{j+1} \frac{1-v_p^{j+1}}{v_0}\right]. \qquad (2.26')$$

The differential form of the relation between χ and g is (ref. 12)

$$\chi = g - (1-v_p)\frac{\partial g}{\partial v_p} \qquad (2.27)$$

$$= g - v_p \frac{\partial g}{\partial (1-v_0)} \qquad (2.27')$$

and χ_0, p_j and p_n are directly related to g, g_j and g_n through the relations (refs 3,5,12),

$$\chi_0 = g_0 - g_1, \qquad (2.28)$$

$$p_j = (j+1)\frac{g_j - g_{j+1}}{g_0 - g_1} \qquad (j = 1, \cdots, n-1), \qquad (2.29)$$

$$p_n = (n+1)\frac{g_n}{g_0 - g_1}. \qquad (2.30)$$

According to the theory of Koningsveld et al., p_j is a function of temperature through g_0. An alternative expression for the concentration-dependence of g was proposed by Koningsveld et al. (refs 19,20) in the closed form,

$$g = \alpha + \frac{\beta}{1 - \gamma v_p} \qquad (2.31)$$

with

$$\beta = \beta_{00} + \beta_{01}/T. \qquad (2.32)$$

Here, α, β_{00}, β_{01} and γ are constants for a given polymer / solvent system. Koningsveld and Kleintjens (ref. 19) remarked that "yet, a closed expression, if available, would seem more elegant and be preferable to a truncated series" and "one would then be inclined to make do with a simple closed expression even if its parameters proved to be void of a physical meaning." If $0<\gamma<1$ holds, $0<\gamma v_p<1$ can be obtained and the second term in the right-hand side of eqn (2.31) expanded as a series function

$$g=\alpha+\beta+\sum_{j=1}^{\infty}\beta(\gamma v_p)^j$$

$$\simeq\alpha+\beta+\sum_{j=1}^{n}\beta(\gamma v_p)^j. \qquad (2.33)$$

Eqn (2.33) reduces to eqn (2.24) by putting $\alpha+\beta=g_0$ and $\beta\gamma^j=g_j$ $(j=1,\cdots,n)$. In this case, χ_0 and p_j can be expressed in terms of α, β and γ by

$$\chi_0=\alpha+\beta(1-\gamma), \qquad (2.28')$$

$$p_j=(j+1)\frac{\beta\gamma^j(1-\gamma)}{\alpha+\beta(1-\gamma)}, \qquad (2.29')$$

$$p_n=(n+1)\frac{\beta\gamma^n}{\alpha+\beta(1-\gamma)}. \qquad (2.30')$$

Note that p_j is sometimes temperature dependent due to the temperature dependences of α, β and γ.

2.3 TWO-PHASE SEPARATION CHARACTERISTICS

Substitution of eqn (2.9) into eqn (1.21) gives (refs 4,6-8),

$$\chi_{00}=\left[\ln\frac{1-v_{p(1)}}{1-v_{p(2)}}+(v_{p(1)}-v_{p(2)})-\left(\frac{v_{p(1)}}{X_{n(1)}}-\frac{v_{p(2)}}{X_{n(2)}}\right)\right]$$

$$/\left[(v_{p(2)}{}^2-v_{p(1)}{}^2)+k'\left(\frac{v_{p(2)}{}^2}{X_{n(2)}}-\frac{v_{p(1)}{}^2}{X_{n(1)}}\right)\right.$$

$$\left.+\sum_{j=1}^{n}p_j\left\{(v_{p(2)}{}^{j+2}-v_{p(1)}{}^{j+2})+k'\left(\frac{v_{p(2)}{}^{j+2}}{X_{n(2)}}-\frac{v_{p(1)}{}^{j+2}}{X_{n(1)}}\right)\right\}\right] \qquad (2.34)$$

and the combination of eqns (1.22) and (2.21) yields

$$\sigma_i\equiv\frac{1}{X_i}\ln\frac{v_{Xi(2)}}{v_{Xi(1)}} \qquad (2.35)$$

$$= \sigma_0 + \sigma_{01}/X_i \tag{2.35'}$$

with

$$\begin{aligned}\sigma_0 = &(v_{p(1)} - v_{p(2)}) - \left(\frac{v_{p(1)}}{X_{n(1)}} - \frac{v_{p(2)}}{X_{n(2)}}\right)\\ &- \chi_{00}\Big[2(v_{p(1)} - v_{p(2)}) - (v_{p(1)}{}^2 - v_{p(2)}{}^2)\\ &\quad + \sum_{j=1}^{n} p_j \Big\{\frac{j+2}{j+1}(v_{p(1)}{}^{j+1} - v_{p(2)}{}^{j+1}) - (v_{p(1)}{}^{j+2} - v_{p(2)}{}^{j+2})\Big\}\Big]\\ &- \chi_{00}k'\Big[\left(\frac{v_{p(1)}}{X_{n(1)}} - \frac{v_{p(2)}}{X_{n(2)}}\right) - \left(\frac{v_{p(1)}{}^2}{X_{n(1)}} - \frac{v_{p(2)}{}^2}{X_{n(2)}}\right)\\ &\quad + \sum_{j=1}^{n} p_j \Big\{\left(\frac{v_{p(1)}{}^{j+1}}{X_{n(1)}} - \frac{v_{p(2)}{}^{j+1}}{X_{n(2)}}\right) - \left(\frac{v_{p(1)}{}^{j+2}}{X_{n(1)}} - \frac{v_{p(2)}{}^{j+2}}{X_{n(2)}}\right)\Big\}\Big]\end{aligned} \tag{2.36}$$

and

$$\sigma_{01} = -k'\chi_{00}\Big\{(v_{p(1)} - v_{p(2)}) + \sum_{j=1}^{n} \frac{p_j}{j+1}(v_{p(1)}{}^{j+1} - v_{p(2)}{}^{j+1})\Big\}. \tag{2.37}$$

If k' and p_j $(j=1,\cdots,n)$ are known, χ_{00}, σ_0 and σ_{01} become functions of four variables, $v_{p(1)}$, $v_{p(2)}$, $X_{n(1)}$ and $X_{n(2)}$ (refs 1,2,7,21,22). According to definition, $v_{p(1)}$, $v_{p(2)}$, $X_{n(1)}$ and $X_{n(2)}$ are expressed in terms of $v_{Xi(1)}$ and $v_{Xi(2)}$ as follows.

$$v_{p(1)} = \sum_{i=1}^{m} v_{Xi(1)}, \tag{2.38a}$$

$$v_{p(2)} = \sum_{i=1}^{m} v_{Xi(2)}, \tag{2.38b}$$

$$X_{n(1)} = \sum_{i=1}^{m} v_{Xi(1)} \Big/ \sum_{i=1}^{m} \frac{v_{Xi(1)}}{X_i} = \sum_{i=1}^{m} g_{(1)}(X_i) \Big/ \sum_{i=1}^{m} \frac{g_{(1)}(X_i)}{X_i}, \tag{2.39a}$$

$$X_{n(2)} = \sum_{i=1}^{m} v_{Xi(2)} \Big/ \sum_{i=1}^{m} \frac{v_{Xi(2)}}{X_i} = \sum_{i=1}^{m} g_{(2)}(X_i) \Big/ \sum_{i=1}^{m} \frac{g_{(2)}(X_i)}{X_i}, \tag{2.39b}$$

where $g_{(1)}(X_i)$ and $g_{(2)}(X_i)$ are the relative amounts of X_i-mer separated into polymer-lean and -rich phases, respectively.

Normalized molecular weight distribution (MWD) of original polymer $g_0(X_i)$ is a summation of $g_{(1)}(X_i)$ and $g_{(2)}(X_i)$:

$$g_0(X_i) = g_{(1)}(X_i) + g_{(2)}(X_i) \qquad (i=1,\cdots,m). \tag{2.40}$$

The weight fraction ρ_s of the polymer in the polymer-lean phase to the total polymer and the fraction ρ_p $(=1-\rho_s)$ of the polymer in the polymer-rich phase are given by eqns (2.41a) and (2.41b).

$$\rho_s = \sum_{i=1}^{m} g_{(1)}(X_i), \tag{2.41a}$$

$$\rho_p = \sum_{i=1}^{m} g_{(2)}(X_i). \tag{2.41b}$$

Using σ_i, the phase volume ratio R, given by

$$R = \frac{V_{(1)}}{V_{(2)}} \tag{2.42}$$

($V_{(1)}$ and $V_{(2)}$, the volumes of the polymer-lean and -rich phases) then $g_0(X_i)$, $g_{(1)}(X_i)$ and $g_{(2)}(X_i)$ are expressed as

$$g_{(1)}(X_i) = \frac{R}{R+\exp(\sigma_i X_i)} g_0(X_i) = f_{(1)}(X_i)\, g_0(X_i), \tag{2.43a}$$

$$g_{(2)}(X_i) = \frac{\exp(\sigma_i X_i)}{R+\exp(\sigma_i X_i)} g_0(X_i) = f_{(2)}(X_i)\, g_0(X_i), \tag{2.43b}$$

where $f_{(1)}(X_i)$ and $f_{(2)}(X_i)$ are

$$f_{(1)}(X_i)\ (= f_{Xi(1)}) = \frac{R}{R+\exp(\sigma_i X_i)}, \tag{2.44a}$$

$$f_{(2)}(X_i)\ (= f_{Xi(2)}) = \frac{\exp(\sigma_i X_i)}{R+\exp(\sigma_i X_i)}. \tag{2.44b}$$

Total volume of the system V is expressed as

$$V = V_{(1)} + V_{(2)} \tag{2.45}$$

$$= V_s + V_p \tag{2.46}$$

where V_s and V_p are the volumes of solvent and polymer, respectively, and combination of eqns (2.42) and (2.45) yields

$$V_{(1)} = \frac{R}{R+1} V, \qquad (2.47a)$$

$$V_{(2)} = \frac{1}{R+1} V. \qquad (2.47b)$$

Using eqns (2.47a) and (2.47b), $v_{X_i(1)}$ and $v_{X_i(2)}$ (or $v_{(1)}(X_i)$ and $v_{(2)}(X_i)$) are represented as

$$v_{X_i(1)} = \frac{V_p g_{(1)}(X_i)}{V_{(1)}} = v_p{}^0 \frac{R+1}{R} g_{(1)}(X_i), \qquad (2.48a)$$

$$v_{X_i(2)} = \frac{V_p g_{(2)}(X_i)}{V_{(2)}} = v_p{}^0 (R+1) g_{(2)}(X_i), \qquad (2.48b)$$

where $v_p{}^0$ is the initial volume fraction of polymer ($= V_p/(V_s+V_p)$: V_s, volume of solvent; V_p, volume of polymer). Utilizing eqns (2.41) and (2.48), eqns (2.49a) and (2.49b) are obtained.

$$v_{p(1)} = v_p{}^0 \frac{R+1}{R} \sum_{i=1}^{m} g_{(1)}(X_i) = v_p{}^0 \frac{R+1}{R} \rho_s, \qquad (2.49a)$$

$$v_{p(2)} = v_p{}^0 (R+1) \sum_{i=1}^{m} g_{(2)}(X_i) = v_p{}^0 (R+1) \rho_p. \qquad (2.49b)$$

A combination of eqns (2.39)-(2.41) gives the following relation:

$$\frac{1}{X_{n(1)}} = \frac{1}{\rho_s} \left\{ \sum_{i=1}^{m} \frac{g_0(X_i)}{X_i} - \sum_{i=1}^{m} \frac{g_{(2)}(X_i)}{X_i} \right\}$$
$$= \frac{1}{\rho_s} \left\{ \frac{1}{X_n{}^0} - \frac{\rho_p}{X_{n(2)}} \right\}. \qquad (2.50)$$

Here, $X_n{}^0$ is the number-average X_i of the original polymer. If ρ_p (accordingly ρ_s) is set as initial condition $\rho_p{}^g$, χ_{00}, σ_0 and σ_{01} are the functions of two variables R^a and $X_{n(2)}{}^a$ (refs 1,2,7,21,22),

$$\chi_{00} = \chi_{00}(R^a, X_{n(2)}{}^a), \qquad (2.51)$$

$$\sigma_0 = \sigma_0(R^a, X_{n(2)}{}^a), \qquad (2.52)$$

$$\sigma_{01} = \sigma_{01}(R^a, X_{n(2)}{}^a), \qquad (2.53)$$

where R^a and $X_{n(2)}{}^a$ are the assumed value of R and $X_{n(2)}$. Combining eqns (2.39b), (2.41b) and (2.42b), ρ_p and $X_{n(2)}$ are finally the functions of σ_0, σ_{01} and R^a. For the sake of convenience, we define the left hand side of eqns (2.54) and (2.55) as A and B, respectively. That is,

$$A \equiv \rho_p(\sigma_0(R^a, X_{n(2)}{}^a), \sigma_{01}(R^a, X_{n(2)}{}^a), R^a) - \rho_p{}^g = 0, \tag{2.54}$$

$$B \equiv X_{n(2)}(\sigma_0(R^a, X_{n(2)}{}^a), \sigma_{01}(R^a, X_{n(2)}{}^a), R^a) - X_{n(2)}{}^a = 0. \tag{2.55}$$

These are the main equations describing the two-phase separation of a quasi-binary system (refs 1,2,7,21,22). By solving the non-linear simultaneous equations (eqns (2.54) and (2.55)), R and $X_{n(2)}$ can be determined. Substituting these two variables into eqns (2.50)-(2.53), it is possible to calculate χ_{00}, σ_0 and σ_{01} and other phase separation characteristics such as $g_{(1)}(X_i)$, $g_{(2)}(X_i)$, $v_{p(1)}$ and $v_{p(2)}$ by use of eqns (2.43a), (2.43b), (2.49a) and (2.49b), respectively (refs 1,2,7,21,22).

In 1968, Kamide et al. (ref. 1) published the first simulation technique for the prediction of phase equilibrium behavior and this was the result of many years' study (1958-1967). They determined the "correct" relationships between σ and R with given $g_0(X_i)$, ρ_p and $v_p{}^0$ (ref. 1) (see, for example, Fig. 2.4). Soon after this initial work, the molecular weight ranges of simulation experiments was widened (ref. 2). Moreover, phase separation characteristics were calculated in the case where χ depends on the polymer concentration (see, eqn (2.3) with $p_1 \neq 0$, $k' = 0$, and $p_2 = \cdots = p_n = 0$) in successive precipitation fractionation (SPF) (ref. 21) and successive solution fractionation (SSF) (ref. 22) (see, section 2.4). In 1981, a more rigorous simulation technique was developed which made it possible to deal with concentration- and molecular weight-dependence χ (see, eqn (2.3) with $p_1 \neq 0$, $k' \neq 0$, and $p_2 = \cdots = p_n = 0$) was established (ref. 7).

Fig. 2.2 shows the main flow chart of the simulation algorithm for phase equilibrium of quasi-binary systems (ref. 7).

1. As prerequisites of computer simulations of two-phase equilibrium (and moreover molecular weight fractionation by solubility difference; see, section 2.4)), (i) $g_0(X_i)$, (ii) k' and p_j ($j = 1, \cdots, n$), (iii) $v_p{}^0$, (iv) ρ_p (and accordingly, ρ_s), (and (v) the way of fractionation in case of molecular weight fractionation; see, section 2.4) are given.

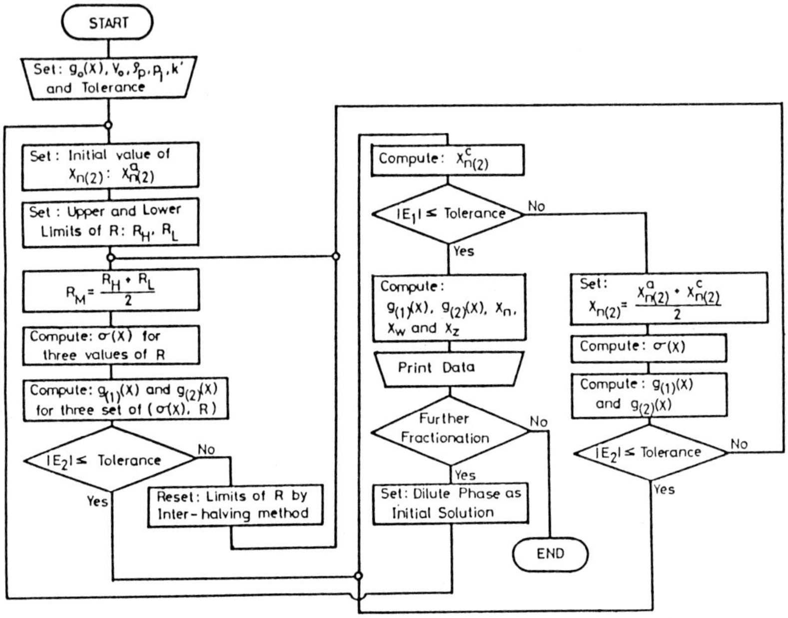

Fig. 2.2 Flow chart of the simulation of phase equilibrium of quasi-binary system.

2. At first $X_{n(2)}{}^a$ and three values of R^a (low, middle, and high values of R^a; R_L, R_M, and R_H (refs 1,2,7,21,22)) were set.
3. For an assumed $X_{n(2)}$,

$$A = A(R^a) = 0 \tag{2.56}$$

can be attained by use of an interhalving method. Thus, R, σ_0 and σ_{01} (accordingly, $g_{(1)}(X_i)$ and $g_{(2)}(X_i)$) are determined for the assumed value of $X_{n(2)}$.
4. When B(given by eqn (2.55)) $\neq 0$ for calculated R, σ_0 and σ_{01}, replace $X_{n(2)}{}^a$ by $(X_{n(2)} + X_{n(2)}{}^a)/2$.
5. Repeating steps 3 and 4 (interhalving method),

$$B = B(X_{n(2)}{}^a) = 0 \tag{2.57}$$

can be solved for variable $X_{n(2)}{}^a$ and therefore simultaneous equations (2.54) and (2.55) are solved in a self-consistent manner and the equilibrium state is finally determined (refs 1,2,7,21,22).

Note that the fraction size is equivalent to the amount of polymer in a phase relative to the original polymer and that, for the polymer remaining in the polymer-rich phase, the fraction size is expressed by ρ_p and for the polymer in the polymer-lean phase fraction size is expressed by ρ_s (see, section 2.4).

In general, four typical molecular weight distribution (MWD) functions were adopted for the original polymer $g_0(X_i)$ (refs 12,23):

(1) Single component (i.e., monodisperse)

(2) Poisson distribution:

$$g_0(X_i) = (X_n^0)^{X_i - 1}\exp(-X_n^0)/(X_i-1)! \tag{2.58}$$

(3) Schulz (ref. 24)-Zimm (ref. 25) (SZ) type distribution with the ratio X_w^0/X_n^0 (X_w^0, the weight-average X_i of the original polymer; denoted as $SZ(X_w^0/X_n^0)$):

$$g_0(X_i) = \frac{y^{h+1}}{\Gamma(h+1)} X_i^h \exp(-yX_i) \tag{2.59}$$

where $$y = h/X_n^0 \tag{2.60}$$

and $$h = \{(X_w^0/X_n^0) - 1\}^{-1}. \tag{2.61}$$

In this case, $X_z^0 - X_w^0 = X_w^0 - X_n^0$ (X_z^0; z-average X_i of $g_0(X_i)$) holds.

(4) Wesslau (ref. 26) (W) type distribution with the ratio X_w^0/X_n^0 (denoted as $W(X_w^0/X_n^0)$):

$$g_0(X_i) = \frac{1}{X_i \beta' (\pi)^{1/2}} \exp[-\frac{1}{\beta'^2}\{\ln(X_i/X^0)\}^2] \tag{2.62}$$

where $$\beta'^2 = 2\ln(X_w^0/X_n^0) \tag{2.63}$$

and $$X^0 = X_w^0/\exp(\beta'^2/4). \tag{2.64}$$

In this case $X_z^0/X_w^0 = X_w^0/X_n^0$ holds.

In actual calculation, it is not necessary to employ the total component number N of the polymer sample. However, when the component number N' is too small, a large error is induced in the values of the distribution breadth ($X_{w(1)}/X_{n(1)}$). All components of the polymers differing in the degree of polymerization X, are then arranged in the order of increasing X and these components are defined as the first, second, third, i-th ··· components from the smallest X, the i-th component being X_i. The number of components N with different X, whose maximum is X_{max} ($\equiv X_m$) and minimum is X_{min} ($\equiv X_1$) are approximated by $X_{max} - X_{min} + 1$. Consequently, calculations are carried out for one [eqn (1.21)] and N [eqn (1.22)] simultaneous equations ($X = X_{min} \sim X_{max}$) (ref. 27). However, the limitations of existing computers mean that it is not always possible to be totally rigorous with the calculation. Actually, the value of X, which is considerably smaller than X_{max}, is taken as the maximum X (referred to as X'_{max}) and the calculation is carried out on only a limited number of components N' ($= X'_{max} - X_{min} + 1$) smaller than N in the range $1 \leq X \leq X'_{max}$. Hereafter N' is termed the "assumed component number". Therefore, the possibility exists for obtaining significantly different results for the true or ideal case (N, $X_{min} \leq X \leq X_{max}$) and actual one simulated (N', $X_{min} \leq X \leq X'_{max}$) (ref. 27).

Kamide and Miyazaki (ref. 27) attempted to determine a minimum reasonable number of components N_m, sufficient for performing the computer experiments with a reasonable degree of accuracy. This is briefly defined as a reasonable polymer component number.

We assume that the MWD of the polymer can be expressed by a SZ or W type distribution, with the ratio of $X_w{}^0/X_n{}^0 = 2$ and 5 (SZ2, SZ5, W2 and W5) and $X_w{}^0 = 150 \sim 3 \times 10^4$ (the suffix 0 means the original polymer).

In the computer experiment, X'_{max} was conveniently expressed by (ref. 27),

$$X'_{max} = 10 \times X_w{}^0 \qquad \text{for the SZ distribution,} \tag{2.65}$$

$$X'_{max} = 20 \times X_w{}^0 \qquad \text{for the W distribution.} \tag{2.66}$$

In eqn (2.65), the weight of the components with $X > 10 \times X_w{}^0$ in the polymer sample is less than 0.1wt%. The distance of the adjacent two Xs, utilized in the calculation was taken as DX ($\equiv X'_{i+1} - X'_i = X'_i - X'_{i-1}$, X'_i is the X value of the i-th component, when all assumed components are numbered in increasing order of X; generally, $X_i \neq X'_i$). The value of N' was then obtained directly as X'_{max}/DX. X'_i is also equivalent to mDX (m = 1,···,N'). The ratio of X_w/X_n of the original polymer, calculated by using only assumed components, is referred to as $X_w{}^{0\prime}/X_n{}^{0\prime}$, in order to distinguish it from the true value.

Preliminary experiments showed that the minimum number of components N_m needed to satisfy the conditions given in eqn (2.67) is 300 for the SZ2 sample and 500 for the W5 sample (ref. 27).

$$X_w^{0\prime}/X_w^0 \leq 1.00 \pm 0.03,$$
$$X_n^{0\prime}/X_n^0 \leq 1.00 \pm 0.03, \quad (2.67)$$
$$(X_w^{0\prime}/X_n^{0\prime})/(X_w^0/X_n^0) \leq 1.00 \pm 0.03.$$

By way of an example we choose a SZ polymer sample with $X_w^0 = 300$ and $X_w^0/X_n^0 = 2$ and a W polymer sample with $X_w^0 = 300$ and $X_w^0/X_n^0 = 5$ and $N' > 300$ ($N = 3000$) for the former and $N' > 1000$ ($N = 6000$) for the latter (ref. 27).

Fig. 2.3 shows the effect of ρ_p on the N' dependence on the σ ($\equiv \sigma_0$, in this case $k' = 0$), R and X_w/X_n of the polymers in both phases (ref. 27). In this figure, $v_p^0 = 1.0\%$ solution of the SZ2 polymer ($X_w^0 = 300$ and $N = 3000$) in a single solvent ($p_J = 0$) was phase-separated. Evidently, $N' > 300$ gives true values for almost all σ, R and $X_{w(2)}/X_{n(2)}$, and these values are virtually equivalent to those obtained for $N' = 3000$. On the other hand, as ρ_p increases, the dependence of N' on $X_{w(1)}/X_{n(1)}$ becomes remarkable. That is, in the region where ρ_p is large, N should be chosen as N'. In conclusion, to calculate X_w/X_n in the two phases within an error of $\pm 3\%$, $N_m = 300$ for $X_{w(2)}/X_{n(2)}$ and $N_m > 1000$ for $X_{w(1)}/X_{n(1)}$ should be employed. The latter N_m value also depends on ρ_p (ref. 27).

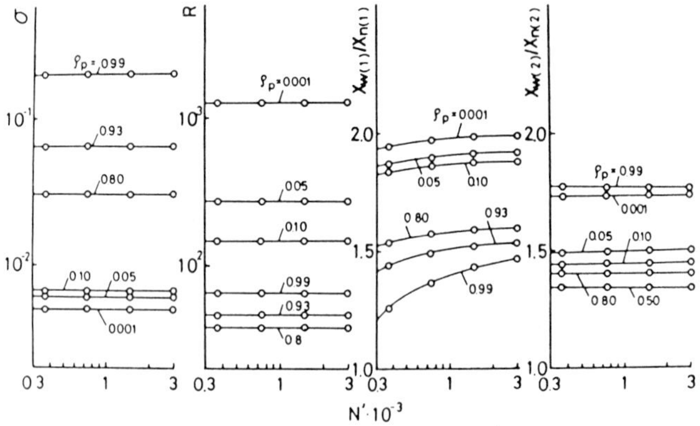

Fig. 2.3 Effect of the fraction size ρ_p on the component number N' dependence on the partition coefficient σ, the volume ratio R, $X_{w(1)}/X_{n(1)}$ and $X_{w(2)}/X_{n(2)}$: Original polymer, Schulz-Zimm type ($X_w^0/X_n^0 = 2$, $X_w^0 = 300$, $N = 3000$); $v_p^0 = 0.01$.

Similar phase separation experiments were carried out on the W5 polymer ($X_w^0 = 300$ and $N = 6000$). The results indicate $N_m = 1500$ for σ, R and $X_{w(2)}/X_{n(2)}$, and $N_m > 3000$ for $X_{w(1)}/X_{n(1)}$ (ref. 27).

In his calculation-fractionation study, Tung (ref. 28) skillfully employed $DX = 1$ in range $1 \leq X \leq 20$, $DX = 3$ in the range $21 \leq X \leq 500$, $DX = 30$ for $501 \leq X \leq 1700$ and $DX = 300$ in the range $1701 \leq X \leq 25700$, for an original polymer ($X_w^0 = 840$). This corresponds to $N' = 300$. Consequently, DX becomes smaller in the lower X region. Thus, as far as the component number is concerned, the reliability of his results may be quite satisfactory. Kamide et al. (ref. 23) employed $N' = 1500 \sim 6000$ (in their very early paper $N' = 800$ was used), which is, except for very special conditions, reasonable to assure the reliability of the results. The validity of the computer simulation data has been experimentally verified by Kamide, Miyazaki and Abe for systems such as polystyrene (PS) / cyclohexane (CH) (ref. 29) and poly(α-methylstyrene) / CH (ref. 30) systems.

When $g_0(X_i)$, k' and p_j are given in advance, there exists explicit but very complicated relationships between v_p^0, ρ_p, σ and R, as the solution theory predicts. In a number of published studies [except those by Tung (ref. 28)] e.g., Kamide et al., (refs 1,2,7,8,21-23,27,29-34) as well as Koningsveld and Staverman (refs 13-17,19,20,35-40), the hypothetical phase separations (or fractionations, see section 2.4) have been carried out either (1) at a constant R value and a given σ value, or (2) at constant v_p^0 and for a given χ (refs 23,41,42). In the former case, v_p^0 as well as ρ_p, was necessarily varied in a very complicated manner and in the latter ρ_p was not kept constant. Fractionation experiments are usually undertaken under the conditions of constant v_p^0 and constant ρ_p, consequently in simulative fractionation where v_p^0 and ρ_p are taken as independent variables these should be carried out under the above mentioned conditions in order that comparisons can be made between experimental and simulation data (see, for example, Table 2.1). Then, it will be useful to establish the relationships among v_p^0, ρ, σ and R. Fig. 2.4 demonstrates such relations for the case, when the solution consisting of the polymer having SZ2 ($X_w^0 = 300$) distribution in a single solvent ($k' = 0$ and $p_j = 0$) was phase-separated (ref. 42).

By inspection, as v_p^0 is reduced keeping ρ_p constant, both σ and R (in particular R) increase progressively. With increasing ρ_p (accordingly, decreasing ρ_s) at a given value for v_p^0, σ

TABLE 2.1
Theoretical studies of the accuracy of the fractionation method by solubility difference.

	Reporter						
	Matsumoto-Ohyanagi	Booth-Beason	Tung	Kamide et al.		Koningsveld et al.	
Theory	FH[a]	FH	FH	Modified FH[b]		Modified FH	
Procedure	SPF[c] and SSF[d]	SPF	SPF	SPF	SSF	SPF	SSF
Original polymer	X_n=500[e]	X_n=50	X_w=50[f]	X_w=300~3000 X_w/X_n=1.05~5 SZ[g] and W[h]	X_w=300 X_w/X_n=2 SZ	X_w=100~1500 X_w/X_n=1.01~2 LN[i] and GE[j]	X_w=100~1500 X_w/X_n=1.01~2 LN and GE
Operating conditions	R=10^2 [k] n_t=8 [l]	R=10^2 n_t=5	v_p^0=0.5 and 2%[m] n_t=10	v_p^0=0.01~1% n_t=4~20 p_1=0 and 0.5[n]	v_p^0=1 and 5% n_t=15 p_1=0 and 0.5	v_p^0=1% n_t=20 p_1=0 and 0.6	v_p^0=1% n_t=20 p_1=0 and 0.6
Number of fractionation runs	2	1	2	17	3	7	5
Analytical methods	S [o]	S	S	S and K [p]	S and K	S	S
Evaluation	q [q]	q	q	E [r] and E'[s]	E and E'	q	q

a) Flory-Huggins, b) concentration dependence of polymer-solvent interaction parameter expressed by p_1 was taken into account for the original FH theory, c) successive precipitation fractionation, d) successive solution fractionation, e) number-average degree of polymerization, f) weight-average degree of polymerization, g) Schulz-Zimm distribution, h) Wesslau distribution, i) log normal distribution, j) generalized exponential distribution, k) volume ratio of polymer-lean phase to polymer-rich phase, l) total number of fractions in a run, m) initial polymer volume fraction, n) p_1 is defined in eq 2, o) Schulz, p) Kamide, q) qualitative, r) E is defined by eqn (2.102), s) E' is defined by eqn (2.107).

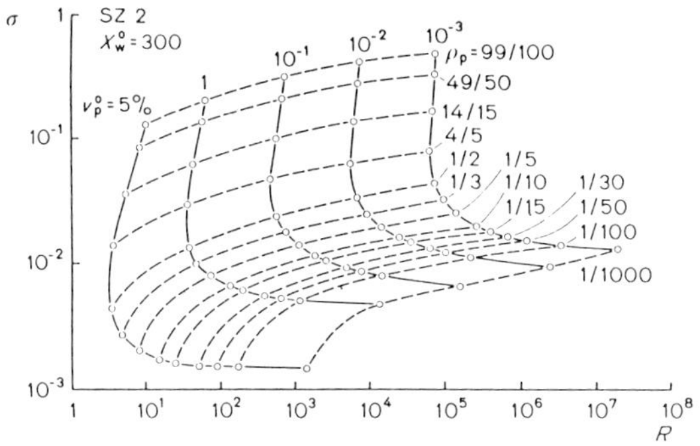

Fig. 2.4 Relations among initial concentration $v_p{}^0$, fraction size ρ_p, partition coefficient σ and phase volume ratio R of the polymer-lean to -rich phase for the first fractionation step: Original polymer, Schulz-Zimm type ($X_w{}^0/X_n{}^0 = 2$, $X_w{}^0 = 300$); $p_j = 0$ ($j = 1, \cdots, n$) and $k' = 0$.

increases monotonically and R decreases markedly where ρ_p is small, then increases very slightly after passing through the minimum value of R, which is located at about $\rho_s = 1/2$. Taking into consideration the fact that in SSF the polydispersity of the fraction always diminishes without limit with a decrease in its amount (ρ_s), it becomes evident that σ rather than R plays the dominant role in controlling the polydispersity of the fraction (ref. 43). It can easily be shown from Fig. 2.4 that despite the theoretical complications, for given external operating conditions, σ together with R is unambiguously determined, i.e., a given pair of $v_p{}^0$ and ρ corresponds rigorously to only one pair of σ and R provided that the molecular weight distribution of the original polymer and the concentration dependence of polymer / solvent interaction parameter are given in advance (ref. 42).

The parameters σ and R can vary within a network structure illustrated in Fig. 2.4 corresponding to a change in $v_p{}^0$, ranging from $10^{-3}\%$ to 5%, and to a change in ρ_p from 1/1000 to 99/100. In other words, varying the values of σ and R outside the network is absolutely impossible from the theoretical view point. This shows that the assumption that σ are independent as assumed R, in the rough approximate calculations published hitherto, are evidently unrealistic (ref. 42).

Inspection of eqns (2.43a) and (2.43b) indicates that the high molecular weight component of $g_0(X_i)$ is easy to separate into polymer-rich phase (PRP) and the low molecular weight component of $g_0(X_i)$ into polymer-lean phase (PLP). Fig. 2.5 shows an example

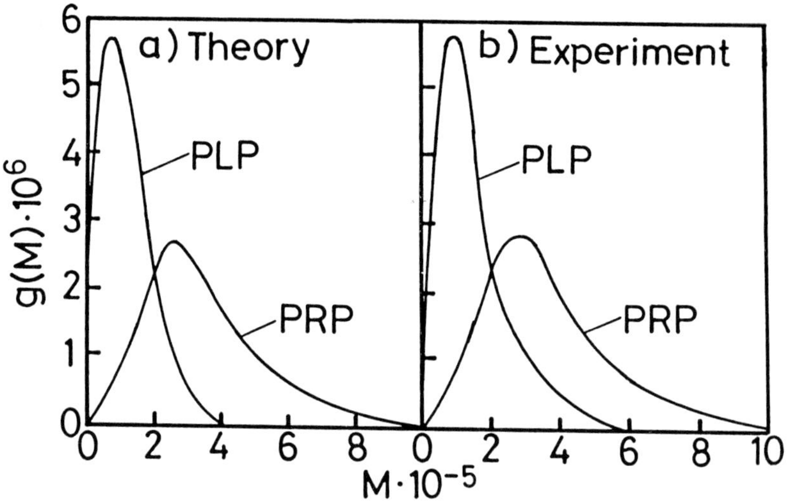

Fig. 2.5 $g_{(1)}(M)$ and $g_{(2)}(M)$ phase-separated from quasi-binary solution consisting of polystyrene ($M_w = 2.39\times10^5$, $M_w/M_n = 2.7$, Schulz-Zimm type distribution) in methylcyclohexane; $v_p{}^0 = 0.94\times10^{-2}$, $\rho_p = 0.52$, $p_1 = 0.7$ and $p_2 = \cdots = p_n = 0$ and $k' = 0$ (theoretical curve).

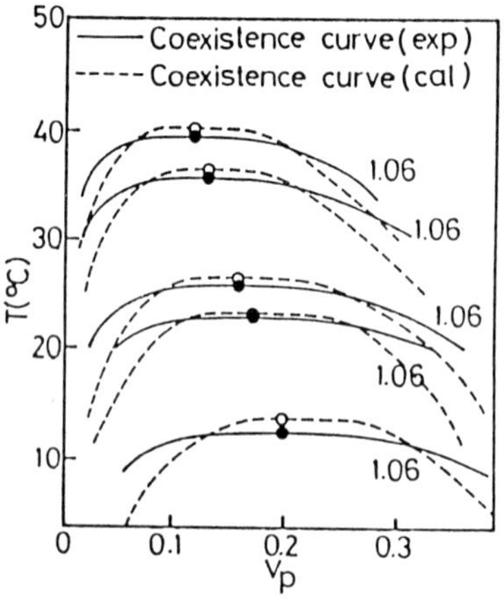

Fig. 2.6 Coexistence curve of polystyrene (Schulz-Zimm type, $M_w = 1.02\times10^4 \sim 4.64\times10^4$, $M_w/M_n = 1.06$) in methylcyclohexane: $v_p{}^0 = 0.117 \sim 0.199$, $p_1 = 0.602$, $p_2 = 0.234$ and $p_3 = \cdots = p_n = 0$ and $k' = 0$; solid line, experimental result; broken line, theoretical curve.

of MWD in both phases (i.e., $g_{(1)}(X_i)$ and $g_{(2)}(X_i)$) when the solutions consisting of polystyrenc (PS; assumed to have SZ type

MWD with the weight-average molecular weight $M_w = 28.9 \times 10^4$ and $M_w/M_n = 2.7$) dissolved in methylcyclohexane (MCH; $v_p{}^0 = 0.94\%$, $k' = 0$, $p_1 = 0.7$ and $p_2 = \cdots = p_n = 0$) is phase-separated ($\rho_p = 0.52$). The experimental MWD in both phases can be reasonably represented by the theory (ref. 44).

Utilizing eqns (2.34), (2.49a) and (2.49b), we can also calculate coexisting curve. Solid lines in Fig. 2.6 show the coexistence curves of the PS / MCH system obtained by Dobashi, Nakata and Kaneko (ref. 45) where the weight-average molecular weights of PS are $M_w = 1.02 \times 10^4$, 1.61×10^4, 2.02×10^4, 3.49×10^4 and 4.64×10^4 and the initial concentrations $v_p{}^0$ are 0.199, 0.1719, 0.159, 0.1293 and 0.117, respectively. Broken lines in Fig. 2.6 are the theoretical curves of the coexistence curves of PS / MCH system with $p_1 = 0.602$, $p_2 = 0.234$ and $k' = 0$ (see, Table 2.24) assumed. Temperature dependent parameters of χ_0 are $a = 0.25$ and $b = 85.05$ (see, eqn (2.7)) evaluated by substituting $\theta = 340.2$K and $\psi = 0.25$ (see, Table 2.24) into eqn (2.217) (or eqns (2.220a) and (2.220b)).

The reliability of the phase equilibrium and fractionation theory has been examined in detail by comparing it with the experimental data obtained by SPF and SSF (see, section 2.4) as well as with the experimental phase separation data of polydisperse polymers in a single solvent. Noda et al. (ref. 30) carried out experiments on both SPF and SSF, using mixtures of two samples of poly(α-methylstyrene) having narrow molecular weight distribution, and compared the resulting data with the theory of Kamide et al. (refs 21-23). The molecular weight distribution of the two samples were sufficiently sharp that the systems could be assumed to be binary mixtures in a single solvent. Consequently, we were able to compare directly, and with clarity, the experimental MWD and true MWD with the theory since the molecular weight distribution of the original binary polymer mixture and fractions are given if the mixed ratios are known (refs 21-23). The molecular characteristics of the samples are listed in Table 2.2.

The binary mixtures used for fractionation experiments, M1, M2 and M3 in Table 2.2, were prepared by mixing two out of four monodisperse polymers. The mixed ratios in weight and the average molecular weights as calculated from the ratio, are also shown in Table 2.2. The samples M1 and M2 are almost identical.

The average polymer-to-solvent molar volume ratio of poly(α-methylstyrene) to cyclohexane X is evaluated by $X = (M/d_p)/(M_s/d_s)$

TABLE 2.2
Molecular characteristics of monodisperse poly(α-methylstyrene) and the binary mixture.

Sample code	$M_n \times 10^{-4}$	$M_v \times 10^{-4}$	$M_w \times 10^{-4}$	X_v[a]	M_w/M_n
a) Monodisperse sample					
α-8	–	20.8		1817	
α-15	38.5	41.0		3582	
α-004	34.2	34.1	34.2	2981	1.00
α-103	50.0	48.7		4255	
b) Binary mixture					
M1: α-8/α-15 (0.514/0.486, w/w)	(27.3)	–	(30.6)	(2633)	(1.12)
M2: α-8/α-15 (0.500/0.500, w/w)	(27.6)	–	(30.9)	(2658)	(1.12)
M3: α-004/α-103(0.501/0.501, w/w)	(40.1)	–	(41.4)	(3602)	(1.03)

[a] $X_v = (M_v/d_p)/(d_s/M_s) = (M_v/1.06)/(0.778/84)$.

where M and M_s and also d_p and d_s are the molecular weights and the densities of the polymer and cyclohexane ($M_s = 84$, $d_p = 1.06$ and $d_s = 0.778$), respectively.

The viscosity-average molecular weight may be satisfactorily used in place of the weight-average molecular weight when the data are compared with the theory since the ratios of the weight-average molecular weight to the number-average molecular weight, M_w/M_n of the fractions are closed to unity as shown in Table 2.2. The weight ratio of two components was determined from the peak area of the Schlieren pattern in sedimentation. Since the ratio of the peak area of two components is not equal to their weight ratio, due to the Johnston-Ogston effect, the weight ratio was determined using the experimental linear relationship between the weight ratio and the peak area ratio obtained by ultracentrifugation of the same polymer mixtures of known composition under the same sedimentation conditions for measuring weight ratio. The value of M_w/M_n was calculated from the weight ratio of two components, neglecting the molecular weight distribution of each component, since the samples were assumed to be monodisperse. The sedimentation experiment was carried out in cyclohexane at 308K with a Beckman Spinco Model E ultracentrifuge at a speed of 59780 rpm.

Sample M1 was dissolved in cyclohexane at 312K. The initial polymer volume fraction of the solution was $0.08_8\%$. The solution was cooled down to the designed phase separation temperature. After keeping the temperature constant for 16 to 24 hours the polymer-rich phase was isolated. The above procedure was carried out successively and 12 fractions were obtained

Successive solution fractionation was carried out using a column, 4cm in inner diameter and 60cm in length, in which about 850g glass beads of 100~150 mesh were packed. The temperature was kept constant by circulating, at constant temperature, water in a water jacket. Sample M2 or M3 was dissolved in cyclohexane whose temperature was about 318K. The solution, whose polymer volume fraction was 0.004, was poured into the column and cooled down to enable the polymer to deposit on the glass beads. Then the column was heated to the designed phase separation temperature. After keeping the temperature constant for about 2 hours, 300ml of the polymer-lean solution was then allowed to flow out by pouring the same amount of cyclohexane into the column. The above procedure was repeated successively.

The fractionation experiments described above were simulated according to the method of Kamide-Sugamiya (refs 21-23). In this simulation, the differential weight distribution of the molecular weight in each component $g_0(X)$, (a component of the binary mixture), was assumed to follow the Poisson distribution (see, eqn (2.58)). The computer simulation was carried out for $p_1 = 0$, 0.6 and 1.2 in SPF so as to clarify the effect of the concentration dependence of χ on the fraction, but only for $p_1 = 0.6$ in SSF, since the value of p_1 has little effect on the simulation result for SSF. To clarify the effect of the initial polymer concentration on SSF, simulations were carried out not only for the experimental initial volume fraction ($v_p^0 = 0.40\%$) but also for $v_p^0 = 0.1$ and 1.0%. Moreover, it should be noted that the fraction size ρ, that is, the weight ratio of each fraction to the original sample, is assumed to be constant in the present simulation program because the improvement of the present program for a variable fraction size exceeds the capacity of the computer. Though not impossible, it is very difficult to keep the fraction size constant in the actual experiment. Thus, the fraction size in the simulation is selected so as to be close to the experimental values in the majority of fractions. The number of fractions for M2 in the simulation of SSF was 12, while the experimental number was 13.

The cumulative weight fraction $I(M_{w,j})$ (or $I(M)$) vs. $M_{w,j}$ (M_w of j-th fraction) (or M) relationship, i.e., molecular weight

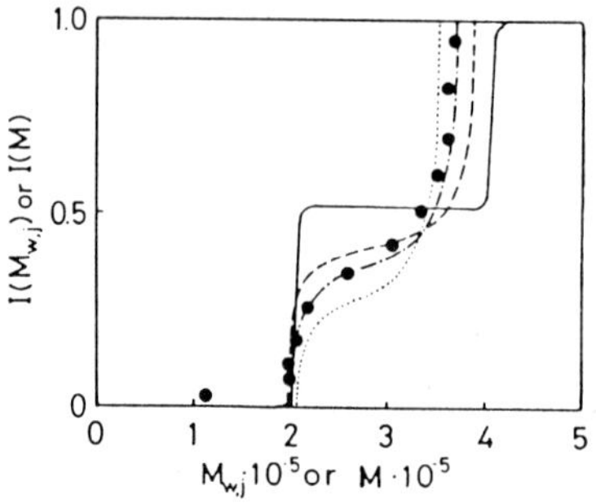

Fig. 2.7 Cumulative weight fraction $I(M_{w,j})$ vs. $M_{w,j}$ relationship of the mixture of poly(α-methylstyrene) ($M_n = 20.8\times10^4$ and 38.5×10^4, 0.514/0.486 (wt/wt)) obtained by SPF: ●, experimental results; full line, "true" $I(M)$; dotted line, theory ($k'=0$ and $p_j=0$); chain line, ($k'=0$ and $p_1=0.6$); broken line, theory ($k'=0$ and $p_1=1.2$).

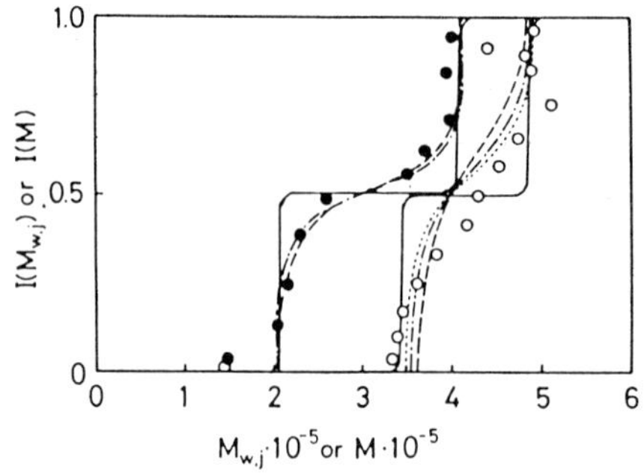

Fig. 2.8 Cummulative weight fraction $I(M_{w,j})$ vs. $M_{w,j}$ relationship of the mixture of poly(α-methylstyrene) obtained by successive solution fractionation: ●, experimental results (mixture of $M_n = 20.8\times10^4$ and 38.5×10^4; 0.500/0.500 (wt/wt)); ○, experimental results (mixture of $M_n = 34.2\times10^4$ and 50.0×10^4; 0.501/0.499 (wt/wt)); full line, "true" $I(M)$; dotted line, theory ($k'=0$ and $p_j=0$); chain line, theory ($k'=0$, $p_1=0.6$ and $v_p^0=0.001$); broken line, theory ($k'=0$, $p_1=0.6$ and $v_p^0=0.01$).

distribution of the binary mixtures calculated according to the method of Schulz (refs 46,47) are shown in Fig. 2.7 and 2.8 where the solid curve denotes the "true" $I(M_{w,j})$ (or $I(M)$) vs. $M_{w,j}$ (or M) relationship calculated on the assumption that the MWD of each component is the Poisson (see, eqn (2.58)) distribution. Here, $I(M_{w,j})$ is given by eqns (2.93) and (2.94) (refs 46,47) in section 2.4.

As shown in Fig. 2.7, the experimental molecular weight distribution obtained from SPF according to the method of Schulz (refs 46,47) is in good agreement with the theoretical curve for

$p_1 = 0.6$ (with $p_2 = \cdots = p_n = 0$) except for a low molecular weight tail part, but deviates from the "true" distribution. The disagreement found in the tail part is understandable and not serious enough to prevent comparison between the experimental results and theory since the low-molecular-weight tail part was not included in the Poisson distribution used in the present computer experiment. Thus, it may be concluded that the simulation result can satisfactorily reproduce the experimental molecular weight distribution if $p_1 = 0.6$ is selected (ref. 32).

In 1944, Flory (ref. 48) defined a parameter ε representing the fractionation efficiency for the molecular weight fractionation (or separation efficiency) based on the solubility difference by

$$\varepsilon = \left[\frac{df_{x(2)}}{d\ln X}\right]_{X=X_p} = \frac{\ln R}{4} \tag{2.68}$$

where $f_{x(2)}$ is the fraction of a given X-mer remaining in the polymer-rich phase and X_p, the value of X at $f_{x(2)} = 0.5$. Eqn (2.68) indicates that a large R, accordingly lower v_p^0, is highly desirable to carry out the fractionation run with high efficiency. Later, Kawai (refs 49,50) showed from some simplified calculations that, in order to make ε large, a smaller ρ_p is also quite effective. On the basis of the results obtained from the systematic computer experiments, Kamide and Nakayama (ref. 51) concluded that phase separation with a large R does not always furnish the narrow MWD and in the case of SPF a decrease in ρ_p unavoidably brings about an increase in the breadth of MWD in the fractions, particularly in smaller ρ_p regions. The validity of the above conclusion was then experimentally ascertained by Kamide and his coworkers (refs 29,52,53). Kamide and his collaborators (refs 51,54) preferred to use a parameter ε', defined by eqn (2.69) in place of ε as a fractionation efficiency parameter.

$$\varepsilon' = \left[\frac{df_{x(2)}}{dX}\right]_{X=X_p} = \frac{\sigma}{4}. \tag{2.69}$$

It should be kept in mind that both ε and ε' are unfortunately inadequate for quantitative discussion as to the ease of separating the components with different X from their mixtures by the fractionation method (ref. 54).

Huggins (ref. 55) defined other parameters, $\varepsilon_{(1)}$ and $\varepsilon_{(2)}$ as efficiency parameters for mixtures of molecules with the same M_w of two components with different X. These parameters are evidently limited in their applicability to only specific binary mixtures he studied (ref. 55).

As is evident from the above discussion, the various efficiency parameters in early studies are only approximate and have a rather ambiguous character, and it is necessary to employ a much more rigorous parameter, like the resolving power R_p, most commonly used and defined in optics (ref. 54). Kamide and Miyazaki (ref. 54) focused attention on this important point, defining at the first step new efficiency parameters such as the resolving power R_p (eqn (2.72)) and then clarifying the effect of $v_p{}^0$ and the total number of fractions on these parameters in the fractionation by solubility difference, using the binary mixture with different X.

Firstly, consider binary mixtures with two different X (X_1 and X_2, $X_1 < X_2$, the suffix indicating the component). Write $W_1{}^0$ for the weight fraction of the component 1 and $W_2{}^0$ ($= 1 - W_1{}^0$) for the weight fraction of the component 2. In the case of equal weight mixtures, we obtain $W_1{}^0 = W_2{}^0 = 0.5$.

A solution of the above binary mixtures is cooled down until phase separation occurs. Here, the fractionating conditions are carefully chosen in such a manner that the weights of the polymers in polymer-lean and -rich phases are equal. The weight fractions of components 1 and 2, dissolved in a polymer-lean phase and a polymer-rich phase, to the total polymer in a starting solution are denoted as $W_{1(1)}$, $W_{2(1)}$, $W_{1(2)}$ and $W_{2(2)}$, respectively (Here, $W_{1(1)} + W_{2(1)} + W_{1(2)} + W_{2(2)} = 1$). Thus, the relation,

$$\frac{\sum_{i=1}^{2} W_{i(1)}}{\sum_{j=1}^{2}\sum_{i=1}^{2} W_{i(j)}} = \rho_s (= 0.5) = 1 - \rho_p \tag{2.70}$$

defines the fractionation conditions (ref. 54). We define the resolving power R_p in the fractionation as the ratio $X_1/(X_2 - X_1)$ of the original polymer, which, under the given conditions, provide fractions having the following heterogeneity.

$$\frac{W_{1(1)}}{\sum_{i=1}^{2} W_{i(1)}} \geq 0.99 \quad \text{or} \quad \frac{W_{2(2)}}{\sum_{i=1}^{2} W_{i(2)}} \geq 0.99. \qquad (2.71)$$

In other words, R_p is defined by (ref. 54),

$$R_p = X_1/(X_2 - X_1). \qquad (2.72)$$

A large R_p value indicates that the fractionation makes possible the separating of each component from the mixture, even if the molar volume ratio of these components X_1 and X_2 are very close to each other. In eqn (2.71), we adopt for convenience a 99% level purity, which may be changed as necessary.

Fig. 2.9 shows the relationships between X_1 and $v_p{}^0$, observed in the phase separation of a 1:1 (by weight) mixture in solvent, yielding a constant R_p. Fig. 2.9a, b and c correspond to the case of $p_1 = 0$, 0.8 and 1.6, respectively. It is noteworthy that for a given R_p, the X_2 value can be clearly determined. With an increase in the X value of the original polymer (X_1 and X_2), the resolving power decreases remarkably, compared at the same $v_p{}^0$. The extent to which R_p is lowered is markedly influenced by the solvent nature (p_1); a small p_1 gives a substantially decreased the value of R_p. For example, in the case of $p_1 = 0$, to obtain

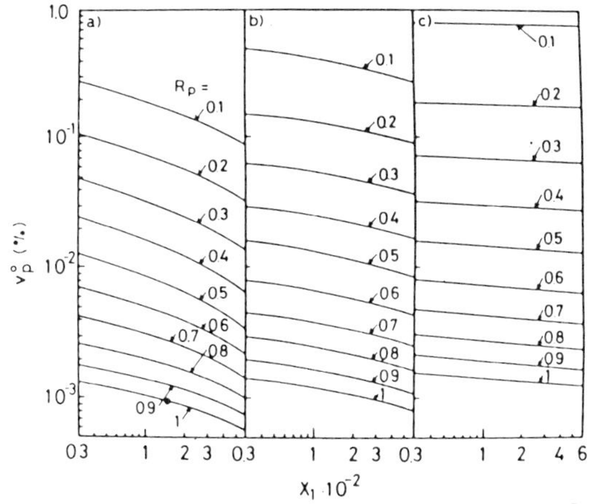

Fig. 2.9 Effect of concentration dependence parameter p_1 on the resolving power $R_p - X_1$ relations: $W_1{}^0 = W_2{}^0 = 0.5$ (binary equal weight mixture); the fraction size, $\rho_p = \rho_s = 0.5$; a) $p_1 = 0$, b) $p_1 = 0.8$ and c) $p_1 = 1.6$ ($k' = 0$ and $p_2 = \cdots = p_n = 0$).

$$\frac{W_{1(1)}}{\sum_{i=1}^{2} W_{i(1)}} = \frac{W_{2(2)}}{\sum_{i=1}^{2} W_{i(2)}} \geq 0.99 \tag{2.73}$$

by a single-phase separation step, from a solution of a binary mixture ($X_1 = 30$, $X_2 = 60$, $W_1^0 = W_2^0 = 0.5$), it is necessary to prepare $1.3 \times 10^{-3}\%$ solution (ref. 54). This value is too low to be practical since the concentration range in which the practical fractionation can be relatively easily carried out is 0.1~several %. The possible combination of X_1 and X_2, constituting a binary mixture, from which the fractions with 99% purity can be separable, is infinite in number; for illustration, when a solution, in which a given binary mixture is dissolved, of a given concentration is to be fractionated, R_p increases significantly with an increase in p_1, indicating that if a solvent with a large p_1 value is employed, high resolving power can be obtained. As is evident from Fig. 2.9, very high resolving power cannot be expected for a single step phase separation (ref. 54).

Kamide, Miyazaki and Abe performed experiments of phase separation and fractionation (SSF and SPF) (see, Fig. 2.22) on solutions of atactic polystyrene (PS) in methylcyclohexane (MCH) or cyclohexane (CH) systems (refs 29,31). For this purpose, the atactic polystyrene manufactured by Asahi Chemical Ind., Tokyo, was employed. The specific material of their study is known commercially as Styron 666. It was purified before use as

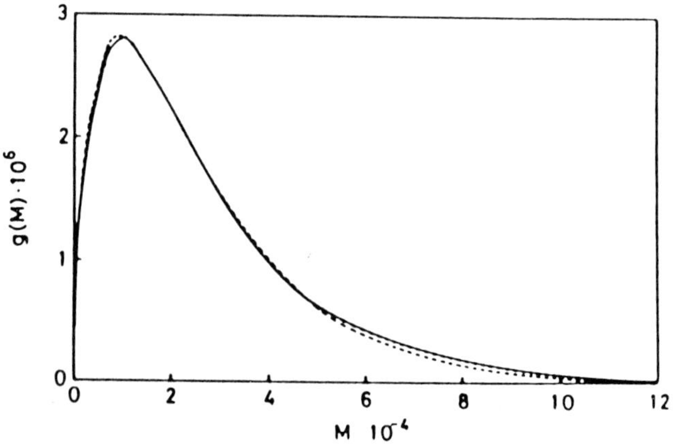

Fig. 2.10 Molecular weight distribution curve g(X) of the polymer sample used for the phase separation experiment: Full line, experimental curve; broken line, Schulz-Zimm distribution curve with $M_w = 24.0 \times 10^4$ and $M_w/M_n = 2.8$.

described in detail before (ref. 52) and was characterized by the following average-molecular weight; the weight-average molecular weight $M_w = 23.2 \times 10^4$ (by light scattering) and 23.9×10^4 (by gel permeation chromatography (GPC)), and the number-average molecular weight $M_n = 8.9 \times 10^4$ (by membrane osmometry) and 8.6×10^4 (by GPC). The molecular weight distribution (MWD) of the polymer estimated by GPC was accurately separated by the Schulz-Zimm distribution with $M_w = 24.0 \times 10^4$ and $M_w/M_n = 2.8$, as shown in Fig. 2.10.

The PS sample was dissolved in MCH at 343K and in CH at 308K to give the starting solution. The solutions prepared by the procedure described above were cooled down at a rate of 4K/h to the designed phase separation temperature T_p. After allowing the solution to stand for 16 to 24h at T_p to reach equilibrium, the volume of the upper phase (i.e., the polymer-lean phase), which was decanted from the lower phase (i.e., the polymer-rich phase), was measured. The remaining phase was weighed. The weight of polymers in both phases was determined after evaporating solvent under reduced pressure. The volume of the polymer-rich phase was calculated from the total weight of the phase and the polymer weight dissolved in it by assuming the additivity rule of polymer and solvent. The volume ratio R, the fraction size ρ, and the polymer volume fraction $v_{p(1)}$ and $v_{p(2)}$ were calculated. M_w and M_w/M_n of the polymers in the two phases were calculated by the GPC elution method (refs 52,53), in which the columns were eluted with tetrahydrofuran at 298K. The width in the GPC curves obtained was corrected, where significant, according to Tung's procedure (ref. 56). The GPC trace was transformed into a MWD curve by the method of Yau and Fleming (ref. 57). From the MWD curve, M_w, M_w/M_n and the standard deviation σ' were determined. The values of σ and ρ_p were elucidated through a direct GPC experiment for polymers in both phases. In this manner it will become possible to verify the real existence of the parameter σ from the experimental view point.

In Fig. 2.11 MWD curve of the polymer remaining in the polymer-rich phase is compared with theoretical curves assuming varying values of the parameter p_1 in eqn (2.4) (here, $k' = 0$ and $p_2 = \cdots = p_n = 0$ are assumed), when atactic PS / MCH solution is cooled to bring about the two-phase separation. As just described before for computer experiment on model polymer solutions ($k' = 0$ and $p_2 = \cdots = p_n = 0$ in eqn (2.4)), the MWD curve markedly depends on the p_1 values chosen (ref. 29). It is generally accepted from these figures that the practical MWD curve of the precipitate for

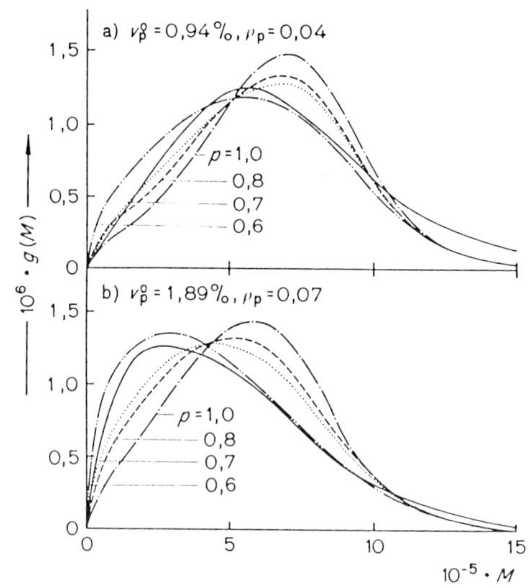

Fig. 2.11 Molecular weight distribution of polymer partitioned into polymer-rich phase $g_{(2)}(M)$ obtained from a solution of polystyrene in methylcyclohexane: Full line, experiment; double chain line, computer simulation with $p_1 = 0.6$; dotted line, computer simulation with $p_1 = 0.7$; broken line, computer simulation with $p_1 = 0.8$; chain line, computer simulation with $p_1 = 1.0$; $k' = 0$ and $p_2 = \cdots = p_n = 0$; a) $v_p^0 = 0.0094$, $\rho_p = 0.04$; b) $v_p^0 = 0.0189$, $\rho_p = 0.07$.

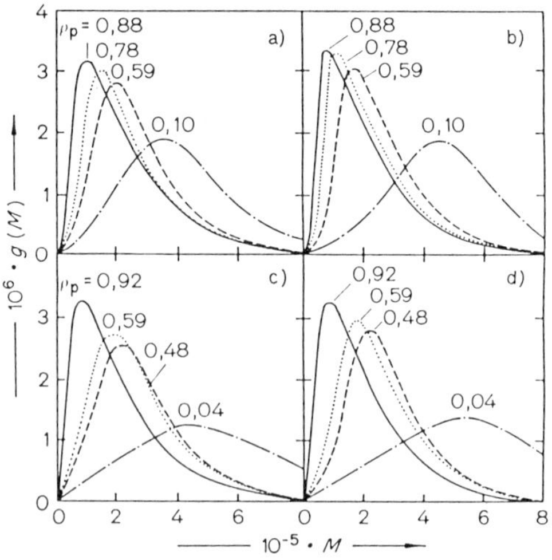

Fig. 2.12 Differential MWD curve of the polymer in polymer-rich phase $g_{(2)}(M)$, precipitated from 0.0047 [a) and b)] and 0.0094 [c) and d)] solutions for different ρ_p: a) and c) experiment for polystyrene in methylcyclohexane; b) and d) computer simulation with parameter $p_1 = 0.7$ (cf. eqn (2.4); $k' = 0$ and $p_2 = \cdots = p_n = 0$); ρ_p is indicated on curves.

PS / MCH can be reasonably approximated by a theoretical curve with $p_1 = 0.7$ (dotted line) (ref. 29). An adequate value of p_1, with which the experimental MWD coincides with the theoretical one, clearly varies in the range $p_1 = 0.6 \sim 0.7$ depending on the operating conditions (v_p^0 and ρ_p) for PS / MCH and PS / cyclohexane (CH) systems. The large discernible solvent effect on the MWD of the fractions is evident and an increase in p_1 value is just equivalent to lowering the v_p^0 (ref. 29).

Fig. 2.12 and Fig. 2.13 show the effect of the ρ_p and ρ_s on the MWD of the polymers in both phases, respectively. As ρ_p decreases, the peak height in the MWD curve decreases, shifting its peak to the large molecular weight region in the polymer-rich phase. This is particularly noticeable in the small ρ_p region (ref. 29). While in the polymer-lean phase, as the fraction size ρ_s decreases the peak height in the MWD curve increases, shifting its peak to the lower molecular weight region. Fig. 2.12 and Fig. 2.13 indicate that the sharpest 1st fraction can be obtained if $\rho_p \to 1.0$ in SPF and $\rho_s \to 0$ in SSF (ref. 29).

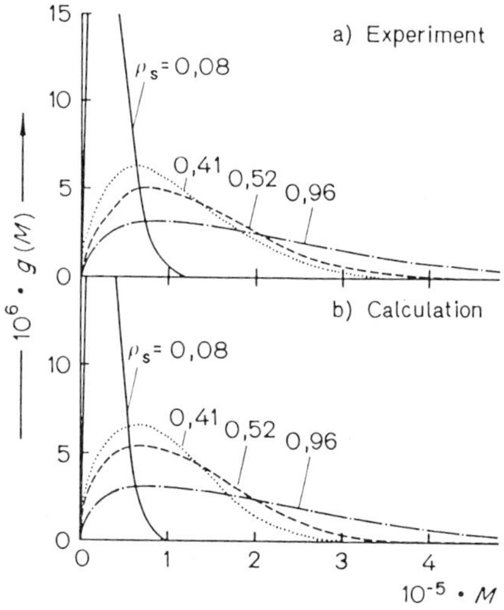

Fig. 2.13 Differential MWD curve $g_{(1)}(M)$ of the polymer in polymer-lean phase obtained from $v_p^0 = 0.0094$ solution for different ρ_s: a) experiment for polystyrene in methylcyclohexane; b) computer simulation with $p_1 = 0.7$ (cf. eqn (2.4); $k' = 0$ and $p_2 = \cdots = p_n = 0$); ρ_s is indicated on curves.

Fig. 2.14 shows the plot of the polymer fraction in the polymer-rich phase $v_{p(2)}$ vs. ρ_p. The broken, chain and full lines mean the theoretical calculations at v_p^0 1.89, 0.94 and 0.47%,

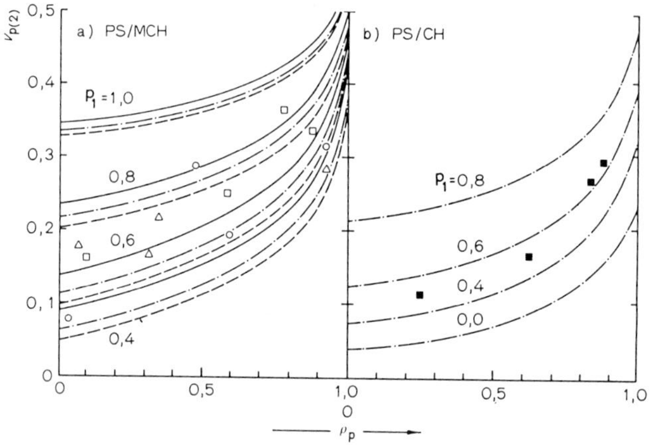

Fig. 2.14 Effect of the relative amount of polymer in polymer-rich phase ρ_p on the polymer volume fraction $v_{p(2)}$ remaining in the polymer-rich phase: Full line, theoretical curve at initial polymer volume fractions $v_p^0 = 0.0047$; chain line, 0.0094; broken line $v_p^0 = 0.0189$; numbers on curves denote the values of p_1 (cf. eqn (2.4); $k' = 0$ and $p_2 = \cdots = p_n = 0$); △, □, ○, experimental data for $v_p^0 = 0.0189$, 0.0094 and 0.0047: a) polystyrene (PS) / methylcyclohexane; b) PS / cyclohexane.

respectively. As ρ_p is increased the value of $v_{p(2)}$ also increases slowly. The magnitude of $v_{p(2)}$ depends largely on p_1 (ref. 29). Although the experimental accuracy of $v_{p(2)}$ is not high, the experimental points scatter around the theoretical curves of $p_1 = 0.6 \sim 0.8$ for PS / MCH and $p_1 = 0.6$ for PS / CH (ref. 29).

The volume ratio R of the two phases decreases drastically as ρ_p increases. Fig. 2.15 is an illustrative case for a 0.94% solution. The difference in the experimental R values and the theoretical curve with $p_1 = 0.7$ for PS / MCH and $p_1 = 0.6$ for PS / CH is within experimental uncertainty in the range $v_p^0 = 0.47$ to 1.89%, irrespective of v_p^0.

The experimental values of $v_{p(1)}$, $v_{p(2)}$, R, σ and M_w/M_n coincide well with theoretical predictions assuming $p_1 = 0.6 \sim 0.7$ for PS / MCH and $p_1 =$ ca.0.6 for PS / CH, respectively (ref. 29). In this way both R and $v_{p(2)}$ depend ordinarily on the nature of solvent. On the contrary, by the use of figures like Figs 2.14 and 2.15 the magnitude of p_1 can be more precisely estimated in such a manner that the experimental point coincides well with the theoretical value at a given ρ_p (ref. 29). Table 2.3 summarizes

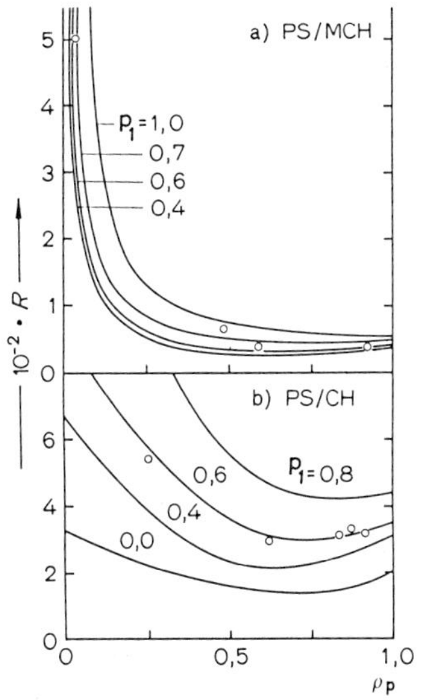

Fig. 2.15 Effect of the relative amount of polymer in polymer-rich phase ρ_p on the volume ratio of two phases R for a) polystyrene (PS) / methylcyclohexane and b) PS / cyclohexane for initial polymer volume fraction $v_p{}^0=0.0094$: Full line, theoretical curves; ○, experimental data for initial polymer volume fraction $v_p{}^0=0.0094$; numbers on curves denote p_1 value ($k'=0$ and $p_2=\cdots=p_n=0$).

TABLE 2.3

Values of p_1 evaluated by separation phenomena and by successive fractionation for the systems polystyrene (PS) / methylcyclohexane (MCH) and PS / cyclohexane (CH).

From[a]		p_1	
		PS/MCH	PS/CH
MWD	(Fig. 2.12)	0.6~0.7	ca. 0.6
$v_{p(2)}$	(Fig. 2.14)	0.66 ± 0.10	0.56 ± 0.05
R	(Fig. 2.15)	0.69 ± 0.10	0.58 ± 0.05
SPF	(Fig. 2.23a)	0.74 ± 0.10	—
SSF	(Fig. 2.23b)	0.72 ± 0.10	—

a) SPF and SSF; successive precipitational and successive solutional fractions, respectively; $v_{p(2)}$, volume fraction of polymer-rich phase; R, volume ratio of polymer-lean phase to polymer-rich phase.

the values of p_1 evaluated. The table includes the values estimated from the plots of X_w/X_n vs. X_w for a series of fractions separated in a successive fractionation run (see, Fig. 2.23). For example, $p_1 = 0.56 \sim 0.60$ was determined for PS / CH, regardless of ρ_p (ref. 29). The reliability of the p_1 values thus estimated can be examined in detail by using other experimental information.

Table 2.3 indicates that at least χ parameter can not be regarded as constant although the accuracy of estimation of p_1 is not so high, and of course, a higher term than p_1 can not be, even roughly, evaluated from $v_{p(2)}$, R, the molecular weight distribution of the polymers partitioned in both phase, and X_w/X_n vs. X_w plot for fractionation run.

In the original Flory-Huggins theory, χ parameter was taken as constant. During the late 1940s and early 1950s, vapor pressure measurements and isothermal distillation equilibration were carried out for some polymer / solvent systems, including polydimethylsiloxane (PDMS) in benzene (ref. 58), Polystyrene (PS) in methylethylketone (MEK) (ref. 59), PS in toluene (ref. 59) and rubber in benzene (ref. 60). Flory constructed plots of χ against v_p for these systems (Fig. 111 of ref. 11) and he stated that "in no other system so far investigated is the agreement so good as for the rubber / benzene" system, "for which χ is remarkably constant over a very wide concentration range", and that "in those cases where either the polymer unit or the solvent possesses a dipole, as in the PDMS / benzene and in PS / MEK systems, χ appears to vary throughout the concentration range". Flory also agreed that the available data is too little to justify generalization.

Surprisingly, Flory's figure (Fig. 111 of ref 11) has been occasionally cited without serious modifications in many text books published later. It should be noted that the experimental data cited in Flory's book is less accurate: p_1 values estimated from Fig. 111 of ref. 11 are 0.15 for PS / MEK, 0.17 for PS / toluene and these values are compared with the values accepted as most probable (0.618 and 0.494), both evaluated by analyzing the critical solution points (see, Table 2.24).

Now, the concentration dependence of χ parameter (eqn (2.1)) can also be estimated by other methods than the phase equilibria as follows:

<u>(a) Vapor pressure, isothermal distillation equilibrium (ref. 58) and vapor sorption isotherm (ref. 61)</u>

$$\ln(p/p_0) = \ln(1 - v_p) + (1 - 1/X_n)v_p + \chi v_p^2 . \qquad (2.74)$$

(b) Membrane osmotic pressure (ref. 58)

$$\pi = (\tilde{R}T/v_1)\{\ln(1-v_p) + (1-1/X_n)v_p + \chi v_p^2\}. \qquad (2.75)$$

Here, v_1 is the molar volume of the solvent.

(c) Ultracentrifuge (sedimentation equilibrium) (ref. 63)

$$M(1-\bar{v}\rho)\omega^2 r = (d\mu/dw)(dn/dr)^{(\omega)}/(dn/dw). \qquad (2.76)$$

$\bar{v}$ is the partial specific volume of the polymer, ω, the rotor speed, ρ, the density of the solution, μ, the chemical potential of the solute (in this case, polymer), w, the weight fraction of polymer in the solution, n, the refractive index and dn/dw, the refractive index increment.

(d) Light scattering (refs 64,65)

$$\frac{Kc}{\Delta R_\theta} = -\frac{1}{\tilde{R}Tc}\left(\frac{\partial \Delta\mu_0}{\partial c}\right)_{T,P} = \frac{1}{M_w p(\theta)} + 2\frac{\bar{v}^2}{N_A V_0}\left(\frac{1}{2} - \chi\right)c \qquad (2.77)$$

with

$$K = \frac{2\pi^2 n^2}{N_A \lambda_0^4}\left(\frac{\partial n}{\partial c}\right)^2. \qquad (2.78)$$

ΔR_θ is the difference of Rayleigh ratio between the solution and the solvent, $p(\theta)$, the scattering function (θ is the scattering angle), c, the polymer concentration (in weight), n, the refractive index of the solution, $(\partial n/\partial c)$, the refractive index increment, N_A, Avogadro number, λ_0, the wave length of the incident light in the solution, $\bar{v}$, the specific volume of the solution and V_0, the molar volume of the solvent.

(e) Temperature dependence of relative amount of precipitate ρ_p, combined with the cloud point curve (simply referred to as the cloud point curve method) (see, section 2.6).

(f) Critical concentration

(i) Comparison of theoretical critical concentration v_p^c(theo) (see eqns (2.121) and (2.122)) with the experimental value v_p^c(exp) [Kamide-Matsuda] (ref. 3).

(ii) Analysis of X_w, X_z, v_p^c and T_c data (see, eqn (2.229) and (2.230)) [Koningsveld-Kleintjens-Shultz] (ref. 16).

Methods d and e will be described in more detail in later sections.

The reliability of p_1 value estimated from the phase separation experiment can be examined by comparing with that by other methods: For the system of PS / CH, Kuwahara et al. (ref. 66) evaluated from a critical miscibility study $g_{00} = 0.5196$, $g_{01} = 67.50$ (see, eqn (2.25)), $g_1 = 0.2369$ and $g_2 = 0.0863$. In the case of $j = 1$, eqn (2.29) is simplified as

$$p_1 = 2(g_1 - g_2)/(g_0 - g_1). \tag{2.79}$$

Substitution of these values into eqns (2.25) and (2.79) yields $p_1 = 0.59$ at 283K and 0.60 at 293K, respectively, which are in good agreement with the values of Table 2.4 by Kamide et al. (refs 29,31). Here, it is of interest to point out that the phase

TABLE 2.4

Parameters a', b', p_1 and p_2 in eqn (2.7) for atactic polystyrene / cyclohexane system (T = 299 K).

Author(s)	Method	a'	b'	p_1	p_2	ref.
Krigbaum and Geymer (1959)	Osmotic pressure	0.2469	76.67	0.630_4	0.480_8	62
Scholte (1970)	Ultra-centrifuge	0.2631	74.31	0.534_4	0.430_4	63
Koningsveld et al. (1970)	Critical point	0.2035	90.50	0.610_6	0.920_7	16
Koningsveld et al. (1970)	Critical point	0.2211	85.31_3	0.622_2	0.289_1	19
Kuwahara et al. (1973)	Threshold cloud point	0.2798	67.50	0.607_3	0.512_1	66
Kamide and Matsuda (1984)	Cloud point	-0.0202_4	158.79	0.643	0.200	4
Kamide and Matsuda (1984)	Critical point	0.23	82.38	0.642	0.190	3

a' and b' (see, eqn (2.7); $k' = 0$, $p_3 = \cdots = p_n = 0$).

separation temperature T_p in the experiments shown in Figs 2.10-2.14 is in the range 283.7 and 297.0K, corresponding to a variation of ρ_p and accordingly, the value of p_1, tabulated in Table 2.4, should be considered as an average quantity over this temperature range. Fortunately, the temperature dependence of χ in the range 273K to 423K is within the experimental uncertainty of determining p_1 (ref. 67). Nakajima et al. (ref. 68) measured osmotic pressure for PS / MEK system and from their data we obtain $p_1 = 0.78_3$ at 283K, 0.78_2 at 298K, 0.74_9 at 313K and 0.73_3 at 328K. Kaneko determined, from osmotic pressure measurements, $p_1 = 0.68_2$ at 293K and 0.70_1 at 333K for the PDMS / chlorobenzene system (ref. 69). These results also imply very small temperature dependence of p_1.

The concentration dependence of the χ parameter has been evaluated for the PS / CH system by many investigations. Table 2.4 collects p_1 and p_2, together with a' and b' (see, eqn (2.7) in section 2.2) established for the system. Fig. 2.16 shows the

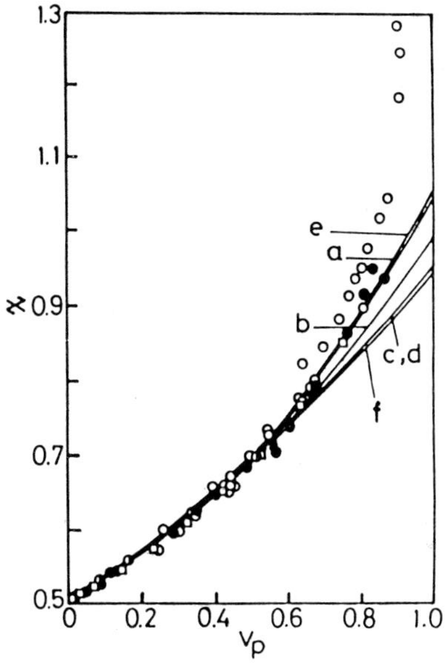

Fig. 2.16 The concentration dependence of the χ parameter for a polystyrene / cyclohexane system: ○, osmotic pressure by Krigbaum-Geymer (ref. 62); ◑, isothermal distillation by Krigbaum-Geymer (ref. 62); ●, vapor pressure by Krigbaum-Geymer (ref. 62); □, ultracentrifuge by Scholte (ref. 63); lines a-f are calculated using p_1 and p_2, obtained by experiments; a, Krigbaum-Geymer (ref. 62); b, Scholte (ref. 63); c, Koningsveld et al. (1970) (ref. 16); d, Koningsveld et al. (1971) (ref. 19); e, Kuwahara et al. (ref. 66); f, Kamide et al. (ref. 4).

concentration dependence of χ parameter for the same system at Flory theta temperature. These kinds of the plots were first produced in 1959 by Krigbaum and Geymer (Fig. 10 of ref. 62) using vapor pressure, isothermal distillation and membrane osmometry and thereafter (in 1970), Koningsveld et al. added the original figure by Krigbaum and Geymer on the experimental data of ultracentrifuge by Scholte and their own data on critical points (Fig. 6 of ref. 16). Fig. 2.16 was constructed by Kamide et al. (ref. 4) using all experimental data available up to 1984 for PS / CH at Flory theta temperature. In this figure, the curves have been calculated from the p_1 and p_2 values which in tern have been evaluated from the CSP data. The experimental data points can be reasonably represented by eqn (2.7), in which terms higher than $v_p{}^2$ are neglected ($k'=0$ and $p_3=\cdots=p_n=0$). That is, in the v_p range of $0\sim0.15$, both p_2 and p_1 are necessary to represent the concentration dependence of χ and in a comparatively dilute range, there is no sharp distinction between χ values reported by the different investigations.

For the PS / CH system, we obtain $a'=0.23$, $b'=81.8$ and $p_1=0.64$ and $p_2=0.20$ as the most probable values. The values of a' and b' estimated from cloud point curve seems less accurate. It will be shown later that the cloud point curve and the critical solution point are sensitively influenced by a combination of p_1 and p_2, and reversely, analysis on the critical and cloud point is the most suitable for determining p_1 and p_2.

Table 2.5 collects p_1 and p_2 values evaluated by various

TABLE 2.5

Parameters p_1 and p_2 in eqn (2.7) for atactic polystyrene / methylcyclohexane system ($T=345K$).

Author(s)	Method	p_1	p_2	ref.
Kamide et al.	Phase separation	$\sim$0.7	–	29
Kamide et al.	Membrane osmometry	0.780	0.329	33
Kamide et al.	Critical point [Dobashi et al.'s data (ref. 70) + Saeki et al.'s data (ref. 71)] by Kamide-Matsuda (KM) method	0.602	0.234	4

methods for the atactic polystyrene / methylcyclohexane system. Dobashi et al. (ref. 45) obtained $p_1 = 0.60_2$ and $p_2 = 0.299$ for the same system by analyzing critical points of nine samples with the Koningsveld and Kleintjens' procedure (ref. 19). Kamide and his coworkers, analyzed Dobashi et al.'s data (9 points) (ref. 70) and Saeki et al.'s data (6 points) (ref. 71) simultaneously by Kamide-Matsuda's method (ref. 3) to estimate p_1 and p_2 as listed in the table. It should be noted that except for the phase equilibrium and cloud point, all methods are limited experimentally to a relatively lower concentration range and do not enables us to evaluate p_2 accurately. The phase equilibrium method is applicable to more concentrated solution, but the experimental accuracy is not high enough to estimate p_2.

The values of p_1 and p_2 for PS and polyethylene (PE) in various single solvents, evaluated from their critical solution points, are summarized in Tables 2.24 and 2.25, respectively.

Now, if one can employ the value of p_1, estimated by other experiments such as osmotic pressure, critical phenomena etc., for a given polymer / solvent system, the MWD of the fraction and any other characteristics of phase separation can be completely calculated by using an electronic computer for the given experimental conditions (the MWD of the original polymer, fractionation scheme, $v_p{}^0$, ρ_p of ρ_s) without any ambiguity (ref. 29).

For a long time it has been considered without any doubt that the solution theory does not permit the exact calculation of σ and that σ is a kind of underestimated parameter (ref. 72): For example, Flory (ref. 11) supported that "even if the calculations were carried out, the numerical value obtained for σ probably would be subject to a considerable error owing to the imperfection of the theory". Okamoto and Sekikawa (ref. 73) were the first who tried to estimate σ experimentally. They derived the equation;

$$1 - R\sigma \int_0^{\infty} \frac{I(X)\exp(-\sigma X)}{\{1 + R\exp(-\sigma X)\}^2} dX = \rho_p \qquad (2.80)$$

from the Flory-Huggins solution theory, where $I(X)$ is the integral size distribution of the original (i.e. unseparated) polymer. Eqn (2.80) was solved with respect to σ for the two cases of phase separation for the system polyethylene / xylene / polyethyleneglycol, by using experimental data on $I(X)$, R and ρ_p. However, the results did not compare the experiment with the theory directly. It should be remembered that I(X) is very difficult to estimate experimentally and the value of I(X) obtained depends on the method utilized. Okamoto and Sekikawa (ref. 73) proposed in their paper the modified Flory phase relation in place of eqn (2.35) as follows:

$$v_{X(2)}/v_{X(1)} = \exp(\sigma_0 X^{\alpha}). \quad (2.81)$$

Two parameters σ_0 and α were chosen by inspection of the agreement between the calculated distribution and the experimental one of the polymer in a polymer-rich phase. By combining eqn (2.35) with eqn (2.1), σ can be expressed by

$$\sigma = \sigma_0 X^{\alpha - 1}. \quad (2.82)$$

On the other hand Breitenbach and Wolf (ref. 74) derived the equation:

$$v_{X(2)}/v_{X(1)} = \exp(\ln C + DX). \quad (2.83)$$

Comparison of eqn (2.35) with eqn (2.83) leads to

$$\sigma = D + (\ln C)/X. \quad (2.84)$$

Eqn (2.84) is principally equivalent to eqn (2.35') derived here.

In 1976 Kamide, Miyazaki and Abe (ref. 29) attempted to make this point clear. Table 2.6 summarizes the value of σ obtained by the experiment on solutions of polystyrene in methylcyclohexane and in cyclohexane as compared with those computed by the theory, in which the molecular weight dependence of χ parameter is neglected (i.e., $k' = 0$ in eqn (2.4)). Evidently, the experimental value for σ deviates appreciably from the theoretical value at $p_1 = 0$, which is based on the original Flory-Huggins theory. According to this, Flory's prediction is correct. However, if the concentration dependence of the polymer / solvent interaction parameter χ, expressed by p_1, is appropriately taken into account, we can calculate the σ value accurately by using the theory. That is, the theoretical value at $p_1 = 0.7$ proves successful.

The last column in Table 2.6 includes the value of σ calculated from $v_{P(2)}$ by:

$$\sigma = (1 - 2/v_{p(2)}) \ln(1 - v_{p(2)}) - 2, \tag{2.85}$$

which was proposed by Scott (ref. 72). Apparently Scott's approximation affords us a larger value than the true one, and consequently it is inapplicable for practical use.

TABLE 2.6
Comparison of the values for the partition coefficient σ obtained by experiment and computer simulation for different initial polymer volume fractions v_p^0 and fraction size ρ_p.

$10^2 \times v_p^0$	ρ_p	$10^3 \times \sigma$			
		experiment	theory Eqn (2.35) (k'=0)		Scott (ref. 72) approx. Eqn (2.85)
Polystyrene / methylcyclohexane:					
			$(p_1=0)$	$(p_1=0.7)$	
0.47	0.88	8.7	9.3	10.8	27.9
	0.78	5.2	5.5	6.5	34.0
	0.59	3.2	2.9	3.7	13.5
	0.10	1.5	1.0	1.4	5.0
0.94	0.92	10.4	10.5	12.6	24.0
	0.59	2.7	2.2	3.1	7.3
	0.48	2.5	1.6	2.4	19.0
	0.004	0.6	0.6	1.0	11.0
1.89	0.92	8.8	7.9	10.1	19.5
	0.35	1.4	0.6	1.4	9.9
	0.31	1.2	0.5	1.3	5.6
	0.07	1.0	0.3	0.8	6.3
Polystyrene / cyclohexane:					
			$(p_1=0)$[a]	$(p_1=0.6)$[a]	
0.94	0.88	5.1	6.2	7.4	20.0
	0.84	4.1	4.8	5.8	15.9
	0.62	2.3	1.9	2.6	5.1
	0.25	1.2	0.8	1.2	2.2

a, p_1 = interaction parameter from eqn (2.3).

The partition coefficient σ determined experimentally for the PS / CH and PS / MCH systems showed a small but significant tendency to decrease with increasing molecular weight, M. For example, among 16 phase-separation experiments for PS / CH and PS / MCH systems, the plots of σ (eqn (2.35)) against the reciprocal molar ratio of X-mer to solvent for 14 experiments exhibit positive slopes, and the slopes of the plots for the rest were positive in the region of small X^{-1} and changed to negative in the region of large X^{-1}. It should be noted that the latter two experiments involved a comparatively large experimental error due to small σ values obtained. Kamide, Miyazaki and Abe tried to attribute the molecular weight dependence of the χ parameter (ref. 8). Koningsveld and Kleintjens (ref. 19) and Kennedy et al. (ref. 20) have pointed out that the molecular weight dependence of the χ parameter influences significantly spinodal and critical miscibility.

Kamide and Miyazaki (ref. 7) carried out a theoretical calculation of the phase separation and successive fractionation on the quasi-binary system, with the molecular weight- and concentration-dependencies of the χ parameter taken into account. They did this so as to get a better understanding of the effects of molecular weight-dependence of the χ parameter on phase separation phenomena. A comparison is made with the experimental data in their previous work (ref. 29).

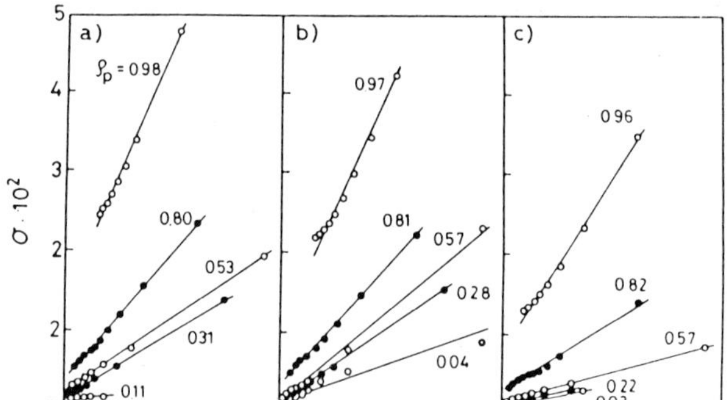

Fig. 2.17 Molecular weight dependence (X = polymer-to-solvent molar volume ratio) of the partition coefficient σ for polystyrene / methylcyclohexane system: a) initial polymer volume fraction $v_p^0 = 0.50 \times 10^{-2}$; b) $v_p^0 = 0.86 \times 10^{-2}$; c) $v_p^0 = 2.0 \times 10^{-2}$; Numbers on curve are the relative weight fraction ρ_p of polymer in the polymer-rich phase. For this purpose, the MWD of polymers partitioned in both phases were evaluated very carefully by high performance liquid chromatography (HPLC) using columns with 12000 theoretical plates, each 50cm in length.

Fig. 2.17 shows empirical relationships between σ and $1/X$ for the PS / MCH system. Throughout the range of v_p^0 investigated, the plot can be reasonably represented by a straight line, whose slope is equal to σ_{01} defined in eqn (2.35). σ_{01} increases with an increase in ρ_p (i.e., with a decrease in T_p) (ref. 8).

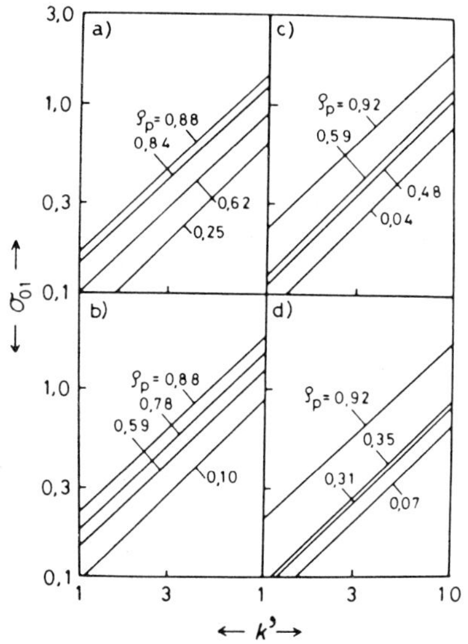

Fig. 2.18 Relationships between σ_{01} and the k' parameter: a) the original polymer (Schulz-Zimm (SZ) distribution, $X_w^0/X_n^0=2.8$, $X_w^0=2110$), $v_p^0=0.94$vol%, $p_1=0.6$; b) the original polymer (SZ distribution $X_w^0/X_n^0=2.8$, $X_w^0=1810$), $v_p^0=0.0047$, $p_1=0.7$; c) the original polymer (SZ distribution, $X_w^0/X_n^0=2.8$, $X_w^0=1810$), $v_p^0=0.0094$, $p_1=0.7$; d) the original polymer (SZ distribution, $X_w^0/X_n^0=2.8$, $X_w^0=1810$), $v_p^0=0.0189$, $p_1=0.7$. The value of ρ_p is indicated on the curve: a) corresponds to polystyrene (PS) / cyclohexane and b)-d) to PS / methylcyclohexane: $p_2=\cdots=p_n=0$.

The theoretical relationships between σ_{01} and the k' parameter at a given ρ_p, estimated from the theoretical σ vs. $1/X$ plots, are shown in Fig. 2.18a for PS / CH and Fig. 2.18b-d for PS / MCH, where the number on each curve denotes the value of ρ_p. Log-log plots of σ_{01} vs. k' can be represented by a straight line throughout the entire range of k' investigated. σ_{01} increases noticeably with increasing k' and ρ_p (ref. 7). The k' parameter was found to be 1.8~7.9 for PS / CH and 3.7~8.7 for PS / MCH. In this manner, we determined k' from the experimentally obtained molecular weight dependence of the partition coefficient σ. Conversely, the molecular weight dependence of σ can, at least partially, be interpreted by the molecular weight dependence of the χ parameter. Obviously, all the k' values estimated by this method are positive (ref. 7).

If the χ parameter is experimentally determined as a function of molecular weight and concentration for a given polymer / solvent system at a given temperature, the data affords us another route for evaluating k'. Scholte (refs 64,65) used the light scattering method to determine the χ parameter for PS samples having different molecular weights in CH. Using his experimental data, (Table 1 of ref. 64 and Tables 1-3 of ref. 65), we determined p_1 and k' for the PS / CH system by the regression method. It should be noted that the k' values above Flory theta temperature are negative.

In order to evaluate k' with two significant figures, we need χ values accurate to four significant digits when the polymer samples have comparatively large M_n as in the case of Scholte's study [$M_n = 49000$ ($X_n = 433$), 154000 ($X_n = 1363$) and 435000 ($X_n = 3550$)]. Such precision seems undoubtedly beyond the accuracy of actual experiment at present.

Fig. 2.19 shows the temperature dependence of k' for the PS / MCH system (ref. 8). In this figure, Kamide et al.' data for PS / MCH (ref. 7) except for $\rho_p \gtrsim 0.88$ and those for PS / CH (ref. 7) are shown for comparison. The k' parameter increases linearly with an increase in 1/T. In other words, this parameter becomes small as the temperature T approaches Flory theta temperature. Kamide and his coworkers demonstrated that for the PS / CH and PS / MCH systems that k' vanished at Flory theta temperatures (307K and 343K, respectively). Therefore, it may be concluded that the molecular weight dependence of χ changes its sign at Flory theta temperature θ and that k' can be expressed by eqn (2.5): k_0 in eqn (2.5) is estimated to be -104.2 for PS / CH system and -132.0 for PS / MCH system.

On the basis of Krigbaum's experiments (ref. 75) on the second virial coefficient A_2 for five polystyrene samples in cyclohexane, Koningsveld and Kleintjens (ref. 19) speculated a change in the sign of molecular weight dependence of the χ parameter in the vicinity of θ temperature and predicted a positive value of k' from a comparison of the data of spinodal with the theory for the PS / CH system. Fig. 2.19 confirms the validity of their prediction. Thus in order to discuss the effect of the molecular weight dependence of the χ parameter on fractionation by solubility, the k' value in the temperature range, wherein the phase separation experiment is carried out,

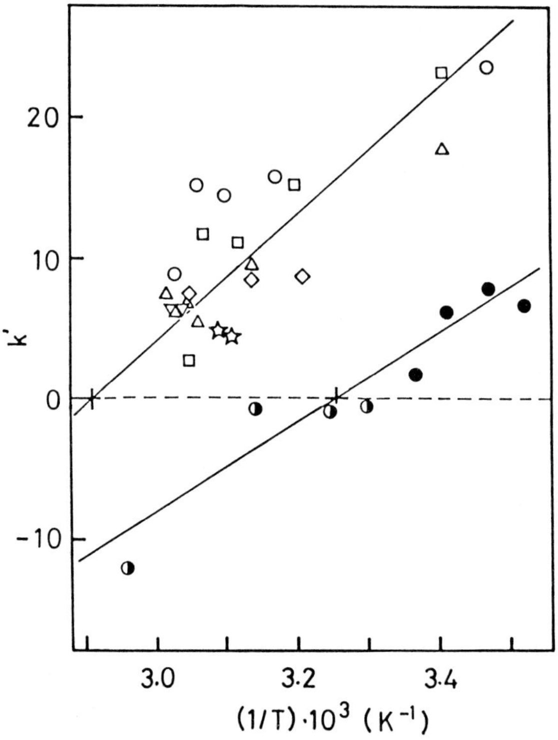

Fig. 2.19 Changes in the k' parameter with temperature T: closed and half-closed mark, polystyrene (PS) / cyclohexane system; ●, Kamide-Miyazaki (ref. 7); ◑, Scholte by light scattering (refs 57,58); open mark, PS / methylcyclohexane system: □, $v_p{}^0 = 0.50\times10^{-2}$; ◇, $v_p{}^0 = 0.47\times10^{-2}$; ○, $v_p{}^0 = 0.86\times10^{-2}$; ☆, $v_p{}^0 = 0.94\times10^{-2}$; △, $v_p{}^0 = 2.0\times10^{-2}$; ▽, $v_p{}^0 = 1.86\times10^{-2}$.

should be utilized in place of that obtained from the χ parameter above the θ temperature.

The breadth of the MWD of polymers in the polymer-rich phase expressed in terms of M_w/M_n is plotted against ρ_p in Fig. 2.20. Open circles are actual experimental data for the PS / MCH system, the broken lines are the theoretical curves at $v_p{}^0 = 0.5\times10^{-2}$, 0.86×10^{-2} and 2.0×10^{-2} for $p_1 = 0.7$ and $k' = 0$ and the full lines are the theoretical relations caluclated for $p_1 = 0.7$ by taking into consideration the complicated changes in k' with phase equilibrium temperature (accordingly ρ_p) (Fig. 2.19). The experimental ρ_p dependence of M_w/M_n in the polymer-rich phase can

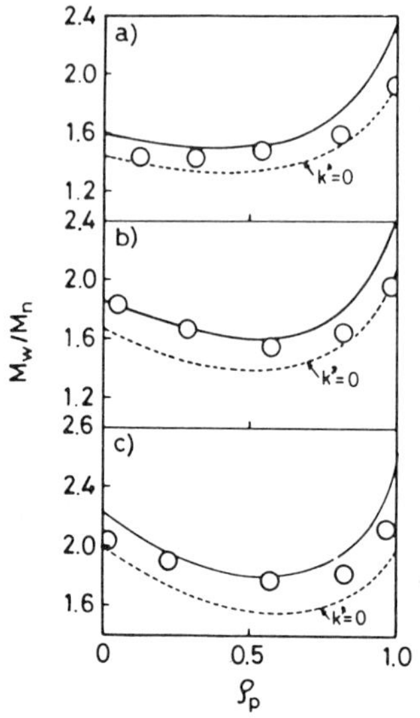

Fig. 2.20 Dependence of the ratio M_w/M_n of polymer in the polymer-rich phase on the relative amount of the polymer partitioned in the polymer-rich phase ρ_p: open circle, experimental data points for polystyrene / methylcyclohexane system; full line, theoretical curve calculated under the same condition as those in actual experiments (in this case, temperature dependence of k' (see, Fig. 2.19) is taken into account); broken line, theoretical curve calculated by assuming k' = 0 under the same conditions as those in actual experiments, except for k'.

be reasonably interpreted by considering the molecular weight and concentration dependencies of the χ parameter (ref. 8).

k' can now be evaluated by the following means:
(1) Comparison of the theoretical and experimental relations between σ and 1/X, (2) osmometric determination of χ as a function of molecular weight and concentration, and (3) comparison of molecular weight distributions (MWD) of the polymer in the two phases with the theoretical curves calculated assuming various values for k'. Kamide and Miyazaki found that the positive molecular weight dependence of the χ parameter (k' > 0) gives rise to a larger polydispersity of the polymers in the polymer-rich and -lean phases, particularly when k' > 3. In the region of low ρ_p (ρ_p is the weight fraction of the polymer in a polymer-rich phase relative to the total polymer dissolved), the ratio of the weight- to number-average of X, X_w/X_n in the polymer-rich phase is remarkably influenced by both p_1 and k'. In the region of high ρ_p, the polydispersity of the polymer in the polymer-lean phase is predominantly controlled by k'. For PS / CH and PS / MCH

systems, k' was found by the following means to lie between 0~10.

Summarizing, the molecular weight dependence of the χ parameter has a minor effect, when compared with its concentration dependence, on the phase separation phenomenon of polymer solutions, but this effect is not always negligible and comes into play with the following characteristics; partition coefficient σ, MWD of the polymer with small ρ_p in the polymer-rich phase

TABLE 2.7
Comparison of the effect of k' and p_1 parameters on phase equilibrium characteristics.

Phase equilibrium characteristics		Comparative role	
Partition coefficient		At limit of $\rho_p \to 0$	$k' < p_1$ [a]
		At limit of $\rho_p \to 1$	$k' > p_1$ [b]
Volume ratio R			$k' \ll p_1$
Polymer volume fraction in polymer-rich phase $v_{p(2)}$			$k' \ll p_1$
Polymer volume fraction in polymer-lean phase $v_{p(1)}$			$k' \simeq p_1$
$M_{w(2)}/M_{n(2)}$		At limit of $\rho_p \to 0$	$k' < p_1$
		At limit of $\rho_p \to 0$ and large p_1	$k' > p_1$
		At limit of $\rho_p \to 0$ and small p_1	$k' < p_1$
		At limit of $\rho_p \to 1$ and large p_1	$k' > p_1$
	SPF	Initial few fractions	$k' < p_1$
		End few fractions	$k' > p_1$
$M_{w(1)}/M_{n(1)}$		At limit of $\rho_s \to 0$	$k' > p_1$
		At limit of $\rho_s \to 1$ and large p_1	$k' > p_1$
		At limit of $\rho_s \to 0$ and small p_1	$k' < p_1$
	SSF	Initial few fractions	$k' > p_1$

[a] p_1 parameter is much more effective than k' parameter.
[b] k' parameter is much more effective than p_1 parameter.

and/or with small ρ_s in the polymer-lean phase, the M_w/M_n vs. M_w relations for the fractions isolated by SPF. It should be noted at the limit $\rho_p \rightarrow 1$, the polymers remaining in the polymer-lean phase are absolutely independent of the p_1 parameter and in the above regions, the k' parameter may play an important role. We can interpret the previously observed small discrepancy between the actual experiments on the PS / CH and PS / MCH systems and the theory, by the molecular weight dependence of the χ parameter. The fraction efficiency decreases with an increase in the molecular weight dependence of the χ parameter.

The effect of the k' and p_1 parameters on the phase equilibrium characteristics is very complicated, but the comparative role of the two parameters is summarized in Table 2.7 (refs 7,29,76).

2.4 FRACTIONATION BASED ON SOLUBILITY DIFFERENCES

2.4.1 Preparative Fractionation

The theory of phase equilibrium shows that the high molecular weight component of hetero-disperse polymers is easily partitioned in polymer-rich phase (see, Fig. 2.5). By utilizing this solubility difference, multicomponent polymers can be fractionated on the basis of the molecular weight. There are two objectives in carrying out fractionation experiments. One is the preparation of monodisperse polymer (preparative fractionation) and the other is the analysis of the MWD of the original polymer (analytical fractionation). Numerous reviews and papers have been published on these topics (refs 23,36,37,72,77-84). Nevertheless, the reliability of fractionation methods including both SPF and SSF, and the operating conditions, have not always been completely established (see, for example, Table 2.1 in Section 2.2).

In a fractionation experiment, a solution of the polymer in a single solvent is sometimes cooled (or nonsolvent is added to it) to allow liquid-liquid separation into polymer-lean and -rich phases. The polymer in the polymer-lean phase is precipitated by addition of a nonsolvent or by evaporation of the solvent (the first fractionation). A large amount of solvent is added to the polymer-rich phase to yield a dilute solution, and the temperature of the solution is lowered (or nonsolvent is added to the solution) again to produce further phase separation. This procedure is repeated to give more fractions. This is a method of fractionation of macromolecules by coacervation and is named

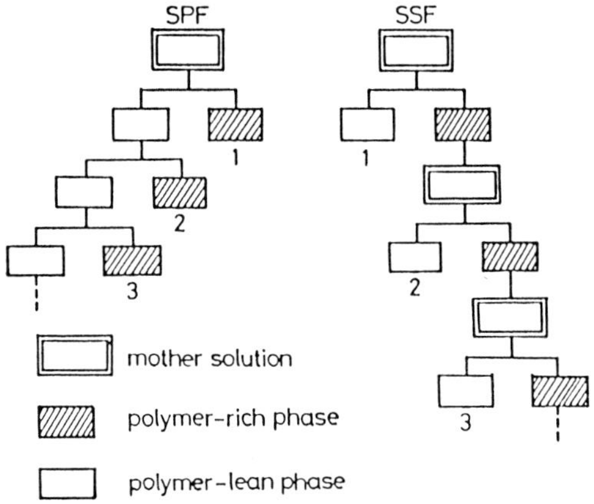

Fig. 2.21 Schematic representation of successive precipitation fractionation (SPF) and successive solution fractionation (SSF): Number denotes fraction number.

successive solution fractionation (SSF). In SPF the polymer-rich phase is separated as the fraction and the principle difference between SPF and SSF is schematically demonstrated in Fig. 2.21. In 1954, Matsumoto and Ohyanagi (ref. 85) made a prediction that SSF has a good possibility to prepare sharp fractions. Since 1972, Kamide et al. (refs 7,8,22,23,27,31-34,86) have studied SSF experimentally and theoretically in a systematic manner.

SPF, as a method of molecular weight fractionation by the solubility difference, has been widely used as a method for studying polymerization and fractionation data, and published work has been accumulated until 1980 (refs 22,86). For the PS / MCH system, the successive fractionation run can be easily performed by cooling the solution to at most 263K because of its high theta temperature (351K).

Fig. 2.22a and b illustrate the MWD curves of some typical fractions obtained by SPF for a 0.94% solution of PS in MCH as well as those obtained by SPF assuming $p_1 = 0.7$ ($k' = 0$ and $p_2 = \cdots = p_n = 0$) from a 0.94% solution ($\rho_p = 1/18$). Fig. 2.22c and d exemplify the MWD curves of fractions isolated by SSF for a 0.94% solution of PS in MCH together with those from the computer simulation with $p_1 = 0.7$ ($k' = 0$ and $p_2 = \cdots = p_n = 0$) from a 0.94% solution ($\rho_s = 1/23$). These figures indicate that the characteristic features observed for SPF and SSF calculations (refs 1,2,22,87) are also maintained qualitatively in practical

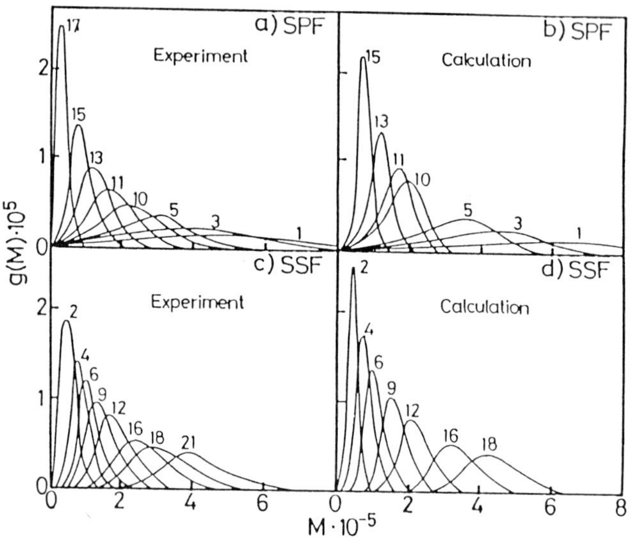

Fig. 2.22 Normalized molecular weight distribution curves of the fractions separated by a successive precipitation fractionation (SPF) [run a) and b)] and a successive solution fractionation (SSF) [run c) and d)] from a $v_p{}^0 = 0.0094$ solution; a) and c) experiments for polystyrene in methylcyclohexane; b) and d) theoretical calculations $p_1 = 0.7$ (cf. eqn (2.4); $k' = 0$ and $p_2 = \cdots = p_n = 0$; fraction size in SPF and SSF, $\rho_p = 1/18$ and $\rho_p = 1/23$, respectively); numbers at curves denote fraction number.

experiment, namely the fractions separated by SSF contain a lower molecular weight material to a less extent, suggesting that the "tail effect" (refs 1,2,22,87) can almost be neglected.

The ratio M_w/M_n was employed as a conventional measure for representing the breadth in MWD of the fractions. Fig. 2.23 demonstrates relationships between M_w/M_n of the fractions obtained by SPF and SSF for the system PS / MCH and M_w of the fractions. Fig. 2.23 also includes, for comparison, the corresponding theoretical curve ($\rho_p = 1/18$ for SPF and $\rho_s = 1/23$ for SSF) attained by assuming various p_1 values ($k' = 0$ and $p_2 = \cdots\cdot = p_n = 0$). In SPF the value of M_w/M_n increases smoothly with increasing M_w. This contrasts sharply with SSF, where M_w/M_n decreases notably approaching a limiting value.

The experimental relationship between M_w/M_n and M_w for a series of fractions obtained by SPF run on the PS / MCH system, reasonably fit the corresponding theoretical curve of $p_1 = 0.6 \sim 0.7$ (Fig. 2.23a) (ref. 29). The experimental points, however, obtained by a successive solution fractionation (SSF) run (hereafter

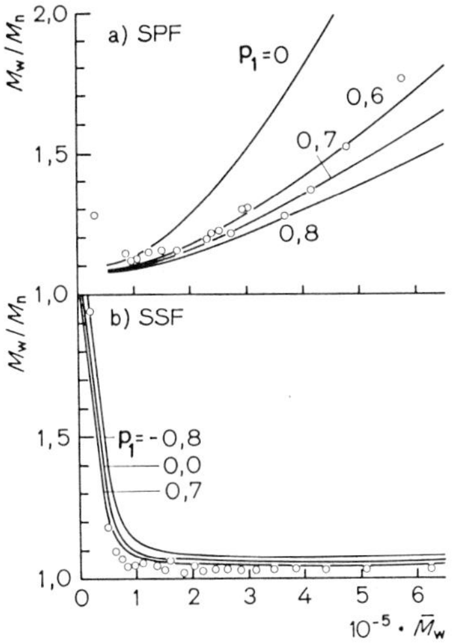

Fig. 2.23 Plot of M_w/M_n vs. M_w of the fractions obtained by a) successive precipitation fractionation (SPF) and b) successive solution fractionation (SSF) on the system polystyrene (PS) / methylcyclohexane: ○, experimental data; full line, theoretical curve; numbers on curves denote p_1 (cf. eqn (2.4); $k'=0$ and $p_2=\cdots=p_n=0$); initial polymer volume fraction, $v_p{}^0=0.0094$; total number of the fractions in a given run $n_t=18$ in SPF and $n_t=23$ in SSF.

referred to as Run 1 (initial polymer volume fraction $v_p{}^0=0.94\%$ and total number of fractions $n_t=23$)) on the same polymer / solvent system represent a significant departure from the theoretical curve of $p_1=0.7$, suggesting that a much larger (and unreal) value of p_1 than 0.7 should be assumed for interpreting these experimental SSF data (Fig. 2.23b). An attempt to explain the M_w/M_n vs. M_w relation in the case of SSF with $p_1=0.7$ met undoubtedly with failure. This was an unresolved problem, toward which further additional experiment and discussion were undertaken as an extension of earlier work (ref. 29).

M_w/M_n in the fraction obtained by SSF is not so sensitive to the p_1 value as compared with SPF. This was exemplified when an original polymer having the Schulz-Zimm distribution ($M_w=2.4\times10^5$, $M_w/M_n=2.8$) was hypothetically dissolved in a 0.94% solution, from which 18 fractions of identical amounts were separated by SPF. The M_w/M_n values of the fraction with $M_w=3\times10^5$ in the p_1 range from 0 to 0.7 are 1.57~1.22, and those values are interpolated from the M_w/M_n-M_w relations for a given series of fractions. By way of contrast, in the case of SSF, the M_w/M_n value of the fraction with $M_w=3\times10^5$ varies from 1.07 to 1.05,

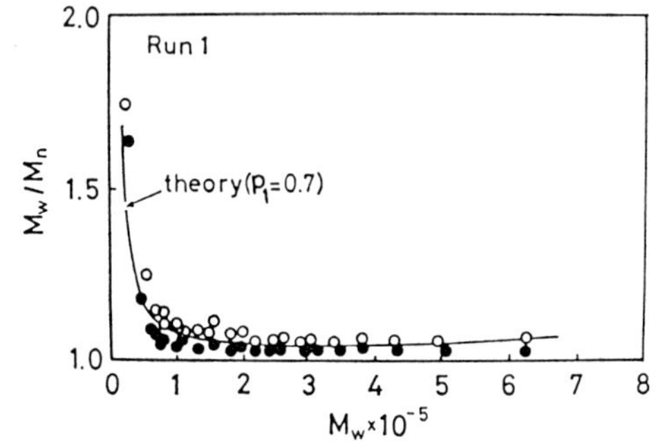

Fig. 2.24 Plot of M_w/M_n vs. M_w of a series of fractions obtained by successive solution fractionation Run 1 ($v_p^0 = 0.0094$, $n_t = 23$) on the system of polystyrene and methylcyclohexane: ○, experimental data estimated by GPC with a high resolution column combination; ●, experimental data estimated by GPC with a low resolution column combination and less accurate h value; full line, theoretical curve ($p_1 = 0.7$, cf. eqn (2.4) $k' = 0$ and $p_2 = \cdots = p_n = 0$).

depending on the variation of p_1 in the range 0~0.7. Thus, if we want to verify the validity of the fractionation theory with actual SSF experiments, we should evaluate M_w/M_n of those fractions obtained with high accuracy. For this purpose, the GPC measurements were carried out again (1) for the same series of fractions, just as those isolated by SSF run (Run 1) and (2) for the fractions which were prepared additionally by two SSF runs under different operating conditions, by using new columns with high resolution, and (3) subsequently the theoretical M_w/M_n-M_w relations of the fractions, which were derived on the bases of the modern theory of molecular weight fractionation, were compared with the experiments (ref. 31).

The results are shown in Fig. 2.24. The experimental data obtained by improved GPC measurements (open mark) are well explained by the theory with $p_1 = 0.7$. If we take properly into consideration the experimental spreads of M_w/M_n data due to inconstancy of ρ_p, the above-mentioned consistency is satisfactory: Theoretical relations between the ratio of the standard deviations of the polymer in both phases to that of the original polymer σ'/σ'_0 and M_w, assuming $p_1 = 0.7$, are graphed in Fig. 2.25, in which values of σ'/σ'_0 in the polymer-rich phase and in the polymer-lean phase for a given separation step are connected by a straight line, and the experimental points are shown by rectangular marks. Here, standard deviation σ' is given

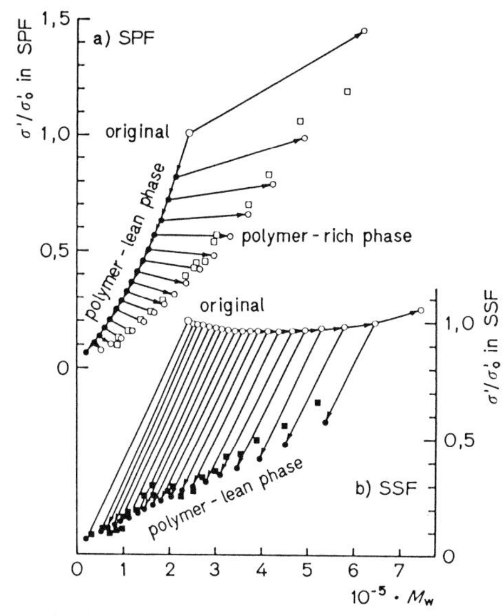

Fig. 2.25 Ratio of the standard deviations σ'/σ_0' (where σ_0' is σ' of the original polymer) in the polymer-rich phase (○) and the polymer-lean phase (●) in a) successive precipitation fractionation (SPF) and b) successive solution fractionation (SSF) obtained by computer simulation with $p_1=0.7$ (cf. eqn (2.4); $k'=0$ and $p_2=\cdots=p_n=0$) and $v_p^0=0.0094$: □ and ■, experimental data for polystyrene / methylcyclohexane.

in the form;

$$\sigma'^2=\int_0^\infty (M-\mu_1)^2 g(M)\,dM = M_w^2\left(\frac{M_z}{M_w}-1\right) \qquad (2.86)$$

where μ_1 is the first moment of the differential weight MWD function g(M) and M_z, z-average molecular weight.

A systematic change in $v_{p(2)}$ and R in a series of SPF or a SSF run is plotted in Figs 2.26 and 2.27, respectively. Full lines stand for the theoretical curves calculated assuming various p_1 values. In this case, $\Sigma\rho_{p,i}$ or $\Sigma\rho_{s,i}$ (where $\rho_{p,i}$ and $\rho_{s,i}$ mean the fraction size at the i-th fractionation step in SPF and SSF, respectively) was taken as the parameter representing the extent of fractionation. As the fractionation proceeds the polymer volume fraction in the consecutive stage of fractionation $v_{p(2)}$ increases appreciably in SPF and decreases gradually in SSF (ref. 29). In contrast to this, R increases rapidly, in particular at a

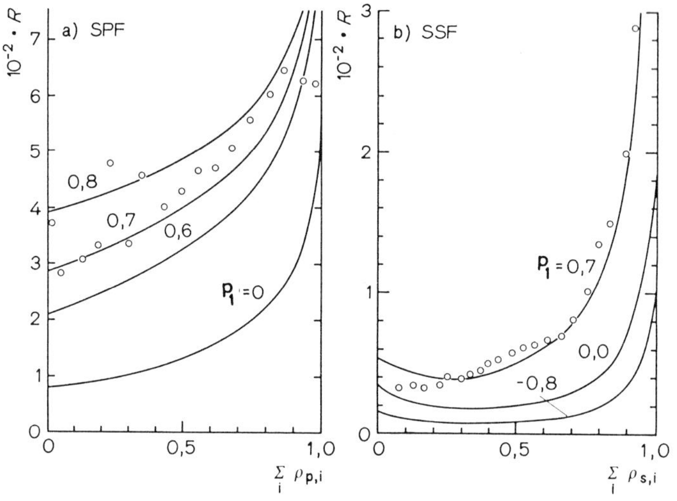

Fig. 2.26 Changes in the volume ratio of the two phases R with the fractionation order in a) successive precipitation fractionation (SPF) and b) successive solution fractionation (SSF) for polystyrene / methylcyclohexane; initial polymer volume fraction, $v_p{}^0 = 0.0094$; total number of fractions in a given run, $n_t = 18$ in SPF and $n_t = 23$ in SSF; full line, computer simulation; numbers at curves denote the values of p_1 (cf. eqn (2.4) with $k' = 0$ and $p_2 = \cdots = p_n = 0$); ○, experimental points.

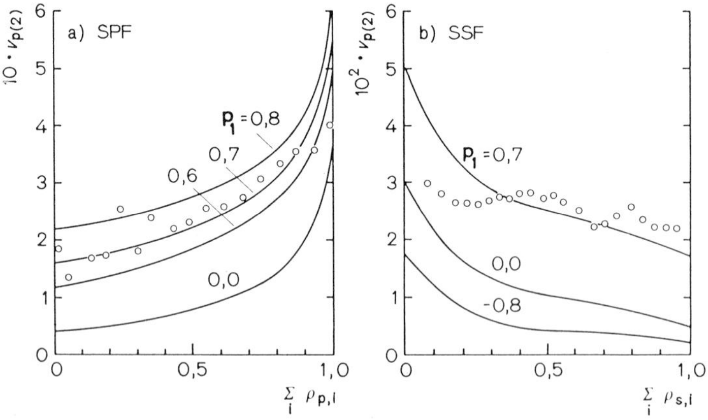

Fig. 2.27 Change in the volume fraction of polymer in the polymer-rich phase $v_{p(2)}$ with the extent of fractionation $\sum \rho_{p,i}$ or $\sum \rho_{s,i}$ in a) successive precipitation fractionation (SPF) and b) successive solution fractionation (SSF) for polystyrene / methylcyclohexane; initial polymer volume fraction $v_p{}^0 = 0.0094$; total number of fractions in a given run, $n_t = 18$ in SPF and $n_t = 23$ in SSF; full line, computer simulation; numbers on curves denote the value of p_1 (cf. eqn (2.4) with $k' = 0$ and $p_2 = \cdots = p_n = 0$); ○, experimental points.

later stage, in both SPF and SSF (ref. 29). Strictly speaking, in view of the theoretical point, the magnitude of R decreases slightly during the initial stages of the fractionation in SSF and, except for the case where the original polymer has a very narrow distribution (ref. 88), no minimum in the R value was, however, observed experimentally. Even in this case the experiments are thoroughly consistent with the theory assuming $p_1 = 0.7$. Experimental difficulties encountered in controlling the relative amount of fractions throughout a given run account for the wide spread of R. The method for estimating p_1 by using the R

TABLE 2.8
Comparison of successive solution fractionation (SSF) with successive precipitational fractionation (SPF).

Parameter	Comparison	Advantage of SSF	ref.
1. Partition coefficient σ	SSF > SPF	yes	88
2. Volume ratio R	SSF ≪ SPF	yes	88
3. Breadth in MWD of the fractions	SSF < SPF (except for extremely low X_w range)	yes	94
4. Operation conditions for controlling fractionation	SSF: ρ SPF: v_p^0	—	93
5. Effect of X_w^0/X_n^0	SSF < SPF	yes	87,88
6. Effect of X_w^0	SSF < SPF	yes	92,94
7. Reverse-order fractionation	only for SPF	yes	94,97
8. Double-peaked MWD in fraction	only for SPF	yes	87,88
9. Upper limit of initial polymer conc. v_p^0	SSF > SPF	yes	42
10. Easiness of phase separation	SSF ≫ SPF	yes	96
11. Accuracy of controlling faction size ρ	SSF ≫ SPF	yes	96
12. Total amount of solvent in a given run	SSF ≫ SPF	no	42
13. Ratio of X_w^0/X_n^0 of the first fraction to X_w^0/X_n^0 of the original polymer	SSF: always less than 1 SPF: always larger than 1	yes	93,94
14. Effect of p_1[a)]	SSF < SPF	—	21,93,96

a) For definition of p_1 see eqn (2.3), $p_2 = \cdots = p_n = 0$.

vs. $\Sigma \rho_{p,i}$ relations is, therefore, susceptible to large errors (see, Table 2.3).

Table 2.8 summarizes a detailed comparison of SSF with SPF, obtained by a series of computer experiments in the previous studies (refs 1,2,21,22,41,47,52,53,76,84,87-97). In SSF the partition coefficient σ is large, while the volume ratio R is small as compared with those in SPF (ref. 88). The value of the polydispersity ratio X_w/X_n converges straightly to unity as ρ_s approaches zero in SSF (ref. 96) and as $v_p{}^0$ reduces to zero in SPF (ref. 76). In addition, the effect of the average degree of polymerization and polydispersity in the original polymer is less remarkable in SSF (refs 88,95), where neither reverse order fractionation nor double peaked MWD in the fraction are detected under any operating conditions. The upper limit of the initial concentration $v_{p,c}{}^0$ is much larger than $v_{p,c}$ in the range $\rho_s < 0.1$ for SSF and it is mainly dependent on $X_w{}^0/X_n{}^0$ of the original polymer. SSF has the advantage of easiness of phase separation and of high accuracy of controlling ρ, e.g. the precise control of the temperature necessary to keep constant ρ in SSF is not as critical as in SPF (ref. 97). Although the total volume of solvent V_t needed for a given SSF run is certainly much larger than that for SPF, the value for the total solvent volume per g of fraction V_t' in SSF necessary to obtain a fraction having a fixed X_w/X_n is, against expectation, smaller if $X_w > X_w{}^0$ (ref. 97).

The SSF technique, employed here, is a kind of coacervate extraction method by solvent, which is somewhat different from the widely accepted concept of SSF, and not much attention had been paid to it until 1953. Although SPF is the most widely used technique for molecular weight fractionation, it is fully realized from a series of studies on SSF and SPF that SSF is preferable to, or at least equivalent to, SPF in every respect, even with respect to the total amount of solvent needed. Separation of fractions as sharp as $X_w/X_n < 1.05$ from a broad original polymer by SSF is by no means unrealistic. The experiments now available are insufficient for a critical test of the SSF method and further systematic study is urgently needed.

The so-called monodisperse polymer samples are useful as standard materials for calibration of GPC, vapor pressure osmometry and, for establishing the structure-property-processibility relations. For example, several kinds of polymer standard materials are now supplied on a commercial basis from a number of organisations, which are produced using an living anionic polymerization technique (PS, polyisoprene and poly-4-hydrofuran) or a preparative GPC technique (PE, isotactic polypropylene, polyvinylchloride) (refs 98-102) besides the fractionation method. The kind of polymers which can be prepared by an ion polymerization technique is relatively limited in number. Moreover, the standard polymer materials, obtained by GPC, have in most cases unsatisfactorily large breadth in their MWD.

Polymer samples with very narrow MWD are achieved by SSF under usual conditions. In Table 2.9 we compiled the number of fractions having M_w/M_n less than 1.10 or 1.05 in that case when the above three SSF runs are applied. Table 2.9 indicates that the agreement between experiment and theory is quite satisfactory, especially for runs 1 and 2, which were performed at relatively low $v_p{}^0$. The usefulness of SSF for this purpose was recently demonstrated by Kamide et al. (ref. 34) for atactic PS, high-density PE and cellulose acetate (CA).

TABLE 2.9
Number of fractions having M_w/M_n less than 1.10 or 1.05, isolated by three successive solution fractionations.

Polydispersity of fractions	Run 1 ($v_p{}^0=0.94\%$, $n_t=23$)		Run 2 ($v_p{}^0=0.94\%$, $n_t=20$)		Run 3 ($v_p{}^0=1.89\%$, $n_t=30$)	
	Exp.	Theory	Exp.	Theory	Exp.	Theory
$M_w/M_n<1.10$	18	19	15	16	16	25
$M_w/M_n<1.05$	7	9	5	7	0	10

A large scale fractionation apparatus, designed in Kamide and his coworkers' laboratory, has been used for preparative SSF and SPF (Fig. 2.28) (ref. 34). In Fig. 2.28, A is a 10 ℓ solvent measuring vessel, B is a 10 ℓ nonsolvent measuring vessel, C is a 20 ℓ five-necked, round bottom fractionation vessel, attached with mechanical sealed motor for agitation, D is a 80 ℓ bath thermostated to $\pm$0.1K in the temperature range 278~423K, E is a 20 ℓ five necked, round bottom storage vessel of the polymer-lean phase, F is a 80 ℓ bath controlled to $\pm$0.5K, ranging from 293~413K, G is a 5 ℓ evaporator made by Tokyo Rika Ltd. (Tokyo), H is a solvent or nonsolvent recovery vessel, I is a low-temperature thermostat bath (Haake Mess-Technik (Berlin) model T-

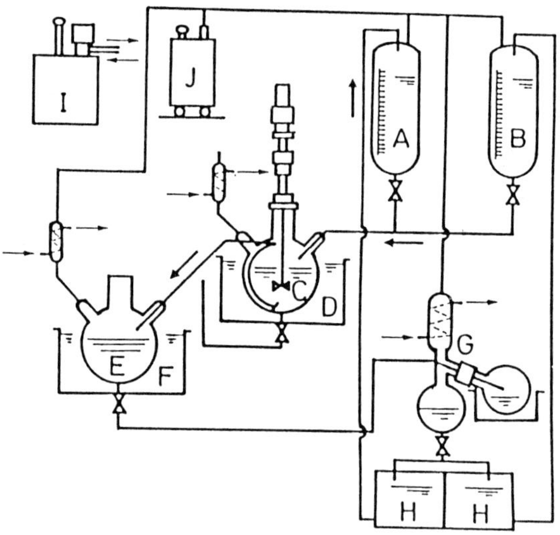

Fig. 2.28 Outline of a large-scale successive solution fractionation apparatus constructed: A and B, solvent and nonsolvent measuring vessel; C, fractionation vessel; D, thermostated bath; E, storage vessel; F, bath; G, evaporator; H, solvent and nonsolvent recovery vessel; I, thermostated bath; J, vacuum pump.

33), supplying cooled water (278∼283K) and J is a vacuum pump. These vessels for measurement, fractionation and storage, and the pipes connecting them, are all made of glass (Shibata Chemical Appl. Mfg. Co. Ltd. (Tokyo)). The leakage of solvent and nonsolvent vapor from the apparatus was completely prevented by allowing water at 278K to flow through a $1 m^2$ heat exchanger installed in the fractionation and storage vessels. All liquids, such as solvent, nonsolvent (if necessary) and polymer-lean phase, were transferred by the pressure generated due to the difference in liquid levels or by the application of a vacuum.

The following four polymers were employed as starting materials. Atactic PS; Styron 666 [M_w (by LS in benzene) $= 23.9 \times 10^4$, M_n (by MO) $= 8.6 \times 10^4$ in toluene] (refs 29,31) and high density PE; Suntec S-360 [M_w (by GPC in trichlorobenzene) $= 13 \times 10^4$, M_n (by GPC in trichlorobenzene) $= 2.9 \times 10^4$, $CH_3/1000C$ $= 3.0$], both manufactured by Asahi Chemical Industry Co. Ltd. (Osaka). CA (the total degree of substitution ⟪F⟫ = 2.92), CA (2.46), CA (1.75) and CA (0.49) were prepared on laboratory scale. CA (2.92); M_w (by LS in dimethylacetamide (DMAc)) $= 23.5 \times 10^4$, $M_n = 5.85 \times 10^4$ combined with acetic acid content (AC) = 61.0wt% (ref. 103),

TABLE 2.10
Operating conditions, M_w/M_n of fractions isolated by SSF.

Experiment No.	polymer	Solvent/Nonsolvent	Temperature /K	Polym. wt.	solution volume (1)	v_p^0	n_t	M_w/M_n	yield (wt%)	ref.
PE Run 1	HDPE[a]	cyclohexane	371~406	—	14.0	0.0108	20	—	96	34
PS Run 1	PS	MCH	293~328	—	2.0	0.0094	14	—	96	34
PS Run 2		MCH	288~328	—	2.0	0.0094	23	1.05~1.2	97	34
PS Run 3		MCH	280~328	—	18.0	0.0094	20	1.05~1.2	94	34
PS Run 4		MCH	283~331	—	18.0	0.0094	30	1.05~1.2	95	34
PS Run 5		MEK[b]/methyl alcohol	308	—	2.0	0.0068	22	1.05~1.2	95	34
MS Run 1	poly(α-MS[c])	CH	—	—	(—)	0.0081	12	1.0 ~1.07	—	30
CA Run 2	CA(2.92)[d]	epichlorohydrin/n-hexane	308	60	(—)	0.0108	13	1.3 ~1.5	—	103
CA Run 1	CA(2.92)	epichlorohydrin/n-hexane	308	—	18.0	0.0050	14	1.3 ~1.5	97	34
CA Run 3	CA(2.46)[e]	acetone/ethyl alcohol	303	—	3.0	0.0210	21	1.2 ~1.5	98	34,104
CA Run 4	CA(2.46)	acetone/ethyl alcohol	303	—	10.0	0.0210	16	1.2 ~1.5	97	34
CA Run 5	CA(1.75)[f]	acetone:water(7:3 vol)/ methyl alcohol	—	—	(3.0)	—	10	1.28~1.36	95	105
CA Run 6	CA(0.49)[g]	water/methyl alcohol	298	—	(3.0)	0.0056	15	1.3	95	106

a HDPE, high density polyethylene; b MEK, methyl ethyl ketone; c MS, methylstyrene; d CA(2.92), cellulose acetate (degree of substitution (DS) = 2.92); e DS = 2.46; f DS = 1.75; g DS = 0.49.

CA(2.46); M_w (by LS in acetone) = 12×10^4, M_n (by MO in acetone) = 5.5×10^4), AC = 55.6wt% (ref. 104), CA(1.75); M_w (by LS in DMAc) = 12×10^4, M_n (by MO in DMAc) = 5.5×10^4, AC = 44.4wt% (ref. 105) and CA(0.49); M_w (by LS in DMAc) = 8.0×10^4, M_n (by MO in DMAc) = 3.9×10^4, AC = 16.1wt% (ref. 106).

The operating conditions employed are summarized, together with the fractionation yield, in Table 2.10. For comparison, small-scale experiments were also carried out: PS Run 1, PS Run 2 and PS Run 5.

The plots of M_w/M_n against M_w, thus estimated for two series of fractions by SSF runs, are shown as open marks in Figs 2.29

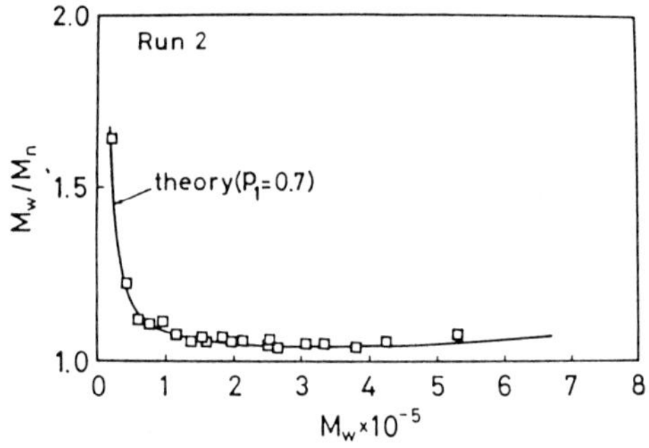

Fig. 2.29 Plot of M_w/M_n vs. M_w of a series of fractions obtained by successive solution fractionation Run 2 ($v_p{}^0 = 0.0094$, $n_t = 20$) on the system of polystyrene and methylcyclohexane: □, experimental data; full line, theoretical curve ($p_1 = 0.7$, cf. eqn (2.4) $k' = 0$ and $p_2 = \cdots = p_n = 0$).

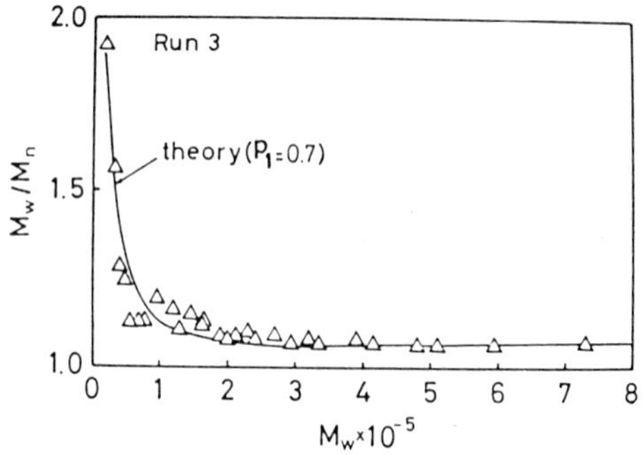

Fig. 2.30 Plot of M_w/M_n vs. M_w of a series of fractions obtained by successive solution fractionation Run 3 ($v_p{}^0 = 0.0189$, $n_t = 30$) on the system of polystyrene and methylcyclohexane: △, experimental data; full line, theoretical curve ($p_1 = 0.7$).

and 2.30, where the corresponding theoretical curves ($p_1 = 0.7$) together with experimental data obtained previously are also included. The data points coincide closely, as expected, with the theoretical curve assuming $p_1 = 0.7$, when the unavoidable variation of the fraction size ρ is considered (i.e., the standard deviation of the fraction size was 0.012 and 0.011 for Runs 2 and 3, respectively) during a given experimental run (ref. 31).

Except for a few initial fractions, a series of fractions having almost the same value of M_w/M_n can be prepared by the SSF technique. Similar results were obtained in other SSF runs, irrespective of the kind of polymers. This is now recognized as a common feature of SSF. About a nine fold scale difference in PS Run 3, compared to PS Run 2, does not yield any significant change in the fractionation characteristics. The theoretical curve

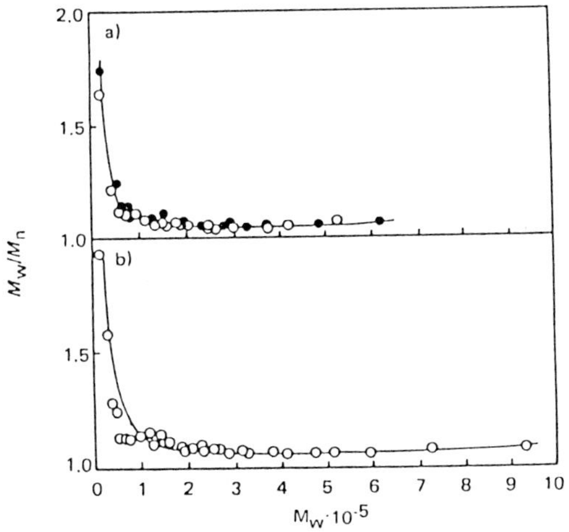

Fig. 2.31 Plot of the M_w/M_n of the fractions against their M_w for polystyrene (PS) Run 2 (closed circle) and PS Run 3 (open circle) [a)] and PS Run 4 (open circle) [b)]: Full lines are the theoretical curves for PS Runs 2 [a)] and 4 [b)].

clearly fits the entire set of data, indicating that these SSF runs were evidently carried out in the equilibrium state. It is of interest to note that from even relatively concentrated solutions the sharp fraction can be isolated by SSF if the total fraction number is large (i.e., $n_t \geq 30$). This trend is self-evident in Fig. 2.31b (PS Run 4). Agreement between the actual experiment and the theory is excellent and for this run a value of 1.07 is obtained for the minimum value of M_w/M_n of the fractions $(M_w/M_n)_{min}$ and about a half of the total fractions have the M_w/M_n value less than 1.1. In addition, the SSF run at higher $v_p{}^0$ and larger n_t yields a series of fractions having larger M_w. Then, without enlargement of the scale we can obtain a larger number of the fractions with reasonably narrow MWD by adopting higher $v_p{}^0$ and larger n_t (ref. 34).

We demonstrate in Fig. 2.32 the theoretical relationship between the value of $(M_w/M_n)_{min}$ or the relative amount of the fractions $Y_{1.1}$, whose M_w/M_n is equal to or less than 1.10, and the total amount of the starting polymers and the operating conditions such as $v_p{}^0$ and n_t. In this figure the MWD of the original polymer was taken as identical with that of the PS used in the experiment and the experimentally determined p_1 value (0.7) for the PS / MCH system was also employed. The curves in the figure are contour lines giving the same value of $(M_w/M_n)_{min}$ or $Y_{1.1}$. Obviously, the large amount of the sharper fractions can be separated by SSF

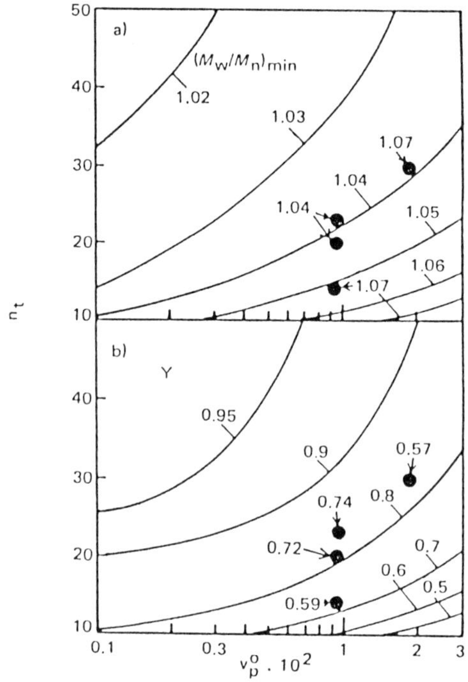

Fig. 2.32 Relationship between $(M_w/M_n)_{min}$ [a)] and $Y_{1.1}$ [b)], $v_p{}^0$ and n_t for polystyrene (PS) / methylcyclohexane system: Full line, theoretical curve; closed circle, experimental data of PS Run 1, 2, 3 and 4.

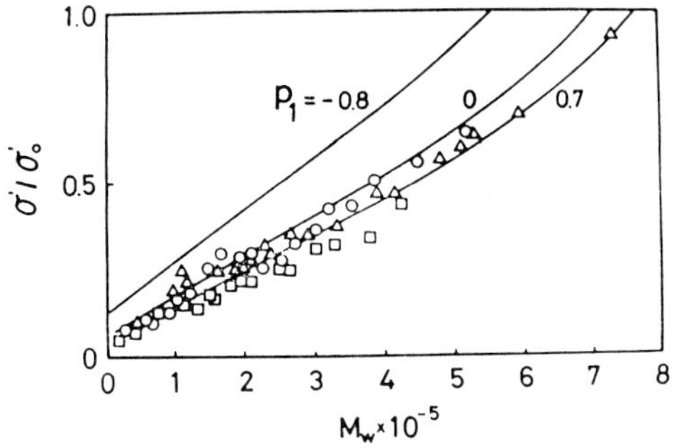

Fig. 2.33 Plot of the ratio of the standard deviation σ' of the fractions obtained by successive solution fractionation runs to that of the original polymer σ_0' vs. their M_w:
○, experimental data of Run 1 ($v_p^0 = 0.0094$, $n_t = 23$);
□, experimental data of Run 2 ($v_p^0 = 0.0094$, $n_t = 20$);
△, experimental data of Run 3 ($v_p^0 = 0.0189$, $n_t = 30$);
full line, theoretical curves p_1 values are indicated on curves (eqn (2.4); $k' = 0$ and $p_2 = \cdots = p_n = 0$).

under the conditions of smaller v_p^0 and larger n_t. The closed marks in the figures are the experimental results. Comparison of the theory with experiment indicates that the latter is only slightly ineffective, owing to experimental difficulty, in particular in a relatively concentrated solution, but the agreement with them is fairly satisfactory.

As a further check, in Fig. 2.33, the ratio of the standard deviations σ' of the fractions, obtained by SSF runs to that of the original polymer σ_0' is employed instead of M_w/M_n as a measure for representing the breadth of the fraction in MWD. For comparison, the figure also gives a family of theoretical curves relating σ'/σ_0' to M_w of the fractions for various p_1 values. Despite the variation of v_p^0 and n_t in Runs 1-3, theoretical σ'/σ_0' increases nearly linearly with increasing M_w. The effect of p_1 on the σ'/σ_0'-M_w relation is significant and the effective value of p_1 which gives the best coincidence to the entire set of experimental data is about 0.7 (ref. 31).

Inspection of Figs 2.29-2.33 leads us to the conclusion that all the experimental data available is thoroughly consistent with the theory developed, and that the conclusions deduced in Fig. 2.23b were collaborated, strongly reflecting the validity of the modern theory of molecular weight fractionation as well as the adequacy of computer simulation techniques developed in our laboratory (ref. 31). Since a significant difference does not exist between theory and experiment on any point, p_1 can be treated as constant throughout fractionation and, moreover, the molecular weight dependence of the partition coefficient σ observed by direct experiment can be reasonably ignored. The reverse, however, is not true and the determination of p_1 from the plot, as in Figs 2.29 and 2.30, may be greatly in error. The conclusions deduced in a series of simulation studies are not so easy to verify by experiment, but can be considered valid. We are now in a position to design the fractionation scheme in a way known in detail and to predict the molecular characteristics of fractions thus obtained.

2.4.2 Refractionation

Relatively low order fractions of SPF have a large tail in low degree of polymerization region and consequently, the MWDs become broad. This long tail is the main reason for low accuracy of SPF as an analytical fractionation and of frequent occurence of reverse-order fractionation (ROF) in SPF, (see, section 2.4.3). In order to improve this shortcoming, relatively low order fractions of SPF are often refractionated into several fractions. Fig. 2.34 shows a few examples of the MWD of refractionated fractions (subfractions) (ref. 1). The first three fractions were refractionated into four subfractions [$v_p{}^0 = 1\%$, $\rho_p = 1/15$, SZ2 (Shulz-Zimm type distribution of $X_w{}^0/X_n{}^0 = 2$) and $X_w{}^0 = 300$]. Here, (j-k) denotes the k-th subfraction of refractionation run for the j-th fraction; for example, the k-th fraction and ℓ-th

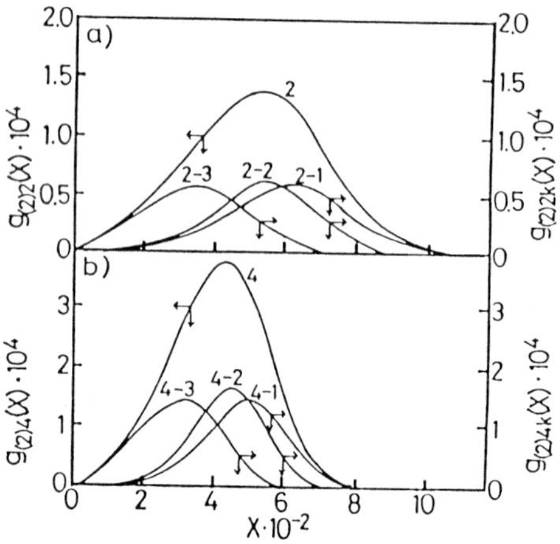

Fig. 2.34 Refractionation calculation of the precipitation fractionation: a) 2-3; b) 4-3.

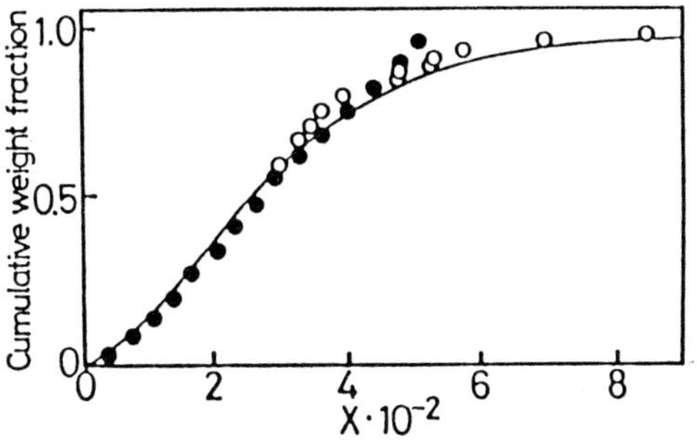

Fig. 2.35 Integral distribution curve estimated from refractionation data by using Schulz's method: ○, plots of the results of computer simulation in which the first four fractions were refractionated into twelve equal subfractions by a computer; ●, plots of single fractionation data.

subfraction are called (k-0) fraction and (k-ℓ) subfraction, respectively. Although the MWD of the (j-1) and (j-2) ($j = 1 \sim 4$) subfractions approaches the normal distribution more closely than that of the j-th fraction, low molecular weight components cannot be removed perfectly. $X_{w(2)}/X_{n(2)}$ of the (j-1) and (j-2) subfractions are smaller than that of the j-th fraction, but $X_{w(2)}/X_{n(2)}$ of the (j-3) subfraction is larger than that of the j-th fraction. $X_{w(2)}/X_{n(2)}$ of the (j-k) subfraction is not always smaller than that of j-th fraction (ref. 1). The cumulative weight fraction distribution of the refractionated material, evaluated by the Schulz's method with $v_p{}^0 = 1.0\%$ (see, eqns (2.93) and (2.94)), approaches the true distribution for the high degree of polymerization region. But the relative error of the refractionation method is larger than that for the simple fractionation evaluated by Schulz's method with $v_p{}^0 = 0.1\%$. The refractionation method is not effective to evaluate the MWD of the original polymer more precisely (Fig. 2.35).

According to the results mentioned above, a simple refractionation of the first few fractions is not sufficient to elucidate accurately the MWD of the original polymer. In order to obtain sharp fractions, however, the refractionation method is preferable to the method of simple SPF from a solution of low initial concentration. Taking into account this point, Kamide et al. (ref. 90) examined three types of successive precipitation refractionation (SPRF), (15-1), (5-3) and (3-5). The procedure of (5-3) refractionation is shown in Fig. 2.36. A $v_p{}^0 = 0.1\%$ solution of the original polymer dissolved in a single solvent, was cooled

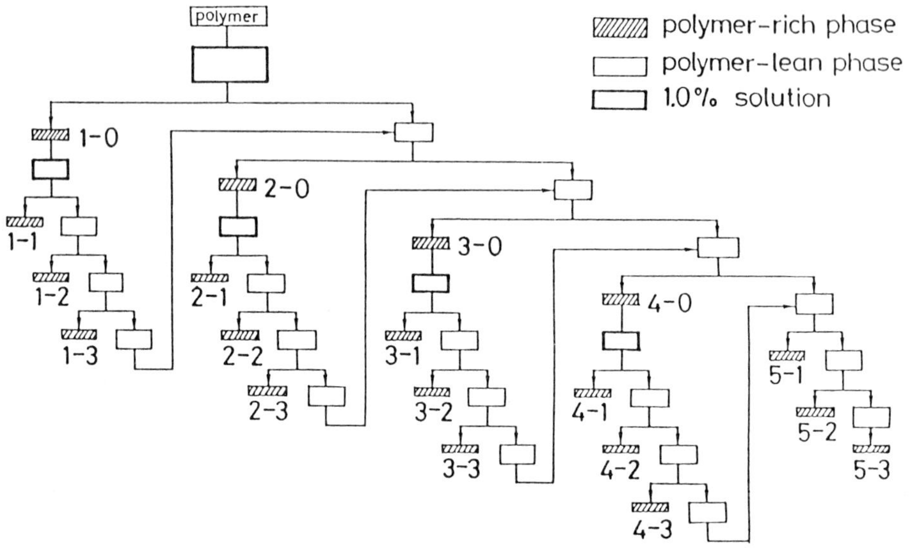

Fig. 2.36 Schematic representation of the (5-3) type successive precipitation refractionation (SPRF).

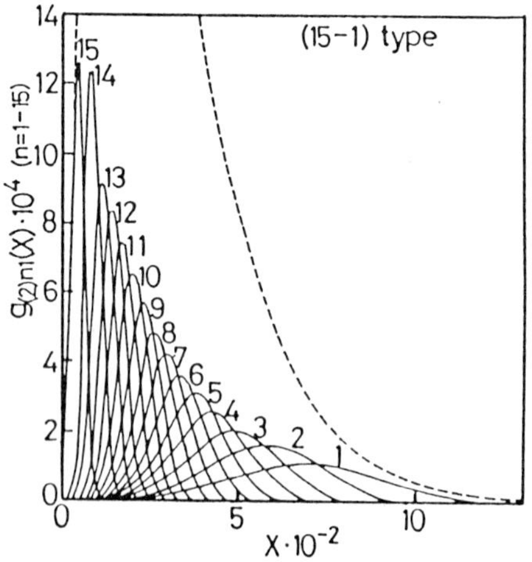

Fig. 2.37 Hypothetical (15-1) type refractionation from a $v_p{}^0 = 0.01$ solution of the polymer having the most probable distribution ($X_n = 150$): Number of subfractions is 15; broken line, original polymer; full line, subfraction.

down in such a manner that an equal amount (i.e., (j+ 1)/ij of the initial polymer in weight) was successively separated by cooling

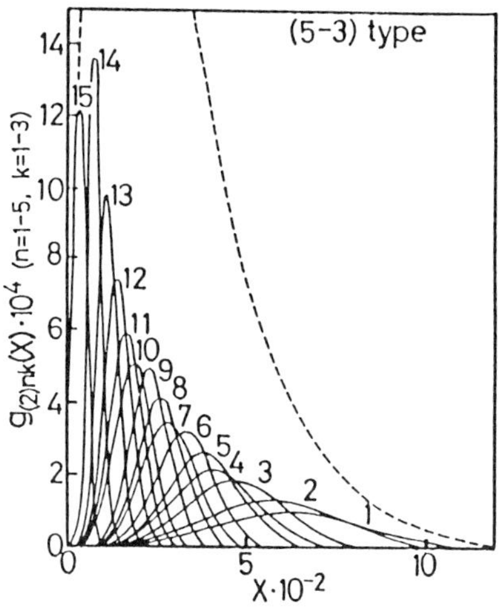

Fig. 2.38 Hypothetical (5-3) type refractionation from a $v_p{}^0 = 0.01$ solution of the polymer having the most probable distribution ($X_n = 150$): Number of subfractions is 15; broken line, original polymer; full line, subfraction.

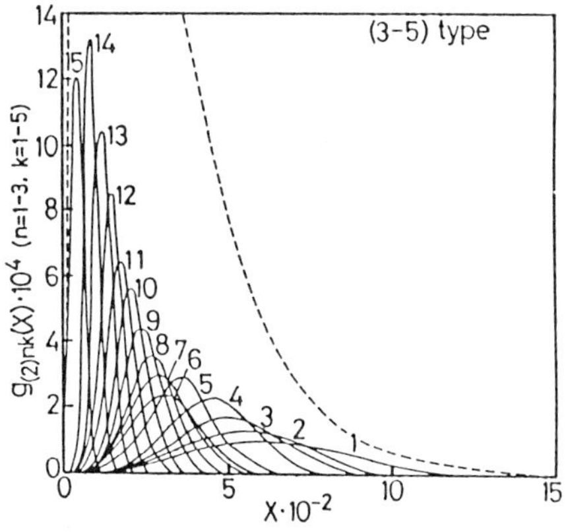

Fig. 2.39 Hypothetical (3-5) type refractionation from a $v_p{}^0 = 0.01$ solution of the polymer having the most probable distribution ($X_n = 150$): Number of subfractions is 15; broken line, original polymer; full line, subfraction.

the solution. The amount of polymer remaining in the polymer-lean phase after the last (i.e., the (k-j)-th) refractionation process for a given fraction was always kept at 1/ij of the total polymer. Figs 2.37-2.39 show the MWDs obtained by the (15-1), (5-3) and (3-5) type refractionations, respectively, together with that of the original polymer (SZ2, $X_w{}^0 = 300$) shown as a broken curve. Other types of SPRF were investigated by Schulz and Dinglinger (ref. 46), Meffroy-Biget (ref. 107), Thurmond and Zimm (ref. 108) and Meyerhoff (ref. 109).

The (5-3) and (3-5) type SPRF methods (ref. 90) were found to be the most efficient techniques to obtain sharp fractions from the whole polymer without laborious and time consuming operations. In general, there are many ways of selecting a pair, i and j, for a given value of the total subfractions. The results of the computer experiments indicate that the pair, i and j, should be chosen in such a way that the i- and j-value do not differ very much in magnitude (ref. 90).

The successive solution refractionation (SSRF) was also examined by Kamide and Yamaguchi (ref. 91). Schematic representation of typical SSRF procedure is illustrated in Fig. 2.40 (ref. 91), where the total number of subfractions, ij, is kept constant (in this case ij = 15). The polymer concentration of the polymer-lean phase is controlled at 1% by evaporating an adequate amount of the solvent in that phase. The first subfraction of the amount 1/ij of the initial polymer by weight is separated from the solution in the polymer-lean phase by lowering the temperature. Solvent is added to the polymer-rich phase to yield a 1.0% solution, which is used as the mother solution for

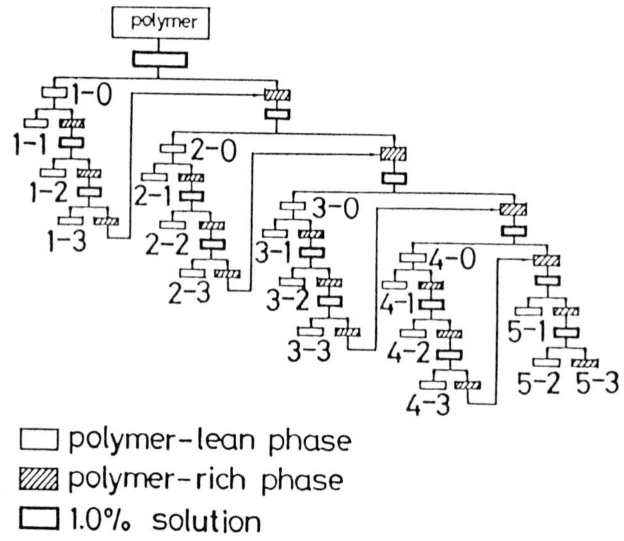

Fig. 2.40 Schematic representation of the (5-3) type successive solution refractionation (SSRF).

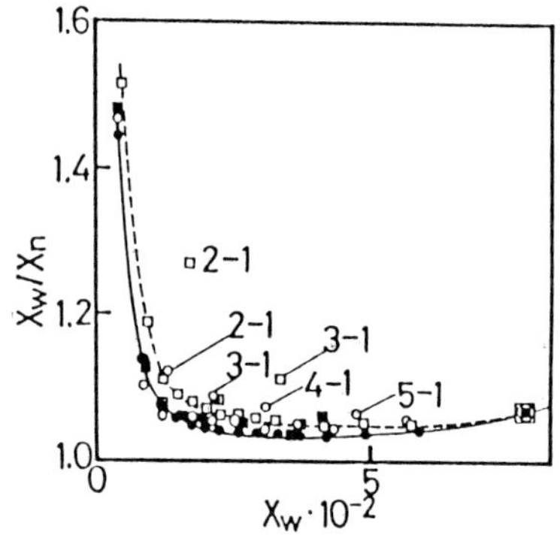

Fig. 2.41 X_w/X_n plotted against X_w for the subfractions obtained under different conditions: ●, (15-1) type; ○, (5-3) type; ■, (3-5) type; □, (0-15) type.

further fractionation. In total, j subfractions of equal amount are refractionated from the first fraction. The polymer-rich phase in the first separating step is employed for successive fractionation. The difference of X_w/X_n vs. X_w relation in (15-1) and (0-15) type fractions is not so large as compared in the case of SPRF (Fig. 2.41). This implies that the refractionation technique is not very effective for separating polymers into very narrow fractions (ref. 91).

SPRF and SSRF are in general not suitable for the purpose of analytical fractionation, but are useful for preparative fractionation (refs 90,91).

2.4.3 Reverse-Order Fractionation

As fractionation proceeds, in other words, as the fractionation order i increases, the average-molecular weight of the i-th fraction obtained by the SPF decreases, irrespective of the nature of the average-molecular weight in the fractionation concerned. However, it is occasionally observed in practical SPF that the average-molecular weight of the (i+1)-th fraction is larger than that in the i-th fraction (ref. 110). This phenomenon is what is commonly termed reverse-order fractionation (ROF). ROF was, up to the mid 1970s, mainly attributed phenomenologically to a too high v_p^0 value, although an explicit explanation has not yet been given for it (refs 111-114). In addition, according to experiments on SPF of polyacrylonitrile by Fujisaki and Kobayashi (refs 111-114), ROF occurs in SPF probably even from a solution of comparatively low v_p^0 in the case of the specific combination of a good solvent and a poor solvent. Matsumoto (ref. 115) predicted after some approximate hypothetical calculations that ROF would be observed, if the fractionation is carried out under conditions where the value of X_a, the value of X corresponding to $f_X(f_{X(2)}) = 0.5$, is adopted as large as possible for the first fraction and as small as possible for the second fraction. Here, f_X denotes the probability of the X_i-mer partitioned in a polymer-rich phase. The computer experiments performed later by Kamide et al. (refs 1,2) indicated that X_a increases eventually with an increase in v_p^0 and with a decrease in p_1 of χ (see, eqn (2.4)) and with a decrease in ρ_p. Accordingly ROF may occur if a very small amount of the first fraction is separated from a dilute solution. However, the computer simulations of Kamide et al. (ref. 2) indicate that Matsumoto's theoretical prediction lacks general character and the parameter X_a cannot be regarded as a reasonable measure of ROF and that he could not exemplify ROF theoretically because of the relatively narrow range of operating conditions adopted (ref. 97).

Kamide and Miyazaki (ref. 97) showed that ROF may occur theoretically in SPF from a dilute solution under operating

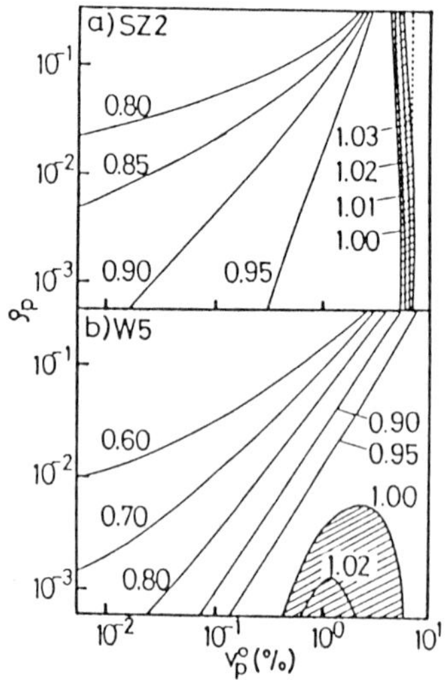

Fig. 2.42 Correlationships between ρ_p and v_p^0 yielding constant $X_{n,2}/X_{n,1}$ in SPF: The original polymer, $X_w^0 = 300$; $p_1 = 0$; numbers on curves denote $X_{n,2}/X_{n,1}$; shadowed region, $X_{n,2}/X_{n,1} > 1$. a) Schulz-Zimm type distribution, $X_w^0/X_n^0 = 2$ (SZ2); b) Wesslau type distribution, $X_w^0/X_n^0 = 5$ (W5).

conditions which are not unusual even when the phase equilibrium is fully achieved. For example, Fig. 2.42 shows the operating conditions v_p^0 and ρ_p, yielding the fractions with a given value of $X_{n,2}/X_{n,1}$ (subscripts 1 and 2 show the order of fractions), which are fractionated by SPF from the two original polymers (the Schulz-Zimm type distribution of $X_w^0/X_n^0=2$ (SZ2) with $X_w^0=300$ and the Wesslau type distribution of $X_w^0/X_n^0=5$ (W5) with $X_w^0=300$) (ref. 97). In Fig. 2.42 the shadowed area represents $X_{n,2}/X_{n,1}\geq 1$. For the SZ type polymer ROF occurs at a very limited range of v_p^0, almost independent of ρ_p (i.e., 4.2~5.2% at $\rho_p=10^{-2}$ and 4.7~6.5% at $\rho_p=10^{-3}$), but for the W type polymer ROF occurs at a relatively wide range of v_p^0 and small ρ_p (ref. 86). Following this, in the former, v_p^0 rather than ρ_p plays an essentially important role in ROF, and in the latter, both ρ_p and v_p^0 contribute effectively to ROF. In the latter, the v_p^0 range yielding ROF becomes wide as ρ_p is small. In this sense, the type of MWD in the original polymer controls predominantly the easiness of occurrence of ROF, and clearly, ROF occurs even if a common value of ρ_p is employed through the first and second fractions. This is inconsistent with predictions by Matsumoto (ref. 115).

Fig. 2.43a shows the plot of $X_{n,i}/X_{n,1}$ against fractionation order for the fractions separated by SPF assuming $p_1=0$ from solutions of v_p^0 ranging from 0.1~7.14% of the polymer having the Schulz-Zimm distribution ($X_w^0/X_n^0=5$ and $X_w^0=300$), where $X_{n,i}$ denotes the value of X_n in the i-th fraction. If v_p^0 exceeds a specific value (in this case, 3.3%), the plot reveals a maximum shifting to the higher order for the higher v_p^0 solution. Obviously ROF occurs at a later stage if the solution is concentrated. A similar behavior was experimentally observed for polyacrylonitrile in aq. nitric acid (60%) / butanol mixture by Fujisaki and Kobayashi (ref. 112). Fig. 2.43b is just an example, which is plotted from their original data (ref. 112) and in this case concentrations given in the figure are calculated according to the definition of v_p^0 in this study and the cumulative weight fraction is employed as the ordinate axis in place of the fractionation order because of inequality in the practical experiments. It should be noticed here that in the strict sense the change in v_p^0 is necessarily accompanied with a small but concurrent change in p_1 and in addition, the systematic variation in the composition of the solvent / nonsolvent mixture during the practical experiment gives rise to the change in p_1 during a SPF run. Then, the effect of p_1 might be contaminated to a small

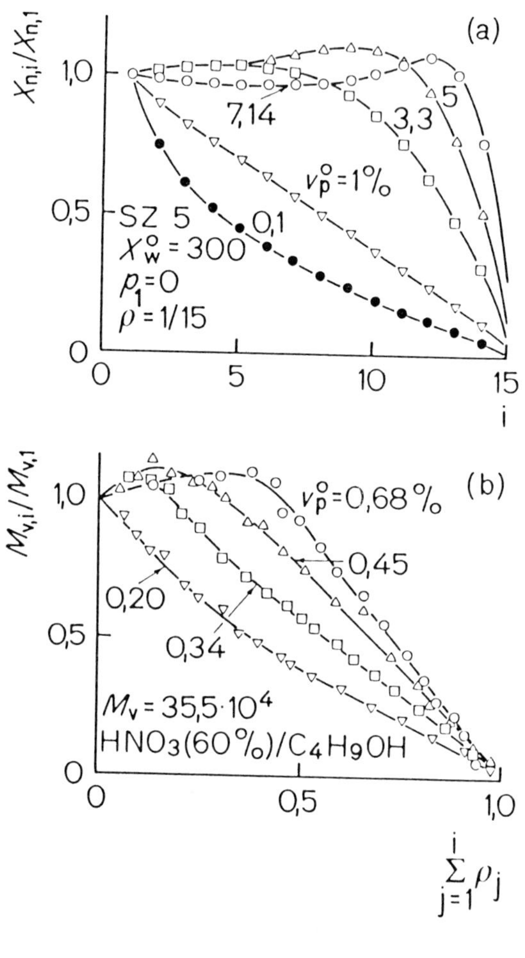

Fig. 2.43 Effect of initial polymer conc. v_p^0 on change in $X_{n,i}/X_{n,1}$ (upper part) and $M_{v,i}/M_{v,1}$ (lower part) with the fractionation order i in successive precipitational fractionation (SPF), $\Sigma \rho_j$; cumulative weight fractionation for relative amount of j-th fraction ρ_j; numbers on curves denote v_p^0: (a) Simulation; The original polymer, Schulz-Zimm distr., $X_w^0/X_n^0 = 5$, $X_w^0 = 300$; $p_1 = p_2 = p_3 = \cdots = 0$, $\rho_P = 1/15$: (b) Experiments by Fujisaki and Kobayashi (ref. 112): The original polymer, unfractionated polyacrylonitrile; the viscosity-average molecular weight $M_v = 3.55 \times 10^5$; solvent / nonsolvent, 60% nitric acid / butanol; v_p^0 was recalculated according to the definition in this study.

extent with the v_p^0 effect in Fig. 2.43b. Generally speaking, when we intend to compare simulations with experiments using

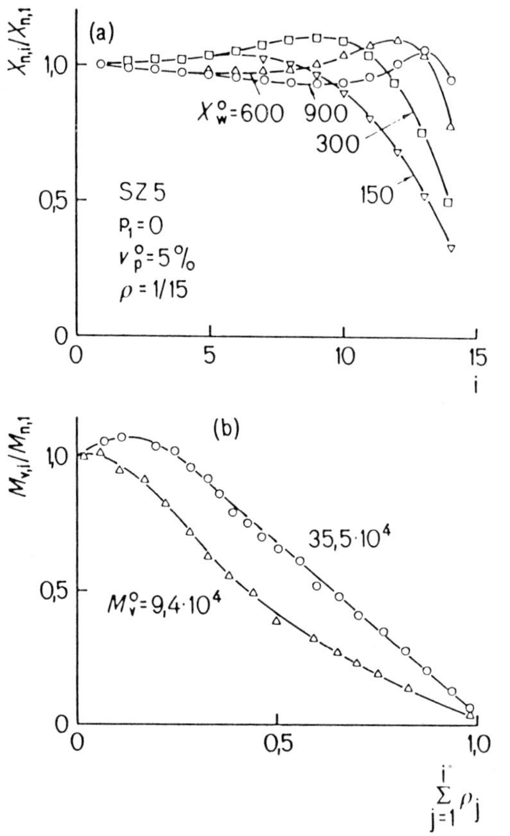

Fig. 2.44 Effect of the average molecular weight of the original polymer on the change in $X_{n,i}/X_{n,1}$ or $M_{v,i}/M_{v,1}$ resp. of the fractions during a successive precipitational fractionation (SPF) run: (a) Simulation; the original polymer, Schulz-Zimm distr., $X_w^0/X_n^0 = 5$, numbers on curves denote X_w^0; $p_1 = p_2 = p_3 = \cdots = 0$, cf. eqn (2.3); initial polymer conc $v_p^0 = 5\%$; amount of fraction $\rho_p = 1/15$: (b) Experiments by Fujisaki and Kobayashi (refs 112,114); the original polymer, whole polyacrylonitrile, numbers on curves denote the viscosity-average molecular weight M_v; $v_p^0 = 0.4\%$; solvent / nonsolvent, 60wt% aq. nitric acid / butanol.

solvent / nonsolvent systems, we cannot avoid encountering the above-mentioned complexities, which make the rigorous and quantitative discussion impossible.

The change in $X_{n,i}/X_{n,1}$ during a SPF run is illustrated in Fig. 2.44a for the case where X_w^0 of the original polymer having Schulz-Zimm distribution ($X_w^0/X_n^0 = 5$) is varied over a wide range.

The peak of the $X_{n,i}/X_{n,1}$ versus i curve shifts to higher i with increasing X_w^0. This means that ROF has a tendency to occur at a later stage of fractionation for the original polymer with the higher molecular weight if the comparison is made for identical operating conditions. This theoretical prediction can be verified by comparing the experimental SPF data on the polyacrylonitrile / aq. nitric acid / butanol system of Fujisaki and Kobayashi (refs 112,114) as shown in Fig. 2.44b.

In the plot of $X_{n,2}/X_{n,1}$ vs. p_1 (not shown here), $X_{n,2}/X_{n,1}$ reveals a maximum at a specific value of p_1 (in this case, $p_1 = -0.2$), suggesting that ROF will occur if p_1 is extremely large or small. This can be explained as follows: With an increase in p_1 the fractionation efficiency becomes higher (ref. 21). Therefore the state in which p_1 is extremely large or small corresponds to one of high or low fractionation efficiency, which does not bring about ROF, because in the former case monodisperse polymers with X_w decreasing consecutively with fractionation order are isolated and in the later case fractions having the same MWD as the original polymer will be separated.

In order to clarify further the role of p_1 in ROF, fifteen equal weight fractions were fractionated by SPF assuming various values of p_1 from a 5.0% solution of the polymer having the Schulz-Zimm distribution ($X_w^0/X_n^0 = 5$ and $X_w^0 = 300$). The plot of $X_{n,i}/X_{n,1}$ against fractionation order i is demonstrated in Fig. 2.45a. As p_1 gets smaller starting from unity, ROF occurs more easily. However, occurrence of ROF again becomes very difficult, if p_1 is highly negative. This suggests that there exists an appropriate range of p_1 values for the ROF phenomena in SPF. In this sense, the specific pair of good and poor solvents, found experimentally by Fujisaki and Kobayashi and suitable for occurrence of ROF, can be recognized. The plot in Fig. 2.45 exhibits a maximum $(X_{n,i}/X_{n,1})_{max}$ at a specific order i, viz. i_{max}. The magnitude of $(X_{n,i}/X_{n,1})_{max}$ has a tendency to have a maximum at a certain p_1 (i.e., in this case, $p_1 \simeq -0.2$), and i_{max} increases continuously with a decrease in p_1. The employment of a solvent having a small p_1 is equivalent with the usage of a high v_p^0 or of a solution of polymer with large X_w^0.

In Fig. 2.45b the ratio of the viscosity-average molecular weight in the i-th and 1st fractions, $M_{v,i}/M_{v,1}$, is plotted against a cumulative weight fraction, as obtained by SPF by using two different binary solvent mixtures (i.e., 60% aq. nitric acid / n-butanol and dimethylsulfoxide / toluene (refs 111,113)). The

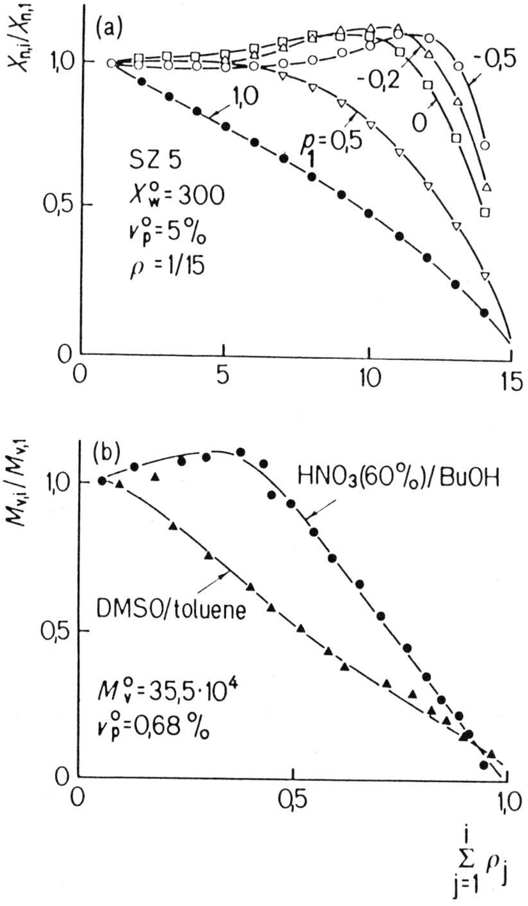

Fig. 2.45 Effect of parameter p_1 (from eqn (2.3)) on change in $X_{n,i}/X_{n,1}$ or $M_{v,i}/M_{v,1}$ of the fractions during a successive precipitational fractionation (SPF) run; $\Sigma \rho_j$ cumulative weight fraction for rel. amount of j-th fraction ρ_j: (a) Simulation; the original polymer, Schulz-Zimm distr., $X_w^0/X_n^0=5$, $X_w^0=300$; initial polymer conc. $v_p^0=5\%$; amount of fraction $\rho_p=1/15$; number on curves denote p_1; $(p_2=p_3=\cdots=0)$: (b) Experiments by Fujisaki and Kobayashi (ref. 113); the original polymer, whole polyacrylonitrile, $M_v=3.55\times10^5$; $v_p^0=0.68\%$; solvent / nonsolvent, 60% aq. nitric acid / butanol (●) and dimethylsulfoxide / toluene(▲).

difference of change in $M_{v,i}/M_{v,1}$ during a given fractionation run using different mixtures can be reasonably interpreted in terms of the difference in solvent nature, expressed by p_1. In this manner, the experimentally unexpected observations on the effects of v_p^0,

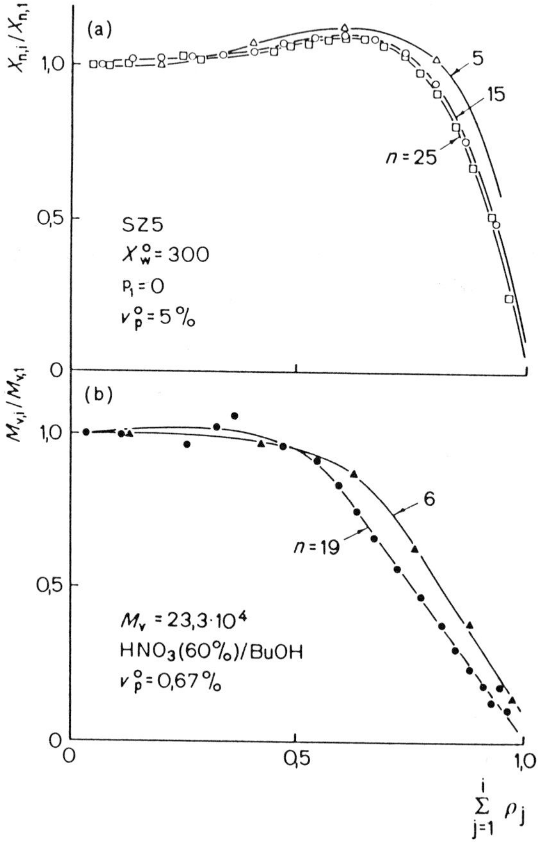

Fig. 2.46 Effect of total number of fractions, n_t, on the change in $X_{n,i}/X_{n,1}$ or $M_{v,i}/M_{v,1}$ of the fractions during a successive precipitation fractionation (SPF) run; $\Sigma\rho_j$ cumulative weight fraction for rel. amount of j-th fraction ρ_j: (a) Simulation; the original polymer, Schulz-Zimm distr., $X_w{}^0/X_n{}^0 = 5$, $X_w{}^0 = 300$; initial polymer conc. $v_p{}^0 = 5\%$; $p_1 = p_2 = p_3 = \cdots = 0$, cf. eqn (2.3); numbers on curves denote n_t: (b) Experiments by Fujisaki and Kobayashi (ref. 112); the original polymer, unfractionated polyacrylonitrile, $M_v = 2.33 \times 10^5$; $v_p{}^0 = 0.6\%$; solvent / nonsolvent, 60wt% aq. nitric acid / butanol; n = 6 (▲) and 19 (●).

$X_w{}^0$ and p_1 on ROF, found in experimental SPF can be well-predicted on the basis of the rigorous fractionation theory. The fractionation efficiency of SPF is very sensitive to p_1 and therefore, it is suggested that ROF will occur favorably if the fractionation efficiency is intermediate between extremely high and low (ref. 21).

The effect of the total number of fractions n_t on the relationship between $X_{n,i}/X_{n,1}$ and cumulative weight fraction $\Sigma \rho_j$ (where ρ_j is the relative amount of the j-th fraction) is demonstrated in Fig. 2.46a, where a 5% solution of the polymer having a Schulz-Zimm distribution ($X_w^0 = 300$, $X_n^0 = 5$) was precipitationally fractionated into 5 to 25 equal fractions. As n_t increases $X_{n,i}/X_{n,1}$ at the later stages decreases significantly, and in consequence, $(X_{n,i}/X_{n,1})_{max}$ has a tendency to become small. These features were also observed by Fujisaki and Kobayashi (ref. 112) in parctical SPF of polyacrylonitrile ($M_v = 23.3 \times 10^4$) solutions in 60% aq. nitric acid / butanol mixture as shown in Fig. 2.46b.

2.4.4 Spencer's Method

In 1948 Spencer (ref. 116) proposed a method of fractionation into summative fractions. A schematic representation is given in Fig. 2.47. Spencer's method is, in principle, an approximate method, based on the assumption that all polymer molecules above a certain degree of polymerization X(S) remain in the polymer-rich phase and all molecules below X(S) remain in the polymer-lean phase, where (S) denotes a given condition of fractionation. The

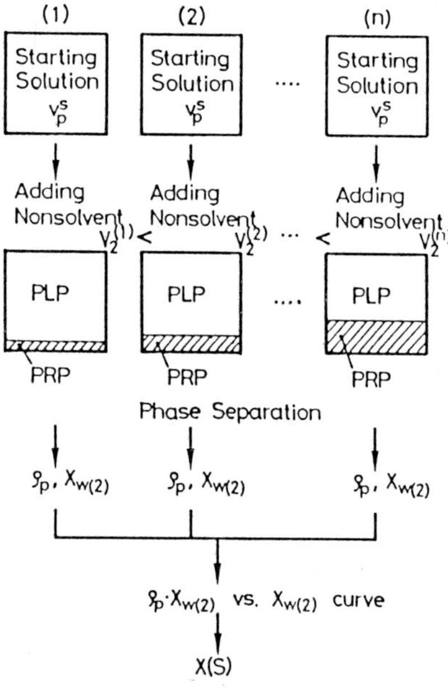

Fig. 2.47 Schematic representation of Spencer's method.

procedures used in this method are, in short, as follows: By measuring the weight fraction of the precipitate, ρ (S) and the weight-average X_i, $X_{w(2)}$ (S) of the precipitated polymer, one may obtain the value of X(S) from the slope of the plot of ρ (S)$X_{w(2)}$(S) against ρ (S). Finally, the plot of $1-\rho$ (S) as a function of X(S) affords directly the integral X_i distribution.

Until now Spencer's method has been examined by Billmeyer and Stockmayer (ref. 117), Matsumoto (ref. 118), Broda et al. (refs 119,120), Okamoto (refs 121,122) and Kamide et al. (ref. 89) in more detail. Kamide et al. pointed out that a rigorous examination can not be achieved without comparing the integral distribution elucidated from the fractionation data with the true distribution. The hypothetical fractionations carried out by Billmeyer and Stockmayer (ref. 117) and Matsumoto (ref. 118) involved approximations which are impractical and not always correct for the sake of simplifying the computation.

Billmeyer and Stockmayer (ref. 117) established, for polymers having several types of distribution, the relationships between ρ (S) and the parameter Z, representing the breadth in the X_i distribution in the precipitate. This parameter Z is given by;

$$Z=\rho(S)(X_{w(2)}(S)-X_w^0)/X_w^0 \qquad (2.87)$$

where X_w^0 is X_w of the initial polymer. In this case, they used the assumption regarding X(S) as described above. Using this parameter, they concluded that Spencer's method can not be expected to disclose any fine details of distribution. However, it is theoretically apparent that the reliability of the integral distribution constructed by Spencer's method can be improved, without limit, by increasing the number of fractions, if the assumption with respect to X(S) is adopted. In addition, as Matsumoto (ref. 118) was the first to indicate, Z is not sufficiently sensitive to express the MWD as expected.

On the basis of the Flory and Huggins theory, the hypothetical fractionation was carried out by Billmeyer and Stockmayer from a dilute solution of a polymer with the exponential MWD at a constant volume ratio R. In their calculations, the partition coefficient σ was assumed to approach zero in the precipitation from an extremely dilute solution. This assumption cannot generally be accepted, since the hypothetical fractionation very accurately computed by Kamide et al. (ref. 1) indicated that for a constant ρ, σ increases with the decreasing initial

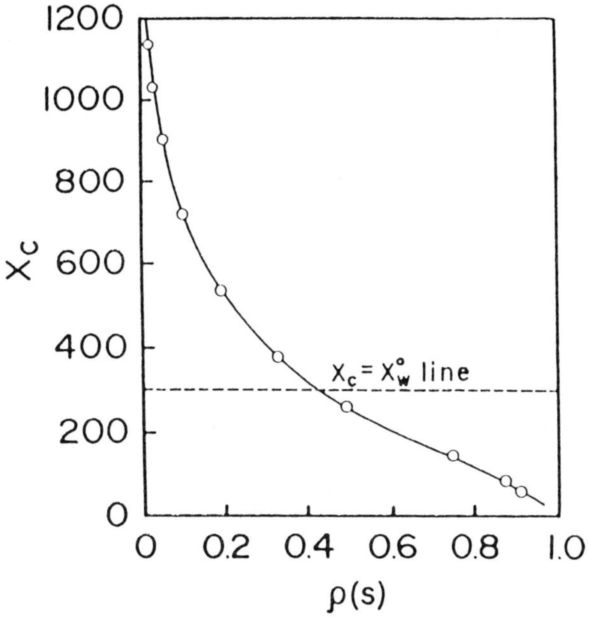

Fig. 2.48 X_c plotted against the amount of fraction ρ (S) (ref. 2). The original polymer, Schulz-Zimm distr., $X_w/X_n=2$, $X_w=300$; $p_j=0$ ($j=1,2,\cdots$), $k'=0$.

concentration $v_p{}^0$. Moreover, Billmeyer and Stockmayer (ref. 117) calculated the value of σ for a constant R, assuming $X_c = X_w{}^0$, where X_c is the value of X at which the proportion of a given X-mer partitioning in the polymer-rich phase, f_X is 0.5. In Fig. 2.48, X_c is plotted against ρ (S) from the data of the fractionation by Kamide et al. (ref. 2). It can be seen that X_c does not necessarily coincide with $X_w{}^0$, but is dependent on the value of ρ (S).

The hypothetical fractionation was attempted by Matsumoto (ref. 118) on the basis of the theory of Flory and Huggins for the heterogeneous polymers of various types of distribution, assuming again a constant R. The fractionation data obtained was treated by the Spencer's method and compared with the true distribution. He came to the conclusion that Spencer's method would give a fairly reliable distribution if the experimental error is sufficiently small. His conclusion is thus contrary to that of Billmeyer and Stockmayer (ref. 117).

It seems very difficult to compare the results calculated by Billmeyer and Stockmayer, and Matsumoto with experimental data, since the fractionation at a constant R can be performed only by changing the initial concentration in a very complicated manner. Thus, it is more desirable to carry out the hypothetical fractionation under the condition that the initial concentration is constant.

According to Okamoto (refs 121,122) ρ (S) is given by the relation;

$$\rho(S) = \int_{X(S)}^{\infty} g(X)\,dX \qquad (2.88)$$

with

$$X(S) = \frac{d\{\rho(S)X_{w(2)}(S)\}/d\rho(S)}{1+\frac{\pi}{6}\left\{\frac{2}{(\ln R(S))^2} - \frac{4}{(\ln R(S))^3}\cdot\frac{d\ln R(S)}{d\ln[d\{\rho(S)X_{w(2)}(S)\}/d\rho(S)]}\right\}} \qquad (2.89)$$

where g(X) represents the normalized differential weight X_i distribution of the initial polymer. Eqn (2.89) was derived from the Flory and Huggins theory using some approximations and is valid under the condition that the order of the magnitude of $|X^n d^n g(X)/dX^n|$ is comparable to or smaller than g(X). The value of X(S) corresponding to a given value of ρ (S) can be calculated precisely from eqns (2.88) and (2.89). Hence, the integral distribution can easily be obtained from the plot of $(1-\rho(S))$ vs. X(S). In this revised version of Spencer's method, an exact value of X(S) can be determined if the data on ρ (S), R(S) and $X_{w(2)}(S)$ is available. The validity of this method has not been confirmed either by the hypothetical fractionations or by experiments.

Kamide et al. (ref. 89) attempted to examine the characteristic features of Spencer's method and Okamoto's method using the hypothetical data of fractionation obtained by an electronic computer for a constant initial concentration. For this purpose, a 1.0% solution of heterogeneous polymers having the most probable distribution with $X_n{}^0 = 150$ in a single solvent was assumed to be cooled. The amount of precipitates ranged from 0.0213 to 0.9182 to obtain eleven fractions.

Fig. 2.49 includes both the plot of ρ (S) against $\rho(S)X_{w(2)}(S)$ and the plot of $(1-\rho(S))$ vs. $(1-\rho(S))X_{w(1)}(S)$. The similar plots in which $X_{w(2)}'(S)$ and $X_{w(1)}'(S)$ were used in place of $X_{w(2)}(S)$ and $X_{w(1)}(S)$, respectively, are shown as the

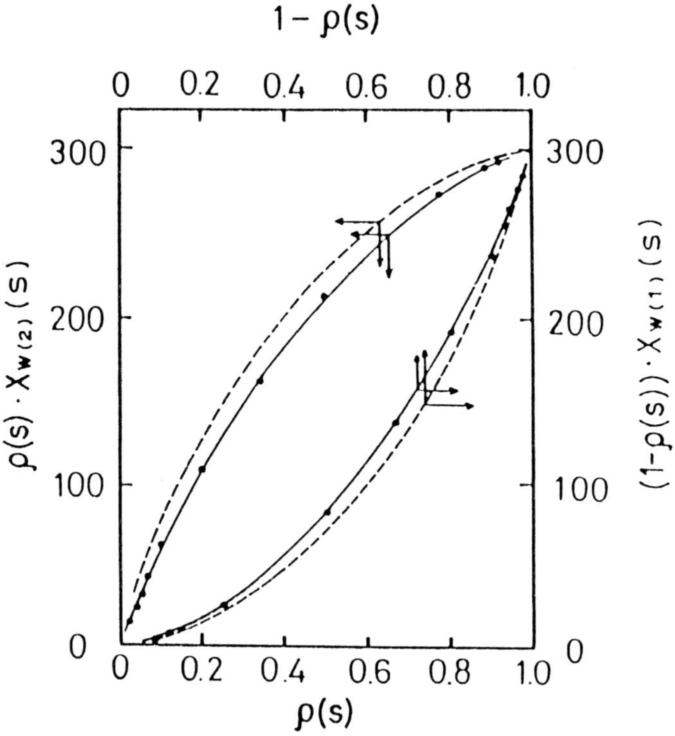

Fig. 2.49 Plot of $\rho(S)X_{w(2)}(S)$ against $\rho(S)$ and plot of $(1-\rho(S))X_{w(1)}$ against $(1-\rho(S))$ (full curves). Similar plots, where $X_{w(2)}'(S)$ and $X_{w(1)}'(S)$ were used in place of $X_{w(2)}(S)$ and $X_{w(1)}(S)$, respectively, are included as dashed curves.

dashed curves in the same figure. In this case, $X_{w(2)}'(S)$ and $X_{w(1)}'(S)$ are given by:

$$\rho(S)X_{w(2)}'(S) = \int_{X(S)}^{\infty} Xg(X)\,dX, \qquad (2.90)$$

$$\rho(S)X_{w(2)}'(S) + (1-\rho(S))X_{w(1)}'(S) = 1. \qquad (2.91)$$

The value of X(S) in eqn (2.90) was calculated using eqn (2.88).

The significant discrepancy between the full curves and the dotted curves in Fig. 2.49 may be mainly attributable to the invalidity of Spencer's assumption on the existence of X(S) described before. It is worth noting, however, that this does not necessarily mean the inapplicability of the method. In fact, a similar situation arises in applying Schulz's method (ref. 77) to the simple successive fractionation.

Another parameter X'(S) defined by

$$\rho(S)X_{w(2)}(S) = \int_{X'(S)}^{\infty} Xg(X)\,dX \tag{2.92}$$

is not generally in agreement with X(S) as was pointed out first by Matsumoto (ref. 118). Table 2.11 illustrates the values of X(S) and X'(S) for various values of ρ (S). Values of X'(S) are always larger than those of X(S) and the difference between them becomes increasingly large as the amount of the precipitate is decreased. It is clear from the above discussions that this disparity of X'(S) from X(S) is most likely the curve for the error involved in Spencer's method. The integral weight distribution obtained by Spencer's method is thus predicted to deviate from the true MWD especially in the region of smaller ρ (S), accordingly, for larger X.

TABLE 2.11
Calculated values of X(S), X'(S) and the supplementary term in eqn (2.89).

ρ (S)	X(S)	X'(S)	The supplementary term in eqn (2.89)
0.9182	70	75	0.302
0.8851	66	92	0.283
0.7557	104	151	0.268
0.4992	167	275	0.159
0.3326	206	376	0.071
0.1981	239	494	−0.054
0.1021	266	640	−0.096
0.0663	275	722	—
0.0519	280	774	−0.106
0.0335	286	861	—
0.0213	290	945	−0.578

Values of X(S) were determined from the slope of the plot of ρ (S) $X_{w(2)}$(S) against ρ (S) given in Fig. 2.49, and plotted as a function of (1 − ρ (S)) in Fig. 2.50. For the sake of comparison, Fig. 2.50 also includes the true MWD (i.e., the most probable distribution) and the integral MWD evaluated by Schulz's method from the SPF data of Kamide et al. (ref. 2). In the simple SPF

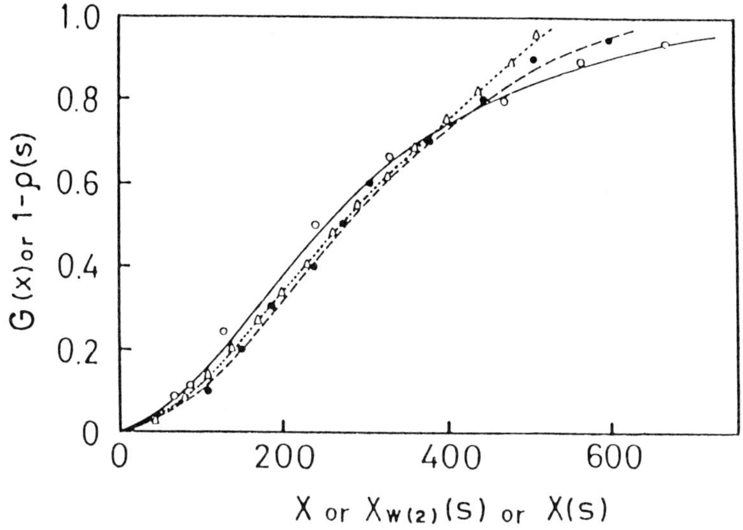

Fig. 2.50 The integral MWD in the initial polymer. Full line, the true distribution (i.e., the Schulz-Zimm distr. $X_w/X_n = 2.0$, $X_w = 300$) (plot of G(X) against X, where $G(X) = \int_0^X g(X)dX$). Open circle, Okamoto's method (plot of circle, Okamoto's method (plot of $(1-\rho(S))$ against X(S)); closed circle, Spencer's method (plot of $(1-\rho(S))$ against $X_{w(2)}(S)$); open triangle, Schulz's method applied on the fractionation data by the simple successive precipitation method (the initial concentration, 1.0%; the total number of fractions, 15 in equal weight).

data cited here, the MWD in the initial polymer and the initial concentration of the solution were entirely the same as those used here and an equal amount (1/15) of fifteen fractions were separated. This kind of fractionation is the typical successive fractionation usually used and the comparison with Spencer's method would be of interest.

Close inspection of Fig. 2.50 leads to the conclusion that Spencer's method has a tendency to underestimate appreciably the amount of the molecular species having smaller and larger X. As compared with Schulz's method, however, Spencer's method furnishes a more reliable distribution in the larger X_i region.

Also in Fig. 2.50 is shown the integral distribution evaluated by Okamoto's method using the data of ρ (S), R(S) and $X_{w(2)}$ (S) obtained by the computer experiments (the original polymer, the most probable MWD with $X_n^0 = 150$; $v_p^0 = 0.01$; ρ (S) $= 0.9182 \sim 0.0213$). The second supplementary term in the denominator in the right-hand side of eqn (2.89) is to be added to the terms given in Spencer's method. As is shown in the fourth column in Table 2.11, the contribution of this term to eqn (2.89) can not be ignored when ρ (S) is near zero or unity. It was also confirmed that the condition necessary for eqn (2.89) to be valid is completely satisfied. Thus, Okamoto's treatment is proved useful for determining the integral MWD with a fairly good accuracy.

Summarizing, the artificial assumption of the existence of the value of X(S) made in Spencer's method can not be justified by the modern solution theory. Nevertheless, Spencer's method offers a convenient and simple technique for determining the integral distribution with fairly good accuracy. In some cases Spencer's method would be more advantageous than Schulz's method (see, eqns (2.93) and (2.94)) applied to the simple SPF data (ref. 89). When the volume ratio R(S) of the polymer-lean phase to the polymer-rich phase is measured together with $X_{w(2)}$(S) and ρ (S), Okamoto's method can be applied and gives the most reliable integral X_i distribution (ref. 89).

2.4.5 Analytical Treatment of Fractionation Data

The molecular weight fractionation by solubility difference has been extensively employed for many years as a principal method for the evaluation of the molecular weight distribution (MWD) of macromolecules, since Schulz's first paper (ref. 123) in 1936 (see, ref. 124 in detail).

Due to the rapid development of gel permeation chromatography (GPC) [for example, separation of dextran by use of dextran gel (1957) (ref. 125), polystyrene (PS) by use of PS soft type gel (1960) (ref. 126) and PS by use of PS hard type gel (1964) (ref. 127)], the frequency of utilization of the fractionation method by the solubility difference has decreased recently as a means of in-plant process control. An exact evaluation of MWD by GPC is possible only in two cases when (1) nearly monodisperse and well-characterized polymer samples are available as standard materials for calibrating the GPC and (2) the hydrodynamic volume universal calibration plot is strictly justified for the polymer type of interest (ref. 47).

Studies (refs 128-132) on the application limit of this universal plot have revealed that the plot was not universal when a second separation mechanism exists in addition to steric exclusion separation and when polymer samples have different molecular geometry and rigidity. It seems sufficient to note that the efficiency of GPC fractionation can not be thoroughly studied without the aid of a complete theory, which has not yet been established (ref. 47).

Fractionation by solubility difference has several merits such as (1) inexpensive equipment, (2) wide choice of solvents, (3) relatively small amount of solvent needed to obtain the samples (as compared with GPC and column fractionation) and (4) definite theoretical background (ref. 47). Some people hold the opinion that analytical use of fractionation by solubility difference is now out of date and meaningless. However, the importance of analytical successive fractionation by solubility difference should never diminish as is evident from the above discussion and even in the future this classical method has, we believe, a good chance of continued use in a research laboratory for analytical purposes (ref. 47).

Using the fractionation data (ρ_p (or ρ_s) and X_w) the MWD of the original polymer can be evaluated by means of the method devised by Schulz (ref. 46) and also that by Kamide et al. (refs 1,47), proposed on the basis of the modern fractionation theory. The principle of these methods can be briefly summarized as follows:

(1) Method of Schulz

The cumulative weight fraction, $I(X_{w,j})$, corresponding to $X_{w,j}$ of the j-th fraction is given by

$$I(X_{w,j}) = \sum_{k=j+1}^{n_t} \sum_{X_i=0}^{\infty} g_k(X_i) + \frac{1}{2} \sum_{X_i=0}^{\infty} g_j(X_i) \quad \text{for SPF,} \qquad (2.93)$$

$$I(X_{w,j}) = \sum_{k=1}^{j-1} \sum_{X_i=0}^{\infty} g_k(X_i) + \frac{1}{2} \sum_{X_i=0}^{\infty} g_j(X_i) \quad \text{for SSF,} \qquad (2.94)$$

where $\sum g_k(X_i)$ denotes the weight fraction of the k-th fraction. If we use X_w for average X_i, the low molecular weight region is also underestimated. As is evident from eqn (2.93) and (2.94), Schulz method assumes that MWD of j-th fraction exists between $X_{w,j-1}$ and $X_{w,j+1}$ (ref. 1). But the true MWD of the j-th fraction does not satisfy this assumption (see, for example, Fig. 2.22).

According to the results of the SPF simulation by Tung (ref. 28), the Schulz method gives fairly good coincidence with the true

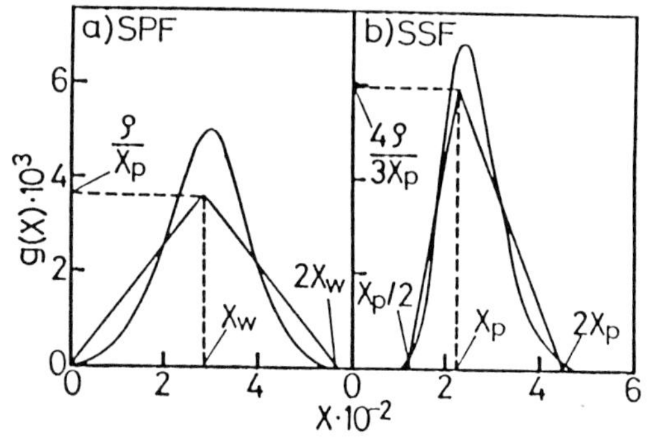

Fig. 2.51 Schematic representation of triangle-shaped molecular weight distribution (MWD): Full line, the true MWD of fraction; a) successive precipitation fractionation (SPF); b) successive solution fractionation (SSF).

distribution (see, Fig. 4 of ref. 28). Strictly speaking, as is shown in Fig. 4 of ref. 36, the MWD calculated by using the Schulz method underestimates the low and high molecular weight regions (ref. 1).

(2) Method of Kamide et al. for SPF and SSF

(a) For SPF, MWD of the fraction separated by SPF can be represented with fairly good accuracy by the equilateral triangle having its peak at X_w in the range $0 \leq X_w \leq 2X_w$ (Fig. 2.51a). The distribution function of the j-th fraction, $g_j(X_i)$ can be expressed as (ref. 1)

$$g_j(X_i) = \frac{\left(\sum_{X_i=0}^{\infty} g_j(X_i)\right) X_i}{X_{w,j}^{\,2}} \qquad \text{for } 0 \leq X_i \leq X_{w,j} \qquad (2.95)$$

and

$$g_j(X_i) = \frac{\left(\sum_{X_i=0}^{\infty} g_j(X_i)\right)(2X_{w,j} - X_i)}{X_{w,j}^{\,2}} \qquad \text{for } X_{w,j} \leq X_i \leq 2X_{w,j} \qquad (2.96)$$

and

$$g_j(X_i) = 0 \qquad \text{for } X_i \geq 2X_{w,j}. \qquad (2.97)$$

Here, $\sum_{X_i=0}^{\infty} g_j(X_i)$ denotes the weight fraction of the j-th fraction to the total weight of polymer, ρ_j. The summation of $g_j(X_i)$ ($\equiv \sum_{j=1}^{n_t} g_j(X_i)$) as given by eqns (2.95)-(2.97) with respect to X_i provides the normalized differential MWD, $g_e(X_i)$, of the original polymer estimated from the fractionation data and its integration yields the cumulative distribution curve, $I_e(X_i)$ (ref. 47).

(b) For SSF an improvement of the original method of Kamide et al., for the case of SPF can be made by introducing modified triangle distribution, ranging from $(1/2)X_p$ to $2X_p$, where X_p is the value of X_i at peak and the triangle has the value of X_w identical with that of the fraction separated by SSF. After a simple calculation, not given here, it can readily be shown that X_p is related to X_w (ref. 22):

$$X_p = \frac{6}{7} X_w . \tag{2.98}$$

The modified triangle distribution is illustrated in Fig. 2.51b. Fig. 2.52 shows the examples of triangle distribution of SPF and SSF.

The triangle has the value of area identical with that of the

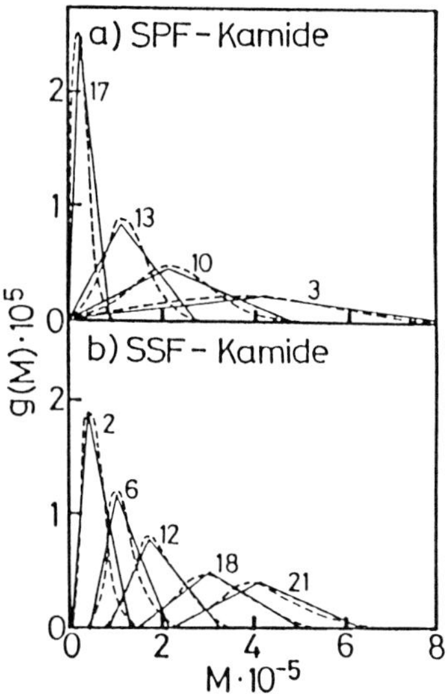

Fig. 2.52 Comparison of the molecular weight distribution (MWD) determined by gel permeation chromatography with triangle-shaped distribution employed in Kamide's method: a) SPF-Kamide; b) SSF-Kamide.

fraction separated (refs 6,9).

In this case, the distribution function of the j-th fraction, $g_j(X_i)$ is given by eqns (2.99)-(2.101):

$$g_j(X_i) = \frac{4(\sum_{X_i=0}^{\infty} g_j(X_i))}{3X_p^2}(2X_i - X_p) \qquad \text{for } X_p/2 < X_i < X_p, \qquad (2.99)$$

$$g_j(X_i) = \frac{4(\sum_{X_i=0}^{\infty} g_j(X_i))}{3X_p^2}(2X_p - X_i) \qquad \text{for } X_p < X_i < 2X_p, \qquad (2.100)$$

$$g_j(X_i) = 0 \qquad \text{for } 0 < X_i < Xp \text{ and } X_i > 2X_p. \qquad (2.101)$$

A parameter E defined by eqn (2.102) is utilized as a measure of the accuracy of the method for estimating the MWD of the original polymer:

$$E = \int |g_0(X) - g_e(X)| dX. \qquad (2.102)$$

Here, $g_0(X)$ is the normalized true differential MWD of the original polymer and $g_e(X)$, the differential MWD curve estimated by the Schulz's method or the Kamide's method.

For convenience, a combination of the fractionation procedure

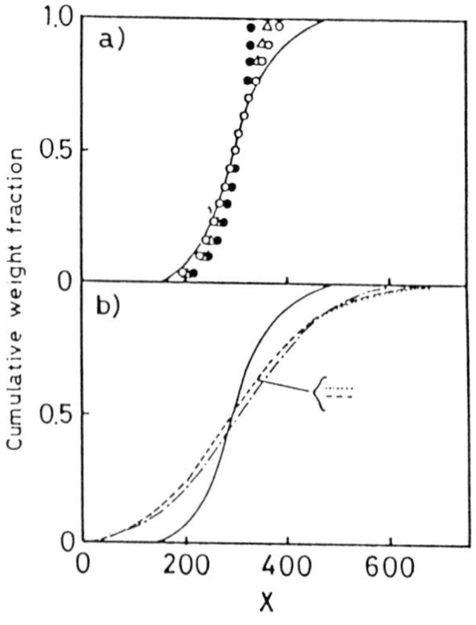

Fig. 2.53 Cumulative weight distribution constructed by the Schulz's method (a) and Kamide's method (b): The original polymer, Schulz-Zimm type distribution ($X_w^0/X_n^0 = 1.05$, $X_w^0 = 300$); $p_1 = 0$, $\rho = 1/15$: $v_p^0 = 1 \times 10^{-2}$ (●, -·-·-); 0.1×10^{-2} (△, -----); 0.01×10^{-2} (○, ·····); true distribution curve (——).

(SPF or SSF) and the analytical method (Schulz or Kamide) will be referred to hereafter as follows; SPF/Schulz (or SSF/Kamide) means that the fractionation data was obtained by SPF (or SSF) and the data was treated by the Schulz's (or Kamide's) method.

The MWD curves of the original polymer, constructed by using the Schulz's method and the Kamide's method from the fractionation data thus obtained, are compared with the true MWD in Fig. 2.53. Inspection of Fig. 2.53 shows that the Schulz's method yields a narrower distribution as compared with the true distribution and that it underestimates the higher region of degree of polymerization X grossly (ref. 21). $I(X)=0.99$ is estimated to be 340 at $v_p{}^0=1\%$ and 410 at $v_p{}^0=0.01\%$ (ref. 93).

Fig. 2.54 shows the relationships between $v_p{}^0$, n_t and E for a polymer with SZ type distribution ($X_w{}^0=300$ and $X_w{}^0/X_n{}^0=2$) (ref. 47). The curves in Fig. 2.54 denote contour lines, relating $v_p{}^0$ to n_t, for a fixed value of E.

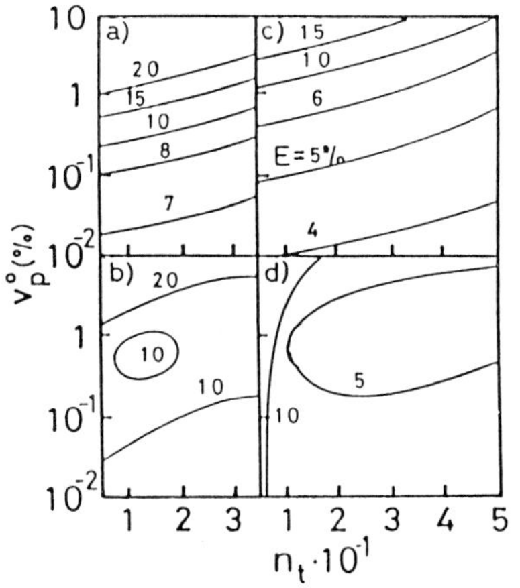

Fig. 2.54 Correlation between the relative error E, $v_p{}^0$ and n_t: The original polymer, Schulz-Zimm type distribution ($X_w{}^0=300$, $X_w{}^0/X_n{}^0=2$); $p_1=0$: a) SPF/Schulz; b) SPF/Kamide; c) SSF/Schulz; d) SSF/Kamide.

The relative error E obtained by Schulz's method approaches closely to zero if $v_p{}^0\rightarrow 0$ and $n_t\rightarrow\infty$ are concurrently realized. In contrast to this, E exhibits a remarkable minimum E_{min} if the Kamide method is employed (in this case, $E_{min}=10\%$ for SPF/Kamide and $E_{min}=4.8\%$ for SSF/Kamide, respectively) (ref. 47). By use of a figure like Fig. 2.54 we can optimize the fractionation procedure for analytical purposes for any polymer sample.

MWD curves of the PS fractions isolated by SPF and SSF from a

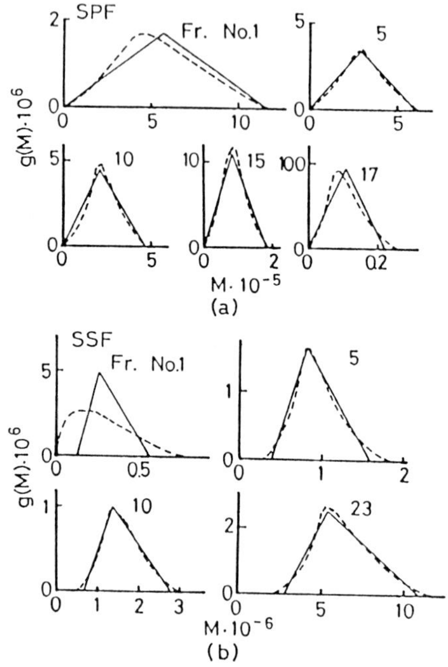

Fig. 2.55 Comparison of the molecular weight distribution (MWD) determined by GPC with the triangle-shaped distribution employed in the Kamide's method for atactic polystyrene fractions; a) successive precipitation fractionation (SPF), $v_p^0 = 0.0094$, $n_t = 18$; b) successive solution fractionation (SSF), $v_p^0 = 0.0094$, $n_t = 23$.

0.94% solution in MCH were determined by GPC. These were compared with the triangle distribution, which was approximated by Kamide et al., as exemplified in Fig. 2.55 (ref. 47). Except for very low order fractions, the triangle distribution employed by Kamide et al., proved successful. Therefore, Fig. 2.55 directly supports the validity of the approximations in Kamide's method.

The maximum and minimum values of the molecular weight, M_{max} and M_{min}, in the distribution for the PS fractions obtained by SPF and SSF were determined conventionally as the values at which the magnitude of $g_e(M)$ of a fraction is equal to 4×10^{-8}. In Fig. 2.56 the plot of M_{min}/M_w and M_{max}/M_w is shown. In this figure the results of the computer experiment for this polymer / solvent system are shown. The agreement shown by M_{min}/M_w and M_{max}/M_w between experiment and theory is fairly satisfactory (ref. 47).

The simplest way of determining X_w and X_n of the whole polymer from the fractionation data is given by;

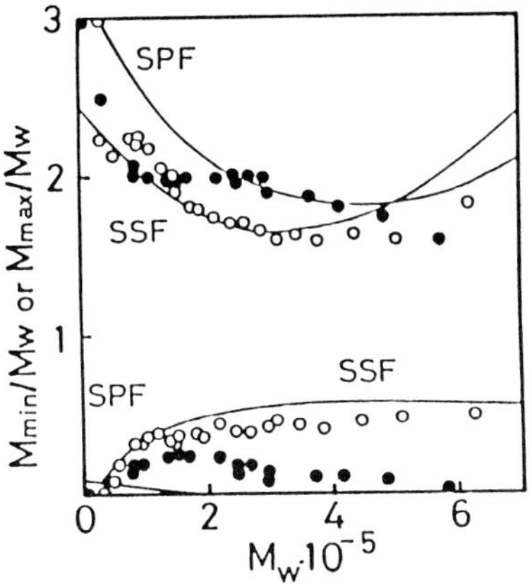

Fig. 2.56 The ratio of M_{max}/M_w (and M_{min}/M_w) plotted as a function of M_w for atactic polystyrene fractions separated by successive solution fractionation (SSF, ○) and successive precipitation fractionation (SPF, ●) experiments: The original polymer, $M_w = 23.9 \times 10^4$ and $M_w/M_n = 2.77$; solvent, methylcyclohexane; $v_p^0 = 0.0094$; $n_t = 23$ in SSF and 18 in SPF: Full line denotes the theoretical curves ($p_1 = 0.7$).

$$X_w = \sum_j \rho_j X_{w,j} \tag{2.103}$$

or

$$X_w = \sum_j \rho_j X_{n,j} \tag{2.103'}$$

and

$$X_n = 1/ \sum_j \frac{\rho_j}{X_{w,j}} \tag{2.104}$$

or

$$X_n = 1/ \sum_j \frac{\rho_j}{X_{n,j}} \tag{2.104'}$$

where ρ_j, $X_{w,j}$ and $X_{n,j}$ are the relative amounts ρ, X_w and X_n of the j-th fraction, respectively. In deriving eqns (2.103) and (2.104) the polydispersity of the fractions is neglected. In consequence, by use of a kind of average degree of polymerization ($X_{w,j}$ or $X_{n,j}$) one cannot calculate both X_w and X_n accurately (ref. 47).

In the strict sense, X_w and X_n of the whole polymer should be calculated from:

$$X_w = \int X\, g_e(X)\, dX, \tag{2.105}$$

$$X_n = 1/\int \frac{g_e(X)}{X} dX. \tag{2.106}$$

The values of M_n, M_w and E, calculated from the actual and hypothetical fractionation data by SPF and SSF on atactic PS in MCH, are compiled in Table 2.12. In this table, E' of the experiment is conveniently defined by:

$$E' = \int |g_G(X) - g_e(X)| dX. \qquad (2.107)$$

Here, $g_G(X)$ is the differential MWD obtained by GPC.

It should be stressed that in the case of computer simulated data the relative error E in eqn (2.102) is the unreducible error inherent in the fractionation procedure/analytical method. In the case of the experimental results and additional error (accompanying the practical determination of X_w and X_n) should be added to the above-mentioned intrinsic error. Moreover, in the hypothetical fractionations it was assumed that equal amounts of the fractions were separated. Then, a variation of the fraction size brings about another kind of error to the value of X_w/X_n. Consequently, the large experimental error makes a detailed comparison with the theory impossible in practice.

The value of M_n obtained from the fractionation data by means of eqn (2.104), using the Schulz's method and the Kamide's method has, even from the theoretical point of view, a large uncertainty compared with the value of M_w and is always appreciably larger than the true value (in this case, 8.6×10^4). The Kamide's method is in principle superior to the Schulz's method for the determination of M_n, because the former takes better account of "the tailing effect" (ref. 47). The hypothetical fractionation data indicates that the Kamide's method without exception affords a more accurate value of M_w/M_n than the Schulz's method does (ref. 47).

The operating conditions employed here, are appropriate for preparative fractionation, since the majority of the fractions isolated by SSF have a polydispersity of $M_w/M_n \leq 1.10$ (for example, 9(8) and 19(20) fractions have $M_w/M_n \leq 1.10$ at $n_t = 14$ and 23 theoretically (experimentally)). As is shown from Fig. 2.57a, b, c and d, where I_e and I_0 or I_G are the cumulative MWD evaluated from the hypothetical fractionation data and that of the original polymer or the cumulative MWD as determined by GPC, the experiment (Fig. 2.57a and c) is in satisfactory agreement with the theory (Fig. 2.57b and d) (ref. 47). SPF/Schulz (Fig. 2.57a and b) is shown to be rather less successful even for estimating the MWD curve of the whole polymer. In contrast to this, analytical SSF (Fig. 2.57c and d) is very suitable for evaluating the total MWD

TABLE 2.12

Value of M_n, M_w and E evaluated by the actual experiment on polystyrene / methylcyclohexane system and the computer simulation with $p_1 = 0.7$ (theory).

Fractionation Procedure	Operating v_p^0 %	Conditions n_t	Analytical method	$M_n \times 10^{-4}$ Exp.	$M_n \times 10^{-4}$ Theory	$M_w \times 10^{-4}$ Exp.	$M_w \times 10^{-4}$ Theory	M_w/M_n Exp.	M_w/M_n Theory	E'[a] Exp.	E[a] Theory
	0.94	13	eqns (2.103) and (2.104)	15.8	11.8	24.7	23.1	1.61	1.96	—	—
	0.94	13	eqns (2.103') and (2.104')	12.1	8.8	17.4	17.4	1.44	1.98	—	—
SPF	0.94	13	Schulz	12.8	11.4	22.6	23.4	1.76	2.03	0.252	0.078
	0.94	13	Kamide	10.9	8.4	24.6	23.2	2.26	2.76	0.181	0.078
	0.94	18	eqns (2.103) and (2.104)	12.5	11.3	23.1	23.1	1.85	2.04	—	—
	0.94	18	eqns (2.103') and (2.104')	10.0	8.8	18.0	17.9	1.80	2.02	—	—
	0.94	18	Schulz	13.1	10.8	23.8	23.7	1.81	2.20	0.222	0.156
	0.94	18	Kamide	9.5	8.7	23.1	24.1	2.43	2.77	0.123	0.103
	0.94	14	eqns (2.103) and (2.104)	10.4	11.5	21.7	23.1	2.10	2.00	—	—
	0.94	14	eqns (2.103') and (2.104')	8.1	8.8	19.9	21.5	2.44	2.44	—	—
SSF	0.94	14	Schulz	10.6	10.6	24.8	23.1	2.34	2.16	0.091	0.068
	0.94	14	Kamide	8.9	9.5	21.9	23.0	2.46	2.41	0.095	0.068
	0.94	23	eqns (2.103) and (2.104)	11.7	10.9	22.1	23.0	1.88	2.11	—	—
	0.94	23	eqns (2.103') and (2.104')	8.9	8.8	20.0	21.9	2.24	2.49	—	—
	0.94	23	Schulz	11.6	10.9	23.8	23.8	2.05	2.17	0.163	0.051
	0.94	23	Kamide	9.7	10.2	22.1	23.1	2.28	2.26	0.102	0.069
GPC	—	—	—	8.6 (8.9)[b]	—	23.9 (23.2)[c]	—	2.27	—	—	—

a) E' and E were defined by eqns (2.107) and (2.102), b) Membrane osmometry, c) Light scattering.

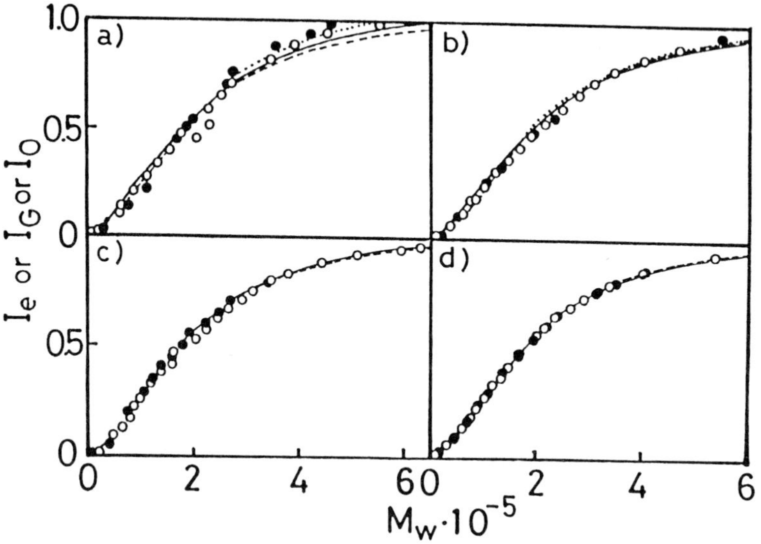

Fig. 2.57 Cumulative weight fraction constructed from the analytical fractionation data (I_e) as compared with the true cumulative weight fraction (I_0) and that estimated by GPC (I_G): a) Successive precipitation fractionation (SPF), experiment; I_G, -----; the Kamide's method ($n_t = 18$), ·····; the Kamide's method ($n_t = 13$), ——; ●, the Schulz's method ($n_t = 13$); ○, the Schulz's method ($n_t = 18$): b) SPF, theory ($p_1 = 0.7$); -----, I_0; other marks have the same meaning as those in Fig. 2.53a: c) Successive solution fractionation (SSF), experiment; I_G, -----; the Kamide's method ($n_t = 14$), ·····; the Kamide's method ($n_t = 23$), ——; ●, the Schulz's method ($n_t = 14$); ○, the Schulz's method ($n_t = 23$): d) SSF, theory ($p_1 = 0.7$); -----, I_0; other marks have the same meaning as those in Fig. 2.53c.

profile (ref. 47). The theoretical calculation indicates that the operating conditions used in Table 2.13 do not fulfill the requirement of $E < 5\%$. Therefore, in order to estimate accurately the M_w/M_n of the whole polymer, much more serious operating conditions should be chosen carefully.

This prediction was accurately examined by computer simulation. Plots of $(M_w/M_n)_t/(M_w/M_n)_0$ vs. v_p^0, thus obtained, are shown in Fig. 2.58, where $(M_w/M_n)_t$ is the M_w/M_n value for the original polymer, evaluated from the SSF data by use of the Schulz's or the Kamide's method and $(M_w/M_n)_0$ is the M_w/M_n value for the original polymer, assumed in advance for simulation (ref. 47). Obviously, as v_p^0 gets small $(M_w/M_n)_t/(M_w/M_n)_0$ approaches unity (ref. 47). The Schulz's method has a tendency to give smaller $(M_w/M_n)_t/(M_w/M_n)_0$ than the Kamide's method, as was found

TABLE 2.13
Parameter used in the computer simulation experiment.

Experiment No	Molecular wt. distribution	$X_w \times 10^{-4}$	X_w/X_n	$v_p{}^0$	n_t	p_1
PS Run 1	Schulz-Zimm	1810	2.8	0.0094	14	0.7
PS Run 2	Schulz-Zimm	1810	2.8	0.0094	20	0.7
PS Run 3	Schulz-Zimm	1810	2.8	0.0094	23	0.7
PS Run 4	Schulz-Zimm	1810	2.8	0.0189	30	0.7
PS Run 5	Schulz-Zimm	2949	2.8	0.0068	22	0.0
PE Run 1	Wesslau	1456	4.6	0.0117	20	0.0
CDA Run 1	Schulz-Zimm	1127	3.0	0.0210	21	0.0
CDA Run 2	Schulz-Zimm	1127	3.0	0.0210	16	0.0
CTA Run 1	Schulz-Zimm	1806	4.0	0.0050	14	0.0

in Table 2.11. If one wishes to estimate M_w/M_n of the original polymer with an error of less than $\pm 5\%$, $v_p{}^0 \leq 5\%$ at $n_t = 20$ or $v_p{}^0 \leq 2\%$ at $n_t = 50$ should be chosen as suitable operating conditions in the case of the Kamide's method (ref. 47). In contrast, if the Schulz's method is employed, the operating conditions of $v_p{}^0 \geq 0.01\%$ and $n_t \leq 50$ do not fulfill the above requirement of $0.95 < (M_w/M_n)_t/(M_w/M_n)_0 < 1.05$.

Unfortunately, the value of X_w/X_n (or M_w/M_n) obtained from the analytical fractionation data by means of eqns (2.105) and

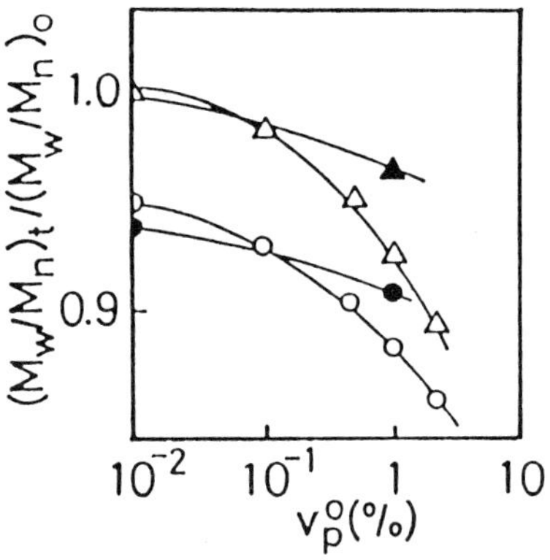

Fig. 2.58 Effect of the initial polymer volume fraction $v_p{}^0$ on the ratio $(M_w/M_n)_t/(M_w/M_n)_0$: Circle, the Schulz's method; triangle, the Kamide's method: Open mark, $n_t = 20$; closed mark, $n_t = 50$: The original polymer, Schulz-Zimm type distribution ($M_w = 23.9 \times 10^4$, $M_w/M_n = 2.77$).

(2.106) or by the Schulz's method has so far been compared directly only with a theoretical value, which has been derived from a knowledge of the polymerization mechanism. The dispersion index has also been correlated with some mechanical properties of polymers. It is obvious from Table 2.13 that much attention should be paid when X_w/X_n of the original polymer is to be determined from the fractionation data.

A combination of SSF and the Kamide's method (SSF/Kamide) is strongly recommended for estimating the MWD of the original polymer.

2.5 SPINODAL CURVE AND CRITICAL SOLUTION POINT

2.5.1 Theoretical background

The critical solution point of the polydisperse polymer in a single solvent was first studied by Stockmayer (ref. 133) who did not consider the concentration dependence of the χ parameter. The most rigorous expressions for critical points were derived by Koningsveld and Staverman (ref. 14), who considered the concentration dependence of the g parameter (eqn (2.24)) along with the polydispersity of the polymer. Further, relations for the spinodal and CSP were studied by Gordon et al. (ref. 38), and the theory was generalized by Koningsveld, Gordon and their collaborators (refs 19,20) by use of a closed form of g parameter (see, eqn (2.31)). The effects of polymer chain branching on CSP were also studied by Kleintjens, Koningsveld and Gordon (ref. 135).

Koningsveld et al. (refs 16,19) applied their theory to estimate the concentration dependence of the g parameter (see, section 2.7). Kamide et al. (ref. 4) also extended the theory to the case in which the polydispersity of a polymer and the concentration- and molecular weight-dependence of χ parameter exist for a quasi-binary mixture and showed the effects of these factors on the CSP in some detail.

The spinodal is determined by eqn (2.108) (refs 14-16,19,20,38)

$$|\Delta G'| = \begin{vmatrix} \Delta \bar{G}'_{11} & \Delta \bar{G}'_{12} & \cdots & \Delta \bar{G}'_{1m} \\ \Delta \bar{G}'_{21} & \Delta \bar{G}'_{22} & \cdots & \Delta \bar{G}'_{2m} \\ \vdots & \vdots & & \vdots \\ \Delta \bar{G}'_{m1} & \Delta \bar{G}'_{m2} & \cdots & \Delta \bar{G}'_{mm} \end{vmatrix} \quad (2.108)$$

and this can be combined with eqn (2.109)

$$|\Delta G''| = \begin{vmatrix} \frac{\partial|\Delta G'|}{\partial v_{x1}} & \frac{\partial|\Delta G'|}{\partial v_{x2}} & \cdots & \frac{\partial|\Delta G'|}{\partial v_{xm}} \\ \bar{\Delta G}'_{21} & \bar{\Delta G}'_{22} & \cdots & \bar{\Delta G}'_{2m} \\ \vdots & \vdots & & \vdots \\ \bar{\Delta G}'_{m1} & \bar{\Delta G}'_{m2} & \cdots & \bar{\Delta G}'_{mm} \end{vmatrix} \qquad (2.109)$$

to obtain the CSP for multicomponent polymer / single solvent system (refs 14-16,19,20,38). In eqns (2.108) and (2.109) $\Delta G'$ is the Gibbs' free energy change of mixing per unit volume, given by

$$\Delta G' = v_0 \left(\frac{\Delta\mu_0}{V_0}\right) + \sum_{i=1}^{m} v_{xi} \left(\frac{\Delta\mu_{xi}}{X_i V_0}\right) \qquad (2.110)$$

and $\bar{\Delta G}'_{ij}$ is defined by eqn (2.111)

$$\bar{\Delta G}'_{ij} = \left(\frac{\partial^2 \Delta G'}{\partial v_{xi} \partial v_{xj}}\right)_{T,P,v_{xk}} \quad (i,j = 1,\cdots,m; \quad k \neq i,j) \qquad (2.111)$$

where v_{xi} is the volume fraction of X_i-mer. The differentiation is carried out at constant temperature and pressure. Determinant expression, eqn (2.108), was first derived by Gibbs for a more general system (ref. 134).

Kamide et al. attempted to solve eqns (2.108) and (2.109) for the case where χ depends on molecular weight as well as concentration, as expressed by eqn (2.4) (ref. 4). Substituting eqns (2.9) and (2.21) into eqn (2.110), we get

$$\Delta G' = \frac{\tilde{R}T}{V_0}\Big[(1-v_p)\ln(1-v_p) + \sum_{i=1}^{m} \frac{v_{xi}\ln v_{xi}}{X_i} + \chi_{00}\{1+\frac{k'}{X_n}(1+X_n-X_w)\}\{1+\sum_{j=1}^{n}\frac{p_j}{j+1}(\sum_{q=0}^{j} v_p{}^q)\}v_p(1-v_p)\Big]. \qquad (2.112)$$

Substitution of eqn (2.112) into eqn (2.111) leads to

$$\frac{V_0}{\tilde{R}T}\Delta \bar{G}'_{ij} = \frac{1}{1-v_p} - \chi_{00}\{1+k'(1+\frac{1}{X_n}-\frac{X_w}{X_n})\}\{2+\sum_{q=1}^{n} p_q(q+2)v_p{}^q\} \equiv M$$

for $i \neq j$ (2.113a)

or

$$\frac{V_0}{\tilde{R}T}\Delta \bar{G}'_{ij} = M + \frac{1}{X_i v_{Xi}} \equiv M + M_i$$

for $i = j$. (2.113b)

M and M_i are defined by eqns (2.113a) and (2.113b) and should be distinguished from the symbol of molecular weight. From eqns (2.108), (2.113a) and (2.113b), we obtain

$$|\Delta G'| = (\frac{\tilde{R}T}{V_0})^m \begin{vmatrix} M+M_1 & M & \cdots & M \\ M & M+M_2 & \cdots & M \\ \vdots & \vdots & & \vdots \\ M & M & \cdots & M+M_m \end{vmatrix}$$

$$= (\frac{\tilde{R}T}{V_0})^m (\prod_{i=1}^{m} M_i)(1+M\sum_{i=1}^{m}\frac{1}{M_i}). \qquad (2.114)$$

Then, we have

$$1+M\sum_{i=1}^{m}\frac{1}{M_i} = 0 \qquad (2.115)$$

for the spinodal. In deriving eqn (2.114) we employed the relation (ref. 4)

$$|D| = \begin{vmatrix} x_1 & a & \cdots & a \\ a & x_2 & \cdots & a \\ \vdots & \vdots & & \vdots \\ a & a & \cdots & x_m \end{vmatrix} = f(a) - a\left.\frac{df(x)}{dx}\right|_{x=a} \qquad (2.116)$$

where

$$f(x) = \prod_{i=1}^{m}(x_i - x). \qquad (2.117)$$

Differentiating eqn (2.114) by v_{Xi}, we get

$$\frac{\partial|\Delta G'|}{\partial v_{Xi}} = (\frac{\tilde{R}T}{V_0})(\prod_{i=1}^{m} M_i)(\frac{\partial M}{\partial v_p}v_p X_w - \frac{X_i}{X_w v_p}) \equiv W_i. \qquad (2.118)$$

Then, from eqns (2.109), (2.113a), (2.113b) and (2.118), and using eqns (2.116) and (2.117) we obtain

$$\left(\frac{V_0}{\tilde{R}T}\right)^m |\Delta G''| = \begin{vmatrix} W_1 & W_2 & W_3 & \cdots & W_m \\ M & M+M_2 & M & \cdots & M \\ M & M & M+M_3 & \cdots & M \\ \vdots & \vdots & \vdots & & \vdots \\ M & M & M & \cdots & M+M_m \end{vmatrix}$$

$$= (-1)^2 W_1 \begin{vmatrix} M+M_2 & M & \cdots & M \\ M & M+M_3 & \cdots & M \\ \vdots & \vdots & & \vdots \\ M & M & \cdots & M+M_m \end{vmatrix}$$

$$+ \sum_{j=2}^{m} W_j (-1)^{j+1} (-1)^j \begin{vmatrix} M & M & & \cdots & & M \\ M & M+M_2 & & \cdots & & M \\ \vdots & \vdots & \ddots & & & \vdots \\ M & M & M+M_{j-1} & M & & \\ M & M & M & M+M_{j+1} & & \\ \vdots & \vdots & & & \ddots & \vdots \\ M & M & & \cdots & & M+M_m \end{vmatrix}$$

$$= -\left(\prod_{i=1}^{m} M_i\right)\left(\frac{M}{M_i}\right)\left(\sum_{i=1}^{m} \frac{W_i}{M_i}\right) = 0. \qquad (2.119)$$

Neutral equilibrium condition is:

$$\sum_{i=1}^{m} \left(\frac{W_i}{M_i}\right) = 0. \qquad (2.120)$$

Rearrangement of eqn (2.115) with the aid of eqns (2.113a) and (2.113b) gives eqn (2.121) (ref. 4).

$$\frac{1}{X_w v_p} + \frac{1}{1-v_p} - \chi_{00}\{1+k'(1+\frac{1}{X_n}-\frac{X_w}{X_n})\}\{2+\sum_{j=1}^{n} p_j (j+2) v_p{}^j\} = 0. \qquad (2.121)$$

Eqn (2.120) can be rewritten, using eqns (2.113a) and (2.113b) in the form (ref. 4):

$$\frac{1}{(1-v_p)^2}-\frac{X_z}{(X_w v_p)^2}-\chi_{00}\{1+k'(1+\frac{1}{X_n}-\frac{X_w}{X_n})\}\{\sum_{j=1}^{n} p_j j(j+2)v_p^{j-1}\}=0. \tag{2.122}$$

Both v_p and χ_{00} at CSP are referred to as v_p^c and χ_{00}^c, and can be calculated by application of eqns (2.121) and (2.122) from X_w and X_z, of the original polymer (X_w^0 and X_z^0), k' and p_j ($j=1,\cdots,n$) using Newton's procedure. In this calculation, as the initial value of v_p^c, the critical value given by the Stockmayer equation (ref. 133),

$$v_p^c=\frac{1}{1+X_w(X_z)^{-1/2}} \tag{2.123}$$

is taken.

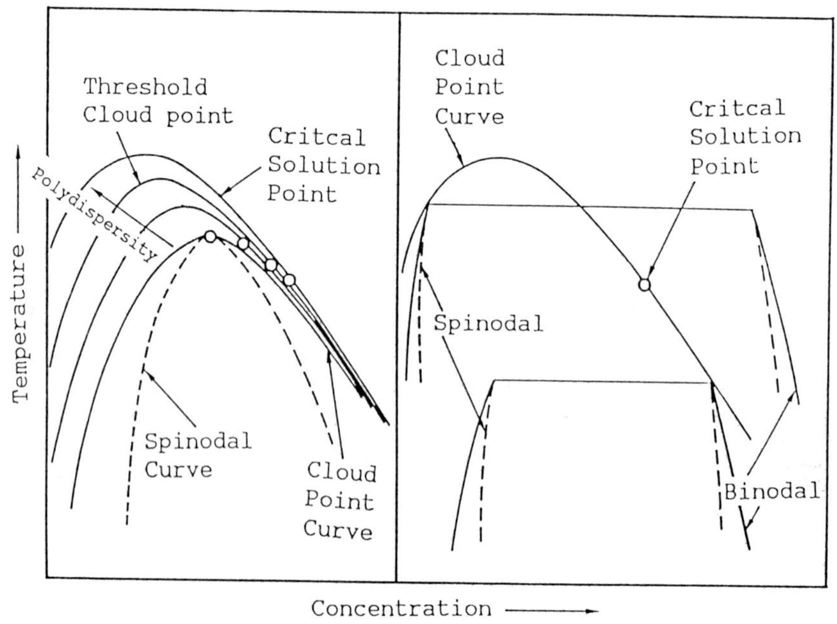

Fig. 2.59 a) Effect of polydispersity of polymer on cloud point curves and b) relations among cloud point curve, binodal and spinodal curve.

Fig. 2.59a illustrates qualitatively the effect of the polymolecularity of the original polymer dissolved in a single solvent on CPC. In the rather concentrated v_p region (i.e., $v_p > v_p^c$), CPC is very insensitive to X_w^0/X_n^0 and in the range of v_p smaller than v_p^c, it shifts to a higher temperature possibly as a result of the precipitation of higher molecular weight components contained in polymers having large polymolecularity. Evidently, the threshold cloud point temperature T_p^{tcp} is much more sensitive than critical temperature T_c to X_w^0/X_n^0. On the contrary, the spinodal curve is independent of polymolecularity as the theory predicts (see, eqn (2.121)).

2.5.2 Experimental data on critical solution points for some polymer / solvent systems

Tables 2.14, 2.15 and 2.16 collect critical solution points determined experimentally for solutions of atactic polystyrene (refs 45,66,70,71,136-143), polyethylene (refs 14,68,144-147) and all cellulose acetate (total degree of substitution, 2.46) (refs 148,149), dissolved in single solvents.

TABLE 2.14
Critical solution point data of quasi-binary system consisting of multicomponent polymer dissolved in a single solvent.

Polystyrene

Solvent	UCSP or LCSP	Number of samples	Method for determining CSP	Polymer Sample $M_w \times 10^{-4}$	Polymer Sample M_w/M_n	Critical solution point $v_p{}^c \times 10^2$	Critical solution point T_c/K	ref.
Methyl ethyl ketone	UCSP	6	tcp[a]	3.70	<1.06	9.01[b]	463.2	136
				9.72	<1.06	7.86[b]	448.8	
				20.0	<1.06	5.92[b]	441.1	
				40.0	<1.06	3.76[b]	434.5	
				67.0	<1.10	2.96[b]	431.8	
				270.0	<1.10	1.34[b]	425.7	
Methyl ethyl ketone	UCSP	6	tcp	11.0	1.1[c]	3.174[b]	446.14	137
				21.2	1.1[c]	2.714[b]	439.52	
				34.2	1.1[c]	2.452[b]	435.76	
				53.3	1.1[c]	2.216[b]	432.29	
				200.0	1.3	1.730[b]	426.10	
				1100.0	1.1[c]	1.307[b]	421.81	
Methyl ethyl ketone	UCSP	4	tcp	9.70	1.1[c]	3.29wt.	445.8	138
				20.0	1.1[c]	2.96wt.	439.2	
				86.0	1.1[c]	2.77wt.	428.0	
				200.0	1.1[c]	2.51wt.	422.9	
Cyclopentane	UCSP	6	tcp	3.70	<1.06	12.64[b]	267.0	136
				9.72	<1.06	9.64	275.2	
				20.9	<1.06	7.23	280.9	
				40.0	<1.06	4.47	284.7	
				67.0	<1.10	3.77	285.9	
				270.0	<1.10	2.14	289.9	

a) tcp = threshould cloud point, b) read from the figure of cloud point curve,
c) assumed value.

TABLE 2.14 (continued-1)

Solvent	UCSP or LCSP	Number of samples	Method for determining CSP	Polymer Sample		Critical solution point		ref.
				$M_w \times 10^4$	M_w/M_n	$v_p{}^c \times 10^2$	T_c/K	
Cyclopentane	LCSP	6	tcp	3.70	<1.06	11.70	455.3	136
				9.72	<1.06	10.60	445.5	
				20.9	<1.06	7.39	440.0	
				40.0	<1.06	4.24	435.4	
				67.0	<1.10	3.69	433.8	
				270.0	<1.10	2.54	429.5	
Cyclohexane	UCSP	9	R[d]	3.54[e]	1.3	17.9wt.	284.60	16
				5.10	1.4	11.34	288.85	
				6.15	1.12	14.25wt.	290.45	
				9.30	1.02	9.02	293.65	
				16.6	1.08	9.90wt.	296.60	
				28.6	1.43	7.60	298.70	
				39.4	1.05	5.29	300.70	
				52.7	1.08	4.87	301.15	
				150.0	1.20	2.14	303.20	
Cyclohexane	UCSP	5	diameter	20.0	<1.02	7.79	296.782	66
				20.6	<1.03	7.81	296.943	
				41.5	<1.06	5.23	299.980	
				68.0	<1.02	4.46	301.027	
				45.0	<1.44	6.77	299.895	
Cyclohexane	UCSP	6	tcp	3.70	<1.06	12.58	285.6	71
				9.72	<1.06	8.35	293.5	
				20.0	<1.06	7.81	297.0	
				40.0	<1.06	5.27	300.3	
				67.0	<1.10	4.47	301.1	
				270.0	<1.10	2.56	304.2	

d) R=phase volume ratio, e) mixture.

TABLE 2.14 (continued-2)

Solvent	UCSP or LCSP	Number of samples	Method for determining CSP	Polymer Sample $M_w \times 10^4$	Polymer Sample M_w/M_n	Critical solution point $v_p^c \times 10^2$	Critical solution point T_c/K	ref.
Cyclohexane	LCSP	6	tcp	3.70	<1.06	9.02	510.9	71
				9.72	<1.06	6.82	502.1	
				20.0	<1.06	5.07	496.9	
				40.0	<1.06	4.15	494.7	
				67.0	<1.10	2.94	491.7	
				270.0	<1.10	1.77	488.6	
Methylcyclohexane	UCSP	9	diameter	1.02	<1.06	19.90	285.71	45,71
				1.61	<1.06	17.19	295.98	
				1.73	<1.06	16.69	296.13	
				2.02	<1.06	15.90	298.95	
				3.94	<1.06	12.93	309.00	
				4.64	<1.06	11.70	312.61	
				10.9	<1.06	8.40	322.71	
				18.1	<1.06	7.01	327.00	
				71.9	<1.06	4.06	334.82	
Methylcyclohexane	UCSP	6	tcp	3.70	<1.06	13.168	309.7	71
				9.72	<1.06	8.673	321.8	
				20.0	<1.06	6.225	327.4	
				40.0	<1.06	4.903	332.7	
				67.0	<1.10	4.412	334.5	
				270.0	<1.10	2.626	339.6	
Methylcyclohexane	UCSP	4	tcp	9.72	1.1	0.25	318.1	139
				41.1	1.1	0.24	344.0	
				86.0	1.1	0.38	346.3	
				180.0	1.1	0.40	348.0	

TABLE 2.14 (continued-3)

Solvent	UCSP or LCSP	Number of samples	Method for determining CSP	Polymer Sample $M_w \times 10^4$	Polymer Sample M_w/M_n	Critical solution point $v_p^c \times 10^2$	Critical solution point T_c/K	ref.
Methylcyclohexane	LCSP	6	tcp	3.70	<1.06	8.56	518.8	71
				9.72	<1.06	5.56	505.9	
				20.0	<1.06	4.82	499.9	
				40.0	<1.06	3.81	494.6	
				67.0	<1.10	2.95	492.3	
				270.0	<1.10	2.11	488.4	
Benzene	LCSP	6	tcp	3.70	<1.06	6.2	538.7	136
				9.72	<1.06	4.1	532.5	
				20.0	<1.06	3.2	530.5	
				40.0	<1.06	2.4	528.3	
				67.0	<1.10	1.8	527.0	
				270.0	<1.10	1.0	525.0	
Toluene	LCSP	6	tcp	3.70	<1.06	7.7	567.2	71
				9.72	<1.06	4.9	559.9	
				20.0	<1.06	4.1	557.2	
				40.0	<1.06	2.7	554.9	
				67.0	<1.10	2.7	553.1	
				270.0	<1.10	1.2	552.0	
Methyl acetate	UCSP	5	tcp	3.70	<1.06	—	266.6	140
				11.0	<1.06	—	284.2	
				20.0	<1.06	—	289.7	
				67.0	<1.15	—	301.5	
				270.0	<1.10	—	311.0	
Methyl acetate	LCSP	5	tcp	3.70	<1.06	—	434.0	140
				11.0	<1.06	—	415.7	
				20.0	<1.06	—	409.9	
				67.0	<1.15	—	398.4	
				270.0	<1.10	—	389.2	

TABLE 2.14 (continued-4)

Solvent	UCSP or LCSP	Number of samples	Method for determining CSP	Polymer Sample $M_w \times 10^4$	M_w/M_n	Critical solution point $v_p{}^c \times 10^2$	T_c/K	ref.
Ethyl acetate	UCSP	5	tcp	3.70	<1.06	—	204.1	140
				11.0	<1.06	—	213.9	
				20.0	<1.06	—	216.5	
				67.0	<1.15	—	222.9	
				270.0	<1.10	—	226.5	
Ethyl acetate	LCSP	5	tcp	3.70	<1.06	—	451.0	140
				11.0	<1.06	—	435.4	
				20.0	<1.06	—	430.6	
				67.0	<1.15	—	421.4	
				270.0	<1.10	—	415.7	
n-Propyl acetate	UCSP	5	tcp	3.70	<1.06	—	—	140
				11.0	<1.06	6.9	183.7	
				20.0	<1.06	6.7	185.5	
				67.0	<1.15	3.9	189.6	
				270.0	<1.10	2.1	191.0	
n-Propyl acetate	LCSP	5	tcp	3.70	<1.06	8.9	484.1	140
				11.0	<1.06	6.3	469.0	
				20.0	<1.06	6.4	464.8	
				67.0	<1.15	5.0	458.2	
				270.0	<1.10	2.5	454.2	
iso-Propyl acetate	UCSP	6	tcp	1.0	<1.06	—	—	140
				3.7	<1.06	11.1	206.6	
				11.0	<1.06	8.10	220.9	
				20.0	<1.06	6.70	225.6	
				67.0	<1.15	4.90	235.5	
				270.0	<1.10	2.70	240.3	

TABLE 2.14 (continued-5)

Solvent	UCSP or LCSP	Number of samples	Method for determining CSP	Polymer Sample $M_w \times 10^4$	Polymer Sample M_w/M_n	Critical solution point $v_p{}^c \times 10^2$	Critical solution point T_c/K	ref.
iso-Propyl acetate	LCSP	6	tcp	1.0	<1.06	10.7	468.5	140
				3.7	<1.06	10.0	436.7	
				11.0	<1.06	7.50	414.2	
				20.0	<1.06	6.60	407.7	
				67.0	<1.15	3.10	395.2	
				270.0	<1.10	2.00	385.9	
Ethyl formate	UCSP	6	tcp	0.22	<1.10	19.4	230	141
				0.40	<1.10	18.7	242	
				1.00	<1.06	16.1	272	
				2.04	<1.06	14.3	294	
				3.70	<1.06	11.2	316	
				9.70	<1.06	—	—	
Dimethoxy methane	LCSP	5	tcp	5.10	<1.06	7.0	413.7	142
				9.72	<1.06	5.6	406.1	
				16.0	<1.06	4.4	401.2	
				41.1	<1.06	3.4	394.7	
				86.0	<1.06	2.6	361.6	
Acetone	UCSP	2	tcp	0.48	<1.06	17.5	222.0	142
				1.03	<1.06	15.0	271.1	
Acetone	LCSP	2	tcp	0.48	<1.06	14.2	464.9	142
				1.03	<1.06	13.0	417.8	
Diphenyl ether	UCSP	1	tcp	1.98	<1.06	9.7	235.6	142
Diphenyl ether	LCSP	3	tcp	0.48	<1.06	12.8	407.3	142
				1.00	<1.06	11.7	353.0	
				1.98	<1.06	10.1	314.5	

TABLE 2.14 (continued-6)

Solvent	UCSP or LCSP	Number of samples	Method for determining CSP	Polymer Sample $M_w \times 10^4$	Polymer Sample M_w/M_n	Critical solution point $v_p^c \times 10^2$	Critical solution point T_c/K	ref.
Trans-decalin	UCSP	6	tcp	3.7	<1.06	14.6	274.5	143
				9.7	<1.06	10.2	281.8	
				20.0	<1.06	8.5	285.0	
				40.0	<1.06	6.0	287.9	
				67.0	<1.10	5.3	289.0	
				270.0	<1.30	3.1	291.3	

TABLE 2.15

Critical solution point data of quasi-binary system consisting of multicomponent polymer dissolved in a single solvent.

Polyethylene

Solvent	UCSP or LCSP	Number of samples	Method for determining CSP	Polymer sample $M_w \times 10^{-4}$	Polymer sample M_w/M_n	Critical Solution point $v_p^c \times 10^2$	Critical Solution point T_c/K	ref.
n-Butyl acetate	UCSP	4	tcp[a]	13.6[b]	1.3[c]	5.190[d]	415	144
				20.0[b]	1.3[c]	4.402[d]	431	
				61.1[b]	1.3[c]	3.758[d]	448	
				64.0[b]	1.3[c] (SSF[e])	3.596[d]	451	
	LCSP	4	tcp	13.6[b]	1.3[c]	5.75[d]	518	144
				20.0[b]	1.3[c]	5.40[d]	507	
				61.1[b]	1.3[c]	4.88[d]	497	
				64.0[b]	1.3[c] (SSF[e])	4.22[d]	490	
n-amyl acetate	UCSP	5	tcp	13.6[b]	—	—	389	144
				20.0[b]	—	—	396	
				61.1[b]	—	—	410	
				64.0[b]	—	—	415	
				175.0[b]	— (SSF[e])	—	421	
n-amyl acetate	LCSP	5	tcp	13.6[b]	—	—	553	144
				20.0[b]	—	—	547	
				61.1[b]	—	—	535	
				64.0[b]	—	—	533	
				175.0[b]	— (SSF[e])	—	528	

a) tcp = threshould cloud point, b) M_n, c) assumed value, d) read from the figure of cloud point curve, e) fractionated by successive solution fractionation.

TABLE 2.15 (continued-1)

Solvent	UCSP or LCSP	Number of samples	Method for determining CSP	Polymer sample $M_w \times 10^{-4}$	Polymer sample M_w/M_n	Critical Solution point $v_p{}^c \times 10^2$	Critical Solution point T_c/K	ref.
Diphenyl ether	UCSP	3	R[f]	5.5	6.4	9.7	410	14
				140.0	1.52	5.0	420.9	
				153.0	12.75	8.2	416.2	
					(W & F[g])			
Diphenyl ether	UCSP	7	tcp	14.3	1.3[c]	6.84[d]	398.2	145
				37.9	1.3[c]	4.89[d]	410.8	
				50.9	1.3[c]	4.22[d]	414.7	
				56.8	1.3[c]	3.77[d]	415.9	
				97.2	1.3[c]	3.26[d]	418.8	
				136.8	1.3[c]	2.48[d]	422.6	
				204.9	1.3[c]	1.70[d]	425.4	
					(column[h])			
Diphenyl	UCSP	4	tcp	50.9	1.3[c]	4.65[d]	383.4	145
				56.8	1.3[c]	4.00[d]	384.8	
				136.8	1.3[c]	2.10[d]	390.2	
				442.1	1.3[c]	0.93[d]	394.3	
					(column)			
Diphenyl methane	UCSP	6	tcp	21.3	1.3[c]	6.74[d]	384.7	145
				34.9	1.3[c]	5.34[d]	390.5	
				56.8	1.3[c]	3.96[d]	396.4	
				76.3	1.3[c]	3.42[d]	397.7	
				97.2	1.3[c]	2.88[d]	400.1	
				136.8	1.3[c]	1.70[d]	402.1	
					(column)			
n-Pentane	LCSP	3	tcp	0.49	1.3[c]	6.59[d]	410.4[i]	146
				12.5	1.3[c]	5.41[d]	386.0[i]	
				14.3	1.3[c]	5.10[d]	382.5[i]	
					(column)			

f) R=phase volume ratio, g) whole and fractionated polymer, h) fractionated by column fractionation (solid extraction), i) read from the figure of cloud point curve.

TABLE 2.15 (continued-2)

Solvent	UCSP or LCSP	Number of samples	Method for determining CSP	Polymer sample $M_w \times 10^{-4}$	Polymer sample M_w/M_n	Critical Solution point $v_p{}^c \times 10^2$	Critical Solution point T_c/K	ref.
n-Hexane	LCSP	4	tcp	34.9	1.3[c]	5.20[d]	433.0[i]	146
				97.2	1.3[c]	3.60[d]	421.8[i]	
				204.9	1.3[c]	2.70[d]	417.2[i]	
				442.1	1.3[c]	2.40[d]	414.2[i]	
					(column)			
n-Heptane	LCSP	4	tcp	76.8	1.3[c]	4.60[d]	463.2[i]	146
				93.5	1.3[c]	2.40[d]	462.0[i]	
				125.0	1.3[c]	2.20[d]	460.3[i]	
				202.0	1.3[c]	1.10[d]	456.7[i]	
					(column)			
n-Octane	LCSP	4	tcp	76.8	1.3[c]	4.36[d]	502.7[i]	146
				93.5	1.3[c]	3.75[d]	501.3[i]	
				125.0	1.3[c]	3.097[d]	499.5[i]	
				202.0	1.3[c]	1.528[d]	495.4[i]	
					(column)			
n-Octyl alcohol	UCSP	4	tcp	41.6	1.3[c]	3.15[d]	423.2	147
				111.0	1.3[c]	2.893[d]	432.8	
				800.0	1.3[c]	2.258[d]	440.2	
				3000.0	1.3[c]	1.598[d]	441.9	
					(W & F)			
	LCSP	4	tcp	41.6	1.3[c]	3.135[d]	627.4	147
				111.0	1.3[c]	2.748[d]	624.3	
				800.0	1.3[c]	2.228[d]	622.6	
				3000.0	1.3[c]	1.323[d]	621.1	
					(W & F)			

TABLE 2.15 (continued-3)

Solvent	UCSP or LCSP	Number of samples	Method for determining CSP	Polymer sample		Critical Solution point		ref.
				$M_w \times 10^{-4}$	M_w/M_n	$v_p{}^c \times 10^2$	T_c/K	
n-Octyl alcohol	UCSP	4	tcp	0.68	1.3[c]	5.261[d]	400.60[i]	68
				2.49	1.3[c]	3.725[d]	425.24[i]	
				7.78	1.3[c]	2.843[d]	437.00[i]	
				16.3	1.3[c] (SPF[j])	1.797[d]	441.48[i]	
n-Decyl alcohol	UCSP	4	tcp	0.68	1.3[c]	6.329[d]	385.58[i]	68
				2.49	1.3[c]	5.127[d]	402.19[i]	
				7.78	1.3[c]	2.943[d]	412.68[i]	
				16.3	1.3[c] (SPF)	1.582[d]	416.23[i]	
n-Lauryl alcohol	UCSP	4	tcp	0.68	1.3[c]	6.678[d]	385.64[i]	68
				2.49	1.3[c]	5.961[d]	388.60[i]	
				7.78	1.3[c]	3.811[d]	397.72[i]	
				16.3	1.3[c] (SPF)	2.541[d]	402.36[i]	
p-tert-Amyl phenol	UCSP	4	tcp	0.68	1.3[c]	6.835[d]	411.40[i]	68
				2.49	1.3[c]	3.797[d]	439.24[i]	
				7.78	1.3[c]	2.057[d]	455.08[i]	
				16.3	1.3[c] (SPF)	1.835[d]	457.48[i]	
p-Octyl phenol	UCSP	4	tcp	0.68	1.3[c]	7.594[d]	397.77[i]	68
				2.49	1.3[c]	5.188[d]	425.48[i]	
				7.78	1.3[c]	2.444[d]	433.64[i]	
				16.3	1.3[c] (SPF)	1.579[d]	435.27[i]	

j) fractionated by successive precipitation fractionation.

TABLE 2.15 (continued-4)

Solvent	UCSP or LCSP	Number of samples	Method for determining CSP	Polymer sample		Critical Solution point		ref.
				$M_w \times 10^{-4}$	M_w/M_n	$v_p^c \times 10^2$	T_c/K	
p-Nonyl phenol	UCSP	4	tcp	0.68	1.3[c]	7.883[d]	394.20[i]	68
				2.49	1.3[c]	7.263[d]	411.15[i]	
				7.78	1.3[c]	5.474[d]	420.13[i]	
				16.3	1.3[c] (SPF)	3.139[d]	426.70[i]	
Anisole	UCSP	4	tcp	0.79	1.3[c]	6.015[d]	372.17[i]	68
				2.49	1.3[c]	3.195[d]	407.86[i]	
				7.78	1.3[c]	2.256[d]	415.75[i]	
				16.3	1.3[c] (SPF)	1.880[d]	418.96[i]	
Benzyl phenyl ether	UCSP	4	tcp	0.68	1.3[c]	3.704[d]	424.39[i]	68
				2.49	1.3[c]	3.259[d]	437.54[i]	
				7.78	1.3[c]	2.593[d]	449.94[i]	
				16.3	1.3[c] (SPF)	1.047[d]	453.00[i]	

TABLE 2.16

Critical solution point data of quasi-binary system consisting of multicomponent polymer dissolved in a single solvent.

Cellulose acetate (total degree of substitution, 2.47)

Solvent	UCSP or LCSP	Number of samples	Method for determining CSP	Polymer Sample		Critical solution point		ref.
				$M_w \times 10^{-4}$	M_w/M_n	$v_p^c \times 10^2$	T_c/K	
Acetone	LCSP	4	tcp	3.76	1.23	0.0330	457.1	148
				7.55	1.26	0.0281	448.0	
				11.1	1.26	0.0264	444.2	
				17.5	1.38	0.0189	440.1	
Methyl ethyl ketone	UCSP	3	tcp	7.55	1.26	0.0314	279.7	149
				11.1	1.26	0.0268	284.5	
				17.5	1.38	0.0229	290.0	
Methyl ethyl ketone	LCSP	3	tcp	7.55	1.26	0.0343	471.5	149
				11.1	1.26	0.0229	465.0	
				17.5	1.38	0.0201	457.9	

2.5.3 Stability conditions of critical solution point

If we attempt to discuss the thermodynamical stability of the critical solution point (CSP) for a system, we should consider the fourth differential of Gibbs' free energy change of mixing to form the system $\delta^4\Delta G$ (refs 150,151). For a relatively simple two components system consisting of a monodisperse polymer dissolved in a single solvent, Kennedy (ref. 152) and Casassa (ref. 153) obtained using $\delta^4\Delta G$ the stability condition of its CSP [see, eqn (23) of ref. 152 and eqn (33) of ref. 153]. Spinodal condition, neutral equilibrium condition, and stability condition for CSP are given by the following equations, respectively (refs 152,153).

$$\left(\frac{\partial^2 \Delta G}{\partial v_X^2}\right)_{T,P} = 0, \tag{2.124}$$

$$\left(\frac{\partial^3 \Delta G}{\partial v_X^3}\right)_{T,P} = 0, \tag{2.125}$$

$$\left(\frac{\partial^4 \Delta G}{\partial v_X^4}\right)_{T,P} > 0. \tag{2.126}$$

Here, v_X is the volume fraction of the monodisperse polymer (X-mer) and ΔG, Gibbs' free energy change of mixing, and T and P, Kelvin temperature and pressure, respectively. According to the Flory-Huggins theory, ΔG is expressed by the following equation

$$\frac{\Delta G}{\tilde{R}T} = (1-v_X)\ln(1-v_X) + \frac{v_X \ln v_X}{X} + \chi v_X(1-v_X) \tag{2.127}$$

where $\tilde{R}$ is gas constant, χ, the thermodynamic interaction parameter between polymer and solvent. When χ is independent of v_X, substitutions of eqn (2.127) into eqns (2.124)-(2.126) yield eqns (2.128)-(2.130), respectively (refs 152,153).

$$\frac{1}{1-v_X} + \frac{1}{Xv_X} - 2\chi = 0, \tag{2.128}$$

$$\frac{1}{(1-v_X)^2} - \frac{1}{Xv_X^2} = 0, \tag{2.129}$$

$$\frac{1}{(1-v_X)^3} - \frac{1}{Xv_X^3} > 0. \tag{2.130}$$

For the system consisting of two monodisperse polymers with the same chemical composition and with different molecular weight dissolved in a single solvent Tompa (ref. 154) proposed, as early as 1949, the equation of condition on coincidence of two plait points (i.e., heterogenous double plait points (HDPP)).

$$3(X_1^2 - 10X_1X_2 + X_2^2)^2 + (X_1 + X_2)(X_1^2 - 106X_1X_2 + X_2^2) - 27X_1X_2 = 0 \qquad (2.131)$$

where X_1 and X_2 are the degree of polymerization of each monodisperse homologues. When both X_1 and X_2 are large, to an approximation made by Chermin (ref. 155) the solution is given by the first term of eqn (2.131), that is

$$X_1^2 - 10X_1X_2 + X_2^2 \simeq 0. \qquad (2.132)$$

From eqn (2.132), one obtains (ref. 155)

$$X_2/X_1 \simeq 9.899 \text{ or } X_2/X_1 \simeq 0.101. \qquad (2.133)$$

Hence, in order to find a ternary critical point in a simple Flory-Huggins system composed of a solvent and two polymer homologues, the ratio of the relative chain lengths should be equivalent with, or larger than 9.899 (ref. 155).

Later, Chermin (ref. 155) succeeded to extend Tompa's original treatment to a more complicated system of two multicomponent polymers with the same chemical composition (i.e., the polymer with a two peaks molecular weight distribution (MWD)) and a single solvent [see, eqn (21) of ref. 155].

$$3(R^2 - 12P)^2 + R(R^2 - 108P) - 27P = 0 \qquad (2.134)$$

where

$$P = M_{w1}M_{w2}\frac{M_{z1} - M_{z2}}{M_{w1} - M_{w2}}, \qquad (2.135a)$$

$$R = M_{z1} + M_{z2}. \qquad (2.135b)$$

Here, M_{w1} and M_{w2} (or M_{z1} and M_{z2}) are the M_w (or M_z) of polydispersed homologues 1 and 2, respectively. It should be pointed out that the stability conditions of CSP, proposed by Tompa (ref. 154) and of course Chermin (ref. 155), were derived indirectly from spinodal and neutral equilibrium conditions and not derived directly from the fourth variable of Gibbs' free energy of mixing (eqn (2.126)) and in this sense seem less rigorous.

Šolc (ref. 156) derived an equation of conditions for a stable CSP, as a specific solution of Korteweg's condition equation [see, eqn (40) of ref. 150]

$$Z_{11}(m^4 Z_{1111} + 4m^3 Z_{1112} + 6m^2 Z_{1122} + 4mZ_{1222} + Z_{2222})$$
$$- 3(m^2 Z_{111} + 2mZ_{112} + Z_{122})^2 < 0 \qquad (2.136)$$

where subscripted Zs are the partial derivatives of ΔG, e.g., $Z_{112} = (\partial^3 \Delta G / \partial^2 v_{X1} \partial v_{X2})$, and m specifies the direction of binodal (and spinodal) at the critical point, $m = (dv_1/dv_2)_c$. From Korteweg's eqns (34) and (35) of ref. 150, m is calculated as

$$m = \frac{Z_{22} - Z_{12}}{Z_{11} - Z_{12}}. \qquad (2.137)$$

Substituting for ΔG, defined by eqn (2.138);

$$\Delta G = \tilde{R}T\left[v_0 \ln v_0 + \frac{v_{X1}}{X_1}\ln v_{X1} + \frac{v_{X2}}{X_2}\ln v_{X2} + \chi v_0 (v_{X1} + v_{X2})\right] \qquad (2.138)$$

into these relations, we have the relation

$$m = \frac{X_1}{X_2}\left(\frac{v_{X1}}{v_{X2}}\right)_c \qquad (2.139)$$

and the instability criterion

$$\frac{2(X_z)^{3/2} + X_z X_{z+1}}{m} - 3(X_z - X_1)^2 < 0 \qquad (2.140)$$

where the known relation

$$v_p{}^c = \frac{1}{1 + X_w (X_z)^{-1/2}} \qquad (2.123)$$

has been used for the critical concentration and X_w, X_z and X_{z+1} are the weight-, z- and (z+1)-averages for the binary polymer mixture (i.e., mixture of X_1-mer and X_2-mer).

Further rearrangement shows that the left-hand side of eqn (2.140) equals $X_z S/m$ where S is defined by eqn (2.141) (ref. 156):

$$S \equiv 3X_z + 2(X_z)^{1/2} - X_{z+1} \qquad (2.141)$$

An attempt is made to derive, directly from $\delta^4 \Delta G > 0$, a discriminant of the criterion for the stability of CSP for quasi-binary systems in a closed form using our rigorous phase-equilibria theory (refs 4,7,8,18).

For simple quasi-binary systems consisting of two single component (i.e., monodisperse) polymers with different X (referred to as X_1 and X_2, respectively) and the same chemical composition dissolved in a single solvent, eqns (2.108) and (2.109) can be given by (ref. 12)

$$|\Delta G'| = \begin{vmatrix} \overline{\Delta G'}_{11} & \overline{\Delta G'}_{12} \\ \overline{\Delta G'}_{21} & \overline{\Delta G'}_{22} \end{vmatrix} = 0, \tag{2.142}$$

$$|\Delta G''| = \begin{vmatrix} \dfrac{\partial |\Delta G'|}{\partial v_{X1}} & \dfrac{\partial |\Delta G'|}{\partial v_{X2}} \\ \overline{\Delta G'}_{21} & \overline{\Delta G'}_{22} \end{vmatrix} = 0. \tag{2.143}$$

$$\Delta G' = (1 - v_p)\left(\frac{\Delta \mu_0}{V_0}\right) + \sum_{i=1}^{2} v_{Xi}\left(\frac{\Delta \mu_{Xi}}{X_i V_0}\right), \tag{2.144}$$

$$\overline{\Delta G'}_{ij} = \left(\frac{\partial^2 \Delta G'}{\partial v_{Xi} \partial v_{Xj}}\right)_{T,P,v_{Xk}} \qquad (i,j = 1,2; \quad k \neq i,j). \tag{2.145}$$

Introducing the following notations,

$$G_i \equiv \left(\frac{\partial \Delta G'}{\partial v_{Xi}}\right)_{T,P,v_{Xk}} \qquad (i = 1 \text{ or } 2, \ k \neq i), \tag{2.146}$$

$$G_{ij} \equiv \overline{\Delta G'}_{ij} \tag{2.147}$$

and using the familiar mathematical formula,

$$\left(\frac{\partial G_1}{\partial v_{X1}}\right)_{G_2} = \left(\frac{\partial G_1}{\partial v_{X1}}\right)_{v_{X2}} - \left(\frac{\partial G_1}{\partial v_{X2}}\right)_{v_{X1}} \frac{\left(\dfrac{\partial G_2}{\partial v_{X1}}\right)_{v_{X2}}}{\left(\dfrac{\partial G_2}{\partial v_{X2}}\right)_{v_{X1}}}$$

$$= \frac{1}{G_{22}}\{G_{11}G_{22} - (G_{12})^2\} = \frac{|\Delta G'|}{G_{22}}. \tag{2.148}$$

Eqn (2.142) can be rewritten as

$$\left(\frac{\partial G_1}{\partial v_{x1}}\right)_{T,P,G_2} = 0. \qquad (2.149)$$

Similarly, $(\partial^2 G_1/\partial v_{x1}^2)_{G2}$ can be given by eqn (2.150):

$$\left(\frac{\partial^2 G_1}{\partial v_{x1}^2}\right)_{G_2} = \left\{\frac{\partial}{\partial v_{x1}}\left(\frac{\partial G}{\partial v_{x1}}\right)_{G_2}\right\}_{G_2}$$

$$= \frac{1}{G_{22}}\left[\left(\frac{\partial|\Delta G'|}{\partial v_{x1}}\right)_{v_{x2}} - \left(\frac{\partial|\Delta G'|}{\partial v_{x2}}\right)_{v_{x1}} \frac{\left(\frac{\partial G_2}{\partial v_{x1}}\right)_{v_{x2}}}{\left(\frac{\partial G_2}{\partial v_{x2}}\right)_{v_{x1}}}\right]$$

$$= \frac{1}{G_{22}^2}\left[\frac{\partial|\Delta G'|}{\partial v_{x1}}G_{22} - \frac{\partial|\Delta G'|}{\partial v_{x2}}G_{21}\right]$$

$$= \frac{|\Delta G''|}{G_{22}}. \qquad (2.150)$$

With the help of eqn (2.150), eqn (2.143) is reduced to a simple form:

$$\left(\frac{\partial^2 G_1}{\partial v_{x1}^2}\right)_{T,P,G_2} = 0. \qquad (2.151)$$

If we take into the consideration eqns (2.149) and (2.151), eqn (2.152) gives the condition of stability of CSP:

$$\left(\frac{\partial^3 G_1}{\partial v_{x1}^3}\right)_{T,P,G_2} > 0. \qquad (2.152)$$

$(\partial^3 G_1/\partial v_{x1}^3)_{G2}$ is given by eqn (2.153):

$$\left(\frac{\partial^3 G_1}{\partial v_{x1}^3}\right)_{G_2} = \frac{1}{G_{22}^2}\left(\frac{\partial|\Delta G''|}{\partial v_{x1}}\right)_{G_2} + |\Delta G''|\left(\frac{\partial}{\partial v_{x1}}\frac{1}{G_{22}^2}\right)_{G_2}$$

$$= \frac{1}{G_{22}^2}\left[\left(\frac{\partial|\Delta G''|}{\partial v_{x1}}\right)_{v_{x2}} - \left(\frac{\partial|\Delta G''|}{\partial v_{x2}}\right)_{v_{x1}} \frac{\left(\frac{\partial G_2}{\partial v_{x1}}\right)_{v_{x2}}}{\left(\frac{\partial G_2}{\partial v_{x2}}\right)_{v_{x1}}}\right]$$

$$= \frac{1}{G_{22}^3}\left[\frac{\partial|\Delta G''|}{\partial v_{x1}}G_{22} - \frac{\partial|\Delta G''|}{\partial v_{x2}}G_{21}\right]. \qquad (2.153)$$

At equilibrium state $G_{22} > 0$ and the combination of eqns (2.152) and (2.153) leads to the following equation as the stability condition of CSP (ref. 157).

$$\begin{vmatrix} \frac{\partial|\Delta G''|}{\partial v_{x1}} & \frac{\partial|\Delta G''|}{\partial v_{x2}} \\ \bar{\Delta G'}_{21} & \bar{\Delta G'}_{22} \end{vmatrix} > 0. \tag{2.154}$$

Criteria of stability of CSP for multicomponent polymer / single solvent system is given by the following inequality (ref. 157).

$$|\Delta G'''| \; (\equiv \delta^4 \Delta G) = \begin{vmatrix} \frac{\partial|\Delta G''|}{\partial v_{x1}} & \frac{\partial|\Delta G''|}{\partial v_{x2}} & \cdots & \frac{\partial|\Delta G''|}{\partial v_{xm}} \\ \bar{\Delta G'}_{21} & \bar{\Delta G'}_{22} & \cdots & \bar{\Delta G'}_{2m} \\ \vdots & \vdots & & \vdots \\ \bar{\Delta G'}_{m1} & \bar{\Delta G'}_{m2} & \cdots & \bar{\Delta G'}_{mm} \end{vmatrix} > 0. \tag{2.155}$$

Eqn (2.154) is the case of $m=2$ in eqn (2.155). Then the validity of eqn (2.155) can be proved by an inductive method. $|\Delta G'''|$ is the fourth variable ($\delta^4\Delta G$) of Gibbs' free energy change and if $|\Delta G'''|>0$, CSP is stable and if $|\Delta G'''|<0$ CSP becomes unstable. If we note that the determinant (eqn (2.155)) for $|\Delta G'''|$ has the same form as that (eqn (2.109)) for $|\Delta G''|$, eqn (2.155) can be written as (ref. 157)

$$|\Delta G'''| = \left(\frac{\tilde{R}T}{V_0}\right)^{m-1} \begin{vmatrix} Z_1 & Z_2 & Z_3 & \cdots & Z_m \\ M & M+M_2 & M & \cdots & M \\ M & M & M+M_3 & \cdots & M \\ \vdots & \vdots & \vdots & & \vdots \\ M & M & M & \cdots & M+M_m \end{vmatrix} \tag{2.156}$$

with

$$Z_j \equiv \left(\frac{\partial|\Delta G'''|}{\partial v_{xj}}\right)_{T,P,v_{xk}} \qquad (k=1,\cdots,m; \quad k\neq j). \tag{2.157}$$

On substitution of eqn (2.119), eqn (2.156) becomes

$$Z_j = -\left(\frac{\tilde{R}T}{V_0}\right)^{m-1} \left(\prod_{i=1}^{m} M_i\right) \frac{M}{M_1} \frac{\partial}{\partial v_{xj}} \left(\sum_{i=1}^{m} \frac{W_i}{M_i}\right)$$

$$= -\left(\frac{\tilde{R}T}{V_0}\right)^{2m-1} \left(\prod_{i=1}^{m} M_i\right)^2 \frac{M}{M_1} \frac{\partial}{\partial v_{xj}} \left[\sum_{i=1}^{m} \left\{\left(\frac{\partial M}{\partial v_p} v_p X_w^0 - \frac{X_i}{v_p X_w^0}\right) / M_i\right\}\right]$$

$$= -\left(\frac{\tilde{R}T}{V_0}\right)^{2m-1} \left(\prod_{i=1}^{m} M_i\right)^2 \frac{M}{M_1} \frac{\partial}{\partial v_{xi}} \left[\frac{\partial M}{\partial v_p} (v_p X_w^0)^2 - X_z^0\right]$$

$$= -\left(\frac{\tilde{R}T}{V_0}\right)^{2m-1}\left(\prod_{i=1}^{m} M_i\right)^2 \frac{M}{M_1}\left[\frac{\partial^2 M}{\partial v_p^2}(v_p X_w^0)^2 + \left\{2\left(\frac{\partial M}{\partial v_p}\right) v_p X_w^0 + \frac{X_z^0}{v_p X_w^0}\right\} X_j - \left(\frac{1}{v_p X_w^0}\right) X_j^2\right]. \quad (2.158)$$

Similarly, after substitution of eqn (2.158) into eqn (2.156), we have (ref. 157)

$$|\Delta G'''| = \left(\frac{\tilde{R}T}{V_0}\right)^{3m-2}\left(\prod_{i=1}^{m} M_i\right)^3 \left(\frac{M}{M_1}\right)^2 \left[\frac{\partial^2 M}{\partial v_p^2}(v_p X_w^0)^3 + 2\frac{\partial M}{\partial v_p}(v_p X_w^0)^2 X_z^0 + (X_z^0)^2 - X_z^0 X_{z+1}^0\right]. \quad (2.159)$$

Combining eqn (2.159) with eqn (2.155), we obtain the stability condition given by

$$\frac{\partial^2 M}{\partial v_p^2}(v_p X_w^0)^3 + 2\frac{\partial M}{\partial v_p}(v_p X_w^0)^2 X_z^0 + (X_z^0)^2 - X_z^0 X_{z+1}^0 > 0. \quad (2.160)$$

When the left-hand side of eqn (2.160) is zero, CSP is meta-stable and when that side is negative, CSP is unstable.
Differentiation of eqn (2.113a) with respect to v_p gives

$$\frac{\partial M}{\partial v_p} = \frac{1}{(1-v_p)^2} - \chi_{00}\left\{1+k'\left(1+\frac{1}{X_n^0}-\frac{X_w^0}{X_n^0}\right)\right\}\left\{\sum_{j=1}^{n} p_j j(j+2) v_p^{j-1}\right\} \quad (2.161)$$

and in a similar way, differentiation of eqn (2.161) with respect to v_p yields

$$\frac{\partial^2 M}{\partial v_p^2} = \frac{2}{(1-v_p)^3} - \chi_{00}\left\{1+k'\left(1+\frac{1}{X_n^0}-\frac{X_w^0}{X_n^0}\right)\right\}\left\{\sum_{j=1}^{n} p_j (j-1) j(j+2) v_p^{j-2}\right\}. \quad (2.162)$$

Substituting eqns (2.161) and (2.162) into eq (2.160), we obtain (ref. 157)

$$2(X_w^0)^3\left(\frac{v_p}{1-v_p}\right)^3 + 2(X_w^0)^2 X_z^0\left(\frac{v_p}{1-v_p}\right)^2 + (X_z^0)^2 - X_z^0 X_{z+1}^0$$
$$- \chi_{00}\left\{1+k'\left(1+\frac{1}{X_n^0}+\frac{X_w^0}{X_n^0}\right)\right\}(X_w^0)^2$$
$$\times \left[\sum_{j=0}^{n} p_j j(j+2)\{(j-1)X_w^0 + 2X_z^0\} v_p^{j+1}\right] > 0 \quad (2.163)$$

where X_{z+1}^0 is the (z+1)-average of X_i. Eqn (2.164) can be readily derived from eqn (2.163) (ref. 157).

$$|\Delta G'''| \sim \frac{2(X_w^0)^3}{(1-v_p)^3} + \frac{2(X_w^0)^2 X_z^0}{v_p(1-v_p)^2} + \frac{(X_z^0)^2 - X_z^0 X_{z+1}^0}{v_p^3} - \chi_{00}\{1+k'(1+\frac{1}{X_n^0} - \frac{X_w^0}{X_n^0})\}(X_w^0)^2 \times [\sum_{j=1}^{n} p_j j(j+2)\{(j-1)X_w^0 + 2X_z^0\} v_p^{j-2}] > 0. \qquad (2.164)$$

In the case of $k' = 0$ ($k_0 = 0$) and $p_j \neq 0$, eqn (2.164) reduces to eqn (2.165).

$$|\Delta G'''| \sim \frac{2(X_w^0)^3}{(1-v_p)^3} + \frac{2(X_w^0)^2 X_z^0}{v_p(1-v_p)^2} + \frac{(X_z^0)^2 - X_z^0 X_{z+1}^0}{v_p^3} - \chi_{00}(X_w^0)^2 [\sum_{j=1}^{n} p_j j(j+2)\{(j-1)X_w^0 + 2X_z^0\} v_p^{j-2}] > 0. \qquad (2.165)$$

However, it should be noted that phase equilibria, especially CPC, of polymer solutions can be reasonably interpreted only when the concentration-dependence parameters of χ, p_1 and p_2 are well taken into consideration (refs 3,5). And in the case of $p_1 \neq 0$, $p_2 \neq 0$ and $p_3 = p_4 = \cdots = p_n = 0$, eqn (2.165) becomes (ref. 157)

$$2(X_w^0)^3(\frac{v_p}{1-v_p})^3 + 2(X_w^0)^2 X_z^0 (\frac{v_p}{1-v_p})^2 + (X_z^0)^2 - X_z^0 X_{z+1}^0 - \chi_{00}(X_w^0)^2\{6p_1(X_z^0)v_p^2 + 8p_2(X_w^0 + 2X_z^0)v_p^3\} > 0. \qquad (2.166)$$

When the discussion is limited to the rather unrealistic, but simple case where χ is independent of the polymer concentration and the polymer molecular weight (i.e., $k'(=k_0) = 0$, $p_1, p_2, \cdots, p_n = 0$), the critical concentration at CSP v_p^c is given by a well-known equation (eqn (2.123)) (ref. 133), and in this case eqn (2.165) (or eqn (2.166)) simply reduces to (ref. 157)

$$2(X_w^0)^3(\frac{v_p}{1-v_p})^3 + 2(X_w^0)^2 X_z^0 (\frac{v_p}{1-v_p})^2 + (X_z^0)^2 - X_z^0 X_{z+1}^0 > 0. \qquad (2.167)$$

Substitution of eqn (2.123) into eqn (2.167) gives eqn (2.168) (ref. 157)

$$3X_z^0 + 2(X_z^0)^{1/2} - X_{z+1}^0 > 0. \tag{2.168}$$

Eqn (2.168) is the stability condition of the system of multicomponent polymers dissolved in a single solvent in the case of $k' = p_j = 0$, but absolutely coincides with the corresponding equation (eqn (2.141), (ref. 156)) derived by Šolc for the system of binary monodisperse homologue dissolved in a single solvent and is rewritten as

$$\frac{X_{z+1}^0}{X_z^0} < 3 + \frac{2}{(X_z^0)^{1/2}}. \tag{2.169}$$

Eqn (2.168) (or eqn (2.169)) indicates that CSP in the system of monodisperse (i.e., single component) polymer in a single solvent is always stable without any reservation if $k' = 0$ and $p_1 = p_2 = \cdots = p_n = 0$.

When the original polymer has the Schulz-Zimm (SZ) type MWD, it follows that

$$X_{z+1}^0 - X_z^0 = X_z^0 - X_w^0 = X_w^0 - X_n^0 \tag{2.170}$$

and the discriminant (eqn (2.168)) can be rewritten as

$$\frac{X_w^0}{X_n^0} > \frac{1 + 2/(X_z^0)^{1/2}}{3 + 4/(X_z^0)^{1/2}}. \tag{2.171}$$

And when the original polymer has Wesslau (W) type MWD, it follows that

$$X_{z+1}^0/X_z^0 = X_z^0/X_w^0 = X_w^0/X_n^0 \tag{2.172}$$

and then, eqn (2.168) becomes

$$\frac{X_w^0}{X_n^0} < 3 + \frac{2}{(X_z^0)^{1/2}}. \tag{2.173}$$

Note that eqn (2.168) is valid irrespective of MWD of the original polymer. Inspection of eqn (2.171) leads to the conclusion that CSP is always stable when χ is independent of the molecular weight and of the concentration and the original polymer has the Schulz-Zimm type distribution. In contrast to this, for the Wesslau type polymer, CSP becomes unstable when X_w^0/X_n^0 exceeds round 3. The discriminant of stability of CSP given by eqn (2.165) is expected to be closely correlated with three-phase equilibria (refs 16,35,158-162) and needs further detailed examination.

Schulz-Zimm type molecular weight distribution (eqns (2.59)-(2.61)) and Wesslau type distribution (eqns (2.62)-(2.64)) were adopted for the original polymer $g_0(X_i)$ with $X_w{}^0 = 300$, 500, 5×10^3 and 3×10^4 and $X_w{}^0/X_n{}^0 = 1.0$ (monodisperse), 1.1, 2, 3, 4, 5, 10 and 20 under the conditions of $k' = 0$ and $p_1 = 0 \sim 1.0$ and $p_2 = p_3 = \cdots = p_n = 0$.

As stability criterion, the combination of χ and v_p, which satisfies $|\Delta G'''| = 0$, and denoted hereafter by $\chi_0{}^{sc}$ and $v_p{}^{sc}$, respectively, were calculated with the aid of eqn (2.164) or eqn (2.165).

Figs 2.60 and 2.61 show the condition of stability of CSP (i.e., $\chi_0{}^{cs}$ vs. $v_p{}^{cs}$ curve) together with spinodal curve and CSP for the above mentioned polymer dissolved in a single solvent. Here, it was assumed that $p_1 = 0.6$ and $p_j = 0$ ($j \geq 2$) and the

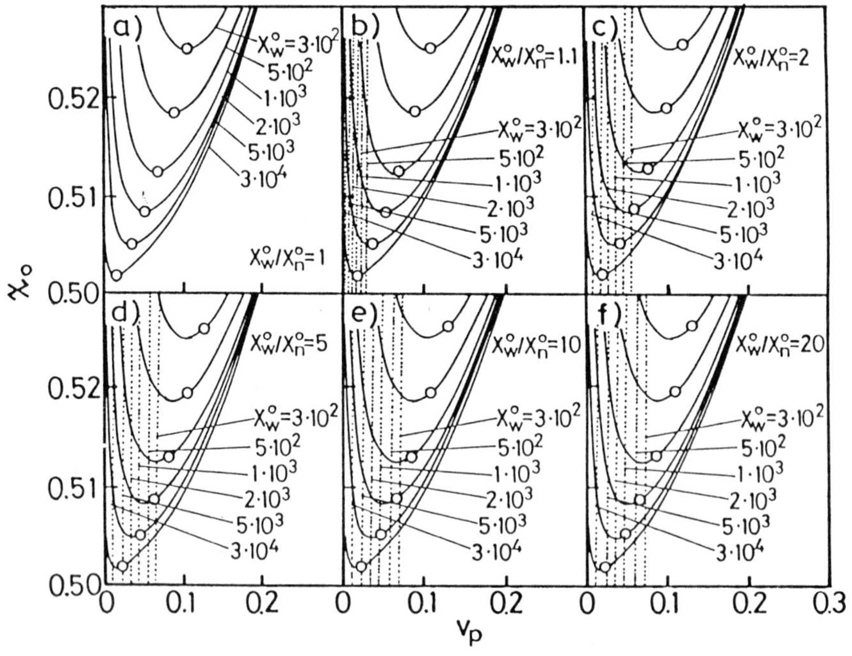

Fig. 2.60 Effect of the weight-average degree of polymerization $X_w{}^0$ on the stability condition (dotted line) of critical solution point (CSP), spinodal curve (solid line) and CSP: Original polymer, Schulz-Zimm type distribution ($X_w{}^0 = 300$, 500, 1×10^3, 2×10^3, 5×10^3 and 3×10^4): a) $X_w{}^0/X_n{}^0 = 1$ (monodisperse); b) $X_w{}^0/X_n{}^0 = 1.1$; c) $X_w{}^0/X_n{}^0 = 2$; d) $X_w{}^0/X_n{}^0 = 5$; e) $X_w{}^0/X_n{}^0 = 10$; f) $X_w{}^0/X_n{}^0 = 20$: $p_1 = 0.6$, $p_2 = \cdots = 0$, $k_0 = 0$; ○, stable CSP.

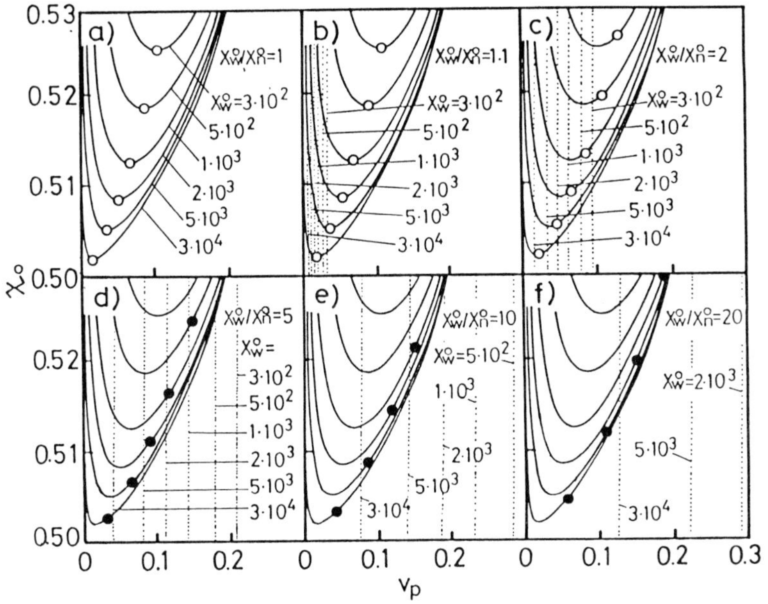

Fig. 2.61 Effect of the weight-average degree of polymerization X_w^0 on the stability condition (dotted line) of critical solution point (CSP), spinodal curve (solid line) and CSP: Original polymer, Wesslau type distribution ($X_w^0=300$, 500, 1×10^3, 2×10^3, 5×10^3 and 3×10^4): a) $X_w^0/X_n^0=1$ (monodisperse); b) $X_w^0/X_n^0=1.1$; c) $X_w^0/X_n^0=2$; d) $X_w^0/X_n^0=5$; e) $X_w^0/X_n^0=10$; f) $X_w^0/X_n^0=20$: $p_1=0.6$, $p_2=\cdots=0$, $k_0=0$; ○, stable CSP; ●, unstable CSP.

stability condition is denoted by the dotted line, which should be compared with CSP. The stability condition line is almost perpendicular to v_p axis (abscissa). For monodisperse polymer solutions with $p_1=0.6$ (Figs 2.60a and 2.61a) CSP is always stable over an entire range of X_w^0 as in the case of monodisperse polymer solutions with $p_1=0$ and $p_j=0$ ($j\geq 2$). The latter was easily verified by eqn (2.168), which holds its validity, irrespective of X_z^0 ($=X_n^0=X_w^0=X_{z+1}^0$), for monodisperse polymer. Generally, the stability limit shifts to a smaller v_p and as previously noted, v_p^c shifts to the same direction with an increase in X_w^0. For the polymer with Schulz-Zimm type distribution (Fig. 2.60), dissolved in a single solvent ($p_1=0.6$), CSP is always located in the right region of the stability condition line, indicating that CSP is always stable over a wide range of X_w^0 ($500\sim3\times10^4$) and X_w^0/X_n^0 ($1.0\sim20$). The stability of CSP for the case of SZ polymer with

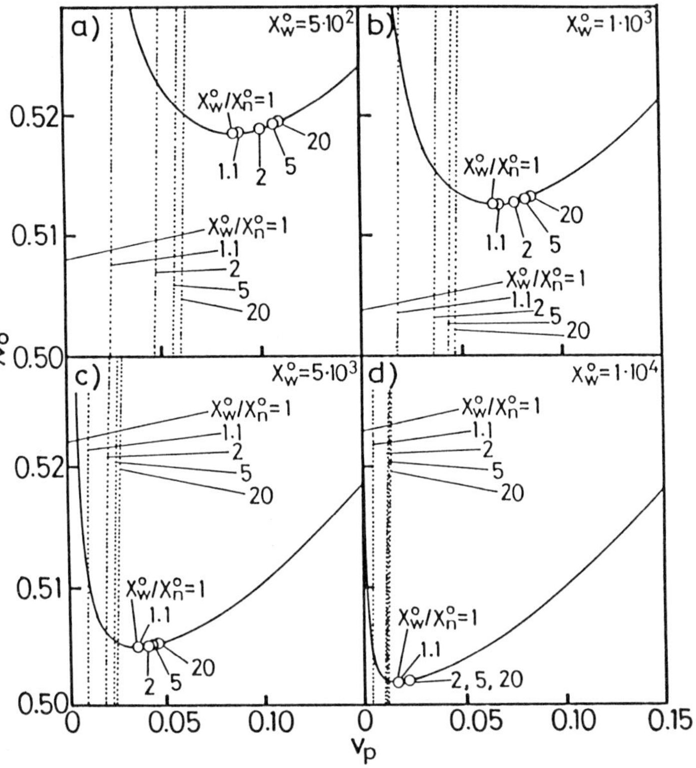

Fig. 2.62 Effect of the polymolecularity X_w^0/X_n^0 on the stability condition (dotted line) of critical solution point (CSP), spinodal curve (solid line) and CSP: Original polymer, Schulz-Zimm type distribution ($X_w^0/X_n^0 = 1$ (monodisperse), 1.1, 1.5, 2, 5, 10, 20): a) $X_w^0 = 500$; b) $X_w^0 = 1\times10^3$; c) $X_w^0 = 5\times10^3$; d) $X_w^0 = 1\times10^4$; $p_1 = \cdots = 0$, $k_0 = 0$; ○, stable CSP.

$p_j = 0$ ($j \geq 1$) was already shown (eqn (2.171)). For the system of the polymer having Wesslau type distribution with less than $X_w^0/X_n^0 = 3$, dissolved in a solvent ($p_1 = 0.6$), CSP is always stable over an entire range of X_w^0, but for the similar system with $X_w^0/X_n^0 > 3$ all CSP lies in the left region of the stability condition, implying that all CSP is unstable, regardless of X_w^0 employed here. Then, we can conclude that the stability of CSP seems to be governed, to the zeroth approximation, by the type of molecular weight distribution of the polymer.

In order to clarify the effect of polymolecularity of the original polymer on the stability of CSP, in Figs 2.62 and 2.63 are shown the stability condition line, CSP and spinodal curve for

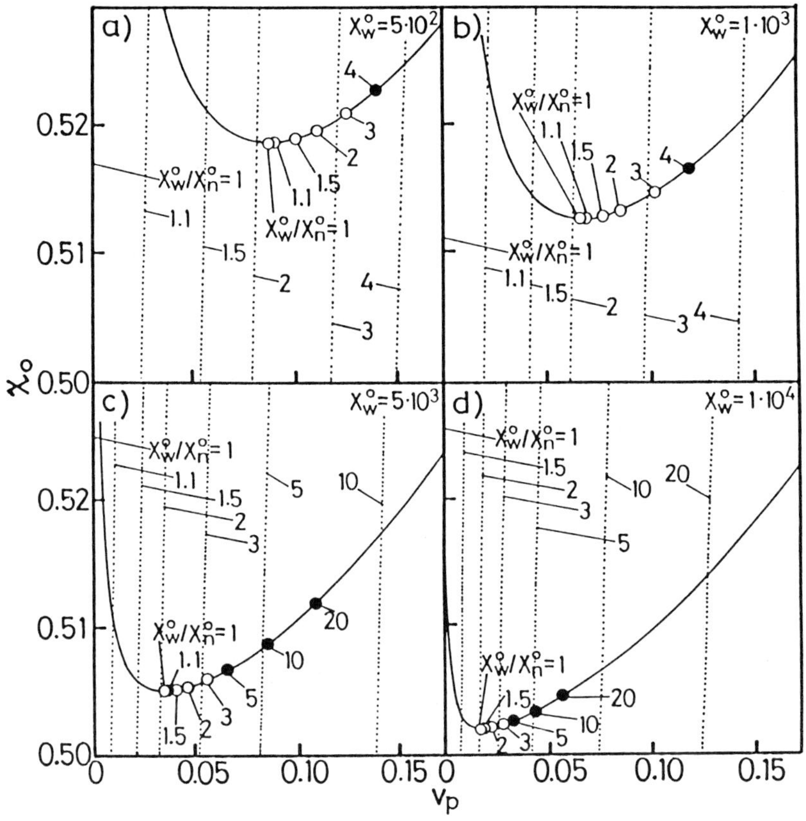

Fig. 2.63 Effect of the polymolecularity $X_w{}^0/X_n{}^0$ on the stability condition (dotted line) of critical solution point (CSP), spinodal curve (solid line) and CSP: Original polymer, Wesslau type distribution ($X_w{}^0/X_n{}^0 = 1$ (monodisperse), 1.1, 1.5, 2, 5, 10, 20): a) $X_w{}^0 = 500$; b) $X_w{}^0 = 1\times10^3$; c) $X_w{}^0 = 5\times10^3$; d) $X_w{}^0 = 1\times10^4$; $p_1 = \cdots = 0$, $k_0 = 0$; ○, stable CSP; ●, unstable CSP.

the system of the polymers with Schulz-Zimm type and Wesslau type distributions, respectively, having the same $X_w{}^0$ and different $X_w{}^0/X_n{}^0$ in the same solvent ($p_1 = 0.6$). It is confirmed again, as eqn (2.173) suggests, that a significant transition of stability of CSP occurs at $X_w{}^0/X_n{}^0$ = around 3, if the original polymer has Wesslau type distribution, regardless of $X_w{}^0$.

Figs 2.64 and 2.65 show the effect of p_1 on the stability condition of CSP for the system of polymers with Schulz-Zimm type

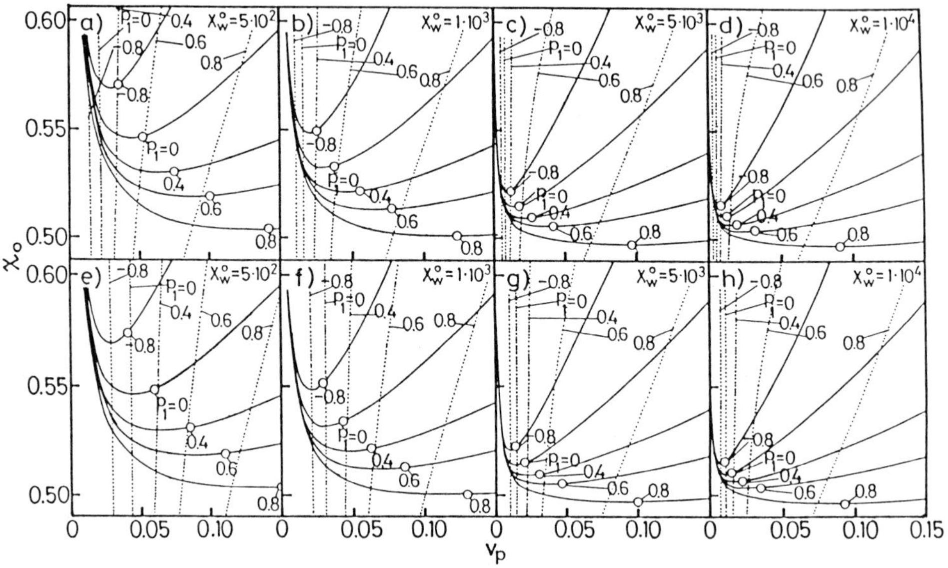

Fig. 2.64 Effect of the 1st order concentration-dependence parameter p_1 of the polymer-solvent interaction χ parameter on the stability condition (dotted line) of critical solution point (CSP), spinodal curve (solid line) and CSP: Original polymer, Schulz-Zimm type distribution (a-d) and Wesslau type distribution (e-h), $X_w^0/X_n^0=2$; a) and e) $X_w^0=500$; b) and f) $X_w^0=1\times10^3$; c) and g) $X_w^0=5\times10^3$; d) and h) $X_w^0=1\times10^4$: $p_1=-0.8$, 0, 0.4, 0.6, 0.8, $p_2=\cdots=0$ and $k_0=0$; ○, stable CSP.

and Wesslau type distributions dissolved in a solvent. As p_1 increases from -0.8 (unrealistically low value) to 0.8, the stability condition line shifts to the direction of larger v_p, with a lowering of its slope. As far as the conditions employed in these two figures ($X_w^0/X_n^0=2$ and $X_w^0=500$, 1×10^3, 5×10^3 and 1×10^4) are concerned, the conclusions obtained in the case of $p_1=0.6$ maintain its validity without any reservation. In the case of Wesslau type distribution with $X_w^0/X_n^0=4$ and $X_w^0=5\times10^3$ and 1×10^4 (Fig. 65g and h), CSP moves from unstable (●) to stable (○) with an increase in p_1. Critical value of p_1, p_1^{cs} is 0.79 for $X_w^0=5\times10^3$ and 0.78 for $X_w^0=1\times10^4$ (ref. 157).

Stability of CSP can also be judged from $|\Delta G'''|$ at CSP (see, eqn (2.165)), then we employ $\delta^4\Delta G/|\delta^4\Delta G(X_w^0/X_n^0=1)|$ at CSP in place of $|\Delta G'''|$ (Here, $|\delta^4\Delta G(X_w^0/X_n^0=1)|$ means absolute value of $\delta^4\Delta G$ for monodisperse polymer having the same X_w^0) and plotted in Figs 2.66-2.68 over a quite wide range of p_1, X_w^0 and X_w^0/X_n^0

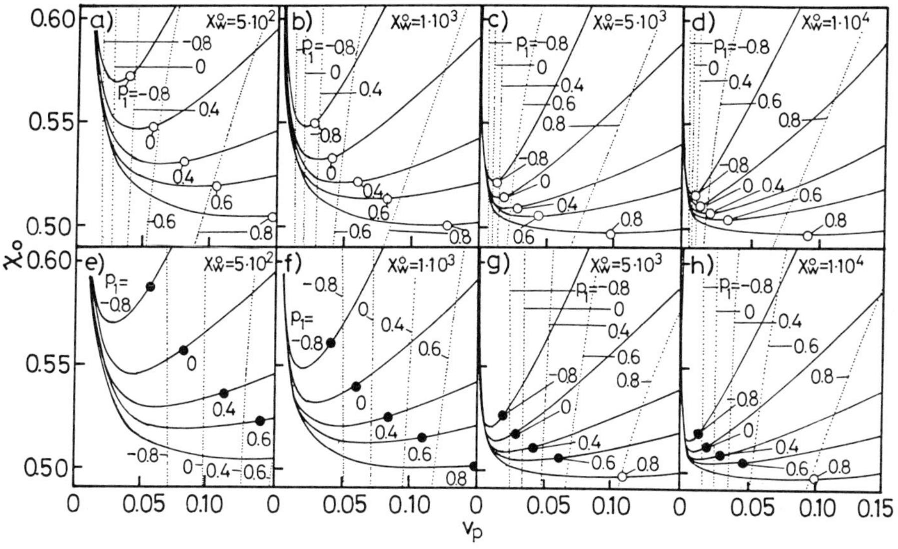

Fig. 2.65 Effect of the 1st order concentration-dependence parameter p_1 of the polymer-solvent interaction χ parameter on the stability condition (dotted line) of critical solution point (CSP), spinodal curve (solid line) and CSP: Original polymer; Schulz-Zimm type distribution (a-d) and Wesslau type distribution (e-h), $X_w^0/X_n^0=4$; a) and e) $X_w^0=500$; b) and f) $X_w^0=1\times10^3$; c) and g) $X_w^0=5\times10^3$; d) and h) $X_w^0=1\times10^4$: $p_1=-0.8$, 0, 0.4, 0.6, 0.8, $p_2=\cdots=0$ and $k_0=0$; ○, stable CSP; ●, unstable CSP.

for polymers with Schulz-Zimm type and Wesslau type distributions. In Schulz-Zimm type polymer solutions $\delta^4\Delta G/|\delta^4\Delta G(X_w^0/X_n^0=1)|$ becomes larger for smaller X_w^0, larger p_1 and larger X_w^0/X_n^0. In contrast to this, in Wesslau-type polymer solutions, $\delta^4\Delta G/|\delta^4\Delta G(X_w^0/X_n^0=1)|$ increases at first shortly with an increase in X_w^0/X_n^0, then remarkably decreases after passing through maximum and finally becomes negative. Initial increase in $\delta^4\Delta G$ becomes significant for larger p_1.

X_w^0/X_n^0, at which $\delta^4\Delta G$ at CSP becomes zero, is nearly constant in the range $p_1\leq0.4$ (probably $p_1\lesssim0.65$), very slightly decreasing with increasing X_w^0, for example from 3.05 ($X_w^0=500$) to 3.005 ($X_w^0=5\times10^4$) at $p_1=0$. For $p_1=0.8$, X_w^0/X_n^0 at $\delta^4\Delta G$(at CSP) $=0$ increases with X_w^0, for example from 3.67 ($X_w^0=500$) to 4.42 ($X_w^0=5\times10^4$).

The effect of p_1 on $\delta^4\Delta G$ at CSP is more systematically demonstrated in Fig. 2.68. For Schulz-Zimm type polymer solutions,

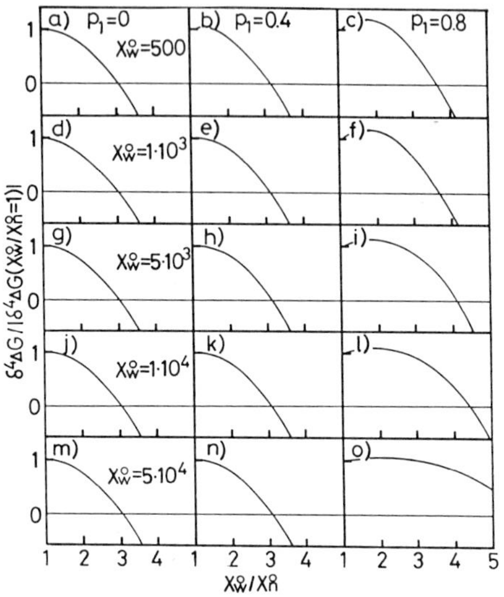

Fig. 2.66 Plots of the ratio of fourth variable of Gibbs' free energy $\delta\,\Delta^4G$ to absolute value of $\delta\,^4\Delta G$ of monodisperse polymer $|\delta\,^4\Delta G(X_w{}^0/X_n{}^0=1)|$ at critical solution point against the polymolecularity $X_w{}^0/X_n{}^0$: Original polymer, Schulz-Zimm type distribution; $X_w{}^0/X_n{}^0=1\sim5$; $X_w{}^0=500$ (a-c), 1×10^3 (d-f), 5×10^3 (g-i), 1×10^4 (j-l), 5×10^4 (m-o): a), d), g), j), m) $p_1=0$; b), e), h), k), n) $p_1=0.4$; c), f), i), l), o) $p_1=0.8$; $p_2=\cdots=0$ and $k_0=0$.

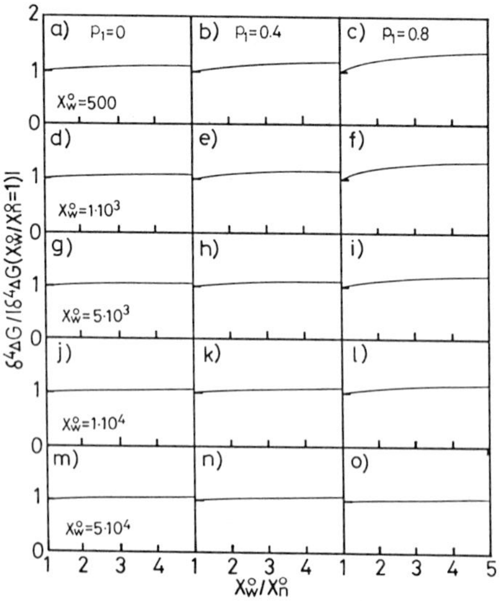

Fig. 2.67 Plots of the ratio of fourth variable of Gibbs' free energy $\delta\,^4\Delta G$ to absolute value of $\delta\,^4\Delta G$ of monodisperse polymer $|\delta\,^4\Delta G(X_w{}^0/X_n{}^0=1)|$ at critical solution point against the polymolecularity $X_w{}^0/X_n{}^0$: Original polymer, Wesslau type distribution; $X_w{}^0/X_n{}^0=1\sim5$; $X_w{}^0=500$ (a-c), 1×10^3 (d-f), 5×10^3 (g-i), 1×10^4 (j-l), 5×10^4 (m-o): a), d), g), j), m) $p_1=0$; b), e), h), k), n) $p_1=0.4$; c), f), i), l), o) $p_1-0.8$; $p_2=\cdots=0$ and $k_0=0$.

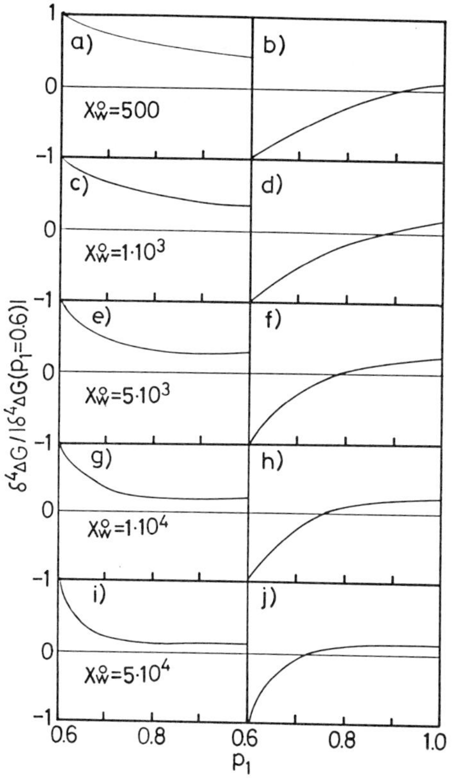

Fig. 2.68 Plots of the ratio of fourth variable of Gibbs' free energy $\delta^4\Delta G$ to absolute value of $\delta^4\Delta G$ at $p_1=0.6$ $|\delta^4\Delta G(p_1=0.6)|$ at critical solution point against the 1st order concentration-dependence parameter p_1 of the polymer-solvent interaction χ parameter: Original polymer, Schulz-Zimm type distribution (a, c, e, g, i) and Wesslau type distribution (b, d, f, h, j); $X_w{}^0/X_n{}^0=4$, and $X_w{}^0=500$ (a and b), $X_w{}^0=1\times10^3$ (c and d), $X_w{}^0=5\times10^3$ (e and f), $X_w{}^0=1\times10^4$ (g and h), $X_w{}^0=5\times10^4$ (i and j), $p_1=0.6\sim1.0$, $p_2=\cdots=0$, $k_0=0$.

$\delta^4\Delta G$ at CSP decreases monotonically from a positive value approaching an asymptotic value (even always positive), depending on $X_w{}^0$ and $X_w{}^0/X_n{}^0$, but for Wesslau type polymer solutions, $\delta^4\Delta G$ at CSP increases from a negative value to a positive value, passing through zero (i.e., the stability condition). For example, p_1 for $\delta^4\Delta G(\text{at CSP})=0$ is 0.926 ($X_w{}^0=500$), 0.873 ($X_w{}^0=1\times10^3$), 0.786 ($X_w{}^0=5\times10^3$), 0.761 ($X_w{}^0=1\times10^4$) and 0.722 ($X_w{}^0=5\times10^4$) for Wesslau type polymer ($X_w{}^0/X_n{}^0=4$) solutions. CSP of the system becomes stable if the concentration dependence of χ is positively larger, even if the polymer has Wesslau type MWD with $X_w{}^0/X_n{}^0$ larger than $3+1/(X_z{}^0)^{1/2}$ (see, eqn (2.172)). Then, it should be

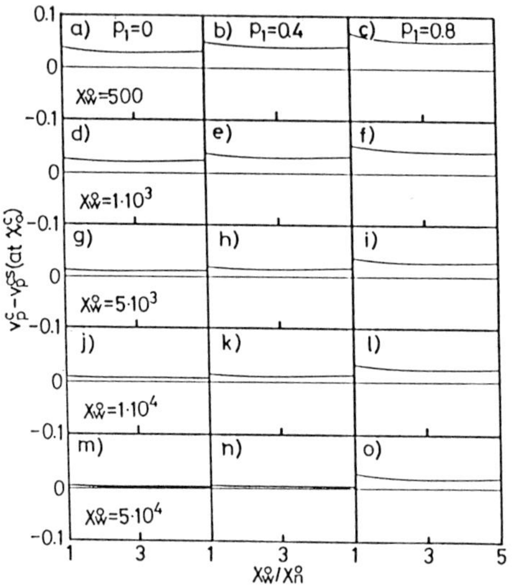

Fig. 2.69 Plots of $v_p^c - v_p^{cs}$(at χ_0^c) against X_w^0/X_n^0; v_p^{cs}(at χ_0^c), v_p of stability condition at χ_0^c: Original polymer, Schulz-Zimm type distribution, $X_w^0/X_n^0 = 1 \sim 5$, $X_w^0 = 500$ (a-c), 1×10^3 (d-f), 5×10^3 (g-i), 1×10^4 (j-l), 5×10^4 (m-o); $p_1 = 0$ (a, d, g, j, m); $p_1 = 0.4$ (b, e, h, k, n); $p_1 = 0.8$ (c, f, i, l, o); $p_2 = \cdots = 0$, $k_0 = 0$.

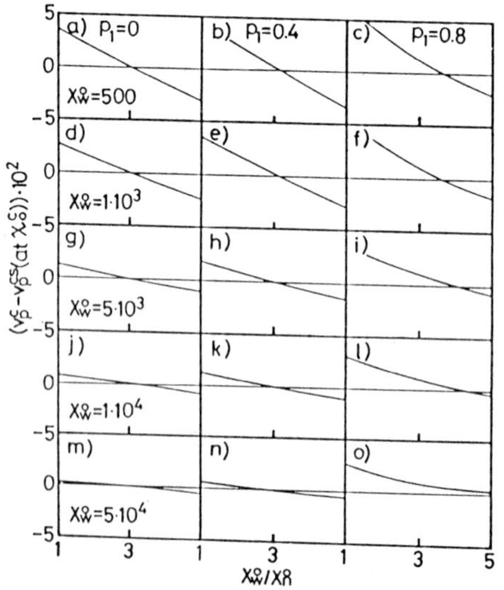

Fig. 2.70 Plots of $v_p^c - v_p^{cs}$(at χ_0^c) against X_w^0/X_n^0; v_p^{cs}(at χ_0^c), v_p of stability condition at χ_0^c: Original polymer, Schulz-Zimm type distribution, $X_w^0/X_n^0 = 1 \sim 5$, $X_w^0 = 500$ (a-c), 1×10^3 (d-f), 5×10^3 (g-i), 1×10^4 (j-l), 5×10^4 (m-o); $p_1 = 0$ (a, d, g, j, m); $p_1 = 0.4$ (b, e, h, k, n); $p_1 = 0.8$ (c, f, i, l, o); $p_2 = \cdots = 0$, $k_0 = 0$.

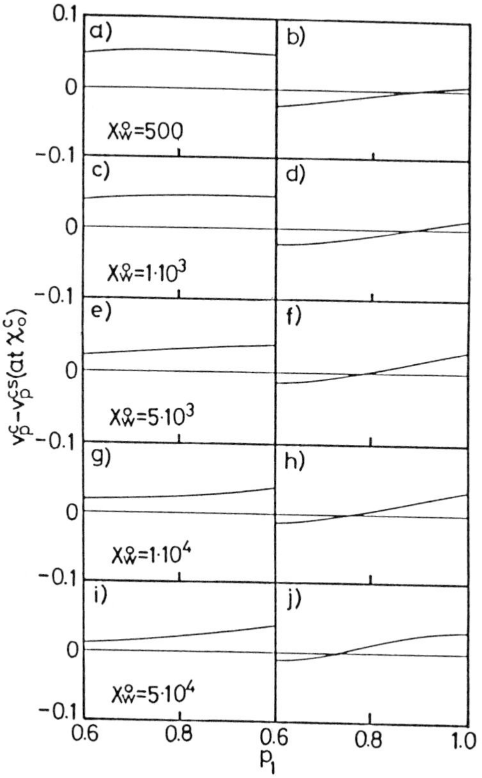

Fig. 2.71 Plots of $v_p^c - v_p^{cs}$(at χ_0^c) against the 1st order concentration dependence parameter p_1 of the polymer-solvent interaction χ parameter; v_p^{cs}(at χ_0^c), v_p of stability condition at χ_0^c: Original polymer, Schulz-Zimm type distribution (b, d, f, h, j), $X_w^0/X_n^0 = 4$ and $X_w^0 = 5600$ (a and b), $X_w^0 = 1\times10^3$ (c and d), $X_w^0 = 5\times10^3$ (e and f), $X_w^0 = 1\times10^4$ (g and h), $X_w^0 = 5\times10^4$ (i and j); $p_1 = 0.6\sim1.0$, $p_2 = \cdots = 0$, $k_0 = 0$.

noted that careless application of eqn (2.172) to an actual polymer / solvent system, in which p_1 is near to the theoretically expected 2/3, if the system is a non-polar polymer / non-polar solvent system, in which the molecular forces acting between the solute and the solvent is only dispersion force, may lead us to erroneous conclusion against the stability of CSP.

We can also use as a parameter of stability criterion the difference between the critical polymer volume fraction v_p^c (v_p at CSP) and v_p^{cs} at χ_0^c (referred to as v_p^{cs}(at CSP); here χ_0^c is χ_0 at CSP). For the polymer with Schulz-Zimm type distribution in a solvent, $v_p^c - v_p^{cs}$(at CSP) is very insensitive to X_w^0/X_n^0, decreasing with increasing X_w^0 (Fig. 2.69), but for the polymer

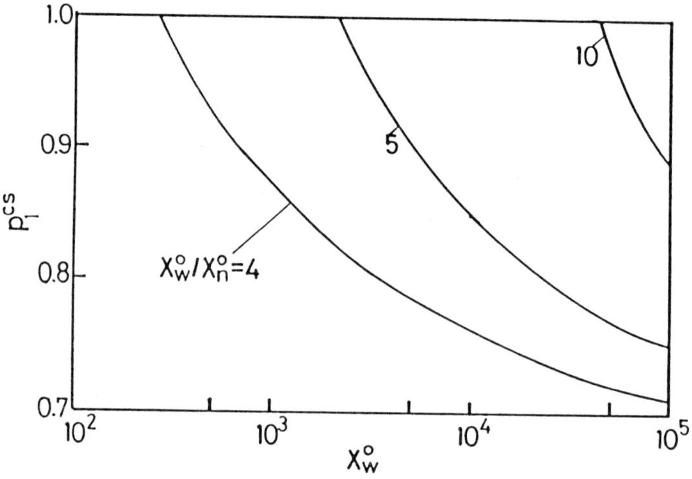

Fig. 2.72 Plot of p_1^{cs} against X_w^0: Original polymer Wesslau type distribution, number on curve denote the polymolecularity X_w^0/X_n^0, $p_2 = \cdots = 0$, and $k_0 = 0$.

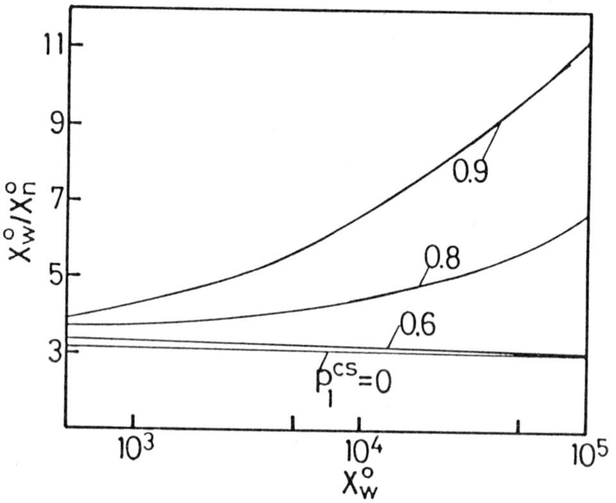

Fig. 2.73 Relations among X_w^0/X_n^0, X_w^0 and p_1^{cs}: Original polymer, Wesslau type distribution.

with Wesslau type distribution dissolved in a solvent, the above difference decreases monotonically with an increase in X_w^0/X_n^0

(Fig. 2.70) and the point at $v_p^{\,c} = v_p^{\,cs}$ (at CSP) is in good agreement with that at $\delta\,{}^{+}\Delta G = 0$. $v_p^{\,c} - v_p^{\,cs}$ (at CSP) increases from a positive (Schulz-Zimm type MWD) or a negative (Wesslau type MWD) value at $p_1 = 0$ (not shown here) with p_1 (Fig. 2.71). For the latter system p_1 at $v_p^{\,c} = v_p^{\,cs}$ (at CSP) is in good agreement with that at $\delta\,{}^{+}\Delta G = 0$.

Fig. 2.72 shows $p_1^{\,cs}$, at which $\delta\,{}^{+}\Delta G$ is zero and above which $\delta\,{}^{+}\Delta G$ becomes positive (i.e., above which CSP is stable), as a function of X_w^0 of the Wesslau type polymer with a given X_w^0/X_n^0. For this system with larger polymer polymolecularity, the stable CSP is realized only at higher X_w^0 and higher p_1 regions.

Fig. 2.73 shows the relation between X_w^0/X_n^0 and X_w^0 for a given p_1 at $\delta\,{}^{+}\Delta G = 0$ for systems of Wesslau type polymer and a single solvent. In the range of $p_1 \leq 0.6$, X_w^0/X_n^0 decreases very slightly with X_w^0 and in the range $p_1 \gtrsim 0.8$, X_w^0/X_n^0 increases with X_w^0.

Fig. 2.74 shows the SC, neutral equilibrium condition and stability criterion, estimated using eqns (2.121), (2.122) and (2.163) for the given polymer (Schulz-Zimm type MWD, $X_w^0/X_n^0 = 2.8$,

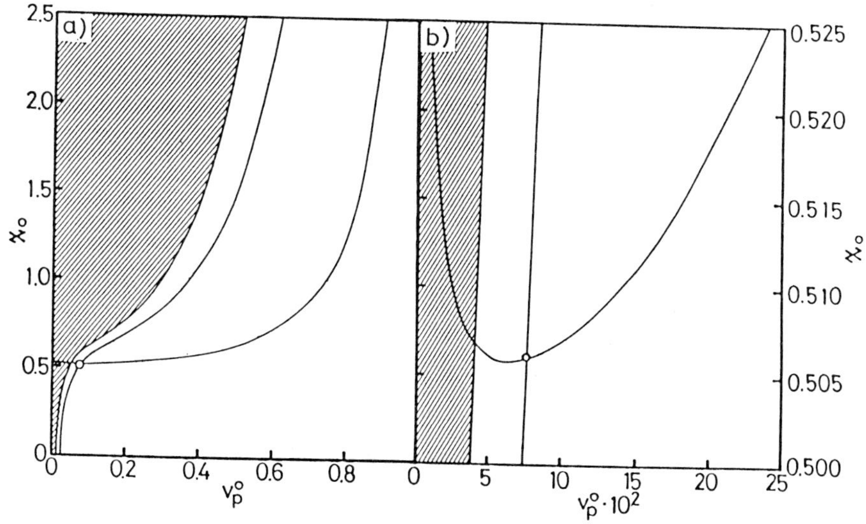

Fig. 2.74 Spinodal curve (spinodal condition), neutral equilibrium and stability conditions of quasi-binary solution: Original polymer, Schulz-Zimm type distribution ($X_w^0 = 2117$ and $X_w^0/X_n^0 = 2.8$); $p_1 = 0.643$, $p_2 = 0.200$ and $k_0 = 0$. Unfilled circle, critical solution point (CSP, $v_p^{\,c} = 0.0766$); shadowed region, unstable area for CSP.

$X_w^0 = 2117$) / solvent ($p_1 = 0.643$, $p_2 = 0.200$ and $k' = 0$) system, which was occasionally used in the previous studies (refs 7,21,22) for computer experiments, and projected on a χ_0-v_p^0 plane. In Fig. 2.74a, v_p^0 covers an entire range ($0 \sim 1$) and accordingly, χ_0 changes from 0 to 2.5. SC reveals a downward curvature (i.e., $\partial^2 \chi_0/\partial (v_p^0)^2 > 0$). The neutral equilibrium condition has an inflection point at $v_p^0 \simeq 0.1$ and its cross point with spinodal gives CSP as shown as an unfilled circle. Since, in the process of derivation of the neutral equilibrium condition (eqn (2.122)), we employed spinodal condition (eqn (2.121)), the neutral equilibrium condition given by eqn (2.122) except CSP should be noted to be meaningless except CSP. In addition, the stability criterion has an inflection point at $v_p^0 \simeq 0.2$. The shadowed area is unstable (i.e., $|\Delta G'''| < 0$) for CSP and unshadowed area is stable for CSP (i.e., $|\Delta G'''| > 0$). For this system, we obtain $v_p^c = 0.0766$, $\chi_0^c = 0.506$ and v_p^{cs} (at χ_0^c) $= 0.040$ to confirm $v_p^c > v_p^{cs}$ (at χ_0^c), indicating that CSP, calculated in the previous studies (ref. 4,6), for the system is unquestionably stable.

Fig. 2.74b shows an expansion of Fig. 2.74a in the narrow area of $0 < v_p^0 < 0.20$ and $0.5 < \chi_0 < 0.525$. Downward convexity of spinodal curve becomes very clear.

Fig. 2.75 shows SC, stability condition, neutral equilibrium condition, all determined in this study as well as the shadow curve and the cloud point curve, estimated for the same polymer / solvent system elsewhere (refs 4,6). The unstable region (i.e., the region of $|\Delta G'''| < 0$) is also shadowed. SC (i.e., χ_0^{SP}-v_p^0 curve, χ_0^{SP} is χ_0 at spinodal condition), estimated using eqn (2.121) crosses with the cloud point curve, determined from eqn (2.191), at CSP as shown by the unfilled circle and they are confirmed to have a common tangent (not shown in the figure).

$$\left(\frac{\partial \chi_0^{SC}}{\partial v_p^0}\right)_{v_p^0 = v_p^c} = \left(\frac{\partial \chi_0^{CPC}}{\partial v_p^0}\right)_{v_p^0 = v_p^c}. \qquad (2.174)$$

In this article, it is shown that the stability condition is strongly governed by the molecular weight distribution of the original polymer; in particular CSP is unstable when the polymer has Wesslau type distribution with $X_w^0/X_n^0 >$ round 3. Instability of CSP is expected not to be characteristic of Wesslau type distribution alone. This is closely correlated with discussion by Tompa (refs 154,158) on three-phase equilibrium, because three-phase equilibrium prerequisites at least instability of CSP in two-phase equilibrium.

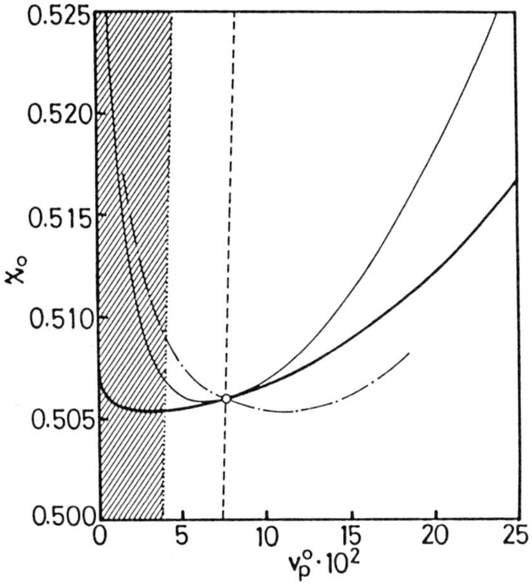

Fig. 2.75 Cloud point curve (broad solid line), spinodal curve (fine solid line), and shadow curve (chain line) together with neutral equilibrium and stability conditions (broken and dotted lines, respectively): Original polymer, Schulz–Zimm type distribution ($X_w^0 = 2117$ and $X_w^0/X_n^0 = 2.8$); $p_1 = 0.643$, $p_2 = 0.200$, and $k_0 = 0$. Unfilled circle, critical solution point (CSP, $v_p^c = 0.0766$); shadowed region, unstable area for CSP.

We consider firstly polymer blends prepared by mixing at various ratios of two polymer fractions (of the same chemical compositions) with extremely sharp Schulz-Zimm type molecular weight distribution ($X_w^0/X_n^0 = 1.03$) and having large different X_w^0 values (i.e., $X_w^0 = 100$ and 4×10^3). Here, the polymer fractions with $X_w^0 = 100$ and 4×10^3 are denoted as P1 and P2, respectively and their weight fractions in the mixture as w_{P1} and w_{P2} ($= 1 - w_{P1}$).

Fig. 2.76 shows the molecular weight distributions of P1 / P2 mixtures with w_{P1} ranging from 0.95 to 0.999. All these mixtures have apparently double-peak distribution. Note that P2 has small X_w^0/X_n^0 (1.03), but has a variety of components ranging between $2.4 \times 10^3 < X < 6 \times 10^3$. The molecular characteristics of the above mixtures are summarized in Table 2.17. Interestingly, X_n^0 can be approximately regarded as constant (97～102) through all mixtures

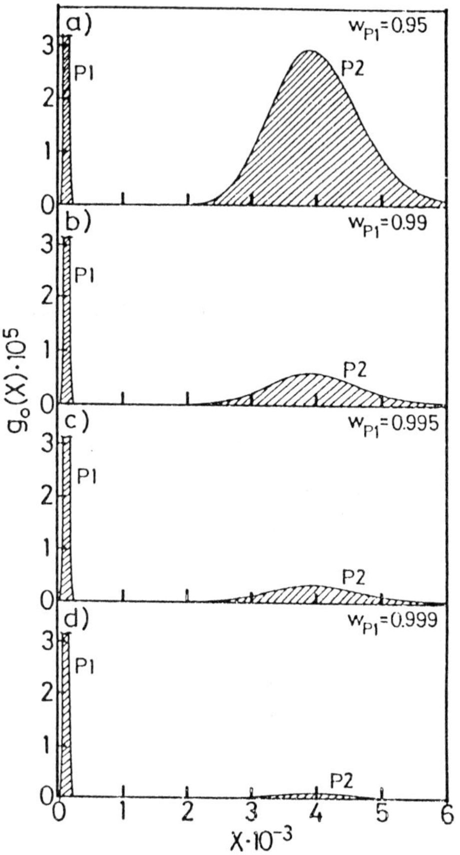

Fig. 2.76 Molecular weight distribution of mixture of two polydisperse polymers belonging to the same chemical homologue: Schulz-Zimm type distribution ($X_w^0/X_n^0=1.03$), $X_w^0=100$ (P1) and 4×10^3 (P2). a) $w_{P1}=0.95$, b) $w_{P1}=0.99$, c) $w_{P1}=0.995$, d) $w_{P1}=0.999$.

TABLE 2.17

Characteristics of the mixture of polymer.

	Average degree of polymerization				Polymolecularity		
w_{P1}	X_n^0	X_w^0	X_z^0	X_{z+1}^0	X_w^0/X_n^0	X_z^0/X_w^0	X_{z+1}^0/X_z^0
0.950	102.1	295.0	2824	4185	2.890	9.573	1.482
0.990	98.04	139.0	1258	3993	1.418	9.050	3.174
0.995	97.56	119.5	774.6	3776	1.225	6.482	4.875
0.999	97.18	103.9	257.4	2647	1.069	2.478	10.28

with a wide range of w_{P1} covering 0.95~0.999. As w_{P1} increases, X_w^0/X_n^0 and X_z^0/X_w^0 decrease and X_{z+1}^0/X_z^0 increases from 1.482 (at $w_{P1}=0.950$) to 10.28 (at $w_{P1}=0.999$) (ref. 157).

Fig. 2.77 shows the spinodal curve, CSP, neutral equilibrium condition and stability criterion for the system of P1 / P2 mixture in a single solvent. For $w_{P1}\geq 0.99$ the mixture has $X_{z+1}^0/X_z^0\geq 3.174$ and as eqn (2.169) predicts, CSP for the system of one of these mixtures in a solvent with $p_1=0$ becomes unstable. CSP for the systems with $p_1=0.6$ is also unstable. Then, we come to the conclusion that the solution of polymer with not only Wesslau type distribution, but also any distribution satisfying eqn (2.164) has unstable CSP (ref. 157).

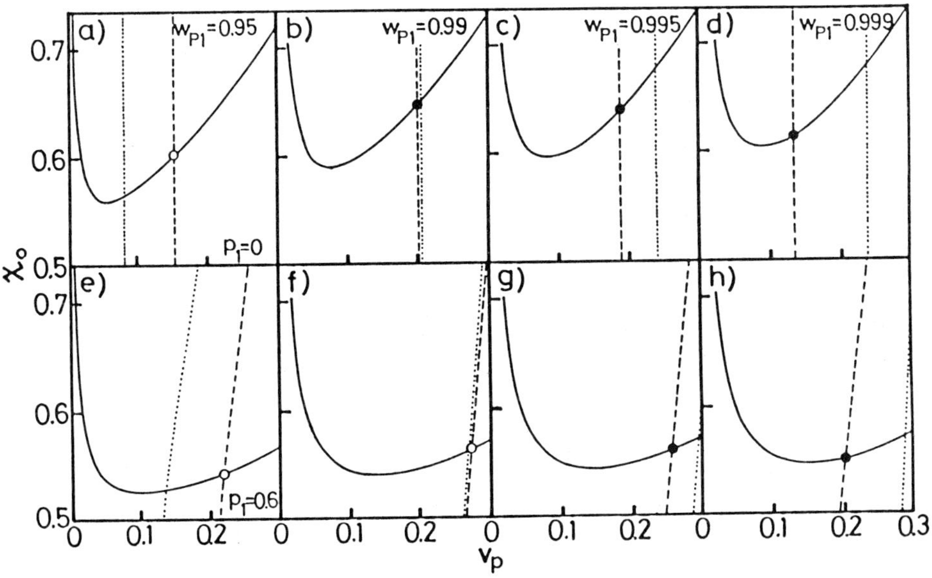

Fig. 2.77 Spinodal curve (solid line), neutral equilibrium condition (broken line), stability condition (dotted line) of critical solution point (CSP), and CSP of quasi-binary solutions: Original polymer, mixture of two Schulz-Zimm type distribution ($X_w^0/X_n^0=1.03$), $X_w^0=100$ (P1) and 4×10^3 (P2); a) and e), $w_{P1}=0.95$, b) and f), $w_{P1}=0.99$, c) and g), $w_{P1}=0.995$, d) and h), $w_{P1}=0.999$; $p_1=0$ (a-d) and $p_1=0.6$ (e-h); ○, stable CSP; ●, unstable CSP.

Three-phase equilibria have been experimentally observed by Koningsveld et al. (refs 16,35,159), Kleintjens et al. (ref. 160), and Einaga and his coworkers (ref. 161) for atactic polystyrene /

cyclohexane system and by Dobashi and Nakata (ref. 162) for atactic polystyrene / methylcyclohexane systems. Fortunately, for these systems p_1 and p_2 have been determined by Kamide and his collaborators (refs 29,31,34) from analysis on CSP and CPC data. Then, we can calculate the stability conditions of CSP using eqn (2.165).

Summarizing, the stability conditions of CSP is derived without approximation for multicomponent polymer / single solvent systems, taking into account the concentration- and the molecular weight-dependence of χ parameter. The equation previously derived by Šolc is confirmed to have a reduced form of general equation of the stability conditions (eqn (2.164)) obtained here. CSP is theoretically expected to be stable for solutions of polymer with larger X_w^0 and smaller X_w^0/X_n^0, dissolved in a solvent having larger p_1 against the polymer. CSP is unstable if X_{z+1}^0/X_z^0 (not X_w^0/X_n^0) is larger than around 3, which is realized for Wesslau-type polymers and for the polymers with double-peak distribution. with a wide range of w_{P1} covering 0.95～0.999. As w_{P1} increases, X_w^0/X_n^0 and X_z^0/X_w^0 decrease and X_{z+1}^0/X_z^0 increases from 1.482 (at $w_{P1}=0.950$) to 10.28 (at $w_{P1}=0.999$) (ref. 157).

2.6 CLOUD POINT CURVE

The experimental cloud point curve (CPC) of a polymer / single solvent system is known to be highly unsymmetrical with respect to polymer concentration, regardless of the polymolecularity of the polymer (refs 163,164). This high degree of dissymmetry in CPC for monodisperse polymer / solvent systems can be explained theoretically in terms of its large molar volume ratio X_1 compared to the solvent. For the strictly monodisperse polymer solution (if available) the coexistence curve should be in excellent agreement with CPC. As early as 1952, Shultz and Flory (ref. 165) determined experimentally the CPC for polydisperse polyisobutyrene / diisobutyl ketone and PS / CH systems, comparing coexistence curve instead of CPC for these systems, using the Flory-Huggins type thermodynamic theory for polymer solutions and assuming a monodisperse polymer / single solvent system and a constant thermodynamic interaction parameter χ over a wide range of concentration (see, Figs 2 and 3 of ref. 165). They noticed that the shapes of the calculated CPC agreed only quantitatively with those observed and the latter was markedly broader and the observed CSP occurred at higher polymer concentrations than those predicted. The significant divergence of the observed CPC from the simple thermodynamic theory may be due to the neglect of (1) the polymolecularity of the polymers and (2) the concentration dependence of the thermodynamic interaction parameter χ , which play an important role in the phase equilibrium of polymer solutions as shown by Kamide and his collaborators (refs 7,8,21,22,29,31,76,93,96). Unfortunately, the effect of the polymolecularity of the polymer and concentration dependence of the χ parameter on CPC and coexistence curve could not be correctly estimated in 1952 due to the lack of theoretical background of phase equilibrium of multicomponent polymer solutions. Shultz and Flory (ref. 165) determined also experimentally the CPC for a mixture of two PS fractions (the relative molar volume ratio of the polymer to the solvent $X = 768$ and 1.095×10^4) dissolved in CH, pointing out that the CPC of this quasi-ternary system is depressed in a higher concentration range than the precipitation threshold point.

Rehage et al. (ref. 166) estimated the cloud point as the cross point of a coexistence curve, obtained for a multicomponent PS / CH system with a given v_p^0 and a v_p^0 axis. Kuwahara et al. (ref. 66) measured directly the CPC of five PS samples dissolved in CH near their CSP, showing experimentally that the polymer volume fractions at the precipitation threshould point (v_p^{tcp}) are significantly smaller than v_p^c even for a solution of "monodisperse" PS characterized by the ratio of the weight- to the number-average molecular weight $M_w/M_n < 1.03$. Koningsveld and Staverman (refs 13,17,35,38,40) applied Rehage et al.'s procedure for estimating the cloud point to their calculations using log-normal, bimodal and exponential type molecular weight distributions for the original polymer.

Up to now two direct methods have been proposed for calculating CPC by Kamide, Matsuda et al. and by Šolc. In this section, an attempt will be made (1) to compare carefully detailed deviation of the two methods and (2) to examine which method gives more accurate CPC and shadow curve.

1) Kamide and Matsuda et al.'s method

In Kamide and Matsuda et al.'s method (refs 4,6), χ parameter is empirically expressed as functions of the polymer volume fraction v_p and the number-average degree of polymerization X_n, as is given by eqn (2.3). The chemical potentials of solvent and the X_i-mer in quasi-binary system, $\Delta\mu_0$ and $\Delta\mu_{Xi}$ are, according to the modified Flory-Huggins solution theory, given by eqns (2.9) and (2.21):

$$\Delta\mu_0 = \tilde{R}T\{\ln(1-v_p) + (1-\frac{1}{X_n})v_p + \chi_{00}(1+\frac{k'}{X_n})(1+\sum_{j=1}^{n} p_j v_p{}^j)v_p{}^2\}, \tag{2.9}$$

$$\Delta\mu_{Xi} = \tilde{R}T[\ln v_{Xi} - (X_i - 1) + X_i(1-\frac{1}{X_n})v_p$$
$$+ X_i(1-v_p)^2\chi_{00}[(1+\frac{k'}{X_n})\{1+\sum_{j=1}^{n}\frac{p_j}{j+1}\{\sum_{q=0}^{j}(q+1)v_p{}^q\}\}$$
$$+ k'(\frac{1}{X_i}-\frac{1}{X_n})\{\frac{1}{1-v_p}+\sum_{j=1}^{n}\frac{p_j}{j+1}(\sum_{q=0}^{j}\frac{v_p{}^q}{1-v_p})\}]]. \tag{2.21}$$

When the quasi-binary system is at constant temperature and constant pressure, in two-phase equilibrium, $\Delta\mu_0$ and $\Delta\mu_{Xi}$ should be subjected to the Gibbs' conditions given by

$$\Delta\mu_{0(1)} = \Delta\mu_{0(2)}, \tag{1.21}$$

$$\Delta\mu_{Xi(1)} = \Delta\mu_{Xi(2)} \qquad (i=1,\cdots,m). \tag{1.22}$$

Here, the suffixes (1) and (2) mean the polymer-lean phase and the polymer-rich phase, respectively, and m is the total number of components consisting of the polymer.

Substitution of eqn (2.9) into eqn (1.21) leads to eqn (2.34).

$$\chi_{00} = [\ln\frac{1-v_{p(1)}}{1-v_{p(2)}} + (v_{p(1)} - v_{p(2)}) - (\frac{v_{p(1)}}{X_{n(1)}} - \frac{v_{p(2)}}{X_{n(2)}})]$$
$$/[(v_{p(2)}{}^2 - v_{p(1)}{}^2) + k'(\frac{v_{p(2)}{}^2}{X_{n(2)}} - \frac{v_{p(1)}{}^2}{X_{n(1)}})$$
$$+ \sum_{j=1}^{n} p_j\{(v_{p(2)}{}^{j+2} - v_{p(1)}{}^{j+2}) + k'(\frac{v_{p(2)}{}^{j+2}}{X_{n(2)}} - \frac{v_{p(1)}{}^{j+2}}{X_{n(1)}})\}]. \tag{2.34}$$

Substituting eqn (2.21) into eqn (1.22) we obtain the partition coefficient σ_i (see, section 2.3),

$$\sigma_i \equiv \frac{1}{X_i}\ln\frac{v_{Xi(2)}}{v_{Xi(1)}} = \sigma_0 + \sigma_{01}/X_i \tag{2.35}$$

with

$$\sigma_0 = (v_{p(1)} - v_{p(2)}) - (\frac{v_{p(1)}}{X_{n(1)}} - \frac{v_{p(2)}}{X_{n(2)}})$$

$$- \chi_{00} [2(v_{p(1)} - v_{p(2)}) - (v_{p(1)}{}^2 - v_{p(2)}{}^2)$$

$$+ \sum_{j=1}^{n} p_j \{\frac{j+2}{j+1}(v_{p(1)}{}^{j+1} - v_{p(2)}{}^{j+1}) - (v_{p(1)}{}^{j+2} - v_{p(2)}{}^{j+2})\}]$$

$$- \chi_{00} k' [(\frac{v_{p(1)}}{X_{n(1)}} - \frac{v_{p(2)}}{X_{n(2)}}) - (\frac{v_{p(1)}{}^2}{X_{n(1)}} - \frac{v_{p(2)}{}^2}{X_{n(2)}})$$

$$+ \sum_{j=1}^{n} p_j \{(\frac{v_{p(1)}{}^{j+1}}{X_{n(1)}} - \frac{v_{p(2)}{}^{j+1}}{X_{n(2)}}) - (\frac{v_{p(1)}{}^{j+2}}{X_{n(1)}} - \frac{v_{p(2)}{}^{j+2}}{X_{n(2)}})\}] \quad (2.36)$$

and

$$\sigma_{01} = -k' \chi_{00} \{(v_{p(1)} - v_{p(2)}) + \sum_{j=1}^{n} \frac{p_j}{j+1}(v_{p(1)}{}^{j+1} - v_{p(2)}{}^{j+1})\}. \quad (2.37)$$

Using eqns (2.42a) and (2.42b), $v_{(1)}(X_i)$ and $v_{(2)}(X_i)$ are given by

$$v_{Xi(1)} = \frac{R+1}{R+\exp(\sigma_i X_i)} v_{Xi}{}^0. \quad (2.175a)$$

$$v_{Xi(2)} = \frac{(R+1)\exp(\sigma_i X_i)}{R+\exp(\sigma_i X_i)} v_{Xi}{}^0. \quad (2.175b)$$

where $v_{Xi}{}^0$ is the volume fraction of X_i-mer in the starting solution ($\equiv v_p{}^0 g_0(X_i)$). If we know $v_p{}^0$ and $g_0(X_i)$ (accordingly $v_{Xi}{}^0$) in advance, $v_{Xi(1)}$ and $v_{Xi(2)}$ can be calculated as functions of $\sigma_0{}^a$, $\sigma_{01}{}^a$ and R^a (see, section 2.3).

$$v_{Xi(1)} = v_{Xi(1)}(\sigma_0{}^a, \sigma_{01}{}^a, R^a), \quad (2.176a)$$

$$v_{Xi(2)} = v_{Xi(2)}(\sigma_0{}^a, \sigma_{01}{}^a, R^a), \quad (2.176b)$$

where $\sigma_0{}^a$ and $\sigma_{01}{}^a$ are the assumed value of σ_0 and σ_{01}. $v_{p(1)}$, $v_{p(2)}$, $X_{n(1)}$ and $X_{n(2)}$ are functions of $\sigma_0{}^a$, $\sigma_{01}{}^a$ and R^a:

$$v_{p(1)} = v_{p(1)}(\sigma_0{}^a, \sigma_{01}{}^a, R^a), \quad (2.177a)$$

$$v_{p(2)} = v_{p(2)}(\sigma_0{}^a, \sigma_{01}{}^a, R^a), \quad (2.177b)$$

$$X_{n(1)} = X_{n(1)}(\sigma_0^a, \sigma_{01}^a, R^a), \tag{2.178a}$$

$$X_{n(2)} = X_{n(2)}(\sigma_0^a, \sigma_{01}^a, R^a). \tag{2.178b}$$

Substituting eqns (2.177a)-(2.178b) into eqns (2.34), (2.36) and (2.37), eqns (2.179), (2.180) and (2.181) can be obtained:

$$\chi_{00} = \chi_{00}(\sigma_0^a, \sigma_{01}^a, R^a), \tag{2.179}$$

$$\sigma_0 = \sigma_0(\sigma_0^a, \sigma_{01}^a, R^a), \tag{2.180}$$

$$\sigma_{01} = \sigma_{01}(\sigma_0^a, \sigma_{01}^a, R^a). \tag{2.181}$$

In the case where the initial polymer volume fraction v_p^0 is lower than the critical polymer concentration v_p^c ($v_p^0 < v_p^c$), which is given as a solution of the simultaneous equations, eqns (2.121) and (2.122), the X_i-mer's volume fraction in the two phases at cloud point, $v_{X_i(1)}$ and $v_{X_i(2)}$, are given by (refs 4,6,167)

$$v_{X_i(1)} = v_{X_i}^0, \tag{2.182a}$$

$$v_{X_i(2)} = v_{X_i}^0 \exp(\sigma_i X_i) \tag{2.182b}$$

and R attains infinity, and the polymer-rich phase becomes "cloud particle" (i.e., $V_{(1)} = V$ and $V_{(2)} = 0$) (refs 4,6,167).

In the case of $v_p^0 > v_p^c$, $R = 0$ (i.e., $V_{(1)} = 0$ and $V_{(2)} = V$) is realized at cloud point and $v_{X_i(1)}$ and $v_{X_i(2)}$ are given by eqns (2.183a) and (2.183b) (refs 4,6,167):

$$v_{X_i(1)} = v_{X_i}^0 \exp(-\sigma_i X_i), \tag{2.183a}$$

$$v_{X_i(2)} = v_{X_i}^0. \tag{2.183b}$$

In this case, the polymer-lean phase becomes "cloud particle".

$v_{X_i}^0$ ($\equiv v_0(X_i)$) is given as $g_0(X_i)/V$ (here, $g_0(X_i)$ is normalized MWD of the original polymer) and then $v_{X_i(1)}$ (eqn (2.182b)) [or $v_{X_i(2)}$ (eqn (2.183a))] is a function of the single variable σ_i at cloud point. $v_{p(1)}$ and $X_{n(1)}$ (or $v_{p(2)}$ and $X_{n(2)}$) are functions of σ_0^a and σ_{01}^a and moreover χ_{00}, σ_0 and σ_{01} (refs 4,6,167):

$$\chi_{00} = \chi_{00}(\sigma_0{}^a, \sigma_{01}{}^a), \tag{2.184}$$

$$\sigma_0 = \sigma_0(\sigma_0{}^a, \sigma_{01}{}^a), \tag{2.185}$$

$$\sigma_{01} = \sigma_{01}(\sigma_0{}^a, \sigma_{01}{}^a). \tag{2.186}$$

True σ_0 and σ_{01} at cloud point (hereafter referred to as $\sigma_0{}^{cp}$ and $\sigma_{01}{}^{cp}$) can be evaluated self-consistently by resolving the simultaneous equations as follows,

$$\Delta\sigma_0 \equiv \sigma_0(\sigma_0{}^a, \sigma_{01}{}^a) - \sigma_0{}^a = 0, \tag{2.187}$$

$$\Delta\sigma_{01} \equiv \sigma_{01}(\sigma_0{}^a, \sigma_{01}{}^a) - \sigma_{01}{}^a = 0. \tag{2.188}$$

Eqns (2.187) and (2.188) are "fundamental equations" for calculation of CPC. Here, $\sigma_0{}^a$ and $\sigma_{01}{}^a$ are the assumed σ_0 and σ_{01}, respectively. Substituting $\sigma_0{}^{cp}$ and $\sigma_{01}{}^{cp}$, obtained from eqns (2.187) and (2.188), into eqn (2.184), we get true χ_{00} at cloud point, $\chi_{00}{}^{cp}$, which can be converted, using eqn (2.6), to T at cloud point, T^{cp}, and finally CPC (T^{cp} vs. $v_p{}^0$ curve) can be calculated.

In the case of $v_p{}^0 < v_p{}^c$, true $v_{Xi(1)}$, $v_{Xi(2)}$, $v_{p(1)}$, $v_{p(2)}$, $X_{n(1)}$ and $X_{n(2)}$ at cloud point, $v_{Xi(1)}{}^{cp}$, $v_{Xi(2)}{}^{cp}$, $v_{p(1)}{}^{cp}$, $v_{p(2)}{}^{cp}$, $X_{n(1)}{}^{cp}$ and $X_{n(2)}{}^{cp}$ can be obtained (refs 4,6,167):

$$v_{Xi(1)}{}^{cp} = v_{Xi}{}^0, \tag{2.182a'}$$

$$v_{Xi(2)}{}^{cp} = v_{Xi}{}^0 \exp(\sigma_i{}^{cp} X_i) \quad (= v_{Xi}{}^{cp}), \tag{2.182b'}$$

$$v_{p(1)}{}^{cp} = v_p{}^0, \tag{2.38a'}$$

$$v_{p(2)}{}^{cp} = \sum_{i=1}^{m} v_{Xi(2)}{}^{cp} \quad (= v_p{}^{cp}), \tag{2.38b'}$$

$$X_{n(1)}{}^{cp} = X_n{}^0, \tag{2.39a'}$$

$$X_{n(2)}{}^{cp} = \sum_{i=1}^{m} v_{Xi(2)}{}^{cp} / \{\sum_{i=1}^{m} (v_{Xi(2)}{}^{cp}/X_i)\} \quad (= X_n{}^{cp}), \tag{2.39b'}$$

where $v_{Xi}{}^{cp}$ $(= v_p{}^{cp}(X_i))$, $v_p{}^{cp}$ and $X_n{}^{cp}$ are v_{Xi}, v_p and X_n of polymer partitioned into "cloud particle" respectively.

In the case of $v_p{}^0 > v_p{}^c$, $v_{Xi(1)}{}^{cp}$, $v_{Xi(2)}{}^{cp}$, $v_{p(1)}{}^{cp}$, $v_{p(2)}{}^{cp}$, $X_{n(1)}{}^{cp}$ and $X_{n(2)}{}^{cp}$ can be obtained (refs 4,6,167):

$$v_{xi(1)}^{cp} = v_{xi}^{0} \exp(-\sigma_i^{cp} X_i) \quad (= v_{xi}^{cp}), \tag{2.183a'}$$

$$v_{xi(2)}^{cp} = v_{xi}^{0}, \tag{2.183b'}$$

$$v_{p(1)}^{cp} = \sum_{i=1}^{m} v_{xi(1)}^{cp} \quad (= v_p^{cp}), \tag{2.38a''}$$

$$v_{p(2)}^{cp} = v_p^{0}, \tag{2.38b''}$$

$$X_{n(1)}^{cp} = \sum_{i=1}^{m} v_{xi(1)}^{cp} / \{\sum_{i=1}^{m} (v_{xi(1)}^{cp}/X_i)\} \quad (= X_n^{cp}), \tag{2.39a''}$$

$$X_{n(2)}^{cp} = X_n^{0}. \tag{2.39b''}$$

Substitution of eqn (2.36) into eqn (2.187) yields eqn (2.189) and substitution of eqn (2.37) into eqn (2.188) gives eqn (2.190) (refs 4,6,167).

$$\begin{aligned}
\Delta\sigma_0 &= \sigma_0(\sigma_0^a, \sigma_{01}^a) - \sigma_0^a \\
&= (v_{p(1)} - v_{p(2)}) - \left(\frac{v_{p(1)}}{X_{n(1)}} - \frac{v_{p(2)}}{X_{n(2)}}\right) \\
&\quad - \chi_{00}\Big[2(v_{p(1)} - v_{p(2)}) - (v_{p(1)}^2 - v_{p(2)}^2) \\
&\qquad + \sum_{j=1}^{n} p_j \Big\{\frac{j+2}{j+1}(v_{p(1)}^{j+1} - v_{p(2)}^{j+1}) - (v_{p(1)}^{j+2} - v_{p(2)}^{j+2})\Big\}\Big] \\
&\quad - \chi_{00} k' \Big[\left(\frac{v_{p(1)}}{X_{n(1)}} - \frac{v_{p(2)}}{X_{n(2)}}\right) - \left(\frac{v_{p(1)}^2}{X_{n(1)}} - \frac{v_{p(2)}^2}{X_{n(2)}}\right) \\
&\qquad + \sum_{j=1}^{n} p_j \Big\{\left(\frac{v_{p(1)}^{j+1}}{X_{n(1)}} - \frac{v_{p(2)}^{j+1}}{X_{n(2)}}\right) - \left(\frac{v_{p(1)}^{j+2}}{X_{n(1)}} - \frac{v_{p(2)}^{j+2}}{X_{n(2)}}\right)\Big\}\Big] - \sigma_0^a = 0,
\end{aligned} \tag{2.189}$$

$$\begin{aligned}
\Delta\sigma_{01} &= \sigma_{01}(\sigma_0^a, \sigma_{01}^a) - \sigma_{01}^a \\
&= -k'\chi_{00}\Big[(v_{p(1)} - v_{p(2)}) \\
&\qquad + \sum_{j=1}^{n} \frac{p_j}{j+1}(v_{p(1)}^{j+1} - v_{p(2)}^{j+1})\Big] - \sigma_{01}^a = 0.
\end{aligned} \tag{2.190}$$

Here, χ_{00} is given by eqn (2.34).

In the case of $k' = 0$ ($k_0 = 0$) and $p_j \neq 0$, we obtain $\sigma_{01} = 0$

(then eqn (2.37) diminishes) and σ_i is independent of X_i and equals to σ (= σ_0). Accordingly, variables in eqn (2.189) reduce to a single variable (σ_0) and eqn (2.189) can be transformed into eqn (2.191), which is a fundamental equation of CPC at $k'=0$ (refs 4,6,167).

$$\Delta\sigma_0 = \sigma_0(\sigma_0{}^a) - \sigma_0{}^a$$

$$= (v_{p(1)} - v_{p(2)}) - \left(\frac{v_{p(1)}}{X_{n(1)}} - \frac{v_{p(2)}}{X_{n(2)}}\right)$$

$$- \chi_{00}\,[2(v_{p(1)} - v_{p(2)}) - (v_{p(1)}{}^2 - v_{p(2)}{}^2)$$

$$+ \sum_{j=1}^{n} p_j \{\frac{j+2}{j+1}(v_{p(1)}{}^{j+1} - v_{p(2)}{}^{j+1}) - (v_{p(1)}{}^{j+2} - v_{p(2)}{}^{j+2})\}] - \sigma_0{}^a = 0. \tag{2.191}$$

χ_{00} (= χ_0 at $k'=0$) in eqn (2.191) can be estimated by putting $k'=0$ in eqn (2.34), that is (refs 4,6,167)

$$\chi_0 = \chi_0(\sigma_a)$$

$$= \frac{\ln\dfrac{1-v_{p(1)}}{1-v_{p(2)}} + (v_{p(1)} - v_{p(2)}) - \left(\dfrac{v_{p(1)}}{X_{n(1)}} - \dfrac{v_{p(2)}}{X_{n(2)}}\right)}{(v_{p(2)}{}^2 - v_{p(1)}{}^2) + \sum_{j=1}^{n} p_j (v_{p(2)}{}^{j+2} - v_{p(1)}{}^{j+2})}. \tag{2.192}$$

In the case of $k'=0$ ($k_0=0$) and $p_j=0$ ($j=1,\cdots,n$), eqn (2.191) can be simplified further to (refs 4,6,167)

$$\Delta\sigma_0 = \sigma_0(\sigma_0{}^a) - \sigma_0{}^a$$

$$= (v_{p(1)} - v_{p(2)}) - \left(\frac{v_{p(1)}}{X_{n(1)}} - \frac{v_{p(2)}}{X_{n(2)}}\right)$$

$$- \chi_{00}\,[2(v_{p(1)} - v_{p(2)}) - (v_{p(1)}{}^2 - v_{p(2)}{}^2)] - \sigma_0{}^a = 0 \tag{2.193}$$

with

$$\chi_0(\sigma_a)\ (= \chi_{00})$$

$$= [\ln\frac{1-v_{p(1)}}{1-v_{p(2)}} + (v_{p(1)} - v_{p(2)}) - \left(\frac{v_{p(1)}}{X_{n(1)}} - \frac{v_{p(2)}}{X_{n(2)}}\right)] / (v_{p(2)}{}^2 - v_{p(1)}{}^2). \tag{2.194}$$

Simultaneous, self-consistent equations (eqns (2.187) and (2.188)), derived by Kamide, Matsuda et al. (refs 4,6,167), can be easily generalized for determination of CPC of the system consisting two multicomponent polymers with different chemical compositions (polymer blend; see, section 4) (ref. 168).

In Kamide et al's calculation, (i) a direct method without any serious assumptions was employed and (ii) not only the phase-diagram, but also σ, and the molecular characteristics ($v^{cp}(X_i)$, X_w^{cp}, $(X_w/X_n)^{cp}$ and σ'^{cp} were calculated. CPC can be estimated from χ_0^{cp}, and the shadow curve from v_p^{cp}. The following procedure was used for computer simulation of CPC (refs 4,6,167).

1. The prerequisites for this simulation are,
   (i) molecular weight distribution $g_0(X)$ and accordingly weight- and number-average X of the original polymer (X_w^0 and X_w^0/X_n^0).
   (ii) polymer volume fraction of the starting solution v_p^0.
   (iii) parameters k_0 and θ in eqn (2.4').
   (iv) concentration-dependence parameters p_j ($j=1,\cdots,n$) in eqn (2.3).
2. At first, the volume fraction of "cloud particle" is calculated by use of eqn (2.182b) (or eqn (2.183a)) with nine sets of $(\sigma_0^a, \sigma_{01}^a)$ (combination of low, middle and high value of $(\sigma_0^a, \sigma_{01}^a)$; $(\sigma_0^L, \sigma_{01}^L)$, $(\sigma_0^L, \sigma_{01}^M)$, $(\sigma_0^L, \sigma_{01}^H)$, $\cdots$, $(\sigma_0^H, \sigma_{01}^H)$). And then $(\Delta\sigma_0, \Delta\sigma_{01})$ is calculated (see, eqns (2.189) and (2.190)). True (σ_0, σ_{01}) should lie in one of the quarter section of (σ_0, σ_{01}) plane ($\sigma_0^L < \sigma_0 < \sigma_0^H$ and $\sigma_{01}^L < \sigma_{01} < \sigma_{01}^H$).
3. Reset the (σ_0, σ_{01}) plane by the quarter section of (σ_0, σ_{01}).
4. Eqns (2.189) and (2.190) are simultaneously solved by use of the interhalving method (reputation of steps 2-3) and true (σ_0, σ_{01}) is determined.
5. Substituting σ_0^{cp} and σ_{01}^{cp} into eqn (2.184), we can evaluate χ_{00}^{cp}. χ_{00}^{cp} can also be calculated by substituting χ_{00}^{cp} into eqn (2.4). Of course, utilizing eqns (2.182), (2.183) and (2.38), we can calculate $v_{(1)}(X_i)$, $v_{(2)}(X_i)$, $v_{p(1)}$ and $v_{p(2)}$.

Fig. 2.78 shows the effect of X_w^0 on CPC, calculated from the solutions of four polymers (SZ, $X_w^0/X_n^0 = 1$, 1.1, 2 and 4) in a single solvent ($p_1 = 0.6$, $p_2 = 0$ and $k_0 = 0$). The unfilled circle in the figure is the critical point and the filled circle, the threshold cloud point. The critical point, independently calculated by a different procedure from CPC, lies just on the

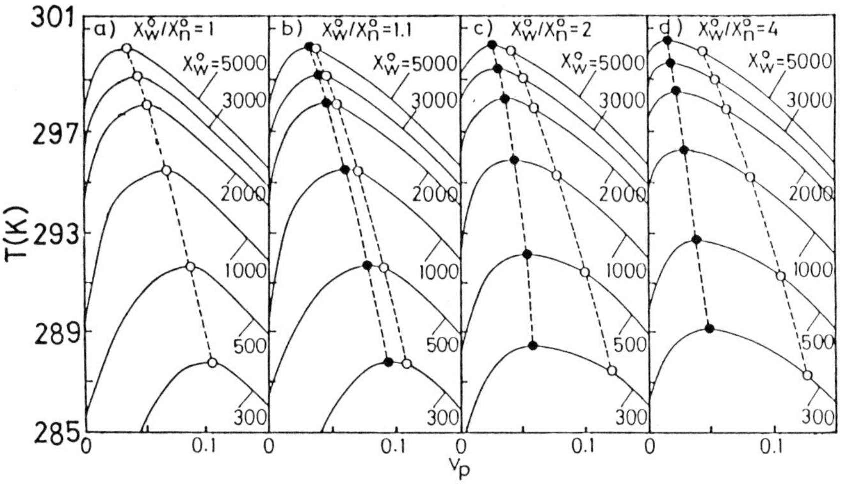

Fig. 2.78 Effect of weight-average molar ratio $X_w{}^0$ of the polymer in a single solvent on cloud point curve ($p_1=0.6$, $p_2=0$ and $k_0=0$): $X_w{}^0/X_n{}^0=1$ a), 1.1 b), 2 c) and 4 d); ○, critical point; ●, precipitation threshold point; number on curve denotes $X_w{}^0$.

CPC. This verifies the reliability of the CPC calculation procedure used here (ref. 4).

A monodisperse polymer solution with low $X_w{}^0$ shows CPC having approximate symmetry, which disappears with increasing $X_w{}^0$ and $X_w{}^0/X_n{}^0$. As $X_w{}^0$ increases, CPC shifts to lower v_p and higher T regions, showing a skewness in the former. The critical points (T_c and $v_p{}^c$) coincides very well as expected from the threshold points ($T_p{}^{tcp}$ and $v_p{}^{tcp}$) for solutions of strictly monodisperse polymers. As the polydispersity of the polymer increases, the CSP shifts to the lower v_p region causing them to differ significantly. The deviation in the critical point from threshold cloud point increases remarkably with an increase in $X_w{}^0/X_n{}^0$ and slightly with a decrease in $X_w{}^0$ (ref. 4). Note that this difference is significant even for a polymer with $X_w{}^0/X_n{}^0=1.1$, which is often regarded as a "monodisperse polymer." Very careful attention should be given to this point in the case of polymers with a low $X_w{}^0$ (ref. 4).

Fig. 2.79 shows the CPC of a PS / CH system. The unfilled circle is the actual experimental data and the full lines, the theoretical curves calculated using the p_1 and p_2 parameters from the literature (Table 2.5). The curve calculated from χ value obtained by Krigbaum and Geymer (ref. 62) can express experimental points in a lower v_p region. The theoretical curve based on the χ

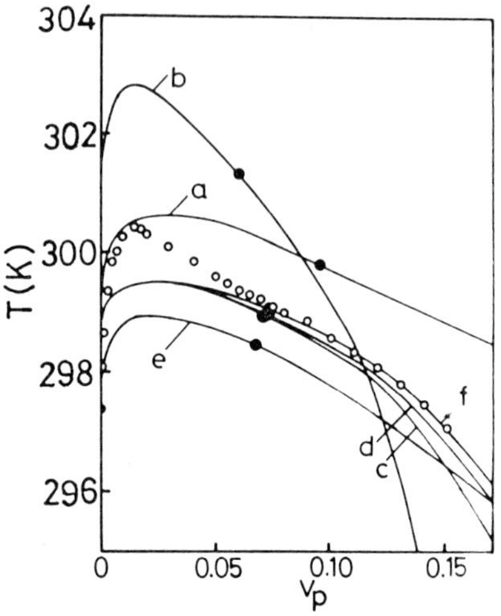

Fig. 2.79 Cloud point curves of polystyrene / cyclohexane system: Full line, theoretical curve calculated using the values of a', b', p_1 and p_2 summarized in Table 2.3. Lines a–f have the same meaning as those in Fig. 2.16. Unfilled circle, experimental data points.

of Scholte (ref. 63) overestimates these points considerably at $v_p < 0.12$ and under estimates them at $v_p > 0.12$. The total shape of CPC, calculated by p_1 and p_2 obtained by Koningsveld et al. (ref. 16) and Kuwahara et al. (ref. 66), using the CSP data of Koningsveld et al.'s method (ref. 16), is similar to the actual experimental points, but their CPCs are underestimated throughout the entire range of v_p by ca.1～1.5K. It is of interest that a very slight change in p_1 and p_2 causes remarkable variation in CPC. p_1 and p_2 should thus be estimated from CPC.

The relationships between ρ_p and phase-separation temperature, T_p were determined for three PS / CH systems with $v_p^0 = 0.6$, 1.184 and 1.737% to an accuracy of ± 0.001 in ρ_p and ± 0.01K in T_p. PS was Styron 666 (Asahi Chem. Industry Co., Ltd.) with $M_w = 23.9 \times 10^4$ and $M_n = 8.6 \times 10^4$ by GPC in tetrahydrofuran and $M_w = 23.2 \times 10^4$ by light scattering in benzene at 298K and $M_n = 8.9 \times 10^4$ by membrane osmometry in toluene at 298K.

Fig. 2.80 shows the relations between ρ_p and T_p, obtained by an actual experiment of the PS / CH system. The relations can be roughly approximated by a part of a circular arc with ρ_p approaching zero at specific temperature T_p^{cp} (i.e., the cloud point). As v_p increases, the ρ_p vs. T_p curve shifts to lower T_p axis concurrently (ref. 4).

Kamide et al. (ref. 4) carried out a computer experiment on

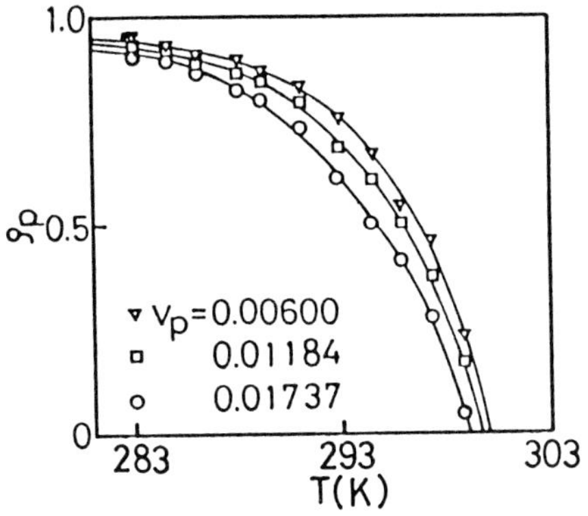

Fig. 2.80 Experimental relations between ρ_p and T for polystyrene ($X_w^0 = 2117$, $X_w^0/X_n^0 = 2.8$) in cyclohexane. The polymer volume fractions of the starting solution v_p^0 are 0.6×10^{-2}(▽), 1.184×10^{-2}(□) and 1.737×10^{-2}(○), respectively.

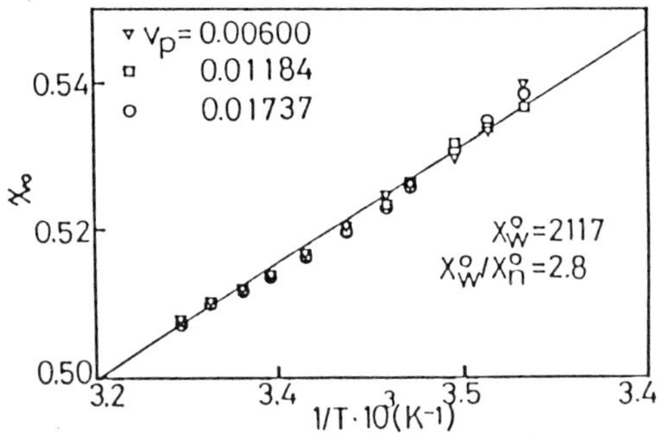

Fig. 2.81 Temperature dependence of χ_{00} in eqn (2.6) for polystyrene ($X_w^0 = 2117$, $X_w^0/X_n^0 = 2.8$) in cyclohexane: $v_p^0 \times 10^2 = 0.60$(▽), 1.184(□) and 1.737(○); full line, theoretical curve calculated by assuming $p_1 = 0.62$, $p_2 = 0.20$ and $k_0 = 0$.

two-phase equilibrium assuming variable p_1 and p_2 to obtain the relations between χ_{00} and T_p by a comparison with the ρ_p-T relations in Fig. 2.81, from which a' and b' were found for a given combination of p_1 and p_2 (see, eqn (2.7)).

The CPC was calculated (T_p^{cp} vs. v_p curve) using a and b obtained in the relations between χ_{00} and 1/T, and then $\delta \equiv \Sigma\ (T_p^{cp}(\mathrm{exp}) - T_p^{cp}(\mathrm{cal}))/N$ (N is the total number of solutions, for which the cloud point was determined) in order to determine a set of p_j ($j = 1,\cdots,n$) to minimize δ (where $N \gg n$). These steps were repeated and a', b' and p_j evaluated where δ is

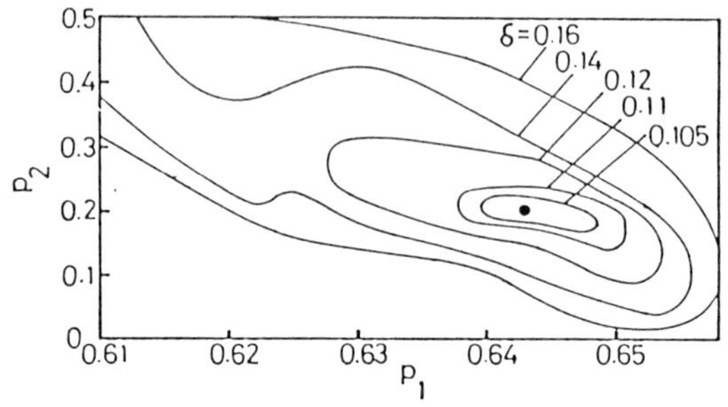

Fig. 2.82 Relations among p_1, p_2 and δ for polystyrene / cyclohexane system: Number denotes δ and the full line is the contour line of the same δ.

below the permissible limit (ref. 4).

Fig. 2.82 shows the relations among p_1, p_2 and δ for the PS / CH system. The most appropriate set of p_1 and p_2 values giving the minimum δ ($\simeq 0.1$), were $p_1 = 0.643$ and $p_2 = 0.200$. These values are very near to that evaluated by Kamide and Matsuda (ref. 3) (see, section 2.7) from these CSP data (refs 16,66,71).

The CPC, calculated using these values of a', b', p_1 and p_2 can express accurately the experimental data, except the threshold point region ($v_p \simeq 0.07$), as shown as the full line in Fig. 2.83 (ref. 4). The maximum deviation of the theoretical T_c from the experiment is about 1K. In the figure the molecular weight

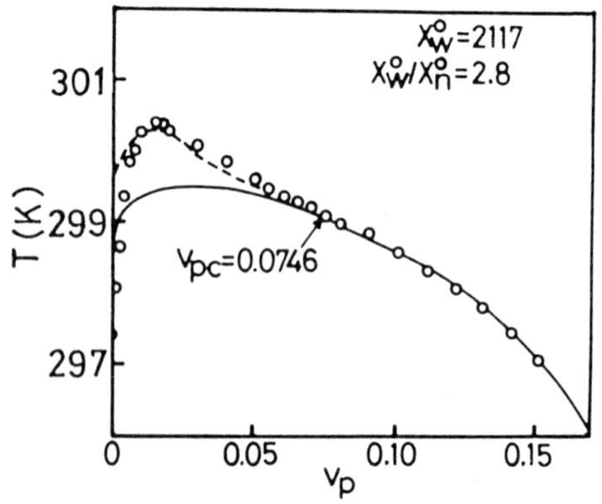

Fig. 2.83 Cloud point curve of polystyrene ($X_w^0 = 2117$, $X_w^0/X_n^0 = 2.8$) in cyclohexane: ○, experimental data; full line, theoretical curve calculated using $p_1 = 0.643$, $p_2 = 0.200$, $k_0 = 0$; broken line, theoretical curve calculated using $p_1 = 0.643$, $p_2 = 0.200$, $\theta = 307.1$K and $k_0 = -104.23$.

dependence of χ parameter was not taken into account. The CPC at a lower v_p region is expected to be significantly influenced by the large molecular weight components and the above inconsistence at a lower v_p region may be due to the neglect of the molecular weight dependence of χ parameter (ref. 4).

Kamide et al. (ref. 8) found that $k_0 = 108$ and $\theta = 307.1$K for the PS / CH system (see, Fig. 2.19). The CPC was calculated using these values together with a', b', p_1 and p_2 values determined before and the result is shown as the broken line in Fig. 2.83. Inspection of Fig. 2.83 leads us to the conclusion that the experimental cloud point data of the polydisperse polymer in a single solvent can be reasonably represented by a theory which takes into consideration (ref. 4) the polydispersity of the polymer and the concentration- and molecular weight-dependence of the χ parameter.

2) Šolc's method

Šolc (ref. 169) derived an equation which allows a direct calculation of CPC (see, eqn (2.203)) for the case of $k' = k_0 = 0$ and $p_j = 0$: When $k' = k_0 = 0$ and $p_j = 0$, eqns (2.9) and (2.21) become

$$\Delta\mu_0 = \tilde{R}T\{\ln(1-v_p) + (1-\frac{1}{X_n})v_p + \chi_{00}v_p^2\}, \qquad (2.9')$$

$$\Delta\mu_{Xi} = \tilde{R}T\{\ln v_{Xi} - (X_i - 1) + X_i(1-\frac{1}{X_n})v_p + X_i\chi_{00}(1-v_p)^2\}. \qquad (2.21')$$

Combining eqns (2.9') and (2.21'), he gives a modified two phase-equilibrium condition, given by

$$\sigma_0 = \ln\frac{1-v_{p(2)}}{1-v_{p(1)}} + 2\chi_{00}(v_{p(2)} - v_{p(1)}) \qquad (2.195)$$

and

$$\frac{\sigma_0(v_{p(1)} + v_{p(2)})}{2} + (v_{p(2)} - v_{p(1)}) + (\frac{v_{p(1)}}{X_{n(1)}} - \frac{v_{p(2)}}{X_{n(2)}})$$

$$+ (1 - \frac{v_{p(1)} + v_{p(2)}}{2})\ln\frac{1-v_{p(2)}}{1-v_{p(1)}} = 0. \qquad (2.196)$$

In the process of calculation of the CPC equation, Šolc represented MWD of the polymer partitioning in "cloud particle" $g^{cp}(X_i)$ as

$$g^{cp}(X_i) \equiv \lim_{Q \to 0} \frac{v^{cp}(X_i)}{v_p^{0}} = g_0(X_i)\exp(K\sigma X_i). \tag{2.197}$$

Here, in the range of $v_p^0 < v_p^c$, $Q = 1/R$, $K = +1$, $v^{cp}(X_i) = v_{X i(2)}{}^{cp}$ and in the range $v_p^0 > v_p^c$, $Q = R$, $K = -1$, $v^{cp}(X_i) = v_{X i(1)}{}^{cp}$. Thus eqn (2.197) is the same as the expression of eqns (2.182b') and (2.183a').

He defined the k-th moment of $g_0(X_i)$ and $g^{cp}(X_i)$, as denoted by μ_k and ν_k, by eqns (2.198) and (2.199), respectively (ref. 169).

$$\mu_k = \int_0^{\infty} X_i^{\,k} g_0(X_i)\, dX_i, \tag{2.198}$$

$$\nu_k = \int_0^{\infty} X_i^{\,k} g^{cp}(X_i)\, dX_i. \tag{2.199}$$

Substituting eqn (2.197) into eqn (2.199) and after carrying out series expansion of the term $\exp(K\sigma X_i)$, we have

$$\nu_k = \int_0^{\infty} X_i^{\,k} g_0(X_i) \sum_{h=0}^{\infty} \frac{(K\sigma X_i)^h}{h!}\, dX_i. \tag{2.199'}$$

Using eqn (2.198), eqn (2.199') can be rewritten in the form (ref. 169),

$$\nu_k = \sum_{h=0}^{\infty} \left[\frac{(K\sigma)^h}{h!}\mu_{k+h}\right] \tag{2.200}$$

$$= \sum_{h=0}^{h_m} \left[\frac{(K\sigma)^h}{h!}\mu_{k+h}\right] \tag{2.200'}$$

where h_m is the minimum h with which we can judge eqn (2.200') to be correct.

$v_{p(1)}$, $v_{p(2)}$, $X_{n(1)}$ and $X_{n(2)}$ can be expressed in terms of v_p^0, ν_0, ν_{-1} and μ_{-1}.

For $v_p^0 < v_p^c$

$$v_{p(1)} = v_p^0 \,(= \mu_0/V), \tag{2.201a}$$

$$v_{p(2)} = v_p^0 \nu_0, \tag{2.201b}$$

$$X_{n(1)} = 1/\mu_{-1}, \tag{2.201c}$$

$$X_{n(2)} = \nu_0/\nu_{-1}. \tag{2.201d}$$

For $v_p^0 > v_p^c$

$$v_{p(1)} = v_p^0 \nu_0, \tag{2.202a}$$

$$v_{p(2)} = v_p^0 \ (= \mu_0/V), \tag{2.202b}$$

$$X_{n(1)} = \nu_0/\nu_{-1}, \tag{2.202c}$$

$$X_{n(2)} = 1/\mu_{-1}. \tag{2.202d}$$

Substituting eqns (2.201a)-(2.201d) or eqns (2.202a)-(2.202d) into eqn (2.196), we obtain (refs 170,171)

$$F(\sigma_0^a) \equiv \frac{K\sigma_0^a(1+\nu_0)}{2} + (\nu_0 - 1) - (\nu_{-1} - \mu_{-1})$$
$$+ \left(\frac{1}{v_p^0} - \frac{1+\nu_0}{2}\right)\ln\frac{1-v_p^0\nu_0}{1-v_p^0} = 0. \tag{2.203}$$

Here, as described before, in the range $v_p^0 < v_p^c$, $K = +1$ and in the range $v_p^0 > v_p^c$, $K = -1$. Since μ_{-1} is known, ν_{-1} and ν_0 are functions of σ_0^a only and eqn (2.203) can be expressed by σ_0^a alone. The value of σ_0^a, which satisfies $F(\sigma_0^a) = 0$ for a given initial polymer volume fraction, is σ_0 at CPC, that is σ_0^{CP}.

Eqn (2.203) is Šolc's CPC equation, which is expressed by ν_{-1} and ν_0, in place of $v_{p(1)}$, $v_{p(2)}$, $X_{n(1)}$ and $X_{n(2)}$, used by Kamide et al.'s method. Substitution of eqns (2.201a)-(2.201d) or eqns (2.202a)-(2.202d) into eqn (2.195) leads to (refs 170,171)

$$2\chi_{00}v_p^0(\nu_0 - 1) = K\sigma_0 + \ln\frac{1-v_p^0}{1-v_p^0\nu_0}. \tag{2.204}$$

By putting σ_0^{CP}, obtained from eqn (2.203), into eqn (2.204), we obtain true χ_{00} at CPC for given v_p^0.

Šolc's CPC equation (eqn (2.203)) and eqn (2.204) are equivalent with Kamide-Matsuda's fundamental equation of CPC (eqn (2.193)) and eqn (2.194) in which $k' = k_0 = 0$ and $p_j = 0$.

Recently, Šolc et al. (ref. 172) derived a generalized "CPC equation" for the quasi-binary system by using the expression of g parameter. Eqns (2.9) and (2.21) can be rewritten as (ref. 14)

$$\Delta\mu_0 = \tilde{R}T\left[\ln(1-v_p) + \left(1-\frac{1}{X_n}\right)v_p + \left\{g - (1-v_p)\frac{\partial g}{\partial v_p}\right\}v_p^2\right], \tag{2.205}$$

$$\Delta \mu_{xi} = \tilde{R}T\left[\ln v_{xi} - (X_i - 1) + X_i\left(1 - \frac{1}{X_n}\right)v_p + X_i\left\{g + v_p\frac{\partial g}{\partial v_p}\right\}(1 - v_p)^2\right]. \tag{2.206}$$

Substituting eqn (2.205) into eqn (1.21), one obtains eqn (2.207)

$$0 = \ln\frac{1 - v_{p(2)}}{1 - v_{p(1)}} + (v_{p(2)} - v_{p(1)}) - \left(\frac{v_{p(2)}}{X_{n(2)}} - \frac{v_{p(1)}}{X_{n(1)}}\right)$$
$$+ \left\{g_{(2)} - (1 - v_{p(2)})\frac{\partial g_{(2)}}{\partial v_p}\right\}v_{p(2)}^{\,2} - \left\{g_{(1)} - (1 - v_{p(1)})\frac{\partial g_{(1)}}{\partial v_p}\right\}v_{p(1)}^{\,2} \equiv G \tag{2.207}$$

where

$$g_{(1)} \equiv g(T, v_{p(1)}), \tag{2.208a}$$

$$g_{(2)} \equiv g(T, v_{p(2)}). \tag{2.208b}$$

And the combination of eqns (2.206), (1.22) and (2.207) yields eqn (2.209)

$$\frac{\sigma_0(v_{p(1)} + v_{p(2)})}{2} + (v_{p(2)} - v_{p(1)}) + \left(\frac{v_{p(1)}}{X_{n(1)}} - \frac{v_{p(2)}}{X_{n(2)}}\right)$$
$$+ \left(1 - \frac{v_{p(1)} + v_{p(2)}}{2}\right)\ln\frac{1 - v_{p(2)}}{1 - v_{p(1)}}$$
$$- (g_{(2)} - g_{(1)})\left[v_{p(1)}v_{p(2)} - \frac{v_{p(1)} + v_{p(2)}}{2}\right]$$
$$+ \frac{v_{p(1)} - v_{p(2)}}{2}\left[\frac{\partial g_{(2)}}{\partial v_p}v_{p(2)}(1 - v_{p(2)}) + \frac{\partial g_{(1)}}{\partial v_p}v_{p(1)}(1 - v_{p(1)})\right] = 0. \tag{2.209}$$

Eqns (2.209) and (2.207) represent a modified two-phase equilibrium condition taking into account the concentration dependence of g.

If we substitute eqns (2.201a)-(2.201d) or eqns (2.202a)-(2.202d) into eqns (2.209) and (2.207), we obtain eqns (2.210) and (2.211), respectively (ref. 172),

$$F_1(\sigma_0^{\,a}) \equiv K\frac{\sigma_0^{\,a}(\nu_0 + 1)}{2} + \nu_0 - 1 + (\mu_{-1} - \nu_{-1})$$

$$+\left(\frac{1}{v_p^0}-\frac{1+\nu_0}{2}\right)\ln\frac{1-v_p^0\nu_0}{1-v_p^0}-(g^*-g)\left[v_p^0-\frac{1+\nu_0}{2}\right]$$

$$+\frac{1-\nu_0}{2}v_p^0\left[\frac{\partial g^*}{\partial v_p}\nu_0(1-v_p^0\nu_0)+\frac{\partial g}{\partial v_p}(1-v_p^0)\right]=0 \qquad (2.210)$$

and

$$2gv_p^0(\nu_0-1)+\ln\frac{1-v_p^0\nu_0}{1-v_p^0}-K\sigma_0^a-(g^*-g)(1-2v_p^0\nu_0)$$

$$-v_p^0\left[\frac{\partial g^*}{\partial v_p}\nu_0(1-v_p^0\nu_0)-\frac{\partial g}{\partial v_p}(1-v_p^0)\right]=0 \qquad (2.211)$$

where for $v_p^0 < v_p^c$, $g = g_{(1)}$, $g^* = g_{(2)}$ and for $v_p^0 > v_p^c$, $g = g_{(2)}$, $g^* = g_{(1)}$.

Eqn (2.210) is Šolc's CPC equation when the concentration dependence of g is taken into consideration. If g is independent of polymer concentration, the last two terms disappear, and eqn (2.210) reduces to eqn (2.203) and eqn (2.211) to eqn (2.204). Eqn (2.210) can be solved definitely if ν_0 and ν_{-1} are evaluated.

The above derivations of eqns (2.189) and (2.190) (in Kamide et al.'s method) seem rather laborious and complicated and it is not so easy to compare the equations in these two methods. This shows that the calculation of CPC is, in spite of an industrial

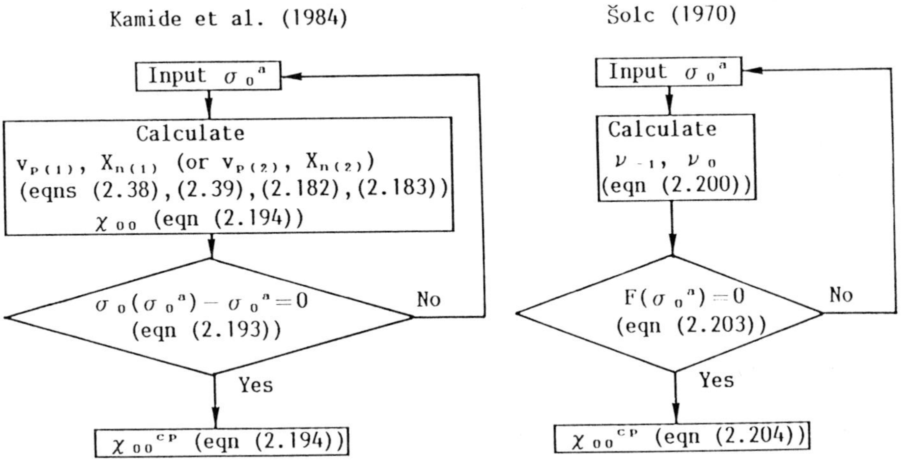

Fig. 2.84 Flow chart of the calculation of CPC by use of the fundamental equations (eqns (2.193) and (2.194)) and CPC equation (eqns (2.203) and (2.204)).

importance, very tedious, unlike the CSP, binodal and spinodal curves.

Fig. 2.84 demonstrates a schematic route of the derivations of the equations. Although both methods are based, as described, on the same theory (i.e., modified Flory-Huggins) combined with Gibbs' two-phase equilibrium conditions (eqns (1.21) and (1.22)), Kamide et al.'s have a more general character than Šolc's in view point of the consideration of molecular weight-dependence of χ parameter (ref. 4).

In Kamide et al.'s method, we can obtain σ_0^{CP} by putting $v_{P(1)}$ and $X_{n(1)}$ (or $v_{P(2)}$ and $X_{n(2)}$) and χ_{00}, all expressed as a function of σ_0^a into fundamental equation. On the other hand, in Šolc's method, ν_{-1} and ν_0 are given as a function of σ_0^a. In addition, in the former method $v_{P(1)}$, $v_{P(2)}$, $X_{n(1)}$ and $X_{n(2)}$ are estimated from the summation of $v_{X_i(1)}$ and $v_{X_i(2)}$, which are calculated from eqn (2.182) or (2.183) with the polymer volume fraction of X_i-mer in the original solution, $v_{X_i}^0$. Šolc's method has a characteristic feature in calculating ν_{-1} and ν_0 from expanding expression of the statistical moment μ_k of the original MWD (see, eqn (2.200)). Here, μ_k can be calculated from the average X_i of the original polymer only. Additionally, the higher-order average X_i for the polymer with Schulz-Zimm type and Wesslau type MWD can be determined from the number- and weight-average X_i, X_n^0 and X_w^0.

Schulz-Zimm type $\quad X_{z+q}^0 = (q+2)X_w^0 - (q+1)X_n^0, \qquad (2.212)$

Wesslau type $\quad X_{z+q}^0 = (X_w^0)^{q+2}/(X_n^0)^{q+1}. \qquad (2.213)$

Therefore, the calculation of ν_k by eqn (2.200) can be readily performed for such polymers. However, for MWD, except for the above distributions, the higher-order average X_i can not be calculated simply from X_n^0 and X_w^0, and in these case, ν_{-1} and ν_0 should be evaluated using eqn (2.199') and this kind of calculation is of course extremely tedious. In contrast to this, Kamide et al.'s method can be applied to any polymer with known MWD.

In Kamide et al.'s method, the factors governing the calculation accuracy are the maximum X_i, X_{max}, and the total number of components N [$N = X_{max}/DX$; DX, the distance of the

TABLE 2.18
Comparison of characteristic features of two direct methods for estimating cloud point curve proposed hitherto for system of multicomponent polymers in a single solvent.

	Šolc (1970)	Šolc et al. (1984)	Kamide et al. (1984)
Theory	modified F-H	modified F-H	modified F-H
• Polydispersity	+	+	+
• χ parameter (or g parameter)			
concentration dependence	—	+	+
molecular weight dependence	—	—	+
Fundamental equation	eqn (2.203) (CPC equation)	eqn (2.210) (CPC equation)	eqns (2.189) and (2.190)
Method of integration	Polynomial method using expansion of exponential (accompanied with much integration)		Direct-integration (Newton-Cotes integration method)

F-H, Flory-Huggins; +, considered; —, not considered.

adjacent two X_i utilized in the calculation [see, eqns (2.214) and (2.215)] and in Šolc's method, h_m in eqn (2.200').

Table 2.18 shows the comparison of characteristic features of the above two direct methods.

In this Chapter, in order to avoid the above mentioned uncertainty of Šolc's method, we also employed our direct integral procedure (see, eqns (2.214) and (2.215)) for estimating v_p^{cp} and X_n^{cp} in Šolc's method (eqns (2.210) and (2.211)) in addition to Šolc's original procedure (infinite series expansion procedure, eqn (2.200')) and compared the two procedures with each other (see, Table 2.20).

The computer experiments were carried out under the conditions as follows: The original polymer was assumed to have the Schulz-Zimm (SZ) type MWD with $X_w^0 = 2117$ and $X_w^0/X_n^0 = 2.8$, which has the same MWD as employed before for computer and actual (atactic polystyrene) experiments of phase-equilibria (ref. 4). The first and second order concentration-dependence parameters were taken as $p_1 = 0.643$ and $p_2 = 0.200$ (this is a combination of p_1 and p_2 experimentally determined for atactic polystyrene / cyclohexane system) (ref. 4) and χ was assumed to be independent of the polymer molecular weight (i.e., $k'(= k_0) = 0$). In Kamide et al.'s method, we assumed that eqn (2.191) was solved when the condition of $|\sigma(\sigma_0^a) - \sigma_0^a| < 0.2 \times 10^{-9}$ was satisfied and in Šolc's method, we assumed that the eqn (2.210) was solved when the condition of $|F(\sigma_0^a)| < 0.2 \times 10^{-9}$ was satisfied. And in Kamide et al.'s method, we performed the calculation of v_p^{cp} and X_n^{cp}, given by the following equations, respectively (refs 4,6,167),

$$v_p^{cp} = \sum_{i=1}^{N} v^{cp}(X_i) \cdot DX, \tag{2.214}$$

$$X_n^{cp} = v_p^{cp} / [\sum_{i=1}^{N} \frac{v^{cp}(X_i)}{X_i} \cdot DX] \tag{2.215}$$

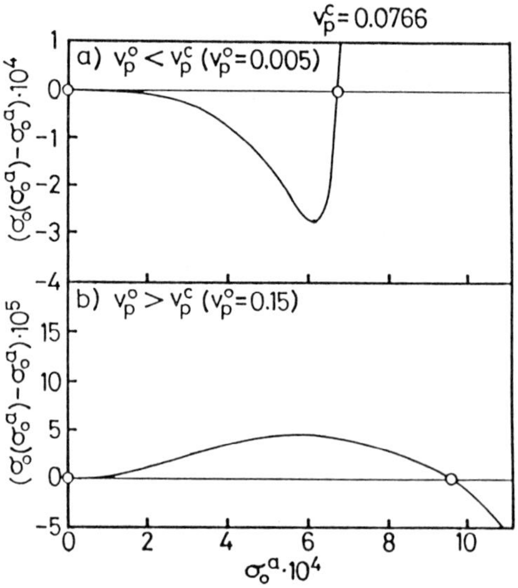

Fig. 2.85 Plots of main equation $\sigma_0(\sigma_0^a) - \sigma_0^a$ vs. σ_0^a: Original polymer, Schulz-Zimm type distribution ($X_w^0 = 2117$ and $X_w^0/X_n^0 = 2.8$); $p_1 = 0.643$, $p_2 = 0.200$ and $k_0 = 0$. $v_p^c = 0.0766$; a) $v_p^0 = 0.005$ and b) $v_p^0 = 0.15$. Open circles correspond to the solution.

under the condition of $X_{max} = 25000$, $DX = 5$ and accordingly, $N = 5000$ which is a reasonable component number for Schulz-Zimm type distribution (ref. 27). In Šolc's method, when $|\nu(h') - \nu(h'-1)|$ was smaller than 1.0×10^{-12}, ν_k was considered to be determined, i.e., $h' = h_m$ (see, eqn (2.200')).

Fig. 2.85 shows the left-hand side of eqn (2.191) plotted as a function of σ_0^a. At the point where the left-hand side term becomes zero (denoted by unfilled circle in the figure), the assumed $\sigma_0(\sigma_0^a)$ coincides with the true σ_0^{cp} and that is the solution of eqn (2.191). From eqns (2.121) and (2.122) we obtain the critical polymer volume fraction $v_p^c = 0.0766$ for the polymer / solvent system. In the range of $v_p^0 < v_p^c$ (for example, $v_p^0 = 0.005$), eqn (2.191) is zero at $\sigma_0^a = 0$ and decreases with an increase in σ_0^a and attains zero again at $\sigma_0^a = 6.8 \times 10^{-4}$ after passing through a minimum (Fig. 2.85a). The latter value among the two solutions ($\sigma_0^a = 6.8 \times 10^{-4}$) has evidently its physical meaning. In the region $v_p^0 > v_p^c$ (for example, $v_p^0 = 0.15$), there

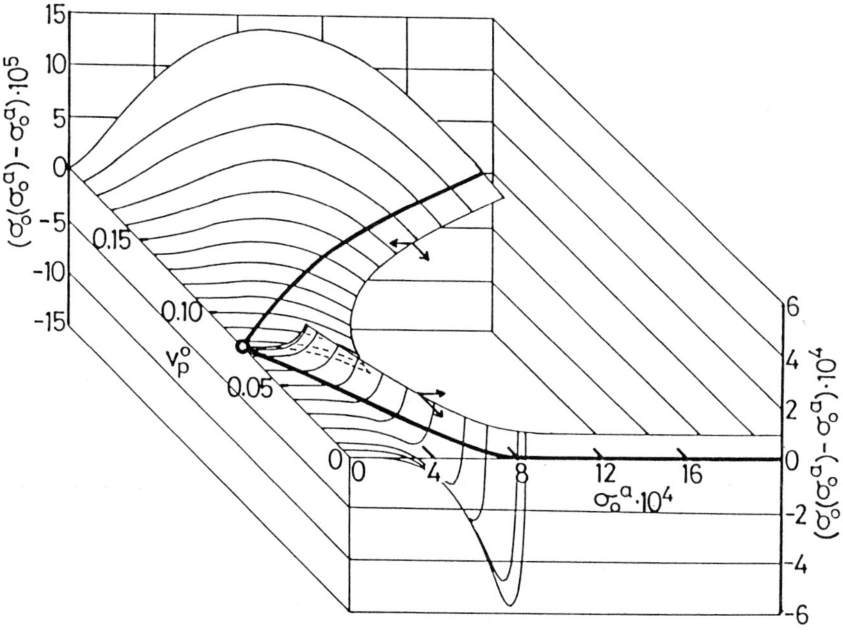

Fig. 2.86 Three dimensional plots of main equation $\sigma_0(\sigma_0^a) - \sigma_0^a$ in σ_0^a-v_p^0 plane: Original polymer, Schulz-Zimm type distribution ($X_w^0 = 2117$ and $X_w^0/X_n^0 = 2.8$); $p_1 = 0.643$, $p_2 = 0.200$ and $k_0 = 0$. Broad solid line, σ_0^{cp} vs. v_p^0 curve and unfilled circle, critical solution point ($v_p^c = 0.0766$).

are two solutions: $\sigma_0{}^a = 0$ and 9.6×10^{-4} (Fig. 2.85b) and of course the latter has its physical meaning (ref. 167).

Fig. 2.86 shows the left-hand side of eqn (2.191) [i.e., $(\sigma_0(\sigma_0{}^a) - \sigma_0{}^a)$], $\sigma_0{}^a$ and $v_p{}^0$ relations. In the figure the axis of ordinate is the difference $\sigma_0(\sigma_0{}^a) - \sigma_0{}^a$ and two abscissas are $\sigma_0{}^a$ and $v_p{}^0$. Here, $v_p{}^0$ was varied from 0 to 0.20 and the unfilled circle is the critical solution point (CSP) for this system. $\sigma_0(\sigma_0{}^a) - \sigma_0{}^a$ at a given $v_p{}^0$ is minimum without exception in the range $v_p{}^0 < v_p{}^c$ and is maximum in the range $v_p{}^0 > v_p{}^c$. Note that in the region near CPC the $\sigma_0(\sigma_0{}^a) - \sigma_0{}^a$ vs. $\sigma_0{}^a$ curve becomes flat and as a result, without magnification, the positions of the minimum and maximum of the curve can not be determined. Although in this case, the error accompanied by the computer simulation becomes sometimes innegligibly large,

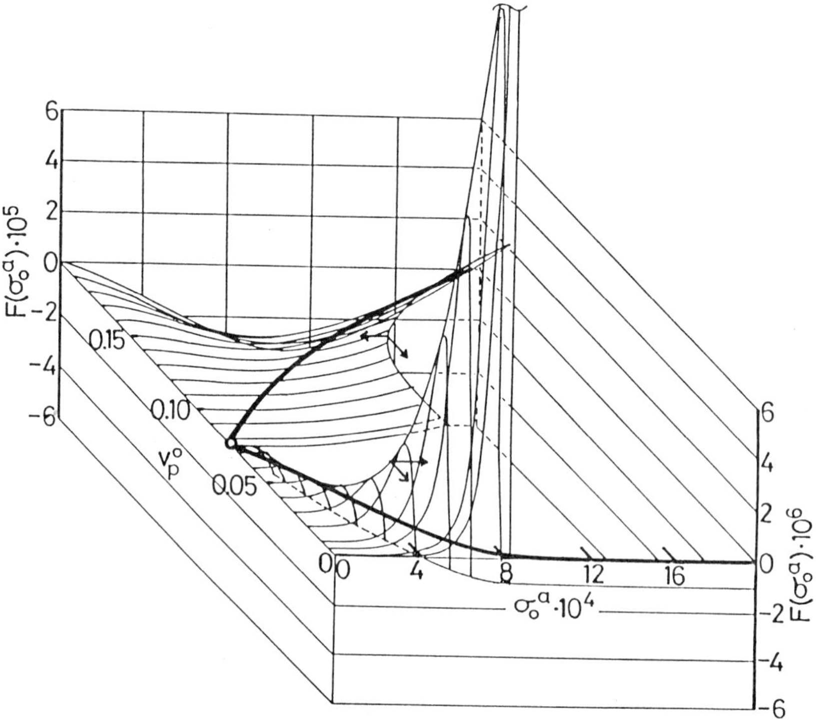

Fig. 2.87 Three dimensional plots of Šolc's CPC equation in $\sigma_0{}^a$-$v_p{}^0$ plane: Original polymer, Schulz-Zimm type distribution ($X_w{}^0 = 2117$ and $X_w{}^0/X_n{}^0 = 2.8$); $p_1 = 0.643$, $p_2 = 0.200$ and $k_0 = 0$. Broad solid line, $\sigma_0{}^{cp}$ vs. $v_p{}^0$ curve and unfilled circle, critical solution point.

depending on the calculation procedure used, the employment of the double precision method enables us ambiguously to determine the solution, which has a physical meaning. The σ_0^{cp} vs. v_p^0 curve is continuous, but the polygonal line whose polygonal point agrees well with CSP, at which the first order differential of σ_0^{cp} with respect to v_p^0, changes its sign from negative to positive (ref. 167).

Fig. 2.87 shows the results obtained by solving Šolc's "CPC equation" (in this case, eqn (2.210)) with direct integration of v_p^{cp} and X_n^{cp} under the same conditions as those in Fig. 2.85 (including $n=2$). In the figure the axis of ordinate $F(\sigma_0^a)$ is defined by the left-hand side of eqn (2.210). As in Fig. 2.85 the unfilled circle is CSP and the bold full line is σ_0^{cp} vs. v_p^0 relation, which is, as expected, in excellent coincidence with that estimated by Kamide et al's method (ref. 167).

Table 2.19 summarizes σ_0^{cp} and χ_0^{cp} values determined over wide range of v_p^0 by Kamide and Matsuda et al.'s method [eqns (2.191) and (2.192) with eqns (2.214) and (2.215)] for the polymer with Schulz-Zimm type distribution (with $X_w^0=2117$ and X_w^0/X_n^0

TABLE 2.19
Kamide's σ_0^{cp} and χ_0^{cp} at various initial concentration v_p^0.

	Kamide, Matsuda et al. (eqn (2.191))			
v_p^0	$\sigma_0^{cp}\times10^3$	$\sigma_0(\sigma_0^{cp})\times10^3$	$\sigma_0(\sigma_0^{cp})-\sigma_0^{cp}$	χ_0^{cp}
0.0005	0.8294649	0.8294649	-0.4482421×10^{-11}	0.5068843
0.0010	0.7894527	0.7894528	0.1415837×10^{-9}	0.5066481
0.0050	0.6765704	0.6765704	0.7176568×10^{-10}	0.5060174
0.0100	0.6083617	0.6083617	0.5377310×10^{-11}	0.5057250
0.0200	0.5090541	0.5090541	0.2419572×10^{-11}	0.5054728
0.0300	0.4213474	0.4213474	0.3975748×10^{-11}	0.5053957
0.0400	0.3351164	0.3351164	-0.3007202×10^{-12}	0.5054140
0.0500	0.2472484	0.2472484	-0.5211310×10^{-13}	0.5055011
0.0600	0.1561916	0.1561916	-0.1052424×10^{-13}	0.5056444
0.0700	0.0608670	0.0608670	-0.3051619×10^{-14}	0.5058373
0.0750	0.0115759	0.0115759	0.1039669×10^{-15}	0.5059510
0.0800	0.0396350	0.0396350	-0.3753067×10^{-15}	0.5060758
0.0900	0.1462143	0.1462143	-0.2102899×10^{-13}	0.5063576
0.0100	0.2596227	0.2596227	-0.1265839×10^{-13}	0.5066815
0.0110	0.3807023	0.3807023	-0.1521097×10^{-12}	0.5070466
0.0120	0.5102899	0.5102899	-0.2331810×10^{-12}	0.5074529
0.0140	0.7984784	0.7984784	-0.1421412×10^{-12}	0.5083891
0.0160	1.1315560	1.1315560	-0.6910338×10^{-12}	0.5094934
0.0180	1.5176370	1.5176370	0.4824771×10^{-12}	0.5107717
0.0200	1.9657260	1.9657260	-0.6988389×10^{-12}	0.5122314
0.0220	2.4858450	2.4858450	-0.1026131×10^{-11}	0.5138814
0.0240	3.0891580	3.0891580	0.4405498×10^{-12}	0.5157317

cf., critical volume fraction of polymer, $v_p^c=0.0766$.

= 2.8) dissolved in a single solvent (p_1 = 0.643 and p_2 = 0.200) and Table 2.20 collects σ_0^{cp} and χ_0^{cp} values estimated by both our direct integral procedures (eqns (2.214) and (2.215)) and the polynomial expansion procedure (eqn (2.200')), which was originally used by Šolc (ref. 172), of Šolc's method (eqns (2.210) and (2.211)) for this polymer / solvent system. $\sigma_0(\sigma_0^{cp}) - \sigma_0^{cp}$ (in Table 2.19) and $F(\sigma_0^{cp})$ (in Table 2.20) are also compiled for the sake of comparison. In Table 2.20, h_m values, defined in eqn (2.200'), of ν_{-1} and ν_0 in Šolc's polynomial expansion procedure are also listed. In Šolc's polynomial expansion procedure, when $v_p^0 \leq 0.001$ and $v_p^0 \geq 0.14$, ν_{-1} and ν_0 could not be solved because of an overflow error inherent to the computer which was caused by large σ_0 value. For example, at $v_p^0 = 0.001$ both ν_{-1} and ν_0 could not converge even at h= 500 in eqn (2.200) and finally overflow error occurred. The same overflow error appeared, for example, $v_p^0 = 0.14$. On the other hand, in Kamide et al.'s method, such an overflow error did not occur under any conditions and accordingly CPC can be calculated over a whole range of v_p^0. Moreover, in Kamide et al.'s method, the fundamental equation (eqn (2.191)) is expressed in terms of the difference between the true σ_0^{cp} and the assumed σ_0^a and then, at the true cloud point, which was independently determined, σ_0^a coincides completely with σ_0^{cp} as the theory requires (ref. 167).

In the range of $0.005 \leq v_p^0 \leq 0.12$, the difference between σ_0^{cp} by Kamide et al.'s method and σ_0^{cp} by Šolc's polynomial expansion method increases remarkably as v_p^0 decreases. For example, at $v_p^0 = 0.005$, Šolc's σ_0^{cp} is about 3.7% smaller than Kamide's σ_0^{cp}. In Fig. 2.88, CPC calculated by both Kamide's method and Šolc's polynomial expansion method are shown as solid and broken lines, respectively: Šolc's CPC is apparently underestimated, compared with CPC by Kamide in the low v_p^0 region. For example, at $v_p^0 = 0.005$, Šolc's $(\chi_0^{cp} - 1/2)$ is 1.22% smaller than Kamide's $(\chi_0^{cp} - 1/2)$. Considering the high accuracy of σ_0^{cp} value by Kamide's method, we can conclude that the Kamide's method yields a more accurate CPC at a low v_p^0 region than the Šolc's original method does. If we adopt the direct integration procedure (eqns (2.214) and (2.215)) in place of the polynomial expansion procedure (eqn (2.200')) in calculating ν_{-1} and ν_0 in Šolc's method, we can avoid the error which is contained in the polynomial expansion method. Truly, the difference between σ_0^{cp} by Šolc's method with a direct integration and σ_0^{cp} by Kamide's method is very small, although the difference amounts to at

TABLE 2.20

Comparison between direct integral procedure's χ_0^{CP} and polynomial expansion procedure's χ_0^{CP} in Šolc's method at various initial concentration v_p^0.

	Šolc (eqns (210) and (2.211))							
	direct integral procedure (eqns (2.214) and (2.215))			polynomial expansion procedure (eqn (2.200'))			h_m(in eqn (2.200'))	
v_p^0	$\sigma_0^{CP}\times10^3$	$F(\sigma_0^{CP})$	χ_0^{CP}	$\sigma_0^{CP}\times10^3$	$F(\sigma_0^{CP})$	χ_0^{CP}	(ν_{-1})	(ν_0)
0.0005	0.8294649	-0.7593535×10^{-10}	0.5068843	—	—	—	—	—
0.0010	0.7894527	-0.1014228×10^{-9}	0.5066481	—	—	—	—	—
0.0050	0.6765704	-0.3773043×10^{-10}	0.5060174	0.6518013	-0.1113997×10^{-11}	0.5059439	86	92
0.0100	0.6083617	-0.1331231×10^{-10}	0.5057250	0.5992794	-0.3376525×10^{-11}	0.5057003	86	92
0.0200	0.5090541	-0.8792652×10^{-11}	0.5054728	0.5075786	0.7825293×10^{-12}	0.5054687	51	82
0.0300	0.4213473	0.1091864×10^{-11}	0.5053957	0.4211462	0.2798365×10^{-12}	0.5053949	35	54
0.0400	0.3351163	0.1809232×10^{-12}	0.5054140	0.3351130	-0.8670955×10^{-13}	0.5054139	25	38
0.0500	0.2472481	0.1244593×10^{-12}	0.5055011	0.2472661	0.2722552×10^{-13}	0.5055011	18	28
0.0600	0.1561907	0.4493058×10^{-14}	0.5056444	0.1562089	-0.4949065×10^{-15}	0.5056445	13	19
0.0700	0.0608585	-0.7216050×10^{-15}	0.5058373	0.0608771	-0.1847419×10^{-15}	0.5058374	8	12
0.0750	0.0113003	0.3773422×10^{-17}	0.5059510	0.0113208	-0.1137062×10^{-17}	0.5059511	5	7
0.0800	0.0396627	-0.4629522×10^{-17}	0.5060758	0.0396428	-0.2542743×10^{-16}	0.5060759	7	10
0.0900	0.1462171	0.1506270×10^{-14}	0.5063577	0.1461947	0.1645313×10^{-14}	0.5063578	13	19
0.0100	0.2596240	0.2325195×10^{-13}	0.5066815	0.2595989	-0.7688479×10^{-14}	0.5066516	19	29
0.0110	0.3807031	-0.6260194×10^{-13}	0.5070467	0.3806752	-0.4512228×10^{-14}	0.5070468	29	46
0.0120	0.5102906	0.1878492×10^{-13}	0.5074529	0.5102594	-0.1049305×10^{-13}	0.5074530	52	83
0.0140	0.7984789	0.1711582×10^{-12}	0.5083891	—	—	—	—	—
0.0160	1.1315570	-0.6531676×10^{-12}	0.5094934	—	—	—	—	—
0.0180	1.5176370	0.2015865×10^{-12}	0.5107717	—	—	—	—	—
0.0200	1.9657260	0.7947712×10^{-12}	0.5122314	—	—	—	—	—
0.0220	2.4858460	-0.1260222×10^{-11}	0.5138814	—	—	—	—	—
0.0240	3.0891590	0.7575863×10^{-14}	0.5157317	—	—	—	—	—

—, calculation is impossible because of overflow.

cf., critical volume fraction of polymer, $v_p^c = 0.0766$.

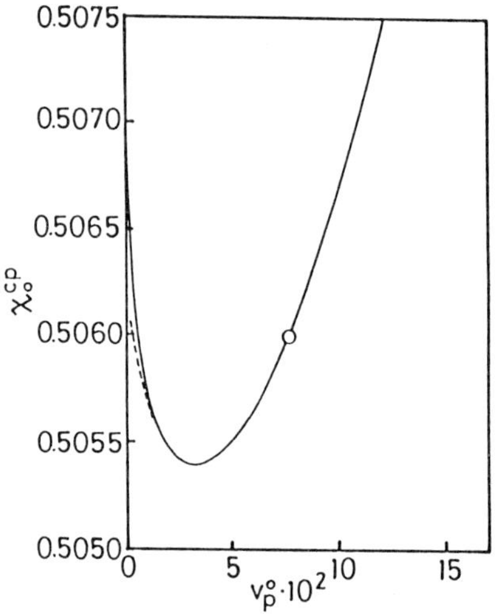

Fig. 2.88 Comparison of cloud point curve obtained by Kamide et al.'s method (solid line) and by Šolc's method (broken line): Original polymer, Schulz-Zimm type distribution ($X_w^0=2117$, $X_w^0/X_n^0=2.8$) and $p_1=0.643$, $p_2=0.200$, $k'=0$. Unfilled circle, critical solution point.

maximum of 3% or less near CSP, and then χ_0^{cp} values by both methods coincide perfectly with each other (ref. 167).

One merit of Šolc's polynomial expansion method is relatively short calculation time. Especially near by CSP, h_m is smaller than 10 and then, calculation time is extremely shortened. For example, in the range of $0.005 \leq v_p^0 \leq 0.12$, calculation time by Šolc's method is less than 1/10 of Kamide's method. But the following points should be noted: Kamide's method gives $v^{cp}(X_i)$, accordingly, X_w^{cp} and X_z^{cp} concurrently, as the result of calculation. In contrast, Šolc's method needs independent calculations of not only $v^{cp}(X_i)$ but also $v_0(X_i)$, if the MWD of polymer partitioned in "cloud particle" is required. Also, when X_w^{cp} and X_z^{cp} are needed, we must calculate progression of ν_1 and ν_2. In addition, if we try to solve skillfully the overflow problem of ν_{-1} and ν_0 for a large σ_0, computation time becomes very long and an accuracy of calculation is unavoidably lowered.

When accurate CPC calculation is needed over a whole range of v_p^0, Kamide and Matsuda et al.'s method is strongly recommended. If the CPC calculation only near by CSP is required, Šolc's method is sufficient for this purpose (ref. 167).

Fig. 2.89 shows the v_p^0 dependence of σ_0^{cp}, the

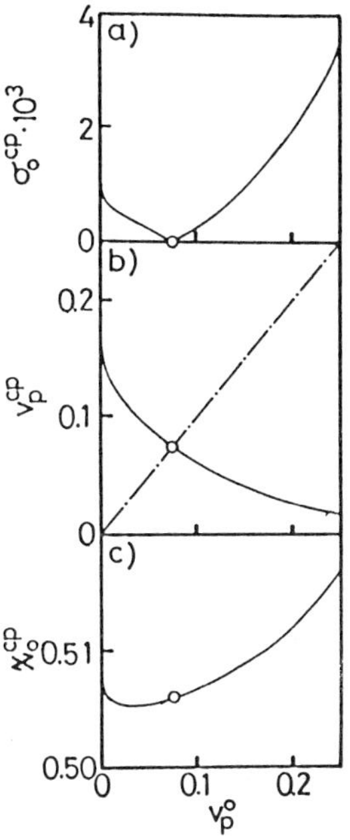

Fig. 2.89 The partition coefficient, the polymer volume fraction and χ_0 parameter at true cloud point, σ_0^{cp}, v_p^{cp} and χ_0^{cp}, plotted as a function of v_p^0: Original polymer, Schulz-Zimm type distribution ($X_w^0=2117$ and $X_w^0/X_n^0=2.8$); $p_1=0.643$, $p_2=0.200$ and $k_0=0$. Open circles correspond to critical solution point ($v_p^c=0.0766$).

concentration of cloud particle v_p^{cp} and χ_0^{cp}, determined by Kamide and Matsuda's method (refs 4,6). Here, the unfilled circle is CSP and the chain line (in Fig. 2.89b) denotes $v_p^{cp}=v_p^0$ line. The σ_0^{cp} vs. v_p^0 curve is continuous (as shown in Fig. 2.85) even at CSP at which σ_0^{cp} becomes zero (Fig. 2.89a). We can not expect any preferential partitioning of polymer molecules based on the difference in the molecular weight. In contrast to this, v_p^{cp} decreases continuously and monotonously with increasing v_p^0 (Fig. 2.89b), and coincides at CSP with v_p^0. χ_0^{cp} vs. v_p^0 relation is also continuous at CSP (Fig. 2.89c).

After estimating shadow curve from Fig. 2.89b and overlapping with CPC (Fig. 2.89c), Fig. 2.90 is obtained. In Fig. 2.90 the chain line is the shadow curve (χ_0^{cp} vs. v_p^{cp} curve), the full line is CPC (χ_0^{cp} vs. v_p^0 curve) and the unfilled circle is CSP. The shadow curve is also continuous at CSP. The CPC and the shadow curve always have downward curvature (i.e., $d^2\chi_0^{cp}/dv_p^2>0$ and $d^2\chi_0^{shadow}/dv_p^2>0$) and any depression around CSP cannot be

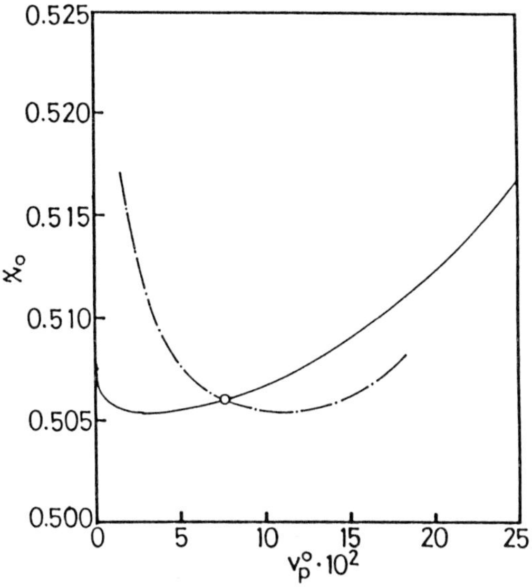

Fig. 2.90 Cloud point curve (solid line) and shadow curve (chain line): Original polymer, Schulz-Zimm type distribution ($X_w^0 = 2117$ and $X_w^0/X_n^0 = 2.8$); $p_1 = 0.643$, $p_2 = 0.200$ and $k_0 = 0$. Open circle, critical solution point.

observed. This is really characteristic of a polymer having Schulz-Zimm type MWD dissolved in a solvent with molecular weight-independent χ . In contrast, CPC has depression at CSP for the same system as shown in Fig. 2.83.

We can easily draw the limiting tie-line (i.e., tie-line for $\chi_0 = \chi_0^{cp}$) in Fig. 2.90 at any initial concentration v_p^0. The polymer concentration on the shadow curve, in other words the concentration of the polymer partitioned in "cloud particle", must be important information enabling us to predict the polymer solid structure formed by the two-phase separation.

Summarizing, both the two direct methods for calculating the cloud point curve (i.e., Kamide et al.'s method and Šolc's method) were shown to be based on the same theoretical principle and it was made clear that the main difference between the two methods is in the procedure of integration of v_p^{cp} and X_n^{cp} and that Kamide et al.'s direct integration method has more general character than the Šolc's polynomial expansion method in the view point of easy and accurate calculation of any type of molecular weight distribution without limitation (ref. 167).

2.7 FLORY TEMPERATURE AND FLORY ENTROPY PARAMETER

2.7.1 Theoretical background

The Flory temperature θ and an entropy parameter ψ_0 (see, eqn (2.216)) were experimentally determined from the temperature dependence of the second virial coefficient A_2; θ as the temperature at which A_2 becomes zero and ψ_0 from the temperature dependence of A_2 at θ (ref. 12),

$$\psi_0 = \left(\frac{V_0}{\bar{v}^2}\right) \theta \left(\frac{\partial A_2}{\partial T}\right)_\theta, \tag{2.216}$$

respectively, and the critical solution point CSP (the temperature T_c and the volume fraction $v_p{}^c$), extrapolating by the Shultz-Flory method for a polymer with an infinite molecular weight (see, eqn (2.227)). Although being only approximate, the Shultz-Flory method has been found to have a rather wide use. For all polymer / solvent systems investigated, it was confirmed that θ temperatures evaluated by the above two procedures agree fairly well with each other, but ψ_0 values estimated by the second virial coefficient method are significantly smaller than those obtained by the critical point method (ref. 75). Stockmayer (ref. 133) pointed out the inadequacy of the basic equations (eqns (2.226) and (2.227)) of the Schulz-Flory theory and proposed alternative equations (eqns (2.123) and (2.224)), rigorously derived for polydisperse polymer solutions. Another factor is the concentration dependence of the χ parameter, which was completely neglected in the Shultz-Flory and Stockmayer methods.

Whether the concentration dependence of the χ parameter and polymolecularity of the polymer should be taken into account to explain the critical point and, in the former case what would be the most reasonable values of p_1, p_2, $\cdots$, p_n, can be decided by comparing the experimental $v_p{}^c$ ($v_p{}^c$(exp)) with theoretical $v_p{}^c$ ($v_p{}^c$(theo)) data [calculated using eqns (2.121) and (2.122) (Kamide-Matsuda), eqn (2.123) (Stockmayer) and eqn (2.226) (Shultz-Flory) (ref. 165)]. The concentration-dependence parameters, evaluated by methods like vapor pressure, isothermal distillation and osmotic pressure, are not accurate enough to calculate $v_p{}^c$.

To solve this problem, Koningsveld et al. (KKS) (ref. 16) and Kamide and Matsuda (KM) (ref. 3) proposed methods for estimating p_1 and p_2 of χ (see, eqn (2.7) with $k_0 = 0$ and $p_3 = \cdots = p_n = 0$)

from the CSP data (T_c and $v_p{}^c$). Using p_1 and p_2 parameters thus estimated and CSP data, the Flory theta temperature θ and the entropy parameter at θ, ψ_0 can also be determined (refs 3,4).

χ_{00} at the critical point ($\chi_{00}{}^c$) is related to T_c, θ and ψ_0 through the relation (ref. 12),

$$\chi_{00}{}^c = (\frac{1}{2} - \psi_0) + \frac{\theta\psi_0}{T_c} \tag{2.217}$$

with

$$\psi_0 = \frac{1}{2} - \chi_{00,s} \tag{2.218}$$

and

$$\theta = \frac{\chi_{00,h}T_c}{\frac{1}{2} - \chi_{00,s}}. \tag{2.219}$$

$\chi_{00,s}$ and $\chi_{00,h}$ are the entropy and enthalpy terms of χ_{00}, respectively (i.e., $\chi_{00} = \chi_{00,s} + \chi_{00,h}$).

Comparing eqn (2.217) with eqn (2.6), we obtain the following relations.

$$a = \frac{1}{2} - \psi_0, \tag{2.220a}$$

$$b = \theta\psi_0. \tag{2.220b}$$

Substitution of eqn (2.220) into eqn (2.8) yields

$$a' = (\frac{1}{2} - \psi_0)(1 + \frac{k_0}{X_n}), \tag{2.221a}$$

$$b' = \theta\psi_0 + \theta(\psi_0 - \frac{1}{4})\frac{2k_0}{X_n}, \tag{2.221b}$$

$$c' = -(\frac{k_0}{X_n})\psi_0\theta^2. \tag{2.221c}$$

The effect of the molecular weight dependence of χ parameter on the critical point can be considered to be relatively small and the critical point is explained in terms of the concentration-dependence parameters alone. In this case ($k' = 0$ and $k_0 = 0$) eqn (2.217) reduces to (refs 3,5)

$$\frac{1}{T_c} = \frac{\chi_0^c}{\theta \psi_0} + \frac{1}{\theta}\left(1 - \frac{1}{2\psi_0}\right). \tag{2.222}$$

Using χ_0^c, calculated from eqns (2.221) and (2.222) and experimental T_c, we can determine θ and ψ_0 from the plot of $1/T_c$ against χ_0^c. This method is hereafter simply referred to as the Kamide-Matsuda method (ref. 3). Note that an accurate values of p_1 and p_2 are necessary to calculate χ_0^c.

Putting $p_j = 0$ (for $j = 1, \cdots, n$) in eqn (2.121), we obtain the Stockmayer equation (eqn (2.123)) (ref. 133),

$$\chi_0^c = \frac{1}{2}\left(\frac{1}{X_w v_p^c} + \frac{1}{1 - v_p^c}\right), \tag{2.223}$$

which was derived rigorously for a multicomponent polymer / single solvent system assuming $p_j = 0$ ($j = 1, \cdots, n$). Substitution of eqn (2.223) into eqn (2.220) yields

$$\chi_0^c = \frac{1}{2}\left(\frac{X_z^{1/2}}{X_w} + \frac{1}{X_w}\right)\left(\frac{X_w}{X_z^{1/2}} + 1\right). \tag{2.224}$$

Combining eqn (2.222) with eqn (2.224), we obtain

$$\frac{1}{T_c} = \frac{1}{\theta \psi_0}\left[\frac{1}{2}\left\{\frac{1}{X_w^{1/2}} + \left(\frac{X_z}{X_w}\right)^{1/2}\right\}\left\{\frac{1}{X_w^{1/2}} + \left(\frac{X_w}{X_z}\right)^{1/2}\right\}\right] + \frac{1}{\theta}\left(1 - \frac{1}{2\psi_0}\right). \tag{2.225}$$

From the plots of T_c against $\left[\frac{1}{2}\left\{\frac{1}{X_w^{1/2}} + \left(\frac{X_z}{X_w}\right)^{1/2}\right\}\left\{\frac{1}{X_w^{1/2}} + \left(\frac{X_w}{X_z}\right)^{1/2}\right\}\right]$, we can estimate θ and ψ_0. This is Stockmayer's method.

When $X_w = X_z$ is assumed (i.e., when the polymer is monodisperse), eqns (2.123) and (2.225) reduce to the well-known equations derived by Shultz and Flory (ref. 165).

$$v_p^c = \frac{1}{1 + X_w^{1/2}} \tag{2.226}$$

and

$$\frac{1}{T_c}=\frac{1}{\theta\,\psi_0}\left[\frac{1}{2}\left(\frac{1}{X_w^{1/2}}+1\right)^2\right]+\frac{1}{\theta}\left(1-\frac{1}{2\psi_0}\right)$$

$$=\frac{1}{\theta\,\psi_0}\left(\frac{1}{X_w^{1/2}}+\frac{1}{2X_w}\right)+\frac{1}{\theta}. \qquad (2.227)$$

Note that eqn (2.227) is strictly valid only for a monodisperse polymer / single solvent system, in which the χ parameter is assumed to be independent of the molecular weight of the polymer and concentration. Slope and intercept of the plots of T_c^{-1} vs. $\left(\frac{1}{X_w^{1/2}}+\frac{1}{2X_w}\right)$ give $1/\theta\,\psi_0$ and $1/\theta$, respectively and then we can estimate ψ_0 and θ from the plots. This method is referred to as the Shultz-Flory (SF) method.

Kamide and Matsuda (ref. 3) showed that the terms more than or equal to v_p^3 can be neglected to make the theoretical critical concentration v_p^c(theo) coincident with the experimental value v_p^c(exp). First calculate χ_0^c and v_p^c(theo) by solving eqns (2.121) and (2.122) with assumed values of p_1 and p_2. Evaluate the square average of the difference between v_p^c(theo) and v_p^c(exp), δ, defined by

$$\delta=\sum_{i=1}^{N}\frac{(v_p^c(\text{exp})-v_p^c(\text{theo}))_i^2}{N_0} \qquad (2.228)$$

(N_0, total number of samples)

for a given combination of p_1 and p_2. Finally determine the most probable p_1 and p_2 as a pair which gives minimum δ. That is, an appropriate choice of $p_1, \cdots, p_n$, v_p^c(theo) will fit v_p^c(exp) for an entire set of data. If v_p^c(exp) and experimental T_c (T_c(exp)) data are available for a polymer / solvent system, we can determine $p_1, \cdots, p_n$ from the plot of v_p^c(exp) vs. v_p^c(theo) and θ and ψ_0 from the $1/T_c-\chi_0^c$ plot concurrently (ref. 165).

Literature data on the upper critical solution points (UCSP) (v_p^c and T_c) of Koningsveld et al. (ref. 16) and of Kuwahara and his collaborators (refs 66,71) and on lower critical solution points (LCSP) of Saeki et al. (ref. 71) are available for determining p_1 and p_2, using the KM method. Here, since Kuwahara et al. and Saeki et al. evaluated only X_w and X_n for these samples, Kamide and Matsuda estimated X_z assuming the Schulz-Zimm type molecular weight distribution for all PS samples.

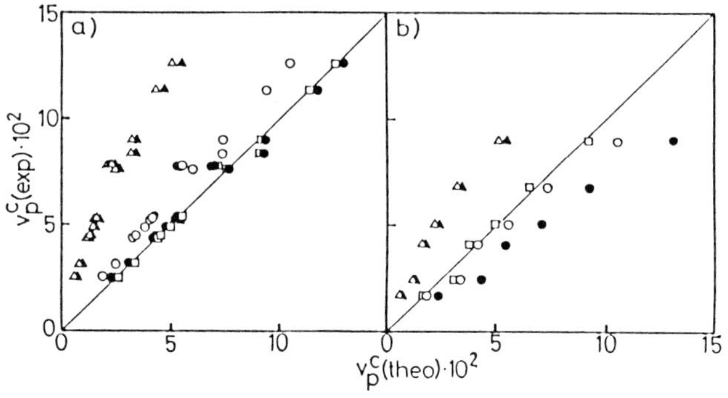

Fig. 2.91 The critical concentration, experimentally determined, $v_p^c(\mathrm{exp})$ plotted against the critical concentration, theoretically calculated, $v_p^c(\mathrm{theo})$ for the UCSP a) and LCSP b) of the of the polystyrene / cyclohexane system: △, Shultz-Flory (eqn (2.223)); ▲, Stockmayer (eqn (2.123)); ○, Kamide-Matsuda (eqns (2.121) and (2.122), $p_1=0.6$, $p_2=0$); ●, Kamide-Matsuda (eqns (2.121) and (2.122), $p_1=0.623$, $p_2=0.290$); □, Kamide-Matsuda (eqns (2.121) and (2.122), $p_1=0.642$, $p_2=0.190$ for UCSP and $p_1=0.602$, $p_2=-0.347$ for LCSP)).

Fig. 2.91 shows the plot of v_p^c (exp) vs. v_p^c (theo) calculated for some typical combinations of p_1 and p_2 for the UCSP and LCSP of a PS / CH system. In this figure the same data as those in Figs 2.92 and 2.93 were utilized. Evidently, the methods of Shultz-Flory (eqn (2.226)) and Stockmayer (eqn (2.123)) can not give reasonable v_p^c and even if p_1 is considered in Kamide-Matsuda method the difference between v_p^c (exp) and v_p^c (theo) remains significant, although improved remarkably.

Fig. 2.92 shows the relations among p_1, p_2 and δ for the PS / CH system. The most reasonable combination of p_1 and p_2, evaluated from Fig. 2.90 is $p_1=0.642$ and $p_2=0.190$ for UCSP and $p_1=0.602$ and $p_2=-0.347$ for LCSP (Tables 2.21 and 2.22) (ref. 3). These values for UCSP should be compared with $p_1=0.643$ and $P_2=0.200$, estimated from the cloud point data by Kamide and Matsuda (ref. 4).

The plots of $1/T_c$ against χ_0^c for the PS / CH system are shown in Fig. 2.93 (see, eqn (2.222)). The results are summarized in Tables 2.21 and 2.22, in which θ and ψ_0 values estimated originally in literature are shown in parenthesis for comparison. Krigbaum and his coworkers (refs 62,75,173), Schulz-Baumann (ref. 174) and Kotera et al. (ref. 175) determined the θ and ψ_0 of UCSP for the PS / CH system from the temperature dependence of the second virial coefficient near θ temperature. The literature data

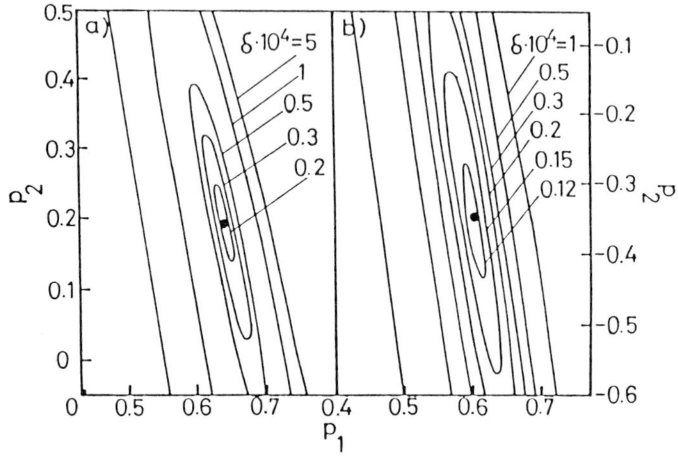

Fig. 2.92 Relations among p_1, p_2 and δ (eqn (2.228)) for the UCSP a) and LCSP b) of the polystyrene / cyclohexane system: Number on curve denotes $\delta \times 10^4$. Filled mark corresponds to the ideal combination of p_1 and p_2 to describe the critical solution point.

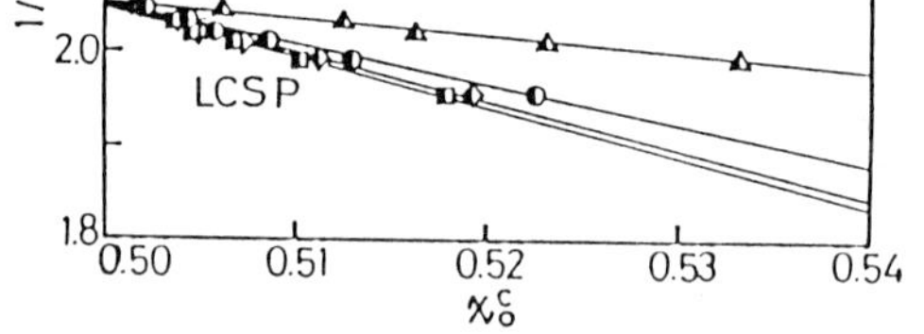

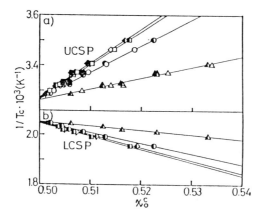

Fig. 2.93 $1/T_c$ plotted against $\chi_0{}^c$ for the UCSP a) and LCSP b) of the polystyrene / cyclohexane system: Triangle, Schulz-Flory plot (eqn (2.227)); circle, Kamide-Matsuda plot (eqns (2.121) and (2.122), $p_1=0.6$, $p_2=0$); rectangle, Kamide-Matsuda plot (eqns (2.121) and (2.122), $p_1=0.623$, $p_2=0.290$ (ref. 3)); diamond, Kamide-Matsuda plot (eqns (2.121) and (2.122), $p_1=0.642$, $p_2=0.190$ for UCSP and $p_1=0.602$, $p_2=-0.347$ for LCSP); unfilled mark, data of Koningsveld et al. (ref. 16); filled mark, data of Kuwahara et al. (ref. 66); half-filled mark, data of Saeki et al. (ref. 71).

show variation in ψ_0 from 0.18 to 0.36 and in θ from 307.0 to 308.4K, as listed in Table 2.21.

TABLE 2.21

Flory θ temperature and entropy parameter ψ_0 of the upper critical solution point for the polystyrene / cyclohexane system.

Method			Concentration dependence of χ p_1	p_2	θ /K	ψ_0
Critical point	Shultz-Flory (eqn (2.227))		0	0	306.2(306.2[a], 307.0[b], 307.2[c], 306.4[d])	0.75(0.78[a], 0.79[b], 1.056[c])
	Stockmayer (eqn (2.225))		0	0	306.2	0.74
	Koningsveld et al. (eqns (2.229) and (2.230))		0.623	0.308	305.2	0.29
			0.623[e]	0.290	305.6	0.30
	Kamide-Matsuda: (eqn (2.222))	Phase separation	0.6[n]	0[n]	306.4	0.35
		Cloud point	0.643[f]	0.200	306.6	0.27
		Critical point	0.623[e]	0.290[e]	305.6	0.27
		Critical point	0.642	0.190	305.1	0.27
Second virial coefficient	Membrane osmometry		—	—	307.6[g]	0.36[g]
			—	—	307.6[h]	0.23[h]
	Light scattering		—	—	308.4[i]	0.19 ± 0.05[i]
			—	—	307.0[j]	0.36[k]
			—	—	307.4[l]	0.18[m]

[a]ref. 62, [b]ref. 71, [c]ref. 165, [d]ref. 62, [e]ref. 19, [f]ref. 4, [g] ref. 75, [h]ref. 62, [i]ref. 173, [j]ref. 174, [k]calculated using $\theta\,(\partial A_2/\partial T)_\theta$ data in ref. 174, [l]ref. 175, [m]calculated using $\theta\,(\partial A_2/\partial T)_\theta$ data in ref. 175, [n]ref. 29.

TABLE 2.22

Flory temperature θ and entropy parameter ψ_0 of the upper critical solution point for the polystyrene / cyclohexane system.

Method	p_1	p_2	θ/K	ψ_0
Critical point				
Shultz-Flory (eqn (2.227))	0	0	486.6(486.0[a])	-1.20(-1.19[a])
Stockmayer (eqn (2.225))	0	0	486.8	-1.20
Koningsveld et al. (eqns (2.229),(2.230))	0.571	-0.047	487.5	-0.61
Kamide-Matsuda: Phase separation	0.60[b]	0[b]	487.6	-0.52
Critical point	0.602	-0.347	487.2	-0.42

[a]ref. 71; [b]ref. 29.

Koningsveld et al. (refs 16,19) defined the polymer-solvent interaction parameter g by eqn (2.24). Eqn (2.122) can be rearranged using eqn (2.24) into (ref. 16)

$$Y \equiv g_1 - g_2 + 4g_2 v_p{}^c$$

$$= \frac{1}{6}\left[\frac{1}{(1+v_p{}^c)^2} - \frac{X_z}{(X_w v_p{}^c)^2}\right] \qquad (2.229)$$

for the case of $k' = 0$ and $n = 2$. Substituting X_w, X_z and $v_p{}^c$ (exp) into eqn (2.229) we obtain the relation between Y and $v_p{}^c$, from which g_1 and g_2 can be determined using curve fitting methods. The value of g_0 depends on the temperature through the relation (ref. 16),

$$g_0 \equiv g_{00} + g_{01}/T_c$$

$$= \frac{1}{2}\left[\frac{1}{1-v_p{}^c} + \frac{1}{X_w v_p{}^c} + 2g_1(1-3v_p{}^c) + 6g_2(1-2v_p{}^c)v_p{}^c\right]. \qquad (2.230)$$

Substitution of X_w, X_z, T_c (exp) data into eqn (2.227) enables us to estimate g_{00} and g_{01} (ref. 16). Comparing eqn (2.219) with eqn (2.28) and eqn (2.25), we obtain

$$\psi_0 = \frac{1}{2} - g_{00} + g_1, \qquad (2.231)$$

$$\theta = g_{01}/\psi_0. \qquad (2.232)$$

This method is referred to as the Koningsveld-Kleintjens-Shultz (KKS) method (ref. 16).

Fig. 2.94 shows the plot of Y in eqn (2.229) vs. $v_p{}^c$ for the literature data on the UCSP and LCSP of a PS / CH system. From this figure, we estimated, using the least-square regression method, g_1 to be 0.214 and g_2, 0.057 for UCSP and $g_1 = 0.1345$ and $g_2 = 0.0079$ for LCSP. The plot for LCSP is not linear because experimental techniques are difficult, thus limiting the accuracy of the g_1 and g_2 values for LCSP. The value of g_0 was calculated by putting the g_1 and g_2 values thus obtained and $v_p{}^c$ into eqn (2.230) and plotting them as a function of $T_c{}^{-1}$ in Fig. 2.95. It was found that $g_{00} = 0.213$ and $g_{01} = 151.7$ at the UCSP and $g_{00} = 1.2447$ and $g_{01} = -297.0$ the LCSP. Using these g parameters,

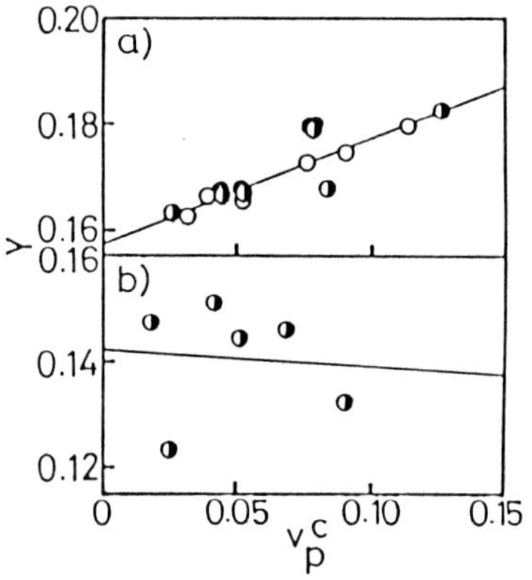

Fig. 2.94 Plot of Y in eqn (2.229) against $v_p{}^c$ for the UCSP a) and LCSP b) of the polystyrene / cyclohexane system: ○, Koningsveld et al. (ref. 16); ●, Kuwahara et al. (ref. 66); ◑, Saeki et al. (ref. 71).

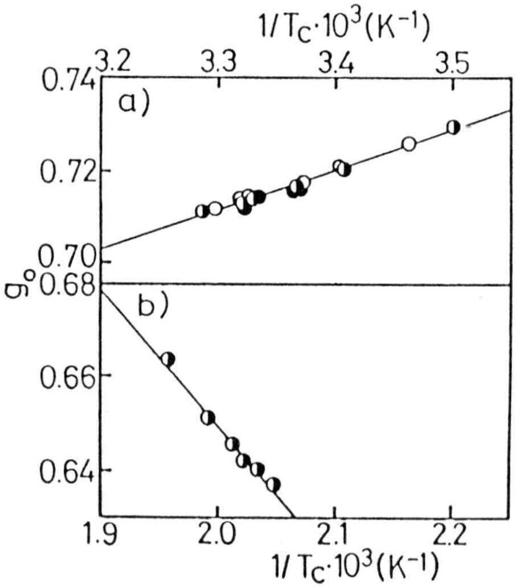

Fig. 2.95 Plot of g_0 in eqn (2.230) versus $1/T_c$ for the UCSP a) and LCSP b) of the polystyrene / cyclohexane system: ○, Koningsveld et al. (ref. 16); ●, Kuwahara et al. (ref. 66); ◑, Saeki et al. (ref. 71).

p_1 and p_2 (both at θ) and θ and ψ_0 were evaluated as summarized in Tables 2.21 and 2.22.

Table 2.23 presents the features of the various methods

TABLE 2.23

Various methods for estimating θ and ψ_0 from critical point data.

Method	Parameter prerequisite for applying method			Parameters deduced from the method
	Molecular characteristics	Critical point data	Concentration dependence of χ	
Schulz-Flory (eqn (2.227))	X_w	T_c	—	θ, ψ_0, v_p^c
Stockmayer (eqn (2.225))	X_w, X_z	T_c	—	θ, ψ_0, v_p^c
Koningsveld et al. (eqns (2.229) and (2.230))	X_w, X_z	T_c, v_p^c	—	θ, ψ_0, g_{00}, g_{01}, g_1, g_2
Kamide-Matsuda (eqns (2.121), (2.122) and (2.222))	X_w, X_z	T_c	p_1, p_2	θ, ψ_0, v_p^c, χ_{00}

proposed so far for estimating θ and ψ_0 from critical point data.

Kamide and Matsuda (ref. 3) applied these two methods to PS / CH and PS / MCH, and concluded that the ψ_0 values obtained by these methods are remarkably smaller than those obtained by the Shultz and Flory (SF) (ref. 165) method which has been widely utilized, and are within the range of variation expect for ψ_0 evaluated from the temperature dependence of the second virial coefficient A_2 (ref. 1). In other words, Kamide and Matsuda clearly established the superiority of the KKS and KM methods over the SF method which assumes $p_1 = p_2 = 0$.

As an extension of their previous study (ref. 3), an attempt was made to apply the above mentioned methods (i.e., KKS, KM and SF) to the literature data on the critical solution points (v_p^c(exp) and T_c(exp)) on PS (refs 16,66,70,71,136,140,142,143) or polyethylene (PE) (refs 14,68,144-147) or cellulose acetate (CA) (refs 148,149,176,177) in a single solvent in order to compare the reliability of these methods and, if possible, to establish the dependence of p_1, p_2, θ and ψ_0 on the nature of the solvent. It should be remembered that except for the work in dimethoxymethane by Siow et al. (ref. 158) and in CH (in part) by Koningsveld and his coworkers (ref. 16), all other data on PS solutions were generated by Kaneko, Kuwahara and their collaborators at Sapporo (refs 66,70,71,136,140,143) using PS samples with $X_w/X_n < 1.10$ (in particular, for samples with the weight-average molecular weight $M_w \times 10^{-4} = 3.7$ to 40, $X_w/X_n < 1.06$) manufactured by Pressure Chemical Co. (Pittsburg, USA). The majority of literature data on PE solutions has been taken from the work of Nakajima and Hamada et al. at Kyoto (refs 146,147) who employed PE fractions isolated by successive precipitation fractionation or the solid extraction (column) method from whole polymers. All the CA data were reported by Suzuki et al. (refs 148,149,176,177) and CA was chosen as a typical polar polymer. The results are collected in Tables 2.24, 2.25 and 2.26. The values in parenthesis are taken from literature.

2.7.2 Polystyrene

For PS / methyl ethyl ketone (LCSP), PS / toluene (LCSP), PS / isopropyl acetate (UCSP), PS / n-propyl acetate (UCSP) and PS / dimethoxy methane (LCSP), the correlation coefficient (r) between Y and v_p^c is low (< 0.55) and in consequence, the reliability of p_1 and p_2 obtained by the KKS method is not high for these systems (Table 2.24). In contrast to this, the

TABLE 2.24

Concentration dependence of χ parameter p_1, p_2, Flory temperature θ and entropy parameter ψ_0 of the critical point for polystyrene / solvent systems.

Polymer: Polystyrene		Methods									
Solvent	UCSP or LCSP	Kamide-Matsuda				Koningsveld et al.				Shultz-Flory	
		p_1	p_2	θ/K	ψ_0	p_1	p_2	θ/K	ψ_0	θ (θ ref.)/K	ψ_0(ψ_0 ref.)
Methyl ethyl ketone	LCSP	0.618	-0.208	423.6	-0.44	0.550	-0.262	423.8	-0.31	423.1(422[a])	-0.63(-0.529[a])
Cyclopentane	UCSP	0.615	0.404	292.1	0.16	0.606	0.497	292.1	0.18	292.7(293[a])	0.53 (0.548[a])
	LCSP	0.631	0.331	428.5	-0.25	0.611	0.468	428	-0.27	427.4(427[a])	-0.81(-0.858[a])
Cyclohexane	UCSP	(0.642	0.190	305.1	0.27[e])	(0.623	0.308	305.2	0.29[e])	306.5(306.2[b])	0.75 (0.78[b])
		—	—	—	—	(0.623	0.290	306.4	0.30[c])	— (307.0[d])	— (0.79[d])
	LCSP	0.638	-0.498	488.6	-0.58	0.621	-0.305	488.3	-0.60	486.8(486.0[d])	-1.21(-1.19[d])
		(0.602	0.347	487.2	-0.42[e])	(0.571	-0.047	487.5	-0.61[e])	— (486.0[e])	— (-1.20[e])
Methylcyclohexane	UCSP	(0.602	0.234	340.2	0.25[e])	(0.602	0.363	339.6	0.27[e])	342.3(344[d])	0.61 (0.56[d])
	LCSP	(0.649	-1.183	487.9	-0.54[e])	(0.643	-1.008	487.8	-0.56[e])	485.1(484[d])	-0.96(-0.94[d])
Toluene	LCSP	0.494	-0.922	550.4	-1.36	0.501	-0.475	550.3	-1.20	549.8(550[d])	-2.02(-1.92[d])
Benzene	LCSP	0.388	-1.781	524.3	-1.81	0.382	-1.655	524.2	-1.82	523.7(523[a])	-2.19(-1.79[a])
Isopropyl acetate	UCSP	0.673	-0.034	240.8	0.11	0.673	-0.082	240.6	0.13	245.3(246[f])	0.29 (0.32[f])
	LCSP	0.839	-2.000	398.1	-0.46	0.773	-1.594	394.6	-0.46	389.5(380[f])	-0.71(-0.46[f])
n-Propyl acetate	UCSP	0.643	-0.018	192.3	0.21	0.623	0.168	192.7	0.22	193.2(193[f])	0.60 (0.63[f])
	LCSP	0.650	-0.202	389.1	-0.25	0.642	-0.083	388.0	-0.24	386.0(—)	-0.61 (—)
trans-Decalin	UCSP	0.630	0.240	292.7	0.33	0.623	0.338	292.5	0.36	293.7(—)	0.95 (—)

[a]ref. 136, [b]ref. 4, [c]ref. 3, [d]ref. 165, [e]ref. 16, [f]ref. 140.

TABLE 2.25

Concentration dependence of χ parameter p_1, p_2, Flory temperature θ and entropy parameter ψ_0 for polyethylene / solvent systems.

Polymer: Polyethylene		Methods									
Solvent	UCSP or LCSP	Kamide-Matsuda				Koningsveld et al.				Shultz-Flory	
		p_1	p_2	θ/K	ψ_0	p_1	p_2	θ/K	ψ_0	θ (θ ref.)/K	ψ_0 (ψ_0 ref.)
n-Butyl acetone	UCSP	2.98	-34.6	453.8	0.89	3.100	-36.6	453.8	0.93	482.1(483[a])	-0.61 (0.65[a])
	LCSP	4.63	-38.0	506.5	-1.15	3.742	-31.59	499.9	-1.27	474.9(471[a])	-1.17(-1.12[a])
n-Pentane	LCSP	1.95	-23.4	366.7	-1.82	4.529	-42.69	376.2	-1.78	351.8(353[b])	-1.06(-1.3[b])
n-Hexane	LCSP	0.662	-6.53	410.4	-0.93	0.675	-6.54	410.3	-0.95	407.3(406.3[b])	-0.98(-1.0[b])
n-Heptane	LCSP	-1.5	12.9	446.5	-1.89	-3.76	34.60	446.5	-2.68	447.1(446.9[b])	-1.16(-1.2[b])
n-Octane	LCSP	-1.0	7.7	482.9	-1.64	-2.873	26.12	481.6	-2.14	484.4(483.0[b])	-1.22(-1.1[b])
n-Octyl alcohol	UCSP	1.7	-33.7	440.9	1.53	3.435	-59.0	436.7	1.52	445.0(444[c])	1.21 (1.16[c])
	LCSP	1.8	-37.0	621.9	-7.40	2.712	-51.43	622.4	-7.13	620.6(621[c])	-5.76(-5.4[c])
	UCSP	0.48	-25.6	446.3	2.59	1.440	-43.79	445.8	2.47	452.9(453.1[d])	1.15(1.15[d])
n-Decyl alcohol	UCSP	-1.82	6.02	425.7	2.82	-2.688	11.24	427.6	2.65	425.5(426.3[d])	1.47 (1.44[d])
n-Lauryl alcohol	UCSP	-0.304	-3.40	409.0	2.47	-0.249	-5.036	409.2	2.39	410.1(410.3[d])	1.67 (1.64[d])
p-tert-Amyl phenol	UCSP	-2.13	2.30	470.2	2.54	-1.923	-6.158	472.0	2.17	472.1(472.2[d])	1.07 (1.07[d])
p-Octyl phenol	UCSP	-1.52	-0.912	446.1	3.26	-2.744	0.997	450.6	2.89	448.2(447.5[d])	1.47 (1.48[d])
p-Nonyl phenol	UCSP	0.175	-2.71	434.1	1.78	-0.095	-1.277	435.7	1.77	435.8(435.4[d])	1.55 (1.62[d])
Anisole	UCSP	1.15	-39.1	422.6	2.61	1.655	-49.63	422.1	2.59	426.3(426.5[d])	1.45 (1.41[d])
Benzyl phenyl ether	UCSP	-0.823	-20.1	461.6	2.95	-1.945	-12.17	464.2	2.74	464.0(464.5[d])	1.32 (1.38[d])
Diphenyl	UCSP	-1.32	9.32	400.5	1.84	-2.064	15.86	401.1	1.95	399.8(400.5[e])	1.19 (1.17[e])
Diphenyl methane	UCSP	-0.89	3.1	414.5	1.61	-1.970	10.66	418.6	1.58	413.7(415.2[e])	1.13 (1.06[e])
Diphenyl ether	UCSP	0.611	-2.483	427.5	1.04	0.400	-1.50	431.4	0.98	431.9(—)	0.96 (—)
	UCSP	-0.31	-1.6	433.0	1.53	-0.698	0.471	435.1	1.49	434.1(436.9[e])	1.12 (1.00[e])

[a]ref. 144, [b]ref. 146, [c]ref. 147, [d]ref. 68, [e]ref. 145.

TABLE 2.26

Concentration dependence of χ parameter p_1, p_2, Flory temperature θ and entropy parameter ψ_0 for cellulose acetate (degree of substitution, DS = 2.46) / solvent systems (ref. 177).

Polymer: Cellulose acetate		Methods									
Solvent	UCSP or LCSP	Kamide-Matsuda				Koningsveld et al.				Shultz-Flory	
		p_1	p_2	θ /K	ψ_0	p_1	p_2	θ /K	ψ_0	θ /K	ψ_0
Acetone	LCSP	0.96	-21.4	433.1	-1.03	0.203	-11.78	429.7	-1.01	427.1	-0.78
Methyl ethyl ketone	UCSP	-0.67	2.52	312.2	0.47	0.47	4.40	313.5	0.46	311.4	0.35
	LCSP	-0.96	7.38	432.0	-0.69	-1.170	9.56	430.7	-0.68	434.6	-0.51

correlation coefficient between $1/T_c$ and $\chi_0{}^c$ (by KM method) is without exception larger than 0.98. Except for PS / toluene (LCSP), PS / benzene (LCSP), PS / isopropyl acetate (LCSP), PS / n-propyl acetate (LCSP), the p_1 value determined for PS solutions by the KM method can be almost regarded as constant (0.638 ± 0.035). This value is near the theoretically expected value (2/3) (eqn (2.13)). The range of variation (± 0.035) in p_1 may contain the solvent effect together with experimental uncertainty. Except for PS / cyclopentane (CP), the p_2 value for LCSP is always negative, lying between -2.0 and -0.202 and, except PS / isopropyl acetate and n-propyl acetate systems, in which p_2 is slightly negative (-0.03 and -0.02), the p_2 value for UCSP is positive (0.297 ± 0.107). This mean value is slightly smaller than the theoretical value ($= 1/2$; see, eqn (2.13)).

The following relations are obtained between p_1, evaluated by the KKS method (p_1(KKS)) and p_1, evaluated by the KM method (p_1(KM)) and between p_2 by the KKS method (p_2(KKS)) and p_2 by the KM method (P_2(KM)):

$$p_1(\mathrm{KKS}) = 0.889 p_1(\mathrm{KM}) + 0.052, \qquad r = 0.9813, \tag{2.233}$$

$$p_2(\mathrm{KKS}) = 0.927 p_2(\mathrm{KM}) + 0.150, \qquad r = 0.9714. \tag{2.234}$$

The Flory temperature, obtained by three methods, is practically independent of the methods employed. ψ_0 is positive for UCSP and negative for LCSP as expected, irrespective of the methods used for evaluation. These are observed not only for PS, but also for PE, strongly suggesting that ψ_0 cannot be regarded as temperature-independent over a wide range of temperatures, which is assumed in the KKS and KM methods and the polymer solution should be athermal at a temperature between UCSP and LCSP. Absolute values of ψ_0, $|\psi_0|$ increases in the order:

$$|\psi_0(\mathrm{KM})| \simeq |\psi_0(\mathrm{KKS})| < |\psi_0(\mathrm{SF})|. \tag{2.235}$$

The fact that $|\psi_0(\mathrm{SF})|$ is larger than $|\psi_0(\mathrm{KM})|$ and $|\psi_0(\mathrm{KKS})|$ is mainly caused from the ignorance of the concentration dependence of χ parameter in the SF method. Among the three methods the following relations holds:

$$\psi_0(\mathrm{KKS}) = 0.983 \psi_0(\mathrm{KM}) + 0.017, \qquad r = 0.9963, \tag{2.236}$$

$$\psi_0(SF) = 1.554\psi_0(KM) + 0.076, \qquad r = 0.9553, \qquad (2.237)$$

$$\psi_0(SF) = 1.578\psi_0(KKS) + 0.094, \qquad r = 0.9571. \qquad (2.238)$$

Even if the solvents have similar molecular weights and almost the same molecular shapes, p_1 and p_2 differ remarkably depending on whether the solvent is aliphatic or aromatic. This point will be discussed in more detail in a later section (section 2.7.4).

Fig. 2.96 shows the plot of p_2 against p_1, both estimated by the KM method, for PS solutions. Here, the unfilled circle and rectangle correspond to UCSP and LCSP, respectively. The point theoretically expected when $A_2 = A_3 = A_4 = 0$ at θ temperature, is denoted as a filled circle. It is obvious that the experimental points for UCSP are not far from the theoretical point. On the other hand, the data points scatter for LCSP, showing negative p_2.

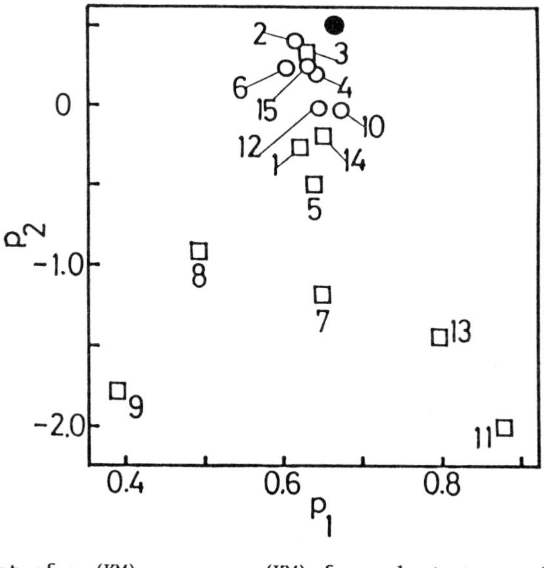

Fig. 2.96 Plot of p_1(KM) versus p_2(KM) for polystyrene solutions; rectangle, LCSP; unfilled circle, UCSP; filled circle, the point theoretically expected when $A_2 = A_3 = A_4 = 0$ at θ: 1, methyl ethyl ketone (LCSP) (ref. 136); 2, cyclopentane (UCSP) (ref. 136); 3, cyclopentane (LCSP) (ref. 136); 4, cyclohexane (UCSP) (refs 16,66,71); 5, cyclohexane (LCSP) (ref. 71); 6, methylcyclohexane (UCSP) (ref. 70,71); 7, methylcyclohexane (LCSP) (ref. 71); 8, toluene (LCSP) (ref. 71); 9, benzene (LCSP) (ref. 136); 10, isopropyl acetate (UCSP) (ref. 140); 11, isopropyl acetate (LCSP) (ref. 140); 12, n-propyl acetate (UCSP) (ref. 140); 13, n-propyl acetate (LCSP) (ref. 140); 14, dimethoxy methane (LCSP) (ref. 142); 15, trans-decalin (UCSP) (ref. 143).

p_1, p_2, θ and ψ_0 values of UCSP for the PS / CH system were calculated from the entire data on v_p^c and T_c, obtained by the three different methods. The data of Kuwahara et al. (ref. 66) by the diameter method gave $p_1 = 0.593$, $p_2 = 0.551$, $\theta = 306.9K$ and $\psi_0 = 0.22$. The threshold cloud point method (ref. 71) gave $p_1 = 0.645$, $p_2 = 0.165$, $\theta = 305.1K$ and $\psi_0 = 0.27$. The phase volume ratio method (ref. 16) yielding $p_1 = 0.631$, $p_2 = 0.221$, $\theta = 305.8K$ and $\psi_0 = 0.27$. Obviously, from the latter two methods, we can obtain the same results, giving p_1 value similar with that theoretical value and also with ψ_0 value (0.264) estimated from the second virial coefficient.

2.7.3 Polyethylene

For PE / p-tert-amyl phenol (UCSP), PE / p-octyl phenol (UCSP), PE / p-nonyl phenol (UCSP), PE / benzyl phenyl ether (UCSP) and PE / diphenyl ether (UCSP), the correlation between Y and v_p^c is low (< 0.5). Even for these systems correlation between $1/T_c$ and χ_0^c in the KM method is larger than 0.97. p_1 and p_2, determined by the KM method, vary in the range of $p_1 = -1.82 \sim 2.89$, $p_2 = -39.1 \sim 9.32$ for UCSP and $p_1 = -1.5 \sim 4.63$, $p_2 = -38.0 \sim 12.9$ for LCSP. Obviously, the range of variation in p_1 and p_2 is much wider for PE than for PS, and only two systems, PE / n-hexane (LCSP) and PE / diphenyl ether (UCSP), have p_1 values in the vicinity of the theoretical value (2/3).

In PE solutions, p_2 often has an unexpectedly large negative value, which brings about larger $|\psi_0(\mathrm{KM})|$ than $|\psi_0(\mathrm{SF})|$. Therefore, if the ψ_0 value, deduced from the temperature dependence of A_2 of PE solutions ($\psi_0(A_2)$) becomes available in the future, we can predict $|\psi_0(A_2)| > |\psi_0(\mathrm{SF})|$.

Among the literature data on PE / solvent systems, those by Koningsveld et al. (ref. 14) for PE / diphenyl ether are believed to have the highest accuracy, because the cloud point curves they constructed for three PE samples and the critical points were determined by using the two-phase volume ratio R, which directly gives the critical point, irrespective of the polymolecularity of the sample. It is well known that PE as polymerized whole polymers have extremely wide molecular weight distributions and PE fractions isolated by successive precipitation fractionation using proper solvent / nonsolvent can never be regarded as monodisperse. Unfortunately, with exception of Koningsveld et al.'s experiments (ref. 14), no experiment was carried out to determine v_p^c and T_c using R. The results, obtained by analyzing Nakajima et al.'s threshold cloud point curves for PE / diphenyl ether system are different remarkably from the data by Koningsveld et al. for the same polymer / solvent system. This means that the threshold cloud points, at least for PE fractions, do not coincide with the critical points with good accuracy and the wide variation of p_1 and p_2 is mainly due to low accuracy of the experiments.

By analyzing Koningsveld et at.'s data (ref. 14) for the PE / diphenyl ether system, we obtain $p_1 = 0.61$, using the KM method, which is close to the theoretical value. Koningsveld et al. (ref. 14) concluded, only considering g_1, that the most reasonable supposition for PE / diphenyl ether system appears to be g_1 (accordingly, p_1) = 0, and δ (in eqn (2.228)), calculated assuming $p_1 = p_2 = 0$ for the same data, is about 60% larger than that obtained using the value of p_1 and p_2, estimated by the KM method for this system ($p_1 = 0.61$ and $p_2 = -2.483$). Similar analysis of

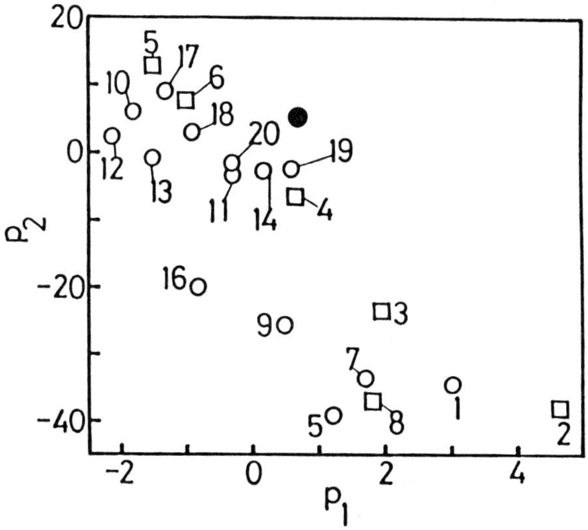

Fig. 2.97 Plot of p_1(KM) versus P_2(KM) for polyethylene solutions; rectangle, LCSP; unfilled circle, UCSP; filled circle, the point theoretically expected when $A_2 = A_3 = A_4 = 0$ at θ: 1, n-butyl acetate (UCSP) (ref. 144); 2, n-butyl acetate (LCSP) (ref. 144); 3, n-pentane (LCSP) (ref. 146); 4, n-hexane (LCSP) (ref. 146); 5, n-heptane (LCSP) (ref. 146); 6, n-octane (LCSP) (ref. 146); 7, n-octyl alcohol (UCSP) (ref. 147); 8, n-octyl alcohol (LCSP) (ref. 147); 9, n-octyl alcohol (UCSP) (ref. 68); 10, n-decyl alcohol (UCSP) (ref. 68); 11, n-lauryl alcohol (UCSP) (ref. 68); 12, p-tert-amyl phenol (UCSP) (ref. 68); 13, p-octyl phenol (UCSP) (ref. 68); 14, p-nonyl-phenol (UCSP) (ref. 68); 15, anisole (UCSP) (ref. 68); 16, benzyl phenyl ether (UCSP) (ref. 68); 17, diphenyl (UCSP) (ref. 145); 18, diphenyl methane (UCSP) (ref. 145); 19, diphenyl ether (UCSP) (ref. 14); 20, diphenyl ether (UCSP) (ref. 145).

Nakajima and his coworkers' data for the same system gave $p_1 = -0.31$ by the KM method. Note that the cloud point curves in Nakajima et al.'s study were constructed from only 4~8 (average 5) different concentration solutions. The large difference in p_1 indicates the experimental difficulty (and accordingly, the experimental uncertainty) contained in determination of the critical point of PE solutions, and as far as PE / diphenyl ether system is concerned, we can obtain $p_1 \simeq 2/3$ as in the case of PS / solvent system.

Fig. 2.97 illustrates the correlation between p_1 and p_2, both estimated by the KM method, for PE solutions. The theoretical point predicted where $A_2 = A_3 = A_4 = 0$ at θ temperature is denoted as a filled circle. It is interesting to note that, unlike PS solutions, p_1 cannot be regarded as nearly constant and p_2 has a tendency to decrease with an increase in p_1, satisfying $p_1 p_2 \lesssim 0$.

Fig. 2.98 shows the correlations between ψ_0 and θ of PS (unfilled mark) and PE (filled mark). For PS solutions, the

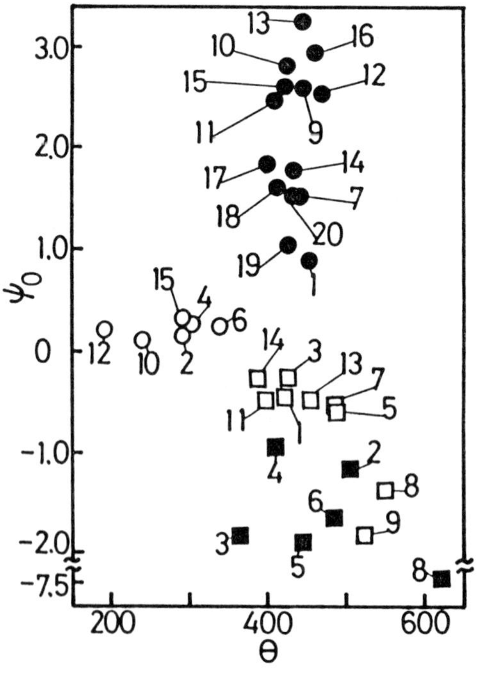

Fig. 2.98 Plot of ψ_0 as a function of θ for polystyrene (unfilled mark) and polyethylene (filled mark) solutions; circle, UCSP; rectangle, LCSP. Number for polymer solutions has the same meaning as those in Figs 2.96 and 2.97.

solvent dependence of ψ_0 is much smaller than that of θ in both ranges of UCSP and LCSP. In contrast to this, for UCSP of PE solutions variation of ψ_0 as compared with the solvent is remarkable as compared with the solvent dependence of θ. The difference in ψ_0 between UCSP and LCSP is very significant in PE solution and the following relations holds:

$$\psi_{0\,PE}^{UCSP} > \psi_{0\,PS}^{UCSP} > \psi_{0\,PS}^{LCSP} > \psi_{0\,PE}^{LCSP}. \quad (2.239)$$

For convenience, we express the Flory temperature evaluated by the KM, KKS and SF methods as θ (KM), θ (KKS) and θ (SF), respectively. For the UCSP of PS / single solvent system, θ (SF) is on the average ca.1.8K higher than θ (KM) and θ (KKS). For LCSP of the same system θ (SF) is 2～3K lower than θ (KS) and θ (KKS). For the PE / single solvent system θ (SF) is ca.3～4K higher than in UCSP and ca.8K lower in LCSP than θ (KM) and θ (KKS). The difference between θ (KM) and θ (KKS) is practically insignificant for all the polymer / solvent systems investigated. In short, θ (KM) $\simeq$ θ (KKS) $<$ θ (SF) for UCSP and θ (KM) $\simeq$ θ (KKS) $>$ θ (SF) for LCSP. The Flory temperature is believed the most reliable when it is determined as the temperature at which the second virial coefficient A_2 by the membrane osmometry or the light scattering method becomes zero (hereafter referred to as θ (A_2)). θ (A_2) was found to be in the range 307.0 and 308.4K (see, Table 2.21) and averaged to 307.6K for UCSP of the PS / CH system. For this system, θ (KM), θ (KKS) and θ (SF) are by 2.5, 2.4 and 1.1K underestimated. θ (A_2) was determined for LCSP of the PS / MCH system to be 340.4 and 341K (Table IV of ref. 3) and averaged to be 340.7K which is 0.5 and 1.1K higher than θ (KM) and θ (KKS) and 1.6K lower than θ (SF). From a theoretical point of view, the KM and KKS methods are superior as compared with the SF method. But, to estimate θ more accurately using the KM or KKS methods, the temperature dependence of p_1 and p_2 and, probably, the molecular weight dependence of ψ_0 should be taken into account.

2.7.4 Cellulose acetate

The θ temperature can be determined independently of the method used. But p_1 and p_2 of CA (the total degree of substitution, DS = 2.46) solutions determined by the Kamide-Matsuda (ref. 3) and Koningsveld et al. (ref. 16) methods do not coincide, because of the low experimental accuracy of v_p^c. The absolute value of ψ_0 determined by the above two methods is larger than that obtained by the Shultz-Flory (ref. 165) method, in which the concentration dependence of χ is ignored, contrarily to the case of polystyrene / solvent systems (ref. 5). In cellulose acetate (DS = 2.46) solutions the solvation phenomena occurs (ref. 178) and the temperature dependence of the solvation possibly affects p_1 and p_2. For the system CA(2.46) / methyl ethyl ketone, θ was determined as 310K from cloud point measurements. The value of θ, defined as the temperature at which the second virial coefficient by the light scattering method vanishes, was found to be 323K (ref. 149). This difference in θ was attributed to association of the dissolved polymer molecules. On the other hand, the weight-average molecular weight M_w of CA samples obtained in methyl ethyl ketone at 323K were in accord with those determined in a good solvent. Then, it was established that methyl ethyl ketone is a theta solvent for CA(2.46) with $\theta = 323$K.

Summarizing, literature data on the cloud point curve and the critical solution point for PS solutions are accurate enough to be analyzed by the KKS and KM methods and p_1 was found to be fairly near to the theoretical value (2/3), expected when $A_2 = A_3 = 0$ at θ. However, p_2 for these polymer / solvent systems deviate occasionally to a large extent, from the theoretical value (1/2), calculated when $A_2 = A_3 = A_4 = 0$ at θ and p_2 is positive for UCSP and negative for LCSP. This suggests that the fourth virial coefficient A_4 does not always become zero even at θ. The corresponding literature data for PE solutions are unfortunately rather qualitative and more reliable data for this polymer are highly anticipated. The entropy at the LCSP region and positive at the UCSP region and an athermal solution will be realized, at least, for these polymers in any solvent at a specific temperature between LCSP and UCSP.

2.7.5 Dissolved state of atactic polystyrene in aromatic solvents

Table 2.24 shows that there is a significant effect of solvent nature on solution properties of PS. For example, for PS / cyclopentane (CP), PS / cyclohexane (CH) and PS / methyl-cyclohexane (MCH) systems, p_1 values for lower and upper critical solution points (LCSP or UCSP) are very similar to each other, but differences in ψ_0 values between CP and CH are much larger than those between CH and MCH for both CSP. From these analyses Kamide et al. (ref. 5) concluded that the skeleton structure of the solvent is a more important factor than the substituent group to the skeleton, covering the thermodynamic interaction between the polymer and solvent, and that even if the solvents have similar molecular weights and almost the same molecular shapes, p_1 and p_2 differ remarkably depending on whether the solvent is aliphatic or aromatic. For aliphatic solvent systems (PS / CH and PS / MCH), p_1 for LCSP was found to be 0.64$\sim$0.65, being near the theoretically expected value (2/3) (eqn (2.13)) for non-polar polymer / non-polar solvent systems, in which the molecular forces acting between the solute and the solvent is only dispersion force. These values also coincide with the averaged p_1 value (0.638$\pm$0.035 (ref. 5)) of twelve PS / solvent systems. In contrast to this, for PS / benzene and PS / toluene systems p_1 for LCSP was 0.39 and 0.49, respectively. These results seems surprising because for these solutions the polymer repeating unit has a similar chemical structure with solvents and minimum deviation, if any, of the free energy change in mixing from that of random mixing athermal solution is expected. In addition to this, if we consider that benzene and toluene are practically non-polar and CH and MCH are less polar, it seems very curious that there are large differences in thermodynamic properties between PS / aromatic and PS / aliphatic solvents.

Kamide et al. (ref. 5) were probably the first to interpret the above findings in terms of the molecular dissolved state of polymer: They measured PS / CH and PS / benzene-d_6 ^{1}H-NMR spectra using a rather less resolving power NMR spectrometer (60MHz JEOL NMR spectrometer type PMX60), finding a significant shift of the proton peak of the phenyl group constituting PS molecules dissolved in benzene-d_6 toward a lower magnetic field. They gave the following explanation for the above experimental findings: The magnetic shielding effect induced by circular electric current is strengthened by planar interaction between the benzene ring of PS and benzene as solvent and this interaction makes benzene a better solvent than CH (ref. 5).

Fujihara (ref. 179) determined using a Ness type (dilution type) constant temperature wall calorimeter the heat of dilution of PS / CH and PS / benzene systems showing that PS / CH is endothermic but PS / benzene is small, but significantly exothermic and as a result, the latter system is expected to be more stable than the former. Fujihara (ref. 179) also speculated in his interesting dissertation the formation of the alternatively over-lapping structure of the phenyl ring of PS and benzene as solvent unfortunately without any experimental direct evidence, which may lead to stability to the system (ref. 179). Except for a few attempts, no specific consideration has been paid up to now to specific interactions between PS molecules and aromatic solvents.

Thereafter Kamide, Matsuda and Kowsaka (ref. 180) challenged this problem more energetically by (1) disclosing the solvent effect of the chemical shifts of protons, assigned to phenyl, methin and methylene of PS, in more detail by measuring more accurately ^{1}H-NMR spectra for atactic PS / CH, atactic PS / benzene, atactic PS / MCH and atactic PS / toluene systems with rexamination of the adequacy of the assignment of each peak in phenyl envelope in observed NMR spectrum, (2) examining the solvent effect of PS / aromatic solvent systems by measuring IR spectra of atactic PS / CH and atactic PS / benzene systems, and (3) evaluating the effect of solvent nature (aliphatic or aromatic) on the degree of solvation by determining the adiabatic compressibility of these systems.

Fig. 2.99a and b show 200MHz ^{1}H-NMR spectra of PS / benzene-d_6 and PS / CH-d_{12} both at 333K and Fig. 2.100a and b show 200MHz ^{1}H-NMR spectra of PS / toluene-d_8 at 333K and PS / MCH-d_{14} at 363K. In these spectra phenyl, aliphatic methin and methylene protons are observed over 6.3～7.2ppm, 1.8～2.2ppm and 1.4～1.7ppm, respectively. The spectra in the phenyl proton region are split into two envelopes. Fig. 2.101a and b demonstrate 400 MHz ^{1}H-NMR spectra of the phenyl proton region of PS / benzene-d_6 and PS / toluene-d_8 at 348K. Harwood and Shepherd (ref. 181)

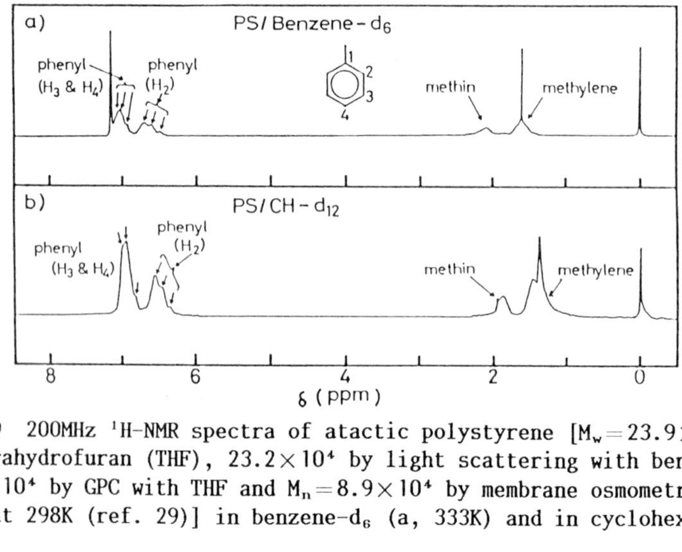

Fig. 2.99 200MHz ^{1}H-NMR spectra of atactic polystyrene [$M_w = 23.9 \times 10^4$ by GPC with tetrahydrofuran (THF), 23.2×10^4 by light scattering with benzene and $Mn = 8.6 \times 10^4$ by GPC with THF and $M_n = 8.9 \times 10^4$ by membrane osmometry with toluene at 298K (ref. 29)] in benzene-d_6 (a, 333K) and in cyclohexane-d_{12} (b, 333K).

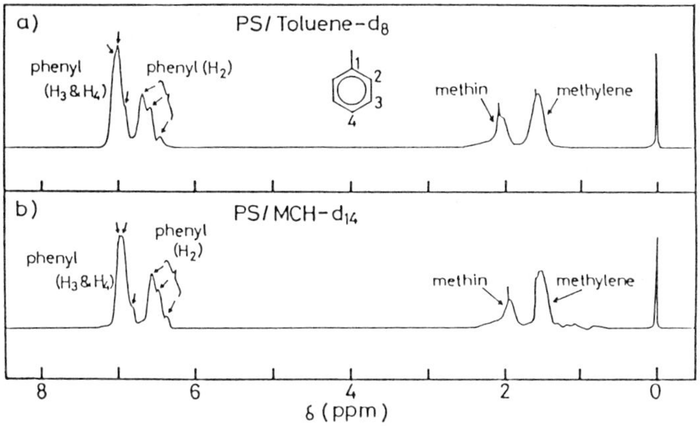

Fig. 2.100 200MHz ^{1}H-NMR spectra of atactic polystyrene [$M_w = 23.9\times10^4$ by GPC with tetrahydrofuran (THF), 23.2×10^4 by light scattering with benzene and $M_n = 8.6\times10^4$ by GPC with THF and $M_n = 8.9\times10^4$ by membrane osmometry with toluene at 298K (ref. 29)] in toluene-d_8 (a, 333K) and in methylcyclohexane-d_{14} (b, 363K).

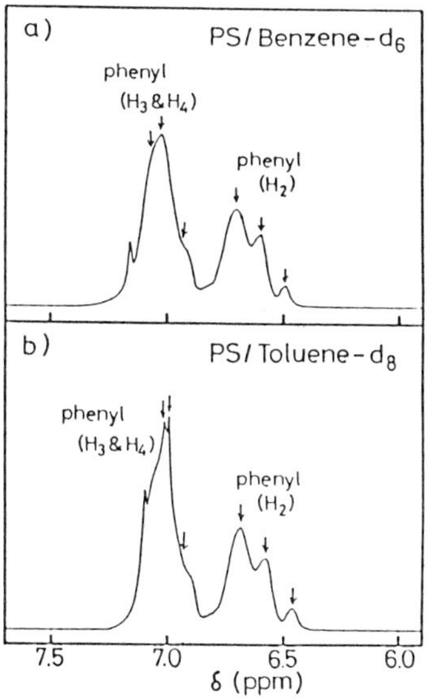

Fig. 2.101 400MHz ^{1}H-NMR spectra of the phenyl proton region of atactic polystyrene in benzene-d_6 (a) and in toluene-d_8 (b), both at 348K.

assigned at a lower magnetic field in two phenyl proton envelopes to H_3-protons and at higher magnetic field in that envelope to H_2-proton. Since the integrated intensity ratio of these two peaks is approximately 3:2, it is clear that the H_4-proton envelope overlaps with the H_3-proton envelope. The validity of the above assignment of the chemical shift of H_4-proton was also confirmed by Carbon-Hydrogen shift correlation NMR spectroscopy (not shown here). In addition, the H_2-proton envelope clearly consists of three components. This indicates that H_2-proton peak are splits according to stereochemical difference. In this case vicinal proton coupling is considered below 10Hz (0.05ppm) (ref. 182) and is of the order of the line width and accordingly, splitting due to this coupling is not observable. Inspection of ^1H-NMR spectra for isotactic and atactic PS reported by Inoue et al. (ref. 183) in tetrachloroethylene at 393K (Fig. 1 of ref. 183) shows that the H_2-proton region of isotactic PS falls fairly well in a region in the H_2-proton peak of atactic PS. Therefore, among the three peaks constituting the H_2-proton envelope, the peak at the lowest magnetic field can be assigned as an iso (mm)-peak (here, m means meso configuration) and the remaining two peaks can be attributed to hetero (mr or rm)- and syndio (rr)-peaks (r is racemo configuration). Similar splitting is also expected to occur in H_3- and H_4-proton envelopes and in fact, more than three peaks are observed, but the degree of splitting is explicitely low. It can be considered that, due to the higher electron density of H_2 proton as compared with H_3 and H_4 protons, the H_2 proton region in ^1H-NMR spectra is sensitive to the configuration of PS, splitting into roughly three parts. H_3- and H_4-envelopes in PS / benzene-d_6 are, more or less, unavoidably overlapping with solvent benzene peaks. In this section, as far as H_3- and H_4-envelopes are concerned the main three peaks observed commonly in all the solvents investigated were analyzed further. Table 2.27 shows the chemical shifts of phenyl, methin and methylene peaks of PS in four solvents at various temperatures. In PS / benzene-d_6, H_3- and H_4-phenyl proton peaks shift by 0.07～0.11ppm to a lower magnetic field and H_2-phenyl proton peaks by 0.16～0.18ppm to a lower magnetic field than those in PS / CH-d_{12}. Similarly, H_3- and H_4-phenyl proton peaks and H_2-phenyl proton peaks in PS / toluene-d_8 shift by 0.08～0.10ppm and by 0.09～0.12ppm, respectively, to a lower magnetic field than those in PS / MCH-d_{14}. Methin and methylene peaks of PS in aromatic solvents are found at a lower magnetic field than those of PS in aliphatic solvents. Differing from aliphatic solvents, aromatic solvents exhibit naturally anti-magneticism without exception and this anti-magnetic field originating from the aromatic solvent also reaches

TABLE 2.27

Temperature dependence of chemical shifts of polystyrene in aromatic and aliphatic solvents.

Solvent	Temp/K	Chemical Shifts (ppm) phenyl(H_3 and H_4)			phenyl(H_2)			methin	methylene
Benzene	293	7.08	7.05	6.95	6.73	6.62	6.48	2.09	1.59
	313	7.07	7.04	6.94	6.72	6.62	6.49	2.09	1.60
	333	7.06	7.04	6.93	6.71	6.62	6.50	2.10	1.61
	348	[7.07]	[7.03]	[6.95]	[6.71]	[6.60]	[6.49]	[2.09]	[1.61]
Cyclohexane	293	7.01	6.94	6.82	6.56	6.43	6.30	1.87	1.51
	303	6.99	6.94	6.82	6.56	6.44	6.30	1.87	1.51
	307.5	7.00	6.94	6.82	6.55	6.44	6.32	1.89	1.48
	313	7.00	6.94	6.81	6.56	6.44	6.33	1.89	1.47
	333[a]	6.99	6.94	6.82	6.55	6.44	6.34	1.89	1.47
	348	[6.99]	[6.94]	[6.82]	[6.68]	[6.57]	[6.35]	[1.92]	[1.48]
(B-CH)[c]	333	0.07	0.10	0.11	0.16	0.18	0.16	0.21	0.13
Toluene	293	7.1[b]	7.0[b]	6.92	6.68	6.58	6.43	2.04	1.57
	313	7.1[b]	7.0[b]	6.92	6.68	6.58	6.44	2.05	1.58
	333[a]	7.07	7.01	6.91	6.69	6.58	6.46	2.06	1.59
	348	[7.1][b]	[7.0][b]	[6.91]	[6.68]	[6.57]	[6.45]	[2.05]	[1.59]
Methylcyclohexane	313	6.99	6.94	6.82	6.55	6.45	6.36	1.90	1.5[b]
	343	6.99	6.95	6.82	6.56	6.45	6.36	1.91	1.5[b]
	351	6.99	6.94	6.82	6.56	6.47	6.36	1.91	1.5[b]
	363[b]	6.99	6.94	6.81	6.57	6.49	6.37	1.93	1.54
(T-MCH)[d]	333,363	0.08	0.07	0.10	0.12	0.09	0.09	0.13	0.05

a, Eliminated solvent peak by pulse sequence; b, overlapped with solvent peak: c, difference of chemical shifts between PS / Benzene and PS / CH; d, difference of chemical shifts between PS / Toluene and PS / Methylcyclohexane; [ ] were measured by JNM-FX400.

tetramethylsilane (TMS). Then, the effects of anti-magneticism of the aromatic solvent on the whole NMR spectra luckily cancel out each other and are neglected here. This suggests strongly that a significant shift of the ^{1}H-peaks of H_2-, H_3- and H_4-phenyl protons, and methin and methylene protons in aromatic solvents to a lower magnetic field as compared with those in aliphatic solvents can only be explained by taking into consideration the specific interaction between phenyl rings of PS and benzene as solvent.

Benzene is a non-polar, but is said to have a partially regular structure in the liquid state due to "$\pi-\pi$" electrostatic interactions originating from overlapping π electron orbitals of benzene molecules. In fact, the molecular orbital calculation (ref. 184) with assumption of C_{2v} symmetry of energy levels of π electrons in benzene, when two neighbouring benzene molecules are packed together like a sheet, shows the splitting of each energy level into two, whose distance from the original state equals the interaction energy between two neighbouring benzene rings and the π electron of overlapping benzene transits naturally to a lower level, which is of course lower than that of a single benzene molecule. Accordingly, liquid benzene may be stabilized by taking, at least partly, the planarly overlapping structure. It is confirmed with experiments of the small-angle X-ray diffraction from liquid benzene that liquid benzene maintains to some extent such a planary overlapping structure observed originally in benzene crystals (ref. 185).

In order to examine the specific interaction between benzene and PS, IR spectra for benzene / CH and PS / CH systems were measured. Theoretically, only four vibrations are expected to be observed at 673(A_{2u}), 1038(E_{1u}), 1486(E_{1u}) and 3063cm^{-1}(E_{1u}), due to the structural symmetry of benzene molecule (D_{6h}) and in fact this is confirmed for gaseous benzene. In the liquid state, the structural symmetry is destroyed partly, resulting in numerous absorption bands such as those at 1448, 1531, 1815, 1960, 3035 and 3095cm^{-1}, which are forbidden in the gas state, become observable (ref. 186). Fig. 2.102 shows the plot of two absorbing wave numbers ν, due to out-of-plane vibrations (1815 and 1960cm^{-1} for pure liquid benzene), as a function of the molar fraction of benzene ϕ_B. All other absorption bands remained constant within experimental uncertainty regardless of ϕ_B. These two absorption peaks are considered to be peaks that are combined in the following ways (ref. 187):

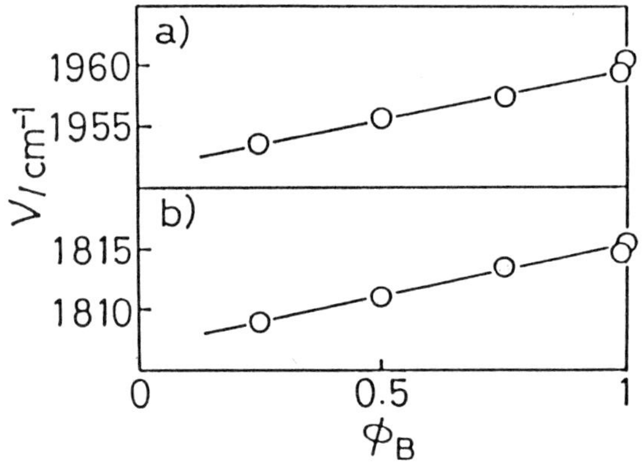

Fig. 2.102 Wavenumbers ν of IR spectrum of liquid benzene / CH system with various molar fraction of benzene ϕ_B.

$$1815\ cm^{-1} \simeq 975\ cm^{-1} + 849\ cm^{-1} \qquad (2.240)$$

and

$$1960\ cm^{-1} = 975\ cm^{-1} + 995\ cm^{-1}. \qquad (2.241)$$

The absorption bands at 849 (E_{1g}), 975 (E_{2u}) and 995cm^{-1} (B_{2g}) are assigned for out-of-plane vibration originally forbidden in IR spectra, and became observable by the existence of an interaction working perpendiculary to the benzene-ring surface. Inspection of Fig. 2.102 leads to the conclusion that the interaction between benzene molecules, which is also expected from analysis of small angle X-ray diffraction experiments of liquid benzene (ref. 185), is weakened by the dilution of benzene with CH.

Fig. 2.103 shows the subtracted IR spectra, which are obtained by subtracting IR spectrum of CH from that of the PS / CH system (weight fraction of polymer, $w_P = 0.1$) (b) and by subtracting IR spectrum of benzene from that of PS / benzene ($w_P = 0.1$) (c), respectively. For the former spectrum, the absorption band at 2927cm^{-1} intrinsic to CH was used as the standard peak so that transmittance at 2927cm^{-1} becomes zero for the subtracted spectrum, and for the latter absorption band 675cm^{-1} was used as the standard peak. In the figure, IR spectra of solid PS are shown for comparison (a). It is interesting to note that the difference spectrum [(PS / CH)-CH] has almost the same IR spectrum of PS, but the difference spectrum [(PS / benzene)

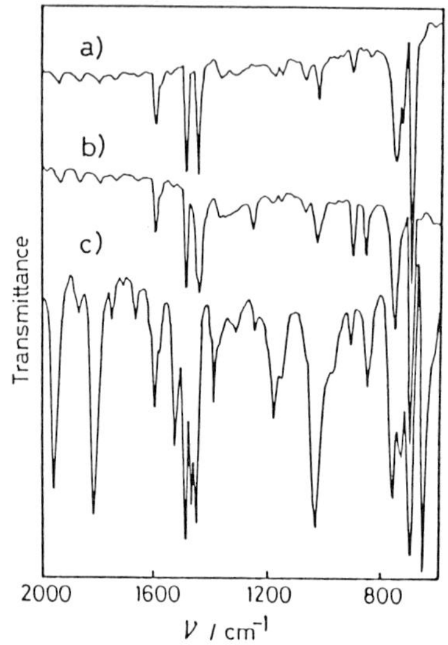

Fig. 2.103 IR spectrum of PS film (a) and subtracted IR spectra of PS from PS / CH (b) and from PS / benzene (c) systems.

– benzene] has a large number of very complicated, additional bands, including out-of-plane vibrations for pure liquid benzene (1815 and 1960 cm^{-1}), besides the original PS bands. This experimental result implies that an interaction between PS molecules and benzene brings about the appearance of new absorption bands, which are observed in neither PS nor benzene, and/or significant shifts of the absorption bands of PS or benzene. In this way, analysis of the IR spectra shows the existence of a specific interaction between PS and benzene.

In contrast to benzene, only a dispersion force is working in CH. CH has molecular shape and molecular size both similar to benzene, but CH has a density smaller by 0.1 (g/cm^3) than benzene at room temperature, indicating that liquid CH is expected to have no regular structure and a larger free volume than benzene.

Possible solvent effects on the polymer proton chemical shifts are the effect due to change in molecular chain conformation with solvent nature and the effect of shielding or deshielding magnetically the solute molecule by the solvent molecule. Hearafter, we discuss the latter effect only.

It can be regarded that liquid CH and methylcyclohexane (MCH) have little magnetic anisotropy and no hydrogen bonding, resulting in neither shielding nor deshielding effects on the solute polymer. In addition, a chemical shift of reference material (in this case, TMS) is influenced by the solvent to the same degree as those of the solute molecule and then an insignificant effect on aliphatic hydrocarbones such as CH and MCH on the PS chemical shift is expected. On the other hand, a significant shift of the chemical shifts of all protons (including H_2-, H_3-, H_4-, methin and methylene protons), of PS dissolved in aromatic solvents, to a lower magnetic field as compared with those in aliphatic solvents indicates some specific interactions between PS and aromatic solvents.

In the phenyl ring of benzene and toluene, placed in the static magnetic field H, the π electron moves circularly along the ring (ring current effect) and as a result the solvents show a strong anti-magnetic effect perpendicular to the ring plane (Fig. 2.104a). The protons in the vicinity of the ring surface shift to a higher magnetic field due to the shielding of H' and protons

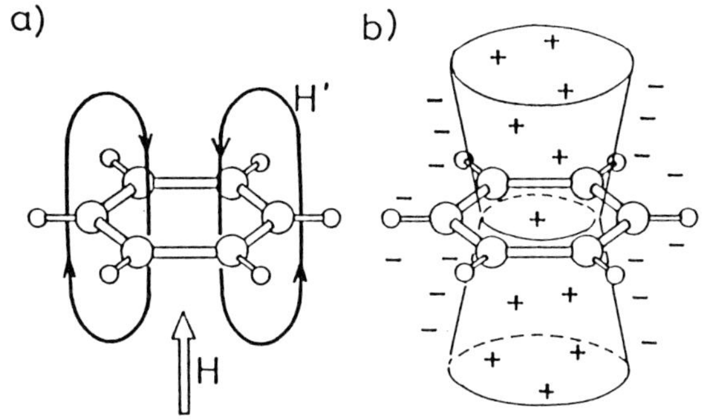

Fig. 2.104 Schematic representation of a phenyl ring under external magnetic field (H); (a) anti-magnetic field (H') of a phenyl ring, (b) shielded area (+) and deshielded area (-) of a phenyl ring.

near the outside of the ring edge are shifted to a lower magnetic field by the deshielding effect (Fig. 2.104b). Here, to the region in which deshielding effect is effective, solvent molecules are not accessible due to steric interference of the proton and in the case when the aromatic solvents are randomly mixed with the solute molecules, a major part of the solute molecules is influenced by the shielding effect, resulting in a higher magnetic field shift of the proton chemical shifts. Conversely, the experimental fact that the proton chemical shifts of PS show lower magnetic field

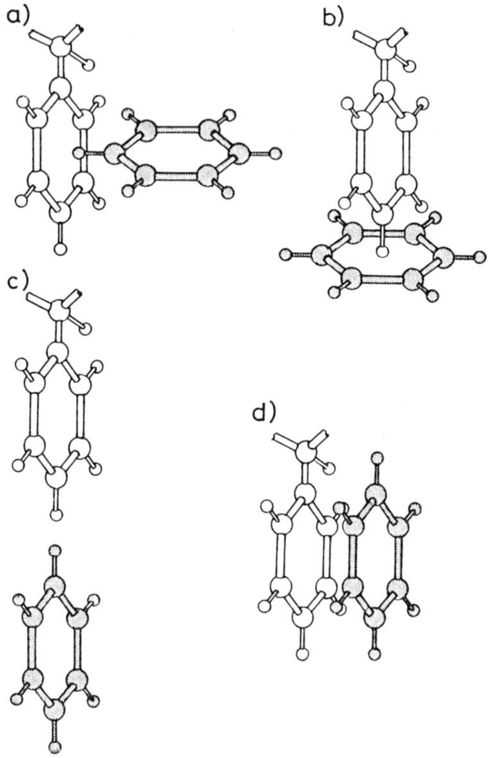

Fig. 2.105 Schematic representation of four types of possible coordination of solvent phenyl ring toward PS phenyl ring.

shift in aromatic solvents suggests that the solvent molecules are selectively coordinated in a manner such that the solvent molecules exert mainly a deshielding effect against PS molecules.

Four types of coordination of the solvent phenyl ring toward PS may be possible as shown in Fig. 2.105. The case when the PS phenyl ring is perpendicular to the solvent phenyl ring (Fig. 2.105a and b), the PS protons are not influenced by deshielding effect. These cases, then, should be discarded for PS / aromatic solvent system. In the case when the PS phenyl ring and the solvent phenyl ring are placed side-by-side on the same plane (Fig. 2.105c), the peak of the protons existing between the two phenyl rings is expected to shift slightly to a lower magnetic field by strengthening the anti-magnetic field between the two phenyl rings, but this coordination type cannot explain the lower magnetic field shift of methin and methylene proton peaks. In the case when the PS phenyl ring is stacked in parallel to the solvent phenyl ring (Fig. 2.105d), the magnetic field H' brought about

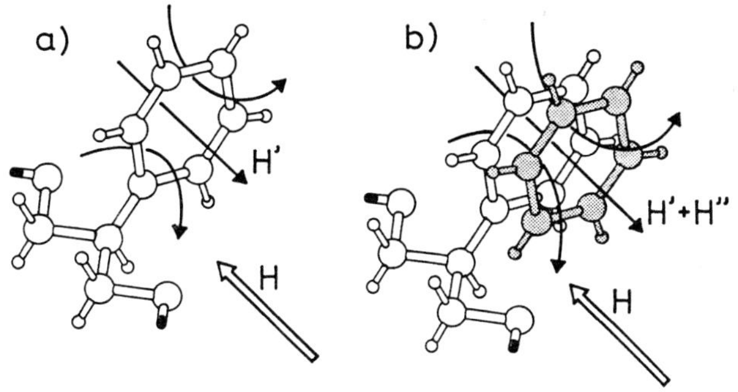

Fig. 2.106 Schematic representation of polystyrene repeating unit under external magnetic field (H); a) polystyrene in aliphatic solvent, b) polystyrene in aromatic solvent.

from circular electric current of the PS phenyl ring and the magnetic field H" from the solvent phenyl ring deshield strongly the PS phenyl protons (Fig. 2.106b), resulting in their chemical shifts to a lower magnetic field and in addition, as Fig. 2.106 clarifies, the magnetic field H' + H" brings about a lower magnetic field shift of methin and methylene peaks. Therefore, among these four coodination types, the experimental results of ^{1}H-NMR spectra can be interpreted only by the case where the PS phenyl ring is stacked in parallel to the solvent phenyl ring. This model may be accepted by speculation that an addition of PS to benzene causes partial destruction of the benzene-benzene planar stacking-in-sheets-structure in benzene liquid and formation of the benzene-PS phenyl ring overlapping structure.

When PS is dissolved in CH, the formation of a regular structure cannot be expected. To PS / toluene and PS / MCH systems, the above discussion will also be applied. Fig. 2.105 illustrates the difference between aromatic and aliphatic solvents in the magnetic interaction against PS. Here H is an external magnetic field. The experimental fact that the differences of the chemical shifts between benzene-d_6 and CH-d_{12} are always larger than those between toluene-d_{12} and MCH-d_{14} (see, Table 2.27) might be interpreted in terms of better molecular symmetry of benzene-d_6, which brings about stronger circular electric current resulting in larger H". The methin and methylene protons are surrounded by the benzene ring of PS and may be also affected by H", resulting in a lower magnetic field shift when dissolved in aromatic solvents.

The ^{1}H- and ^{13}C-NMR spin lattice relaxation time T_1 as well

TABLE 2.28
^{1}H- and ^{13}C-NMR spin lattice relaxation time T_1 and the chemical shift of solvent proton for cyclohexane, benzene and their solutions of polystyrene ($M_w = 1.00 \times 10^4$) at 313K.

sample	^{1}H T_1(s)	^{1}H δ (ppm)^a	^{13}C T_1(s)	^{13}C δ (ppm)^a
CH	4.07	1.44	19.89	27.27
CH with PS	3.85	1.43	17.09	27.61
Benzene	6.55	7.15	27.14	128.58
Benzene with PS	6.34	7.15	24.59	128.48

a, TMS(=0 ppm) internal reference.

as the chemical shift of solvent proton for CH, benzene and their solutions of a monodisperse PS with $M_w = 1.00 \times 10^4$ (concentration, 0.1g PS / 1.0ml solvent) are shown in Table 2.28. It is evident that some liquid structures of CH and benzene are slightly destroyed by the addition of PS molecules. But the degree of destruction is comparatively small. For example, Yamashiki et al. (ref. 188) have observed that T_1 of water (1.6sec) decreased, by the addition of sodium hydroxide to produce 10wt% aq. sodium hydroxide, to 0.98sec and also decreased, by the addition of cellobiose to produce 10% solution, to 1.0sec.

Fig. 2.107a and b show the concentration (c) dependence of the density ρ, and the sound velocity V of PS / CH at 313K and PS / benzene at 293K. Both ρ and V increase linearly with an increase in the polymer concentration c (g/cm^3).

Adiabatic compressibility β, calculated through Laplace's equation (ref. 189)

$$\beta = \frac{1}{\rho V^2} \tag{2.242}$$

from ρ and V values, is plotted against c in Fig. 2.107c. β decreases in linear proportion to c.

The gram of solvated solvent molecules per gram of polymer n is given by (ref. 190)

$$n = \left(1 - \frac{\beta}{\beta_0}\right) \frac{100\rho - c}{c} \tag{2.243}$$

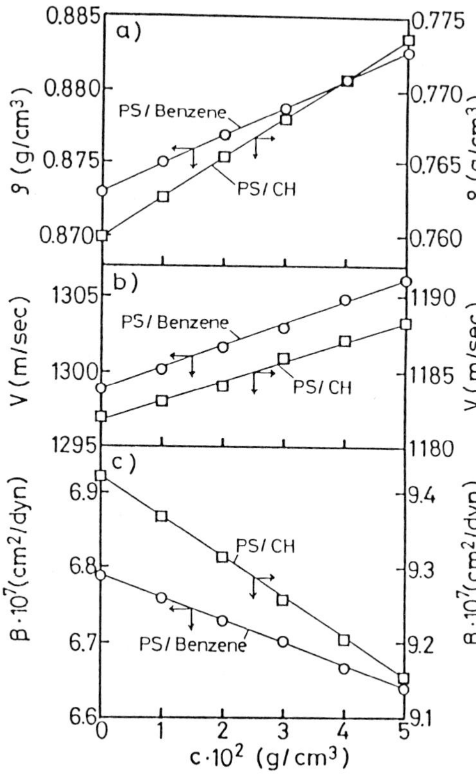

Fig. 2.107 Plots of density ρ (g/cm^3), sound velosity V(m/sec), and adiabatic compressibility β (cm^2/dyn) vs. concentration c(g/cm^3) of polystyrene / benzene (○, 293K) and polystyrene / cyclohexane (□, 313K) solutions.

where β_0 is β of the pure solvent. n can be readily transformed into the number of solvated solvent molecules per repeating unit of the polymer (S) by

$$S = \frac{m_p}{m_s} n. \qquad (2.244)$$

Here, m_p and m_s are the molecular weights of the polymer repeating unit and the solvent, respectively. Fig. 2.108 shows the concentration dependence of S for PS / benzene at 293K and PS / CH system at 313K. S is roughly constant, independent of c. S at infinite dilution (S_0) was found to be 0.49 for PS / benzene and 0.44 for PS / CH, indicating that the degree of solvation is very low but is slightly larger in benzene than in CH. The experimental error of sonic velocity and density measurements are estimated to be $\pm$ 1m/s and 5×10^{-4}g/cm^3, respectively, and for PS solutions S

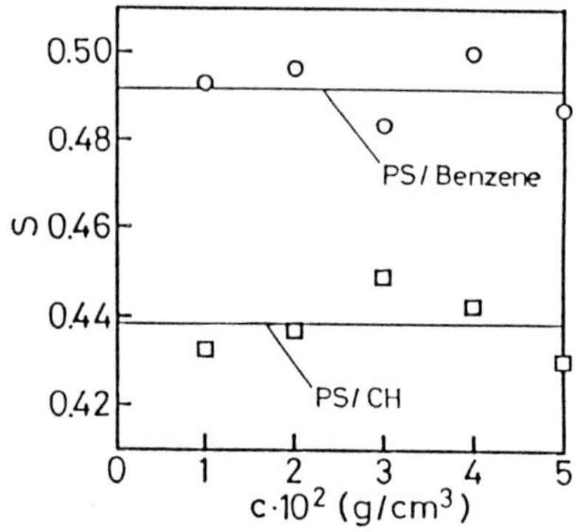

Fig. 2.108 Number of solvated solvent molecules per repeating unit of polystyrene for polystyrene / benzene (○, at 293K) and polystyrene / cyclohexane (□, 313K) with various concentration c(g/cm³).

and accordingly S_0 can be determined with a relative error of ± 0.17. Therefore, the values estimated here ($S_0 = 0.49$ for PS / benzene and $S_0 = 0.44$ for PS / CH) indicate that the solvation occurs, although to less extent, significantly. These S_0 values should be compared with those (2.0~3.0) determined for cellulose acetates dissolved in polar solvents (ref. 178).

2.8 FLORY ENTHALPY PARAMETER

The pair interaction Flory enthalpy parameter κ is defined by the van Laar-Scatchard type relation (ref. 11):

$$\kappa = \frac{\Delta H_0}{\tilde{R} T v_p^{\,2}} \qquad (2.245)$$

where ΔH_0 is the partial molar heat of dilution with respect to the solvent.

Calorimetry allows direct determination of ΔH_0, and thus κ. The calorimetric experiments, made by Fujishiro and his students, showed that κ was not independent of T and v_p, which was firstly assumed in the original Flory-Huggins theory, but that it is a purely phenomenological parameter, which depends on both T and v_p (ref. 191).

$$\kappa = \kappa_0 + \kappa_1 v_p + \kappa_2 v_p^{\,2} + \cdots \qquad (2.246)$$

κ_0 in eqn (2.246) is given by

$$\kappa_0 = \lim_{v_p \to 0} \frac{\Delta H_0}{\tilde{R}Tv_p^{\,2}} \tag{2.247}$$

and κ_1, κ_2, ··· are the 1st, 2nd, ··· order concentration-dependence parameters of κ. Theoretically, κ or simply κ_0 can be evaluated by various methods, directly or indirectly as will be described later. In other words, if good agreement between κ or κ_0 values evaluated by various methods is confirmed, the theory, on which the principles of the methods are based, is considered thoroughly acceptable to explain all the thermodynamic properties of polymer solutions (ref. 192).

κ_0 can be determined by the following methods:

(a) The temperature dependence of the chemical potential of the solvent, estimated from vapor pressure and osmotic pressure, through use of the relations (ref. 11),

$$\kappa_0 = \left(\frac{1}{\tilde{R}Tv_p^{\,2}}\right)\left(\frac{\partial(\Delta\mu_0/T)}{\partial(1/T)}\right)_{P,v_p} \tag{2.248}$$

where $\Delta\mu_0$ is given by eqn (1.1). The partial differentiation of $\Delta\mu_0/T$ with respect to $1/T$ is carried out under constant pressure and constant composition except the polymer.

(b) The critical phenomena (critical solution temperature T_c and critical polymer concentration $v_p^{\,c}$). κ_0 is related to the Flory theta temperature θ and Flory entropy parameter ψ_0 through the definition of θ (ref. 11).

$$\kappa_0 = \theta\,\psi_0/T \tag{2.249}$$

with

$$\psi_0 = \lim_{v_p^0 \to 0} \frac{\Delta S_0 - \Delta S_0^{\,comb}}{\tilde{R}Tv_p^{\,2}}$$

$$= \lim_{v_p^0 \to 0} \frac{\Delta\mu_0 - \Delta H_0 - T\Delta S_0^{\,comb}}{\tilde{R}Tv_p^{\,2}}. \tag{2.250}$$

ΔS_0 is the partial molar entropy of dilution and ΔS_0^{comb}, the combinatorial entropy term.

θ and ψ_0 can be evaluated from T_c (and v_p^c) for a series of solutions of polymers having different molecular weights by the method described in section 2.7 [SF (ref. 165), Stockmayer (ref. 133), KM (refs 3,5) and KKS (ref. 16)]. In KKS notation, κ_0 is given as g_{01}/T.

(c) Temperature dependence of A_2 by membrane osmometry or light scattering measured in the vicinity of the θ temperature (eqn (2.216)) via θ and ψ_0 (ref. 11). It is also possible to evaluated κ_0 by putting θ and ψ_0 into eqn (2.249).

(d) Calorimetry

The heat of dilution ΔH and accordingly ΔH_0 ($\equiv (\partial \Delta H/\partial n_0)_{T,P}$; n_0 is the number of moles of the solvent) can be directly measured by calorimetry.

Fig. 2.109 demonstrates schematic routes of determination for κ_0.

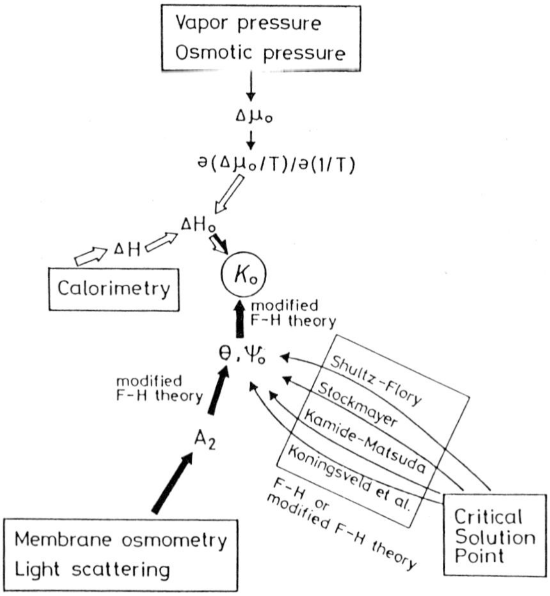

Fig. 2.109 Routes of calculation of the Flory enthalpy parameter κ_0 from experimental data of vapor pressure, osmotic pressure, light scattering and critical solution points.

The thermodynamic properties of atactic PS / CH system have hitherto been most thoroughly investigated and the literature data available for this system were chosen for estimating κ_0. As a supplement the literature data for atactic PS / trans-decalin system were also analyzed.

For atactic PS / CH system, the membrane osmotic pressure (MO) and light scattering (LS) measurements have been carried out to estimate A_2 over a temperature range including θ temperature by Krigbaum (MO) (ref. 75), Krigbaum and Geymer (MO) (ref. 62), Outer et al. (LS) (ref. 193), Krigbaum and Carpenter (LS) (ref. 173), Schulz and Baumann (LS) (ref. 174), Kotera et al. (LS) (ref. 175), Miyaki-Fujita (LS) (refs 194,195) and Tong et al. (LS) (ref. 196).

The results of the present analysis with data taken from the literature are assembled in Table 2.29. Here, we have constructed the plots of A_2 against T using the literature data, from which $(\partial A_2/\partial T)_\theta$ was estimated. In the table κ_0 at 308 K (i.e., the most probable θ temperature) is demonstrated (ref. 192).

Fig. 2.110 shows κ_0 at θ estimated by the A_2 method for atactic PS / CH system, as a function of M_w^{-1} or M_n^{-1}. In the figure the κ_0 value (0.32), obtained from the temperature dependence of the chemical potential by Scholte (ref. 63), is shown for comparison.

Least-square straight lines a and b in the figure pass through the observed data points of Krigbaum (ref. 75) (except for sample code H-2-8) and Krigbaum and Geymer (ref. 62), respectively. From these lines, we can evaluate κ_0 at an infinite molecular weight to be 0.28 from Krigbaum's data and 0.27 from Krigbaum and Geymer's data (0.28, on average). The κ_0 values obtained from Krigbaum's and also by Krigbaum and Geymer's data have a small but significant tendency to decrease with an increase in M_n.

It is probably true to say that as early as the mid 1950s membrane osmometry was established as an accurate and reliable method for the estimation of A_2, but in the 1950s, the light scattering technique was embryonic, not being as precise as osmometry as noticed by Krigbaum and Carpenter (ref. 173). Therefore, the LS data for that period may contain large experimental uncertainty. However, an inspection of Fig. 2.109 leads us to the conclusion that κ_0 values estimated from the classical LS A_2 data by Outer et al., Krigbaum-Carpenter, Schulz-Baumann and Kotera et al. decrease rapidly with an increase in M_w

TABLE 2.29

Evaluation of Flory κ_0 parameter from the temperature dependence of the second virial coefficient A_2 by membrane osmometry and light scattering for atactic polystyrene / cyclohexane system.

Method	Reporter (Year)	Sample code	$M_w(M_n)$ $\times 10^{-4}$	Temp.range K	$(\partial A_2/\partial T)_\theta$ $\times 10^6 cm^3 mol/g\ K$	θ K	κ_0 at 308K
Membrane osmometry	Krigbaum (1954)	L5-6	(5.05)		9.53	305.8	0.37
		HA-9	(12.5)		8.72	305.8	0.34
		M2-2	(35.9)	303～323	7.84	305.6	0.30
		H1-4	(56.6)		7.22	305.4	0.28
		H2-8	(20.3)		9.14	307.6	0.36
	Krigbaum-Geymer (1959)	F-I	(7.2)	313,323	8.9	307.2	0.35
		F-II	(44.0)	303,313,323	7.2	307.9	0.28
Light scattering	Outer et al. (1950)	B-2	163	300～314.5	4.25	308	0.17
	Krigbaum-Carpenter (1955)	HB2-3		320	4.76	308.4	0.19
	Schulz-Baumann (1961)	—	17	—	10	307	0.39
	Kotera-Saito-Fujisaki (1963)	SMI	30(15)	303.2～323.2	6.76	307.4	0.26

TABLE 2.29 (continued)

Method	Reporter (Year)	Sample code	$M_w(M_n)$ $\times 10^{-4}$	Temp.range K	$(\partial A_2/\partial T)_\theta$ $\times 10^6 cm^3 mol/g\ K$	θ K	κ_0 at 308K
Light Scattering	Miyaki-Fujita (1981)	IK1500-1	5680	306～323	4.62	307.9	0.18
		IK1500-2	3900	306～323	5.34	307.9	0.21
		BK2500-1	3190	306～328	5.86	307.8	0.23
		BK2500-2	2320	306～323	5.95	307.8	0.23
		BK2500-3	1500	305～328	6.62	308.3	0.26
		BK2500-4	876	305～328	5.58	308.3	0.22
		C1	647	304,328	6.01	308.2	0.24
		F450	480	303～328	5.43	308.9	0.21
		F288	308	305～328	5.91	307.3	0.23
		F126	134	302～328	7.04	307.3	0.27
	Tong et al. (1984)	F1	1.00	307.7K～			
		F2	2.15	a value			
		F4	4.36	about 1K	8.4	307.7	0.30
		F20	18.1	below the			
		F40	49.8	cloud point			

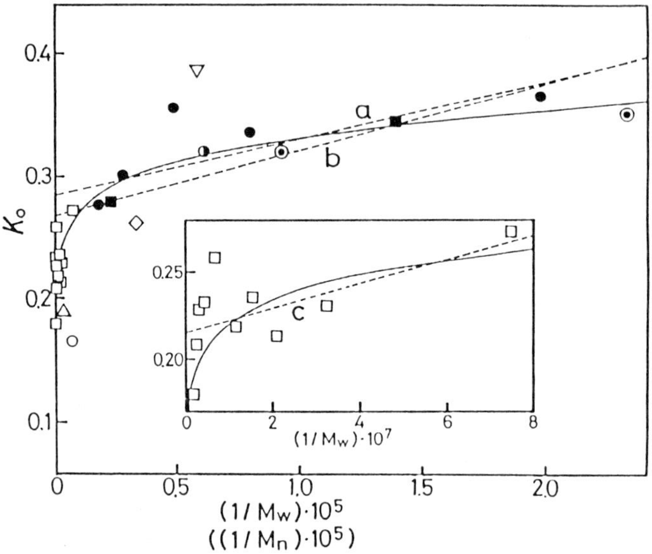

Fig. 2.110 Molecular weight dependence of Flory enthalpy parameter at infinite dilution κ_0, evaluated by the temperature dependence of the chemical potential and of the second virial coefficient A_2 in vicinity to theta temperature and by calorimetry for atactic polystyrene / cyclohexane system: ●, Krigbaum (ref. 75) (membrane osmometry (MO)); ■, Krigbaum-Geymer (ref. 62) (MO); ◑, Scholte (ref. 63) (ultracentrifuge); △, Krigbaum-Carpenter (ref. 173) (light scattering (LS)); ▽, Schulz-Baumann (ref. 174) (LS); ◇, Kotera et al. (ref. 175) (LS); ○, Outer et al. (ref. 193) (LS); □, Miyaki (ref. 195) (LS); ⊙, Fujihara (ref. 179) (calorimetry). Solid line, master curve of all data; broken line a, drawn pass through the data points of Krigbaum; b, Krigbaum-Carpenter. Inserted figure, enlarged plots of Miyaki et al. data. Broken line c, pass through the data points of Miyaki et al.

and for PS samples with $M_w > 10^6$ κ_0 was found to be below 0.20, which is significantly lower than those estimated by MO ($M_n = 5.50\times 10^4 \sim 56.6\times 10^4$) by Outer et al. and do not deviate strongly from a common relation obtained by classical MO and recent LS measurements.

Miyaki and Fujita (refs 194,195) carried out a tremendously elaborate work in measuring A_2 for ten atactic PS samples with very high molecular weights ($M_w = 1.34\times 10^6 \sim 5.68\times 10^7$) in CH at three to seven different temperatures ranging from 309.2 to 328.2 K. Except for a fraction (Sample code IK1500-1, $M_w = 5.68\times 10^6$), the almost constant value of κ_0 ($= 0.23\pm 0.01$, at 308K) was obtained over a wide range of M_w ($8.76\times 10^5 \sim 6.47\times 10^6$) as shown in Table 2.29. The figure inserted in Fig. 2.110 also shows the enlarged plots in the lower M_w^{-1} region. κ_0 values thus determined by Miyaki and Fujita scatter fairly widely, probably due to experimental difficulty of LS measurements of solutions of polymers with $M_w \sim 10^7$. $\kappa_0 = 0.22$ at 308 K was obtained for $M_w = \infty$ from line c in the figure, constructed by the least-squares method. Tong et al. (ref. 196) determined A_2 of five monodisperse PS samples in the temperature range from 307.7K to about 1K above the cloud point and they found A_2 for monodisperse PS in CH below θ to be independent of M, within the ranges of T and M studied. Since the A_2 values were not described in their paper, we estimated $(\partial A_2/\partial T)_\theta$ from the initial slope of the plot of $-A_2$ vs. T^{-1} (Fig. 4 of ref. 196), to be 0.30.

Amaya and Fujishiro (ref. 197) reported the heat of dilution ΔH, obtained with a calorimetry specially designed and constructed, for five CH solutions of an atactic PS (unfractionated sample, $X = 275$) at 303.2K. From Table I of ref. 197 we can calculate κ_0 at 298.2K to be 6.0. This value is unbelievably large. In this case the measurements were carried out almost 4~5K below the θ temperature and Amaya and Fujishiro (ref. 197) pointed out that it may be reasonable to suppose that in the solution at 303.2K there remain more or less aggregations of polymers. Schulz and Horbach (ref. 198) measured ΔH directly for nine atactic PS fractions (the viscosity-average molecular weight $M_v = 400 \sim 1.7 \times 10^4$) in CH at 296.2K and for two PS samples ($M_v = 1200$ and 1.1×10^4) they obtained ΔH_0 as a function of the polymer concentration v_p. The value of κ, calculated by eqn (2.245) from the experimental ΔH_0 (Table 7 of ref. 198), indicated a strong negative molecular weight dependence. From these data, however, we cannot evaluate an accurate κ or κ_0 at the infinite molecular weight in a θ solvent ($\simeq 307.5$K). For example, we obtained $\kappa_0 = 0.9$ for PS with $M_v = 1.1 \times 10^4$, but this value is almost 2.5 times larger than that (0.4) estimated from Fujihara's data (ref. 179) for monodisperse PS with $M_w = 1.02 \times 10^4$ at 307.2K. The difference of temperature in measurements provides a partial explanation of this large difference. From Fig. 4 of ref. 198 for atactic PS ($M_w = 6.8 \times 10^4$) in CH ($w_p = 0.1$) [cited from Cantow's data (ref. 199)] it is shown that the κ value at θ is nearly half of that at 297.2K.

By overcoming the numerous experimental difficulties encountered with calorimetry Fujihara (ref. 179) succeeded in measuring directly ΔH_0 of four monodisperse PS samples ($M_w = 6.2 \times 10^3 \sim 1.07 \times 10^5$) in CH at 307.2K. According to him, the

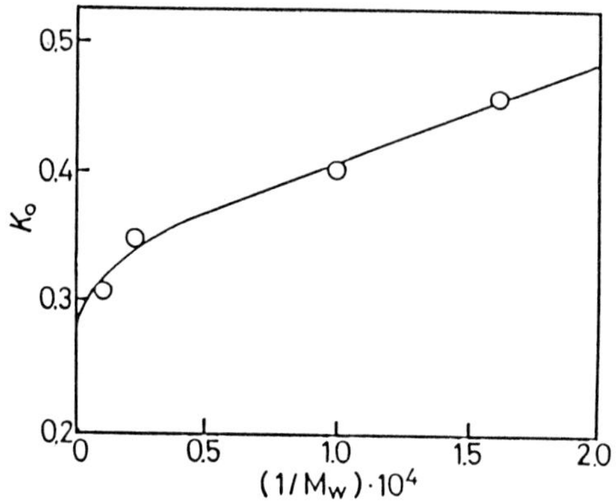

Fig. 2.111 Plots of Flory enthalpy parameter at infinite dilution κ_0, evaluated directly by Fujihara (ref. 179) using calorimetry, atactic polystyrene / cyclohexane system at 308.2K.

PS / CH system is a large endothermic solution, decreasing ΔH_0 with increasing M_w. First we constructed the plot of κ against v_p for each sample from the data in Table I of ref. 199 and estimated κ_0 as an intercept of the above plot. Fig. 2.111 shows the molecular weight dependence of κ_0 determined thus. It is noticeable that κ_0 decreases with increasing M_w. The data points, except for the highest M_w sample, can be reasonably represented by a straight line, yielding $\kappa_0 = 0.33$ at $M_w = \infty$ (i.e., $\kappa_{00} = 0.33$). If all the data points are taken into consideration for the extrapolation procedure, we obtain $\kappa_0 = 0.28$ at $M_w = \infty$ (i.e., $\kappa_{00} = 0.28$).

The molecular weight range covered by calorimetry is lower than that for MO and LS experiments, to ease the experimental measurements. For the two highest M_w samples κ_0 values are plotted in Fig. 2.110. It is noted that the chemical potential method and the calorimetry give almost the same κ_0 values as those determined from A_2 with the same M_w polymer. The chemical potential method (ref. 63) for $M_n = 15.4\times 10^4$ gives $\kappa_0 = 0.32$ and by calorimetry (ref. 179) for $M_w = 10.7\times 10^4$ $\kappa_0 = 0.31$, which should be compared with 0.34 by A_2 method (Krigbaum (ref. 75)) for $M_w = 12.5\times 10^4$ and 0.32 by A_2 method (Tong et al. (ref. 196)) for $M_w = 1\times 10^4 \sim 18.1\times 10^4$.

Fig. 2.110 indicates that with the exception of the points by Schulz and Baumann (ref. 174) and by Krigbaum (ref. 75) (sample code H-2-8), all the data points estimated by numerous researchers

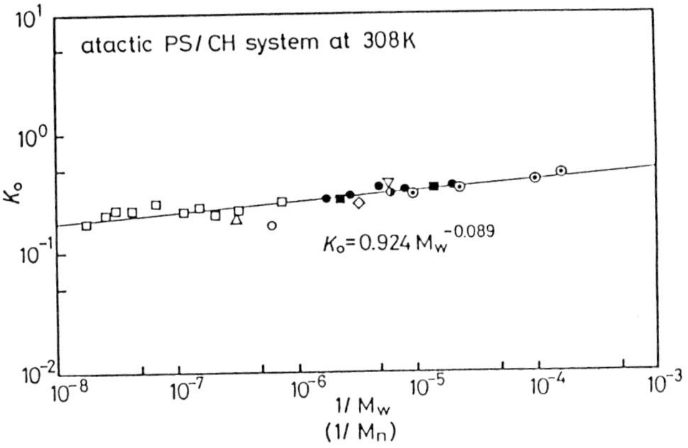

Fig. 2.112 Log-log plot of Flory enthalpy parameter at infinite dilution κ_0, evaluated by the temperature dependence of the chemical potential and second virial coefficient A_2 in the vicinity of the theta temperature and by calorimetry versus the reverse of weight- (or number-) average molecular weight M_w (M_n) for the atactic polystyrene / cyclohexane system: ●, Krigbaum (membrane osmometry (MO)); ■, Krigbaum-Geymer (MO); ◑, Scholte (ultracentrifuge); △, Krigbaum-Carpenter (light-scattering (LS)); ▽, Schulz-Baumann (LS); ◇, Kotera et al. (LS); ○, Outer et al. (LS); □, Miyaki-Fujita (LS); ⊙, Fujihara (calorimetry). Solid line, the equation $\kappa_0 = 0.92\, M_{w(n)}^{-0.089}$.

using various methods seem to be represented by a single master curve, which very gradually decreases with M_w above $M_w > 10^6$. In order to obtain a better understanding of the molecular weight dependence of κ_0, a log-log plot of κ_0 against M_w^{-1} or M_n^{-1} for atactic PS / CH systems at 307.2K is shown in Fig. 2.112. All data points available yielded a straight line given by (ref. 192)

$$\kappa_0 (\equiv \psi_0) = 0.924\, M_w \text{ (or } M_n)^{-0.089}. \tag{2.251}$$

Eqn (2.251) is valid over the entire molecular range which is accessible experimentally from 6.2×10^3 to 5.680×10^7. Eqn (2.251) is represented by the full line in Fig. 2.112. Therefore, we can conclude that the most probable κ_0 value is, in a strict sense, dependent on M_w (or M_n), irrespective of the method employed and that if eqn (2.251) can be expanded its applicability to $M_w = \infty$, κ_0 at the infinite molecular weight may be zero. This is an experimental indication that both the randomness in the mixing of a polymer and solvent and the spacial homogeneity of the polymer segment density in solution are expected to be realized in dilute

TABLE 2.30
The Flory θ temperature and entropy parameter κ_0 at infinite dilution for the upper critical solution point of the atactic polystyrene / trans-decalin system.

Method		θ /K	κ_0 at θ
Critical point			
	Shultz-Flory	293.7	0.95
	Kamide-Matsuda	292.7	0.33
Second virial coefficient			
	Light scattering	296.8	0.32

solutions of polymer with infinitely large molecular weight (i.e., $\Delta S_0 = \Delta S_0^{comb}$). The fact that methods (1)-(4) give essentially identical κ_0 values within ± 0.02 for a given PS sample in CH, strongly supports the validity of the modified Flory-Huggins theory (ref. 192).

For the atactic PS / trans-decalin system, the upper critical solution points and the temperature dependence of the second virial coefficient have been reported in literature. Nakata et al. (ref. 143) determined v_p^c and T_c for four monodisperse PS samples ($M_w = 3.7\times 10^4 \sim 270\times 10^4$) in trans-decalin. In section 2.7, we estimated θ and ψ_0 from the above data according to the Shultz-Flory, Kamide-Matsuda methods. The results are shown in Table 2.30, where κ_0 values at θ calculated from ψ_0 are shown.

Inagaki and his coworkers (ref. 200) measured the light scattering second virial coefficient A_2 of four well-fractionated PS samples ($M_w = 14.3\times 10^4 \sim 112\times 10^4$) in trans-decalin over temperatures ranging from 291.2 to 349.2K and evaluated θ and $(\partial A_2/\partial T)_\theta$ (Table III of ref. 200), from which we calculated ψ_0 and κ_0 at θ assuming $(V_1/\bar{v}^2) = 183.82 g^2 cm^3 mol^{-1}$ (V_1 is the molar volume of the solvent and $\bar{v}$ is the specific volume of the polymer), regardless of temperature. The results are summarized in Table 2.31. From the plot of κ_0 vs. $1/M_w$ (Fig. 2.113), κ_0 at the infinite molecular weight was found to be 0.32 (Table 2.30). κ_0 can also be experimentally represented by (ref. 192)

$$\kappa_0 = 31.5\ M_w^{-0.312}. \qquad (2.252)$$

The κ_0 value from the critical point by the Kamide-Matsuda treatment (0.33) for the atactic PS / trans-decalin system agrees

TABLE 2.31
Evaluation of Flory κ_0 parameter from the temperature dependence of the second virial coefficient A_2 by light scattering for atactic polystyrene / trans-decalin system.

Method	Reporter (Year)	Sample code	M_w $\times 10^{-4}$	Temp. K	$(\partial A_2/\partial T)_\theta$ $\times 10^6$ $cm^3 mol/g\ K$	θ K	κ_0 at θ*
Light scattering	Inagaki et al. (1966)	G-2	14.3	291.2~343.2	1.43	296.8	0.78
		J-4	22.2	295.5~302.1	1.14	296.8	0.62
		J-10	51.3	295.2~300.1	1.11	296.7	0.61
		P-4	112	295.2~302.8	0.69	296.8	0.38

*, $(V_1/\bar{v}^2) = 183.82$.

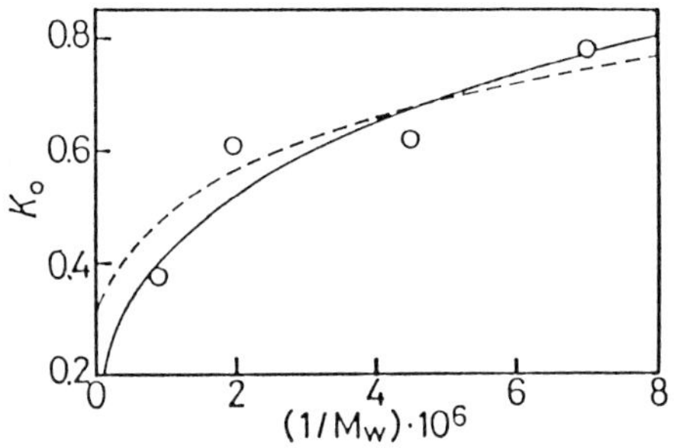

Fig. 2.113 The molecular weight dependence of Flory enthalpy parameter at infinite dilution κ_0, evaluated by Inagaki et al. from the temperature dependence of the second virial coefficient A_2, for the atactic polystyrene / trans-decalin system.

fairly well with that from $(\partial A_2/\partial T)_\theta$ (0.32) because the molecular weight ranges of these two sets of data overlap fairly well.

Summarizing, Kamide et al. (ref. 192) clarified the reliability of κ_0 in Flory theta solvent experimentally estimated and a molecular weight dependence of κ_0 for typical non-polar polymer / non-polar solvent systems. Then, we come to the conclusion that the modified Flory-Huggins theory, derived by introducing the molecular weight- and the concentration-dependencies of the χ parameter into the original theory, can be regarded as valid and sufficiently accurate to express any thermodynamic property of a polymer solution.

2.9 EXPERIMENTAL DETERMINATION OF PHASE DIAGRAM ON POLYMER SOLUTIONS

Cloud point is determined the most commonly by visual observation (i.e., naked eye observation) or from the pattern of a He-Ne transmitted laser beam (ref. 201), or by differential thermal analysis (refs 138,139,202,203). The last method is unavoidably less accurate.

Coexistence curve (i.e., binodal curve) is evaluated from direct or indirect analysis of the two coexisting phases: Direct method, separation of the two phases and determination of volume and weight of each phase, precipitation of polymer or evaporation of solvent and determination of polymer weight (ref. 29); indirect method, measurement of refractive index difference between each phase and pure solvent, and determination of volume and weight of the two phases (refs 204,205).

Critical point is determined as the solution, in which two-phase volume ratio R becomes unity [phase volume method] (refs 35,206), or by extrapolation of the curve of unit phase volume ratio to the cloud point curve [diameter method] (ref. 66), or as the maximum and minimum points of the cloud point temperature vs. composition plot [maximum or minimum precipitation temperature method], or as the composition of the solution, at which pore shape of the membrane, cast from the solution, changes from circular to noncircular [membrane method] (ref. 207). Diameter method and maximum or minimum precipitation temperature method can only be applicable to nearly monodisperse polymer solution.

Spinodal curve is very difficult to determine experimentally. Spinodal point is determined by light scattering measurements at temperatures slightly above the miscibility gap (ref. 65): The scattered light intensity $I(\theta)$ is measured as a function of the scattered angle θ and extrapolated to $\theta = 0$. If $1/I(0)$ is plotted vs. $1/T$, extrapolation to $1/I(0) = 0$ yields spinodal temperature for the polymer solution. Pulse induced critical scattering (PICS) method is based on this principle, although not yet widely popularized. In PICS the relative intensities of scattered laser light are measured over a period of seconds after a fast temperature step into the region of Debye critical opalescence (ref. 208). Another method proposed is the differential scanning calorimetry (ref. 203).

REFERENCES

1 K. Kamide, T. Ogawa, M. Sanada and M. Matsumoto, Kobunshi Kagaku, 25 (1968) 440.
2 K. Kamide, T. Ogawa and M. Matsumoto, Kobunshi Kagaku, 25 (1968) 788.
3 K. Kamide and S. Matsuda, Polym. J., 16 (1984) 825.
4 K. Kamide, S. Matsuda, T. Dobashi and M. Kaneko, Polym. J., 16 (1984) 839.
5 K. Kamide, S. Matsuda and M. Saito, Polym. J., 17 (1985) 1013.
6 S. Matsuda, Polym. J., 18 (1986) 981.
7 K. Kamide and Y. Miyazaki, Polym. J., 13 (1981) 325.
8 K. Kamide, T. Abe and Y. Miyazaki, Polym. J., 14 (1982) 355.
9 D. Patterson and G. Delmas, Trans. Faraday Soc., 65 (1969) 708.
10 P.J. Flory, Discuss. Faraday Soc., 49 (1970) 7.
11 P.J. Flory, Principle of Polymer Chemistry, Cornell Univ. Press, Ithaca, NY, 1953.
12 M. Kurata, Thermodynamics of Polymer Solutions, Harwood Academic Pubs. Chur, London, NY, 1982, Chapter 2.
13 R. Koningsveld and A.J. Staverman, J. Polym. Sci., A-2, 6 (1968) 305.
14 R. Koningsveld and A.J. Staverman, J. Polym. Sci., A-2, 6 (1968) 325.
15 G. Rehage and R. Koningsveld, Polymer Letter, 6 (1968) 421.
16 R. Koningsveld, L.A. Kleintjens and A.R. Shultz, J. Polym. Sci., A-2, 8 (1970) 1261.
17 R. Koningsveld and A.J. Staverman, J. Polym. Sci., A-2, 6 (1968) 349.
18 S. Matsuda, Polym. J., 18 (1986) 993.
19 R. Koningsveld and L.A. Kleintjens, Macromolecules, 5 (1971) 637.
20 J.W. Kennedy, M. Gordon and R. Koningsveld, J. Polym. Sci., C, 39 (1972) 43.
21 K. Kamide and K. Sugamiya, Makromol. Chem., 139 (1970) 197.
22 K. Kamide and K. Sugamiya, Makromol. Chem., 156 (1972) 259.
23 K. Kamide, in L.H. Tung (Ed.), Fractionation of Synthetic Polymers, Marcel Dekker Inc., New York, 1977, Chapter 2.
24 G.V. Schulz, Z. Physik. Chem., B43 (1939) 25.
25 B.H. Zimm, J. Chem. Phys., 16 (1948) 1099.
26 H. Wesslau, Makromol. Chem., 20 (1956) 111.
27 K. Kamide and Y. Miyazaki, Polym. J., 12 (1980) 205.
28 L.H. Tung, J. Polym. Sci., 61 (1962) 449.
29 K. Kamide, Y. Miyazaki and T. Abe, Makromol. Chem., 177 (1976) 485.
30 I. Noda, H. Ishizawa, Y. Miyazaki and K. Kamide, Polym. J., 12 (1980) 87.
31 K. Kamide, Y. Miyazaki and T. Abe, Polym. J., 9 (1977) 395.
32 K. Kamide and S. Matsuda, Netsu Sokutei, 13 (1986) 173.
33 K. Kamide, K. Sugamiya, T. Kawai and Y. Miyazaki, Polym. J., 12 (1980) 67.

34 K. Kamide, Y. Miyazaki and T. Abe, Brit. Polym. J., 13 (1981) 168.
35 R. Koningsveld and A.J. Staverman, Kolloid-Z. Z. Polym., 218 (1967) 114.
36 R. Koningsveld and A.J. Staverman, J. Polym. Sci., A-2, 6 (1968) 367.
37 R. Koningsveld and A.J. Staverman, J. Polym. Sci., A-2, 6 (1968) 383.
38 M. Gordon, H.A.G. Chermin and R. Koningsveld, Macromolecules, 2 (1969) 107.
39 R. Koningsveld, Adv. Polym. Sci., 7 (1970) 1.
40 R. Koningsveld, W.H. Stockmayer and J.W. Kennedy, L.A. Kleintjens, Macromolecules, 7 (1974) 731.
41 K. Kamide, Molecular Weight Fractionation on the Basis of Solubility, IUPAC Macromolecular Chemistry-8, 1973, 147.
42 K. Kamide and Y. Miyazaki, Makromol. Chem., 176 (1975) 3453.
43 K. Kamide and Y. Miyazaki, Makromol. Chem., 176 (1975) 1029.
44 K. Kamide, in Soc. Polym. Sci., Japan (Ed.), Polymer Solutions (Polymer Experiments, vol. 12) Kyoritu Pub. Co., Tokyo, 1982, Section 1.3.2.
45 T. Dobashi, M. Nakata and M. Kaneko, J. Chem. Phys., 72 (1980) 6692.
46 G.V. Schulz and A. Dinglinger, Z. Phys. Chem. B43 (1939) 47.
47 Y. Miyazaki and K. Kamide, Polym. J., 9 (1977) 61.
48 P.J. Flory, J. Chem. Phys., 12 (1944) 425.
49 T. Kawai, Kobunshi Kagaku, 12 (1955) 63.
50 T. Kawai, Kobunshi Kagaku, 12 (1955) 71.
51 K. Kamide and C. Nakayama, Makromol. Chem., 129 (1969) 289.
52 K. Kamide, K. Sugamiya, T. Ogawa, C. Nakayama and N. Baba, Makromol. Chem., 135 (1970) 23.
53 K. Kamide, K. Sugamiya, T. Terakawa and T. Hara, Makromol. Chem., 156 (1972) 287.
54 K. Kamide and Y. Miyazaki, Polym. J., 12 (1980) 153.
55 M.L. Huggins, J. Polym. Sci., A-2, 5 (1967) 1221.
56 L.H. Tung, J. Appl. Polym. Sci., 10 (1960) 375.
57 W.W. Yau and S.W. Fleming, J. Appl. Polym. Sci., 12 (1968) 2111.
58 M.J. Newing, Trans. Faraday Soc., 46 (1950) 613.
59 C.E.H. Bawn, R.F.J. Freeman and A.R. Kamaliddin, Trans. Faraday Soc., 46 (1950) 677.
60 G. Gee and W.J.C. Orr, Trans. Faraday Soc., 42 (1946) 507.
61 See, for example, S. Saeki, J.C. Holste and D.C. Bonner, J. Polym. Sci., Phys. Ed., 19 (1981) 307.
62 W.R. Krigbaum and D.O. Geymer, J. Am. Chem. Soc., 81 (1959) 1859.
63 Th.G. Scholte, J. Polym. Sci., A-2, 8 (1970) 841.
64 Th.G. Scholte, Eur. Polym. J., 6 (1970) 1063.
65 Th.G. Scholte, J. Polym. Sci., A-2, 9 (1971) 1553.
66 N. Kuwahara, M. Nakata and M. Kaneko, Polymer, 14 (1973) 415.
67 N. Kuwahara, S. Saeki, S. Konno and M. Kaneko, Polymer, 15 (1974) 66.

68 A. Nakajima, H. Fujiwara and F. Hamada, J. Polym. Sci., A-2, 4 (1966) 507.
69 M. Kaneko and N. Kuwahara, Asahi Glass Gijutu Shoreikai Hokoku, 13 (1967) 377.
70 T. Dobasi, M. Nakata and M. Kaneko, J. Chem. Phys, 72 (1980) 6685.
71 S. Saeki, N. Kuwahara, S. Konno and M. Kaneko, Macromolecules, 6 (1973) 246.
72 R.L. Scott, J. Chem. Phys., 13 (1945) 178.
73 H. Okamoto and K. Sekikawa, J. Polym. Sci., 55 (1961) 597.
74 J.W. Breitenbach and B.A. Wolf, Makromol. Chem., 108 (1967) 263.
75 W.R. Krigbaum, J. Am. Chem. Soc., 76 (1954) 3758.
76 K. Kamide, Y. Miyazaki and K. Yamaguchi, Makromol. Chem., 173 (1973) 157.
77 G.V. Schulz, Z. Physik. Chem., B47 (1940) 155.
78 L.H. Cragg and H. Hammerschlag, Chem. Rev., 39 (1946) 79.
79 R.W. Hall, in P.W. Allen (Ed.), Techniques of Polymer Characterization, Butterworths, London, 1959, Chapter 2.
80 G.M. Guzman, in J.C. Robb and F.W. Peaker (Ed.), Progress in High Polymers, Heywood & Co., 1961, Vol. 1, p113
81 M.J.R. Cantow (Ed.), Polymer fractionation, Academic Press, NY, 1967.
82 T. Kawai (Ed.), Polymer Engineering, Vol. 4, Chijin Shokan, Tokyo, 1967.
83 L.H. Tung (Ed.), Fractionation of Synthetic Polymers, Marcel Dekker Inc., NY, 1977.
84 K. Kamide, Kobunshi Ronbunshu, 31 (1974) 147.
85 M. Matsumoto and Y. Ohyanagi, Kobunsi Kagaku, 11 (1954) 7.
86 K. Kamide, Pure and Appl. Chem., Makromol. Chem., 8 (1972) 147.
87 K. Kamide, K. Yamaguchi and Y. Miyazaki, Makromol. Chem., 173 (1973) 133.
88 K. Kamide and Y. Miyazaki, Makromol. Chem., 176 (1975) 1051.
89 K. Kamide, T. Ogawa and C. Nakayama, Makromol. Chem., 132 (1970) 65.
90 K. Kamide, T. Ogawa and C. Nakayama, Makromol. Chem., 135 (1970) 9.
91 K. Kamide and K. Yamaguchi, Makromol. Chem., 167 (1973) 287.
92 K. Kamide, Y. Miyazaki and K. Sugamiya, Makromol. Chem., 173 (1973) 113.
93 K. Kamide, Y. Miyazaki and K. Yamaguchi, Makromol. Chem., 173 (1973) 175.
94 K. Kamide and Y. Miyazaki, Makromol. Chem., 176 (1975) 1029.
95 K. Kamide and Y. Miyazaki, Makromol. Chem., 176 (1975) 1427.
96 K. Kamide and Y. Miyazaki, Makromol. Chem., 176 (1975) 1447.
97 K. Kamide and Y. Miyazaki, Makromol. Chem., 176 (1975) 2393.
98 United States Department of Commerce. National Bureau of Standards, Washington, D.C., 20234, U.S.A.
99 Pressure Chemical Co., 3419-25 Swallman Street, Pittsburgh, U.S.A.

100 Rubber and Plastic Research Association, Shawburh, Shrewsburg, Shropshire, England.
101 National Physical Laboratory, Teddington, Middlesex, TW110LW, UK, England.
102 Toyo Soda Manufacturing Co., Ltd., Toso Building, 1-7-7, Akasaka, Minato-ku, Tokyo 107, Japan.
103 K. Kamide, Y. Miyazaki and T. Abe, Polym. J., 11 (1979) 523.
104 K. Kamide, T. Terakawa and Y. Miyazaki, 11 (1979) 285.
105 M. Saito, Polym. J., 15 (1983) 249.
106 K. Kamide, M. Saito and T. Abe, Polym. J., 13 (1981) 421.
107 A.M. Meffroy-Biget, Ampt. Rend., 240 (1955) 1707.
108 C.D. Thurmond and B.H. Zimm, J. Polym. Sci., 8 (1952) 477.
109 G. Meyerhoff, Makromol. Chem., 12 (1945) 45.
110 A. Kotera, in M.J.R. Cantow (Ed.), Polymer Fractionation, Academic Press, New York, 1967.
111 Y. Fujisaki and H. Kobayashi, Kobunshi Kagaku, 18 (1961) 305.
112 Y. Fujisaki and H. Kobayashi, Kobunshi Kagaku, 18 (1961) 312.
113 Y. Fujisaki and H. Kobayashi, Kobunshi Kagaku, 19 (1961) 49.
114 Y. Fujisaki and H. Kobayashi, Kobunshi Kagaku, 19 (1961) 69.
115 M. Matsumoto, in A. Kotera (Ed.), Polymer Chemistry (Experimental Chemistry vol. 8) Maruzen Pub. Co., Tokyo, 1957.
116 R.S. Spencer, J. Polym. Sci., 4 (1948) 606.
117 F.W. Billmeyer, Jr. and W. H. Stockmayer, J. Polym. Sci., 5 (1949) 121.
118 M. Matsumoto, Kobunshi Kagaku, 11 (1954) 182.
119 A. Broda, T. Niwinska and S. Polowinski, J. Polym. Sci., 22 (1958) 343.
120 A. Broda, B. Bawronska, T. Niwinska and S. Polowinski, J. Polym. Sci., 29 (1958) 183.
121 H. Okamoto, J. Polym. Sci., 41 (1959) 535.
122 H. Okamoto, J. Phys. Soc. Japan, 14 (1959) 1388.
123 G.V. Schulz, Z. Phys. Chem., B32 (1936) 27.
124 K. Kamide and S. Matsuda, "Fractionation Method," Part C, Chapter I in "Molecular Weight Determination," John Wiley & Sons Inc. Publishers, in press.
125 J. Porath and P. Flodin, Nature, 183 (1957) 1657.
126 M.F. Vaugham, Nature, 188 (1960) 55.
127 J.C. Moore, J. Polym. Sci., A2 (1964) 835.
128 D. Berek, D. Bakos, T. Bleha and L. Soltes, Makromol. Chem., 176 (1975) 391.
129 J.V. Dawkins and M. Hemming, Makromol. Chem., 176 (1975) 1777.
130 J.V. Dawkins and M. Hemming, Makromol. Chem., 176 (1975) 1795.
131 J.V. Dawkins and M. Hemming, Makromol. Chem., 176 (1975) 1815.
132 M.R. Ambler and D. McIntyre, J. Polym. Sci., Polym. Lett., 13 (1975) 589.
133 W.H. Stockmayer, J. Chem. Phys., 17 (1949) 588.

134 J.W. Gibbs, The Scientific Papers of J. Willard Gibbs, Ph. D., LL. D., Vol. I, Thermodynamics, p65, Dover Pub., (1961), unaltered republication, originally published by Longmans, Green, and Co., in 1906.
135 L.A. Kleintjens, R. Koningsveld and M. Gordon, Macromolecules, 13 (1980) 303.
136 S. Saeki, N. Kuwahara, S. Konno and M. Kaneko, Macromolecules, 6 (1973) 589.
137 A. Nakajima, F. Hamada, K. Yasue and K. Fujisawa, Makromol. Chem., 175 (1974) 177.
138 Y. Baba, Y. Fujita and A. Kagemoto, Makromol. Chem., 164 (1973) 349.
139 K. Kagemoto and Y. Baba, Kobunshi Kagaku, 28 (1971) 784.
140 S. Saeki, S. Konno, N. Kuwahara, M. Nakata and M. Kaneko, Macromolecules, 7 (1974) 521.
141 S. Konno, S. Saeki, N. Kuwahara, M. Nakata and M. Kaneko, Macromolecules, 8 (1975) 799.
142 K.S. Siow, G. Delmas and D. Patterson, Macromolecules, 5 (1972) 29.
143 M. Nakata, S. Higashida, N. Kuwahara, S. Saeki and M. Kaneko, J. Chem. Phys., 64 (1976) 1022.
144 N. Kuwahara, S. Saeki, T. Chiba and M. Kaneko, Polymer, 15 (1974) 777.
145 A. Nakajima, F. Hamada and S. Hayashi, J. Polym. Sci., C, 15 (1966) 285.
146 F. Hamada, K. Fujisawa and A. Nakajima, Polym. J., 4 (1973) 316.
147 Y. Muraoka, H. Inagaki and H. Suzuki, Br, Polym. J., 15, (1983) 110.
148 H. Suzuki, K. Kamide and M. Saito, Eur. Polym. J., 18 (1980) 123.
149 H. Suzuki, Y. Muraoka, M. Saito and K. Kamide, Eur. Polym. J., 18 (1982) 831.
150 D.J. Korteweg, Sitzungsber Kaizerlichen Akad. Wiss., Math. Naturwiss. Kl. 98, 2 Abt. A (1889) 1154.
151 D.J. Korteweg, Archs néerl Sci., 24 (1891) 295.
152 J.W. Kennedy, J. Polym. Sci., Part C, 39 (1972) 71.
153 E.F. Casassa, Macromolecules, 8 (1975) 242.
154 H. Tompa, Trans Faraday Soc., 45 (1949) 1142.
155 H.A.G. Chermin, Br. Polym. J., 9 (1977) 195.
156 K. Šolc, Macromolecules, 16 (1983) 236.
157 K. Kamide, S. Matsuda and H. Shirataki, Eur. Polym. J., to be published.
158 H. Tompa, Polymer Solutions, Butterworths Scientific Publications, London, 1956.
159 R. Koningsveld and A.J. Staverman, Kolloid- Z. Z. Polym. 210 (1966) 151.
160 L.A. Kleintjens, H.M. Schoffeleers and L. Domingo, Br. Polym. J., 8 (1976) 29.
161 Y. Einaga, Y. Nakamura and H. Fujita, Macromolecules, 20 (1987) 1083.

162 T. Dobashi and M. Nakata, J. Chem. Phys., 84 (1986) 5775.
163 J.N. Bronsted and K. Volquartz, Trans. Faraday Soc., 35 (1939) 576.
164 R.B. Richards, Trans. Faraday Soc., 42 (1946) 10, 20.
165 A.R. Shultz and P.J. Flory, J. Am. Chem. Soc., 74 (1952) 4760.
166 G. Rehage, D. Moller and O. Ernst, Makromol. Chem., 38 (1965) 232.
167 K. Kamide, S. Matsuda and H. Shirataki, Eur. Polym. J., 26 (1990) 379.
168 H. Shirataki, S. Matsuda and K. Kamide, Brit. Polym. J., to be published.
169 K. Šolc, Collect. Czech. Chem. Commun., 34 (1969) 992.
170 K. Šolc, Macromolecules, 3 (1970) 665.
171 K. Šolc, J. Polym. Sci., Polym. Phys. Ed., 12 (1974) 555.
172 K. Šolc, L.A. Kleintjens and R. Koningsveld, Macromolecules, 17 (1984) 573.
173 W.R. Krigbaum and D.K. Carpenter, J. Phys. Chem., 59 (1959) 1166.
174 G.V. Schulz and H. Baumann, Makromol. Chem., 60 (1963) 120.
175 A. Kotera, T. Saito and N. Fujisaki, Rep. Prog. Polym. Phys. J., Ⅵ (1963) 9.
176 H. Suzuki, Y. Muraoka, M. Saito and K. Kamide, Brit. Polym. J., 14 (1982) 23.
177 K. Kamide and M. Saito, Adv. Polym. Sci., 83 (1987) 5.
178 K. Kamide and M. Saito, Eur. Polym. J., 20 (1984) 963.
179 I. Fujihara, Ph.D. Dissertation, 1979, Osaka City Univ.
180 K. Kamide, S. Matsuda and K. Kowsaka, Polym. J., 20 (1988) 231.
181 H.J. Harwood and L. Shepherd, unpublished results cited in D.Y. Yoon and P.J. Flory, Macromolecules, 10 (1977) 532.
182 See, for example, R. Abraham and P. Loftus, Proton and Carbon-13 NMR Spectroscopy, John Wiley & Sons, Inc., New York, N. Y., 1978, Chapter Ⅲ and Ⅵ.
183 Y. Inoue, A. Nishioka and R. Chujo, Makromol. Chem., 156 (1972) 207.
184 D. Roberts, Note on Molecular Orbital Calculation, W.A. Benjamin Inc., New York, N. Y., 1962; Y. Yukawa, R. Mikawa and K. Itoh, Exercises in Molecular Orbital Calculations, Hirokawa Pub., Co., Tokyo, 1965.
185 See, for example, A.H. Narten, J. Chem. Phys., 48 (1968) 1630.
186 C. Bailey, C. Ingold, H. Poole and C. Wilson, J. Chem. Soc., 222 (1946).
187 R. Mair and G. Hornig, J. Chem. Phys., 17 (1949) 1236.
188 T. Yamashiki, K. Kamide, K. Okajima, K. Kowsaka, T. Matsui and H. Fukase, Polym. J., 20 (1988) 447.
189 L.D. Landau and E.M. Lifshitz, Fluid Dynamics, Pergamon Press, 1954.
190 A. Passynsky, Acta Physicochim. U.S.S.R., 22 (1974) 137.

191 See, for example, A. Kagemoto, S. Murakami and R. Fujishiro, Bull. Chem. Soc. Jpn., 39 (1966) 15.
192 K. Kamide, S. Matsuda and M. Saito, Polym. J., 20 (1988) 31.
193 P. Outer, C.I. Carr and B.H. Zimm, J. Chem. Phys., 18 (1950) 830.
194 Y. Miyaki and H. Fujita, Macromolecules, 14 (1981) 742.
195 Y. Miyaki, Ph. D. Dissertation, 1981, Osaka Univ.
196 Z. Tong, S. Ohashi, Y. Einaga and H. Fujita, Polym. J., 15 (1983) 835.
197 K. Amaya and R. Fujishiro, Bull. Chem. Soc. Jpn., 22 (1959) 377.
198 G.V. Schulz and A. Horbach, Z. Phys. Chem. Neue Folge, 22 (1959) 377.
199 H.J. Cantow, Z. Phys. Chem. Neue. Folge, 7 (1956) 58.
200 H. Inagaki, H. Suzuki, M. Fujii and T. Matsuo, J. Phys. Chem., 70 (1966) 1718.
201 S.P. Lee, W. Tscharnuter, B. Chu and N. Kuwahara, J. Chem. Phys., 57 (1972) 4240.
202 Y. Baba and A. Kagemoto, Kobunshi Ronbunshu, 31 (1974) 446.
203 P.T. van Emmerik and C.A. Smolders, Eur. Polym. J., 9 (1973) 293.
204 R. Koningsveld, Thesis, 1967, Leiden.
205 M. Nakata, N. Kuwahara and M. Kaneko, J. Chem. Phys., 62 (1975) 4278.
206 R. Koningsveld and A.J. Staverman, J. Polym. Sci., C16 (1967) 1775.
207 K. Kamide, H. Iijima and M. Iwata, unpublished results.
208 K.W. Derham, J. Goldbrough and M. Gordon, Pure & Appl. Chem., 38 (1974) 97.

Chapter 3

QUASI-TERNARY SOLUTION CONSISTING OF MULTICOMPONENT POLYMER DISSOLVED IN A BINARY SOLVENT MIXTURE

3.1 CHEMICAL POTENTIAL

The two-phase equilibrium phenomena of a quasi-ternary system consisting of multicomponent polymers, all belonging to chemical homologue, solvent and nonsolvent are used not only for molecular weight fractionation by solubility difference, but also extensively for many industrial processes, such as the wet-spinning of regenerated cellulose and synthetic fibers, solvent-casting of micro-porous membranes and polymer coating. The quasi-ternary system thus warrants study so as to establish some scientific basis for it.

The phase separation phenomena of a single component polymer / solvent / nonsolvent system (i.e., rigorous ternary system) have been extensively studied from in the late 1940s to 1950s. Even for this relatively simple system, all two-phase equilibrium calculations at constant temperature and pressure carried out in the literature are far from rigorous and one based on the following crude assumptions: (a) a solvent mixture can be approximated as a "single solvent" (Flory (ref. 1), Scott (refs 2,3)), (b) a polymer molecular weight is infinite (Scott (refs 2,3), Nakagaki-Sunada (ref. 4)), (c) among three thermodynamic interaction parameters (χ_{12}, solvent 1 (ordinarily, good solvent)-solvent 2 (ordinarily, nonsolvent); χ_{13}, solvent 1-polymer; χ_{23}, solvent 2-polymer), $\chi_{12}=\chi_{13}$ and $\chi_{23}=0$ hold (Tompa (ref. 5), Nakagaki and Sunada (ref. 4)) and (d) any polymer does not exist in a polymer-lean phase (Krigbaum-Carpenter (ref. 6), Suh-Liou (ref. 7)). Assumptions (a)-(d) are not generally acceptable and hence the conclusions obtained by them cannot be regarded as having general applicability.

Theoretical study of the phase equilibrium of multicomponent polymer / solvent / nonsolvent system has been carried out only on a small scale since Münster's first work on a quasi-ternary system (ref. 8). Okamoto (ref. 9) calculated the threshold point of the same system and Koningsveld (ref. 10) demonstrated, on the basis of the Flory-Huggins high polymer solution theory, some coexistence curves, using rigorous calculation technique, but he showed none of the details for these phase characteristics.

In spite of its scientific and industrial importance, the study of the phase equilibrium of the quasi-ternary system, is still at a very primitive stage. This prompted Kamide and his co-workers to study, as an extension of the phase equilibrium theory of the quasi-binary system (see, Chapter 2), applying the simulation techniques that were used for the quasi-ternary system (refs 11-14). Based on this theory, a series of computer experiments were carried out on quasi-ternary system to study the effects of (a) three thermodynamic interaction parameters (ref. 11) (b) ρ_P and v_P^0 and (c) the average molecular weight and MWD of the original polymer (ref. 13), on the phase separation characteristics, and the results are compared with those of quasi-binary systems consisting of multicomponent polymers consisting of polymers and a single solvent. Aminavhabi and Munk (ref. 15) and Altena and Smolders (ref. 16) showed that the magnitude of χ_{12} and its concentration dependence influence thermodynamic properties, such as phase separation characteristics, of monodisperse polymer in a binary solvent system.

In this Chapter, we generalize a theory for phase equilibrium of a quasi-ternary system studied in the previous papers (refs 11-13) to the case where the three χ parameters (not only χ_{12} but also χ_{13} and χ_{23}) depend significantly on the concentration and explore the effects of this concentration dependence on the two-phase equilibrium characteristics.

On the basis of the Flory-Huggins theory (ref. 17), Kamide et al. (refs 11-13) proposed a theory of a quasi-ternary system, assuming that χ_{12}, χ_{13} and χ_{23} are independent of the polymer molecular weight and concentration. Chemical potentials of solvent 1, 2 and X_i-mer ($\Delta\mu_1$, $\Delta\mu_2$ and $\Delta\mu_{X_i}$, respectively) were given by following equations.

$$\Delta\mu_1 = \tilde{R}T\{\ln v_1 + (1-\frac{1}{X_n})v_P$$

$$+\chi_{12}v_2(1-v_1)+\chi_{13}v_P(1-v_P)-\chi_{23}v_2v_P\}, \qquad (3.1)$$

$$\Delta\mu_2 = \tilde{R}T\{\ln v_2 + (1-\frac{1}{X_n})v_P$$

$$+\chi_{12}v_1(1-v_2)+\chi_{23}v_P(1-v_2)-\chi_{13}v_1v_P\}, \qquad (3.2)$$

$$\Delta\mu_{xi} = \tilde{R}T[\ln v_{xi} - (X_i - 1) + X_i(1 - \frac{1}{X_n})v_p + X_i\{\chi_{13}v_1(1 - v_p)$$

$$+ \chi_{13}v_2(1 - v_p) - \chi_{12}v_1v_2\}] \qquad (i = 1, \cdots, m), \qquad (3.3)$$

where X_i is the molar volume ratio of the i-th polymer to solvent 1 or 2, v_1 and v_2, the volume fraction of solvent 1 and 2. We assume that (a) the molar volume of solvent 1 is the same as that of solvent 2, (b) solvent 1, 2 and the polymer are volumetrically additive and (c) the densities of solvent 1, 2 and the polymer are the same (ref. 11). These assumptions do not lessen the validity of the theory. For the quasi-binary system consisting of multicomponent polymers dissolved in a single solvent, we considered concentration dependence of χ parameter, as follows (refs 18-21):

$$\chi = \chi_0(1 + \sum_{j=1}^{n} p_j v_p^j). \qquad (3.4)$$

This is a special case of eqn (2.4) (k' = 0). From the analogy to the quasi-binary system, concentration dependence of χ_{12}, χ_{13} and χ_{23} for the quasi-ternary system should be written as (ref. 14)

$$\chi_{12} = \chi_{12}^0(1 + \sum_{s=1}^{n_s} p_{12,s}v_p^s) \times \{1 + \sum_{u=1}^{n_u}(p_{1,u}v_1^u + p_{2,u}v_2^u)\}, \qquad (3.5)$$

$$\chi_{13} = \chi_{13}^0(1 + \sum_{q=1}^{n_q} p_{13,q}v_p^q), \qquad (3.6)$$

$$\chi_{23} = \chi_{23}^0(1 + \sum_{r=1}^{n_r} p_{23,r}v_p^r), \qquad (3.7)$$

where χ_{12}^0, χ_{13}^0 and χ_{23}^0 are parameters independent of concentration and the degree of polymerization ($\simeq X_i$) and are dependent on temperature only. $p_{12,s}$, $p_{13,q}$ and $p_{23,r}$ are independent of X_i and temperature. $p_{1,u}$ and $p_{2,u}$ are the solvent composition-dependence parameters. Here, we neglected the theoretical and experimental possibility of the dependence on composition of binary solvent mixture.

$$\chi_{12} = \chi_{12}^{0}(1 + \sum_{s=1}^{n_s} p_{12,s} v_p^{s}). \tag{3.5'}$$

Gibbs' free energy of mixing of these solvents and a polymer, ΔG can be divided into four parts; Gibbs' free energy of ideal solution, ΔG^{id}, the excess free energy of solvent 1-2, ΔG_{12}^{E}, the excess free energy of solvent 1-polymer, ΔG_{13}^{E} and the excess free energy of solvent 2-polymer, ΔG_{23}^{E}:

$$\Delta G = \Delta G^{id} + \Delta G_{12}^{E} + \Delta G_{13}^{E} + \Delta G_{23}^{E}. \tag{3.8}$$

In the same way as the quasi-binary system (see, eqn (2.15)), eqns (3.6) and (3.7) are combined with the original equations of the excess chemical potential of solvent 1 and 2 ($\partial \Delta G_{13}^{E}/\partial N_1$ and $\partial \Delta G_{23}^{E}/\partial N_2$), to give

$$\frac{\partial \Delta G_{13}^{E}}{\partial N_1} = \tilde{R}T\chi_{13}^{0}(1 + \sum_{q=1}^{n_q} p_{13,q} v_p^{q}) v_p (1 - v_1), \tag{3.9}$$

$$\frac{\partial \Delta G_{23}^{E}}{\partial N_2} = \tilde{R}T\chi_{23}^{0}(1 + \sum_{r=1}^{n_r} p_{23,r} v_p^{r}) v_p (1 - v_2). \tag{3.10}$$

Here, N_1 and N_2 are number of solvent 1 and 2, respectively. Following the procedure of quasi-binary system, ΔG_{13}^{E} and ΔG_{23}^{E} for the quasi-ternary system can be given by integration of eqns (3.9) and (3.10), respectively (ref. 14).

$$\Delta G_{13}^{E} = \int_0^{N_1} \left(\frac{\partial \Delta G_{13}^{E}}{\partial N_1}\right) dN_1$$

$$= \tilde{R}T(N_1 + N_2 + \sum_{i=1}^{m} X_i N_{Xi}) \times [\chi_{13}^{0}\{1 + \sum_{q=1}^{n_q} \frac{p_{13,q}}{q+1} \left(\frac{v_p}{v_2 + v_p}\right)^{q}$$

$$\times \frac{1 - (v_2 + v_p)^{q+1}}{v_1}\}], \tag{3.11}$$

$$\Delta G_{23}^{E} = \int_0^{N_2} \left(\frac{\partial \Delta G_{23}^{E}}{\partial N_2}\right) dN_2$$

$$= \tilde{R}T(N_1 + N_2 + \sum_{i=1}^{m} X_i N_{Xi}) \times [\chi_{23}{}^0 \{1 + \sum_{r=1}^{n_r} \frac{p_{23,r}}{r+1} (\frac{v_p}{v_2 + v_p})^r$$

$$\times \frac{1 - (v_2 + v_p)^{r+1}}{v_2}\}] . \qquad (3.12)$$

Eqn (3.5') was put directly into a $\Delta G_{12}{}^E$ equation (eqn (3.13)) in order to satisfy symmetry with respect to the exchange of solvents 1 and 2,

$$\Delta G_{12}{}^E \equiv \tilde{R}TL\chi_{12} v_1 v_2 \qquad (3.13)$$

$$= \tilde{R}TL\chi_{12}{}^0 (1 + \sum_{s=1}^{n_s} p_{12,s} v_p{}^s) v_1 v_2 \qquad (3.13')$$

where $L = N_1 + N_2 + \Sigma X_i N_{Xi}$ and N_{Xi} is the number of X_i-mer. ΔG^{id} is expressed by eqn (3.14) (ref. 22)

$$\Delta G^{id} = \tilde{R}TL(N_1 + N_2 + \sum_{i=1}^{m} X_i N_{Xi})$$

$$\times [v_1 \ln v_1 + v_2 \ln v_2 + \sum_{i=1}^{m} \frac{v_{Xi}}{X_i} \ln v_{Xi}] . \qquad (3.14)$$

Substituting eqns (3.11)-(3.14) into eqn (3.8), ΔG is given by eqn (3.15)

$$\Delta G = \tilde{R}TL(N_1 + N_2 + \sum_{i=1}^{m} X_i N_{Xi})$$

$$\times [v_1 \ln v_1 + v_2 \ln v_2 + \sum_{i=1}^{m} \frac{v_{Xi}}{X_i} \ln v_{Xi} + \chi_{12}{}^0 (1 + \sum_{s=1}^{n_s} p_{12,s} v_p{}^s) v_1 v_2$$

$$+ \chi_{13}{}^0 \{1 + \sum_{q=1}^{n_q} \frac{p_{13,q}}{q+1} (\frac{v_p}{v_2 + v_p})^q \times \frac{1 - (v_2 + v_p)^{q+1}}{v_1}\} v_1 v_p$$

$$+ \chi_{23}{}^0 \{1 + \sum_{r=1}^{n_r} \frac{p_{23,r}}{r+1} (\frac{v_p}{v_1 + v_p})^r \times \frac{1 - (v_1 + v_p)^{r+1}}{v_2}\} v_2 v_p] . \qquad (3.15)$$

Differentiations of ΔG by N_1, N_2 and N_{Xi} give $\Delta\mu_1$, $\Delta\mu_2$ and

$\Delta\mu_{xi}$ (i= 1,⋯,m) as shown in eqns (3.16), (3.17) and (3.18), respectively (ref. 14).

$$\Delta\mu_1 = \tilde{R}T[\ln v_1 - (1-\frac{1}{X_n})v_p$$

$$+ \chi_{12}^0\{v_2(1-v_1) + \sum_{s=1}^{n_s} p_{12,s}v_2v_p^s(1-(s+1)v_1)\}$$

$$+ \chi_{13}^0v_p(1-v_1)\{1+\sum_{q=1}^{n_q} p_{13,q}v_p^q\} - \chi_{23}^0v_2v_p\{1+\sum_{r=1}^{n_r} p_{23,r}v_p^r$$

$$+ \sum_{r=1}^{n_r} p_{23,r}(\frac{r}{r+1})\frac{v_p^r}{(1-v_2)^{r+1}}(\frac{1-(v_1+v_p)^{r+1}}{v_2})\}], \qquad (3.16)$$

$$\Delta\mu_2 = \tilde{R}T[\ln v_2 - (1-\frac{1}{X_n})v_p$$

$$+ \chi_{12}^0\{v_1(1-v_2) + \sum_{s=1}^{n_s} p_{12,s}v_1v_p^s(1-(s+1)v_2)\}$$

$$+ \chi_{23}^0v_p(1-v_2)\{1+\sum_{r=1}^{n_r} p_{23,r}v_p^r\} - \chi_{13}^0v_1v_p\{1+\sum_{q=1}^{n_q} p_{13,q}v_p^q$$

$$+ \sum_{q=1}^{n_q} p_{13,q}(\frac{q}{q+1})\frac{v_p^q}{(1-v_1)^{q+1}}(\frac{1-(v_2+v_p)^{q+1}}{v_1})\}], \qquad (3.17)$$

$$\Delta\mu_{xi} = \tilde{R}T[\ln v_{xi} - (X_i-1) + X_i(1-\frac{1}{X_n})v_p$$

$$- X_i\chi_{12}^0v_1v_2\{1-\sum_{s=1}^{n_s} p_{12,s}sv_p^{s-1}(1-\frac{s+1}{s})\}$$

$$+ X_i\chi_{13}^0v_1\{1-v_p(1+\sum_{q=1}^{n_q} p_{13,q}v_p^q)$$

$$+ \sum_{q=1}^{n_q} p_{13,q}(1-(\frac{q}{q+1})\frac{v_p}{1-v_1})(\frac{v_p}{1-v_1})^q(\frac{1-(v_2+v_p)^{q+1}}{v_1})\}$$

$$+ X_i \chi_{23}{}^0 v_2 \{1 - v_p (1 + \sum_{r=1}^{n_r} p_{23,r} v_p{}^r)$$

$$+ \sum_{r=1}^{n_r} p_{23,r} (1 - (\frac{r}{r+1}) \frac{v_p}{1-v_p}) (\frac{v_p}{1-v_p})^r (\frac{1-(v_1+v_p)^{r+1}}{v_2}) \}]$$

$$(i = 1, \cdots, m) . \qquad (3.18)$$

The Gibbs-Duhem relation holds for $\Delta\mu_1$, $\Delta\mu_2$ and $\Delta\mu_{xi}$ $(i = 1, \cdots, m)$ (refs 11,14)

$$N_1 d(\Delta\mu_1) + N_2 d(\Delta\mu_2) + \sum_{i=1}^{m} N_{xi} d(\Delta\mu_{xi}) = 0. \qquad (3.19)$$

3.2 TWO-PHASE SEPARATION CHARACTERISTICS

3.2.1 Theoretical background

Conditions for two-phase equilibrium of the quasi-ternary system at a constant temperature and pressure are given by eqns (3.20) - (3.22)

$$\Delta\mu_{1(1)} = \Delta\mu_{1(2)}, \qquad (3.20)$$

$$\Delta\mu_{2(1)} = \Delta\mu_{2(2)}, \qquad (3.21)$$

$$\Delta\mu_{xi(1)} = \Delta\mu_{xi(2)}, \qquad (3.22)$$

where subscripts (1) and (2) mean the same as those in eqns (1.21) and (1.22). The partition coefficient σ is defined by:

$$\sigma = \frac{1}{X_i} \ln \frac{v_{xi(2)}}{v_{xi(1)}}. \qquad (3.23)$$

Combining of eqns (3.18), (3.22) and (3.23) gives

$$\sigma = (v_{p(1)} - v_{p(2)}) + (\frac{v_{p(2)}}{X_{n(2)}} - \frac{v_{p(1)}}{X_{n(1)}})$$

$$+ \chi_{12}{}^0 [(v_{1(2)} v_{2(2)} - v_{1(1)} v_{2(1)})$$

$$-\sum_{s=1}^{n_s} p_{12,s}s\{v_{1(2)}v_{2(2)}v_{p(2)}{}^{s-1}(1-\frac{s+1}{s}v_{p(2)})$$

$$-v_{1(1)}v_{2(1)}v_{p(1)}{}^{s-1}(1-\frac{s+1}{s}v_{p(1)})\}]$$

$$+\chi_{13}{}^{0}[(v_{1(2)}-v_{1(1)})-(v_{1(2)}v_{p(2)}-v_{1(1)}v_{p(1)})$$

$$-\sum_{q=1}^{n_q} p_{13,q}(v_{1(2)}v_{p(2)}{}^{q+1}-v_{1(1)}v_{p(1)}{}^{q+1})$$

$$+\sum_{q=1}^{n_q}\frac{p_{13,q}}{q+1}\{v_{1(2)}(q+1-\frac{qv_{p(2)}}{1-v_{1(2)}})$$

$$\times(\frac{v_{p(2)}}{1-v_{1(2)}})^{q}\frac{1-(v_{2(2)}+v_{p(2)})^{q+1}}{v_{1(2)}}$$

$$-v_{1(1)}(q+1-\frac{qv_{p(1)}}{1-v_{1(1)}})(\frac{v_{p(1)}}{1-v_{1(1)}})^{q}$$

$$\times\frac{1-(v_{2(1)}+v_{p(1)})^{q+1}}{v_{1(1)}}\}]$$

$$+\chi_{23}{}^{0}[(v_{2(2)}-v_{2(1)})-(v_{2(2)}v_{p(2)}-v_{2(1)}v_{p(1)})$$

$$-\sum_{r=1}^{n_r} p_{23,r}(v_{2(2)}v_{p(2)}{}^{r+1}-v_{2(1)}v_{p(1)}{}^{r+1})$$

$$+\sum_{r=1}^{n_r}\frac{p_{23,r}}{r+1}\{v_{2(2)}(r+1-\frac{rv_{p(2)}}{1-v_{2(2)}})$$

$$\times(\frac{v_{p(2)}}{1-v_{2(2)}})^{r}\frac{1-(v_{1(2)}+v_{p(2)})^{r+1}}{v_{2(2)}}$$

$$-v_{2(1)}(r+1-\frac{rv_{p(1)}}{1-v_{2(1)}})(\frac{v_{p(1)}}{1-v_{2(1)}})^{r}$$

$$\times\frac{1-(v_{1(1)}+v_{p(1)})^{r+1}}{v_{2(1)}}\}]. \qquad (3.24)$$

Substitution of eqn (3.16) into eqn (3.20) yields eqn (3.25) (the lefthand side of eqn (3.25) is labeled A):

$$A \equiv \ln\frac{v_{1(2)}}{v_{1(1)}} + (v_{p(2)} - v_{p(1)}) + (\frac{v_{p(2)}}{X_{n(2)}} - \frac{v_{p(1)}}{X_{n(1)}})$$

$$+ \chi_{12}^{0}[(v_{2(2)} - v_{2(1)}) - (v_{1(2)}v_{2(2)} - v_{1(1)}v_{2(1)})$$

$$- \sum_{s=1}^{n_s} p_{12,s}\{(v_{2(2)}v_{p(2)}{}^{s} - v_{2(1)}v_{p(1)}{}^{s})$$

$$- (s+1)(v_{1(2)}v_{2(2)}v_{p(2)}{}^{s} - v_{1(1)}v_{2(1)}v_{p(1)}{}^{s})\}]$$

$$+ \chi_{13}^{0}[(v_{p(2)} - v_{p(1)}) - (v_{1(2)}v_{p(2)} - v_{1(1)}v_{p(1)})$$

$$- \sum_{q=1}^{n_q} p_{13,q}\{v_{p(2)}{}^{q+1}(1 - v_{1(2)}) - v_{p(1)}{}^{q+1}(1 - v_{1(1)})\}]$$

$$+ \chi_{23}^{0}[(v_{2(2)}v_{p(2)} - v_{2(1)}v_{p(1)})$$

$$- \sum_{r=1}^{n_r} p_{23,r}(v_{2(2)}v_{p(2)}{}^{r+1} - v_{2(1)}v_{p(1)}{}^{r+1})$$

$$+ \sum_{r=1}^{n_r} p_{23,r}(\frac{r}{r+1})\{v_{2(2)}(\frac{v_{p(2)}}{1 - v_{2(2)}})^{r+1}$$

$$\times \frac{1 - (v_{1(2)} + v_{p(2)})^{r+1}}{v_{2(2)}} - v_{2(1)}(\frac{v_{p(1)}}{1 - v_{p(1)}})^{r+1}$$

$$\times \frac{1 - (v_{1(1)} + v_{p(1)})^{r+1}}{v_{2(1)}}\}] = 0. \qquad (3.25)$$

Substitution of eqn (3.17) into eqn (3.21) gives eqn (3.26) (the lefthand side of eqn (3.26) is labeled B):

$$B \equiv \ln\frac{v_{2(2)}}{v_{2(1)}} + (v_{p(2)} - v_{p(1)}) - (\frac{v_{p(2)}}{X_{n(2)}} - \frac{v_{p(1)}}{X_{n(1)}})$$

$$+ \chi_{12}^{0}[(v_{1(2)} - v_{1(1)}) - (v_{1(2)}v_{2(2)} - v_{1(1)}v_{2(1)})$$

$$+\sum_{s=1}^{n_s} p_{12,s}\{(v_{1(2)}v_{p(2)}{}^{s}-v_{1(1)}v_{p(1)}{}^{s})$$

$$-(s+1)(v_{1(2)}v_{2(2)}v_{p(2)}{}^{s}-v_{1(1)}v_{2(1)}v_{p(1)}{}^{s}\}]$$

$$+\chi_{23}{}^{0}[(v_{p(2)}-v_{p(1)})-(v_{2(2)}v_{p(2)}-v_{2(1)}v_{p(1)})$$

$$+\sum_{r=1}^{n_r} p_{23,r}\{v_{p(2)}{}^{r+1}(1-v_{2(2)})-v_{p(1)}{}^{r+1}(1-v_{2(1)})\}]$$

$$-\chi_{13}{}^{0}[(v_{1(2)}v_{p(2)}-v_{1(1)}v_{p(1)})$$

$$+\sum_{q=1}^{n_q} p_{13,q}(v_{1(2)}v_{p(2)}{}^{q+1}-v_{1(1)}v_{p(1)}{}^{q+1})$$

$$+\sum_{q=1}^{n_q} p_{13,q}\left(\frac{q}{q+1}\right)\{v_{1(2)}\left(\frac{v_{p(2)}}{1-v_{p(2)}}\right)^{q+1}$$

$$\times\frac{1-(v_{2(2)}+v_{p(2)})^{q+1}}{v_{1(2)}}-v_{1(1)}\left(\frac{v_{p(1)}}{1-v_{p(1)}}\right)^{q+1}$$

$$\times\frac{1-(v_{2(1)}+v_{p(1)})^{q+1}}{v_{1(1)}}\}]=0. \qquad (3.26)$$

If both (a) $\chi_{12}{}^{0}$, $\chi_{13}{}^{0}$ and $\chi_{23}{}^{0}$ and (b) $p_{12,s}$, $p_{13,q}$ and $p_{23,r}$ ($s=1,\cdots,n_s$; $q=1,\cdots,n_q$; $r=1,\cdots,n_r$) are given in advance, σ, A and B are functions having six variables, $X_{n(1)}$, $X_{n(2)}$, $v_{2(1)}$, $v_{2(2)}$, $v_{p(1)}$ and $v_{p(2)}$.

If one dissolves a polymer in solvent (solvent 1) and then adds nonsolvent (solvent 2), two-phase separation finally occurs as is shown in Fig. 3.1. If volumes of solvent 1, 2 and polymer are $V_1{}^0$, V_2 and $V_3{}^0$, respectively, the volume of the starting solution V_0 is given by $V_1{}^0+V_3{}^0$ and the starting concentration $v_p{}^s$ is given by

$$v_p{}^s=\frac{V_3{}^0}{V_0}=\frac{V_3{}^0}{V_1{}^0+V_3{}^0}. \qquad (3.27)$$

With two-phase equilibrium, the total volume of the system is $V_1{}^0+V_2+V_3{}^0$ (= V) and the initial concentration $v_p{}^0$ is expressed by

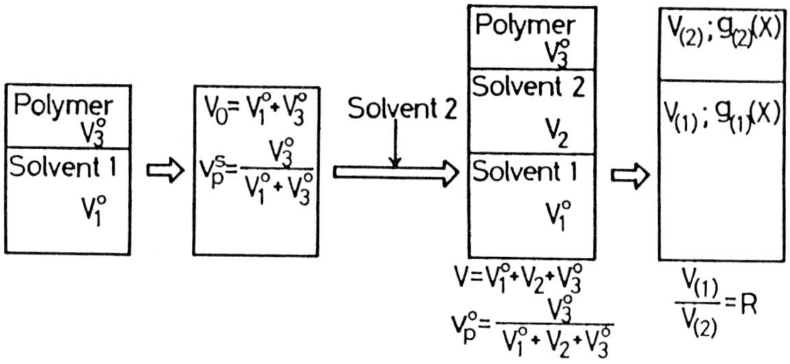

Fig. 3.1 Phase equilibrium experiment for a quasi-ternary system; $V_1{}^0$, volume of solvent 1; V_2, volume of solvent 2; $V_3{}^0$, volume of polymer; $V_0=V_1{}^0+V_3{}^0$; $v_p{}^s$, polymer volume fraction of the starting solution; V, total volume of the solution at phase equilibrium; $v_p{}^0$, polymer volume fraction of the solution at phase equilibrium; R, phase volume ratio, $V_{(1)}$, volume of polymer-lean phase; $V_{(2)}$, volume of polymer-rich phase; $g_{(1)}(X)$, MWD of the polymer in the polymer-lean phase; $g_{(2)}(X)$, MWD of the polymer in the polymer-rich phase.

$$v_p{}^0=\frac{V_3{}^0}{V}=\frac{V_3{}^0}{V_1{}^0+V_2+V_3{}^0}. \tag{3.28}$$

As is in the case of the quasi-binary system, $g_0(X_i)$ is a summation of $g_{(1)}(X_i)$ and $g_{(2)}(X_i)$ (see, eqn (2.40)):

$$g_0(X_i)=g_{(1)}(X_i)+g_{(2)}(X_i). \tag{3.29}$$

The weight fraction ρ_s of the polymer in the polymer-lean phase to the total polymer and the fraction $\rho_p(=1-\rho_s)$ of the polymer in the polymer-rich phase are given by eqns (2.30a) and (2.30b) (see, eqn (2.41)).

$$\rho_s=\sum_{i=1}^{m} g_{(1)}(X_i), \tag{3.30a}$$

$$\rho_p=\sum_{i=1}^{m} g_{(2)}(X_i). \tag{3.30b}$$

Using σ, R and $g_0(X_i)$, $g_{(1)}(X_i)$ and $g_{(2)}(X_i)$ are expressed as (see, eqn (2.42)):

$$g_{(1)}(X_i) = \frac{R}{R + \exp(\sigma X_i)} g_0(X_i), \tag{3.31a}$$

$$g_{(2)}(X_i) = \frac{\exp(\sigma x_i)}{R + \exp(\sigma X_i)} g_0(X_i). \tag{3.31b}$$

$v_{p(1)}$ and $v_{p(2)}$ are finally expressed by (refs 11,14)

$$v_{p(1)} = v_p{}^s \frac{R + 1 - Rv_{2(1)} - v_{2(2)}}{R} \rho_s, \tag{3.32a}$$

$$v_{p(2)} = v_p{}^s (R + 1 - Rv_{2(1)} - v_{2(2)}) \rho_p \tag{3.32b}$$

and $X_{n(2)}$ can also be expressed by eqn (2.45). If ρ_p is set as the initial condition $(= \rho_p{}^g;$ given $\rho_p)$, σ, A and B become the functions of four variables $v_{2(1)}$, $v_{2(2)}$, R_a and $X_{n(2)}{}^a$ (refs 11,14).

$$\sigma = \sigma(v_{2(1)}, V_{2(2)}, R^a, X_{n(2)}{}^a), \tag{3.33}$$

$$A = A(v_{2(1)}, v_{2(2)}, R^a, X_{n(2)}{}^a) = 0, \tag{3.34}$$

$$B = B(v_{2(1)}, v_{2(2)}, R^a, X_{n(2)}{}^a) = 0, \tag{3.35}$$

where R^a and $X_{n(2)}{}^a$ are the assumed values of R and $X_{n(2)}$. $X_{n(2)}$ finally becomes the function of σ and R^a. We define C and D by eqns (3.36) and (3.37), respectively.

$$C \equiv \rho_p(\sigma(v_{2(1)}, v_{2(2)}, R, X_{n(2)}{}^a), R_a) - \rho_p{}^g = 0, \tag{3.36}$$

$$D \equiv X_{n(2)}(\sigma(v_{2(1)}, v_{2(2)}, R^a, X_{n(2)}{}^a), R^a) - X_{n(2)}{}^a = 0. \tag{3.37}$$

By solving non-linear simultaneous eqns (3.34)-(3.37), $v_{2(1)}$, $v_{2(2)}$, R and $X_{n(2)}$ are determined. Substituting these four values into eqn (3.33), we can calculate σ and other phase separation characteristics.

Computer experiments were carried out according to the procedure, established by Kamide et al. (refs 11-14).

1. As prerequisites, (a) $\chi_{12}{}^0$, $\chi_{13}{}^0$ and $\chi_{23}{}^0$, (b) $p_{12,s}$, $p_{13,q}$ and $p_{23,r}$ $(s = 1, \cdots, n_s;$ $q = 1, \cdots, n_q;$ $r = 1, \cdots, n_r)$, (c) $V_1{}^0$, $V_3{}^0$, $g_0(X_i)$ $(i = 1, \cdots, m)$ and $\rho_p (= 1 - \rho_p)$ are given.
2. At first, $X_{n(2)}{}^a$ and three values of R^a (low, middle and high

value of R^a; R_L, R_M and R_H, respectively) are assumed. True R should be between R_L and R_M or R_M and R_H.

3. For the assumed $X_{n(2)}$ and R_a, simultaneous eqns (3.38) and (3.39),

$$A = A(v_{2(1)}, v_{2(2)}) = 0, \tag{3.38}$$

$$B = B(v_{2(1)}, v_{2(2)}) = 0 \tag{3.39}$$

are solved by using the two variables Newton method.

4. Substituting the three sets of $(v_{2(1)}, v_{2(2)})$ which correspond to R_L, R_M and R_H, into eqn (3.36), we get C_L, C_M and C_H. Reset (R_L, R_H) by (R_L, R_M) for $C_L \times C_M < 0$ and by (R_M, R_H) for $C_M \times C_H < 0$. σ_L, σ_M and σ_H are also obtained.
5. Using the interhalving method (repetition of step 3-4),

$$C = (R^a) = 0 \tag{3.40}$$

can be attained. Thus, R and σ (accordingly, $g_{(1)}(X_i)$ and $g_{(2)}(X_i)$) are determined for the assumed value of $X_{n(2)}$.

6. When D (given by eqn (3.37)) $\neq 0$ for calculated σ and R, replace $X_{n(2)}{}^a$ by $(X_{n(2)} + X_{n(2)}{}^a)/2$.
7. Repeating step 3-6 (interhalving method), the following equation can be solved for $X_{n(2)}{}^a$;

$$D = D(X_{n(2)}{}^a) = 0 \tag{3.41}$$

and therefore an equilibrium state is finally determined.

8. When a constant initial polymer volume fraction ($v_p{}^0$ = constant, $v_p{}^{0g}$) is needed, we solve the following equation

$$v_p{}^0(v_{2(1)}, V_{2(2)}, R^a, X_{n(2)}{}^a, V_1{}^0) - v_p{}^{0g} = 0 \tag{3.42}$$

by replacing given $V_1{}^0$ by $V_1{}^0$, satisfying eqn (3.42), using the interhalving method (repetition of step 1-7).

9. Compute other phase separation characteristics (a) $V_?$, V, $V_{(1)}$ and $V_{(2)}$, (b) $v_{1(1)}$, $v_{1(2)}$, $v_{p(1)}$ and $v_{p(2)}$ and (c) $X_{n(1)}$, $X_{w(1)}$, $X_{w(2)}$ and so on.

This simulation procedure is superior to others (refs 1-7,16) in that:

(i) The MWD of polymers in both phases can be obtained directly, in addition to the volume fractions of solvents and a polymer.

(ii) ρ_p can be determined. For example, Altena and Smolders (ref. 16) did not write clearly about ρ_p in their computer experiments on the phase separation of monodisperse polymer / binary solvent mixture.

(iii) Calculations under constant v_p^0 are feasible. Constant v_p^0 is important to compare the quasi-ternary system with the quasi-binary system. Fig. 3.2 shows the main flow chart for simulation.

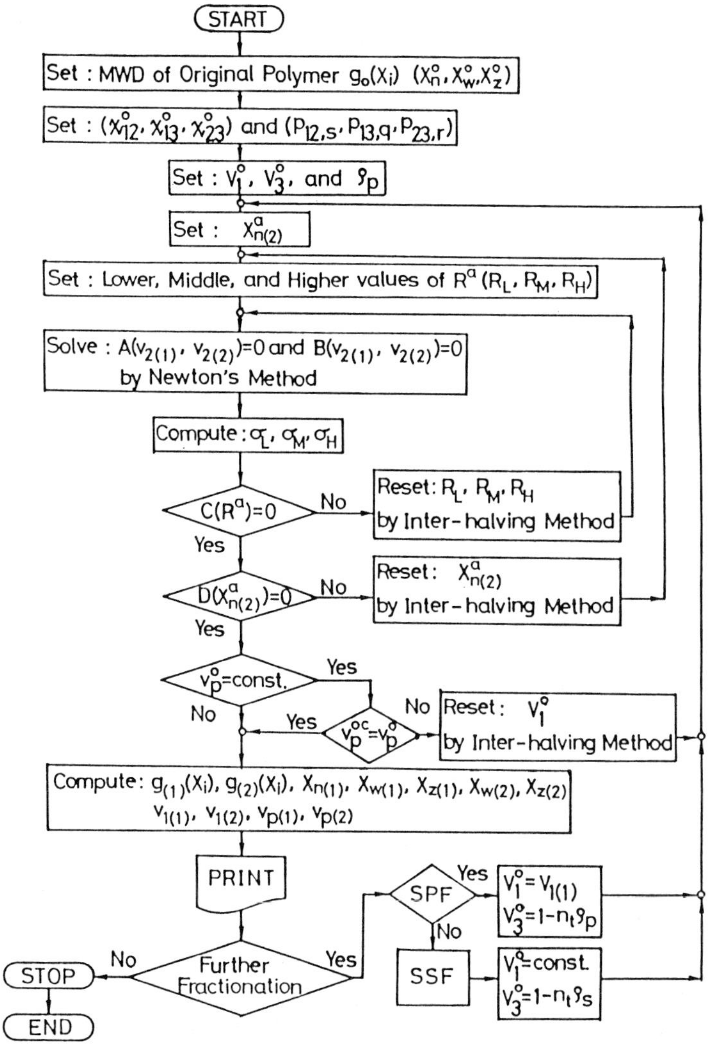

Fig. 3.2 Flow chart of the simulation.

The original polymer was assumed to have the Schulz-Zimm (SZ) type MWD (see, eqns (2.59)-(2.61)) with $X_w{}^0 = 300$ and $X_w{}^0/X_n{}^0 = 2$. Kamide et al. (ref. 13) ascertained that the conclusion obtained for the SZ type polymer with $X_w{}^0 = 300$ has a very general character applicable for common polymer solutions. The calculations were made under the following conditions: (a) $\chi_{12}{}^0 = 0.5$, $\chi_{13}{}^0 = 0.2$, and $\chi_{23}{}^0 = 1.0$, (b) $p_{12,1} = -1.0 \sim 2.0$, $p_{13,1} = -0.6 \sim 0.6$ and $p_{23,1} = -0.15 \sim 0.15$ ($p_{12,s} = p_{13,q} = p_{23,r} = 0$ for $s,q,r \geq 2$), (c) $\rho_p = 1/100 \sim 99/100$, $V_3{}^0 = 1$ and $v_p{}^0 = 0.005$. For comparison, calculations of a quasi-binary system, under conditions $p_1 = -0.6$, 0 and 0.6 ($p_j = 0$, $j \geq 2$), $V_1{}^0 = 1$ and $v_p{}^0 = 0.005$, were also carried out. When coexisting curves were calculated, we put $v_p{}^s = 0.01$ in place of $v_p{}^0 = 0.005$. Eqns (3.40)-(3.42) were considered to be solved, if $|C(R^a)| < E_1$, $|D(X_{n(2)}{}^a)| < E_2$, and $|(v_p{}^0 - v_p{}^{0g})/v_p{}^{0g}| < E_3$ were satisfied. E_1, E_2 and E_3 are allowance errors. Here, $E_1 = 0.001$, $E_2 = 0.01$ and $E_3 = 0.001$ are adopted.

3.2.2 Suitable choice of solvent 1 and solvent 2

In actual experiments, we often find that not just any combination of a solvent and a nonsolvent brings about two-phase equilibrium at a given temperature under a given pressure and that the proper choice of solvent (i.e., solvent 1) and the nonsolvent (i.e., solvent 2) is very important (ref. 11). For example, consider a case in which a solvent power of solvent 1 is weak (i.e., χ_{13} is relatively large) and solvent 1 and 2 are immiscible (i.e., χ_{12} is larger than 2.0), the polymer dissolved in solvent 1 does not completely precipitate on adding an infinite amount of solvent 2 (ref. 11).

In Fig. 3.3, the shadowed area denotes the possible combination of χ_{12}, χ_{13} and χ_{23}, by which phase separation might occur under the given conditions. When $\rho_p = 1/15$ and $v_p{}^s = 0.01$ (Fig. 3.3a), χ_{23} are at least larger than 0.65. Otherwise, no phase separation will occur due to insufficient precipitation capacity of solvent 2. The choice of solvent 1 and 2 is easier if χ_{12} is larger (i.e., solvent 1 has less thermodynamic affinity toward solvent 2). That is, as χ_{13} increases with a decrease in solvent power, the maximum χ_{23} value attainable becomes smaller. In this case, only a nonsolvent with weak precipitation power should be allowed to combine with solvent 1. As χ_{13} becomes smaller (i.e., the solvent power increases), phase separation will occur over a wide range of χ_{23}, from weak to strong precipitation capacity. When $\rho_s = 1/15$ (Fig. 3.3b), the

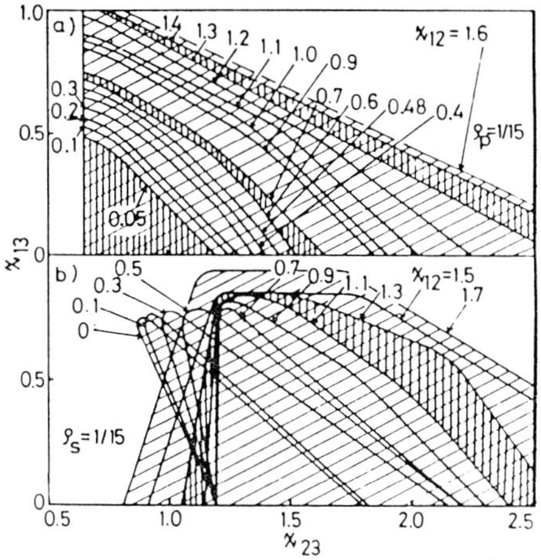

Fig. 3.3 Allowable combination of three χ parameters, χ_{12}, χ_{13} and χ_{23} for given conditions: The original polymer, $X_w^0 = 300$, $X_w^0/X_n^0 = 2.0$; $v_p^s = 0.01$; $\rho_p = 1/15$ in a), $\rho_s = 1/15$ in b); shadowed area denotes allowed area, $p_{12,s} = p_{13,q} = p_{23,r} = 0$.

lower limit of χ_{23}, especially in the higher χ_{12} range, is much larger than 0.8. As expected intuitively, when either the solvent power is strong (smaller χ_{13}) or the precipitation power is weak (smaller χ_{23}), the solution will not separate into the two liquid phases. At $\chi_{12} = 0.7$, the theoretically possible lower limit of χ_{23} is 1.2, regardless of χ_{12} and this value is a maximum of the lower limits permissible for all other χ_{12} values. As is the case of $\rho_p = 1/15$, large χ_{12} value affords large permissible χ_{13} and χ_{23} (ref. 11).

Fig. 3.4 illustrates the effects of χ_{12}, χ_{13} and χ_{23} on the relation between ρ_p and total volume V. In this case, $V_1^0 = 100$ and $V_3^0 = 1$ (accordingly, $v_p^s = 0.01$) are assumed. As χ_{12} increases, the amount of solvent 2 needed for a given ρ_p increases. But all V vs. ρ_p relations are superposable by shifting themselves along the horizontal axes. The effect of χ_{13} on this relation is just the reverse of that of χ_{12}. As χ_{23} decreases, the maximum ρ_p value decreases and V becomes large. To precipitate a given relative amount of polymer using as small an amount of solvent 2 as possible, it is necessary to use a combination of solvent 1 and 2 having small χ_{12}, large χ_{13} and large χ_{23}. When the mutual miscibility of solvents 1 and 2 is bad and the solvent power of solvent 1 is large and the precipitation capacity of solvent 2 is weak, we should add a large amount of

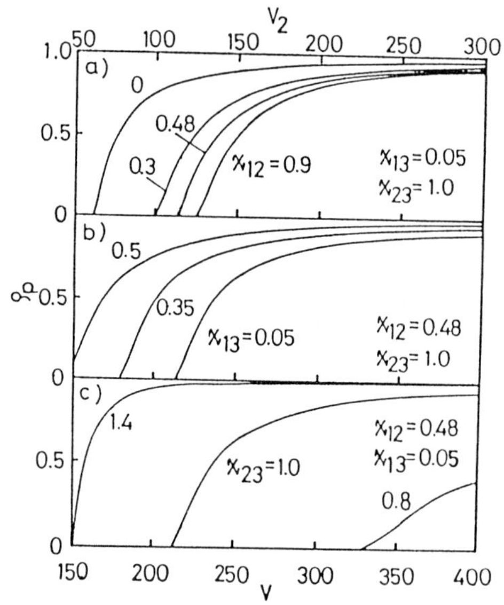

Fig. 3.4 Plot of the relative amount of the polymer in the polymer-rich phase ρ_P against the total volume V and the volume of solvent 2: The original polymer, Schulz-Zimm distr., $X_w{}^0=300$, $X_w{}^0/X_n{}^0=2.0$; $v_p{}^s=0.01$; a) $\chi_{13}=0.05$, $\chi_{23}=1.0$; b) $\chi_{12}=0.48$, $\chi_{23}=1.0$; c) $\chi_{12}=0.48$, $\chi_{13}=0.05$; χ_{13} is shown on the curves.

solvent 2 to the quasi-binary solution of the polymer and solvent 1, in order to bring about phase separation. Among three χ parameters, χ_{23} plays an important role in controlling V (ref. 11). In a quasi-ternary system, ρ_P is controlled by adding solvent 2 to the quasi-binary system (see, Fig. 3.4). As a result, the total volume increases unavoidably for larger ρ_P, approaching the experimental limit accessible, if the same solvent 1 / solvent 2 pair is employed. For a quasi-ternary system, there are numerous combinations of solvent 1 and 2, enabling us to separate a polymer-rich phase effectively over a wide range of ρ_P. Fig. 3.4 demonstrates that when we try to carry out a molecular weight fractionation based on the phase equilibrium phenomena of the quasi-ternary system, solvent 2 should be chosen carefully. For SPF, a small χ_{23} (and, if possible, large χ_{13}) is favorable and for SSF, a large χ_{23} is desirable for keeping total volume as small as possible. It is very rare case that when a drop of solvent 2 is added to the polymer / solvent 1 system, the solution becomes instantly turbid, indicating the occurrence of a phase separation. Usually, after a measurably large amount of solvent 2

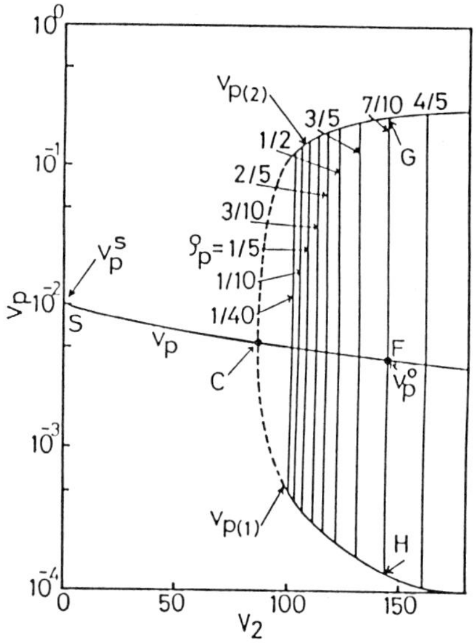

Fig. 3.5 Change in the polymer volume fraction v_p with the addition of solvent 2 (volume of solvent 2, V_2) in the phase equilibrium: v_p^s, the polymer volume fraction of the starting solution S; F, see the text. The original polymer, Schulz-Zimm distr., $X_w^0=300$, $X_w^0/X_n^0=2.0$; $v_p^s=0.01$; $\chi_{12}=0.48$, $\chi_{13}=0.20$ and $\chi_{23}=1.00$.

is added, the cloud point of the solution (point C in Fig. 3.5) is observed. Further addition of solvent 2 rapidly increases the relative amount of the polymer-rich phase (R), as shown in Fig. 3.5. In this figure, the point S is the composition of a starting solution whose polymer volume fraction is v_p^s and by adding solvent 2, the polymer volume fraction v_p changes along the full line leading to the cloud point C. Further addition of solvent 2 after the cloud point makes ρ_p large and, for example, at point F phase equilibrium is attained. A fine line passing through F is a tie-line connecting the polymer-rich phase G and the polymer-lean phase H. The polymer volume fraction at F is defined as an initial polymer volume fraction (i.e., initial "concentration") and is denoted by v_p^0. In the small ρ_p region, an approximation of $v_{p(1)}=0$ should not be employed when the detailed phase separation characteristics are to be evaluated.

Fig. 3.6 shows the effects of χ_{12}, χ_{13} and χ_{23} on $v_{p(1)}$. The broken lines in the figures are the limit of the phase separation, occurring theoretically under given conditions. The full lines give the same $v_{p(1)}$ value. Regardless of χ_{12} and χ_{13}, $v_{p(1)}$ is strongly controlled by χ_{23} (ref. 11).

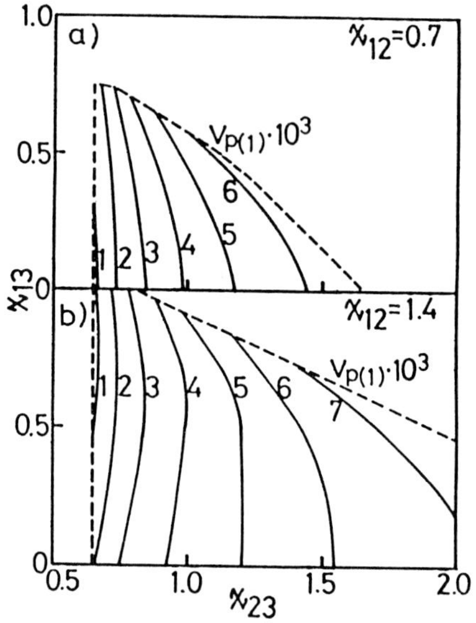

Fig. 3.6 Relationships among χ_{13}, χ_{23} and the polymer volume fraction in a polymer-lean phase $v_{p(1)}$ at given χ_{12} [0.7 in a) and 1.4 in b)]: In the area surrounded by broken lines, a phase equilibrium under given conditions is theoretically possible. The original polymer, Schulz-Zimm distr., $X_w^0 = 300$, $X_w^0/X_n^0 = 2.0$; $v_p^s = 0.01$; $\rho_p = 1/15$; $p_{12,s} = p_{13,q} = p_{23,r} = 0$.

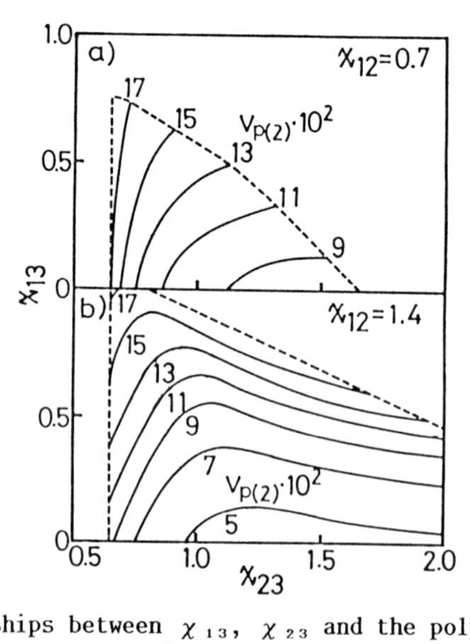

Fig. 3.7 Relationships between χ_{13}, χ_{23} and the polymer volume fraction in a polymer-rich phase $v_{p(2)}$ at given χ_{12} [0.7 in a) and 1.4 in b)]: In the area surrounded by broken line, a phase equilibrium can theoretically occur under given conditions. The original polymer, Schulz-Zimm distr., $X_w^0 = 300$, $X_w^0/X_n^0 = 2.0$; $v_p^s = 0.01$; $\rho_p = 1/15$; $v_{p(2)}$ is denoted on the curve; $p_{12,s} = p_{13,q} = p_{23,r} = 0$.

Similar effects of χ_{13} and χ_{23} on $v_{P(2)}$ are shown in Fig. 3.7. In the range of $\chi_{23} < 1.0$, $v_{P(2)}$ is mainly determined by χ_{23} and in the range of $\chi_{23} > 1.0$, $v_{P(2)}$ is strongly χ_{13}-dependent. The broken line shows the limit within which the phase separation under given conditions is theoretically possible.

From Fig. 3.7, it is concluded that $v_{P(2)}$ becomes larger when the mutual miscibility of solvent 1 and 2 is high and the affinity of solvent 1 to the polymer is weak and the precipitation capacity of solvent 2 is weak.

Fig. 3.8 shows the effects of χ_{12}, χ_{13} and χ_{23} on σ and R. Fig. 3.8a and c correspond to the case of $\chi_{12} = 0.7$ ($\rho_P = 1/15$) and Fig. 3.8b and d to $\chi_{12} = 1.4$ ($\rho_P = 1/15$). In the former case,

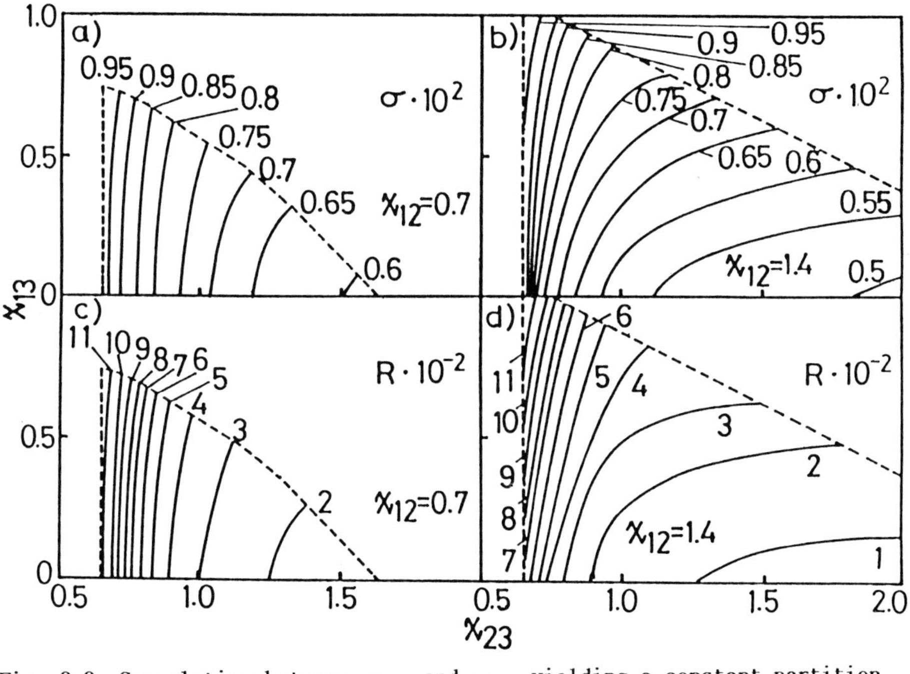

Fig. 3.8 Correlation between χ_{13} and χ_{23} yielding a constant partition coefficient σ or phase volume ratio R at given χ_{12} [0.7 in a) and c), 1.4 in b) and d)]: The original polymer, Schulz-Zimm distr., $X_w^0 = 300$, $X_w^0/X_n^0 = 2.0$; $\rho_P = 1/15$; $v_P^s = 0.01$; $p_{12,s} = p_{13,q} = p_{23,r} = 0$.

both σ and R are predominantly controlled by χ_{23}, regardless of χ_{13}. Smaller χ_{23} yields larger σ and larger R, and hence, the careful choice of a not-so poor solvent is strongly recommended for effective fractionation. In the latter case ($\chi_{12} = 1.4$ and

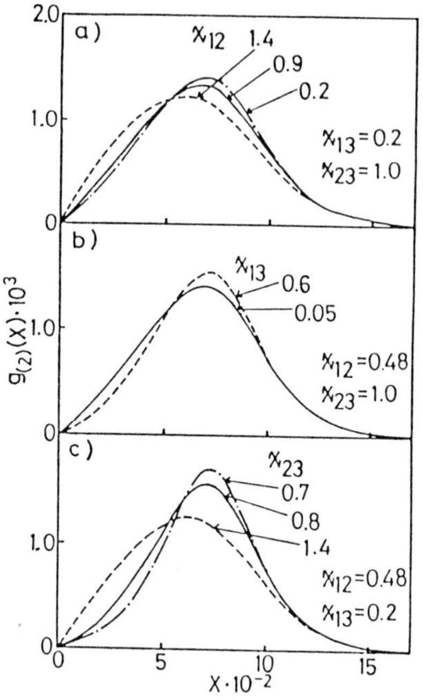

Fig. 3.9 Effect of χ_{12}, χ_{13} and χ_{23} parameters on the normalized molecular weight distribution curve of the polymer-rich phase $g_{(2)}(X)$, separated from a solution ($v_p{}^s = 0.01$) of the polymer with Schulz-Zimm distr., ($X_w{}^0 = 300$, $X_w{}^0/X_n{}^0 = 2.0$): $\rho_p = 1/15$; a) $\chi_{13} = 0.2$, $\chi_{23} = 1.0$; b) $\chi_{12} = 0.48$, $\chi_{23} = 1.0$; c) $\chi_{12} = 0.48$, $\chi_{13} = 0.2$; $p_{12,s} = p_{13,q} = p_{23,r} = 0$.

$\rho_p = 1/15$), in the range of $\chi_{23} < 1.0$ both σ and R are χ_{23}-dependent, but in the range of $\chi_{23} > 1.0$ σ together with R depend on χ_{23} as well as χ_{13}. To make σ and R as large as possible, it is better to choose solvent 1 and 2 with large χ_{13} and small χ_{23} (i.e., not a very good solvent and a not-so poor solvent). The use of solvent 2 with extremely strong precipitation capacity should be avoided from the standpoint of separation efficiency (ref. 11).

Fig. 3.9 shows the effects of χ_{12}, χ_{13} and χ_{23} on the normalized MWD curve $g_{(2)}(X)$ of the polymer in a polymer-rich phase at $\rho_p = 1/15$. A combination of solvent 1 and 2 with smaller χ_{12}, larger χ_{13} and smaller χ_{23} yields a polymer with a sharp distribution still in a polymer-rich phase. The effects of χ_{13} on $g_{(2)}(X)$ are negligible, but χ_{23} has a powerful effect on $g_{(2)}(X)$ (ref. 11).

Fig. 3.10 shows the effects of χ_{13} and χ_{23} on the ratio $X_{w(2)}/X_{n(2)}$ and the standard deviation of $g_{(2)}(X)$, $\sigma'_{(2)}$. Both

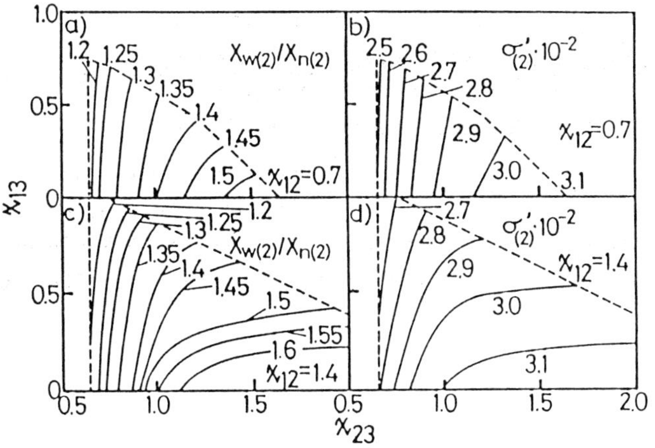

Fig. 3.10 Correlation between χ_{13} and χ_{23} yielding constant $X_{w(2)}/X_{n(2)}$ or the standard deviation $\sigma_{(2)}$ in the polymer-rich phase at given χ_{12} [0.7 in a) and b), 1.4 in c) and d)]: The original polymer, Schulz-Zimm distr., $X_w^0 = 300$, $X_w^0/X_n^0 = 2.0$, $\sigma_0' = 212.2$; $\rho_P = 1/15$, $v_P^S = 0.01$.

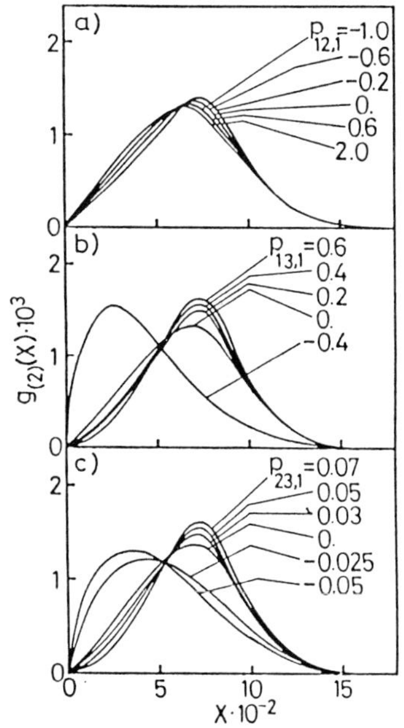

Fig. 3.11 Effects of the 1st order concentration-dependence parameters $p_{12,1}$, $p_{13,1}$ and $p_{23,1}$ on the normalized molecular weight distribution of the polymer partitioned in polymer-rich phase $g_{(2)}$: Original polymer, Schulz-Zimm type distribution ($X_w^0 = 300$, $X_w^0/X_n^0 = 2$); $\chi_{12}^0 = 0.5$, $\chi_{13}^0 = 0.2$ and $\chi_{23}^0 = 1.0$, $\rho_P = 1/15$ and $v_P^0 = 0.005$. a) $p_{13,1} = p_{23,1} = 0$, b) $p_{12,1} = p_{23,1} = 0$, c) $p_{12,1} = p_{13,1} = 0$.

parameters represent the breadth of the MWD of the polymer in a polymer-rich phase. In the region where χ_{23} is small, the polydispersity of the polymer remaining in a polymer-rich phase is χ_{23}-dependent, being smaller as χ_{23} decreases. In preparing polymer fractions by isolating the polymers from a polymer-rich phase to be separated from a quasi-ternary system, a not-so poor solvent should be used as solvent 2.

Fig. 3.11 shows the effects of $p_{12,1}$, $p_{13,1}$ and $p_{23,1}$ on the normalized $g_{(2)}(X)$ at $\rho_p = 1/15$. $p_{12,1}$ was found to have a small but still significant effect on $g_{(2)}(X)$ and the breadth in $g_{(2)}(X)$ attains minimum at $p_{12,1} = -0.2$ when $p_{13,1} = p_{23,1} = 0$. The polydispersity of the polymer in a polymer-rich phase becomes lower with an increase in $p_{13,1}$ or $p_{23,1}$, at least in the range of $p_{13,1} \leq 0.6$ and $p_{23,1} \leq 0.07$. Particularly, a small change in $p_{23,1}$ brings about a large change in $g_{(2)}(X)$ (ref. 14). The effect of the concentration dependence of χ parameters on $g_{(2)}(X)$ decreases in the following order: $p_{23,1} > p_{13,1} > p_{12,1}$.

3.2.3 Role of initial concentration and relative amount of polymers partitioned in two phases

Fig. 3.12 shows the relationship among σ, R and v_p^0. The full line in the figure corresponds to a given combination of three χ parameters. Fig. 3.12 also shows the data (as the broken line) for quasi-binary solutions, in which the molecular characteristics of the original polymer are the same as those used for quasi-ternary solutions and the number on each line is a p_1 parameter defined in eqn (2.3). The relations between χ_0 and v_p^0 for quasi-binary solutions have already been discussed in detail by Kamide and Sugamiya (ref. 23). The σ as well as R decreases with an increase in v_p^0. As χ_{12} and χ_{23} become smaller and χ_{13} larger, both σ and R increase. For any combination of χ_{12}, χ_{13} and χ_{23}, the same σ vs. v_p^0 relations as those for quasi-binary solutions assuming $p_1 = 0$ can not be obtained. That is, for given v_p^0, the quasi-ternary system always affords smaller σ and R than does the quasi-binary system. The lines for the quasi-binary system with $p_1 = 0$ can be regarded as asymptotic lines attainable by the quasi-ternary system. The higher fractionation efficiency for a quasi-binary system was experimentally demonstrated by Kamide et al. (ref. 24) using PS (see, section 2.4).

Fig. 3.13 shows the effects of ρ_p on σ and R for the quasi-ternary (full line) and -binary (broken line) solutions of $v_p^0 = 0.005$ of the polymer having the SZ distribution (X_w^0 300 and

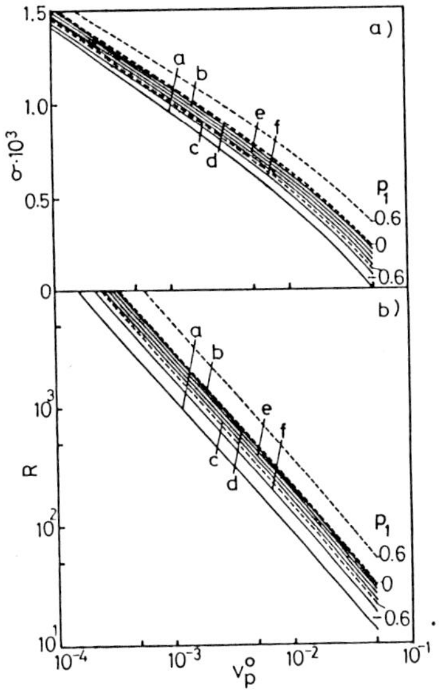

Fig. 3.12 Effects of the χ_{12}, χ_{13} and χ_{23} parameters on the σ or R vs. the initial polymer volume fraction $v_p{}^0$ relationship: The original polymer, Schulz-Zimm distr., $X_w{}^0 = 300$, $X_w{}^0/X_{n0} = 2.0$; $\rho_p = 1/15$; a, $\chi_{12} = 1.30$, $\chi_{13} = 0.20$, $\chi_{23} = 1.0$; b, $\chi_{12} = 0.05$, $\chi_{13} = 0.20$, $\chi_{23} = 1.0$; c, $\chi_{12} = 0.48$, $\chi_{13} = 0.05$, $\chi_{23} = 1.0$; d, $\chi_{12} = 0.48$, $\chi_{13} = 0.50$, $\chi_{23} = 1.0$; e, $\chi_{12} = 0.48$, $\chi_{13} = 0.20$, $\chi_{23} = 1.0$; f, $\chi_{12} = 0.48$, $\chi_{13} = 0.20$, $\chi_{23} = 1.4$: $p_{12,s} = p_{13,q} = p_{23,r} = 0$.

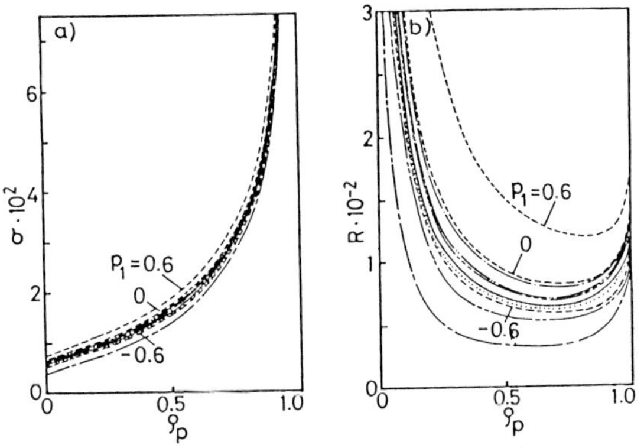

Fig. 3.13 Effect of the χ_{12}, χ_{13} and χ_{23} parameters on the relationship between the partition coefficient σ or the phase volume ratio R and the relative amount of polymer partitioned in a polymer-rich phase ρ_p: The original polymer, Schulz-Zimm distr., $X_w{}^0 = 300$, $X_w{}^0/X_{n0} = 2.0$; $v_p{}^0 = 0.005$; ———, $\chi_{12} = 0.48$, $\chi_{13} = 0.20$, $\chi_{23} = 1.0$; ————, $\chi_{12} = 0$, $\chi_{13} = 0.2$, $\chi_{23} = 1.0$; – – – –, $\chi_{12} = 0.9$, $\chi_{13} = 0.2$, $\chi_{23} = 1.0$; – – –, $\chi_{12} = 1.4$, $\chi_{13} = 0.2$, $\chi_{23} = 1.0$; —·—, $\chi_{12} = 0.48$, $\chi_{13} = 0.5$, $\chi_{23} = 1.0$; —··—, $\chi_{12} = 0.48$, $\chi_{13} = 0.2$, $\chi_{23} = 0.8$; ······, $\chi_{12} = 0.48$, $\chi_{13} = 0.2$, $\chi_{23} = 1.3$.

$X_n^0 = 150$). As ρ_p increases from zero to unity, σ increases monotonically, but R decreases first and increases after passing through a minimum (ρ_{min}). As described before, σ and R, both calculated for the quasi-ternary solution, do not exceed, by any combination of the three χ parameters, those for the quasi-binary solution with $p_1 = 0$. A decrease in χ_{12} and χ_{23} and an increase in χ_{13} bring about an increase in σ and R.

In Figs 3.14 and 3.15, the effects of $p_{12,1}$, $p_{13,1}$ and $p_{23,1}$ on the σ (or R) vs. v_p^0 relations at $\rho_p = 1/15$ are shown.

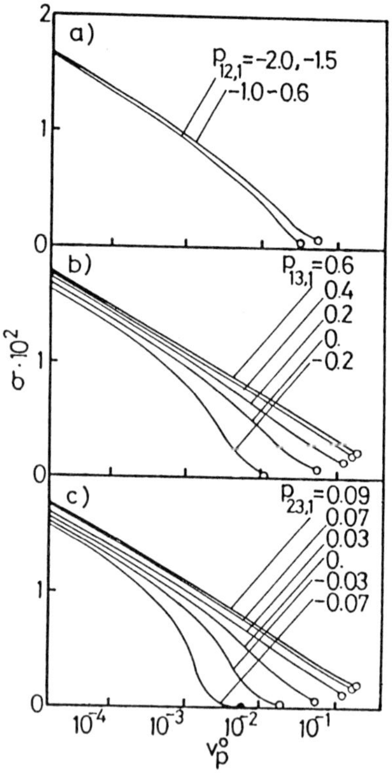

Fig. 3.14 Effect of the 1st order concentration-dependence parameter $p_{12,1}$, $p_{13,1}$ and $p_{23,1}$ on the relationship between partition coefficient σ and v_p^0: Original polymer, Schulz-Zimm type distribution ($X_w^0 = 300$, $X_w^0/X_n^0 = 2$); $\chi_{12}^0 = 0.5$, $\chi_{13}^0 = 0.2$ and $\chi_{23}^0 = 1.0$, $\rho_p = 1/15$. a) $p_{13,1} = p_{23,1} = 0$, b) $p_{12,1} = p_{23,1} = 0$, c) $p_{12,1} = p_{13,1} = 0$. Unfilled circle denotes $v_{p,c}^0$.

The unfilled circle indicates the point at which both σ and R reach a minimum. Above v_p^0 at minimum, two-phase separation under given conditions becomes impossible and this v_p^0 was defined as "critical point" by Kamide et al. (ref. 25) for quasi-binary system and referred to as $v_{p,c}^0$. Note that $v_{p,c}^0$ does not mean the

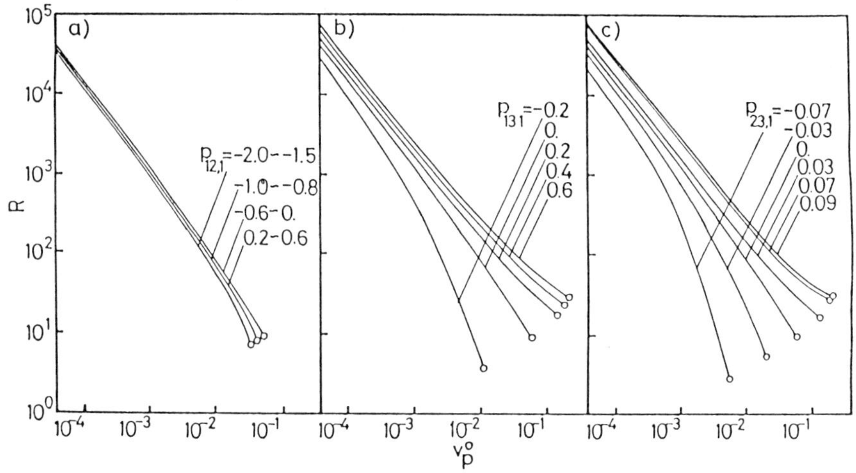

Fig. 3.15 Effect of $p_{12,1}$, $p_{13,1}$ and $p_{23,1}$ on the relationship between the two phase volume ratio R and v_p^0: Original polymer, Schulz-Zimm type distribution ($X_w^0=300$, $X_w^0/X_n^0=2$); $\chi_{12}^0=0.5$, $\chi_{13}^0=0.2$ and $\chi_{23}^0=1.0$, $\rho_p=1/15$. a) $p_{13,1}=p_{23,1}=0$, b) $p_{12,1}=p_{23,1}=0$, c) $p_{12,1}=p_{13,1}=0$. Unfilled circle denotes $v_{p,c}^0$.

critical solution point and the unfilled circle approaches the critical solution point (CSP) in the limit of $\rho_p \rightarrow 0$ (not shown in the figure). Volume fractions of solvent 1, 2 and the polymer at CSP (v_1^c, v_2^c and v_p^c, respectively) can be calculated from the spinodal and neutral-equilibrium conditions according to the method proposed by Kamide and Matsuda (ref. 26) (see, section 3.3). (v_1^c, v_2^c, v_p^c) is (0.4880, 0.4609, 0.0511) under the conditions of (χ_{12}^0, χ_{13}^0, χ_{23}^0) = (0.5, 0.2, 1.0) with $p_{12,1}=p_{13,1}=p_{23,1}=0$. When $p_{12,1}=p_{13,1}=p_{23,1}=0$, $v_{p,c}^0$ at $\rho_p=1/15$ is ca.0.05 and is close to v_p^c. Both σ and R decrease with an increase in v_p^0. Generally, σ and R, at given ρ_p, increase with increasing $p_{13,1}$ and $p_{23,1}$. The effects of $p_{12,1}$ on the σ (or R) vs. v_p^0 curve are small (ref. 14).

The effects of $p_{12,1}$, $p_{13,1}$ and $p_{23,1}$ on the relations between σ (or R) and ρ_p are shown in Fig. 3.16. In the figure, results of quasi-binary system with $p_1=-0.6$, 0 and 0.6 ($p_j=0$, $j\geq 2$) are also shown as a broken line (refs 24,27-34). In the case of $p_{12,1}=p_{13,1}=p_{23,1}=0$, σ and R for a quasi-ternary system is smaller than those for a quasi-binary system with $p_1=0$ for any combination of χ_{12}^0, χ_{13}^0 and χ_{23}^0 (ref. 11). When $p_{13,1}\geq 0.2$ (with $p_{12,1}=p_{23,1}=0$) or $p_{23,1}\geq 0.03$ (with $p_{12,1}=p_{23,1}=0$), σ (and R) for the quasi-ternary system exceeds σ (and R) for the

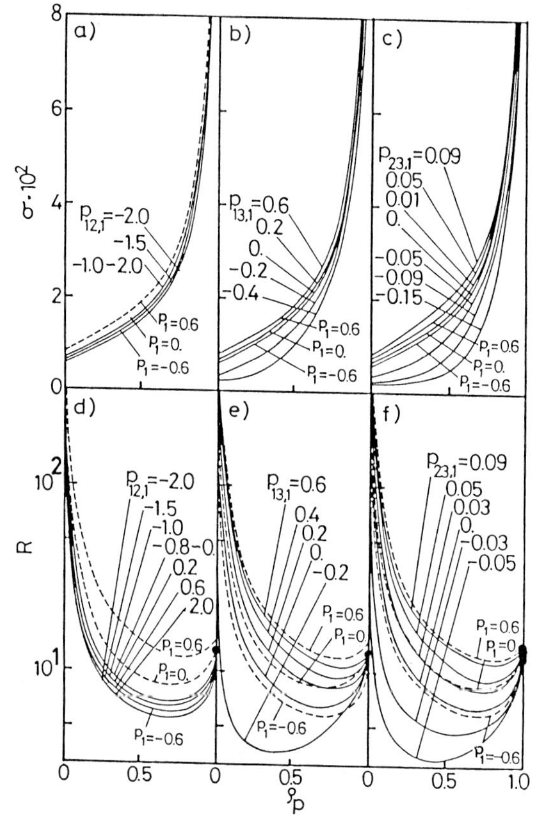

Fig. 3.16 Effect of $p_{12,1}$, $p_{13,1}$ and $p_{23,1}$ on the relationship between partition coefficient σ or the phase volume ratio R and ρ_p: Original polymer, Schulz-Zimm type distribution ($X_w^0=300$, $X_w^0/X_n^0=2$); $\chi_{12}^0=0.5$, $\chi_{13}^0=0.2$ and $\chi_{23}^0=1.0$, $v_p^0=0.005$. a) and d), $p_{13,1}=p_{23,1}=0$; b) and e), $p_{12,1}=p_{23,1}=0$; c) and f), $p_{12,1}=p_{13,1}=0$. The broken lines are the results for a quasi-binary system. p_1 is denoted on the curve. Filled circles denote R at $\rho_p=1.0$.

quasi-binary system with $p_1=0$. Specially, σ and R for the quasi-ternary system with $p_{23,1}=0.09$ ($p_{12,1}=p_{13,1}=0$) are larger than σ and R for the quasi-binary system with $p_1=0$. As a consequence, under specific conditions, a better fractionation efficiency is expected to be obtained for the quasi-ternary solutions than for quasi-binary solutions. This prediction has, unfortunately, not yet been experimentally confirmed and clearly more detailed experimental examination is called for. The effect of $p_{12,1}$ on σ (or R) is not as large.

We denote extrapolated values of σ, R and the volume fractions of each phase for $\rho_p \to 1$ with an asterisk as σ^*, R^* ($v_{1(1)}^*$, $v_{2(1)}^*$, $v_{p(1)}^*$) and ($v_{1(2)}^*$, $v_{2(2)}^*$, $v_{p(2)}^*$). $v_{p(2)}^*$ can be calculated by the following equation (ref. 14).

$$\ln(1-v_{p(2)}^*)+v_{p(2)}^*-\frac{v_{p(2)}^*}{X_n^0}$$

$$+\chi_{23}^0[v_{p(2)}^{*2}+\sum_{r=1}^{n_r}p_{23,r}v_{p(2)}^{*r+2}]=0. \qquad (3.43)$$

and R* is expressed by

$$R^*=\frac{v_{p(2)}^*}{v_p^0}-1. \qquad (3.44)$$

As is known from eqns (3.43) and (3.44), $v_{p(2)}^*$ and R^* are independent of χ_{12}^0, χ_{13}^0, $p_{12,s}$ and $p_{13,q}$ ($s=1,\cdots,n_s$; $q=1,\cdots n_q$) and determined by χ_{23}^0 and $p_{23,r}$ ($r=1,\cdots n_r$). The extrapolated values of R for $\rho_p \to 1$ are given by filled circles in Fig. 3.16d, e and f and are consistent with R* calculated from eqn (3.43). From the combination of $v_{Xi(1)}=0$ ($i=1,\cdots,m$) for

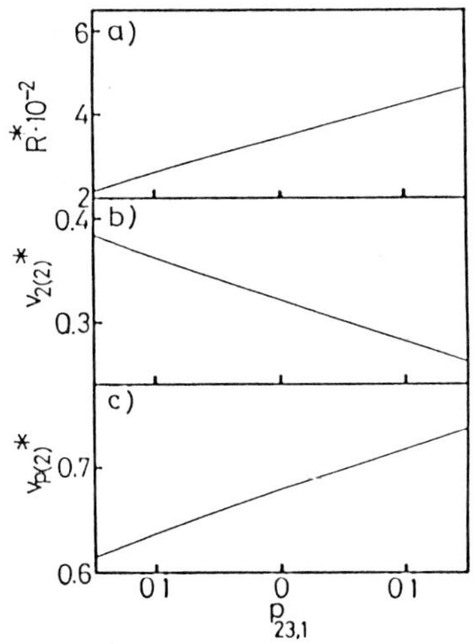

Fig. 3.17 Effect of $p_{23,1}$ on the R, $v_{2(2)}$ and $v_{p(2)}$ for $\rho_p=1.0$ (R^*, $v^*_{2(2)}$ and $v^*_{(2)}$, respectively): Original polymer, Schulz-Zimm type distribution ($X_w^0=300$, $X_w^0/X_n^0=2$); $\chi_{12}^0=0.5$, $\chi_{13}^0=0.2$ and χ_{23}^0 1.0, v_p^0 0.005.

$\rho_p = 1$ and the definition of σ (eqn (3.23)), σ for $\rho_p = 1$ (σ*) becomes infinite.

Fig. 3.17 shows the effect of $p_{23,1}$ on R*, $v_{2(2)}$* and $v_{p(2)}$*. With an increase in $p_{23,1}$, R* and $v_{2(2)}$* increase and $v_{p(2)}$* decreases. We can evaluate χ_{23}^0 and $p_{23,r}$ from R* experimentally determined.

Fig. 3.18 shows the effect of v_p^0 on the relationship between σ (or R) and ρ_p. In this case, $\chi_{12} = 0.48$, $\chi_{13} = 0.2$ and $\chi_{23} = 1.3$ were assumed. The characteristic feature of the σ (or R) vs. ρ_p relationship in Fig. 3.13 holds its validity over a wide range of v_p^0, from 1×10^{-5} to 1×10^{-2}. That is, as ρ_p increases, σ as well as $d\sigma/d\rho_p$ increases monotonically, but R decreases first and increases after passing through the minimum $\rho_{p,min}$ at all the v_p^0 values investigated. $\rho_{p,min}$ is almost constant, irrespective of v_p^0.

For any value of ρ_p, both σ and R increases with v_p^0 and for any value of v_p^0, σ increases with ρ_p, but R is the minimum at $\rho_p \simeq 0.7$. This confirms the general characteristics in Fig. 3.18. Fig. 3.19 shows the relationships among v_p^0, ρ_p, σ and R obtained at phase equilibrium. As Kamide and his coworkers pointed

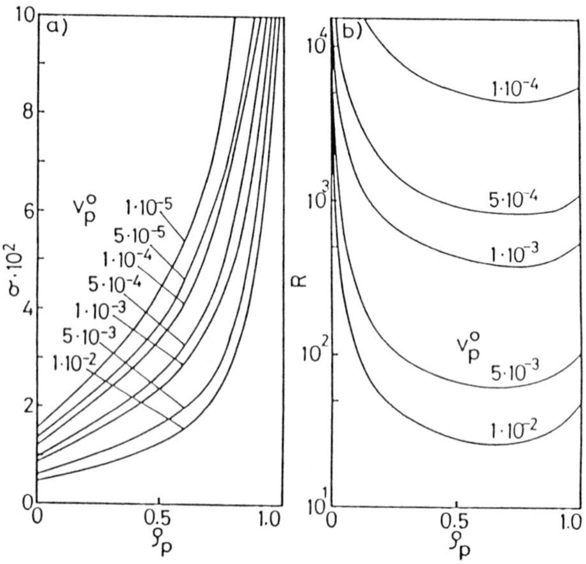

Fig. 3.18 Effect of the initial polymer volume fraction v_p^0 on the relationship between the partition coefficient σ or the phase volume ratio R and the relative amount of the polymer in a polymer-rich phase ρ_p: The original polymer, Schulz-Zimm distr., $X_w^0 = 300$, $X_w^0/X_n^0 = 2.0$; $\chi_{12} = 0.48$, $\chi_{13} = 0.20$, $\chi_{23} = 1.30$.

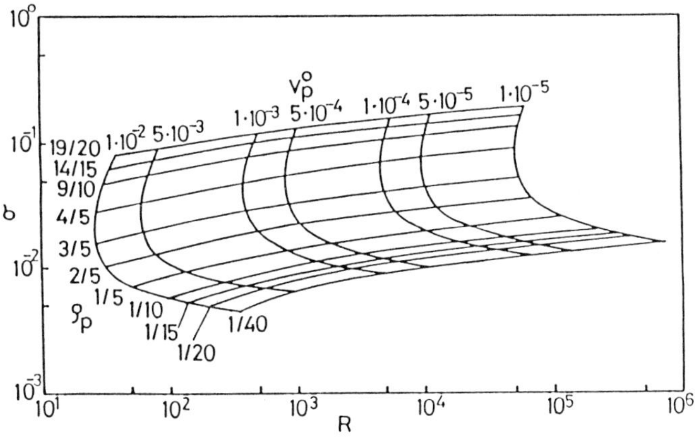

Fig. 3.19 Relationships among the initial polymer fraction $v_p{}^0$, the relative amount of the polymer in a polymer-rich phase ρ_p, the partition coefficient σ and the phase volume ratio R for quasi-ternary polymer solution: The original polymer, Schulz-Zimm distr., $X_w{}^0=300$, $X_w{}^0/X_n{}^0=2.0$; $\chi_{12}=0.48$, $\chi_{13}=0.20$, $\chi_{23}=1.30$.

out for a quasi-binary system (ref. 35), under given external operating conditions ($v_p{}^0$ and ρ_p), σ together with R is unambiguously determined. In other words, a given pair of $v_p{}^0$ and ρ_p corresponds rigorously to a specialized pair of σ and R as independent separation variables even for a quasi-ternary system, provided the molecular characteristics of the original polymer and three χ parameters are given in advance.

It should be remarked that σ and R may vary within a network structure in Fig. 3.19 corresponding to a change in $v_p{}^0$ ranging from 1×10^{-5} to 1×10^{-2} and to a change in ρ_p from 1/40 to 19/20. As $v_p{}^0$ reduces with keeping ρ_p constant, both σ and R (particularly, R) increase progressively and with increasing ρ_p at given $v_p{}^0$, σ increases monotonically and R decreases in a small ρ_p region, then increases slightly after passing through the minimum.

Fig. 3.20a exemplifies the normalized MWD curves $g_{(2)}(X)$ remaining in the polymer-rich phase, which is phase-separated, keeping $\rho_p=1/15$, from the solutions of various concentrations $v_p{}^0$. $g_{(2)}(X)$ for the polymers isolated from a comparatively concentrated solution shows a peak at a lower X (and this X is designed as X_p) and approximated as X_p for the original polymer. As $v_p{}^0$ decreases, the peak of the MWD curve moves from a lower to a higher X region, without changing practically the upper and

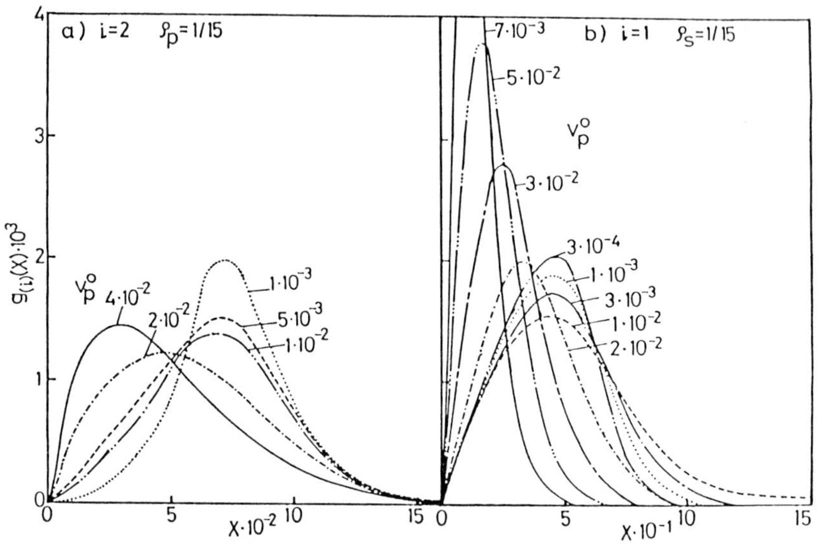

Fig. 3.20 Differential MWD curve $g_{(i)}(X)$ of the polymers in a polymer-rich phase ($i=2$) or in a polymer-lean phase ($i=1$), separated at $\rho_p=1/15$ (a)) or 14/15 (i.e., $\rho_s=1/15$) (b)) from different $v_p{}^0$ quasi-ternary solutions (numbers at curves): The original polymer, Schulz-Zimm distr., $X_w{}^0=300$, $X_w{}^0/X_n{}^0=2.0$; $\chi_{12}=0.48$, $\chi_{13}=0.20$, $\chi_{23}=1.30$.

lower limits of MWD, and this shift, results in a sharpening of a MWD.

Fig. 3.20b shows a normalized MWD curve $g_{(1)}(X)$ of some polymers partitioned in the polymer-lean phase, which is phase-separated, keeping $\rho_s=1/15$ (i.e., $\rho_p=14/15$), from the solutions of various of $v_p{}^0$. The higher X region in MWD is notably changed by $v_p{}^0$ and as $v_p{}^0$ is lowered, X_p of $g_{(1)}(X)$ decreases with a remarkable peak sharpening. The shift of X_p and peak sharpening contributes in an reverse manner to MWD and in this case, the effect of the latter is significantly larger on MWD than that of the former, resulting in a decrease in the breadth of the MWD of the polymer with a decrease in $v_p{}^0$. A similar dependence of MWD on $v_p{}^0$ was observed for multicomponent polymer / single solvent systems.

In Fig. 3.21, the X_p-$v_p{}^0$ relations, obtained at constant ρ_p or ρ_s (1/15) for various combinations of χ_{12}, χ_{13} and χ_{23} are shown. Similar relations for quasi-binary solutions with various values of the concentration-dependence parameter p_1 in eqn (2.3) (with $p_2=\cdots=p_n=0$) are also shown as full lines for comparison. All the data for the quasi-ternary solutions yield a

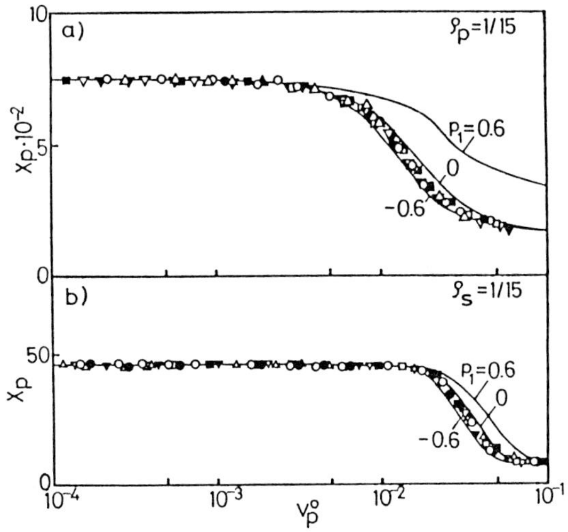

Fig. 3.21 Dependence of X_p for polymers in a polymer-rich phase (a)) or in a polymer-lean phase (b)) on the initial polymer volume fraction $v_p{}^0$: The original polymer, Schulz-Zimm distr., $X_w{}^0=300$, $X_w{}^0/X_n{}^0=2.0$; ○ $\chi_{12}=0.48$, $\chi_{13}=0.20$, $\chi_{23}=1.00$; ● $\chi_{12}=0.48$, $\chi_{13}=0.20$, $\chi_{23}=1.2$; △ $\chi_{12}=0.48$, $\chi_{13}=0.20$, $\chi_{23}=0.80$; □ $\chi_{12}=0.48$, $\chi_{13}=0.05$, $\chi_{23}=1.00$; ■ $\chi_{12}=0.48$, $\chi_{13}=0.50$, $\chi_{23}=1.00$; ▽ $\chi_{12}=0.20$, $\chi_{13}=0.20$, $\chi_{23}=1.00$; ▼ $\chi_{12}=1.30$, $\chi_{13}=0.20$, $\chi_{23}=1.00$. Full lines are for quasi-binary solutions and the concentration dependence of χ parameter, p_1 is denoted on the curve.

single master curve, irrespective of the combination of χ_{12}, χ_{13} and χ_{23} values, lying between those for the quasi-binary solutions with $p_1=0$ and -0.6. These plots demonstrate an asymptotic approach of X_p to a maximum value at lower $v_p{}^0$ and to a minimum at higher $v_p{}^0$. X_p can be regarded as roughly constant in the range of $v_p{}^0 < 2\times 10^{-3}$ for the polymer in the polymer-rich phase and of $v_p{}^0 < 2\times 10^{-2}$ for the polymer in the polymer-lean phase and X_p transits rapidly in a comparatively narrow $v_p{}^0$ range (ref. 12).

Fig. 3.22 shows the relation between X_w/X_n of the polymer in the polymer-lean and -rich phases and $v_p{}^0$ for a given combination of three χ parameters when a constant $\rho_p (\equiv 1-\rho_s)$ of the polymers is isolated. The MWD of the polymer in the polymer-lean phase becomes sharp without limit as ρ_s decreases and the effect of ρ_s predominates over $v_p{}^0$ in the polydispersity of the polymer-lean phase. As ρ_p decreases, the $v_p{}^0$ dependence of X_w/X_n of the polymer in the polymer-rich phase becomes remarkable, as expected. The ρ_p value, at which the minimum X_w/X_n is obtained at given $v_p{}^0$ and $\rho_{p,min}$, depends on $v_p{}^0$ and becomes smaller as $v_p{}^0$ diminishes.

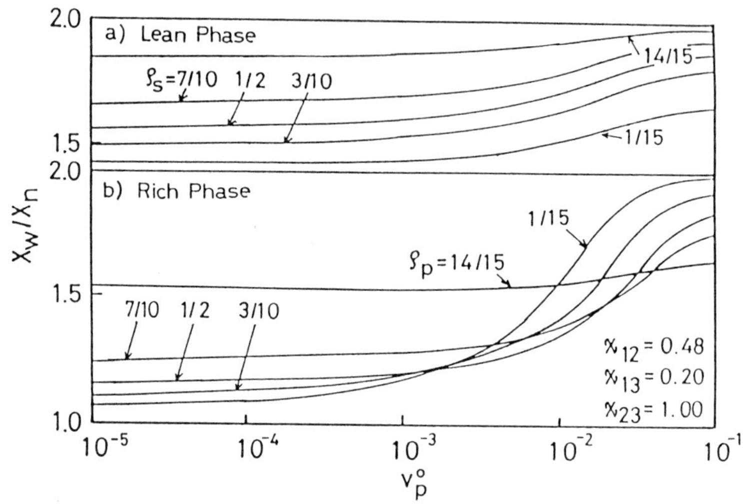

Fig. 3.22 Effect of the relative amounts of polymers in a polymer-lean phase, ρ_s, (number on curve in a)) and in a polymer-rich phase, ρ_p, (number on curve in b)) on the relationship between X_w/X_n of the polymers and the initial polymer volume fraction v_p^0: The original polymer, Schulz-Zimm distr., $X_w^0=300$, $X_w^0/X_n^0=2.0$; $\chi_{12}=0.48$, $\chi_{13}=0.20$, $\chi_{23}=1.00$.

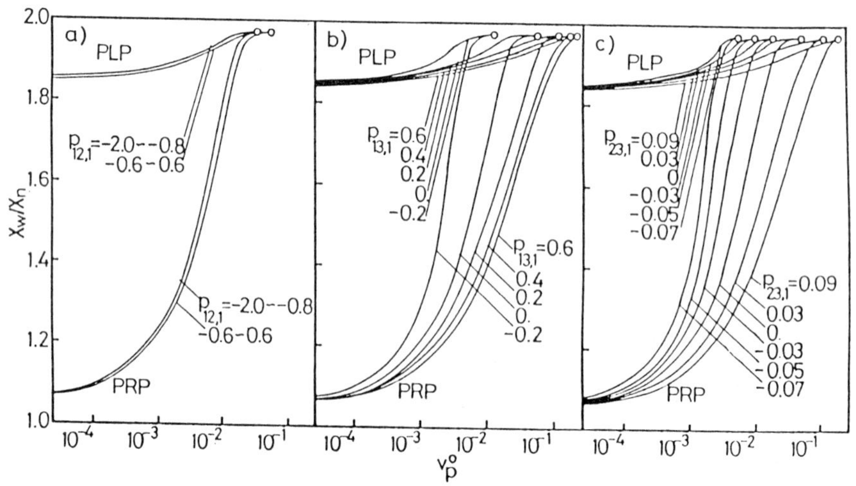

Fig. 3.23 Effect of the concentration-dependence parameters $p_{12,1}$, $p_{13,1}$ and $p_{23,1}$ on the relationship between X_w/X_n of the polymers in the polymer-lean phase (PLP) or the polymer-rich phase (PRP) and v_p^0: Original polymer, Schulz-Zimm type distribution ($X_w^0=300$, $X_w^0/X_n^0=2$); $\chi_{12}^0=0.5$, $\chi_{13}^0=0.2$ and $\chi_{23}^0=1.0$, $\rho_p=1/15$. a) $p_{13,1}=p_{23,1}=0$, b) $p_{12,1}=p_{13,1}=0$. Unfilled circles denote $v_{p,c}^0$.

The effects of $p_{12,1}$, $p_{13,1}$ and $p_{23,1}$ on the relationship between X_w/X_n and v_p^0 (for $\rho_p = 1/15$) are shown in Fig. 3.23. With an increase in $p_{13,1}$ and $p_{23,1}$, X_w/X_n decreases abruptly. X_w/X_n of the polymer in polymer-lean and -rich phases ($X_{w(1)}/X_{n(1)}$ and $X_{w(2)}/X_{n(2)}$) increases with an increase in v_p^0 and coincides with each other for $v_{p,c}^0$. As ρ_p decreases from 1/15 to zero, $v_{p,c}^0$ approaches the critical polymer concentration.

Fig. 3.24 shows the normalized MWD curve of the polymers in the polymer-rich and -lean phases, $g_{(2)}(X)$ and $g_{(1)}(X)$, isolated from a quasi-ternary solution. With a decrease in ρ_p, a remarkable increase in its content in a larger X region and a significant lowering of the peak height are observed in both $g_{(2)}(X)$ and $g_{(1)}(X)$ and in the former, X_p shifts, noticeably to a larger X region, but remains constant in the latter, roughly identical with X_p of the original polymer (ref. 12).

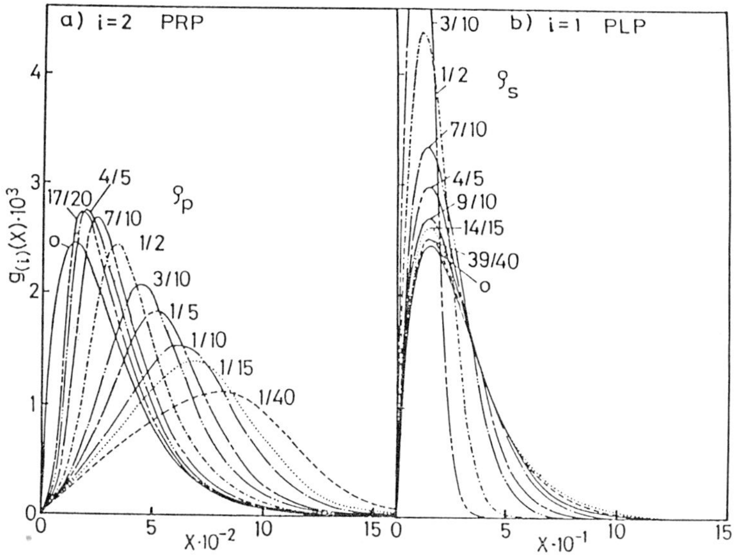

Fig. 3.24 Effect of relative amounts of polymers in a polymer-rich phase ρ_p (a)) and in a polymer-lean phase ρ_s (b)) on the MWD curve $g_{(i)}(X)$ (i = 1 and 2) of the polymers: The original polymer, Schulz-Zimm distr., $X_w^0 = 300$, $X_w^0/X_n^0 = 2.0$; the initial polymer volume fraction $v_p^0 = 0.005$; $\chi_{12} = 0.48$, $\chi_{13} = 0.20$, $\chi_{23} = 1.00$; ρ_p and ρ_s are denoted on the curve.

Fig. 3.25 shows the effects of χ_{12}, χ_{13} and χ_{23} on the relationships between X_w/X_n of the polymer dissolving in the polymer-rich phase and those remaining in the polymer-lean phase

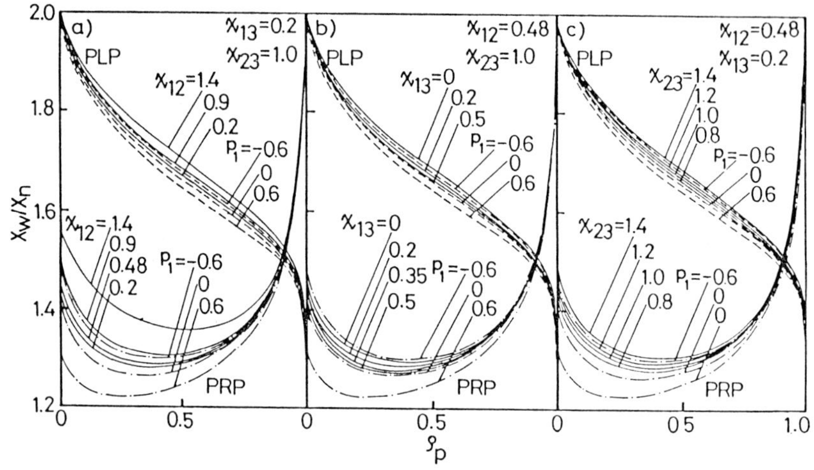

Fig. 3.25 Effects of the parameters χ_{12} (a)), χ_{13} (b)) and χ_{23} (c)) on the relationship between X_w/X_n of the polymers in a polymer-rich phase ρ_p: The original polymer, Schulz-Zimm distr., $X_w^0=300$, $X_w^0/X_n^0=2.0$; a) χ_{12} is denoted on the curve, $\chi_{13}=0.2$, $\chi_{23}=1.0$; b) $\chi_{12}=0.48$, χ_{13} is denoted on the curve, $\chi_{23}=1.0$; c) $\chi_{12}=0.48$, $\chi_{13}=0.20$ and χ_{23} is denoted on the curve; broken line, quasi-binary solution; the concentration dependence of χ parameter p_1 is denoted on the curve.

and ρ_p for $v_p^0=0.005$. Fig. 3.25 also shows the data for a quasi-binary system. X_w/X_n of the polymer in the polymer-rich phase decreases first and then increases after passing through a minimum of X_w/X_n ($(X_w/X_n)_{min}$). ρ_p, yielding $(X_w/X_n)_{min}$, which is defined as $\rho_{p,min}$. In contrast to this, X_w/X_n of the polymer in the polymer-lean phase decreases monotonically with a decrease in ρ_s, approaching 1.0 at a limit of $\rho_s=0$ (i.e., $\rho_p=1$). Fig. 3.25 indicates that characteristics observed in a quasi-binary system hold even in a quasi-ternary (and probably much higher order) system. The X_w/X_n of the polymers in both phases is without exception smaller in a quasi-binary system, suggesting that fractionation using a quasi-binary system is much more effective than using a quasi-ternary system (ref. 12).

Fig. 3.26 shows the effects of $p_{12,1}$, $p_{13,1}$ and $p_{23,1}$ on the relationship between the ratio X_w/X_n of the two phases and ρ_p. In this figure the results for quasi-binary solutions with various p_1 are shown as broken lines for comparison. For any combination of $p_{12,1}$, $p_{13,1}$ and $p_{23,1}$, X_w/X_n of the polymer in the polymer-lean phase, $X_{w(1)}/X_{n(1)}$ decreases monotonically from 2.0 to 1.0 with an

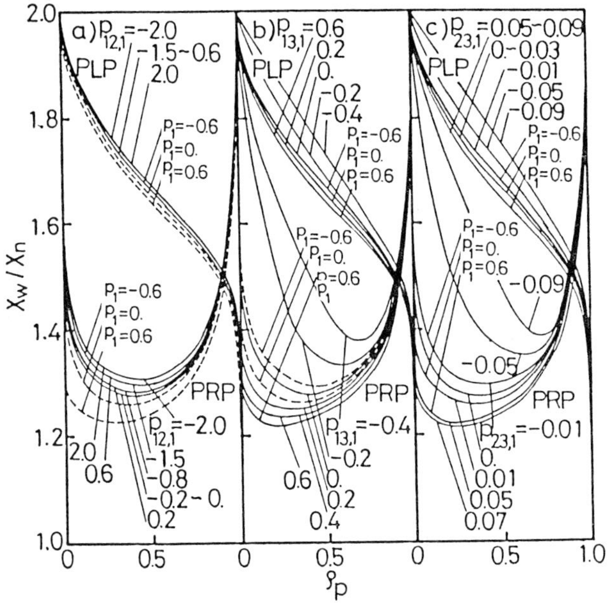

Fig. 3.26 Effect of the concentration-dependence parameters $p_{12,1}$, $p_{13,1}$ and $p_{23,1}$ on the relationship between X_w/X_n of the polymers in polymer-lean phase (PLP) or polymer-rich phase (PRP) and ρ_p: Original polymer, Schulz-Zimm type distribution ($X_w^0 = 300$, $X_w^0/X_n^0 = 2$); $\chi_{12}^0 = 0.5$, $\chi_{13}^0 = 0.2$ and $\chi_{23}^0 = 1.0$; $v_p^0 = 0.005$. a) $p_{13,1} = p_{23,1} = 0$, b) $p_{12,1} = p_{13,1} = 0$. The broken lines are results for a quasi-binary system. p_1 is denoted on the curve.

increase in ρ_p. X_w/X_n of the polymer in a polymer-rich phase $X_{w(2)}/X_{n(2)}$ reveals minimum at a specific ρ_p, approaching with X_w^0/X_n^0 at $\rho_p = 1.0$. Over a whole range of ρ_p, $X_{w(2)}/X_{n(2)}$ attains minimum at $p_{12,1} = -0.2$ when $\chi_{12}^0 = 0.5$, $\chi_{13}^0 = 0.2$ and $\chi_{23}^0 = 1.0$ and becomes smaller for a larger $p_{13,1}$ and $p_{23,1}$. Note that the relationship between $X_{w(2)}/X_{n(2)}$ and ρ_p for the quasi-ternary systems agrees well with that of the quasi-binary systems with $p_1 = 0.6$. For example, when $p_{13,1} = 0.6$. (with $p_{12,1} = p_{23,1} = 0$) or $p_{23,1} = 0.05$ (with $p_{12,1} = p_{13,1} = 0$) for quasi-ternary systems (SZ polymer $X_w^0 = 300$, $X_w^0/X_n^0 = 2$, $v_p^0 = 0.005$), the $X_{w(2)}/X_{n(2)}$ vs. ρ_p relationship almost coincides with that for quasi-binary systems (SZ polymer, $X_w^0 = 300$, $X_w^0/X_n^0 = 2$, $v_p^0 = 0.005$). Combination of solvent 1 and 2 with larger values of $p_{13,1}$ and $p_{23,1}$ (in this case, $p_{13,1} \geq 0.6$ and $p_{23,1} \geq 0.05$, respectively) will afford us a polymer having a narrower MWD than that obtained from quasi-binary systems with $p_1 = 0.6$ in polymer fractionation experiments (ref. 14). It has been shown that the effects of χ_{13}^0

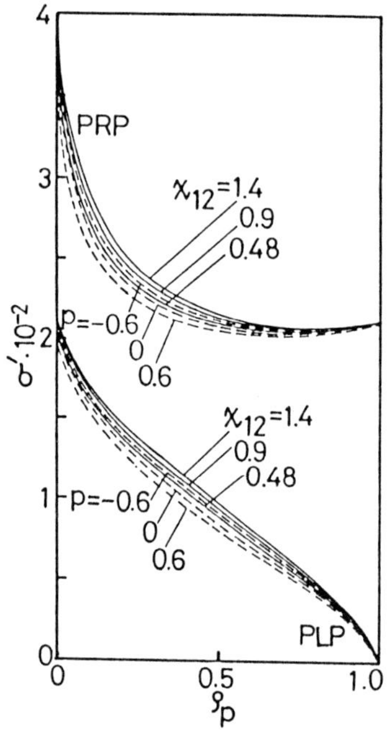

Fig. 3.27 The standard deviation σ' of the polymers in a polymer-rich phase and in a polymer-lean phase separated from a quasi-ternary solution (full line) and a quasi-binary solution (broken line), plotted vs. the relative amount of polymers in a polymer-rich phase ρ_p: The original polymer, Schulz-Zimm distr., $X_w^0=300$, $X_w^0/X_n^0=2.0$; χ_{12}, number on the curve, $\chi_{13}=0.20$, $\chi_{23}=1.00$; initial polymer volume faction $v_p^0=0.005$; for quasi-binary solution, the concentration dependence of χ parameter p_1, number on the curve.

on $g_{(2)}(X)$ is small but χ_{23}^0 has a large effect on $g_{(2)}(X)$ (ref. 12). Fig. 3.26 indicates that $p_{23,1}$ has a strong influence on $g_{(2)}(X)$, accordingly $X_{w(2)}/X_{n(2)}$. In other words, a suitable choice for solvent 2 (i.e., nonsolvent) is required to separate polymers with sharp MWD from quasi-ternary systems.

Fig. 3.27 shows the standard deviation σ' of the polymers in both phases as a function of ρ_p for quasi-ternary and -binary systems. In this case, for the former system, various combinations of χ_{12}, χ_{13} and χ_{23} are assumed. For the same value of ρ_p, the combination of these χ parameters controls, to some extent, the polydispersity of the polymers. A decrease in χ_{12} improves the separation efficiency. It should be noted that σ', a much more adequate parameter representing the polydispersity than the ratio X_w/X_n, is always larger in the polymer-rich phase than in the polymer-lean phase. This clearly indicates the superiority of the solution method over the precipitation method, even for quasi-

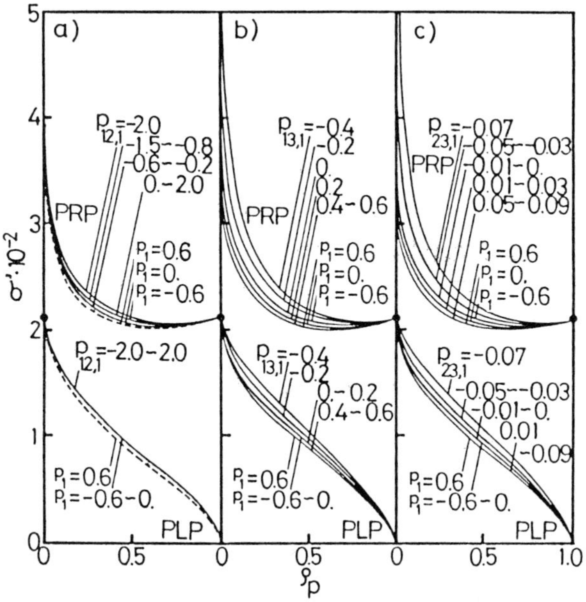

Fig. 3.28 Effects of the concentration-dependence parameters $p_{12,1}$, $p_{13,1}$ and $p_{23,1}$ on the relationship between σ' of the polymers in polymer-lean phase (PLP) or polymer-rich phase (PRP) and ρ_p: Original polymer, Schulz-Zimm type distribution ($X_w^0 = 300$, $X_w^0/X_n^0 = 2$); $\chi_{12}^0 = 0.5$, $\chi_{13}^0 = 0.2$ and $\chi_{23}^0 = 1.0$, $v_p^0 = 0.005$. a) $p_{13,1} = p_{23,1} = 0$, b) $p_{12,1} = p_{13,1} = 0$. The broken lines are the results for a quasi-binary system. p_1 is denoted on the curve.

ternary solutions as well as quasi-binary solutions (see, section 3.6).

Fig. 3.28 shows the effects of $p_{12,1}$, $p_{13,1}$ and $p_{23,1}$ on σ' vs. ρ_p curve. The figure includes results (refs 24,29-34,36) for a quasi-binary system as the broken line for comparison. The filled circle denotes the value for the original polymer (σ'_0). As well as a $X_{w(2)}/X_{n(2)}$ vs. ρ_p curve, the $\sigma'_{(2)}$ vs. ρ_p curve has a minimum at some ρ_p. $\sigma'_{(1)}$ and $\sigma'_{(2)}$ gradually decrease with an increase in $p_{13,1}$ and $p_{23,1}$. $\sigma'_{(2)}$ of the quasi-ternary system with $p_{13,1} = 0.4 \sim 0.6$ ($p_{12,1} = p_{23,1} = 0$) or $p_{23,1} = 0.05 \sim 0.09$ ($p_{12,1} = p_{13,1} = 0$) agrees fairly well with $\sigma'_{(2)}$ of the quasi-binary system with $p_1 = 0.6$. Similar behaviour is observed for $\sigma'_{(1)}$.

Figs 3.29 and 3.30 depict the relationship among polymer volume fractions, $v_{p(2)}$, in the polymer-rich phase and v_p^0 and ρ_p. Under the same operating conditions, i.e., χ_{12}, χ_{13}, χ_{23} and ρ_p ($= 1/15$) (or $v_p^0 = 0.005$), $v_{p(2)}$ decreases with an

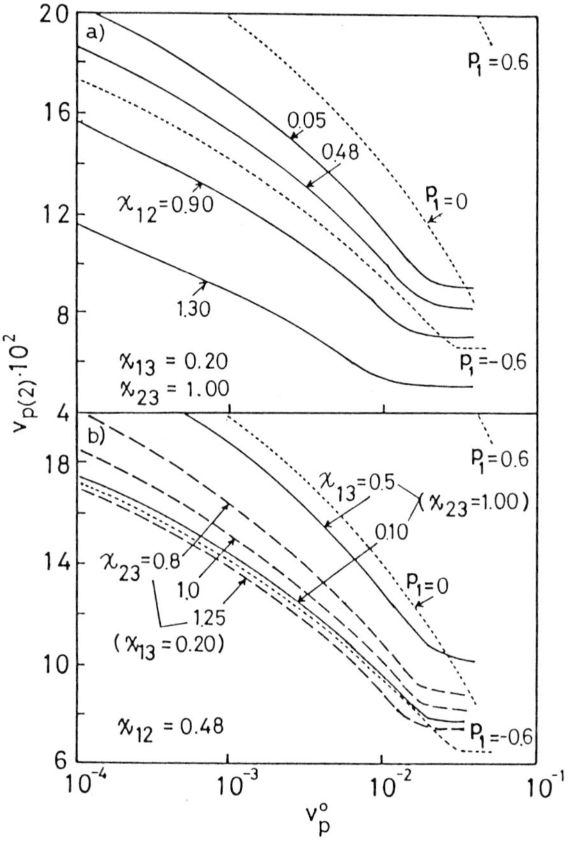

Fig. 3.29 Plot of the polymer volume fraction in a polymer-rich phase $v_{p(2)}$ versus the initial polymer volume fraction v_p^0 for quasi-ternary (full and broken lines) and quasi-binary (dotted line) solutions with a constant relative amount of the polymer ρ_p: The original polymer, Schulz-Zimm distr., $X_w^0 = 300$, $X_w^0/X_n^0 = 2.0$; for a quasi-ternary solution, a) χ_{12}, numbers on the curves, $\chi_{13} = 0.20$, $\chi_{23} = 1.00$; b) full line, $\chi_{12} = 0.48$, χ_{13}, numbers on the curve, $\chi_{23} = 1.00$; broken line, $\chi_{12} = 0.48$, $\chi_{13} = 0.20$, χ_{23}, numbers on the curve. For a quasi-binary solution, the concentration dependence of χ parameter, p_1 is denoted on the curve.

increase in v_p^0 and a decrease in ρ_p. Unlike $v_{p(1)}$, $v_{p(2)}$ is evidently dependent on the χ parameters and $v_{p(2)}$ (and accordingly, $v_{p(2)} - v_{p(1)}$) is larger for a smaller χ_{12}, larger χ_{13} and smaller χ_{23}. The quantity $v_{p(2)} - v_{p(1)}$ is a measure of the ease of separating polymers (ref. 37). The dotted lines in $v_{p(2)}$ of a quasi-ternary solution can never exceed that of the quasi-binary solution with $p_1 = 0$, suggesting that an effective separation of the two phases for quasi-ternary systems are comparatively difficult compared to that for quasi-binary system.

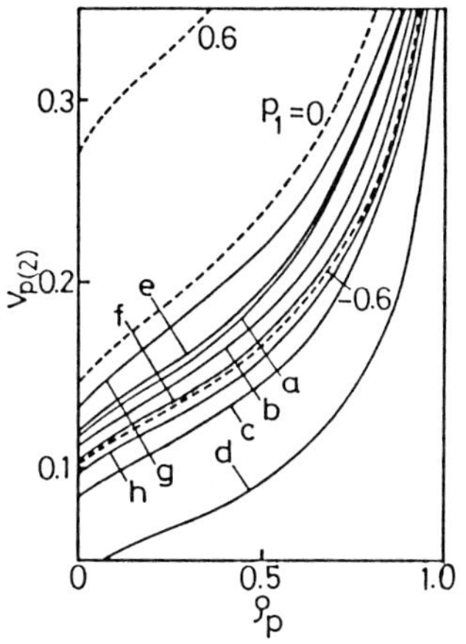

Fig. 3.30 The polymer volume fraction in a polymer-rich phase $v_{p(2)}$ plotted as a function of the relative amount of the polymer in a polymer-rich phase ρ_p: The original polymer, Schulz-Zimm distr., $X_w^0=300$, $X_w^0/X_n^0=2.0$; initial polymer volume fraction $v_p^0=0.005$; full line, quasi-ternary solution; a, $\chi_{12}=0$, $\chi_{13}=0.20$, $\chi_{23}=1.0$; b, $\chi_{12}=0.48$, $\chi_{13}=0.20$, $\chi_{23}=1.0$; c, $\chi_{12}=0.90$, $\chi_{13}=0.20$, $\chi_{23}=1.0$; d, $\chi_{12}=1.4$, $\chi_{13}=0.2$, $\chi_{23}=1.0$; e, $\chi_{12}=0.48$, $\chi_{13}=0.20$, $\chi_{23}=0.80$; f, $\chi_{12}=0.48$, $\chi_{13}=0.20$, $\chi_{23}=1.30$; g, $\chi_{12}=0.48$, $\chi_{13}=0.50$, $\chi_{23}=1.00$; h, $\chi_{12}=0.48$, $\chi_{13}=0$, $\chi_{23}=1.0$.

Figs 3.31 and 3.32 show the effects of $p_{12,1}$, $p_{13,1}$ and $p_{23,1}$ on the $v_{p(2)}$ vs. v_p^0 curve and $v_{p(2)}$ vs. ρ_p curve, respectively. The effect of $p_{12,1}$ on $v_{p(2)}$ is small but not negligible and $v_{p(2)}$ for given ρ_p shows a maximum at $p_{12,1}$, ranging between -0.6 and 0. The effects of $p_{13,1}$ and $p_{23,1}$ are remarkable and $v_{p(2)}$ increases abruptly with an increase in $p_{13,1}$ and $p_{23,1}$. Especially, $v_{p(2)}$ increases suddenly with a small increase in $p_{23,1}$. $v_{p(2)}$ vs. v_p^0 curve has a minimum at the point corresponding to the maximum of $V_{(2)}$ vs. v_p^0 curve. $v_{p(2)}$ increases with an increase in ρ_p, and coincides with $v_{p(2)}^*$ at $\rho_p=1$ (see, Fig. 3.17). By using eqn (2.49b), the relationship between R and ρ_p is expressed by (ref. 14):

$$R=\frac{v_{p(2)}}{v_p^0\rho_p}-1. \tag{3.45}$$

In the range of a small ρ_p (i.e., $\rho_p<0.3$)), R decreases abruptly with an increase in ρ_p because $v_{p(2)}$ changes slowly with

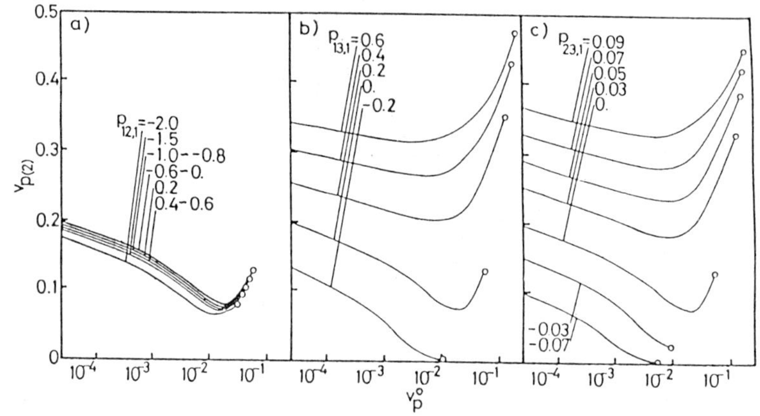

Fig. 3.31 Effects of the concentration-dependence parameters $p_{12,1}$, $p_{13,1}$ and $p_{23,1}$ on the relations between polymer volume fraction in polymer-rich phase $v_{p(2)}$ and $v_p{}^0$: Original polymer, Schulz-Zimm type distribution ($X_w{}^0=300$, $X_w{}^0/X_n{}^0=2$); $\chi_{12}{}^0=0.5$, $\chi_{13}{}^0=0.2$, and $\chi_{23}{}^0=1.0$; $\rho_p=1/15$.
a) $p_{13,1}=p_{23,1}=0$, b) $p_{12,1}=p_{23,1}=0$, c) $p_{12,1}=p_{13,1}=0$. Unfilled circles denote $v_{p,c}{}^0$.

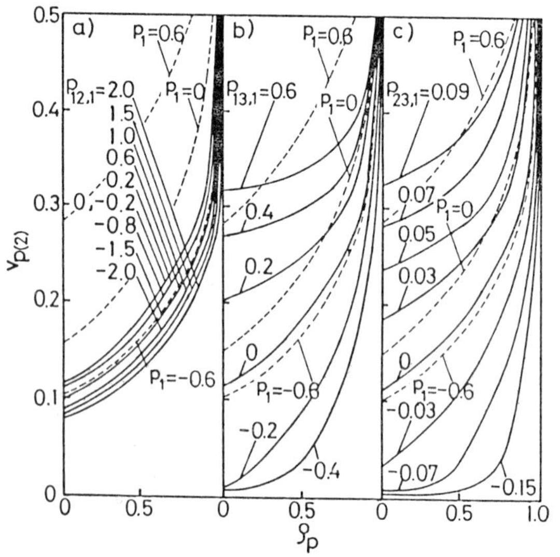

Fig. 3.32 Effects of the concentration-dependence parameters $p_{12,1}$, $p_{13,1}$ and $p_{23,1}$ on the relationship between polymer volume fractions in the polymer-rich phase $v_{p(2)}$ and ρ_p: Original polymer, Schulz-Zimm type distribution ($X_w{}^0=300$, $X_w{}^0/X_n{}^0=2$); $\chi_{12}{}^0=0.5$, $\chi_{13}{}^0=0.2$ and $\chi_{23}{}^0=1.0$, $v_p{}^0=0.005$. a) $p_{13,1}=p_{23,1}=0$, b) $p_{12,1}=p_{23,1}=0$, c) $p_{12,1}=p_{13,1}=0$. The broken lines are results for quasi-binary systems. p_1 is denoted on the curve.

ρ_p. In contrast to this, in the large ρ_p region, R increases gradually with ρ_p because of the rapid change in $v_{p(2)}$ with ρ_p.

Eqn (2.44a) can be rewritten as follows.

$$v_{p(1)} = v_p^0 (1-\rho_p)(1+\frac{1}{R}). \qquad (3.46)$$

If $1/R \ll 1$, $v_{p(1)}$ is independent of χ_{12}^0, χ_{13}^0, χ_{23}^0, $p_{12,s}$, $p_{13,q}$ and $p_{23,r}$ and dependent on ρ_p and v_p^0 (see, for example, Fig. 19 of ref. 12).

3.2.4 Coexistence curve

Fig. 3.33 shows the effects of the χ parameters on the phase diagram. A polymer-lean phase lies almost on the solvent 1-solvent 2 axis because $v_{p(1)} \simeq 0$. The coexistence curve of a polymer-rich phase side is approximately parallel to the polymer-solvent 1 axis. The tie-line connecting the two coexistence phases is, interestingly, parallel to the polymer-solvent 2 axis, particularly for larger χ_{12} and smaller χ_{13}. That is, the volume fractions of solvent 1 in both phases are approximately equal, irrespective of ρ_p. An increase in χ_{12} and a decrease in χ_{13} and χ_{23} shift the coexistence curve of a polymer-rich phase to a large v_2 region (ref. 11). The shape of the coexistence curves is not significantly affected by the χ_{12} and χ_{23} parameters. The role of χ_{13} at this point cannot be neglected.

Fig. 3.34 shows the effect of the concentration dependence of three thermodynamic interaction parameters on the coexistence curve. Phase separation is obtained under the same conditions as those in Fig. 3.33. In the figure tie lines for $\rho_p = 1/100$ are shown by the solid lines, which can be accurately approximated with the tie-line for $\rho_p = 0$ (limiting tie-line), and the line of $v_p^s = 0.01$ is shown by the broken line. The cross point of the limiting tie-line and the line of $v_p^s = 0.01$ can be regarded as a cloud point. When the concentration dependencies of three χ parameters are neglected (refs 11-13), the coexistence curve shifted to a direction of decreasing v_2 (i.e., decreasing nonsolvent) with a decrease in χ_{12}^0 and an increase in χ_{13}^0 and χ_{23}^0. With an increase in $p_{12,1}$ from -2.0 to 0.6, a similar shift in the coexistence curve was observed, but $v_{p(2)}$ changes a little, showing a maximum at $p_{12,1} = -0.2$ (ref. 14). With an increase in $p_{13,1}$ and $p_{23,1}$, the limiting tie-line approaches to v_2-v_p axis and $v_{p(2)}$ increases drastically (ref. 14).

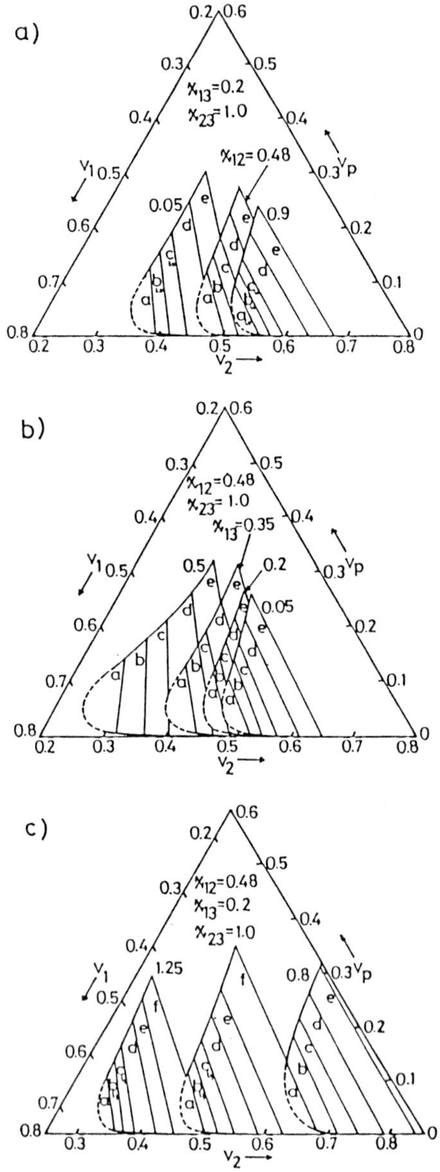

Fig. 3.33 Effects of χ_{12}, χ_{13} and χ_{23} parameters on the phase diagram of a quasi-ternary system of a polymer (Schulz-Zimm distr., $X_w^0 = 300$, $X_w^0/X_n^0 = 2.0$) dissolved in binary mixture: $v_p^S = 0.01$; a) $\chi_{13} = 0.2$, $\chi_{23} = 1.0$; b) $\chi_{12} = 0.48$, $\chi_{23} = 1.0$; c) $\chi_{12} = 0.48$, $\chi_{13} = 0.2$; a, $\rho_p = 1/15$; b, $\rho_p = 3/10$; c, $\rho_p = 1/2$; d, $\rho_p = 7/10$; e, $\rho_p = 4/5$; f, $\rho_p = 14/15$. $p_{12,s} = p_{13,q} = p_{23,r} = 0$.

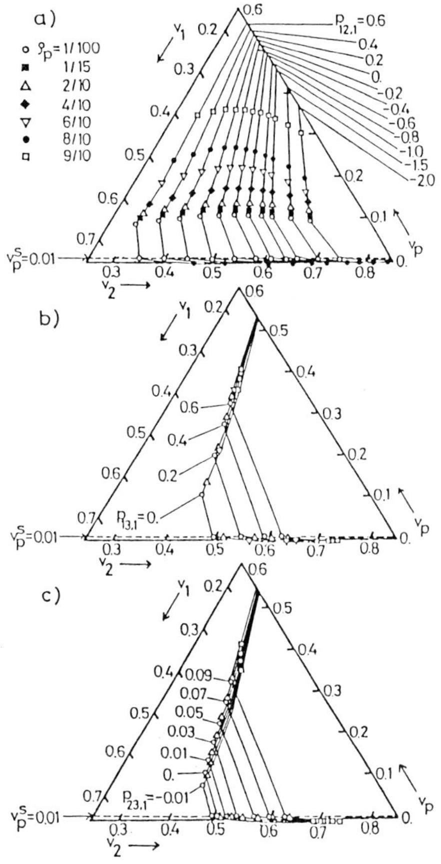

Fig. 3.34 Effects of the concentration-dependence parameters $p_{12,1}$, $p_{13,1}$ and $p_{23,1}$ on the coexistence curves and the tie-lines for a quasi-ternary system consisting of multicomponent polymers (Schulz-Zimm distribution, $X_w^0 = 300$, $X_w^0/X_n^0 = 2$) in binary solvent mixture: $\chi_{12}^0 = 0.5$, $\chi_{13}^0 = 0.2$ and $\chi_{23}^0 = 1.0$; $v_p^s = 0.01$. a) $p_{13,1} = p_{23,1} = 0$, b) $p_{12,1} = p_{23,1} = 0$, c) $p_{12,1} = p_{13,1} = 0$. ○ $\rho_p = 1/100$, ■ $\rho_p = 1/15$, △ $\rho_p = 2/10$, ◆ $\rho_p = 4/10$, ▽ $\rho_p = 6/10$, ● $\rho_p = 8/10$, □ $\rho_p = 9/10$.

Fig. 3.35 shows the relationship among χ_{12}, χ_{13} and χ_{23}, yielding the same phase separation characteristics as those obtained for quasi-binary solutions with fixed p_1 values (see, eqn (2.4), with $k' = 0$ and $p_2 = \cdots = p_n = 0$). Quasi-ternary solutions having small χ_{12}, large χ_{13} and large χ_{23} can be reasonably approximated with a quasi-binary solution with large negative p_1 (ref. 12). A combination of small χ_{12} and large χ_{13} or that

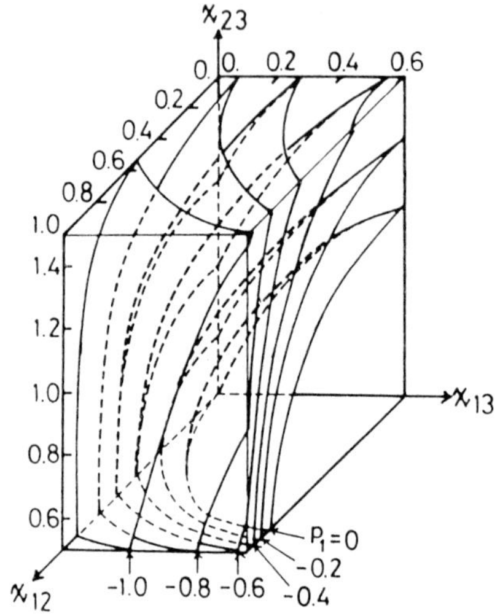

Fig. 3.35 Relationships between χ_{12}, χ_{13} and χ_{23}, yielding the same phase separation characteristics as those for a quasi-binary solution with a concentration dependence of the χ parameter p_1 (number on the curve): The original polymer, Schulz-Zimm distr., $X_w{}^0=300$, $X_w{}^0/X_n{}^0=2.0$; the initial polymer volume fraction $v_p{}^0=0.005$; relative amount of the polymer-rich phase $\rho_p=1/15$; $p_{12,s}=p_{13,q}=p_{23,r}=0$.

of large χ_{12} and small χ_{13} corresponds to fixed p_1, if χ_{23} is maintained constant. This relationship between χ_{12} and χ_{13} belongs to a family of hyperbolas. A suitable relationship between χ_{13} and χ_{23}, corresponding to a quasi-binary solution with fixed p_1, is almost χ_{13}-dependent, especially for small p_1. Large χ_{23} is combined with large χ_{12} at constant χ_{13} if the quasi-ternary solution is approximated with the quasi-binary solution with fixed p_1. Conclusively, a quasi-ternary polymer solution gives smaller σ, smaller R and broader MWD for polymers partitioned in the two phases, compared to a quasi-binary polymer solution in which a polymer having the same molecular characteristics as that used in a quasi-ternary solution is dissolved (ref. 12). This strongly suggests that the fundamental characteristics observed for the two-phase equilibrium of a quasi-binary polymer solution can be generalized quantitatively to a quasi-ternary polymer solution by assuming negative p_1, which is unrealistic for a quasi-binary solution (ref. 12). A quasi-ternary solution is inferior in separation efficiency, when the solution is used for fractionation.

3.2.5 Effects of average molecular weight and molecular weight distribution of the original polymer

In this chapter, hitherto, only a polymer having an SZ distribution with $X_w^0/X_n^0 = 2$ and $X_w^0 = 300$ was used and hence the influence of the average molecular weight and MWD of the original polymer was not clarified. For solutions consisting of multicomponent polymer in a single solvent, the important role of molecular characteristics in the phase separation phenomena has already been shown by Kamide and his collaborators (refs 25,38-40).

Fig. 3.36 shows the effects of X_w^0 of the original polymer having an SZ distribution ($X_w^0/X_n^0 = 2$) on the phase diagram such as coexistence curves and the tie-lines. A solid line is used for $X_w^0 = 1\times 10^3$, and a broken line, for $X_w^0 = 150$. With an increase in X_w^0, (1) the branch of the coexistence curve at the polymer-rich

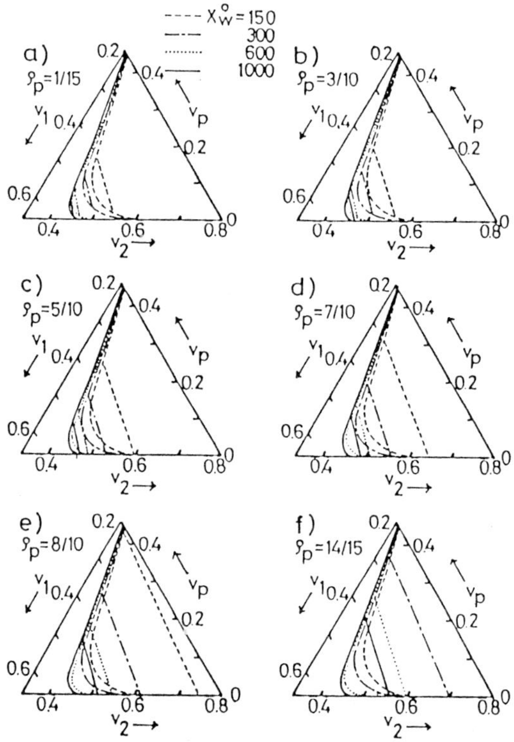

Fig. 3.36 Effect of the weight-average molar volume ratio X_w^0 of the original polymer on the phase diagram of quasi-ternary polymer solutions: The original polymer, Schulz-Zimm distribution, $X_w^0/X_n^0 = 2.0$; $v_P^S = 0.01$; a) $\rho_P = 1/15$, b) 3/10, c) 5/10, d) 7/10, e) 8/10, f) 14/15; $\chi_{12} = 0.48$, $\chi_{13} = 0.20$, $\chi_{23} = 1.20$; $X_w^0 = 150$ (broken line), 300 (chain line), 600 (dotted line), 1000 (full line).

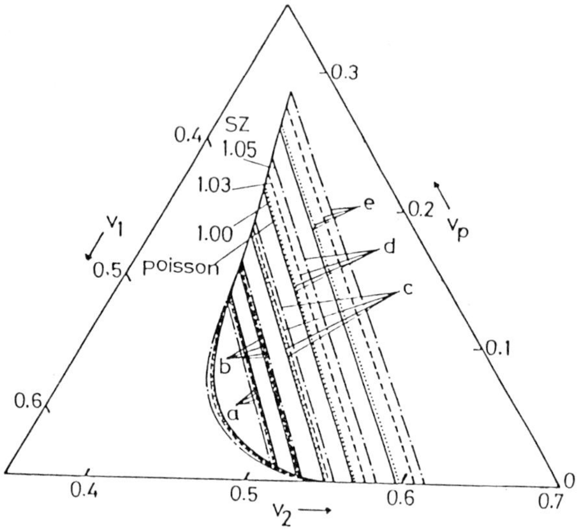

Fig. 3.37 Effect of molecular weight distribution on the phase diagram of solutions of a polymer dissolved in a binary solvent mixture: The original polymer, $X_w^0 = 300$, monodisperse (full line), Poisson distribution (dotted line), Schulz-Zimm distribution, $X_w^0/X_n^0 = 1.03$ (broken line), Schulz-Zimm distribution, $X_w^0/X_n^0 = 1.05$ (chain line); $v_p^s = 0.01$; $\chi_{12} = 0.48$, $\chi_{13} = 0.20$, $\chi_{23} = 1.00$; a) $\rho_p = 1/15$, b) 3/10, c) 1/2, d) 7/10, e) 14/15.

phase side shifts to a lower v_p region, (2) the peak point (or plait point) of the coexistence curve changes its composition, i.e., increasing v_1 and decreasing v_p and v_2 and (3) the distance of the tie lines corresponding to given ρ_p decreases drastically. (2) and (3) are consistent with a fact that the solution of the polymer with larger X_w tends to separate into two phases with smaller amounts of solvent 2. At given ρ_p, the polymer volume fraction in a polymer-rich phase $v_{p(2)}$ decreases with increasing X_w^0 (ref. 13).

Fig. 3.37 shows the effects of X_w^0/X_n^0 on the phase diagrams calculated under the same conditions: $X_w^0 = 300$, $\chi_{12} = 0.48$, $\chi_{13} = 0.20$, $\chi_{23} = 1.00$ and $v_p^s = 0.01$. Fig. 3.37 shows phase diagrams for a monodisperse polymer, a polymer with Poisson distribution ($X_w^0/X_n^0 = 1.003$), and a polymer having an SZ distribution (with $X_w^0/X_n^0 = 1.03$ and 1.05; SZ1.03 and SZ1.05, respectively) are shown. All of these polymers belong to the category of so-called "monodisperse" polymers. The coexistence curves for these polymers are not easily distinguishable from each other, particularly in the large ρ_p region. Accordingly, experiments with polymers having X_w/X_n of 1.03～1.05 as monodisperse polymers for studying phase separation phenomena may be theoretically acceptable except at small ρ_p, provided only the coexistence curve is being considered. But, the effect of X_w^0/X_n^0 on the tie-line must never be neglected even if X_w^0/X_n^0 is less than 1.05 (ref. 13). There is a significant difference in the tie-line between the strictly monodisperse polymer and polymers with $X_w^0/X_n^0 = 1.03$～1.05. Unfortunately, this point has been overlooked hitherto and "monodisperse" polymers have been employed in actual experiments to obtain phase diagrams of "truly monodisperse" polymers.

Fig. 3.38a and c show the effects of X_w^0 on the σ vs. v_p^0 and R vs. v_p^0 relationships, respectively. Evidently, σ decreases gradually with increasing v_p^0 throughout the entire range of v_p^0. At a same value of v_p^0, an increase in X_w^0 by a factor of ten is accompanied by a substantial decrease in σ by a factor of ten. A single master curve is obtained by shifting concurrently the σ - v_p^0 curves along the σ -axis and the v_p^0-axis. The value of R is approximately linear to $-\ln v_p^0$ in the range of $R > 20$. The slope of the plot is estimated to be 1.15, irrespective of X_w^0. These

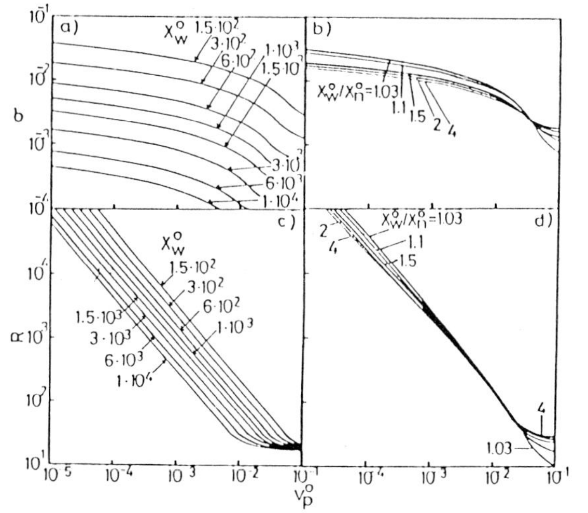

Fig. 3.38 Effect of the weight-average molar volume ratio X_w^0 and molecular weight distribution of the original polymer on the relationship between the partition coefficient σ or phase volume ratio R and the initial polymer volume fraction v_p^0: The original polymer, Schulz-Zimm distribution; a) and c), $X_w^0/X_n^0 = 2.0$; b) and d), $X_w^0 = 300$; $\chi_{12} = 0.48$, $\chi_{13} = 0.20$, $\chi_{23} = 1.0$; $\rho_p = 1/15$.

findings are the same as those observed for the quasi-binary solutions by Kamide et al. (refs 27,28). The volume fraction v_p^0, above which the linearity of the plot ceases to hold, is roughly in agreement with the critical volume fraction.

The effect of X_w^0/X_n^0 on the correlation of σ and R with v_p^0 is displayed graphically in Fig. 3.38b and d. Inversion of the effect of X_w^0/X_n^0 is observed at $v_p^0 \simeq 0.05$ for σ and $v_p^0 \simeq 0.02$ for R, above which σ as well as R is larger for the polymer with smaller X_w^0/X_n^0, suggesting that fractionation is more effective for a polymer with a sharp MWD, at least, in a low v_p^0 range.

Fig. 3.39 shows plots for σ and R as a function of X_w^0 and X_w^0/X_n^0 for constant ρ_p. Both σ and R decrease rapidly, firstly with an increase in X_w^0, approaching asymptotic values depending in ρ_p. This feature is much more pronounced with σ. The sign of $(\partial \sigma / \partial (X_w^0/X_n^0))$ changes at $\rho_p = 4/10$. R vs. X_w^0/X_n^0 curves for various ρ_p values form an envelope line representing minimum R, R_{min} at given X_w^0/X_n^0 and the ρ_p value corresponding to R_{min} depends markedly on X_w^0/X_n^0.

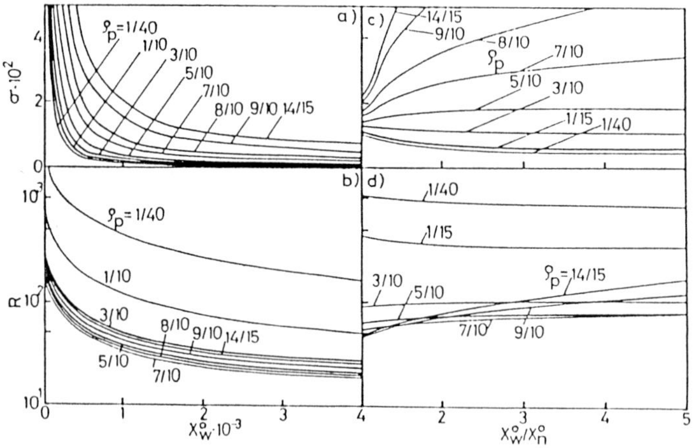

Fig. 3.39 Effect of the relative amount of polymer dissolved in a polymer-rich phase ρ_p on the dependence of the partition coefficient σ and phase ratio R on the weight-average molar volume ratio X_w^0 and X_w^0/X_n^0 of the original polymer: The original polymer, Schulz-Zimm distr., $X_w^0/X_n^0 = 2.0$ [a) and b)], $X_w^0 = 300$, [c) and d)], $v_p^0 = 0.005$; $\chi_{12} = 0.48$, $\chi_{13} = 0.20$, $\chi_{23} = 1.00$.

Fig. 3.40a and b illustrate the effect of X_w^0 and X_w^0/X_n^0 on the normalized MWD, $g_{(2)}(X)$ of the polymer ($\rho_p = 1/15$) in the polymer-rich phase from a quasi-ternary solution of $v_p^0 = 0.005$.

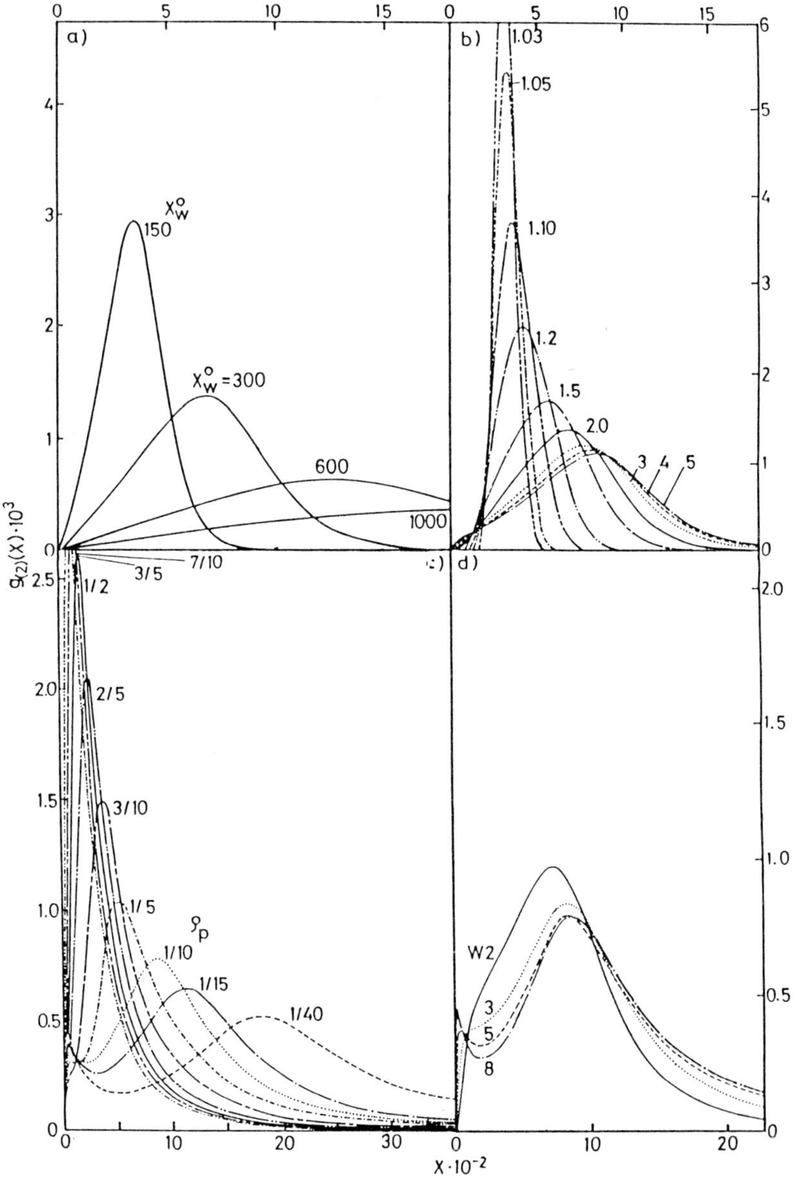

Fig. 3.40 The molecular weight distribution curves, $g_{(2)}(X)$, of the polymer in a polymer-rich phase separated from a quasi-ternary polymer solution: The original polymer, Schulz-Zimm distr. [a) and b)] and Wesslau distr. [c) and d)]; $v_p^0=0.005$; $\chi_{12}=0.48$, $\chi_{13}=0.20$ and $\chi_{23}=1.00$ [a) and b)] or 1.30 [c) and d)]; $\rho_p=1/15$ [a), b) and d)] or 1/2−1/40 [c)]; a) $X_w^0/X_n^0=2.0$, X_w^0 is denoted on the curve; b) $X_w^0=300$, X_w^0/X_n^0 is denoted on the curve; c) $X_w^0=300$, $X_w^0/X_n^0=5.0$, ρ_p is denoted on the curve; d) $X_w^0=300$, X_w^0/X_n^0 is denoted on the curve.

$g_{(2)}(X)$ becomes broader (i.e., the content of the larger X component increases) with increasing X_w^0 and X_w^0/X_n^0, and its peak becomes lower, but the ratio (X_w/X_n) of the polymers in polymer-rich phase becomes smaller.

Fig. 3.40c and d show that when the polymer-rich phase is separated at small ρ_p from the solution of the original polymer having a broad MWD, another peak appears near $X=0$ and with decreasing ρ_p and increasing X_w^0/X_n^0, its height increases. The possibility of the appearance of double peaks for polymers in a polymer-rich phase has already been noted for quasi-binary solutions (ref. 39). The discussions described can be applied to all quasi-ternary systems.

Fig. 3.41 shows the effects of X_w^0 on the relationships between X_w/X_n of polymers in the polymer-rich phase and polymer-

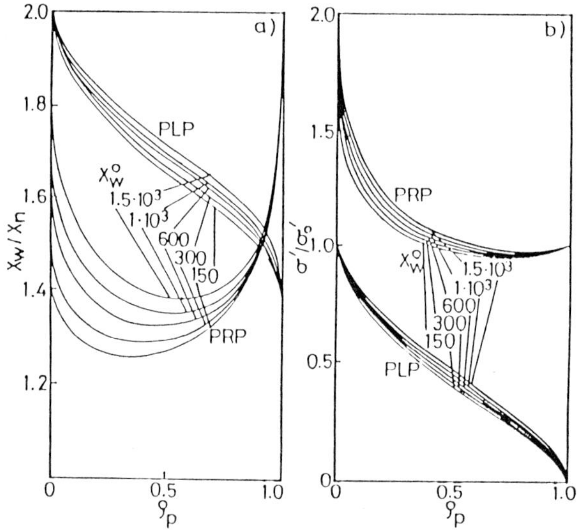

Fig. 3.41 Effect of the weight-average molar volume ratio X_w^0 of the original polymer on the relationship between X_w/X_n or σ'/σ_0' and the relative amount of the polymer in a polymer-rich phase: The original polymer, Schulz-Zimm distr., $X_w^0/X_n^0=2.0$, X_w^0 is denoted on the curve; $v_p^0=0.005$; $\chi_{12}=0.48$, $\chi_{13}=0.20$, $\chi_{23}=1.00$; PLP, polymer-lean phase; PRP, polymer-rich phase.

lean phase or on the ratio of the standard deviation of $g_{(1)}(X)$ and $g_{(2)}(X)$ to that of $g_0(X)$, σ'/σ'_0 and ρ_p. This figure was obtained for polymers with $X_w^0=150\sim1.5\times10^3$ having the SZ distribution, $X_w^0/X_n^0=2.0$, assuming $\chi_{12}=0.48$, $\chi_{13}=0.2$, $\chi_{23}=1.0$ and $v_p^0=0.005$. As X_w^0 increases, X_w/X_n of the polymer in the polymer-rich phase increases significantly and the minimum

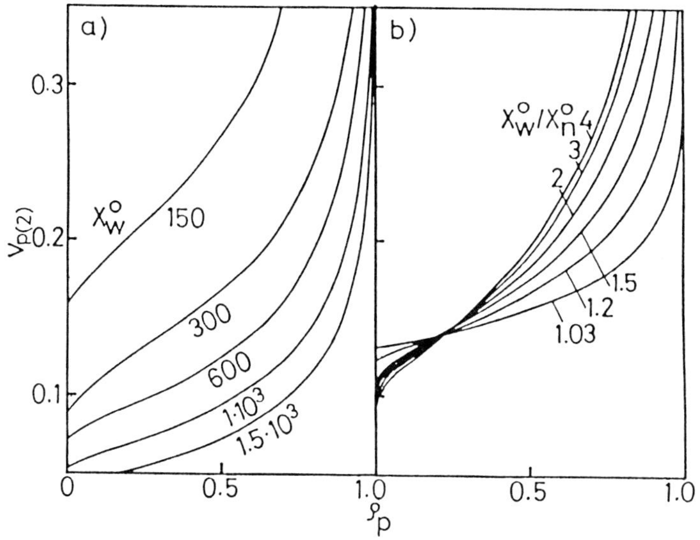

Fig. 3.42 Change in the polymer volume fraction of a polymer-rich phase $v_{p(2)}$ with a relative amount of polymer ρ_p: The original polymer, Schulz-Zimm distribution; a) X_w^0 is denoted on the curve; $X_w^0/X_n^0=2.0$; b) $X_w^0=300$, X_w^0/X_n^0 is denoted on the curve; the polymer volume fraction of the initial solution $v_p^0=0.005$; $\chi_{12}=0.48$, $\chi_{13}=0.20$, $\chi_{23}=1.30$ (a)) or 1.00 (b)).

value of X_w/X_n, $(X_w/X_n)_{min}$, increases, shifting $\rho_{p,min}$ towards a higher ρ_p region. An increase in X_w^0 exerts an influence, entirely analogous to that of increasing v_p^0 (ref. 41) on the relationship between X_w/X_n or σ'/σ_0' and ρ_p. σ'/σ_0 for the polymer-rich phase exhibits a minimum in a comparatively higher ρ_p region ($\rho_p>0.5$) and σ'_{min} appears independent of X_w^0. X_w/X_n of the polymer-rich phase is smaller than that of the polymer-lean phase, except for the large ρ_p region. This does not indicate that the polydispersity of the polymer in a polymer-rich phase is lower than that in the polymer-lean phase. σ', the most reasonable parameter representing polydispersity, is much smaller in the polymer-lean phase than in the polymer-rich phase over the entire range of ρ_p.

Fig. 3.42 shows the effects of X_w and X_w/X_n on the $v_{p(2)}-\rho_p$ relations. In this case, the solutions ($v_p^0=0.005$) of SZ polymer with $X_w^0/X_n^0=2$ (Fig. 3.42a) and $X_w^0=300$ (Fig. 3.42b) were phase-separated at $\chi_{12}=0.48$, $\chi_{13}=0.2$ and $\chi_{23}=1.3$. The polymer

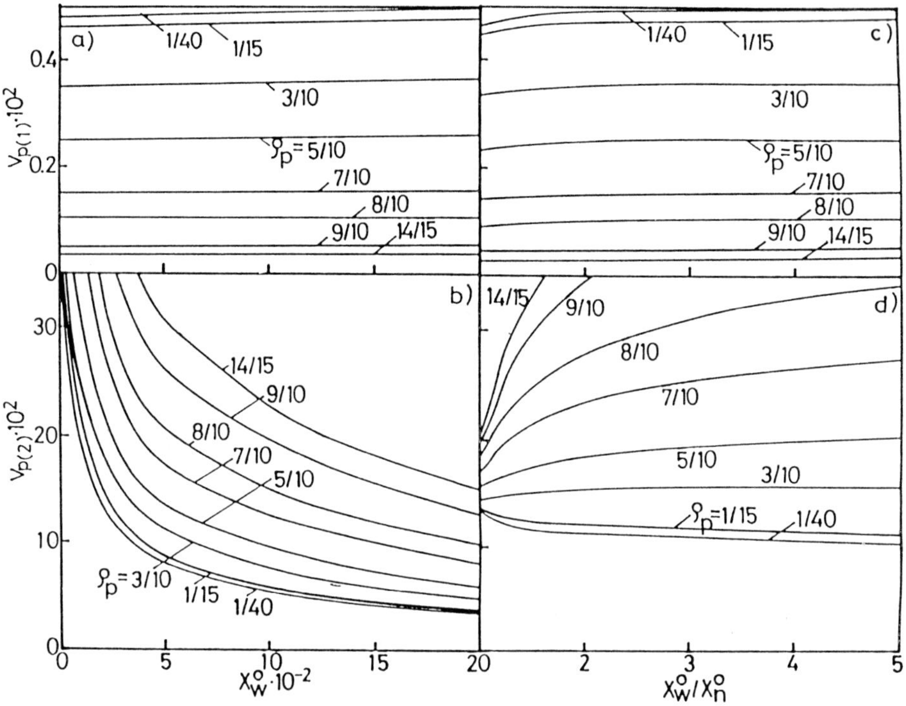

Fig. 3.43 Relationships among polymer volume fractions in polymer-lean and -rich phase, $v_{p(1)}$ and $v_{p(2)}$, the relative amount of a polymer in a polymer-rich phase ρ_p, X_w^0 and X_w^0/X_n^0 of the original polymer: The original polymer, Schulz-Zimm distr., $X_w^0/X_n^0=2.0$ [a) and b)], $X_w^0=300$ [c and d)]; $v_p^0=0.005$; $\chi_{12}=0.48$, $\chi_{13}=0.20$, $\chi_{23}=1.00$.

volume fraction in the polymer-rich phase $v_{p(2)}$ increases with ρ_p and decreases with X_w^0. The effect of X_w^0/X_n^0 on $v_{p(2)}$ changes its sign at $\rho_p=0.24$, above which $v_{p(2)}$ increases with X_w^0/X_n^0.

Fig. 3.43 shows the effect of ρ_p on the polymer volume fraction in the polymer-lean phase $v_{p(1)}$ (or $v_{p(2)}$) vs. X_w^0 (or X_w^0/X_n^0) relations. As expected, $v_{p(1)}$ is nearly independent of X_w^0 irrespective of ρ_p. In contrast, $v_{p(2)}$ decreases remarkably with an increase in X_w^0 and is independent of X_w^0/X_n^0 at $\rho_p=0.24$. For $\rho_p>0.24$, $v_{p(2)}$ increases with X_w^0/X_n^0, but at $\rho_p<0.24$, $v_{p(2)}$ gradually decreases with X_w^0/X_n^0. The effect of X_w^0 and X_w^0/X_n^0 of the original polymer on the phase separation parameters for quasi-ternary solutions are summarized in Table 3.1.

TABLE 3.1

Effect of X_w^0 and X_w^0/X_n^0 on phase separation parameters for quasi-ternary solutions

Parameters	Effect	
	X_w^0	X_w^0/X_n^0
Coexistence curve	Shift to lower v_2[a]	Shift to lower v_2
Tie line	Shift to lower v_2	Shift to lower v_2 at small ρ_p
		Shift to higher v_2 at large ρ_p
Total volume V	−[b]	− at small ρ_p
		+[c]at large ρ_p
Composition V_1/V_2	+	+ at large ρ_p
		− at large ρ_p
Partition coefficient σ	−	− at small v_p^0
		+ at large v_p^0
		− at small ρ_p
		+ at large ρ_p
Phase volume ratio R	−	− at small v_p^0
		+ at large v_p^0
		− at small ρ_p
		+ at large ρ_p
Breadth in mol. wt. distr. of polymer in PRP	+	+
Double peaked MWD of polymer in PRP		+ Wesslau at small ρ_p
ρ_p giving minimum X_w/X_n of polymer in PRP	+	~0[d]
$(X_w/X_n)_{min}$ in PRP	+	+
Polymer volume fraction $v_{p(2)}$ in PRP	−	+ (Effect of small region) p_1

[a]Parameter shift to lower or higher v_2 with an increase in X_w^0 or X_w^0/X_n^0; [b]−, decrease; [c]+, increase; [d]~0, almost insensitive; PRP, polymer-rich phase.

Fig. 3.44 shows plots for M_w/M_n vs. M_w for SSF runs (ref. 12). The unfilled circles are a series of PS fractions ($n_t = 22$), separated successively using methylethylketone as the solvent and methanol as the nonsolvent ($v_p^0 = 6.15 \times 10^{-3}$) (ref. 24) (see, section 2.4). The solid line is a theoretical curve calculated assuming $\chi_{12} = 0.2$, $\chi_{13} = 0.05$, $\chi_{23} = 0.80$ and $\rho_p \simeq 1/n_t$ ($= 1/22$). The coincidence of the experimental data with theoretical calculation is satisfactory. However, note that the

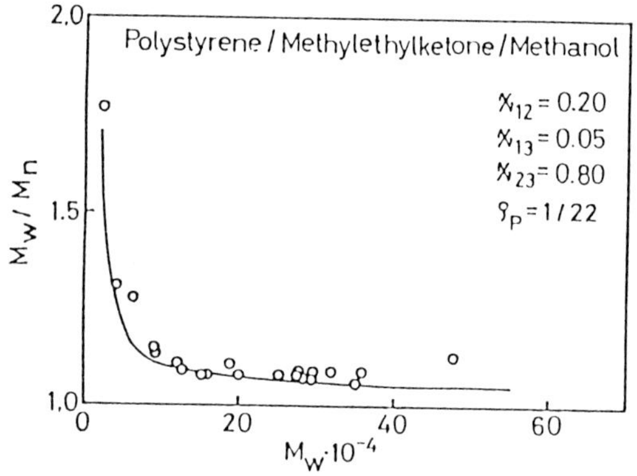

Fig. 3.44 The ratio M_w/M_n plotted against M_w of fractions separated by successive solution fractionation from a quasi-ternary system: ○, experimental data on polystyrene / methylethylketone / methanol; full line, theoretical curve ($\chi_{12}=0.20$, $\chi_{13}=0.05$, $\chi_{23}=0.80$ and $\rho_p=1/22$); $v_p^0=6.15\times10^{-3}$, $p_{12,s}=p_{13,q}=p_{23,r}=0$.

three χ_{ij} parameters are not experimentally determined and this figure does not mean a direct comparison between the actual experiment and the computer experiment based on the theory of phase separation of quasi-ternary solutions. This problem is still open for further research.

3.3 SPINODAL CURVE AND CRITICAL SOLUTION POINT

In 1949, Scott (ref. 3) derived equations giving the critical solution point (CSP) of a ternary system consisting of monodisperse polymer, solvent and nonsolvent. Two equations derived by Scott for spinodal and so-called neutral equilibrium conditions (i.e., eqns (25) and (26) in ref. 3) are as will be discussed in more detail, not thermodynamically consistent and in particular the latter equation is incorrect (ref. 26). Kurata (ref. 22) proposed general and strict equations of CSP of the mixed solvent system including a ternary system, consisting of the monodisperse polymer dissolved in a binary solvent mixture. All the studies made by Tompa (ref. 5), Bamford and Tompa (ref. 42), Okamoto (ref. 9) and Nakagaki and Sunada (ref. 43), who put their theoretical starting point on the Scott equations, are obviously not theoretically rigorous. Of course, no theoretical study has ever been published on CPC and CSP of quasi-ternary system.

In this section, an attempt is made to establish a theory of

CSP of a quasi-ternary system, studying the effects of χ_{12}, χ_{13} and χ_{23} and that of $X_w{}^0$ and $X_w{}^0/X_n{}^0$ of the original polymer on CSP and to propose a method for calculating indirectly CPC for the system, as intercepts of the line of constant starting concentration $v_p{}^s$ (see, eqn (3.27)) with a coexistence curve theoretically obtained (ref. 26).

The spinodal condition for quasi-ternary system is given by (ref. 26)

$$|\Delta G'| \equiv \begin{vmatrix} \Delta\bar{G}'_{NN} & \Delta\bar{G}'_{N1} & \cdots & \Delta\bar{G}'_{Nm} \\ \Delta\bar{G}'_{1N} & \Delta\bar{G}'_{11} & \cdots & \Delta\bar{G}'_{1m} \\ \Delta\bar{G}'_{2N} & \Delta\bar{G}'_{21} & \cdots & \Delta\bar{G}'_{2m} \\ \vdots & \vdots & & \vdots \\ \Delta\bar{G}'_{mN} & \Delta\bar{G}'_{m1} & \cdots & \Delta\bar{G}'_{mm} \end{vmatrix} = 0 \qquad (3.47)$$

and the Gibbs' free energy change of mixing per unit volume, $\Delta G'$ is given by (ref. 26)

$$\Delta G' = v_1 \left(\frac{\Delta\mu_1}{V_0}\right) + v_2 \left(\frac{\Delta\mu_2}{V_0}\right) + \sum_{i=1}^{m} v_{Xi} \left(\frac{\Delta\mu_{Xi}}{X_j V_0}\right) \qquad (3.48)$$

and $\Delta G'_{ij}$ is defined by eqn (3.49):

$$\Delta G'_{ij} = \left(\frac{\partial^2 \Delta G'}{\partial v_{Xi} \partial v_{Xj}}\right)_{T,P,v_{Xk}} \qquad (3.49)$$

$$(i,j = N,1,2 \cdots,m; \quad k \neq i,j).$$

V_0 is the molar volume of solvent 1 and 2. In eqns (3.47) and (3.49), the suffix N denotes nonsolvent (in this case, solvent 2) in order to distinguish the component of the polymer (i = 2).

The neutral equilibrium condition is expressed by eqn (3.50) (ref. 26).

$$|\Delta G''| \equiv \begin{vmatrix} \frac{\partial|\Delta G'|}{\partial v_2} & \frac{\partial|\Delta G'|}{\partial v_{X1}} & \frac{\partial|\Delta G'|}{\partial v_{X2}} & \cdots & \frac{\partial|\Delta G'|}{\partial v_{Xm}} \\ \Delta\bar{G}'_{1N} & \Delta\bar{G}'_{11} & \Delta\bar{G}'_{12} & \cdots & \Delta\bar{G}'_{1m} \\ \Delta\bar{G}'_{2N} & \Delta\bar{G}'_{21} & \Delta\bar{G}'_{22} & \cdots & \Delta\bar{G}'_{2m} \\ \vdots & \vdots & \vdots & & \vdots \\ \Delta\bar{G}'_{mN} & \Delta\bar{G}'_{m1} & \Delta\bar{G}'_{m2} & \cdots & \Delta\bar{G}'_{mm} \end{vmatrix} = 0. \qquad (3.50)$$

Substitution of eqns (3.1)-(3.3) into eqn (3.48) yields

$$\Delta \bar{G}' = \left(\frac{\tilde{R}T}{V_0}\right) [v_1 \ln v_1 + v_2 \ln v_2 + \sum_{i=1}^{m} \frac{v_{Xi}}{V_0} \ln v_{Xi}$$

$$+ \chi_{12} v_1 v_2 + \chi_{13} v_1 v_p + \chi_{23} v_2 v_p] . \tag{3.51}$$

From eqns (3.49) and (3.51), we obtain

$$\left(\frac{V_0}{\tilde{R}T}\right) \Delta G'_{ij} = \frac{1}{v_1} - 2\chi_{13} \equiv M \qquad \text{(for } i \neq j, i \text{ and } j \neq N), \tag{3.52a}$$

$$\left(\frac{V_0}{\tilde{R}T}\right) \Delta G'_{ij} = M + \chi_{13} + \chi_{23} - \chi_{12} \equiv M + K$$

$$\text{(for } i \neq j,\ i \text{ or } j = N), \tag{3.52b}$$

$$\left(\frac{V_0}{\tilde{R}T}\right) \Delta G'_{ij} = M + \frac{1}{v_2} + 2(\chi_{13} - \chi_{12}) \equiv M + U$$

$$\text{(for } i = j = N), \tag{3.52c}$$

$$\left(\frac{V_0}{\tilde{R}T}\right) \Delta G'_{ij} = M + \frac{1}{X_i v_{Xi}} \equiv M + M_i \qquad \text{(for } i = j \neq N). \tag{3.52d}$$

Substituting eqns (3.52a)-(3.52d) into eqn (3.47), we obtain

$$|\Delta G'| = \left(\frac{\tilde{R}T}{V_0}\right)^{m+1} \begin{vmatrix} M+U & M+K & M+K & \cdots & M+K \\ M+K & M+M_1 & M & \cdots & M \\ M+K & M & M+M_2 & \cdots & M \\ \vdots & \vdots & \vdots & & \vdots \\ M+K & M & M & \cdots & M+M_m \end{vmatrix}$$

$$= \left(\frac{\tilde{R}T}{V_0}\right)^{m+1} \left(\prod_{j=1}^{m} M_j\right) [(M+U)\{1 + M\sum_{j=1}^{m} \frac{1}{M_j}\} - (M+K)^2 \sum_{j=1}^{m} \frac{1}{M_j}] = 0. \tag{3.53}$$

Then, the spinodal condition is (ref. 26):

$$(M+U)\{1 + M\sum_{j=1}^{m} \frac{1}{M_j}\} - (M+K)^2 \sum_{j=1}^{m} \frac{1}{M_j} = 0. \tag{3.54}$$

Partial differentiation of eqn (3.53) with v_{xi} $(i=N,1,2,\cdots,m;$ $v_{xN}=v_2)$ yields

$$\left(\frac{\partial|\Delta G'|}{\partial v_{xi}}\right)_{T,P,v_{xk}} = \left(\frac{\tilde{R}T}{V_0}\right)^{m+1}\left(\prod_{j=1}^{m} M_j\right)\frac{\partial}{\partial v_{xi}}\left\{(M+U)(1+Mv_pX_w^{\,0}) - (M+K)^2 v_pX_w^{\,0}\right\}_{T,P,v_{xk}} \equiv W_i \qquad (3.55)$$

$(k\neq i,\ k=N,1,2,\cdots,m;\quad v_{xN}=v_2)$.

Substituting eqns (3.52a)-(3.52d) and (3.55) into eqn (3.50), we obtain eqn (3.56),

$$\left(\frac{V_0}{\tilde{R}T}\right)^{m+1}|\Delta G''| = \begin{vmatrix} W_N & W_1 & W_2 & \cdots & W_m \\ M+K & M+M_1 & M & \cdots & M \\ M+K & M & M+M_2 & \cdots & M \\ \vdots & \vdots & \vdots & & \vdots \\ M+K & M & M & \cdots & M+M_m \end{vmatrix}$$

$$= \left(\prod_{j=1}^{m} M_j\right)\left[W_N\left(1+M\sum_{j=1}^{m}\frac{1}{M_j}\right) - (M+K)\sum_{j=1}^{m}\frac{W_j}{M_j}\right]. \qquad (3.56)$$

Neutral equilibrium condition is given by (ref. 26)

$$W_N\left(1+M\sum_{j=1}^{m}\frac{1}{M_j}\right) - (M+K)\sum_{j=1}^{m}\frac{W_j}{M_j} = 0. \qquad (3.57)$$

Combining eqn (3.54) with eqns (3.52a)-(3.52b), we obtain eqn (3.58) (put the lefthand side of eqn (3.58) with A):

$$A\equiv\left(\frac{1}{v_1}+\frac{1}{v_2}-2\chi_{12}\right)\left(\frac{1}{v_pX_w^{\,0}}+\frac{1}{v_1}-2\chi_{13}\right) - \left(\frac{1}{v_1}+\chi_{23}-\chi_{13}-\chi_{12}\right)^2 = 0. \qquad (3.58)$$

Eqn (3.59) can be derived from eqn (3.51) considering eqns (3.52a)-(3.52d) and eqn (3.55) (put the lefthand side of eqn (3.59) with B).

$$B \equiv \left[\frac{1}{v_p X_w^0}\left(\frac{1}{v_1^2}-\frac{1}{v_2^2}\right)+\frac{1}{v_1^2}\left(\frac{1}{v_2}-2\chi_{23}\right)\right.$$

$$-\frac{1}{v_2^2}\left(\frac{1}{v_1}-2\chi_{13}\right)\Big]\left(\frac{1}{v_p X_w^0}+\frac{1}{v_1}-2\chi_{13}\right)$$

$$-\left(\frac{1}{v_1}+\chi_{23}-\chi_{13}-\chi_{12}\right)\Big[\frac{1}{v_1^2}\left(\frac{1}{v_2}-2\chi_{23}\right)$$

$$+\frac{1}{v_1^2 v_p X_w^0}+\frac{X_z^0}{v_p X_w^0}\Big\{\frac{1}{v_1 v_2}-\frac{2\chi_{13}}{v_2}-\frac{2\chi_{23}}{v_1}$$

$$\left.+2(\chi_{12}\chi_{13}+\chi_{13}\chi_{23}+\chi_{23}\chi_{12})-(\chi_{12}^2+\chi_{13}^2+\chi_{23}^2)\Big\}\right]=0. \tag{3.59}$$

Note that the following relation holds

$$v_1+v_2+v_3=1. \tag{3.60}$$

If χ_{12}, χ_{13}, χ_{23}, X_w^0 and X_z^0 are known in advance, both A and B are functions of v_1 and v_p (ref. 26):

$$A=A(v_1,\ v_p)=0, \tag{3.61}$$

$$B=B(v_1,\ v_p)=0. \tag{3.62}$$

For given v_p, $A=A(v_1)=0$ is obtained and v_1 can be evaluated using the single-variable Newton method and gives a spinodal curve. By solving the simultaneous equations eqns (3.61) and (3.62), using the two-variables Newton method, the compositions v_1^c and v_p^c (and v_2^c) at CSP can be evaluated. Eqns (3.58) and (3.59) are evidently symmetrical with respect to the exchange of solvent 1 and 2, respectively (ref. 26).

Fig. 3.45 shows a coexistence curve by a full line for some quasi-ternary solutions having v_p^s ranging from $10^{-5}\sim 0.7$. Here, a solid line linking filled and unfilled marks is a "limiting tie-line" and a chain line is a constant v_p^s line and the cross point of constant v_p^s line and a coexistence curve correspond to a cloud point demonstrated as a filled mark in the figure (ref. 26). An unfilled circle is the composition of the polymer-lean ($v_p^0 > v_p^c$) or -rich ($v_p^0 < v_p^c$) phase which is separated from the mother solution at the cloud point. A cloud point curve for the system

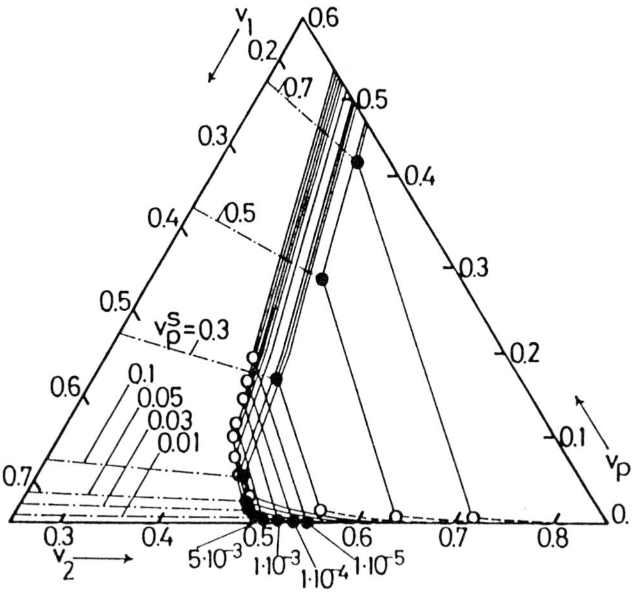

Fig. 3.45 Binodal curve (solid line) and cloud point (filled circle) separated from the original solution of a quasi-ternary system with the starting polymer volume fraction $v_p{}^s = 1\times10^{-5}\sim0.7$: Original polymer, Schulz-Zimm type distribution ($X_w{}^0 = 300$, $X_w{}^0/X_n{}^0 = 2$); $\chi_{12} = 0.5$, $\chi_{13} = 0.2$, $\chi_{23} = 1.0$; chain line, constant $v_p{}^s$ line; filled and unfilled circle are the coexisting phases; $p_{12,s} = p_{13,q} = p_{23,r} = 0$.

can be obtained by connecting the filled marks and a so-called shadow curve by connecting unfilled marks.

Fig. 3.46 exemplifies CPC, constructed in the same manner as shown in Fig. 3.46, the spinodal curve and CSP for quasi-ternary solutions. As theory requires, the cloud point and spinodal curves intercept beautifully on CSP (ref. 26). This strongly implies that both the present theory and the simulation technique proposed here are reasonable and unconditionally acceptable.

Fig. 3.47 shows the effects of the polydispersity of the polymer on the spinodal curve and CSP for quasi-ternary systems containing a polymer with SZ type distribution ($X_w{}^0 = 300$). As expected from eqn (3.58), the spinodal curve is absolutely independent of $X_w{}^0/X_n{}^0$, depending only on $X_w{}^0$ and three χ parameters. CSP is only slightly influenced by $X_w{}^0/X_n{}^0$. It is interesting to note that CSP of a monodisperse polymer does not

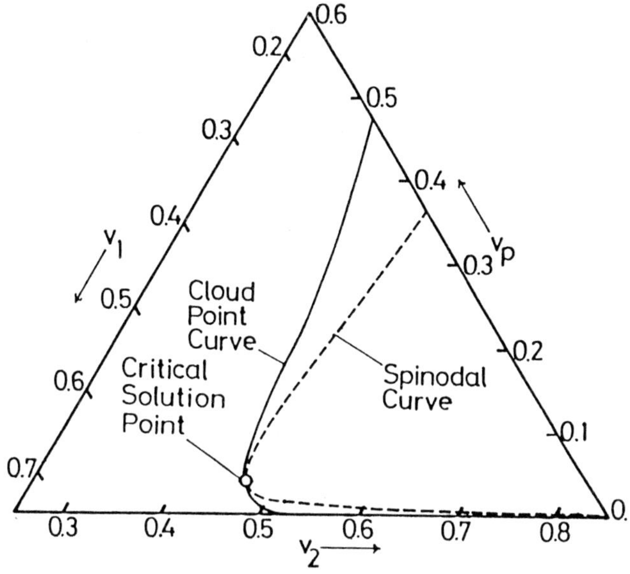

Fig. 3.46 Cloud point curve (full line), constructed from Fig. 3.45, spinodal curve (broken line) and critical solution point (unfilled circle) of a quasi-ternary system: Original polymer, Schulz-Zimm type distribution ($X_w{}^0 = 300$, $X_w{}^0/X_n{}^0 = 2$); $\chi_{12} = 0.5$, $\chi_{13} = 0.2$, $\chi_{23} = 1.0$; $p_{12,s} = p_{13,q} = p_{23,r} = 0$.

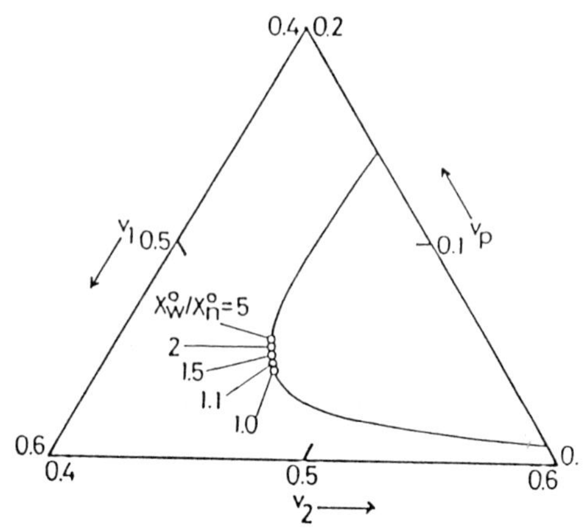

Fig. 3.47 Effect of $X_w{}^0/X_n{}^0$ of the original polymer on the spinodal curve and critical solution point of quasi-ternary solutions: Original polymer, Schulz-Zimm type distribution ($X_w{}^0 = 300$); $\chi_{12} = 0.5$, $\chi_{13} = 0.2$, $\chi_{23} = 1.0$; $p_{12,s} = p_{13,q} = p_{23,r} = 0$.

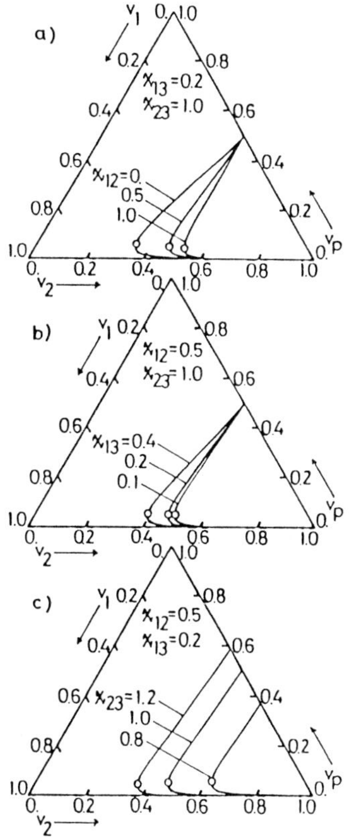

Fig. 3.48 Effects of χ_{12}, χ_{13} and χ_{23} on the spinodal curve and CSP of quasi-ternary systems consisting of multicomponent polymers with Schulz-Zimm type distribution ($X_w{}^0=300$, $X_w{}^0/X_n{}^0=2$) in a binary solvent mixture: a) $\chi_{13}=0.2$, $\chi_{23}=1.0$; b) $\chi_{12}=0.5$, $\chi_{23}=1.0$; c) $\chi_{12}=0.5$, $\chi_{23}=0.2$; $p_{12,s}=p_{13,q}=p_{23,r}=0$.

locate on the peak (maximum v_1 point; $v_1{}^P$, $v_2{}^P$, $v_p{}^P$) of spinodal curve, but shifts to the higher v_p side. In this case, ($v_1{}^c$, $v_2{}^c$, $v_p{}^c$) = (0.4920, 0.4661, 0.0419) and ($v_1{}^P$, $v_2{}^P$, $v_p{}^P$) = (0.4927, 0.4705, 0.0368).

Figs. 3.48a–c show the effects of χ_{12}, χ_{13} and χ_{23} on the spinodal curve and critical point. With a decrease in χ_{12} and with an increase in χ_{13} and χ_{23}, the spinodal curve and accordingly CSP shifts in a direction of decreasing content of solvent 2 (nonsolvent). χ_{23} has the strongest effect on the spinodal curve and χ_{12} is second. At a point of $v_1=0$ (i.e., the case of polymer-solvent 2 mixture) the spinodal curve is almost independent of χ_{12} and χ_{13} and is governed by χ_{23} alone, because the composition of this point can be given by solving equation;

$$v_p + \frac{v_2}{X_w^0} - 2\chi_{23} v_2 v_p = 0 \tag{3.63}$$

and at the limit of $X_w^0 \to \infty$ eqn (3.63) reduces to

$$v_p = 1 - \frac{1}{2\chi_{23}} \tag{3.64}$$

and

$$v_2 = \frac{1}{2\chi_{23}}. \tag{3.65}$$

Eqn (3.63) can be readily derived from eqn (3.58).

Fig. 3.49 shows the effects of χ_{12}, χ_{13} and χ_{23} on v_1^c, v_2^c and v_p^c for quasi-ternary solutions of a polymer with SZ type distribution ($X_w^0 = 300$, $X_w^0/X_n^0 = 2$). For smaller χ_{12} and larger χ_{13}, both v_1^c and v_p^c are larger and v_2^c is smaller. The effects of χ_{12} on change in v_1^c, v_2^c and v_p^c are smaller than those of χ_{13} and χ_{23}.

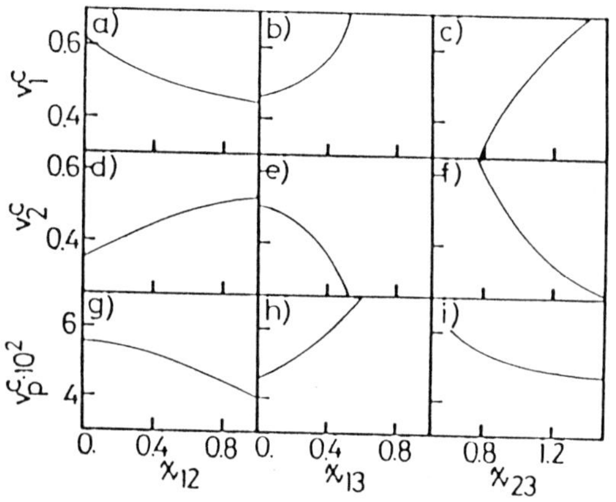

Fig. 3.49 Plots of v_1^c, v_2^c and v_p^c against χ_{12}, χ_{13} and χ_{23} of quasi-ternary solutions: Original polymer, Schulz-Zimm type distribution ($X_w^0 = 300$, $X_w^0/X_n^0 = 2$); a), d), and g) $\chi_{13} = 0.2$, $\chi_{23} = 10$; b), e), and h) $\chi_{12} = 0.5$, $\chi_{23} = 1.0$; c), f), and i) $\chi_{12} = 0.5$, $\chi_{13} = 0.2$; $p_{12,s} = p_{13,q} = p_{23,r} = 0$.

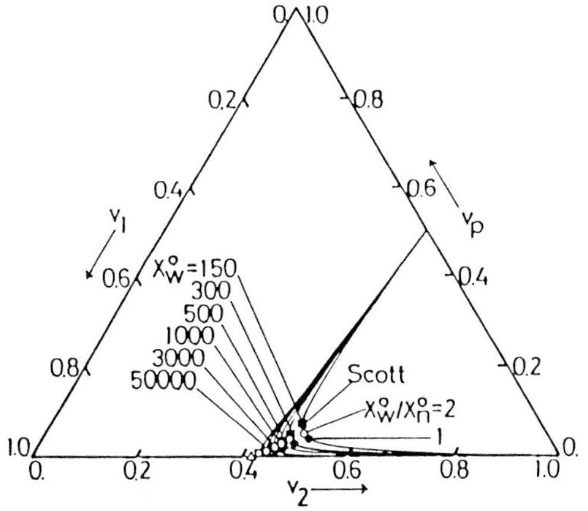

Fig. 3.50 Effects of the average molecular weight of the original polymer on the spinodal curve and CSP of quasi-ternary solutions: Original polymer, Schulz-Zimm type distribution ($X_w{}^0/X_n{}^0 = 2$, unfilled circle), monodisperse (filled circle); $\chi_{12} = 0.5$, $\chi_{13} = 0.2$, $\chi_{23} = 1.0$: Filled rectangle, CSP of ternary solutions (Scott); $p_{12,s} = p_{13,q} = p_{23,r} = 0$.

Fig. 3.50 shows the effects of $X_w{}^0$ on the spinodal curve and CSP for quasi-ternary solutions of polymers with SZ type distribution ($X_w{}^0/X_n{}^0 = 2$, unfilled circle) and a monodisperse polymer (filled circle). For comparison, CSP for a monodisperse polymer in a binary solvent mixture was calculated using Scott's equations (eqns 26 and 27 of ref. 3) and is shown as filled rectangles. CSP calculated by Scott's method is located on the spinodal curve, but it is significantly different from the true critical point obtained using the method proposed here, and shifted to a higher polymer volume fraction range. In other words, Scott's critical point for monodisperse polymer / solvent 1 / solvent 2 system corresponds to the true critical point for multicomponent polymers with relatively large heterogeneity / solvent 1 / solvent 2 system. With an increase in $X_w{}^0$, CSP changes roughly linearly in the phase diagram, approaching a point on the v_2-axis at limit of $X_w{}^0 \rightarrow \infty$. This point is denoted as the unfilled triangle in the figure, giving the solvent composition of the Flory's theta solvent for given values of χ_{12}, χ_{13} and χ_{23} at a given temperature. In other words, we can predict theoretically the solvent composition of the theta solvent mixture from three χ values (see, section 3.4) (ref. 26).

Fig. 3.51a-c illustrate the effects of $X_w{}^0$ on CSP ($v_1{}^c$, $v_2{}^c$ and $v_p{}^c$) for quasi-ternary solutions ($X_w{}^0/X_n{}^0 = 1$, 2, 4: solid

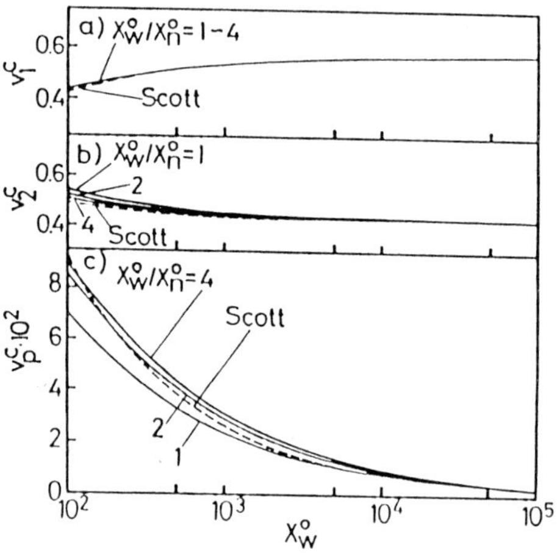

Fig. 3.51 Plots of the composition of critical solution points $v_1{}^c$, $v_2{}^c$ and $v_p{}^c$ versus $X_w{}^0$ for quasi-ternary solutions: Original polymer, Schulz-Zimm type distribution ($X_w{}^0/X_n{}^0=1$, 2 and 4); $\chi_{12}=0.5$, $\chi_{13}=0.2$, $\chi_{23}=1.0$: Broken line, $v_1{}^c$, $v_2{}^c$ and $v_p{}^c(v_p{}^c)$ versus $X_w{}^0(X)$ of ternary solutions (Scott); $p_{12,s}=p_{13,q}=p_{23,r}=0$.

line) and CSP obtained from Scott's equations (broken line). Scott's $v_p{}^c$ is 20~40% higher than the correct $v_p{}^c$ in the range of $X_w{}^0=100\sim300$, evaluated by Kamide-Matsuda's method. With an increase in $X_w{}^0$, $v_1{}^c$ increases and $v_2{}^c$ decreases approaching asymptotic values. This point gives the composition of the Flory solvent. $v_p{}^c$ decreases gradually approaching zero at infinite $X_w{}^0$. $v_p{}^c$ for ternary solutions is significantly smaller than $v_p{}^c$ for binary solutions. When solvent 2 is added to polymer solutions with a starting concentration $v_p{}^s$, CSP can be realized only at a specific $v_p{}^s$ (denoted by $v_p{}^{sc}$; see, Fig. 3.46). $v_p{}^{sc}$ is rather comparable with $v_p{}^c$ for binary solutions. In order to determine the composition of the Flory solvent, it is necessary to use polymers with $X_w{}^0$ larger than at least 1×10^3 and desirably 1×10^4. $v_1{}^c$ is practically independent of $X_w{}^0/X_n{}^0$ of the polymer, in $X_w{}^0/X_n{}^0$ range investigated. As $X_w{}^0/X_n{}^0$ increases, both $v_2{}^c$ and $v_p{}^c$ decrease significantly and the effects of $X_w{}^0/X_n{}^0$ on $v_2{}^c$ and $v_p{}^c$ become quite remarkable in a lower $X_w{}^0$ range. The corresponding values estimated by Scott's equations deviate noticeably from true values for polymers with lower $X_w{}^0$ and larger $X_w{}^0/X_n{}^0$. These deviations are apparently in the same direction of increasing $X_w{}^0/X_n{}^0$.

3.4 FLORY SOLVENT COMPOSITION

Inspection of Figs 3.50 and 3.51a-c lead us to the conclusion that, regardless of the MWD of the original polymer, at the limit of $X_w{}^0 \to \infty$ (and also, $X_z{}^0 \to \infty$), $v_p{}^c$ reduces to zero concurrently with $v_1{}^c$ and $v_2{}^c$ Flory solvent compositions, $v_1{}^F$ and $v_2{}^F$, respectively. In the case of $X_w{}^0 \to \infty$, it follows that $v_1{}^c \to v_1{}^F$, $v_2{}^c \to v_2{}^F$ and $v_p{}^c \to 0$ and eqn (3.58) reduces to eqn (3.66),

$$\left(\frac{1}{v_1}+\frac{1}{v_2}-2\chi_{12}\right)\left(\frac{1}{v_1}-2\chi_{13}\right)-\left(\frac{1}{v_1}+\chi_{23}-\chi_{13}-\chi_{12}\right)^2=0. \qquad (3.66)$$

In deriving eqn (3.66) we assumed $v_p{}^c$ approaches zero at an order of $(X_w{}^0)^{-a}$ $(0<a<1)$ with increasing $X_w{}^0$: $v_p{}^c \sim O((X_w{}^0)^{-a})$. Accordingly, $1/(v_p{}^c X_w{}^0)$ is to the order of $(X_w{}^0)^{a-1}$ (i.e., $1/(v_p{}^c X_w{}^0) \sim O((X_w{}^0)^{a-1})$ and at the limit of $X_w{}^0 \to \infty$ $1/(v_p{}^c X_w{}^0)$ approaches zero (ref. 44). Eqn (3.59) can be rewritten by dividing both sides by $X_z{}^0$ as

$$\left[\frac{1}{v_p X_w{}^0 (X_z{}^0)^{1/2}}\left(\frac{1}{v_1{}^2}-\frac{1}{v_2{}^2}\right)+\frac{1}{v_1{}^2}\left(\frac{1}{v_2}-2\chi_{23}\right)\frac{1}{(X_z{}^0)^{1/2}}\right.$$
$$\left.-\frac{1}{v_2{}^2}\left(\frac{1}{v_1}-2\chi_{13}\right)\frac{1}{(X_z{}^0)^{1/2}}\right]$$
$$\times\left[\frac{1}{v_p X_w{}^0 (X_z{}^0)^{1/2}}+\frac{1}{v_1 (X_z{}^0)^{1/2}}-\frac{2\chi_{13}}{(X_z{}^0)^{1/2}}\right]$$
$$-\left(\frac{1}{v_1}+\chi_{23}-\chi_{13}-\chi_{12}\right)\left[\frac{1}{v_1{}^2}\left(\frac{1}{v_2}-2\chi_{23}\right)\frac{1}{X_z{}^0}+\frac{1}{v_1{}^2 v_p X_w{}^0 X_z{}^0}\right.$$
$$+\frac{1}{v_p X_w{}^0}\left\{\frac{1}{v_1 v_2}-\frac{2\chi_{13}}{v_2}-\frac{2\chi_{23}}{v_1}+2(\chi_{12}\chi_{13}+\chi_{13}\chi_{23}+\chi_{23}\chi_{12})\right.$$
$$\left.\left.-(\chi_{12}{}^2+\chi_{13}{}^2+\chi_{23}{}^2)\right\}\right]=0. \qquad (3.67)$$

The lefthand side of eqn (3.67) reduces to zero at limit of $X_w{}^0 \to \infty$ regardless of v_1 and v_2. Therefore, $v_1{}^c$ and $v_2{}^c$ at a limit of infinite $X_w{}^0$ (in other words, $v_1{}^F$ and $v_2{}^F$) can be calculated by solving eqn (3.66) alone for a given combination of χ_{12}, χ_{13} and χ_{23} under the condition that $0<v_1<1$ and $0<v_2<1$ (ref. 44).

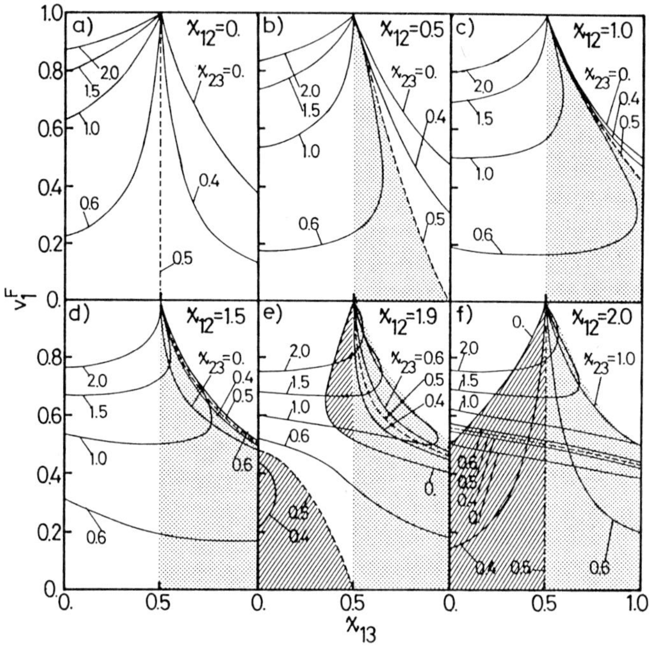

Fig. 3.52 Relationship between Flory solvent composition v_1^F ($=1-v_2^F$) and χ_{13}: a) $\chi_{12}=0$, b) $\chi_{12}=0.5$, c) $\chi_{12}=1.0$, d) $\chi_{12}=1.5$, e) $\chi_{12}=1.9$, f) $\chi_{12}=2.0$; number on curve means χ_{23}; $p_{12,s}=p_{13,q}=p_{23,r}=0$.

Fig. 3.52a-f show the relationship between v_1^F and χ_{13} for fixed values of χ_{12} and χ_{23}. In the figure the broken line denotes the $\chi_{23}=0.5$. For $\chi_{12}=0$ (Fig. 3.52a), the v_1^F decreases rapidly with increasing χ_{13} in the range of $\chi_{23}<0.5$, but it increases with increasing χ_{13} if $\chi_{23}>0.5$. For a specific combination of χ_{12} and χ_{23} ($\chi_{12}=0$ and $\chi_{23}=0.5$), $\partial v_1^F/\partial \chi_{13}$ changes its sign at $\chi_{13}=0.5$. With $\chi_{12}=0$ and $\chi_{13}=\chi_{23}=0.5$, solvent 1 and 2 are thermodynamically identical and then, the system can be obtained over the whole range of the composition of v_1. Even in the case of $\chi_{13}>0.5$, there exists a region in which v_1^F decreases with an increase in χ_{13} (for example, $\chi_{12}=0.5$ and $\chi_{23}=0.6$). Generally, this region affords us two Flory's compositions for this polymer / solvents system as denoted by shadowed area in the figure (ref. 44).

The existence of two Flory solvent compositions for a given polymer / solvent / nonsolvent system has not been demonstrated experimentally, but this phenomenon probably is closely correlated with the occurrence of cosolvency (see, section 2.5). Bamford and

Tompa (ref. 42) reported for cellulose acetate / chloroform / ethylacetate system the occurrence of cosolvency and two critical solution points. Of course, two critical solution points, corresponding to the infinite polymer, yield two Flory solvent compositions.

As χ_{12} increases beyond 1.0, the area giving two Flory solvent compositions becomes remarkably wide (Fig. 3.52c-f). For example, in the case of $\chi_{12}=1.0$ and $\chi_{23}=0.6$ (Fig. 3.52c), v_1^F diminishes with increasing χ_{13} even for $\chi_{13}<0.5$. This means that there is a possibility that as the power of solvent 1 decreases the composition of solvent 1 (good solvent) for the Flory solvent decreases unexpectedly. Fig. 3.52e shows that mixing two good solvents with $\chi_{13}<0.5$ and $\chi_{23}<0.5$ is predicted theoretically to sometimes give a Flory solvent. The area giving a Flory solvent is denoted by hatch marks. Inspection of Fig. 3.52a-f also reveals that in this case, mixing two good solvents gives two Flory solvent mixtures (ref. 44). This is just the reverse of the cosolvency phenomena (i.e., cononsolvency), but unfortunately it has not yet been confirmed by experiments (ref. 44). In the case of $\chi_{12}=2.0$ (Fig. 3.52f), all combinations of χ_{13} and χ_{23}, except for $\chi_{13}<0.5$ and $\chi_{23}>0.5$ (or $\chi_{13}>0.5$ and $\chi_{23}<0.5$), have two Flory solvent compositions.

Fig. 3.53a-f show the relationship among v_1^F, χ_{13} and χ_{23}, constructed using the same data as employed in Fig. 3.52a-f, for a fixed χ_{12} value. The curved surface surrounded by bold full lines corresponds to v_1^F for a given combination of three χ parameters. Reflecting on the symmetry of eqn (3.66), the axis determined as an interception of two planes; $\chi_{13}=\chi_{23}$ and $v_1^F=0.5$ is two-fold axis (not shown in the figure). Any combination of solvent 1 and solvent 2 for a given polymer having three χ parameters, whose values lie in shadowed area in Fig. 3.53, always affords us two Flory solvent compositions (ref. 44). It is clear that for $\chi_{12}=1.9\sim2.0$ two Flory solvents compositions are expected to exist in the range of $\chi_{13}<0.5$ and $\chi_{23}<0.5$: For example, for $(\chi_{12}, \chi_{13}, \chi_{23})=(1.9, 0, 0.4)$ we obtain $v_1^F=0.2$ and 0.45 and for $(\chi_{12}, \chi_{13}, \chi_{23})=(1.9, 0.4, 0)$ we obtain $v_1^F=0.55$ and 0.8, respectively.

Fig. 3.54 shows the region giving two Flory solvent compositions in χ_{12}-χ_{13}-χ_{23} space. In the figure, the region is shadowed and has a mirror symmetry with respect to a $\chi_{13}=\chi_{23}$ plane, as requested by eqn (3.62). As well as the region of $\chi_{13}>0.5$ and $\chi_{23}>0.5$, relating to cosolvency, the region of

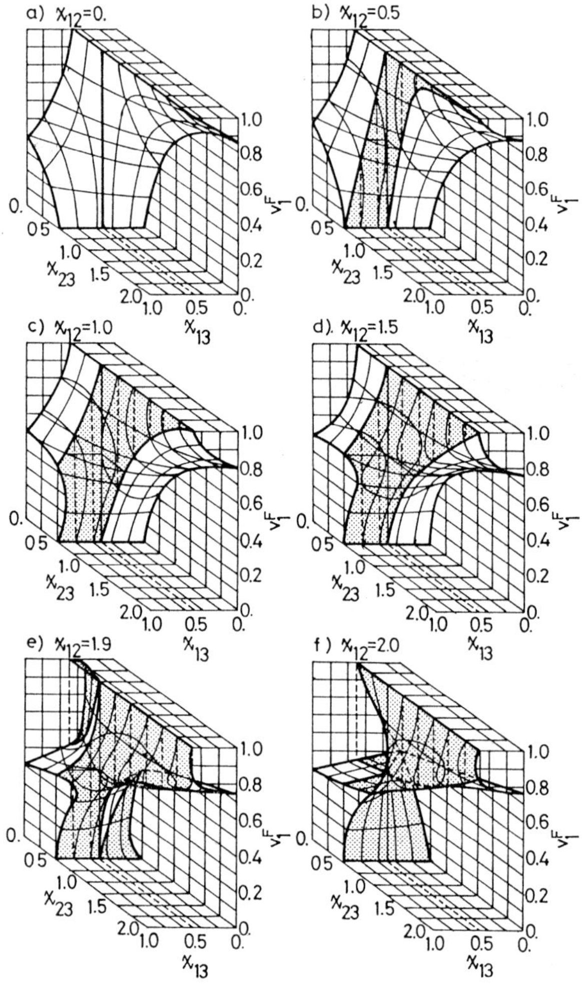

Fig. 3.53 Relationship among Flory solvent composition $v_1{}^F(=1-v_2{}^F)$, χ_{13} and χ_{23}: a) $\chi_{12}=0$, b) $\chi_{12}=0.5$, c) $\chi_1=1.0$, d) $\chi_{12}=1.5$, e) $\chi_{12}=1.9$, f) $\chi_{12}=2.0$; shadowed area has two Flory solvent compositions; $p_{12,s}=p_{13,q}=p_{23,r}=0$.

$\chi_{13}<0.5$ and $\chi_{23}<0.5$, relating to cononsolvency increases its area remarkably with increasing χ_{12} (ref. 45).

In summary, (1) Flory solvent composition can be theoretically calculated regardless of the polymolecularity of the polymer sample for any polymer / solvent 1 / solvent 2 system, if χ_{12}, χ_{13} and χ_{23} are known in advance for the system. (2) Two Flory solvent compositions exist for a specific combination of solvents 1 and 2 (i.e., for specific values of χ_{12}, χ_{13} and χ_{33}). This prediction has not yet been experimentally confirmed.

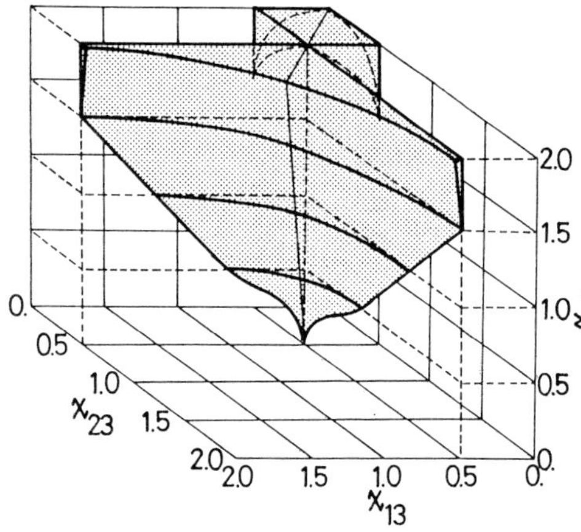

Fig. 3.54 The region giving two Flory solvent compositions in $\chi_{12}-\chi_{13}-\chi_{23}$ space; $p_{12,s}=p_{13,q}=p_{23,r}=0$.

(3) When $\chi_{12}>1.5$ (i.e., when the thermodynamic property of two solvents are greatly different), mixing two good solvents is expected theoretically to give a Flory solvent. This prediction has also not yet been experimentally ascertained.

3.5 COSOLVENCY

3.5.1 Cosolvency

Even if two solvents are poor against a given polymer when they are used separately, a binary mixture consisting of these two solvents sometimes dissolves the polymer. In this sense the binary mixture acts as a good solvent. This phenomenon is conventionally called cosolvency, firstly discovered in 1920s expeimentally for cellulose nitrate solution systems (ref. 46). Thereafter, cosolvency has been observed for numerous polymer / two-nonsolvents mixture systems including cellulose acetate (CA) / chloroform / ethyl alcohol (ref. 47), PS / acetone / n-propyl laurate (ref. 48), PS / MCH / acetone (ref. 49), PS / acetone / cyclohexanol (ref. 50), PS / MCH / diethyl ether (ref. 50), PS / acetone / diethyl ether (refs 51,52), poly(methyl methacrylate) (PMMA) / benzyl alcohol / sec-butyl chloride (ref. 53), poly(p-tert butyl phenyl)methacrylate / acetone / CH (ref. 54), PMMA / CCl_4 / tert-butyl chloride (ref. 55) and PMMA / CCl_4 / butyl chloride (refs 55,56).

Tompa (ref. 57) predicted without showing detailed derivations that for infinite molecular weight polymers, cosolvency occurs if the following conditions are satisfied.

$$2(\chi_{12}\chi_{13}+\chi_{13}\chi_{23}+\chi_{23}\chi_{12})-(\chi_{12}^2+\chi_{13}^2+\chi_{23}^2)>0, \quad (3.68)$$

$$2>\chi_{12}>\chi_{13}+\chi_{23}-1+(2\chi_{13}-1)^{1/2}(2\chi_{23}-1)^{1/2}. \quad (3.69)$$

Tompa explained cosolvency as follows; when the cohesive energy density of the polymer lies between that of two solvents, the solvent mixture may have on average the cohesive energy density near to that of the polymer and in consequence the binary mixture of two nonsolvents dissolves the polymer (ref. 57).

The second virial coefficient A_2 measured by the light scattering and the limiting viscosity number $[\eta]$ have been investigated in correlation to cosolvency phenomena (refs 49-56,58-63). A_2 and $[\eta]$ were shown to exhibit maximum at a specific composition of binary mixture and these were discussed in terms of selective adsorption coefficient λ^* (ref. 64). The solubility parameter δ (refs 49,50), $\Delta G^E{}_{12}$ (eqn (3.13), refs 49,58), the long range intermolecular interaction parameter B' (ref. 55) in the two-parameter theory estimated by Stockmayer-Fixman plot (ref. 65), coil size (refs 49,50,53) and the linear expansion factor α (ref. 66) were also studied as a function of binary mixture composition. In particular, Katime and Ochoa (ref. 56) attempted to explain cosolvency by using the solubility parameter defined by Koenhen and Smolders (ref. 67). It should be noted that physical quantities such as A_2, $[\eta]$, λ^*, δ, $\Delta G^E{}_{12}$, B' and α do not have "the relation of cause and effect" against occurrence of cosolvency, but may be indirectly correlated with cosolvency.

This section intends to disclose the thermodynamical conditions of the occurrence of cosolvency using the Kamide-Matsuda theory (refs 11-14,26) of phase equilibria of polydisperse polymer / solvent 1 / solvent 2 (i.e., quasi-ternary) systems and to demonstrate theoretically that cosolvency is a very generally observable phenomena for quasi-ternary systems.

An equation for SC at $v_1=0$ is obtained from eqn (3.58) as eqn (3.63) (refs 26,45). SP at $v_1=0$ is evidently independent of χ_{12} and χ_{13} and is determined by χ_{23} alone.

The compositions of SP ($v_2^{(2)}$, $v_p^{(2)}$) at $v_1=0$ are given by solving eqn (3.58) as

$$v_p^{(2)}=\frac{1}{2}[1-\frac{1}{2\chi_{23}}+\frac{1}{2\chi_{23}X_w^0}+\{(1-\frac{1}{2\chi_{23}}+\frac{1}{2\chi_{23}X_w^0})^2-\frac{2}{\chi_{23}X_w^0}\}^{1/2}] \quad (3.70a)$$

with

$$v_2^{(2)} = 1 - v_p^{(2)}. \tag{3.70b}$$

At $v_2 = 0$, SC and SP are independent of χ_{12} and χ_{23} and a single function of χ_{13} in the form (ref. 45):

$$v_p + \frac{v_1}{X_w^0} - 2\chi_{13} v_1 v_p = 0. \tag{3.71}$$

By solving eqn (3.71), the composition of SP ($v_1^{(1)}$, $v_p^{(1)}$) at $v_2 = 0$ is given by eqn (3.72)

$$v_p^{(1)} = \frac{1}{2}[1 - \frac{1}{2\chi_{13}} + \frac{1}{2\chi_{13}X_w^0} + \{(1 - \frac{1}{2\chi_{13}} + \frac{1}{2\chi_{13}X_w^0})^2 - \frac{2}{\chi_{13}X_w^0}\}^{1/2}] \tag{3.72a}$$

with

$$v_1^{(1)} = 1 - v_p^{(1)}. \tag{3.72b}$$

Then eqns (3.70a) and (3.72a) give $v_p^{(1)}$ and $v_p^{(2)}$, separately and cosolvency occurs only when $v_p^{(1)}$ and $v_p^{(2)}$, evaluated thus, satisfy the following two inequalities concurrently (ref. 45).

$$0 < v_p^{(1)} < 1 \quad \text{and} \quad 0 < v_p^{(2)} < 1. \tag{3.73}$$

Without complete satisfaction of eqn (3.73), cosolvency never occurs. In this sense, the reverse is rigorously true.

In other words, a necessary condition of cosolvency can also be obtained by putting X_w^0 values into eqns (3.70a) and (3.72a), combined with eqn (3.73). If an additional rough approximation of $X_w^0 \to \infty$ is applied to eqn (3.73) the necessary condition reduces to the well-known inequalities,

$$\chi_{13} > 0.5 \quad \text{and} \quad \chi_{23} > 0.5. \tag{3.74}$$

Moreover, when two CSP are obtained from eqns (3.58) and (3.59), the system has two-phase equilibria regions and unavoidably cosolvency occurs. Then, existence of two CSP is a sufficient condition (ref. 45).

Fig. 3.55a-d illustrate the effect of X_w^0 on CPC, SC and CSP. In the figure, the solid and broken lines are CPC and SC,

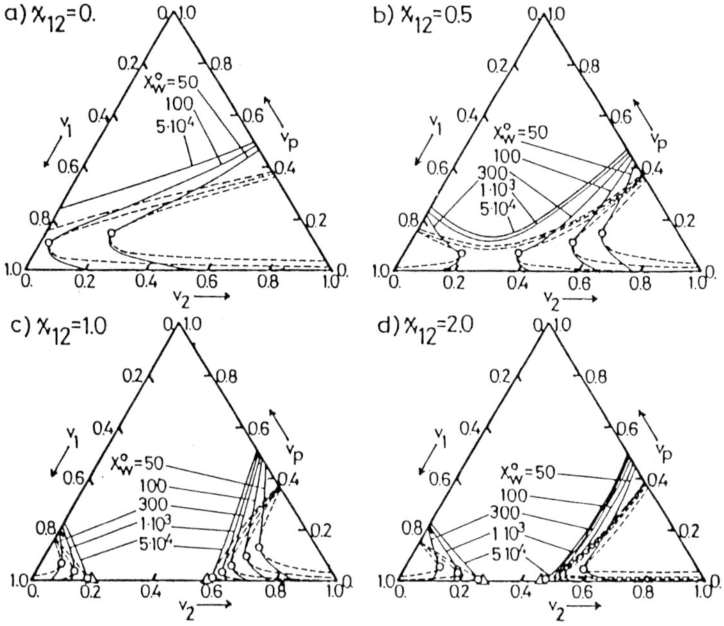

Fig. 3.55 Effect of the weight-average degree of polymerization of the original polymer $X_w{}^0$ on the cosolvency of quasi-ternary solution systems consisting of a multicomponent polymer (Schulz-Zimm distribution, $X_w{}^0/X_n{}^0=2$) dissolved in a binary mixture: a) $\chi_{12}=0$, $\chi_{13}=0.6$, $\chi_{23}=0.8$; b) $\chi_{12}=0.5$, $\chi_{13}=0.6$, $\chi_{23}=0.8$; c) $\chi_{12}=1.0$, $\chi_{13}=0.6$, $\chi_{23}=0.8$; d) $\chi_{12}=2.0$, $\chi_{13}=0.6$, $\chi_{23}=0.8$; full line, cloud point curve; broken line, spinodal curve; unfilled circle, critical solution point; unfilled triangle, Flory solvent; $p_{12,s}=p_{13,q}=p_{23,r}=0$.

respectively, and CSP is shown as a unfilled circle. In the case of $\chi_{12}=1.0$ and 2.0 with $\chi_{13}=0.6$ and $\chi_{23}=0.8$ (Fig. 3.55c and d), cosolvency occurs over an entire range of $X_w{}^0$ and the systems have two Flory solvent compositions (ref. 44): The volume fraction of solvent 1 at Flory solvent, $v_1{}^F=0.4248$ and 0.8024 for $\chi_{12}=1.0$ and $v_1{}^F=0.5281$ and 0.7283 for $\chi_{12}=2.0$ (for these cases, $\chi_{13}=0.6$ and $\chi_{23}=0.8$). At a limit of $X_w{}^0=\infty$, the two CSPs coincide with their Flory solvent compositions. In the case of $\chi_{12}=0.5$, Flory solvent composition does not exist, but in the range of relatively smaller $X_w{}^0$ ($=300$), two CSPs are observed and consequently cosolvency occurs. Two two-phase equilibria regions overlap heavily for larger $X_w{}^0$ polymer solutions.

In the case of $\chi_{12}=0$ (Fig. 3.55a), solvent 1 having $\chi_{13}=0.6$ behaves as good solvent for the smaller $X_w{}^0$ (for example, 50 and 100). For larger $X_w{}^0$, solvent 1 with $\chi_{13}=0.6$

becomes, of course, poorer and no cosolvency occurs.

Even in the case where two two-phase equilibria regions overlap as demonstrated in Figs 3.55a ($X_w{}^0 = 5\times 10^4$) and 3.55b ($X_w{}^0 = 1\times 10^3$ and 5×10^4), a one-phase region exists at higher v_p. This region can also be, in a broad sense, regarded as that of cosolvent, if the region maintains the liquid phase. But, in this article, we deal only with the case when two-phase equilibria regions are observed. Tompa (ref. 57) predicted that for $\chi_{12} = 2.0$, solvents 1 and 2 become critical consolute states and for $\chi_{12} > 2.0$, a three-phase separation region exists. A sign of the occurrence of three-phase separation can be observed at $\chi_{12} = 2.0$. In Fig. 3.55d ($\chi_{12} = 2.0$) CSP exists on CPC at solvent 2 side for $X_w{}^0 = 50$ and 5×10^4, but CSP does not exist on CPC on

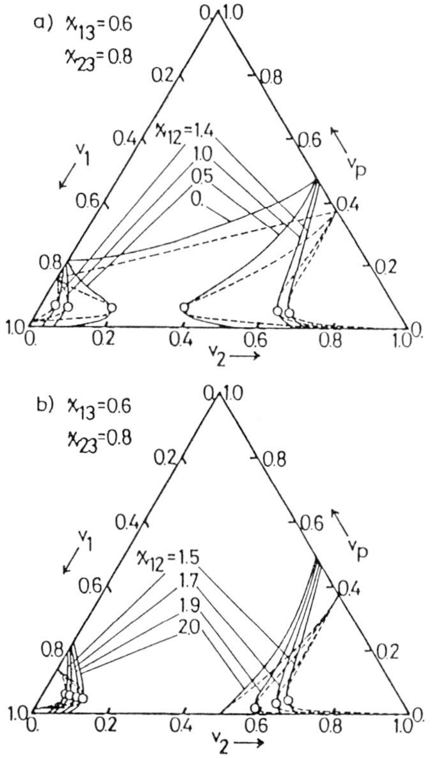

Fig. 3.56 Effect of χ_{12} on the cosolvency of quasi-ternary solution systems consisting of a multicomponent polymer in a binary solution: Original polymer, Schulz-Zimm distribution, $X_w{}^0 = 300$, $X_w{}^0/X_n{}^0 = 2$; $\chi_{13} = 0.6$ and $\chi_{23} = 0.8$; full line, cloud point curve; broken line, spinodal curve; unfilled circle, critical solution point; $p_{12,s} = p_{13,q} = p_{23,r} = 0$.

the solvent 2 side for $X_w{}^0 = 100$, 300 and 1×10^3. This singular property is considered to be closely correlated with the three-phase separation phenomena.

Fig. 3.56 shows the effect of χ_{12} on cosolvency for quasi-ternary systems consisting of the polymer (SZ, $X_w{}^0 = 300$) in a binary solvent mixture ($\chi_{13} = 0.6$ and $\chi_{23} = 0.8$). In the case of $\chi_{12} = 0$, no cosolvency occurs. SC at $\chi_{12} = 0$ can be accurately represented by a straight line connecting the two points: $(0, v_2{}^{(2)}, v_p{}^{(2)})$ (SP given by eqn (3.63)) and $(v_1{}^{(1)}, 0, v_p{}^{(1)})$ (SP given by eqn (3.71)). Cosolvency occurs at $\chi_{12} = 0.5$ and the region of cosolvency becomes wide as χ_{12} increases, approaching maximum at $\chi_{12} = 1.4$. Further increases in χ_{12} beyond 1.4, makes the cosolvency region smaller again, increasing the two-phase equilibria region. The two $v_p{}^c$ decrease slowly, but monotonically with an increase in χ_{12}.

3.5.2 Cononsolvency

As Tompa (ref. 57) predicted as early as 1956, the theory indicates that a mixture of solvent 1 having $\chi_{13} < 0.5$ and solvent 2 having $\chi_{23} < 0.5$ (in other words, both solvents are good solvents when used separately) may sometimes be nonsolvent for the

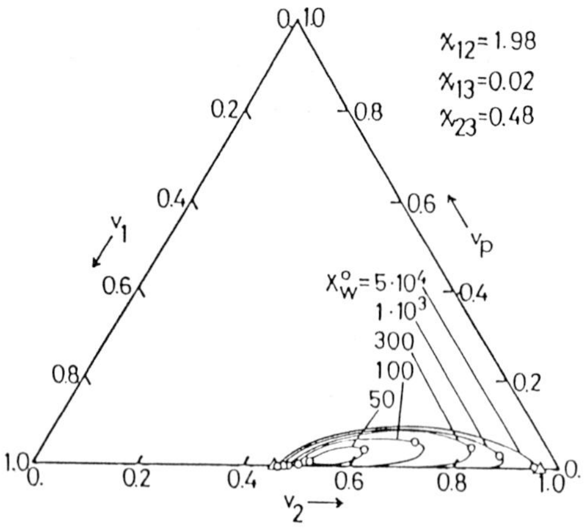

Fig. 3.57 Effect of the weight-average degree of polymerization of the original polymer $X_w{}^0$ on the cononsolvency of quasi-ternary solution systems consisting of a multicomponent polymer (Schulz-Zimm distribution, $X_w{}^0/X_n{}^0 = 2$) in a binary solvent mixture: $\chi_{12} = 1.98$, $\chi_{13} = 0.02$, $\chi_{23} = 0.48$; full line, spinodal curve; unfilled circle, critical solution point; unfilled triangle, Flory solvent; $p_{12,s} = p_{13,q} = p_{23,r} = 0$.

polymer. Unfortunately, this phenomenon has not yet been experimentally confirmed. The phenomenon is just the reverse of cosolvency, and we define this hereafter as cononsolvency. Cononsolvency is apparently accompanied by two CSPs as with the case of cosolvency.

Fig. 3.57 shows the effect of X_w^0 on cononsolvency for quasi-ternary systems ($\chi_{12}=1.98$, $\chi_{13}=0.02$ and $\chi_{23}=0.48$). With an increase in X_w^0 the two-phase equilibrium region expands its space. The two CSPs, denoted as unfilled circles in the figure, reduce to two Flory solvent compositions at a limit of $X_w^0=\infty$. Note that there are cases when cosolvency occurs, even if the system has no Flory solvent compositions (see, for example, Figs 3.55b and 3.56b). The system, in which cononsolvency occurs, always has two Flory solvent compositions.

3.6 FRACTIONATION BASED ON SOLUBILITY DIFFERENCES

3.6.1 Introduction

Since late 1960s, Kamide and coworkers (refs 24,27-34,68-70) and Koningsveld et al. (refs 71-77) have studied, theoretically and experimentally, molecular weight fractionation by solubility differences, such as successive precipitation fractionation (SPF) and successive solution fractionation (SSF), for quasi-binary solutions consisting of multicomponent polymer / single solvent systems, as was discussed in section 2.4. The former group strongly recommended SSF rather than SPF, for isolating fractions with narrower molecular weight distribution. Then, SPF and/or SSF, accompanied actually by changing the temperature of multicomponent polymer dissolved in a single solvent, should be the best, in the sense that they have a sound theoretical basis. Unfortunately, SPF and SSF using quasi-binary solutions have not widely been applied mainly due to the experimental difficulty of finding a suitable solvent (see, Table 3.3). Accordingly, in practice the fractionations have been carried out generally by adding nonsolvent to polymer / solvent systems (refs 24,78-87). No theoretical study of fractionation was made for the quasi-ternary system itself. An exception is Kamide and Matsuda's preliminary study (ref. 11), in which an example of an SPF calculation which was compared with actual SPF consisting of atactic polystyrene, methyl ethyl ketone, and methanol (see, Fig. 3.44). Even at present the importance of molecular weight fractionation by the solubility difference is not diminished, in particular for preparative purposes, in spite of the wide and rapid popularization of various chromatographic methods. Recent remarkable progress in the thermodynamics of phase-equilibria of multicomponent polymer solutions enables us to discuss quantitatively molecular weight fractionation using quasi-ternary solutions consisting of multicomponent polymer in a binary (solvent and nonsolvent) mixture, which have for many years been considered too complicated systems to analyze theoretically.

In this section, we generalize the theory of phase equilibria of quasi-ternary solutions, previously established by Kamide, Matsuda et al., (refs 11-14) (section 3.1 and 3.2) to the case of SPF and SSF of quasi-ternary solutions and to carry out systematic computer experiments, based on the theory and its simulation technique, in order to clarify the effects of the following factors on SPF and SSF (ref. 88): (1) three thermodynamic χ parameters between solvent 1 (ordinary, good solvent)-solvent 2 (ordinary, poor solvent), solvent 1-polymer, and solvent 2-polymer (as denoted by χ_{12}, χ_{13} and χ_{23}, respectively), (2) the starting solution concentration v_p^s ($\equiv V_1^0/(V_1^0+V_3^0)$, V_1^0 and V_3^0 are the volumes of solvent and polymer) and (3) total number of fractions in a run n_t.

The theory of phase equilibria of quasi-ternary solutions, established by Kamide, Matsuda and Miyazaki (refs 11-14), and the computer simulation technique, based on it, are used as basic tenets. Both the theory and the calculation programme, corresponding only to a single step of phase separation, were generalized to the case of successive fractionations.

Three χ parameters are in general given as functions of the polymer concentration and the solvent composition as follows (see, section 3.1) (ref. 14):

$$\chi_{12}=\chi_{12}^0\left(1+\sum_{s=1}^{n_s} p_{12,s}v_p^s\right)\left\{1+\sum_{u=1}^{n_u}(p_{1,u}v_1^u+p_{2,u}v_2^u)\right\}, \tag{3.5}$$

$$\chi_{13}=\chi_{13}^0\left(1+\sum_{q=1}^{n_q} p_{13,q}v_p^q\right), \tag{3.6}$$

$$\chi_{23}=\chi_{23}^0\left(1+\sum_{r=1}^{n_r} p_{23,r}v_p^r\right). \tag{3.7}$$

In this section, effects of χ_{12}^0, χ_{13}^0 and χ_{23}^0 on the

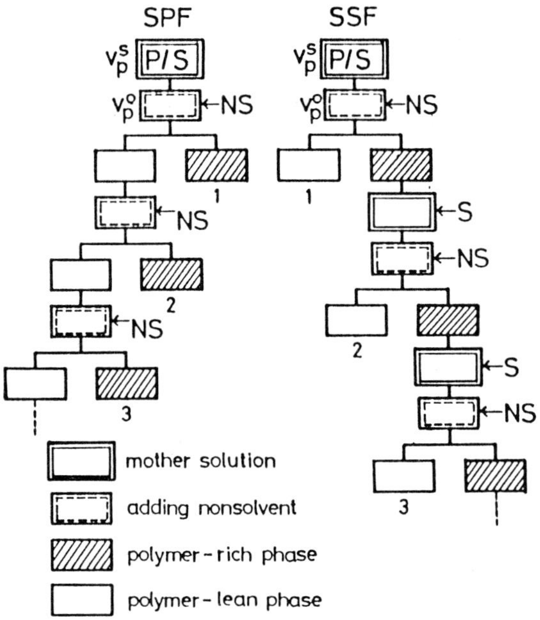

Fig. 3.58 Schematic representation of successive precipitation and solution fractionation (SPF and SSF).

fractionation characteristics will be investigated, assuming all other parameters in eqns (3.1)-(3.3) equal to zero.

Fig. 3.58 shows schematic procedures of SPF and SSF using quasi-ternary solutions. First, we dissolve a given amount of the polymer sample with known molecular weight distribution in solvent 1 (with known χ_{13}^0) to make a starting solution, with a given polymer volume fraction of v_p^s ($\equiv V_3^0/(V_1^0+V_3^0)$, V_1^0 and V_3^0, volume of solvent 1 and polymer, respectively), consisting of polymer and a single solvent. To the solution, we add nonsolvent (solvent 2) to bring about occurrence of two liquid-phase separation under the predetermined conditions of the relative amount of polymer partitioning in a polymer-rich phase ρ_p or that of polymer remaining in a polymer-lean phase ρ_s ($=1-\rho_p$) and χ_{12}^0, χ_{13}^0 and χ_{23}^0. If necessary, we can control the polymer volume fraction v_p^0 ($\equiv V_3^0/(V_1^0+V_2+V_3^0)$, V_2 is the volume of nonsolvent) at the step of phase separation by changing v_p^s (and accordingly, V_2). In SPF, we separate the polymer-rich phase as the first fraction and use the remaining polymer-lean phase as the starting solution again. The same process is then repeated n_t times. In SSF, we separate the polymer-lean phase as the first fraction and add solvent to the polymer-rich phase to give a solution with the same total volume V as the first cycle and use this solution as the starting solution. This process is then repeated n_t times.

The original polymer was assumed to have the SZ type molecular weight distribution $g_0(X)$ with $X_w^0 = 300$ and $X_n^0 = 150$ (the superscript 0 denotes the original polymer). The number of components in the original polymer was taken as 1500 (ref. 32). The computer experiments were carried out under the conditions of $\chi_{12}^0 = 0.2 \sim 1.3$, $\chi_{13}^0 = 0 \sim 0.5$, $\chi_{23}^0 = 0.8 \sim 1.5$, $v_p^s = 1 \times 10^{-5} \sim 5 \times 10^{-2}$ and $n_t = 5 \sim 50$.

Using the fractionation data on the fraction size ρ ($= \rho_p$ for SPF or $= \rho_s$ for SSF) or n_t (if equal size of the fractions is separated, $n_t = 1/\rho$) and X_w of a series of fractions, isolated by SPF or SSF, the molecular weight distribution (MWD) of the original polymer was calculated by the method of Miyazaki and Kamide (ref. 70). They demonstrated that their method affords, without exception, a more accurate value of the ratio M_w/M_n of the

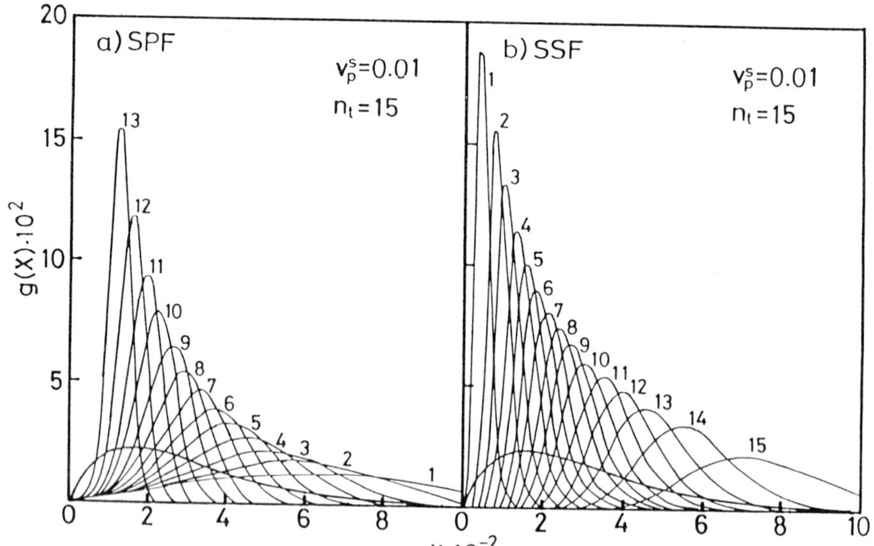

Fig. 3.59 Normalized molecular weight distribution curves of the fractions separated by a successive precipitation fractionation (SPF) run (a) and successive solution fractionation (SSF) run (b): The original polymer, Schulz-Zimm type distribution, $X_w^0/X_n^0 = 2$, $X_w^0 = 300$; $\chi_{12}^0 = 1.3$, $\chi_{13}^0 = 0.2$, $\chi_{23}^0 = 1.0$; starting polymer volume fraction $v_p^s = 0.01$; total number of fractions in a run, $n_t = 15$. The numbers on the curve denote the order of the fraction and the bold curve denotes the original polymer.

original polymer than the Schulz method (ref. 89) does.

The Kamide and Miyazaki method is expressed by eqns (2.76)-(2.78) for SPF (Fig. 2.47a) and is expressed by eqns (2.76')-(2.78') for SSF (Fig. 2.47b). A parameter E defined by eqn (2.80) is also utilized as a measure of the accuracy of the method.

3.6.2 The molecular weight distribution of SPF and SSF fraction

Fig. 3.59a and b show the molecular weight distribution (MWD) of a series of fractions isolated by SPF or SSF runs from quasi-ternary solutions consisting of the same polymer and the same combination of solvent and nonsolvent (i.e., the same χ_{12}^0, χ_{13}^0 and χ_{23}^0) under the same operating conditions (v_p^s and n_t). As previously noted for a quasi-binary system (section 2.4) (refs 24,27-34,68-70), the SPF fractions have extremely small X components (i.e., the so-called "tailing effect"), and in contrast, the SSF fractions do not reveal a tailing effect. Note that in this SPF run the last step fractionation (in this case, $j=14$ for $n_t=15$) was impossible (ref. 88).

3.6.3 Choice of solvent and nonsolvent

Fig. 3.60a and b illustrate the range of three χ parameters suitable for SPF or SSF of the system of a given polymer dissolved in a binary solvent mixture under given operating conditions (i.e., $v_p^s=0.01$ and $n_t=15$). Here, the shadowed area is an allowable combination of three χ parameters for the first step of phase separation (ref. 11) at ρ_p (or ρ_s) $=1/n_t$ and the numbers in circle mean the maximum order of fractions j_{max}, isolated by SPF or SSF under given operating conditions.

If three χ parameters are chosen in SSF so that the first step fractionation can be carried out, the fractionation up to the last step seems unquestionably possible with the same three χ values as in the first step. It can be shown in SPF that j_{max} is not so sensitive to χ_{13}^0 and χ_{12}^0 (see, Fig. 3.64). In contrast, in SPF j_{max} is most dependent on χ_{23}^0 and increases rapidly and then decreases remarkably with increase in χ_{23}^0, after passing through a maximum (in this case, $n_t=15$), suggesting that the choice of the poor solvent determines the extent of fractionation practically carried out in SPF and a not too weak and not too strong poor solvent is strongly recommended from this point of view. Note, however, that the too strong poor solvent yields the broad MWD fractions in SPF (see, Fig. 3.64). In contrast to SPF, SSF can be performed up to the last step (i.e., $j=n_t$) using any

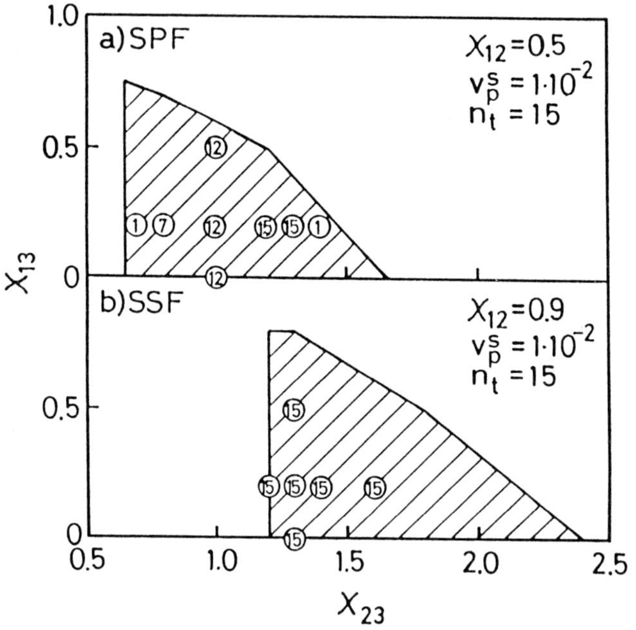

Fig. 3.60 Allowable combination of three χ parameters, $\chi_{12}{}^0$, $\chi_{13}{}^0$ and $\chi_{23}{}^0$ for given conditions of successive precipitation fractionation (SPF) [a)] and successive solution fractionation [b)]: The original polymer, Schulz-Zimm type distribution, $X_w{}^0/X_n{}^0=2$, $X_w{}^0=300$; starting polymer volume fraction $v_p{}^s=0.01$; total number of fractions in a run, $n_t=15$; shadowed area is allowable combination for single step of phase separation and number in circle means the maximum fractionation step attained in the SPF run, which was first designed as $n_t=15$.

three χ parameters which allow the first step phase separation. Then, for success in an SSF run, it is sufficient to determine only an adequate combination of three χ parameters fulfilling the requirements at the first step (ref. 88).

In the first step of SSF, the predominantly larger portion of higher molecular weight components in the original polymer is partitioned into a polymer-rich phase if a suitable combination of a solvent and a nonsolvent is chosen. As SSF proceeds, the lower molecular weight components are preferentially extracted as fractions from the original polymer, and as a result the higher molecular weight components precipitated as a polymer-rich phase increase gradually, with the total volume V keeping constant throughout SSF. Then, the polymer-rich phase, diluted with solvent to give the j-th starting solution at the j-th step, can be readily phase-separated again by addition of a relatively small

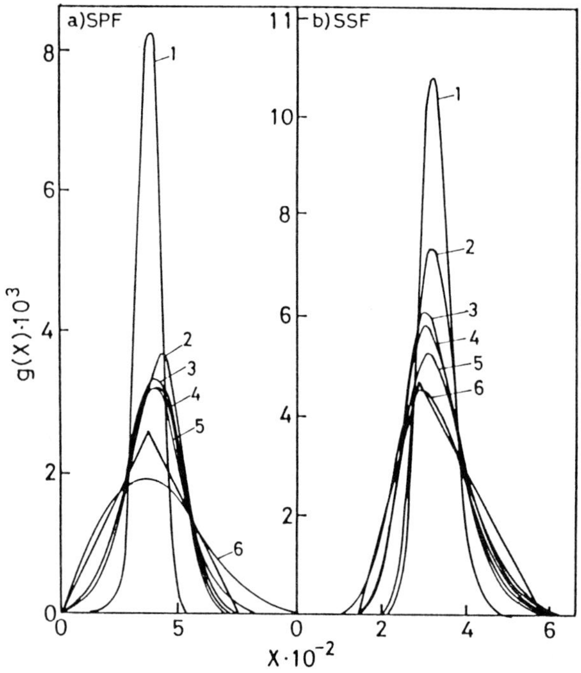

Fig. 3.61 Differential molecular weight distribution curve g(X) of fractions separated by successive precipitation fractionation (SPF) [a)] and successive solution fractionation (SSF) [b)] under various conditions: The original polymer, Schulz-Zimm type distribution, $X_w{}^0/X_n{}^0=2$, $X_w{}^0=300$; the fractions having similar X_w are chosen.

a) 1, $\chi_{12}{}^0=0.5$, $\chi_{13}{}^0=0.2$, $\chi_{23}{}^0=1.0$, $j=5$ ($n_t=15$), $X_w=369.2$, $v_p{}^s=1\times10^{-5}$, $X_p=380.9$ ($X_p/X_w=1.031$);

2, $\chi_{12}{}^0=0.5$, $\chi_{13}{}^0=0.2$, $\chi_{23}{}^0=1.0$, $j=15$ ($n_t=50$), $X_w=391.9$, $v_p{}^s=0.01$, $X_p=440.4$ ($X_p/X_w=1.124$);

3, $\chi_{12}{}^0=0.5$, $\chi_{13}{}^0=0.2$, $\chi_{23}{}^0=1.0$, $j=5$ ($n_t=15$), $X_w=384.7$, $v_p{}^s=0.01$, $X_p=410.7$ ($X_p/X_w=1.068$);

4, $\chi_{12}{}^0=0.5$, $\chi_{13}{}^0=0.2$, $\chi_{23}{}^0=1.0$, $j=2$ ($n_t=5$), $X_w=387.5$, $v_p{}^s=0.01$, $X_p=410.7$ ($X_p/X_w=1.060$);

5, $\chi_{12}{}^0=1.3$, $\chi_{13}{}^0=0.2$, $\chi_{23}{}^0=1.0$, $j=5$ ($n_t=15$), $X_w=385.9$, $v_p{}^s=0.01$, $X_p=422.6$ ($X_p/X_w=1.095$);

6, $\chi_{12}{}^0=0.5$, $\chi_{13}{}^0=0.2$, $\chi_{23}{}^0=1.0$, $j=5$ ($n_t=15$), $X_w=393.4$, $v_p{}^s=0.05$, $X_p=369.0$ ($X_p/X_w=0.938$);

b) 1, $\chi_{12}{}^0=0.9$, $\chi_{13}{}^0=0.2$, $\chi_{23}{}^0=1.3$, $j=10$ ($n_t=15$), $X_w=324.4$, $v_p{}^s=1\times10^{-5}$, $X_p=319.5$ ($X_p/X_w=0.985$);

2, $\chi_{12}{}^0=0.9$, $\chi_{13}{}^0=0.2$, $\chi_{23}{}^0=1.3$, $j=33$ ($n_t=50$), $X_w=340.8$, $v_p{}^s=0.01$, $X_p=314.8$ ($X_p/X_w=0.923$);

3, $\chi_{12}{}^0=1.3$, $\chi_{13}{}^0=0.2$, $\chi_{23}{}^0=1.3$, $j=10$ ($n_t=15$), $X_w=327.1$, $v_p{}^s=0.01$, $X_p=305.3$ ($X_p/X_w=0.933$);

4, $\chi_{12}{}^0=0.5$, $\chi_{13}{}^0=0.2$, $\chi_{23}{}^0=1.3$, $j=10$ ($n_t=15$), $X_w=326.8$, $v_p{}^s=0.01$, $X_p=300.6$ ($X_p/X_w=0.919$);

5, $\chi_{12}{}^0=0.5$, $\chi_{13}{}^0=0.2$, $\chi_{23}{}^0=1.3$, $j=10$ ($n_t=15$), $X_w=328.0$, $v_p{}^s=0.05$, $X_p=291.1$ ($X_p/X_w=0.919$);

6, $\chi_{12}{}^0=0.5$, $\chi_{13}{}^0=0.2$, $\chi_{23}{}^0=1.0$, $j=7$ ($n_t=10$), $X_w=337.0$, $v_p{}^s=0.01$, $X_p=310.1$ ($X_p/X_w=0.920$).

X_p, X value at peak of MWD.

amount of nonsolvent and the complete fractionation run up to the final step is usually achievable if the first step fractionation is possible for a given solvent / nonsolvent system. In contrast, as an SPF run progresses, the lower molecular weight components increase in the polymer-rich phase and v_p^0 decreases remarkably, especially at the later steps.

Figs 3.61a and b show MWD of some SPF and SSF fractions, with nearly the same X_w (380$\pm$ 10 in SPF and 330$\pm$ 15 in SSF), arbitrarily chosen among the numerous computer experimental data obtained under a very wide range of operating conditions, such as v_p^s, χ_{12}^0, χ_{13}^0, χ_{23}^0 and n_t. Evidently, MWD of the fractions has the common features characteristic of SPF or SSF, by which they are separated: SPF fractions are symmetrical in the MWD curve, ranging approximately over $0\sim 2X_w$ and X at peak, X_p, coincides roughly with X_w ($X_p/X_w = 1.00\pm 0.07$). Then, the SPF fractions, isolated from the quasi-ternary system, exhibit a MWD which obeys eqns (2.95)-(2.97). SSF fractions have an

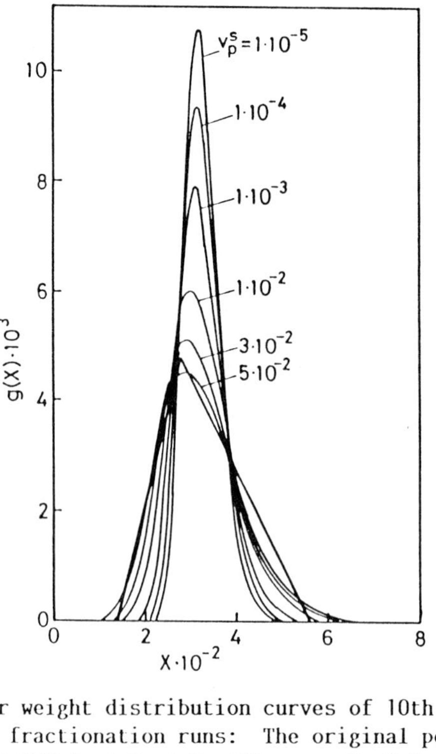

Fig. 3.62 Molecular weight distribution curves of 10th fractions isolated by successive solution fractionation runs: The original polymer, Schulz-Zimm type distribution, $X_w^0/X_n^0 = 2.0$, $X_w^0 = 300$; $\chi_{12}^0 = 0.9$, $\chi_{13}^0 = 0.2$ and $\chi_{23}^0 = 1.3$; $n_t = 15$ and $v_p^s = 1\times 10^{-5}\sim 5\times 10^{-2}$.

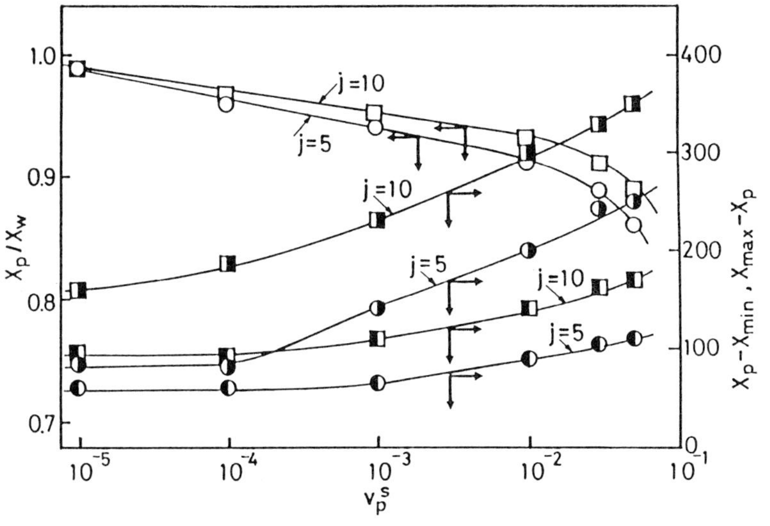

Fig. 3.63 Effect of the starting concentration v_p^s on X_p/X_w, X_p-X_{min}, and $X_{max}-X_p$; X_p, X value at peak, X_{min} and X_{max}, minimum and maximum value of X in the MWD for a given fraction defined conventionally as the values at which MWD is almost equal to 0: The original polymer, Schulz-Zimm type distribution, $X_w^0/X_n^0=2.0$, $X_w^0=300$; $\chi_{12}^0=0.9$, $\chi_{13}^0=0.2$ and $\chi_{23}^0=1.3$; $j=5$ and 10 ($n_t=15$).

unsymmetrical MWD curve, covering very roughly from $(1/2)X_w$ to $2X_w$. The ratio X_p/X_w ranges for $v_p^s=0.01\sim0.05$ from 0.92 to 0.93, which is a little larger than 6/7. The MWD of SSF fractions, obtained under conventional conditions, easily employable in practice, from quasi-ternary system, can also be represented by eqns (2.99)-(2.101), originally proposed for those from a quasi-binary system. Strictly speaking, however, the shape of the MWD of SSF fractions varies depending on v_p^s (ref. 88).

Fig. 3.62 shows the effect of polymer concentration v_p^s on MWD of SSF 10th fractions, separated under the conditions of $\chi_{12}^0=0.9$, $\chi_{13}^0=0.2$, $\chi_{23}^0=1.3$ and $n_t=15$. The upper limit of MWD, expressed by X_{max}, decreases and the lower limit of MWD, X_{min}, increases as v_p^s decreases. Fig. 3.63 demonstrates the effect of v_p^s on the X_p/X_w, $X_{max}-X_p$ and X_p-X_{min} for the 5th and 10th fractions separated by SSF. As v_p^s increases, X_p/X_w increases and $X_{max}-X_p$ and X_p-X_{min} decrease resulting in sharp SSF fractions with a more symmetrical distribution without limit (Fig. 3.62).

Fig. 3.64a-f show the effect of χ_{12}^0, χ_{13}^0 and χ_{23}^0 on the relations between X_w/X_n and X_w for a series of fractions, isolated by SPF (a, c and e) and SSF (b, d and f) under the conditions of $v_p^s = 0.01$ and $n_t = 15$. As observed in the case of the quasi-binary system (refs 32,68), with increase in X_w, X_w/X_n in a SPF run increases significantly and that, in a SSF run, decreases first remarkably in the lower M_w region, and then remains approximately constant. In other words, as the fractionation proceeds, X_w/X_n continues to decrease without limit in SPF, but changes little in SSF, except for the first few fractions, approaching an asymptotic value, which is much less than 1.2. The black circle in the figure is the polymer in a polymer-rich phase at the final separation

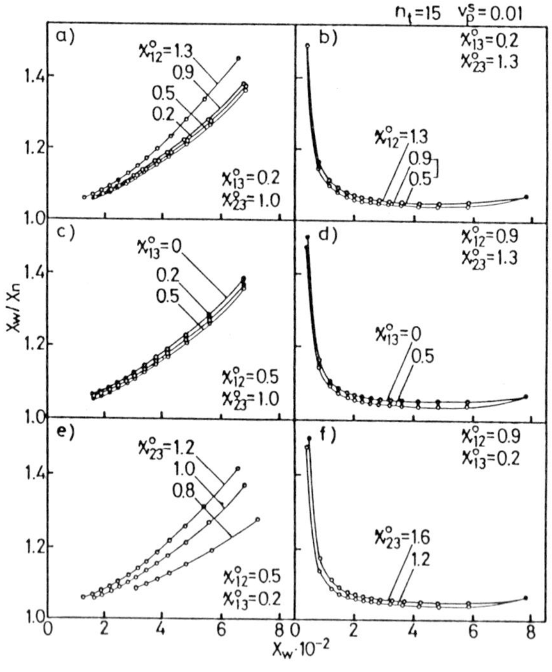

Fig. 3.64 Effects of three thermodynamic interaction parameters χ_{12}^0, χ_{13}^0 and χ_{23}^0 on the relationship between X_w/X_n and X_w for series of fractions isolated by SPF [a), c), e)] and SSF [b), d), f)] runs: The original polymer, Schulz-Zimm type distribution, $X_w^0/X_n^0 = 2$, $X_w^0 = 300$; starting polymer volume fraction, $v_p^s = 0.01$; total number of fractions in a run, $n_t = 15$. For SPF, a) $\chi_{13}^0 = 0.2$, $\chi_{23}^0 = 1.0$, c) $\chi_{12}^0 = 0.5$, $\chi_{23}^0 = 1.0$, e) $\chi_{12}^0 = 0.5$, $\chi_{13}^0 = 0.2$ and for SSF, b) $\chi_{13}^0 = 0.2$, $\chi_{23}^0 = 1.3$, d) $\chi_{12}^0 = 0.9$, $\chi_{23}^0 = 1.3$, f) $\chi_{12}^0 = 0.9$, $\chi_{13}^0 = 0.2$; closed circle, counter part polymer at final fractionation step.

step, for which X_w/X_n is a little larger than that of the counterpart polymer in a polymer-lean phase, coexisting to the polymer-rich phase in question. In some computer experiments, it becomes impossible to pursue the fractionation further, especially in SPF. Evidently, by using SPF we cannot obtain a series of fractions with the same X_w/X_n. In contrast, SSF proves very powerful for preparative purpose even in quasi-ternary systems, giving fractions with $X_w/X_n \lesssim 1.1$ over a wide range of X_w ($100 \lesssim X_w \lesssim 800$). As is known well, it is very important to prepare a series of polymer samples, for which X_w/X_n is almost constant, even if not unity, for establishing structure-property relationships. It is interesting to note that X_w of the first fraction by SPF is considerably lower than that of the last fraction by SSF under the same fractionation conditions except fractionation scheme (ref. 88).

In the range $0.2 \lesssim \chi_{12}^0 \lesssim 0.9$, the effect of χ_{12}^0 on X_w/X_n of the fractions, isolated by SPF is extremely small and can be regarded as insignificant. In the strict sense, X_w/X_n attains a minimum at a given fractionation step. In the Fig. 3.64, v_p^s at first step was kept constant (0.01), then with increasing χ_{12}^0 the total volume V increases and in consequence v_p^0 decreases. A decrease in v_p^0 brings about an increase in the partition coefficient σ and the phase volume ratio R ($\equiv V_{(1)}/V_{(2)}$, $V_{(1)}$ and $V_{(2)}$ are the volumes of the polymer-lean and rich phases, respectively) (refs 12,14), i.e., an increase in fractionation (separation) efficiency. On the other hand, larger χ_{12}^0 causes essentially lower separation efficiency (refs 11,12,14). For example, at constant v_p^0 X_w/X_n increases with χ_{12}^0 (see, Fig. 14a of ref. 12). In consequence, X_w/X_n of the fractions in a SPF run becomes minimum at a specific value (in this case, 0.5) of χ_{12}^0. An increase in χ_{13}^0 and a decrease in χ_{23}^0 brings about a decrease in X_w/X_n of the SPF fractions. Particularly, the effect of χ_{23}^0 is remarkable. This indicates that, in order to isolate a series of sharp fractions by SPF, the choice of mild nonsolvent is very important as pointed out by Kamide and Matsuda in their study on phase equilibria of quasi-ternary systems (see, Fig. 13 of ref. 11). A small, but significant tendency of lowering X_w/X_n of fractions with decrease in χ_{12}^0 and χ_{23}^0 and an increase in χ_{13}^0 is observed also in SSF, but the effect of these three χ parameters on the fraction polydispersity is small as compared with that in SPF. We can conclude that X_w/X_n of SSF fractions is almost independent of χ_{12}^0, χ_{13}^0 and χ_{23}^0 (ref. 88).

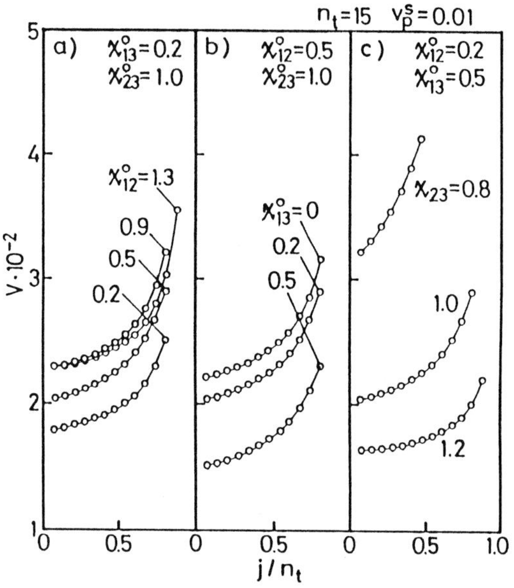

Fig. 3.65 Effect of three χ parameters on change in total volume V ($\equiv V_1{}^0 + V_2 + V_3{}^0$) of solutions in equilibrium with the extent of fraction in successive precipitation fractionation (SPF) runs: The original polymer, Schulz-Zimm type distribution, $X_w{}^0/X_n{}^0 = 2$, $X_w{}^0 = 300$; starting polymer volume fraction, $v_p{}^s = 0.01$; total number of fractions in a run, $n_t = 15$. a) $\chi_{13}{}^0 = 0.2$, $\chi_{23}{}^0 = 1.0$, b) $\chi_{12}{}^0 = 0.5$, $\chi_{23}{}^0 = 1.0$, c) $\chi_{12}{}^0 = 0.5$, $\chi_{13}{}^0 = 0.2$

Fig. 3.65a-c show the effect of $\chi_{12}{}^0$, $\chi_{13}{}^0$ and $\chi_{23}{}^0$ on the relations between the total volume V and the extent of the SPF run j/n_t (j is the order of fractionation step). Fig. 3.66a-c and d-f show the effect of $\chi_{12}{}^0$, $\chi_{13}{}^0$ and $\chi_{23}{}^0$ on the relations between the polymer volume fractions in polymer-lean and -rich phases, $v_{p(1)}$ and $v_{p(2)}$, and the extent of the SPF run j/n_t, respectively. As the SPF run proceeds $v_{p(2)}$ (and V) increases almost exponentially, but $v_{p(1)}$ decreases roughly linearly.

An increase in $\chi_{12}{}^0$ and a decrease in $\chi_{13}{}^0$ and $\chi_{23}{}^0$ bring about an increase in V and a decrease in $v_{p(1)}$. With a decrease in $\chi_{12}{}^0$ and $\chi_{23}{}^0$ and with an increase in $\chi_{13}{}^0$, $v_{p(2)}$ increases. It was often observed in the computer experiments that the fractionation run becomes impossible when V is large [$\gtrsim 500$; with V_0 ($\equiv V_1{}^0 + V_3{}^0$) = 100] and $v_{p(2)} \gtrsim 0.35$. Whether this is due to the theoretical requirement or not (i.e., only a problem of the computer program employed here) remained unclear.

Fig. 3.67a-c show the effect of $\chi_{12}{}^0$, $\chi_{13}{}^0$ and $\chi_{23}{}^0$ on the relations between X_w/X_n of the fractions, separated by the SPF

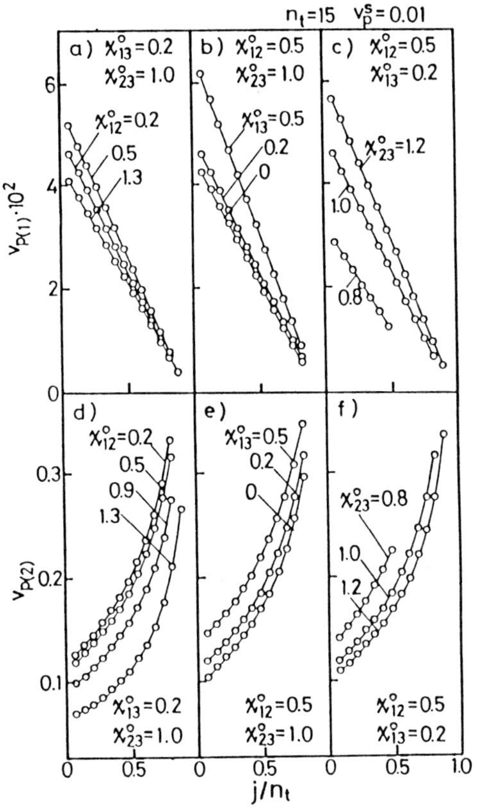

Fig. 3.66 Change in the volume fractions of polymers in the polymer-lean phase $v_{p(1)}$ and in the polymer-rich phase $v_{p(2)}$ with the extent of fractionation j/n_t in successive precipitation fractionation: The original polymer, Schulz-Zimm type distribution, $X_w^0/X_n^0 = 2$, $X_w^0 = 300$; starting polymer volume fraction, $v_p^s = 0.01$; total number of fractions in a run, $n_t = 15$. a) and d) $\chi_{13}^0 = 0.2$, $\chi_{23}^0 = 1.0$; b) and e) $\chi_{12}^0 = 0.5$, $\chi_{23}^0 = 1.0$; c) and f) $\chi_{12}^0 = 0.5$, $\chi_{13}^0 = 0.2$.

run, and their X_w under the conditions of constant v_p^0 [the polymer volume fraction at which the two-phase separation occurs (i.e., the initial polymer volume fraction)] = 0.005, in place of the starting polymer volume fraction v_p^s. In the figure is included as a broken line the results of SPF for the quasi-binary systems, where the concentration-dependence χ parameter (p_1) was neglected (refs 29,35). At constant v_p^0, unlike constant v_p^s, X_w/X_n monotonically decreases with decreasing χ_{12}^0. Smaller X_w/X_n of the SPF fractions is obtained by SPF for smaller χ_{12}^0, larger χ_{13}^0 and smaller χ_{23}^0. Nevertheless X_w/X_n of these fractions never becomes smaller than that by quasi-binary solutions. This is

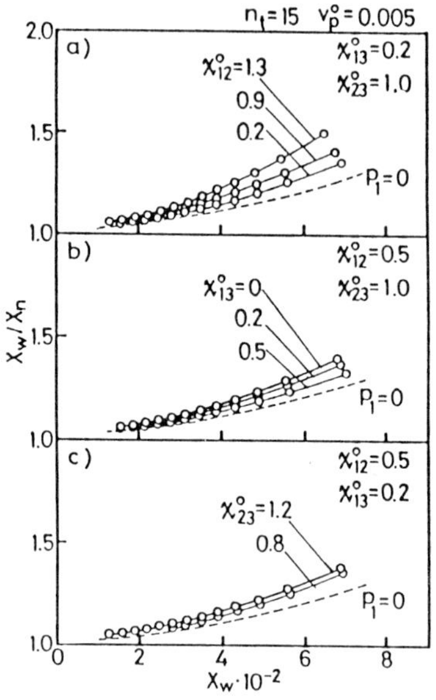

Fig. 3.67 Effect of three χ parameters on the relationship between the polydispersity of fractions X_w/X_n and their X_w in successive precipitation fractionation (SPF) runs: The original polymer, Schulz-Zimm type distribution, $X_w^0/X_n^0 = 2$, $X_w^0 = 300$; initial polymer volume fraction $v_p^0 = 0.005$; total number of fractions, $n_t = 15$. Broken line, the results of SPF by a quasi-binary system, where the concentration dependence of the χ parameter was neglected (i.e., $p_1 = 0$) and the initial polymer volume fraction v_p^0 was taken as 0.01. a) $\chi_{13}^0 = 0.2$, $\chi_{23}^0 = 1.0$; b) $\chi_{12}^0 = 0.5$, $\chi_{23}^0 = 1.0$; c) $\chi_{12}^0 = 0.5$, $\chi_{13}^0 = 0.2$.

in good agreement with the previous conclusion that the separation efficiency of the first fraction by quasi-ternary solution never exceeds, by any possible combination of three χ parameters, that of the first fraction by quasi-binary solution (ref. 88). Note that if the concentration dependence of these χ parameters is taken into account, the situation is changed as previously described (section 3.2, Fig. 3.26) (ref. 14).

3.6.4 Role of operating conditions

Fig 3.68a and b show the effect of v_p^s on the relations between X_w/X_n and X_w of a series of fractions, isolated by SPF (a) and SSF (b) run under the conditions of $n_t = 15$, $\chi_{12}^0 = 0.5$, $\chi_{13}^0 = 0.2$ and $\chi_{23}^0 = 1.0$ for SPF and $\chi_{12}^0 = 0.9$, $\chi_{13}^0 = 0.2$ and $\chi_{23}^0 = 1.3$ for SSF. Obviously, from higher v_p^s solutions, fractions with larger X_w/X_n (and accordingly, lower separation efficiency) are isolated both by SPF and SSF. However, the v_p^s

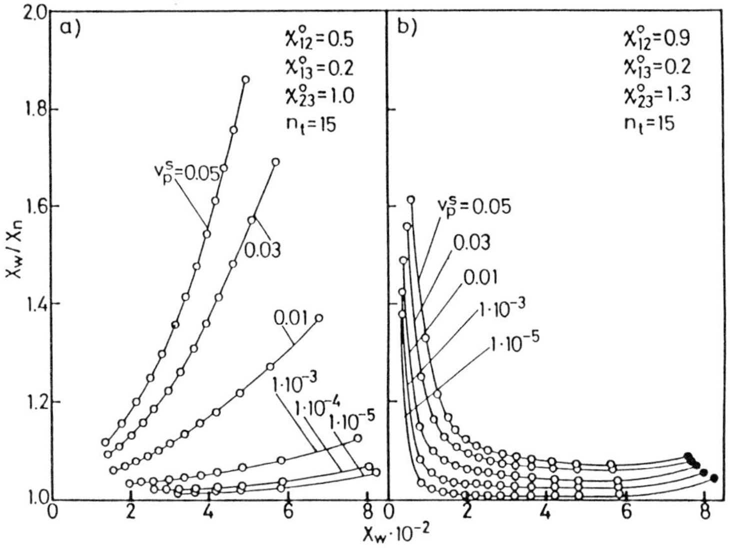

Fig. 3.68 Effect of starting polymer volume fraction v_p^s on relationship of X_w/X_n of the fractions and their X_w in successive precipitation fractionation (SPF) [a)] and successive solution fractionation (SSF) [b)] runs: The original polymer, Schulz-Zimm type distribution, $X_w^0/X_n^0=2$, $X_w^0=300$; $\chi_{12}^0=0.5$, $\chi_{13}^0=0.2$, $\chi_{23}^0=1.0$ in SPF and $\chi_{12}^0=0.9$, $\chi_{13}^0=0.2$, $\chi_{23}^0=1.3$ in SSF; total number of fractions in a run, $n_t=15$.

dependence of X_w/X_n is remarkable in SPF, but is moderate in SSF. For example, X_w and X_w/X_n of the first SPF fraction are 500 and 1.87 for $v_p^s=0.05$ and 830 and 1.06 for $v_p^s=1\times10^{-5}$, respectively. The corresponding values of the first SSF fraction are 70 and 1.62 for $v_p^s=0.05$ and 40 and 1.38 for $v_p^s=1\times10^{-5}$, respectively and (X_w, X_w/X_n) of the final SSF fractions are (750, 1.1) for $v_p^s=0.05$ and (830, 1.05) for $v_p^s=1\times10^{-5}$. SSF enables us to isolate, over a wide v_p^s range, the fractions with larger X_w and sharp MWD (ref. 88). When the SPF fractions are separated from relatively concentrated solution (~0.05, in this case), V can be regarded as almost constant (see, Fig. 3.70).

Fig. 3.69 shows the maximum fractionation step j_{max} practically attainable in computer experiments of SPF and SSF runs, plotted as a function of v_p^s. It is interesting to note that in this case j_{max} for SSF is absolutely independent of v_p^s, indicating that the combination of three χ parameters adequate for the first step phase separation enables us to continue the SSF

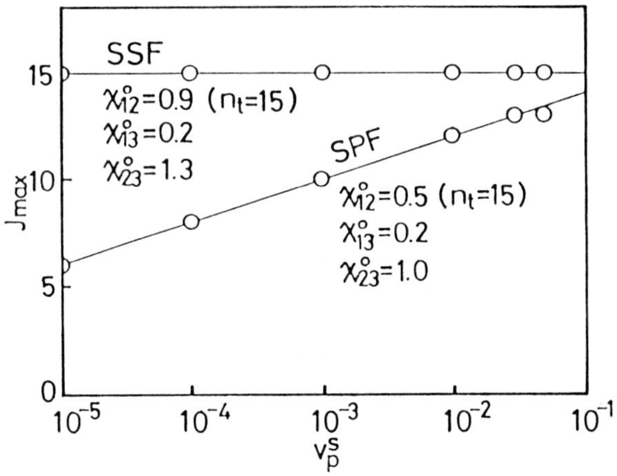

Fig. 3.69 Starting polymer volume fraction v_p^s dependence on the maximum fractionation step j_{max} practically attainable in SPF and SSF runs: The original polymer, Schulz-Zimm type distribution, $X_w^0/X_n^0 = 2$, $X_w^0 = 300$: $\chi_{12}^0 = 0.5$, $\chi_{13}^0 = 0.2$, $\chi_{23}^0 = 1.0$ in SPF and $\chi_{12}^0 = 0.9$, $\chi_{13}^0 = 0.2$, $\chi_{23}^0 = 1.3$ in SSF; $n_t = 15$.

run up to the last step and in other words, such a combination, originally found for a specific v_p^s, can also be applied to a wide range of v_p^s, being very insensitive to v_p^s. In contrast to this, in SPF j_{max} depends strongly on v_p^s, increasing linearly with log v_p^s. Then, we can predict that at lower v_p^s complete fractionation (i.e., fractionation up to the last step, $j = n_t$) is very difficult.

Figs 3.70 and 3.71a and b show the effects of starting concentration v_p^s on the relations between the total volume V and the extent of fractionation j/n_t, and those between the polymer volume fractions in a polymer-lean or -rich phase $v_{p(1)}$ or $v_{p(2)}$ and j/n_t, respectively in a SPF run ($n_t = 15$, $\chi_{12}^0 = 0.5$, $\chi_{13}^0 = 0.2$ and $\chi_{23}^0 = 1.0$). The effect of v_p^s on V becomes notable in the smaller v_p^s range and is negligibly small at $v_p^s = 0.05$. In other words, in the latter case V remained nearly constant throughout a SPF run. The effect of v_p^s on $v_{p(1)}$ is larger for larger v_p^s, but interestingly, $\partial v_{p(2)}/\partial (j/n_t)$ is very insensitive to v_p^s.

Fig. 3.72a and b demonstrate the effect of the total number of fractions n_t in a SPF and a SSF run, on the relations between X_w/X_n of the fractions, isolated by a run, and their X_w. In this

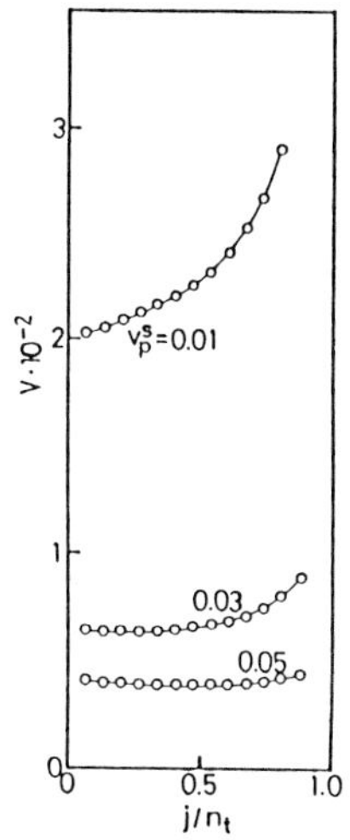

Fig. 3.70 Effect of starting polymer volume fraction $v_p{}^s$ on the relations the between the total volume of the solution V and the extent of SPF fractionation j/n_t: The original polymer, Schulz-Zimm type distribution, $X_w{}^0/X_n{}^0=2$, $X_w{}^0=300$; $\chi_{12}{}^0=0.5$, $\chi_{13}{}^0=0.2$, $\chi_{23}{}^0=1.0$; $n_t=15$. $v_p{}^s$ is denoted on the curve.

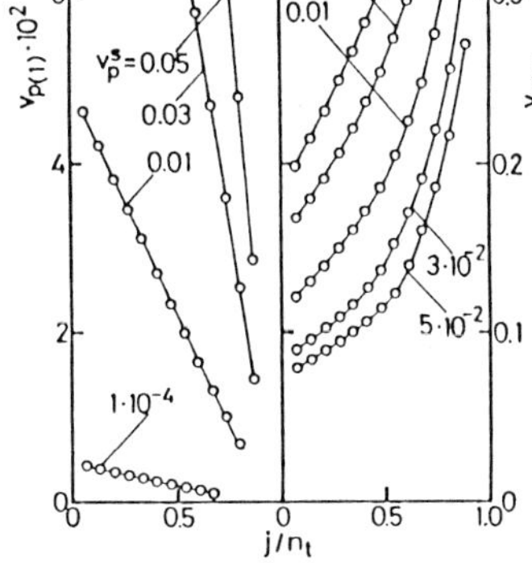

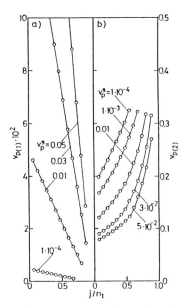

Fig. 3.71 Effect of starting polymer volume fraction $v_p{}^s$ on the relationship between the polymer volume fraction in polymer-lean and -rich phases, $v_{p(1)}$ and $v_{p(2)}$, in an SPF run: The original polymer, Schulz-Zimm type distribution, $X_w{}^0/X_n{}^0=2$, $X_w{}^0=300$; $\chi_{12}{}^0=0.5$, $\chi_{13}{}^0=0.2$, $\chi_{23}{}^0=1.0$; $n_t=15$. $v_p{}^s$ is denoted on the curve.

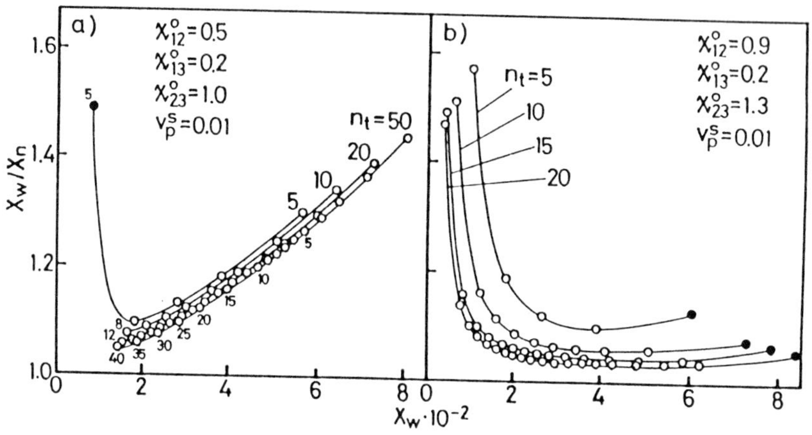

Fig. 3.72 Effect of the total number of fractions in a run n_t on the relationship between X_w/X_n and X_w of the fractions obtained by successive precipitation fractionation (SPF) [a)] and successive solution fractionation (SSF) [b)]: The original polymer, Schulz-Zimm type distribution, $X_w{}^0/X_n{}^0=2$, $X_w{}^0=300$; $\chi_{12}{}^0=0.5$, $\chi_{13}{}^0=0.2$, $\chi_{23}{}^0=1.0$ in SPF and $\chi_{12}{}^0=0.9$, $\chi_{13}{}^0=0.2$, $\chi_{23}{}^0=1.3$ in SSF; $v_p{}^s=0.01$. n_t is denoted on the curve; closed circles denotes the counterpart polymer at the final fractionation step.

case, all conditions, except for n_t, are kept constant. Generally, as n_t increases X_w/X_n of the fractions with a given X_w decreases: In SPF the X_w/X_n vs. X_w relations are not significantly influenced by n_t, if $n_t \gtrsim 10$, suggesting that $n_t=10$ is sufficient to obtain the fractions as sharp as possible. X_w of the first fraction is influenced in SPF by n_t and 50 should be taken as n_t in order to isolate the fraction with large X_w. Contrary to this, in SSF X_w/X_n and X_w are shifted to the origin of the coordinates. X_w/X_n of the fraction with a given X_w decreases without limit to unity with an increase in n_t. These features were already observed for quasi-binary solutions (refs 68,90) (Fig. 2.23).

Fig. 3.73 shows a plot of $(X_w/X_n)_{min}$ vs. n_t in SPF and SSF runs. $(X_w/X_n)_{min}$ decreases with increasing n_t in SPF and SSF, and n_t dependence of $(X_w/X_n)_{min}$ is considerable in relatively small n_t (~ 10) region. In this case, $(X_w/X_n)_{min}$ in SSF is always smaller than $(X_w/X_n)_{min}$ in SPF, compared at the same $v_p{}^s$ and the same n_t, but the difference in $(X_w/X_n)_{min}$ between SPF and SSF under the same conditions ($v_p{}^s$ and n_t) becomes larger with an increase in n_t and this difference cannot be cancelled by choosing an adequate combination of three χ parameters (ref. 88).

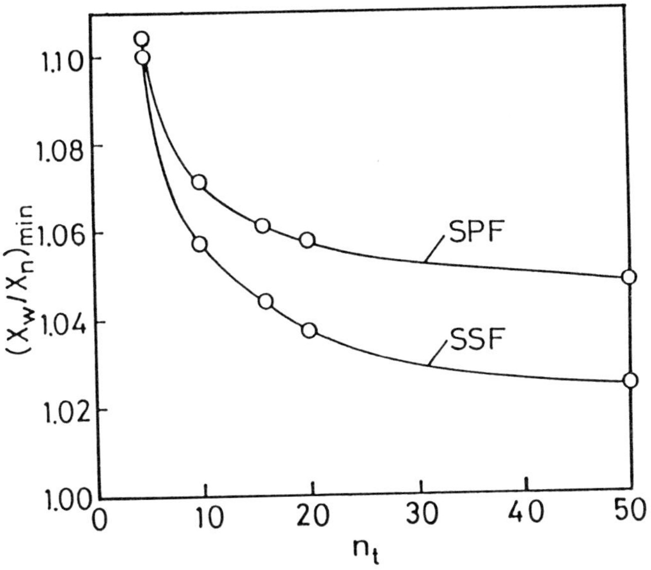

Fig. 3.73 Relationship between $(X_w/X_n)_{min}$ and n_t for successive precipitation fractionation (SPF) and successive solution fractionation (SSF): The original polymer, Schulz-Zimm type distribution, $X_w^0/X_n^0=2$, $X_w^0=300$; $\chi_{12}^0=0.5$, $\chi_{13}^0=0.2$, $\chi_{23}^0=1.0$ in SPF and $\chi_{12}^0=0.9$, $\chi_{13}^0=0.2$, $\chi_{23}^0=1.3$ in SSF; $v_p^s=0.01$.

Fig. 3.74 shows the effect of v_p^s and n_t on the minimum value of X_w/X_n of the fractions, obtained by a given fractionation run, $(X_w/X_n)_{min}$, and the relative amount of fractions $Y_{1.1}$, for which X_w/X_n is equal to or less than 1.1. For a given combination of v_p^s and n_t, SSF always affords us smaller $(X_w/X_n)_{min}$ than SPF and it is expected to isolate by SSF the fractions with $(X_w/X_n)_{min}=1.02\sim1.04$ from solutions ($v_p^s=0.001\sim0.01$) of the polymer with Schulz-Zimm type distribution ($X_w^0=300$ and $X_w^0/X_n^0=2.0$) under the operating conditions of $n_t=10\sim30$. This value of $(X_w/X_n)_{min}$ should be compared with $1.04\sim1.06$ for SPF fractions produced under the same conditions. In SPF, $(X_w/X_n)_{min}$ is dependent predominantly on v_p^s and, in SSF, $(X_w/X_n)_{min}$ is sensitive to n_t if $n_t<10$ and is governed by v_p^s in the range of $n_t>10$. $Y_{1.1}$ of SPF is determined only by v_p^s if $n_t>15$ and $v_p^s>10^{-3}$ and v_p^s dependence of $Y_{1.1}$ is less remarkable in SSF as compared with SPF. It is obvious that $Y_{1.1}>0.6$ cannot be achieved by SPF, but $Y_{1.1}\simeq0.90\sim0.98$ is easily attainable employing conventional conditions in SSF. Then Fig. 3.74 exhibits unquestionable

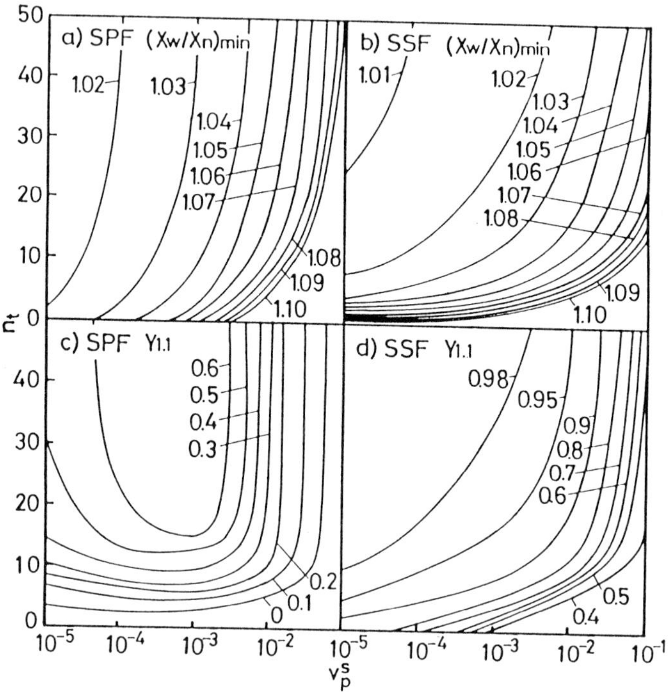

Fig. 3.74 Relationship between the minimum value of X_w/X_n obtained by a given fractionation run $(X_w/X_n)_{min}$ [a) and b)], and the relative amount of fractions $Y_{1.1}$ [c) and d)] whose X_w/X_n is equal to or less than 1.10, $v_p{}^s$ and n_t: The original polymer, Schulz-Zimm type distribution, $X_w{}^0/X_n{}^0=2$, $X_w{}^0=300$; $\chi_{12}{}^0=0.5$, $\chi_{13}{}^0=0.2$, $\chi_{23}{}^0=1.0$ for SPF and $\chi_{12}{}^0=0.9$, $\chi_{13}{}^0=0.2$, $\chi_{23}{}^0=1.3$ for SSF; $v_p{}^s=0.01$.

superiority of SSF over SPF for preparative purposes (ref. 88).

Fig. 3.75 shows the relationships between the standard deviation of MWD of the polymer dissolved in a given mother solution (hereafter referred to as σ'_0) and that of the fractions ($\rho=1/15$), separated from the solutions, for each fractionation step in the case of SPF and SSF, respectively. In this figure values of σ'_0 and the corresponding values of σ' for a given separation step (closed circle for SPF and open circle for SSF) are connected by a straight line, with an arrow. In SPF, the polymolecularity, expressed in terms of σ', of the polymers in the mother solution at a given step (j) [i.e., the polymer-lean phase at an immediately previous step (j-1) ($j\geq 2$) or the starting solution (when $j=1$)] drastically diminishes as the fractionation

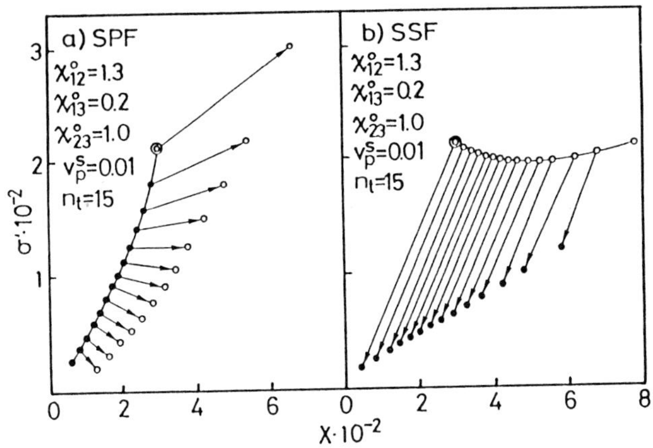

Fig. 3.75 Standard deviation σ ' in the polymer-rich phase (open circle) and the polymer-lean phase (filled circle) in a) successive precipitation fractionation (SPF) run and b) successive solution fractionation (SSF) run, both obtained by computer experiments under the following conditions: The original polymer, Schulz-Zimm type distribution, $X_w{}^0/X_n{}^0=2$, $X_w{}^0=300$, $\sigma_0{}'=212.2$ (denoted by ◎); $\chi_{12}{}^0=1.3$, $\chi_{13}{}^0=0.2$, $\chi_{23}{}^0=1.0$; starting polymer volume fraction $v_p{}^s=0.01$; total number of fractions in a run, $n_t=15$.

proceeds. In this case ($n_t=15$, $v_p{}^s=0.01$), at $j\leq 4$ the polymer remaining in the polymer-rich phase has always larger σ ' than that in the mother solution at the step j, from which two-phase separation occurred. In addition, SPF yields often fractions (in this case, first and second fractions) having broader MWD than the original polymer. σ ' of polymers in the polymer-rich phase and in the mother solution agrees well at some specific step (in this case, $j=5$). At a further step (i.e., $j\geq 6$) the reversion of σ ' between the polymer in polymer-rich phase and the mother solution is observed. In SSF, σ ' of the polymer dissolved in the mother solution remains roughly constant, which is the same as that of the original polymer, throughout the fractionation process. In the strict sense, σ ' decreases very slightly, passes through a minimum and then increases as fractionation progresses. SSF always yields fractions having much smaller values of σ ' than the corresponding σ ' values in the mother solution. Note that the slope of the connecting line between the polymer-lean phase and the mother solution at a given step is almost a constant during an entire SSF run. The above mentioned characteristics were theoretically and experimentally confirmed for a quasi-binary solution (ref. 35) (Fig. 2.25).

3.6.5 Analytical fractionation

Fig. 3.76 shows the cumulative weight distribution of the degree of polymerization X, $I_e(M)$ of the original polymer (in this case, SZ, $X_w^0=300$, $X_w^0/X_n^0=2$), estimated by means of SPF/Kamide and SSF/Kamide. Here, the true I(M), $I_0(M)$, is shown as a full line. SPF overestimates I(M) (and accordingly, $g_0(X_i)$) in the lower X (i.e., $X<X_w^0$) region and underestimates it in the larger X (i.e., $X>X_w^0$) region. In contrast, SSF gives a reasonable cumulative distribution curve very similar to the true distribution curve, although in the range of $X<2X_w^0$ SSF has a tendency to overestimate slightly.

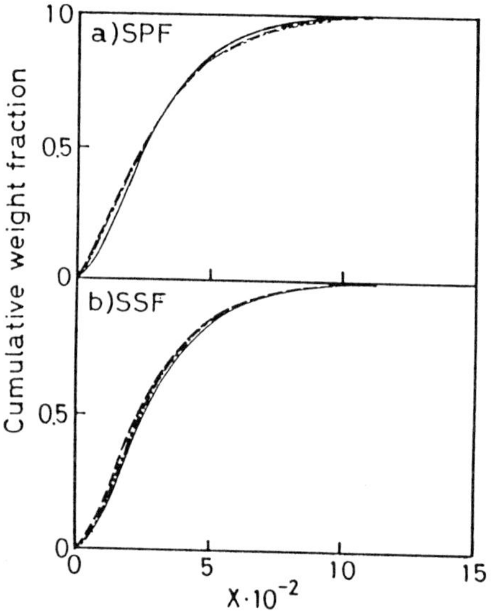

Fig. 3.76 Cumulative weight fraction, constructed from the analytical fractionation data, I_e as compared with the true cumulative weight fraction I_0: The original polymer, Schulz-Zimm type distribution, $X_w^0/X_n^0=2$, $X_w^0=300$; a) $\chi_{12}^0=0.5$, $\chi_{13}^0=0.2$, $\chi_{23}^0=1.0$ for successive precipitation fractionation (SPF); b) $\chi_{12}^0=0.9$, $\chi_{13}^0=0.2$, $\chi_{23}^0=1.3$ for successive solution fractionation (SSF); $v_p^s=0.01$; chain line, $n_t=5$ and broken line, $n_t=15$.

Fig. 3.77 shows the effects of χ_{12}^0, χ_{13}^0 and χ_{23}^0 on the relative error E, defined by eqn (2.102), in the estimated MWD of the original polymer by the SPF/Kamide and SSF/Kamide methods. E decreases under the following conditions: In SPF, larger χ_{12}^0, smaller χ_{13}^0 and smaller χ_{23}^0; in SSF, smaller χ_{12}^0, smaller

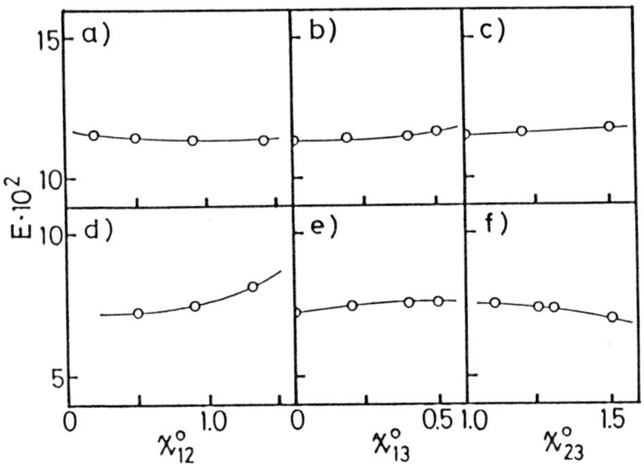

Fig. 3.77 Effect of three thermodynamic interaction parameters χ_{12}^0, χ_{13}^0 and χ_{23}^0 on the relative error E (defined by eqn (2.102)): Original polymer Schulz-Zimm type distribution; $X_w^0/X_n^0=2$, $X_w^0=300$; $v_p^s=0.01$; $n_t=15$. a), b), c), successive precipitation fractionation (SPF); d), e), f), successive solution fractionation (SSF). a) $\chi_{13}^0=0.2$, $\chi_{23}^0=1.0$; b) $\chi_{12}^0=0.5$, $\chi_{23}^0=1.0$; c) $\chi_{12}^0=0.5$, $\chi_{13}^0=0.2$; d) $\chi_{13}^0=0.2$, $\chi_{23}^0=1.3$; e) $\chi_{12}^0=0.9$, $\chi_{23}^0=1.3$; f) $\chi_{12}^0=0.9$, $\chi_{13}^0=0.2$.

TABLE 3.2
Relative error E in evaluation of MWD of the original polymer.

Fractionation Process	Analytical / Method	Relative Error (E)	
		Polymer/Solvent	Polymer/Solvent/Nonsolvent
	SPF/Kamide	~0.15	~0.12
	SSF/Kamide	~0.09	0.07~0.08

χ_{13}^0 and larger χ_{23}^0. the SPF/Kamide gives a larger E value ($\simeq 0.12$) as compared with the SSF/Kamide ($\simeq 0.07$), indicating that the SSF/Kamide method is strongly recommended for analytical purposes (ref. 88).

Table 3.2 lists the estimated range of relative error E when SPF and SSF are applied to quasi-binary (polymer-solvent) and quasi-ternary (polymer / solvent / nonsolvent) systems and the fractionation data are analyzed by the Kamide's method. For both systems, the SSF/Kamide method has an advantage over the the SPF/Kamide method and the quasi-ternary system is, surprisingly, more adequate for estimation of MWD of the original polymer. Then, we can conclude that the SSF/Kamide method applied to the system of polymer / solvent / nonsolvent is the best choice for this purpose.

Summarizing, the numerous characteristic features observed by Kamide et al. for quasi-binary systems are expected (and partly confirmed by the computer experiments) to be found also for quasi-ternary systems. A newly encountered problem in the latter system is an adequate selection of the two solvents. The criterion of choosing of solvents and nonsolvents adequate for separation of sharp fractions from a quasi-ternary system by SPF is as follows: good mutual solubility of a solvent and a nonsolvent (small χ_{12}^0), a nonsolvent with a low precipitation power (weak nonsolvent, small χ_{23}^0) and a solvent with a low dissolving power against polymer (large χ_{13}^0). In particular, χ_{23}^0 plays an important role. In contrast, SSF does not need any careful selection of solvents and nonsolvents, because the effect of these three χ parameters is ordinarily small and often even negligible. In contrast to SPF, SSF can be performed up to the last step (i.e., $j=n_t$) using three χ parameters which allow the first step phase separation. Then, in this sense in SSF it is sufficient to determine only an adequate combination of three χ parameters fulfilling the requirements at the first step. The fractionation efficiency of SPF is strongly dependent on v_p^s and rather insensitive to n_t if $n_t \gtrsim 10$, but that of SSF is less strongly influenced by v_p^s and remarkably depends on n_t. SSF provides a series of sharp fractions with $X_w/X_n \lesssim 1.1$ over a wide range of X_w. For analytical fractionation, the ternary system is superior over a binary system if the SSF/Kamide method is applied and for the former system also the SSF/Kamide method gives a more accurate MWD curve of the original polymer than the SPF/Kamide method does.

Sufficient theoretical guide line useful in practical fractionation were obtained in this section. Now, if three χ parameters are determined in advance, we can design operating conditions of fractionation with the help of computer experiments. Here, the possible concentration dependencies of three χ parameters were neglected and if this dependence must be taken into account for phase equilibrium, the theory presented this section can be readily generalized. We expect that even in that case, a not very different result as in the case for a quasi-binary system will be obtained.

3.6.6 Experimental examples of molecular weight fractionation using quasi-ternary solutions

Adequate combinations of solvent and nonsolvent and the operating conditions in successive weight fractionation runs are carefully selected from literature. The criterion employed is as follows:

(1) The total number of fractions in a run, $n_t > 10$.

(2) The ratio of the maximum average molecular weight to the minimum molecular weight of a series of fractions, fractionated in a run, at least, larger than 10 (in SPF) or 7 (in SSF).

(3) The starting polymer concentration $v_p{}^s$ and phase separation temperature are clearly described.

The results are summarized in Table 3.3, in which the literature for quasi-binary systems are also included.

TABLE 3.3
Adequate combination of solvent and nonsolvent and the operating conditions in molecular weight fractionation runs using quasi-ternary systems.

Polymer	Method of fract.	Solvent	Nonsolvent	Temp. /℃	$v_p^s \times 10^2$ /g/cm³	n_t	$M_w \times 10^{-4}$	M_w/M_n	Method for determining M_w(or M_n)	ref.
Polybutadiene										
(98%cis)	SPF	Toluene	Methanol	25	0.1	10	5~100	—	MO[a]	91
(98%cis)	SPF	Chloroform	Methanol	25	0.1	10	5~100	—	MO	91
(95%cis)	SPF	Benzene	Methanol	22.8	1.0	12	4~37	1.4~1.6	MO, LS[b]	92
(71%trans)	SPF	Benzene	Acetone	25	1.0	12	230~800	—	LS	93
Polybutadiene	SSF	Benzene	n-Pentane	80~	0.9	10	3~76	1.3~2.0	[η][c], MO	94
Polystyrene	SPF	Butanone	Methanol	25	1.0	6	25~100	—	MO	95
	SPF	Butanone	Ethanol	20	1.0	6~9	3~1100	—	[η]	96
	SPF	Ethyl acetate	Ethanol	20	1.0	12	3~55	—	MO	97
	SSF	Methylcyclohexane	—	20~55	0.94	14	2~50	<1.1	GPC[d]	24,29
	SSF	Methylcyclohexane	—	15~55	0.94	23	2~63	<1.1	GPC	24,29
	SSF	Methylcyclohexane	—	7~55	0.94	20	2~53	<1.1	GPC	24,29
	SSF	Methylcyclohexane	—	10~58	1.89	30	2~93	<1.1	GPC	24,29
	SSF	2-Butanone	Methanol	35	0.68	22	2~48	<1.1	GPC	24
Polyethylene										
(HDPE[e])	SPF	Xylene	n-Propanol	95	0.84	20	1~65	—	[η]	98
(HDPE)	SPF	Xylene	Triethylene glycol	130	1.0	17	~65	—	MO	99
(HDPE)	SPF	p-Xylene	PEG[f] 300	125	1.0	12	4~27	—	[η]	100
(HDPE)	SSF	Xylene	PEG 400	120	0.5	11	1~55	—	[η]	101
(HDPE)	SPF	2-Ethyl-1-hexanol /decalin (85/15)	—	20 ~133	0.5 ~0.7	5 ~13	0.4 ~285	—	[η]	102

a) Membrane osmometry, b) Light scattering, c) Viscometer, d) Gel permeation chromatography, e) High density polyethylene, f) Poly(ethylene glycol).

TABLE 3.3 (continued-1)

Polymer	Method of fract.	Solvent	Nonsolvent	Temp. /℃	v_p^s $\times 10^2$ /g/cm^3	n_t	$M_w \times 10^{-4}$	M_w/M_n	Method for determining M_w (or M_n)	ref.
(HDPE)	SSF	Cyclohexane	—	98~133	1.1	20	2~45	1.3	GPC	24
(LDPE[g])	SPF	Toluene	n-Propanol	80	2.0	7~10		~5	MO	103
(LDPE)	SPF	p-Xylene	n-Amyl-alcohol	110	1.1	19	~22	—	MO	104
(LDPE)	SPF	Xylene	PEG 300	80	0.5~1.0	10~14	—	—	[η]	105
Polypropylene	SSF	Xylene	PEG 200	134	1.6	16	1~132	—	[η]	106
	SSF	Tetralin	PEG 200	140	0.1	15	1~300	—	[η]	107
Poly(methyl methacrylate)	SPF	Benzene	Methanol	20~30	3	11	41~337	—	LS	108
	SPF	Benzene	n-Hexane	40	0.5	15	5~128	1.2~2.5	MO, LS, [η]	109
	SPF	Benzene	n-Hexane	25	1.0	19	~41	—	MO	110
Poly(vinyl acetate)	SPF	Acetone	Methanol /Water(1/1)	30	6.5	37	3~163	—	MO	111
	SPF	Acetone	Water	30	3	5	2~43	—	MO	112
	SPF	Acetone	Water	30	5	11	2~43	—	MO	112
	Extraction	Ethylacetate	Petroleum ether	25	0.75	23	2~40	—	[η]	113
Poly(vinyl alcohol)	SSF	Water	Methyl-acetate /Methanol(3/1)	50	0.79	20	1~30	—	[η]	113

g) Low density polyethylene.

TABLE 3.3 (continued-2)

Polymer	Method of fract.	Solvent	Nonsolvent	Temp. /℃	$v_p^s \times 10^2$ /g/cm³	n_t	$M_w \times 10^{-4}$	M_w/M_n	Method for determining M_w (or M_n)	ref.
PAN[h] (mm[i]=25%)	SPF	DMSO[j]	Toluene	35	1.0	18	5~72	—	[η]	80
PAN (mm=25%)	SPF	DMSO	Toluene	35	1.0	19	6~99	—	[η]	80
PAN (mm=25%)	SSF	DMSO	Toluene	45	1.0	18	2~52	1.2~1.4	MO, LS	114
PAN (mm=52%)	SSF	DMSO	Toluene	30	2.0	2~78		—	LS	115
PAN/MA[k] (92/8)	SSF	DMSO	Toluene	30	1.0	16	2~53	1.2~1.4	MO, LS	116
PAN/VDC[l] (58/42)	SSF	DMSO	Toluene	30	3.9	20	2~50	—	LS	117
Nylon 6	SPF	Phenol/Tetra-chloroethane (2/3, w/w)	n-Heptane	20	0.6	14	0.6~6.3	1.3~1.7	MO, EGA[m] GPC	118
PET[n]	SPF	Phenol/Tetra-chloroethane (2/3, w/w)	n-Heptane	20	0.3	30	1.9~6.2	—	MO	119
Segment-polyurethane	SPF	DMAc[o]	Ethylether /n-Heptane (1/1)	20	0.8	23	4~189	—	LS	120
	SSF	DMAc	Methanol	25	2.5	23	3~35	—	LS	121
Cellulose	SPF	Cuprammonium-Hydroxide	Propanol	—	—	4	0.51 ~6.81[p]	—	[η]	122

h) Polyacrylonitrile, i) Isotactic triad, j) Dimethyl surfoxide, k) Methylacrylate, l) Vinylidene chloride, m) End group analysis, n) Poly(ethylene telephthalate), o) Dimethylacetoamide, p) Value of [η].

TABLE 3.3 (continued-3)

Polymer	Method of fract.	Solvent	Nonsolvent	Temp. /℃	v_p^s ×10² /g/cm³	n_t	M_w×10⁻⁴	M_w/M_n	Method for determining M_w(or M_n)	ref.
Cellulose acetate										
DS[q]=0.49	SSF	Water	Methanol	25	1.0	15	1~15	1.28~1.31	MO, LS	84
DS=1.75	SSF	Acetone/water (7/3)	Water	—	—	10	2~13	1.28~1.39	MO, LS	85
DS=2.46	SSF	Acetone	Water	30	2.1	21	1~27	1.25~1.5	MO, LS	82
DS=2.46	SSF	Acetone	Ethanol	30	2.1	16	2~35	1.2~1.4	GPC	24
DS=2.46	SSF	Acetone	Ethanol	30	2.1	21	2~37	1.2~1.5	GPC	24
DS=2.92	SSF	1-chloro-2,3-epoxypropane	n-Hexane	35	1.0	13	2~69	1.4~1.5	MO, LS	83
Cellulose nitrate	SPF	Acetone/Water (91/9)	Water	25	0.5	12	~74	—	[η]	123
	SPF	Acetone/Water (98/2)	Water	25	0.125	14	~190	—	[η]	124
						18	~190	—	[η]	124
						19	~190	—	[η]	124
	SPF	Acetone	Water	21	0.20	12	1~41	—	[η]	125
	SPF	Acetone	Water	21	0.20	12	6~107	—	[η]	125
	SPF	Acetone	Water	30	1.0	19	~31	—	[η]	126
	SPF	Acetone	Hexane	25	0.8	5	~48	—	[η]	127
(2-Hydroxy propyl) cellulose	SPF	Acetone	Methanol /Water (85/15)	—	—	8	4~85	1.2~1.7	GPC	128
Amylose	SPF	DMSO	Ethanol	25	0.5	8	22~305	—	LS	129
	SPF	DMSO	Ethanol	4	0.5	10	16~229	—	LS	130

q) Degree of substitution.

TABLE 3.3 (continued-4)

Polymer	Method of fract.	Solvent	Nonsolvent	Temp. /℃	$v_p^s \times 10^2$ /g/cm^3	n_t	$M_w \times 10^{-4}$	M_w/M_n	Method for determining M_w(or M_n)	ref.
Amylose	SPF	Nitromethane	Methanol	25	0.35	13	12~480	—	LS	131
acetate	SPF	Nitromethane	Methanol	30	0.2	16	13~103	—	LS	132
	SPF	DMSO	Butanol	25	0.5	12	15~311	—	LS	133

REFERENCES

1 P.J. Flory, J. Chem. Phys., 12 (1944) 425.
2 R.L. Scott, J. Chem. Phys., 13 (1945) 178.
3 R.L. Scott, J. Chem. Phys., 17 (1949) 268.
4 M. Nakagaki and H. Sunada, Yakugaku Zasshi, 83 (1963) 1147.
5 H. Tompa, Trans. Faraday Soc., 45 (1949) 1142.
6 W.R. Krigbaum and D.K. Carpenter, J. Polym. Sci., 14 (1954) 241.
7 K.W. Suh and D.W. Liou, J. Polym. Sci., A-2, 6 (1968) 813.
8 A. Münster, J. Polym. Sci., 5 (1949) 333.
9 H. Okamoto, J. Polym. Sci., 33 (1958) 507.
10 R. Koningsveld, Discuss. Faraday Soc., 49 (1970) 144.
11 K. Kamide, S. Matsuda and Y. Miyazaki, Polym. J., 16 (1984) 479.
12 K. Kamide and S. Matsuda, Polym. J., 16 (1984) 515.
13 K. Kamide and S. Matsuda, Polym. J., 16 (1984) 591.
14 S. Matsuda, Polym. J., 18 (1986) 993.
15 T.M. Aminavhavi and P. Munk, Macromolecules, 12 (1979) 607.
16 F.W. Altena and C.A. Smolders, Macromolecules, 15 (1982) 1941.
17 P.J. Flory, Principle of Polymer Chemistry, Cornell Univ. Press, Ithaca, NY, 1953.
18 K. Kamide and S. Matsuda, Polym. J., 16 (1984) 825.
19 K. Kamide, S. Matsuda, T. Dobashi and M. Kaneko, Polym. J., 16 (1984) 839.
20 K. Kamide, S. Matsuda and M. Saito, Polym. J., 17 (1985) 1013.
21 S. Matsuda, Polym. J., 18 (1986) 981.
22 M. Kurata, Thermodynamics of Polymer Solutions, Harwood Academic Pubs. Chur, London, NY, 1982, Chapter 2.
23 K. Kamide and K. Sugamiya, Makromol. Chem., 156 (1972) 259.
24 K. Kamide, Y. Miyazaki and T. Abe, Brit. Polym. J., 13 (1981) 168.
25 K. Kamide, Y. Miyazaki and K. Sugamiya, Makromol. Chem., 173 (1973) 113.
26 K. Kamide and S. Matsuda, Polym. J., 18 (1986) 347.
27 K. Kamide, T. Ogawa, M. Sanada and M. Matumoto, Kobunshi Kagaku, 25 (1968) 440.
28 K. Kamide, T. Ogawa and M. Matsumoto, Kobunshi Kagaku, 25 (1968) 788.
29 K. Kamide, Y. Miyazaki and T. Abe, Polym. J., 9 (1977) 395.
30 I. Noda, H. Ishizawa, Y. Miyazaki and K. Kamide, Polym. J., 12 (1980) 87.
31 K. Kamide, K. Sugamiya, T. Kawai and Y. Miyazaki, Polym. J., 12 (1980) 67.
32 K. Kamide and Y. Miyazaki, Polym. J., 12 (1980) 205.
33 K. Kamide and Y. Miyazaki, Polym. J., 13 (1981) 325.
34 K. Kamide, T. Abe and Y. Miyazaki, Polym. J., 14 (1982) 355.
35 K. Kamide, Y. Miyazaki and T. Abe, Makromol. Chem., 177 (1976) 485.
36 K. Kamide, Molecular Weight Fractionation on the Basis of Solubility, IUPAC Macromolecular Chemistry-8, (1973) 147.

37 K. Kamide and Y. Miyazaki, Makromol. Chem., 176 (1975) 1447.
38 K. Kamide and Y. Miyazaki, Makromol. Chem., 176 (1975) 1051.
39 K. Kamide, K. Yamaguchi and Y. Miyazaki, Makromol. Chem., 173 (1973) 133.
40 K. Kamide and Y. Miyazaki, Makromol. Chem., 176 (1975) 1029.
41 K. Kamide and K. Sugamiya, Makromol. Chem., 139 (1970) 197.
42 C.H. Bamford and H. Tompa, Trans. Faraday Soc., 46 310(1950).
43 M. Nakagaki and H. Sunada, Yakugaku Zasshi, 86 (1966) 336.
44 S. Matsuda and K. Kamide, Polym. J., 19 (1987) 203.
45 S. Matsuda and K. Kamide, Polym. J., 19 (1987) 211.
46 F.D. Miles, Cellulose Nitrate, Imperial Chem. Industries Ltd., Oliver and Boyd, Edinburgh, London, 1955, Chapter V.
47 A. Dobry, J. Chim. Phys., 35 (1938) 387.
48 J.N. Bronsted and K. Volqvartz, Trans. Faraday Soc., 36 (1940) 619.
49 J.M.G. Cowie and J.T. McCrindle, Eur. Polym. J., 8 (1972) 1185.
50 J.M.G. Cowie and J.T. McCrindle, Eur. Polym. J., 8 (1972) 1325.
51 J.M.G. Cowie and I.J. McEven, Macromolecules, 3 (1974) 291.
52 J.M.G. Cowie and I.J. McEven, J. Chem. Soc., Faraday Trans. 1, 70 (1974) 171.
53 I. Katime, R. Valenciano and J.M. Teijon, Eur. Polym. J., 15 (1979) 261.
54 L. Gragallo, D. Radic and I.A. Katime, Eur. Polym. J., 17 (1981) 439.
55 I. Katime, J.R. Ochoa and L.C. Cesteros, Eur. Polym. J., 19 (1983) 1167.
56 I. Katime and J.R. Ochoa, Eur. Polym. J., 20 (1984) 99.
57 H. Tompa, Polymer Solutions, Butterworths Scientific Publications, London, 1956, Chapter 7.
58 A. Dondos and D. Patterson, J. Polym. Sci., A-2, 7 (1969) 209.
59 P. Munk, M.T. Abizaoude and M.E. Halbrook, J. Polym. Sci., Polym. Phys. Ed., 16 (1978) 105.
60 M.G. Prolongo, R.M. Masegosa, I. Hernandez-Feures and A. Horta, Macromolecules, 14 (1981) 1526.
61 A. Horta and I. Fernandez-Pierola, Macromolecules, 14 (1981) 1519.
62 R.M. Masegosa, M.G. Prolongo, I. Hernandez-Feures and A. Horta, Macromolecules, 17 (1984) 1181.
63 M.G. Prolongo, R.M. Masegosa, I. Hernandez-Feures and A. Horta, Polymer, 25 (1984) 1307.
64 B.E. Read, Trans. Faraday Soc., 56 (1960) 382.
65 W.H. Stockmayer and M. Fixman, J. Polym. Sci., C, 1 (1963) 137.
66 M. Nakata and N. Numasawa, Macromolecules, 18 (1985) 1736.
67 D.M. Koenhen and C.A. Smolders, J. Appl. Polym. Sci., 19 (1975) 1163.
68 K. Kamide, Fractionation of Synthetic Polymers, L. H. Tung ed., Marcel Dekker Inc., New York, 1977, Chapter 2.

69 K. Kamide and Y. Miyazaki, Polym. J., 12 (1980) 153.
70 Y. Miyazaki and K. Kamide, Polym. J., 9 (1977) 61.
71 R. Koningsveld and A.J. Staverman, Kolloid-Z. Z. Polym., 218 (1967) 114.
72 R. Koningsveld and A.J. Staverman, J. Polym. Sci., A-2, 6 (1968) 305.
73 R. Koningsveld and A.J. Staverman, J. Polym. Sci., A-2, 6 (1968) 367.
74 R. Koningsveld and A.J. Staverman, J. Polym. Sci., A-2, 6 (1968) 383.
75 R. Koningsveld, Adv. Polym. Sci., 7 (1970) 1.
76 M. Gordon, H.A.G. Chermin and R. Koningsveld, Macromolecules, 2 (1969) 107.
77 R. Koningsveld, W.H. Stockmayer and J.W. Kennedy, L.A. Kleintjens, Macromolecules, 7 (1974) 731.
78 Y. Fujisaki and H. Kobayashi, Kobunshi Kagaku, 18 (1961) 305, 312 and Kobunshi Kagaku, 19 (1962) 49, 69.
79 Y. Fujisaki and H. Kobayashi, Kobunshi Kagaku, 18 (1961) 312.
80 Y. Fujisaki and H. Kobayashi, Kobunshi Kagaku, 19 (1962) 49.
81 Y. Fujisaki and H. Kobayashi, Kobunshi Kagaku, 19 (1962) 64.
82 K. Kamide, T. Terakawa and Y. Miyazaki, Polym. J., 11 (1979) 285.
83 K. Kamide, Y. Miyazaki and T. Abe, Polym. J., 11 (1979) 523.
84 K. Kamide, M. Saito and T. Abe, Polym. J., 13 (1981) 421.
85 M. Saito, Polym. J., 15 (1983) 249.
86 J. Bandrup and E.H. Immergut, Ed., Polymer Handbook, 2nd Ed., John Wiley & Sons Inc., New York, 1974.
87 R. Koningsveld, Discuss. Faraday Soc., 49 (1970) 144.
88 K. Kamide, S. Matsuda, H. Shirataki, and Y. Miyazaki, Eur. Polym. J., 25 (1989) 1153.
89 G.V. Schulz and A. Dinglinger, Z. Phys. Chem., B43 (1939) 47.
90 K. Kamide and Y. Miyazaki, Makromol. Chem., 176 (1975) 3453.
91 F. Dunusso, G. Moraglio and G. Gianott, J. Polym. Sci., 51 (1974) 475.
92 M. Abe, Y. Murakami and H. Fujita, J. Appl. Polym. Sci., 9 (1965) 2549.
93 R.L. Cleland, J. Polym. Sci., 27 (1958) 349.
94 D.J. Harmon, J. Polym. Sci., C-8 (1965) 243.
95 G.M. Guzman, J. Polym. Sci., 19 (1956) 519.
96 J.P. Bianchi, F.P. Price and B.H. Zimm, J. Polym. Sci., 25 (1957) 27.
97 H.P. Frank and J.W. Breitenbach, J. Polym. Sci., 6 (1951) 609.
98 R.S. Aries and A.P. Sachs, J. Polym. Sci., 21 (1956) 551.
99 L.H. Tung, J. Polym. Sci., 24 (1957) 333.
100 H.J.L. Schunrmans, J. Polym. Sci., 57 (1962) 557.
101 H. Okamoto, J. Polym. Sci., A-2 (1964) 3451.
102 H. Wesslau, Makromol. Chem., 20 (1956) 111.
103 K. Ueberreiter, H.J. Onthmann and G. Sorge, Makromol. Chem., 8 (1952) 21.
104 Q.A. Trementozze, J. Polym. Sci., 23 (1957) 887.

105 C. Mussa, J. Polym. Sci., 28 (1958) 587.
106 T.E. Davis and R.L. Tobias, J. Polym. Sci., 50 (1961) 227.
107 O. Redlich, A.L. Jacobson and W.H. McFadden, J. Polym. Sci., A-1 (1963) 393.
108 S.N. Chimai, J.D. Mallack, A.L. Resnik and R.J. Samuels, J. Polym. Sci., 17 (1955) 391.
109 S. Krause and E. Cohen-Ginsberg, Polymer, 3 (1962) 565.
110 W.R. Moore and R.J. Fort, J. Polym. Sci., A-1 (1963) 929.
111 R.H. Wagner, J. Polym. Sci., 2 (1947) 21.
112 A. Nakajima and I. Sakurada, Kobunshi Kagaku, 11 (1954) 110.
113 A. Beresniewiez, J. Polym. Sci., 35 (1959) 321.
114 K. Kamide, Y. Miyazaki and H. Kobayashi, Polym. J., 17 (1985) 607.
115 K. Kamide, H. Yamazaki and Y. Miyazaki, Polym. J., 18 (1986) 819
116 K. Kamide, Y. Miyazaki and H. Kobayashi, Polym. J., 14 (1982) 591.
117 K. Kamide, Y. Miyazaki and H. Yamazaki, Polym. J., 18 (1986) 645.
118 K. Kamide and Y. Miyazaki, Kobunshi Ronbunshu, 35 (1978) 467.
119 K. Kamide, Y. Miyazaki and H. Kobayashi, Polym. J., 9 (1977) 317.
120 Y. Miyazaki and K. Kamide, Kobunshi Ronbunshu, 44 (1987) 1.
121 K. Kamide, A. Kiguchi and Y. Miyazaki, Polym. J., 18 (1986) 919.
122 H. Sihtola, E. Kaila and L. Laamanen, J. Polym. Sci., 23 (1957) 809.
123 R.L. Mitchell, Ind. Eng. Chem., 45 (1953) 2526.
124 T.E. Timell, Ind. Eng. Chem., 47 (1955) 2166.
125 M. Mark-Figini, Makromol. Chem., 32 (1959) 233.
126 K. Kamide, T. Shiomi, H. Ohkawa and K. Kaneko, Kobunshi Kagaku, 22 (1965) 785.
127 K. Kamide, T. Okada and K. Kaneko, Polym. J., 10 (1978) 547.
128 S.N. Bhadani, S.L. Tseng and D.G. Gray, Makromol. Chem., 184 (1983) 1727.
129 J.M.G. Cowie, Makromol. Chem., 42 (1961) 230.
130 W. Banks and C.T. Greenwood, Makromol. Chem., 67 (1963) 49.
131 J.M.G. Cowie, J. Polym. Sci., 49 (1961) 455.
132 R.S. Patel and R.D. Patel, Makromol. Chem., 90 (1966) 262.
133 W. Banks, C.T. Greenwood and D.J. Hourston, Trans. Faraday Soc., 64 (1968) 363.

Chapter 4

QUASI-BINARY SOLUTION CONSISTING OF MULTICOMPONENT POLYMER 1 AND MULTICOMPONENT POLYMER 2 SYSTEM

4.1 INTRODUCTION

Scott (ref. 1) is probably first who carried out theoretical study on phase equilibria of polymer solutions consisting of two kinds of polymer with different chemical compositions (polymer 1 and polymer 2 each) only (without solvent) (i.e., quasi-binary polymer mixture). He derived, based on Flory-Huggins solution theory, the relations giving the chemical potentials of monodisperse polymer 1 and of monodisperse polymer 2, $\Delta\mu_X$ and $\Delta\mu_Y$ (ref. 1):

$$\Delta\mu_X = \tilde{R}T[\ln v_1 + (1 - X/Y)v_2 + X\chi_{12}v_2^2], \tag{4.1}$$

$$\Delta\mu_Y = \tilde{R}T[\ln v_2 + (1 - Y/X)v_1 + Y\chi_{12}v_1^2], \tag{4.2}$$

where $\tilde{R}$ is the gas constant, T, Kelvin temperature, χ_{12}, the thermodynamic interaction parameter between polymers 1 and 2, X and Y, the DP (in the strict sense, the molecular volume ratio of the polymer and the lattice unit (= the polymer segment)), v_1 and v_2, the volume fraction of polymers 1 and 2 ($v_1 + v_2 = 1$). The spinodal curve (SC) and the neutral equilibrium condition are given as $\partial\Delta\mu_X/\partial v_1 = 0$ and $\partial^2\Delta\mu_X/\partial v_1^2 = 0$ (or $\partial\Delta\mu_Y/\partial v_2 = 0$ and $\partial^2\Delta\mu_Y/\partial v_2^2 = 0$) and finally we obtain

$$Xv_1 + Yv_2 - 2XY\chi_{12}v_1v_2 = 0 \tag{4.3}$$

for SC and

$$X - Y + 2XY\chi_{12}(v_1 - v_2) = 0 \tag{4.4}$$

for a neutral equilibrium condition. At critical solution point (CSP), both eqns (4.3) and (4.4) should be satisfied concurrently and χ_{12}, v_1 and v_2 at CSP (i.e., χ_{12}^c, v_1^c and v_2^c) are given by eqns (4.5)-(4.7).

$$\chi_{12}{}^{c} = (1/2)\,(X^{-1/2} + Y^{-1/2})^{2}, \tag{4.5}$$

$$v_1{}^{c} = Y^{1/2}/(X^{1/2} + Y^{1/2}) \tag{4.6}$$

$$v_2{}^{c} = X^{1/2}/(X^{1/2} + Y^{1/2}). \tag{4.7}$$

Scott (ref. 1) predicted that χ_{12} for these systems is smaller by several digits than those for a mixture of two low molecular weight liquids ($\chi_{12} \simeq 2.0$) and those for polymer / solvent system ($\chi_{12} \simeq 0.5$). An attempt to generalize CSP equations (eqns (4.5)-(4.7)) for the two monodisperse polymers mixture to the case of multicomponent polymer 1 / multicomponent polymer 2 system was made by Koningsveld, Chermin and Gordon (ref. 2). They showed the equations of spinodal and neutral equilibrium conditions (eqns 12 and 13 of ref. 2) for systems of multicomponent polymer 1 / multicomponent polymer 2 / single solvent. As recently Kamide et al. pointed out (ref. 3), they referred the equation of spinodal (eqn 12 of ref. 2) originally derived by Chermin (unpublished (up to now) work) and they did not show the detailed mathematical derivation of the equation of neutral the equilibrium condition (eqn 13 of ref. 2). They described that the spinodal condition for multicomponent polymer 1 / multicomponent polymer 2 system was derived as eqn 23 of ref. 2 after multiplying eqn 12 of ref. 2 by v_0 and reducing v_0 to zero, but the detailed derivation, including the treatment of the solvent-polymer 1 interaction parameter χ_{01} and the solvent-polymer 2 interaction parameter χ_{02} was not given. In addition, it is very difficult to consider that by application of the above-mentioned procedure to eqn 12 of ref. 2, the spinodal condition for multicomponent polymer 1 / multicomponent polymer 2 / single solvent system can be simply reduced to that for multicomponent polymer 1 / multicomponent polymer 2 system. We believe that for this purpose (i.e., for the derivation of a spinodal condition for a quasi-binary polymer mixture) we should define first Gibbs' free energy change of mixing, ΔG (see, eqn (4.11)) and derive the equation straightforwardly from determinant (eqn (4.16)) constructed using the ΔG. Koningsveld et al. described that the equation of the neutral equilibrium condition (eqn 24 of ref. 2) for two different multicomponent polymers was derived using an analogous method with a spinodal condition. The equation should be rigorously derived from the determinant (eqn (4.19)) of the neutral equilibrium condition. They demonstrated few and not systematic examples of calculation using their theoretical equations (eqns 23 and 24 of ref. 2).

On the other hand, Koningsveld et al. (refs 2,4) also demonstrated a few examples of calculated CPC for multicomponent polymer 1 / multicomponent polymer 2 systems (see, Fig. 2 of ref. 2 and Fig. 4 of ref. 4) with a consideration of the concentration dependence of χ_{12}. However, in their papers (refs 2,4) only the incomplete process of derivation of SC equation was described and in addition any derivation of CPC was not given and after the appearance of their papers no publication by them could be found concerning the theoretical derivations of SC and CPC for polymer blend.

On the basis of the Koningsveld et al.'s theoretical equations (eqns 23 and 24 of ref. 2), McMaster (ref. 5) calculated SC and CSP of the system of two chemically different single component (and multicomponent) polymers and also applying nonlinear optimization procedure (eqn 34 of ref. 5) to the equations of Gibbs' phase equilibria conditions (eqns 32 and 33 of ref. 5) of two different single component polymers.

McMaster (ref. 5), based on the Flory's state equation, derived $\Delta\mu_X$ and $\Delta\mu_Y$ for monodisperse polymer 1 / monodisperse polymer 2 system (eqns 22 and 23 of ref. 5) and calculated a binodal curve (BC) using the equations of $\Delta\mu_X$ and $\Delta\mu_Y$ derived thus. In this case, BC agrees absolutely with CPC because of the monodispersity of polymer 1 and polymer 2. His procedure of calculation of BC is simply as follows: Well-known Gibbs' two-phase equilibrium conditions under constant temperature and constant pressure, are given by the following simultaneous relations:

$$F_1 \equiv (\Delta\mu_{X(1)} - \Delta\mu_{X(2)})/k_BT = 0, \tag{4.8}$$

$$F_2 \equiv (\Delta\mu_{Y(1)} - \Delta\mu_{Y(2)})/k_BT = 0. \tag{4.9}$$

Here, the suffixes (1) and (2) attached to the chemical potentials denote coexisting two phases, respectively, k_B, Boltzmann constant, T, Kelvin temperature. Next, for a given initial polymer volume fraction of one phase (for example (1) phase) $v_{1(1)}$, calculate a function A, defined by

$$A \equiv \frac{F_1{}^2 + F_2{}^2}{v_{1(1)} - v_{1(2)}} \tag{4.10}$$

as a function of $v_{1(2)}$. The true value of $v_{1(2)}$ can be obtained at $A=0$ and thus, various reasonable combinations of $v_{1(1)}$ and $v_{1(2)}$ (i.e., BC) can be estimated (ref. 5). This method, as evident from eqn (4.10), avoids the trivial solution, $v_{1(1)}=v_{1(2)}$. Needless to mention, this method is applicable to estimate CPC only for a monodisperse polymer 1 / monodisperse polymer 2 system and not for the quasi-binary system consisting of at least one polydisperse polymer, such as multicomponent polymer 1 / multicomponent polymer 2 system due to incoincidence between CPC and BC. Accordingly, the calculations shown by McMaster are severely limited to a monodisperse polymer 1 / monodisperse polymer 2 system. Then, it is very clear from the above description that the discussions made hitherto on CPC for a polymer 1 / polymer 2 system are extremely insufficient and any valuable theoretical consideration can not be added to the actual system.

In this Chapter, considering the concentration (v_1 or v_2) dependence of χ_{12}, we intend to derive the equations of SP, CSP and CPC for multicomponent polymer 1 / multicomponent polymer 2 system in the manner that the thermodynamical requirements are strictly satisfied, and to examine the effect of the molecular weight distributions (MWD), the average DP of the two original polymers, and the concentration dependence of χ_{12} in the above quasi-binary polymer mixture on SP, CSP and CPC (refs 6,7).

4.2 THEORETICAL BACKGROUND OF CRITICAL SOLUTION POINT AND SPINODAL CURVE

According to the Flory-Huggins polymer solution theory, the change in mean molar Gibbs' free energy of mixing ΔG is given by

$$\Delta G = \tilde{R}TL\left[\sum_{i=1}^{m_1} \frac{v_{Xi}}{X_i} \ln v_{Xi} + \sum_{j=1}^{m_2} \frac{v_{Yj}}{Y_j} \ln v_{Yj} + \chi_{12} v_1 v_2\right] \tag{4.11}$$

where L is the total number of lattice ($\equiv \sum_i X_i N_{Xi} + \sum Y_j N_{Yj}$; N_{Xi} and N_{Yj} are the numbers of X_i-mer of polymer 1 and of Y_j-mer of polymer 2, respectively), m_1 and m_2 the total numbers of the components consisting of polymer 1 and polymer 2, v_{Xi}, the volume fraction of X_i-mer of multicomponent polymer 1, v_{Yj}, the volume fraction of Y_j-mer of multicomponent polymer 2, v_1 and v_2, the total volume fractions of polymer 1 and polymer 2 as defined by the relations;

$$v_1 = \sum_{i=1}^{m_1} v_{Xi}, \tag{4.12}$$

$$v_2 = \sum_{j=1}^{m_2} v_{Yj}. \tag{4.13}$$

The first and second terms in the right-hand side of eqn (4.11) are the combinatory terms and the third term in the same side is the term relating to the mutual thermodynamic interaction.

Differentiations of eqn (4.11) with respect to N_{Xi} and N_{YJ} give the chemical potentials of X_i-mer and Y_J-mer, $\Delta\mu_{Xi}$ and $\Delta\mu_{YJ}$ in the forms;

$$\Delta\mu_{Xi} = \tilde{R}T\,[\ln v_{Xi} - (X_i - 1) + X_i\,(1 - \frac{1}{X_n})\,v_1 + X_i\,(1 - \frac{1}{Y_n})\,v_2 + X_i\,\chi_{12}v_2{}^2], \tag{4.14}$$

$$\Delta\mu_{Yi} = \tilde{R}T\,[\ln v_{YJ} - (Y_J - 1) + Y_J\,(1 - \frac{1}{X_n})\,v_1 + Y_J\,(1 - \frac{1}{Y_n})\,v_2 + Y_J\,\chi_{12}v_1{}^2], \tag{4.15}$$

where X_n and Y_n are the number-average X_i and Y_J (i.e., the number-average DP of the original polymers). In deriving eqns (4.14) and (4.15), we assumed that (1) the molar volume of the segment of polymer 1 is the same as that of polymer 2, (2) polymer 1 and polymer 2 are volumetrically additive, and (3) the densities of polymer 1 and polymer 2 are the same (= unity). Note that eqns (4.14) and (4.15) are symmetric with respect to exchange of polymer 1 and polymer 2. When both polymer 1 and polymer 2 are monodisperse (i.e., single component each), eqns (4.14) and (4.15) straightforwardly reduce to eqns (4.1) and (4.2), respectively.

The thermodynamical requirement of SC is that the second variation of Gibbs' free energy change of mixing is always zero and this requirement can be described for multicomponent polymer 1 / multicomponent polymer 2 system as a spinodal condition of $(m_1 - 1 + m_2) \times (m_1 - 1 + m_2)$ determinant (eqn (4.16)) (ref. 6).

$$
|\Delta G'| = \begin{vmatrix}
\bar{\Delta G'}_{X_2X_2} & \bar{\Delta G'}_{X_2X_3} & \cdots & \bar{\Delta G'}_{X_2X_{m1}} & \bar{\Delta G'}_{X_2Y_1} & \bar{\Delta G'}_{X_2Y_2} & \cdots & \bar{\Delta G'}_{X_2Y_{m2}} \\
\bar{\Delta G'}_{X_3X_2} & \bar{\Delta G'}_{X_3X_3} & \cdots & \bar{\Delta G'}_{X_3X_{m1}} & \bar{\Delta G'}_{X_3Y_1} & \bar{\Delta G'}_{X_3Y_2} & \cdots & \bar{\Delta G'}_{X_3Y_{m2}} \\
\vdots & \vdots & & \vdots & \vdots & \vdots & & \vdots \\
\bar{\Delta G'}_{X_{m1}X_2} & \bar{\Delta G'}_{X_{m1}X_3} & \cdots & \bar{\Delta G'}_{X_{m1}X_{m1}} & \bar{\Delta G'}_{X_{m1}Y_1} & \bar{\Delta G'}_{X_{m1}Y_2} & \cdots & \bar{\Delta G'}_{X_{m1}Y_m} \\
\bar{\Delta G'}_{Y_1X_2} & \bar{\Delta G'}_{Y_1X_3} & \cdots & \bar{\Delta G'}_{Y_1X_{m1}} & \bar{\Delta G'}_{Y_1Y_1} & \bar{\Delta G'}_{Y_1Y_2} & \cdots & \bar{\Delta G'}_{Y_1Y_{m2}} \\
\bar{\Delta G'}_{Y_2X_2} & \bar{\Delta G'}_{Y_2X_3} & \cdots & \bar{\Delta G'}_{Y_2X_{m1}} & \bar{\Delta G'}_{Y_2Y_1} & \bar{\Delta G'}_{Y_2Y_2} & \cdots & \bar{\Delta G'}_{Y_2Y_{m2}} \\
\vdots & \vdots & & \vdots & \vdots & \vdots & & \vdots \\
\bar{\Delta G'}_{Y_{m2}X_2} & \bar{\Delta G'}_{Y_{m2}X_3} & \cdots & \bar{\Delta G'}_{Y_{m2}X_{m1}} & \bar{\Delta G'}_{Y_{m2}Y_1} & \bar{\Delta G'}_{Y_{m2}Y_2} & \cdots & \bar{\Delta G'}_{Y_{m2}Y_{m2}}
\end{vmatrix}
$$

$$= 0. \qquad (4.16)$$

Here, we employed the Gibbs' free energy change of mixing per unit volume of the solution $\Delta G'$ defined by

$$\Delta G' = \sum_{i=1}^{m_1} v_{X_i} \left(\frac{\Delta \mu_{X_i}}{X_i V_0}\right) + \sum_{j=1}^{m_2} v_{Y_j} \left(\frac{\Delta \mu_{Y_j}}{Y_j V_0}\right). \qquad (4.17)$$

V_0 is the molar volume of the polymer segment, $\bar{\Delta G'}_{X_iY_j}$ is the second order partial differential of $\Delta G'$ with respect to volume fraction;

$$\bar{\Delta G'}_{X_iY_j} = \left(\frac{\partial^2 \Delta G'}{\partial v_{X_i} \partial v_{Y_j}}\right)_{T,P,v_k}$$
$$(i = 2,3,\cdots,m_1,\ \ j = 1,2,\cdots,m_2;\ \ k \neq X_i, Y_j). \qquad (4.18)$$

As compared with the determinant giving the spinodal condition for multicomponent polymer 1 / multicomponent polymer 2 / single solvent system (eqn 10 of ref. 3), the term $\bar{\Delta G'}_{X_1X_1}$ is dropped out and the matrix starts with X_2 in eqn (4.16), making the calculation somewhat complicated.

At CSP, in addition to eqn (4.16), it is simultaneously necessary that the third variation of Gibbs' free energy change be also zero, in other words, the neutral equilibrium condition (eqn (4.19)) should be satisfied.

$$|\Delta\bar{G}''| = \begin{vmatrix} \frac{\partial|\Delta G'|}{\partial v_{X2}} & \frac{\partial|\Delta G'|}{\partial v_{X3}} & \cdots & \frac{\partial|\Delta G'|}{\partial v_{Xm1}} & \frac{\partial|\Delta G'|}{\partial v_{Y1}} & \frac{\partial|\Delta G'|}{\partial v_{Y2}} & \cdots & \frac{\partial|\Delta G'|}{\partial v_{Ym2}} \\ \Delta\bar{G}'_{X3X2} & \Delta\bar{G}'_{X3X3} & \cdots & \Delta\bar{G}'_{X3Xm1} & \Delta\bar{G}'_{X3Y1} & \Delta\bar{G}'_{X3Y2} & \cdots & \Delta\bar{G}'_{X3Ym2} \\ \vdots & \vdots & & \vdots & \vdots & \vdots & & \vdots \\ \Delta\bar{G}'_{Xm1X2} & \Delta\bar{G}'_{Xm1X3} & \cdots & \Delta\bar{G}'_{Xm1Xm1} & \Delta\bar{G}'_{Xm1Y1} & \Delta\bar{G}'_{Xm1Y2} & \cdots & \Delta\bar{G}'_{Xm1Ym2} \\ \Delta\bar{G}'_{Y1X2} & \Delta\bar{G}'_{Y1X3} & \cdots & \Delta\bar{G}'_{Y1Xm1} & \Delta\bar{G}'_{Y1Y1} & \Delta\bar{G}'_{Y1Y2} & \cdots & \Delta\bar{G}'_{Y1Ym2} \\ \Delta\bar{G}'_{Y2X2} & \Delta\bar{G}'_{Y2X3} & \cdots & \Delta\bar{G}'_{Y2Xm1} & \Delta\bar{G}'_{Y2Y1} & \Delta\bar{G}'_{Y2Y2} & \cdots & \Delta\bar{G}'_{Y2Ym2} \\ \vdots & \vdots & & \vdots & \vdots & \vdots & & \vdots \\ \Delta\bar{G}'_{Ym2X2} & \Delta\bar{G}'_{Ym2X3} & \cdots & \Delta\bar{G}'_{Ym2Xm1} & \Delta\bar{G}'_{Ym2Y1} & \Delta\bar{G}'_{Ym2Y2} & \cdots & \Delta\bar{G}'_{Ym2Ym2} \end{vmatrix}$$

$$= 0. \qquad (4.19)$$

In eqn (4.19), the term of partial differential of $\Delta\bar{G}'$ with respect to v_{X1} is dropped out and this makes the calculation tedious.

Substitution of eqns (4.14) and (4.15) into eqn (4.17) leads to

$$\Delta G' = \left(\frac{\tilde{R}T}{V_0}\right)\left[\sum_{i=1}^{m_1}\frac{v_{Xi}\ln v_{Xi}}{X_i} + \sum_{j=1}^{m_2}\frac{v_{Yj}\ln v_{Yj}}{Y_j} + \chi_{12}v_1v_2\right]. \qquad (4.20)$$

From eqns (4.18) and (4.20), we obtain (ref. 6)

$$\left(\frac{V_0}{\tilde{R}T}\right)\Delta\bar{G}'_{kl} = \frac{1}{X_1 v_{X1}} \equiv M \quad (\text{for } k \neq l \ (k,l) = (X_i, X_j)), \qquad (4.21a)$$

$$\left(\frac{V_0}{\tilde{R}T}\right)\Delta\bar{G}'_{kl} = \frac{1}{X_1 v_{X1}} - 2\chi_{12} \equiv N \quad (\text{for } k \neq l \ (k,l) = (Y_i, Y_j)), \qquad (4.21b)$$

$$\left(\frac{V_0}{\tilde{R}T}\right)\Delta\bar{G}'_{kl} = \frac{1}{X_1 v_{X1}} \equiv M \quad (\text{for } k \neq l \ (k,l) = (X_i, Y_j) \text{ or } (Y_i, X_j)), \qquad (4.21c)$$

$$\left(\frac{V_0}{\tilde{R}T}\right)\Delta\bar{G}'_{kl} = \frac{1}{X_1 v_{X1}} + \frac{1}{X_i v_{Xi}} \equiv M + M_i \quad (\text{for } k = l = X_i), \qquad (4.21d)$$

$$\left(\frac{V_0}{\tilde{R}T}\right)\Delta\bar{G}'_{kl} = \frac{1}{X_1 v_{X1}} - 2\chi_{12} + \frac{1}{Y_j v_{Yj}} \equiv N + N_j \quad (\text{for } k = l = Y_j). \qquad (4.21e)$$

Combining eqns (4.21a)-(4.21e) with eqn (4.16), we obtain eqn (4.22);

$$
|\Delta G'| = \left(\frac{\tilde{RT}}{V_0}\right)^{m1+m2-1}
\left|\begin{array}{cccc|cccc}
M+M_2 & M & \cdots & M & M & M & \cdots & M \\
M & M+M_3 & \cdots & M & M & M & \cdots & M \\
\vdots & \vdots & & \vdots & \vdots & \vdots & & \vdots \\
M & M & \cdots & M+M_{m1} & M & M & \cdots & M \\
\hline
M & M & \cdots & M & N+N_1 & N & \cdots & N \\
M & M & \cdots & M & N & N+N_2 & \cdots & N \\
\vdots & \vdots & & \vdots & \vdots & \vdots & & \vdots \\
M & M & \cdots & M & N & N & \cdots & N+N_{m2}
\end{array}\right| .
\tag{4.22}
$$

$|\Delta G'|$ given by eqn (4.22) can be expanded, according to addition rule,

$$
\left(\frac{V_0}{\tilde{RT}}\right)^{m1+m2-1}|\Delta G'| =
\left|\begin{array}{cccc|cccc}
M_2 & 0 & \cdots & 0 & M & M & \cdots & M \\
0 & M_3 & & 0 & M & M & \cdots & M \\
\vdots & & \ddots & \vdots & \vdots & \vdots & & \vdots \\
0 & & \cdots & M_{m1} & M & M & \cdots & M \\
\hline
0 & 0 & \cdots & 0 & N+N_1 & N & & N \\
0 & 0 & \cdots & 0 & N & N+N_2 & \cdots & N \\
\vdots & \vdots & & \vdots & \vdots & \vdots & \ddots & \vdots \\
0 & 0 & \cdots & 0 & N & N & \cdots & N+N_{m2}
\end{array}\right|
$$

$$
+\sum_{i=2}^{m_1}
\left|\begin{array}{cccc|c|ccc|ccc}
M+M_2 & M & \cdots & M & \boxed{M} & M & \cdots & M & M & \cdots & M \\
M & M+M_3 & \cdots & M & M & M & \cdots & M & M & \cdots & M \\
\vdots & \vdots & \ddots & \vdots & \vdots & \vdots & & \vdots & \vdots & & \vdots \\
M & M & \cdots & M+M_{i-1} & M & M & \cdots & M & M & \cdots & M \\
M & M & \cdots & M & M & M & \cdots & M & M & \cdots & M \\
\vdots & \vdots & & \vdots & \vdots & M+M_{i+1} & & \vdots & \vdots & & \vdots \\
\vdots & \vdots & & \vdots & \vdots & \vdots & \ddots & \vdots & \vdots & \ddots & \vdots \\
M & M & \cdots & M & M & M & \cdots & M+M_{m1} & M & \cdots & M \\
\hline
M & M & \cdots & M & M & M & \cdots & M & N+N_1 & \cdots & M \\
\vdots & \vdots & & \vdots & \vdots & \vdots & \ddots & \vdots & \vdots & \ddots & \vdots \\
M & M & \cdots & M & M & M & \cdots & M & N & \cdots & N+N_{m2}
\end{array}\right| .
\tag{4.23}
$$

Note that all elements at the i-th column in the (i+1)-th term determinant of the right-hand side of eqn (4.23) are M (i = 1, ···,

m_1). After transferring M at (i,i) element to the position of (1,1) element, the (i+1)-th term can be rewritten as

$$
(i+1)\text{-th term} = (-1)^{2i-4}\left|\begin{array}{cccc|ccc}
M & M & \cdots & M & M & \cdots & M \\
M & M+M_2 & & \vdots & \vdots & & \vdots \\
\vdots & \vdots & \ddots & & & & \\
M & M & [i] & & & & \\
\vdots & \vdots & \ddots & \vdots & \vdots & & \vdots \\
M & M & \cdots & M+M_{m1} & M & \cdots & M \\
\hline
M & M & \cdots & M & N+N_1 & \cdots & N \\
\vdots & \vdots & & \vdots & \vdots & \ddots & \vdots \\
M & M & \cdots & M & N & \cdots & N+N_{m2}
\end{array}\right|
$$

$$
= (-1)^{2i-4}\left|\begin{array}{cccc|ccc}
M & 0 & \cdots & 0 & 0 & \cdots & 0 \\
M & M_2 & & \vdots & \vdots & & \vdots \\
\vdots & \vdots & \ddots & & & & \\
M & 0 & [i] & & & & \\
\vdots & \vdots & \ddots & \vdots & \vdots & & \vdots \\
M & 0 & \cdots & M_{m1} & 0 & \cdots & 0 \\
\hline
M & 0 & \cdots & 0 & N-M+N_1 & \cdots & N \\
\vdots & \vdots & & \vdots & \vdots & \ddots & \vdots \\
M & 0 & \cdots & 0 & N-M & \cdots & N+N_{m2}
\end{array}\right| . \tag{4.24}
$$

Here [i] denotes that (i,i) element is transferred to (1,1) component.

By expanding the first and (i+1)-th terms by Laplace expansion, eqn (4.23) can be rewritten in the form,

$$
\left(\frac{V_0}{RT}\right)^{m1+m2-1}|\Delta G'| = \left|\begin{array}{cccc}
M_2 & 0 & \cdots & 0 \\
0 & M_3 & & \vdots \\
\vdots & & \ddots & \vdots \\
0 & & \cdots & M_{m1}
\end{array}\right| \cdot \left|\begin{array}{cccc}
N+N_1 & N & \cdots & N \\
N & N+N_2 & & \vdots \\
 & & \ddots & \\
N & & \cdots & N+N_{m2}
\end{array}\right|
$$

$$
+\sum_{i=2}^{m_1}(-1)^{2i-4}\left|\begin{array}{cccc}
M & 0 & \cdots & 0 \\
M & M_2 & \cdots & \vdots \\
\vdots & \vdots & \ddots & \\
 & & [i] & \\
\vdots & \vdots & \ddots & \vdots \\
M & 0 & \cdots & M_{m1}
\end{array}\right| \cdot \left|\begin{array}{ccc}
N-M+N_1 & \cdots & N-M \\
\vdots & \ddots & \vdots \\
N-M & \cdots & N-M+N_{m2}
\end{array}\right|
$$

$$=\frac{1}{M}\left(\prod_{i=1}^{m_1} M_i\right)\left(\prod_{j=1}^{m_2} N_j\right)\left[\left(1+N\sum_{j=1}^{m_2}\frac{1}{N_j}\right)\right.$$

$$\left.+\sum_{i=2}^{m_1}\frac{M}{MM_i}\left(\prod_{i=1}^{m_1} M_i\right)\left(\prod_{j=1}^{m_2} N_j\right)\left\{1+(N-M)\sum_{j=1}^{m_2}\frac{1}{N_j}\right\}\right]$$

$$=\frac{1}{M}\left(\prod_{i=1}^{m_1} M_i\right)\left(\prod_{j=1}^{m_2} N_j\right)\left[1+N\sum_{j=1}^{m_2}\frac{1}{N_j}\right.$$

$$\left.+\sum_{i=2}^{m_1}\frac{M}{M_i}\left\{1+(N-M)\sum_{j=1}^{m_2}\frac{1}{N_j}\right\}\right]$$

$$=\left(\prod_{i=1}^{m_1} M_i\right)\left(\prod_{j=1}^{m_2} N_j\right)\left[\sum_{i=1}^{m_1}\frac{1}{M_i}+\sum_{j=1}^{m_2}\frac{1}{N_j}+(N-M)\left(\sum_{i=1}^{m_1}\frac{1}{M_i}\right)\left(\sum_{j=1}^{m_2}\frac{1}{N_j}\right)\right]=0. \qquad (4.25)$$

Eqn (4.25) coincides with eqn (4.22')

$$|\Delta G'|=\left(\frac{\tilde{R}T}{V_0}\right)^{m1+m2-1}\left(\prod_{i=1}^{m_1} M_i\right)\left(\prod_{j=1}^{m_2} N_j\right)$$

$$\times\left[\sum_{i=1}^{m_1}\frac{1}{M_i}+\sum_{j=1}^{m_2}\frac{1}{N_j}+(N-M)\left(\sum_{i=1}^{m_1}\frac{1}{M_i}\right)\left(\sum_{j=1}^{m_2}\frac{1}{N_j}\right)\right]=0. \qquad (4.22')$$

A similar combination of eqns (4.21a)-(4.21e) with eqn (4.19) yields eqn (4.26), which can also be calculated as eqn (4.26')

$$|\Delta G''|=\left(\frac{\tilde{R}T}{V_0}\right)^{m1+m2-1}\begin{vmatrix} W_{X2} & W_{X3} & \cdots & W_{Xm1} & W_{Y1} & W_{Y2} & \cdots & W_{Ym2} \\ M & M+M_3 & \cdots & M & M & M & \cdots & M \\ \vdots & \vdots & & \vdots & \vdots & \vdots & & \vdots \\ M & M & \cdots & M+M_{m1} & M & M & \cdots & M \\ M & M & \cdots & M & N+N_1 & N & \cdots & N \\ M & M & \cdots & M & N & N+N_2 & \cdots & N \\ \vdots & \vdots & & \vdots & \vdots & \vdots & & \vdots \\ M & M & \cdots & M & N & N & \cdots & N+N_{m2} \end{vmatrix}. \qquad (4.26)$$

Transfer of component after cofactor decomposition of eqn (4.26) leads to

$$
\left(\frac{V_0}{\underset{\sim}{RT}}\right)^{m1+m2-2}|\Delta G''| = (-1)^2 W_{X2}
\left|\begin{array}{cccc:cccc}
M+M_3 & \cdots & & M & M & \cdots & & M \\
\vdots & & \ddots & \vdots & \vdots & & & \vdots \\
M & \cdots & & M+M_{m1} & M & & \cdots & M \\
\hdashline
M & \cdots & & M & N+N_1 & & \cdots & N \\
\vdots & & & \vdots & \vdots & \ddots & & \vdots \\
M & \cdots & & M & N & & \cdots & N+N_{m2}
\end{array}\right|
$$

$$
+\sum_{i=3}^{m_1} W_{Xi}(-1)^{i}(-1)^{i-3}
\left|\begin{array}{cccccc:ccc}
M & M & \cdots & & & M & M & \cdots & M \\
M & M+M_3 & & & & \vdots & \vdots & & \vdots \\
\vdots & & \ddots & & & & & & \\
 & & & [i] & & & & & \\
 & & & & \ddots & & & & \\
M & M & \cdots & & & M+M_{m1} & M & \cdots & M \\
\hdashline
M & M & \cdots & & & M & N+N_1 & \cdots & N \\
\vdots & & & & & \vdots & \vdots & \ddots & \vdots \\
M & M & \cdots & & & M & N & \cdots & N+N_{m2}
\end{array}\right|
$$

$$
+\sum_{j=1}^{m_2} W_{Yj}(-1)^{m1+j}(-1)^{m1+j-3}
\left|\begin{array}{cccc:ccccc}
M & M & \cdots & M & M & \cdots & & & M \\
M & M+M_3 & & \vdots & \vdots & & & & \vdots \\
\vdots & & \ddots & & & & & & \\
M & \cdots & & M+M_{m1} & M & \cdots & & & M \\
\hdashline
M & \cdots & & M & N+N_1 & \cdots & & & N \\
\vdots & & & \vdots & \vdots & \ddots & & & \vdots \\
 & & & & & & [j] & & \\
 & & & & & & & \ddots & \\
M & \cdots & & M & N & \cdots & & & N+N_{m2}
\end{array}\right| .
\tag{4.27}
$$

After applying addition rule and Laplace expansion, eqn (4.27) is rewritten as

$$
\left(\frac{V_0}{\underset{\sim}{RT}}\right)^{m1+m2-2}|\Delta G''| = \frac{1}{M_2}\left(\prod_{i=1}^{m_1} M_i\right)\left(\prod_{j=1}^{m_2} N_j\right)\left[-\frac{W_{X2}}{M_2}-\frac{W_{X2}(N-M)}{M_2}\left(\sum_{j=1}^{m_2}\frac{1}{N_j}\right)\right.
$$

$$
\left.-\sum_{i=3}^{m_1}\frac{W_{Xi}}{M_i}\left\{1+(N-M)\sum_{j=1}^{m_2}\frac{1}{N_j}\right\}-\sum_{j=1}^{m_2}\frac{W_{Yj}}{N_j}\right] . \tag{4.28}
$$

Eqn (4.22') (spinodal condition) can be transformed into

$$(N-M)\sum_{j=1}^{m_2}\frac{1}{N_j}=-\left[\left(\sum_{i=1}^{m_1}\frac{1}{M_i}\right)+\left(\sum_{j=1}^{m_2}\frac{1}{N_j}\right)\right]\left(\sum_{i=1}^{m_1}\frac{1}{M_i}\right)^{-1}. \tag{4.29}$$

Substitution of eqn (4.29) into eqn (4.28) gives

$$\left(\frac{V_0}{\tilde{RT}}\right)^{m1+m2-2}|\Delta G''|=\frac{1}{M_2}\left(\prod_{i=1}^{m_1}M_i\right)\left(\prod_{j=1}^{m_2}N_j\right)\left(\sum_{j=1}^{m_2}\frac{1}{N_j}\right)\left[\left(\sum_{i=1}^{m_1}\frac{W_{Xi}}{M_i}\right)\left(\sum_{i=1}^{m_1}\frac{1}{M_i}\right)^{-1}\right.$$

$$\left.-\left(\sum_{j=1}^{m_2}\frac{W_{Yj}}{N_j}\right)\left(\sum_{j=1}^{m_2}\frac{1}{N_j}\right)^{-1}\right]. \tag{4.30}$$

Then, it is proved that the condition of $|\Delta G''|=0$ is represented by eqn (4.26').

$$|\Delta G''|=\left(\frac{\tilde{RT}}{V_0}\right)^{m1+m2-2}\frac{1}{M_2}\left(\prod_{i=1}^{m_1}M_i\right)\left(\prod_{j=1}^{m_2}N_j\right)\left(\sum_{j=1}^{m_2}\frac{1}{N_j}\right)$$

$$\times\left[\left(\sum_{j=1}^{m_2}\frac{W_{Xi}}{N_j}\right)\left(\sum_{j=1}^{m_2}\frac{1}{N_j}\right)^{-1}-\left(\sum_{i=1}^{m_1}\frac{W_{Yj}}{M_i}\right)\left(\sum_{i=1}^{m_1}\frac{1}{M_i}\right)^{-1}\right]=0. \tag{4.26'}$$

Here, W_{Xi} and W_{Yj} are given by

$$W_{Xi}\equiv\left(\frac{\partial|\Delta G'|}{\partial v_{Xi}}\right)_{T,P,v_k}\quad (k\neq X_i,\ k=X_1,\cdots,X_{m1},Y_1,\cdots,Y_{m2}) \tag{4.31a}$$

and

$$W_{Yj}\equiv\left(\frac{\partial|\Delta G'|}{\partial v_{Yj}}\right)_{T,P,v_k}\quad (k\neq Y_j,\ k=X_1,\cdots,X_{m1},Y_1,\cdots,Y_{m2}). \tag{4.31b}$$

From eqns (4.22') and (4.26'), the equations of spinodal and neutral equilibrium conditions are derived in the form of eqns (4.32) and (4.33), respectively (ref. 6).

$$\sum_{i=1}^{m_1}\frac{1}{M_i}+\sum_{j=1}^{m_2}\frac{1}{N_j}+(N-M)\left(\sum_{i=1}^{m_1}\frac{1}{M_i}\right)\left(\sum_{j=1}^{m_2}\frac{1}{N_j}\right)=0, \tag{4.32}$$

$$\left(\sum_{i=1}^{m_1}\frac{W_{Xi}}{M_i}\right)\left(\sum_{i=1}^{m_1}\frac{1}{M_i}\right)^{-1}-\left(\sum_{j=1}^{m_2}\frac{W_{Yj}}{N_j}\right)\left(\sum_{j=1}^{m_2}\frac{1}{N_j}\right)^{-1}=0. \tag{4.33}$$

All parameters in eqns (4.32) and (4.33) can be expressed, by using eqns (4.21a)-(4.21e), (4.31a) and (4.31b), in terms of the experimentally determinable physical quantities such as the volume fraction, the average DP and χ_{12}.

From eqns (4.21a)-(4.21e), we obtain

$$N-M=-2\chi_{12}, \tag{4.34}$$

$$\sum_{i=1}^{m_1}\frac{1}{M_i}=\sum_{i=1}^{m_1}X_iV_{Xi}=v_1X_w{}^0, \tag{4.35}$$

$$\sum_{j=1}^{m_2}\frac{1}{N_j}=\sum_{j=1}^{m_2}Y_jV_{Yj}=v_2Y_w{}^0. \tag{4.36}$$

Substitution of eqns (4.34)-(4.36) into eqn (4.32) yields directly eqn (4.37).

$$\frac{1}{v_1X_w{}^0}+\frac{1}{v_2Y_w{}^0}-2\chi_{12}=0. \tag{4.37}$$

On the other hand, substituting eqn (4.22') into eqn (4.31a) and by using eqn (4.32), we obtain

$$W_{Xi}=\left(\frac{\tilde{R}T}{V_0}\right)^{m1+m2-1}\frac{\partial}{\partial v_{Xi}}\left[\left(\prod_{i=1}^{m_1}M_i\right)\left(\prod_{j=1}^{m_2}N_j\right)\right.$$

$$\left.\times\left\{\sum_{i=1}^{m_1}\frac{1}{M_i}+\sum_{j=1}^{m_2}\frac{1}{N_j}+(N-M)\left(\sum_{i=1}^{m_1}\frac{1}{M_i}\right)\left(\sum_{j=1}^{m_2}\frac{1}{N_j}\right)\right\}\right]$$

$$=\left(\frac{\tilde{R}T}{V_0}\right)^{m1+m2-1}\left(\prod_{i=1}^{m_1}M_i\right)\left(\prod_{j=1}^{m_2}N_j\right)$$

$$\times\frac{\partial}{\partial v_{Xi}}\left[\sum_{i=1}^{m_1}\frac{1}{M_i}+\sum_{j=1}^{m_2}\frac{1}{N_j}+(N-M)\left(\sum_{i=1}^{m_1}\frac{1}{M_i}\right)\left(\sum_{j=1}^{m_2}\frac{1}{N_j}\right)\right]. \tag{4.38}$$

And from eqns (4.21d) and (4.21e), we obtain

$$\frac{\partial}{\partial v_{Xi}}\left(\sum_{i=1}^{m_1}\frac{1}{M_i}\right)=\frac{\partial}{\partial v_{Xi}}\left(\sum_{i=1}^{m_1}X_iv_{Xi}\right)=-X_1+X_i, \tag{4.39}$$

$$\frac{\partial}{\partial v_{Xi}}\left(\sum_{j=1}^{m_2}\frac{1}{N_j}\right)=\frac{\partial}{\partial v_{Xi}}\left(\sum_{i=1}^{m_1}Y_j v_{Yj}\right)=0. \tag{4.40}$$

After the combination of eqns (4.39) and (4.40) with eqn (4.38), we can derive eqn (4.41)

$$W_{Xi}=\left(\frac{\tilde{R}T}{V_0}\right)^{m1+m2-1}\left(\prod_{i=1}^{m_1}M_i\right)\left(\prod_{j=1}^{m_2}N_j\right)(-X_1+X_i)\left[1+(N-M)\left(\sum_{j=1}^{m_2}\frac{1}{N_j}\right)\right]. \tag{4.41}$$

In a similar manner, W_{Yj} is given by

$$W_{Yj}=\left(\frac{\tilde{R}T}{V_0}\right)^{m1+m2-1}\left(\prod_{i=1}^{m_1}M_i\right)\left(\prod_{j=1}^{m_2}N_j\right)$$

$$\times\left[-X_1\left\{1+(N-M)\left(\sum_{j=1}^{m_2}\frac{1}{N_j}\right)\right\}+Y_j\left\{1+(N-M)\left(\sum_{i=1}^{m_1}\frac{1}{M_i}\right)\right\}\right]. \tag{4.42}$$

From eqn (4.41), we obtain

$$\sum_{i=1}^{m_1}\frac{W_{Xi}}{M_i}=\left(\frac{\tilde{R}T}{V_0}\right)^{m1+m2-1}\left(\prod_{i=1}^{m_1}M_i\right)\left(\prod_{j=1}^{m_2}N_j\right)$$

$$\times\left(-X_1\sum_{i=1}^{m_1}\frac{1}{M_i}+\sum_{i=1}^{m_1}\frac{X_i}{M_i}\right)\left[1+(N-M)\left(\sum_{j=1}^{m_2}\frac{1}{N_j}\right)\right]. \tag{4.43}$$

From eqn (4.42), we obtain

$$\sum_{j=1}^{m_2}\frac{W_{Yj}}{N_j}=\left(\frac{\tilde{R}T}{V_0}\right)^{m1+m2-1}\left(\prod_{i=1}^{m_1}M_i\right)\left(\prod_{j=1}^{m_2}N_j\right)$$

$$\times\left[-X_1\left\{1+(N-M)\left(\sum_{j=1}^{m_2}\frac{1}{N_j}\right)\right\}\left(\sum_{j=1}^{m_2}\frac{1}{N_j}\right)\right.$$

$$\left.+\left\{1+(N-M)\left(\sum_{i=1}^{m_1}\frac{1}{M_i}\right)\right\}\left(\sum_{j=1}^{m_2}\frac{Y_j}{N_j}\right)\right]. \tag{4.44}$$

By combining eqns (4.43) and (4.44) with eqn (4.33), we can derive eqn (4.45).

$$\frac{X_z{}^0}{(v_1 X_w{}^0)^2} - \frac{Y_z{}^0}{(v_2 Y_w{}^0)^2} = 0. \tag{4.45}$$

In eqns (4.37) and (4.45), $X_w{}^0$ and $Y_w{}^0$ are the weight-average DP of the two original polymers and $X_z{}^0$ and $Y_z{}^0$ are the corresponding z-average quantities.

By solving simultaneous equations eqns (4.37) and (4.45) and $v_1 + v_2 = 1$ with given $X_w{}^0$, $Y_w{}^0$, $X_z{}^0$ and $Y_z{}^0$ of the original polymers, we obtain eqns (4.46)-(4.48), from which we can calculate analytically $\chi_{12}{}^c$, $v_1{}^c$ and $v_2{}^c$.

$$\chi_{12}{}^c = \frac{1}{2X_w{}^0 Y_w{}^0}\left(\frac{X_w{}^0}{(X_z{}^0)^{1/2}} + \frac{Y_w{}^0}{(Y_z{}^0)^{1/2}}\right)[(X_z{}^0)^{1/2} + (Y_z{}^0)^{1/2}], \tag{4.46}$$

$$v_1{}^c = \left[\frac{Y_w{}^0}{(Y_z{}^0)^{1/2}}\right] / \left[\frac{X_w{}^0}{(X_z{}^0)^{1/2}} + \frac{Y_w{}^0}{(Y_z{}^0)^{1/2}}\right], \tag{4.47}$$

$$v_2{}^c = \left[\frac{X_w{}^0}{(X_z{}^0)^{1/2}}\right] / \left[\frac{X_w{}^0}{(X_z{}^0)^{1/2}} + \frac{Y_w{}^0}{(Y_z{}^0)^{1/2}}\right]. \tag{4.48}$$

In the case when both polymer 1 and polymer 2 are single component each, eqns (4.37), (4.45), (4.46)-(4.48) reduce to the well-known equations (eqns (4.3)-(4.7)), originally derived by Scott (ref. 1).

Next, we take into consideration the concentration (i.e., in this case compositional) dependence of χ_{12}, given in the form,

$$\chi_{12} = \chi_{12}{}^0 \left[1 + \sum_{u=1}^{n_u} (p_{1,u} v_1{}^u + p_{2,u} v_2{}^u)\right]. \tag{4.49}$$

Eqn (4.49) is symmetrical with respect to the exchange of polymer 1 and polymer 2. $\chi_{12}{}^0$ in eqn (4.49) is a parameter, independent of v_1 and v_2 and inversely proportional to T. $p_{1,u}$ and $p_{2,u}$ are concentration-dependence parameters. After combining eqn (4.11) with eqn (4.49), we can obtain $\Delta\mu_{Xi}$ and $\Delta\mu_{Yj}$ in the case when χ_{12} is concentration dependent in a similar manner with the derivation of eqns (4.14) and (4.15).

$$\Delta\mu_{Xi} = \left(\frac{\partial \Delta G}{\partial N_{Xi}}\right)_{T,P,N_{Xk}} \qquad (k = 1, \cdots, m_1; \quad k \neq i)$$

$$= \tilde{R}T[\ln v_{Xi} - (X_i - 1) + X_i(1 - \frac{1}{X_n})v_1 + X_i(1 - \frac{1}{Y_n})v_2$$

$$+ \chi_{12}{}^0 X_i v_2{}^2 \{1 + \sum_{u=1}^{n_u} (p_{1,u} v_1{}^u + p_{2,u} v_2{}^u)$$

$$+ \sum_{u=1}^{n_u} u(p_{1,u} v_1{}^{u-1} - p_{2,u} v_2{}^{u-1}) v_1\}] \quad (i = 1, \cdots, m_1), \tag{4.50}$$

$$\Delta\mu_{Yj} = (\frac{\partial \Delta G}{\partial N_{Yj}})_{T,P,N_{Yk}} \qquad (k = 1, \cdots, m_2; \quad k \neq j)$$

$$= \tilde{R}T[\ln v_{Yj} - (Y_j - 1) + Y_j(1 - \frac{1}{X_n})v_1 + Y_j(1 - \frac{1}{Y_n})v_2$$

$$+ \chi_{12}{}^0 Y_j v_1{}^2 \{1 + \sum_{u=1}^{n_u} (p_{1,u} v_1{}^u + p_{2,u} v_2{}^u)$$

$$+ \sum_{u=1}^{n_u} u(p_{2,u} v_2{}^{u-1} - p_{1,u} v_1{}^{u-1}) v_2\}] \quad (j = 1, \cdots, m_2). \tag{4.51}$$

Substitution of eqns (4.50) and (4.51) into eqn (4.17) yields $\Delta G'$ in the form;

$$\Delta G' = (\frac{\tilde{R}T}{V_0}) [\sum_{i=1}^{m_1} \frac{v_{Xi}}{X_i} \ln v_{Xi} + \sum_{j=1}^{m_2} \frac{v_{Yj}}{Y_j} \ln v_{Yj}$$

$$+ \chi_{12}{}^0 \{1 + \sum_{u=1}^{n_u} (p_{1,u} v_1{}^u + p_{2,u} v_2{}^u)\} v_1 v_2]$$

$$= (\frac{\tilde{R}T}{V_0}) [\sum_{i=1}^{m_1} \frac{v_{Xi}}{X_i} \ln v_{Xi} + \sum_{j=1}^{m_2} \frac{v_{Yj}}{Y_j} \ln v_{Yj} + \chi_{12} v_1 v_2]. \tag{4.52}$$

From eqns (4.52) and eqn (4.18), we obtain

$$(\frac{V_0}{\tilde{R}T}) \Delta \bar{G}'_{kl} = \frac{1}{X_l v_{Xl}} \equiv M \qquad (\text{for } k \neq l \ (k,l) = (X_i, X_j)), \tag{4.53a}$$

$$(\frac{\tilde{V}_0}{RT})\,\Delta\bar{G}'_{kl} = \frac{1}{X_1 v_{X1}} - 2\chi_{12} + 2(1-2v_2)\frac{\partial \chi_{12}}{\partial v_2} + v_2(1-v_2)\frac{\partial^2 \chi_{12}}{\partial v_2^2} \equiv N,$$

(for $k \neq l$ $(k,l) = (Y_i, Y_j)$) (4.53b)

$$(\frac{\tilde{V}_0}{RT})\,\Delta\bar{G}'_{kl} = \frac{1}{X_1 v_{X1}} \equiv M \quad (\text{for } k \neq l \ (k,l) = (X_i, Y_j) \text{ or } (Y_i, X_j)), \qquad (4.53c)$$

$$(\frac{\tilde{V}_0}{RT})\,\Delta\bar{G}'_{kl} = \frac{1}{X_1 v_{X1}} + \frac{1}{X_i v_{Xi}} \equiv M + M_i \qquad (\text{for } k = l = X_i), \qquad (4.53d)$$

$$(\frac{\tilde{V}_0}{RT})\,\Delta\bar{G}'_{kl} = \frac{1}{X_1 v_{X1}} - 2\chi_{12} + \frac{1}{Y_j v_{Yj}}$$

$$+ 2(1-2v_2)\frac{\partial \chi_{12}}{\partial v_2} + v_2(1-v_2)\frac{\partial^2 \chi_{12}}{\partial v_2^2} \equiv N + N_j$$

(for $k = l = Y_j$). (4.53e)

Substitution of eqn (4.22') into eqns (4.31a) and (4.31b) gives W_{Xi} and W_{Yj}, respectively;

$$W_{Xi} \equiv (\frac{\partial |\Delta G'|}{\partial v_{Xi}})_{T,P,v_k} = (\frac{\tilde{R}T}{V_0})^{m1+m2-1} (\prod_{i=1}^{m_1} M_i)(\prod_{j=1}^{m_2} N_j)$$

$$\times (-X_1 + X_i)\,[1 + (N-M)\sum_{j=1}^{m_2} \frac{1}{N_j}] \qquad (4.54a)$$

$(k \neq X_i,\ k = X_1, \cdots, X_{m1}, Y_1, \cdots, Y_{m2})$,

$$W_{Yj} \equiv (\frac{\partial |\Delta G'|}{\partial v_{Yj}})_{T,P,v_k} = (\frac{\tilde{R}T}{V_0})^{m1+m2-1} (\prod_{i=1}^{m_1} M_i)(\prod_{j=1}^{m_2} N_j)$$

$$\times [-X_1\{(1 + (N-M)\sum_{j=1}^{m_2} \frac{1}{N_j}\}$$

$$+ Y_j\{1 + (N-M)\sum_{i=1}^{m_1} \frac{1}{M_i}\} + \frac{\partial (N-M)}{\partial v_{Yj}} (\sum_{i=1}^{m_1} \frac{1}{M_i})(\sum_{j=1}^{m_2} \frac{1}{N_j})] \qquad (4.54b)$$

$(k \neq Y_j,\ k = X_1, \cdots, X_{m1}, Y_1, \cdots, Y_{m2})$.

By substituting eqns (4.53a)-(4.53e) into eqn (4.32) and substituting eqns (4.53a)-(4.53e) and eqns (4.54a)-(4.54b) into eqn (4.33), we obtain the equations of spinodal and neutral equilibrium conditions (eqns (4.55) and (4.56)) (ref. 6).

$$\frac{1}{v_1 X_w^0} + \frac{1}{v_2 Y_w^0} - 2\chi_{12} + 2(v_1 - v_2)\frac{\partial \chi_{12}}{\partial v_2} + v_1 v_2 \frac{\partial^2 \chi_{12}}{\partial v_2^2} = 0, \tag{4.55}$$

$$\frac{X_z^0}{(v_1 X_w^0)^2} - \frac{Y_z^0}{(v_2 Y_w^0)^2} - 6\frac{\partial \chi_{12}}{\partial v_2} + 3(v_1 - v_2)\frac{\partial^2 \chi_{12}}{\partial v_2^2} + v_1 v_2 \frac{\partial^3 \chi_{12}}{\partial v_2^3} = 0. \tag{4.56}$$

Eqns (4.55) and (4.56) coincide absolutely with Koningsveld-Chermin-Gordon's equations (eqns 23 and 24 of ref. 2). Note that in Koningsveld et al.'s method the spinodal conditions for the quasi-binary system of polymer 1 and polymer 2 was derived, without detailed derivation, by reducing the solvent volume fraction in the spinodal condition for the quasi-ternary system of polymer 1, polymer 2 and solvent to zero. In contrast to this, eqns (4.55) and (4.56) are derived by Kamide, Shirataki and Matsuda in a very orthodox manner from the second variation of Gibbs' free energy change for the polymer 1 / polymer 2 system (ref. 6).

We can calculate the differential term of χ_{12} in eqns (4.55) and (4.56) with an aid of eqn (4.49) to obtain the following equations;

$$\frac{1}{v_1 X_w^0} + \frac{1}{v_2 Y_w^0} - 2\chi_{12}^0 \left[1 + \sum_{u=1}^{n_u} (p_{1,u} v_1^u + p_{2,u} v_2^u)\right]$$
$$- 2(v_1 - v_2)\chi_{12}^0 \sum_{u=1}^{n_u} u(p_{1,u} v_1^{u-1} - p_{2,u} v_2^{u-1})$$
$$+ v_1 v_2 \chi_{12}^0 \sum_{u=1}^{n_u} u(u-1)(p_{1,u} v_1^{u-2} + p_{2,u} v_2^{u-2}) = 0, \tag{4.57}$$

$$\frac{X_z^0}{(v_1 X_w^0)^2} - \frac{Y_z^0}{(v_2 Y_w^0)^2} + 6\chi_{12}^0 \sum_{u=1}^{n_u} u(p_{1,u} v_1^{u-1} - p_{2,u} v_2^{u-1})$$
$$+ 3(v_1 - v_2)\chi_{12}^0 \sum_{u=1}^{n_u} u(u-1)(p_{1,u} v_1^{u-2} + p_{2,u} v_2^{u-2})$$

$$-v_1 v_2 \chi_{12}^0 \sum_{u=1}^{n_u} u(u-1)(u-2)(p_{1,u} v_1^{u-3} - p_{2,u} v_2^{u-3}) = 0. \tag{4.58}$$

Note that eqns (4.57) and (4.58) (and eqns (4.55) and (4.56)) are symmetrical with respect to exchange of polymer 1 and polymer 2. In the case when all the concentration-dependence parameters are zero (i.e., $p_{1,t} = p_{2,t} = 0$; $t = 1, \cdots, n_t$), eqn (4.57) reduces to eqn (4.37), and eqn (4.58) to eqn (4.45). We can calculate SC from eqn (4.57) and CSP by solving simultaneous equations eqns (4.57) and (4.58).

Then, SC (i.e., χ_{12}^0 vs. v_1 (or v_2) relation) can be calculated, using eqn (4.57), from X_w^0, Y_w^0 and the concentration-dependence parameters. CSP (i.e., v_1^c (or v_2^c) and χ_{12}^c) can also be determined, using eqns (4.57) and (4.58), from X_w^0, X_z^0, Y_w^0, Y_z^0 and the concentration-dependence parameters.

Koningsveld et al. (ref. 4) expressed the Gibbs' free energy change of mixing in terms of g function as follows:

$$\Delta G = \tilde{R}TL\left[\sum_{i=1}^{m_1} \frac{v_{Xi}}{X_i} \ln v_{Xi} + \sum_{j=1}^{m_2} \frac{v_{Yj}}{Y_j} \ln v_{Yj} + g v_1 v_2\right] \tag{4.11'}$$

with

$$g = \sum_{u=0}^{n_u} g_u v_2^u = g_0 + g_1 v_2 + g_2 v_2^2 + \cdots + g_{nu} v_2^{nu}. \tag{4.59}$$

Here, g_0 is a parameter depending on T alone and g_1, g_2, $\cdots$, g_{nt} are concentration dependent, but T- and v_2-independent parameters. Substituting eqn (4.59) into χ_{12} in eqn (4.55) and eqn (4.56), we obtain the spinodal and neutral equilibrium conditions, given in the framework of Koningsveld et al.'s expression in the forms:

$$\frac{1}{v_1 X_w^0} + \frac{1}{v_2 Y_w^0} - 2\sum_{u=0}^{n_u} g_u v_2^u + 2(v_1 - v_2)\sum_{u=1}^{n_u} u g_u v_2^{u-1}$$

$$+ v_1 v_2 \sum_{u=1}^{n_u} u(u-1) g_u v_2^{u-2} = 0, \tag{4.60}$$

$$\frac{1}{(v_1X_w{}^0)^2}+\frac{1}{(v_2Y_w{}^0)^2}-6\sum_{u=1}^{n_u}ug_uv_2{}^{u-1}+3(v_1-v_2)\sum_{u=1}^{n_u}u(u-1)g_uv_2{}^{u-2}$$

$$+v_1v_2\sum_{u=1}^{n_u}u(u-1)(u-2)g_uv_2{}^{u-3}=0. \qquad (4.61)$$

Using these equations, we can also determine SC and CSP in the same manner as in the case of using eqns (4.57) and (4.58). As far as we know nobody has examined Koningsveld et al.'s computer experiments on SC and CSP for polymer 1 / polymer 2 system. This might be due to the fact that eqns (4.60) and (4.61) have not been given in literature by Koningsveld et al. (refs 2,4).

Differing from χ_{12}, given by eqn (4.49), g is expanded into a series of v_2 alone and then is not symmetrical with respect to the exchange of polymer 1 and polymer 2. In addition, it is assumed that the expanded terms in eqn (4.59) are independent of temperature. This is equivalent to an assumption that the third, fourth and higher order virial coefficients are absolutely temperature-independent. Therefore, in Koningsveld et al.'s theory it can be said that the concentration dependence of g was simplified under the sacrifice of physical strictness.

It follows

$$\chi_{12}=g. \qquad (4.62)$$

Although χ_{12} is given an as expanded expression of v_1 and v_2, g is an expanded expression of v_2 alone. Then, in order to make a comparison of the corresponding terms in eqns (4.49) and (4.59) easier, eqn (4.49) is rewritten in advance by putting $p_{1,1}=p_{1,2}=\cdots=0$ as

$$\chi_{12}=\chi_{12}{}^0(1+p_{2,1}v_2+p_{2,2}v_2{}^2+\cdots). \qquad (4.63)$$

Then by comparing eqn (4.59) with eqn (4.63), we obtain

$$\chi_{12}{}^0=g_0, \qquad (4.64)$$

$$p_{2,1}=\frac{g_1}{\chi_{12}{}^0}=\frac{g_1}{g_0}, \qquad (4.65)$$

$$p_{2,2}=\frac{g_2}{\chi_{12}{}^0}=\frac{g_2}{g_0}. \tag{4.66}$$

Firstly, $p_{2,1}$ and $p_{2,2}$ can be determined by substituting the values of g_1 and g_2 employed by Koningsveld et al. and the value of g_0 at their CSP (denoted by $g_0{}^c$) into eqns (4.65) and (4.66) and by putting $p_{2,1}$ and $p_{2,2}$ values obtained thus, together with $p_{1,1}=p_{1,2}=0$, into eqns (4.57) and (4.58) we can reproduce, by means of our theory, the computer calculations on SC and CSP for the same system as used by Koningsveld et al.

4.3 THEORETICAL BACKGROUND OF CLOUD POINT CURVE

Gibbs' two-phase equilibrium conditions under a constant temperature and a constant pressure ($dT=0$, $dp=0$; p is pressure) are given by the following equations.

$$\Delta\mu_{Xi(1)}=\Delta\mu_{Xi(2)} \qquad (i=1,\cdots,m_1), \tag{4.67}$$

$$\Delta\mu_{Yj(1)}=\Delta\mu_{Yj(2)} \qquad (j=1,\cdots,m_2). \tag{4.68}$$

Here, we define the first phase as the polymer 1-rich phase (i.e., polymer 2-lean phase), represented by the suffix (1), and in a similar manner, we define the second phase as the polymer 1 lean phase (i.e., polymer 2-rich phase), representing it by the suffix (2). Principally, by solving m_1+m_2 simultaneous equations (i.e., m_1, number of eqn (4.67) and m_2, that of eqn (4.68)), the two-phase equilibrium can be completely calculated.

The partition coefficients of X_i-mer and Y_j-mer, σ_{Xi} and σ_{Yj} are defined by eqns (4.69) and (4.70), respectively (ref. 7).

$$\sigma_{Xi}\equiv\frac{1}{X_i}\ln\frac{v_{Xi}\,(\text{polymer 1-rich phase})}{v_{Xi}\,(\text{polymer 1-lean phase})}, \tag{4.69}$$

$$\sigma_{Yj}\equiv\frac{1}{Y_j}\ln\frac{v_{Yj}\,(\text{polymer 2-rich phase})}{v_{Yj}\,(\text{polymer 2-lean phase})}. \tag{4.70}$$

Differing from a polymer / single solvent system, the polymer 1 / polymer 2 system has a characteristic feature which prerequisites two kinds of the partition coefficients as defined by eqns (4.69) and (4.70), in order to describe thermodynamically the two-phase equilibrium.

Substitution of $\Delta\mu_{Xi}$ given by eqn (4.50) into eqn (4.67) and combination of the equation derived thus with eqn (4.69) lead to

$$\sigma_{Xi} = \left(1-\frac{1}{X_{n(2)}}\right)v_{1(2)} + \left(1-\frac{1}{Y_{n(2)}}\right)v_{2(2)}$$

$$-\left(1-\frac{1}{X_{n(1)}}\right)v_{1(1)} - \left(1-\frac{1}{Y_{n(1)}}\right)v_{2(1)}$$

$$+\chi_{12}{}^{0}\,[v_{2(2)}{}^{2}\,\{1+\sum_{u=1}^{n_u}(p_{1,u}v_{1(2)}{}^{u}+p_{2,u}v_{2(2)}{}^{u})$$

$$+\sum_{u=1}^{n_u}u(p_{1,u}v_{1(2)}{}^{u-1}-p_{2,u}v_{2(2)}{}^{u-1})v_{1(2)}\}$$

$$-v_{2(1)}{}^{2}\,\{1+\sum_{u=1}^{n_u}(p_{1,u}v_{1(1)}{}^{u}+p_{2,u}v_{2(1)}{}^{u})$$

$$+\sum_{u=1}^{n_u}u(p_{1,u}v_{1(1)}{}^{u-1}-p_{2,u}v_{2(1)}{}^{u-1})v_{1(1)}\}]. \qquad (4.71)$$

Similarly, from eqns (4.51), (4.68) and (4.70), we obtain

$$\sigma_{Yj} = \left(1-\frac{1}{X_{n(1)}}\right)v_{1(1)} + \left(1-\frac{1}{Y_{n(1)}}\right)v_{2(1)}$$

$$-\left(1-\frac{1}{X_{n(2)}}\right)v_{1(2)} - \left(1-\frac{1}{Y_{n(2)}}\right)v_{2(2)}$$

$$+\chi_{12}{}^{0}\,[v_{1(1)}{}^{2}\,\{1+\sum_{u=1}^{n_u}(p_{1,u}v_{1(1)}{}^{u}+p_{2,u}v_{2(1)}{}^{u})$$

$$+\sum_{u=1}^{n_u}u(p_{2,u}v_{2(1)}{}^{u-1}-p_{1,u}v_{1(1)}{}^{u-1})v_{2(1)}\}$$

$$-v_{1(2)}{}^{2}\,\{1+\sum_{u=1}^{n_u}(p_{1,u}v_{1(2)}{}^{u}+p_{2,u}v_{2(2)}{}^{u})$$

$$+\sum_{u=1}^{n_u} u(p_{2,u}v_{2(2)}^{u-1} - p_{1,u}v_{1(2)}^{u-1})v_{2(2)}\}]. \quad (4.72)$$

Eqns (4.71) and (4.72) indicate that σ_{Xi} is independent of X_i and σ_{Yj} is independent of Y_j and then σ_{Xi} and σ_{Yj} are hereafter simply denoted by σ_X and σ_Y, respectively.

Eqns (4.71) and (4.72) are rewritten as

$$\chi_{12}^0 \equiv \chi_{12}^0{}_{,X}$$

$$= [\sigma_X - (1-\frac{1}{X_{n(2)}})v_{1(2)} - (1-\frac{1}{Y_{n(2)}})v_{2(2)}$$

$$+ (1-\frac{1}{X_{n(1)}})v_{1(1)} + (1-\frac{1}{Y_{n(1)}})v_{2(1)}]$$

$$/ [v_{2(2)}^2\{1+\sum_{u=1}^{n_u}(p_{1,u}v_{1(2)}^u + p_{2,u}v_{2(2)}^u)$$

$$+\sum_{u=1}^{n_u} u(p_{1,u}v_{1(2)}^{u-1} - p_{2,u}v_{2(2)}^{u-1})v_{1(2)}\}$$

$$- v_{2(1)}^2\{1+\sum_{u=1}^{n_u}(p_{1,u}v_{1(1)}^u + p_{2,u}v_{2(1)}^u)$$

$$+\sum_{u=1}^{n_u} u(p_{1,u}v_{1(1)}^{u-1} - p_{2,u}v_{2(1)}^{u-1})v_{1(1)}\}]. \quad (4.73)$$

and

$$\chi_{12}^0 \equiv \chi_{12}^0{}_{,Y}$$

$$= [\sigma_Y - (1-\frac{1}{X_{n(1)}})v_{1(1)} - (1-\frac{1}{Y_{n(1)}})v_{2(1)}$$

$$+ (1-\frac{1}{X_{n(2)}})v_{1(2)} + (1-\frac{1}{Y_{n(2)}})v_{2(2)}]$$

$$/ [v_{1(1)}^2\{1+\sum_{u=1}^{n_u}(p_{1,u}v_{1(1)}^u + p_{2,u}v_{2(1)}^u)$$

$$+\sum_{u=1}^{n_u} u(p_{2,u}v_{2(1)}^{u-1} - p_{1,u}v_{1(1)}^{u-1})v_{2(1)}\}$$

$$-v_{1(2)}^2\{1+\sum_{u=1}^{n_u}(p_{1,u}v_{1(2)}^{u}+p_{2,u}v_{2(2)}^{u})$$

$$+\sum_{u=1}^{n_u} u(p_{2,u}v_{2(2)}^{u-1} - p_{1,u}v_{1(2)}^{u-1})v_{2(2)}\}]. \qquad (4.74)$$

Note that for the polymer 1 / polymer 2 system there are two kinds of χ_{12}^0; $\chi_{12}^0{}_{,X}$ (eqn (4.73)) and $\chi_{12}^0{}_{,Y}$ (eqn (4.74)). Since χ_{12}^0 is a function of temperature alone, these two kinds of χ_{12}^0 should be the same when two-phase equilibrium is completely attained, that is

$$F \equiv \chi_{12}^0{}_{,X} - \chi_{12}^0{}_{,Y} = 0. \qquad (4.75)$$

Accordingly, the conditions, under which eqn (4.75) is realized, are considered as an alternative expression of the Gibbs' two-phase equilibrium conditions, under which eqns (4.67) and (4.68) are satisfied. In this sense, eqn (4.75) is a fundamental equation for calculating two-phase equilibrium of a multicomponent polymer 1 / multicomponent polymer 2 system.

If $p_{1,u}$ and $p_{2,u}$ are given in advance, $\chi_{12}^0{}_{,X}$ in eqn (4.73) becomes a function of seven variables such as σ_X, $X_{n(1)}$, $X_{n(2)}$, $Y_{n(1)}$, $Y_{n(2)}$, $v_{1(1)}$ $(=1-v_{2(1)})$ and $v_{1(2)}$ $(=1-v_{2(2)})$. In the same manner, for a given combination of $p_{1,u}$ and $p_{2,u}$, $\chi_{12}^0{}_{,Y}$ in eqn (4.74) is a function of seven variables such as σ_Y, $X_{n(1)}$, $X_{n(2)}$, $Y_{n(1)}$, $Y_{n(2)}$, $v_{1(1)}$ and $v_{1(2)}$. Then, F defined by eqn (4.75) becomes a function of 8 variables including σ_X, σ_Y, $X_{n(1)}$, $X_{n(2)}$, $Y_{n(1)}$, $Y_{n(2)}$, $v_{1(1)}$ and $v_{1(2)}$.

Here, note that $X_{n(1)}$, $X_{n(2)}$, $Y_{n(1)}$, $Y_{n(2)}$, $v_{1(1)}$ and $v_{1(2)}$ are given by

$$X_{n(k)} = \sum_{i=1}^{m_1} v_{Xi(k)} \Big/ \sum_{i=1}^{m_1}(v_{Xi(k)}/X_i), \qquad (4.76a)$$

$$Y_{n(k)} = \sum_{j=1}^{m_2} v_{Yj(k)} \Big/ \sum_{j=1}^{m_2}(v_{Yj(k)}/Y_j), \qquad (4.76b)$$

$$v_{1(k)} = \sum_{i=1}^{m_1} v_{Xi(k)}, \tag{4.76c}$$

$$v_{2(k)} = \sum_{j=1}^{m_2} v_{Yj(k)} \tag{4.76d}$$

$$(k = 1,\ 2)$$

and eight variables for F is reduced to 6 variables of σ_X, σ_Y, $v_{Xi(1)}$, $v_{Xi(2)}$, $v_{Yj(1)}$ and $v_{Yj(2)}$.

By definition, $v_{Xi(1)}$, $v_{Xi(2)}$, $v_{Yj(1)}$ and $v_{Yj(2)}$ are given by the relations,

$$v_{Xi(1)} = \frac{V_1{}^0 g_{(1)}(X_i)}{V_{(1)}}, \tag{4.77a}$$

$$v_{Xi(2)} = \frac{V_1{}^0 g_{(2)}(X_i)}{V_{(2)}}, \tag{4.77b}$$

$$v_{Yj(1)} = \frac{V_2{}^0 g_{(1)}(Y_j)}{V_{(1)}}, \tag{4.78a}$$

$$v_{Yj(2)} = \frac{V_2{}^0 g_{(2)}(Y_j)}{V_{(2)}}. \tag{4.78b}$$

Here, $V_1{}^0$ and $V_2{}^0$ are the volumes of polymer 1 and polymer 2, respectively, $g_{(1)}(X_i)$ (or $g_{(2)}(X_i)$) and $g_{(1)}(Y_j)$ (or $g_{(2)}(Y_j)$) are the weight fractions of X_i-mer of polymer 1 and Y_j-mer of polymer 2, partitioned in the first phase (or the second phase), respectively and these are related to the distribution of DP of the original polymers, $g_0(X_i)$ and $g_0(Y_j)$, through eqns (4.79) and (4.80).

$$g_0(X_i) = g_{(1)}(X_i) + g_{(2)}(X_i), \tag{4.79}$$

$$g_0(Y_j) = g_{(1)}(Y_j) + g_{(2)}(Y_j). \tag{4.80}$$

Substitution of eqns (4.77a) and (4.77b) into eqn (4.69) and substitution of eqns (4.78a) and (4.78b) into eqn (4.70) lead to eqns (4.81) and (4.82), respectively.

$$\sigma_{Xi} = \frac{1}{X_i} \ln \frac{g_{(1)}(X_i)}{R \cdot g_{(2)}(X_i)}, \tag{4.81}$$

$$\sigma_{Yj} = \frac{1}{Y_j} \ln \frac{R \cdot g_{(2)}(Y_j)}{g_{(1)}(Y_j)}. \tag{4.82}$$

Combining eqns (4.79) and (4.81), we obtain

$$g_{(1)}(X_i) = \frac{R \cdot \exp(\sigma_{Xi} X_i)}{1 + R \cdot \exp(\sigma_{Xi} X_i)} g_0(X_i), \tag{4.83a}$$

$$g_{(2)}(X_i) = \frac{1}{1 + R \cdot \exp(\sigma_{Xi} X_i)} g_0(X_i). \tag{4.83b}$$

From eqns (4.80) and (4.82), it follows

$$g_{(1)}(Y_j) = \frac{R}{R + \exp(\sigma_{Yj} Y_j)} g_0(Y_j), \tag{4.84a}$$

$$g_{(2)}(Y_j) = \frac{\exp(\sigma_{Yj} Y_j)}{R + \exp(\sigma_{Yj} Y_j)} g_0(Y_j). \tag{4.84b}$$

By the definition of R and V, $V_{(1)}$ and $V_{(2)}$ are

$$V_{(1)} = \frac{R}{R+1} V, \tag{4.85a}$$

$$V_{(2)} = \frac{1}{R+1} V. \tag{4.85b}$$

Substitution of eqns (4.83a), (4.83b) and (4.85a) into eqns (4.77a) and (4.77b) then yield

$$v_{Xi(1)} = \frac{(R+1)\exp(\sigma_{Xi} X_i)}{1 + R \cdot \exp(\sigma_{Xi} X_i)} \left(\frac{V_1{}^0}{V}\right) g_0(X_i), \tag{4.86a}$$

$$v_{Xi(2)} = \frac{R+1}{1 + R \cdot \exp(\sigma_{Xi} X_i)} \left(\frac{V_1{}^0}{V}\right) g_0(X_i). \tag{4.86b}$$

Similarly from eqns (4.84a), (4.84b), (4.85b), (4.78a) and (4.78b), we obtain

$$v_{Yj(1)} = \frac{(R+1)}{R+\exp(\sigma_{Yj}Y_j)} \left(\frac{V_2{}^0}{V}\right) g_0(Y_j) , \tag{4.87a}$$

$$v_{Yj(2)} = \frac{(R+1)\exp(\sigma_{Yj}Y_j)}{R+\exp(\sigma_{xi}X_i)} \left(\frac{V_1{}^0}{V}\right) g_0(Y_j) . \tag{4.87b}$$

Considering the following relations

$$\frac{V_1{}^0}{V} g_0(X_i) = v_1{}^0 g_0(X_i) = v_0(X_i) , \tag{4.88}$$

$$\frac{V_2{}^0}{V} g_0(Y_j) = v_2{}^0 g_0(Y_j) = v_0(Y_j) . \tag{4.89}$$

We can derive eqns (4.90a)-(4.91b) from eqns (4.86a)-(4.87b).

$$v_{Xi(1)} (= v_{(1)}(X_i)) = \frac{(R+1)\exp(\sigma_X X_i)}{1+R\cdot\exp(\sigma_X X_i)} v_0(X_i) , \tag{4.90a}$$

$$v_{Xi(2)} (= v_{(2)}(X_i)) = \frac{(R+1)}{1+R\cdot\exp(\sigma_X X_i)} v_0(X_i) \tag{4.90b}$$

and

$$v_{Yj(1)} (= v_{(1)}(Y_j)) = \frac{(R+1)}{R+\exp(\sigma_Y Y_j)} v_0(Y_j) , \tag{4.91a}$$

$$v_{Yj(2)} (= v_{(2)}(Y_j)) = \frac{(R+1)\exp(\sigma_Y Y_j)}{R+\exp(\sigma_Y Y_j)} v_0(Y_j) \tag{4.91b}$$

where R is the ratio of the volume of the first phase $V_{(1)}$ to the volume of the second phase $V_{(2)}$,

$$R = \frac{V_{(1)}}{V_{(2)}} \tag{4.92}$$

and $v_0(X_i)$ and $v_0(Y_j)$ are the volume fractions of X_i-mer in original polymer 1 and of Y_j-mer in original polymer 2, respectively. The volume fractions of the original polymers 1 and 2, $v_1{}^0$ and $v_2{}^0$ $(= 1 - v_1{}^0)$ (the initial concentrations) are then given by

$$v_1{}^0 = \sum_{i=1}^{m_1} v_0(X_i), \tag{4.93}$$

$$v_2{}^0 = \sum_{j=1}^{m_2} v_0(Y_j). \tag{4.94}$$

If $v_0(X_i)$ and $v_0(Y_j)$ are given in advance (from $v_1{}^0$, $v_2{}^0$ and the molecular weight distributions of the original polymers), 6 variables in F function reduces furthermore to 3 variables of σ_X, σ_Y and R,

$$F = F(\sigma_X, \sigma_Y, R)$$

$$= \chi_{12}{}^0{}_{,X}(\sigma_X, \sigma_Y, R) - \chi_{12}{}^0{}_{,Y}(\sigma_X, \sigma_Y, R). \tag{4.95}$$

The cloud point is defined as the temperature at which the cloud particle with an infinitely small volume appears from the solution with a given initial concentration (refs 8,9). When the volume fractions of polymer 1 and polymer 2 at CSP are denoted by $v_1{}^c$ and $v_2{}^c$, the cloud particles are the second phase (i.e., polymer 1-lean phase = polymer 2-rich phase) itself for $v_2{}^0 < v_2{}^c$ $(v_1{}^0 > v_1{}^c)$ and the first phase (i.e., polymer 1-rich phase = polymer 2-lean phase) for $v_2{}^0 > v_2{}^c$ $(v_1{}^0 < v_1{}^c)$. Accordingly, we obtain the following relations.

(i) $v_2{}^0 < v_2{}^c$ $(v_1{}^0 > v_1{}^c)$

$$V_{(1)} = V, \quad V_{(2)} = 0, \quad R = V_{(1)}/V_{(2)} = \infty \tag{4.96}$$

and

(ii) $v_2{}^0 > v_2{}^c$ $(v_1{}^0 < v_1{}^c)$

$$V_{(1)} = 0, \quad V_{(2)} = V, \quad R = V_{(1)}/V_{(2)} = 0. \tag{4.97}$$

Here V is the total volume $(= V_{(1)} + V_{(2)} = V_1{}^0 + V_2{}^0$; $V_1{}^0$ and $V_2{}^0$, the volume of polymer 1 and polymer 2).

Using eqns (4.96) and (4.97), eqns (4.90a), (4.90b), (4.91a) and (4.91b) can be transformed into

(i) $v_2{}^0 < v_2{}^c$ $(v_1{}^0 > v_1{}^c)$

$$v_{(1)}{}^{CP}(X_i) = v_0(X_i)\,, \tag{4.98a}$$

$$v_{(2)}{}^{CP}(X_i) = v_0(X_i)\exp(-\sigma_X{}^{CP}X_i)\,, \tag{4.98b}$$

$$v_{(1)}{}^{CP}(Y_j) = v_0(Y_j)\,, \tag{4.99a}$$

$$v_{(2)}{}^{CP}(Y_j) = v_0(Y_j)\exp(\sigma_Y{}^{CP}Y_j)\,, \tag{4.99b}$$

(ii) $v_2{}^0 > v_2{}^c$ $(v_1{}^0 < v_1{}^c)$

$$v_{(1)}{}^{CP}(X_i) = v_0(X_i)\exp(\sigma_X{}^{CP}X_i)\,, \tag{4.100a}$$

$$v_{(2)}{}^{CP}(X_i) = v_0(X_i)\,, \tag{4.100b}$$

$$v_{(1)}{}^{CP}(Y_j) = v_0(Y_j)\exp(-\sigma_Y{}^{CP}Y_j)\,, \tag{4.101a}$$

$$v_{(2)}{}^{CP}(Y_j) = v_0(Y_j)\,. \tag{4.101b}$$

The superscript CP means the cloud point. As a consequence, F reduces, at the cloud point, to a function of two variables of σ_X and σ_Y. Here, the definition of the volume fraction gives the following relations.

$$\sum_{i=1}^{m_1} v_{(1)}{}^{CP}(X_i) + \sum_{j=1}^{m_2} v_{(1)}{}^{CP}(Y_j) = 1, \tag{4.102a}$$

$$\sum_{i=1}^{m_1} v_{(2)}{}^{CP}(X_i) + \sum_{j=1}^{m_2} v_{(2)}{}^{CP}(Y_j) = 1. \tag{4.102b}$$

In the case of $v_2{}^0 < v_2{}^c$, substituting eqn (4.102b) with eqns (4.98b) and (4.99b), we obtain

$$\sum_{i=1}^{m_1} v_0(X_i)\exp(-\sigma_X{}^{CP}X_i) + \sum_{j=1}^{m_2} v_0(Y_j)\exp(\sigma_Y{}^{CP}Y_j) = 1 \tag{4.103}$$

and in the case of $v_2{}^0 > v_2{}^c$, from eqns (4.100a), (4.101a) and (4.102a), we obtain

$$\sum_{i=1}^{m_1} v_0(X_i)\exp(\sigma_X{}^{CP}X_i) + \sum_{j=1}^{m_2} v_0(Y_j)\exp(-\sigma_Y{}^{CP}Y_j) = 1. \tag{4.104}$$

$$F \equiv \chi_{12}{}^0{}_{,X} - \chi_{12}{}^0{}_{,Y} \quad [\text{eqn } (4.75)]$$

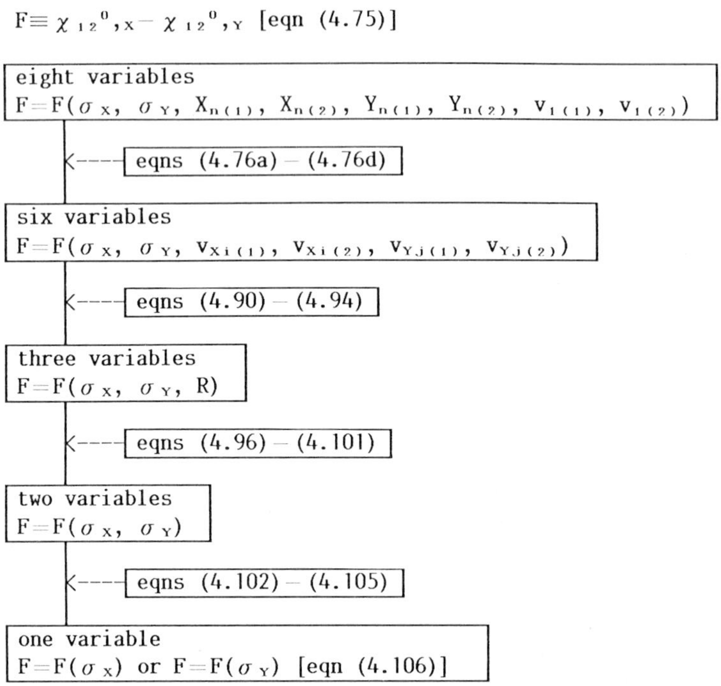

Fig. 4.1 Schematic route of derivation of eqn (4.106) from fundamental equation (eqn (4.75)).

Although both equations (eqns (4.103) and (4.104)) are not given in the closed form, σ_Y can be unambiguously determined for a given σ_X. That is, σ_Y is a function of σ_X and the reverse is true.

$$\sigma_X = \sigma_X(\sigma_Y) \quad (\text{or } \sigma_Y = \sigma_Y(\sigma_X)). \tag{4.105}$$

And ultimately F becomes a function of the simple variable (σ_X or σ_Y),

$$F = F(\sigma_X)$$

$$= \chi_{12}{}^0{}_{,X}(\sigma_X) - \chi_{12}{}^0{}_{,Y}(\sigma_X). \tag{4.106}$$

Eqn (4.106) is a fundamental equation for calculating CPC of polymer 1 / polymer 2 system: By dissolving eqn (4.106), we can determine $\sigma_X{}^{CP}$ and putting $\sigma_X{}^{CP}$ thus determined into eqn (4.103) or (4.104), we can estimate σ_Y^{CP}. Substitution of σ_X^{CP} into the equation,

$$\chi_{12}^{0}{}_{,X} = \chi_{12}^{0}{}_{,X}(\sigma_X) \qquad (\text{or } \chi_{12}^{0}{}_{,Y} = \chi_{12}^{0}{}_{,Y}(\sigma_X)) \tag{4.107}$$

enables us to evaluate χ_{12}^{0} at cloud point ($= \chi_{12}^{CP}$). Of course, by use of eqns (4.98a)-(4.99b) (or eqns (4.100a)-(4.101b)) $v_{X_i}{}^{CP}{}_{(k)}$ and $v_{Y_j}{}^{CP}{}_{(k)}$ ($k = 1$ and 2) can be obtained and furthermore, by putting these variables into eqns (4.76a)-(4.76d), we obtain $X_n{}^{CP}{}_{(k)}$, $Y_n{}^{CP}{}_{(k)}$, $v_1{}^{CP}{}_{(k)}$ and $v_2{}^{CP}{}_{(k)}$ ($k = 1$ and 2). CPC is determined by solving eqn (4.106) over a wide range of $v_1{}^{0}$ and the shadow curve is calculated from $v_1{}^{CP}{}_{(k)}$ (or $v_2{}^{CP}{}_{(k)}$ ($k = 1$ and 2)) for a wide range of $v_1{}^{0}$.

Since the above discussion seems rather complicated, the schematic route of derivation of eqn (4.106) from eqn (4.75) is demonstrated in Fig. 4.1.

4.4 EFFECT OF POLYMER CHARACTERISTICS ON TWO-PHASE EQUILIBRIUM OF MULTICOMPONENT POLYMER 1 AND MULTICOMPONENT POLYMER 2

4.4.1 Spinodal curve and critical solution point (ref. 6)

Two χ_{12}^{0}, satisfying the spinodal and neutral equilibrium conditions, as denoted by $\chi_{12}^{0}{}_{SP}$ and $\chi_{12}^{0}{}_{NE}$, respectively, are given , by rewriting eqns (4.57) and (4.58), in the forms

$$\chi_{12}^{0}{}_{SP} = \left(\frac{1}{v_1 X_w^{0}} + \frac{1}{v_2 Y_w^{0}}\right) \Big/ \Big[\, 2\{1 + \sum_{u=1}^{n_u} (p_{1,u} v_1{}^{u} + p_{2,u} v_2{}^{u})\}$$

$$+ 2(v_1 - v_2) \sum_{u=1}^{n_u} u(p_{1,u} v_1{}^{u-1} - p_{2,u} v_2{}^{u-1})$$

$$- v_1 v_2 \sum_{u=1}^{n_u} u(u-1)(p_{1,u} v_1{}^{u-2} + p_{2,u} v_2{}^{u-2})\Big], \tag{4.108}$$

$$\chi_{12}^{0}{}_{NE} = \left\{\frac{Y_z{}^{0}}{(v_2 Y_w{}^{0})^2} - \frac{X_z{}^{0}}{(v_1 X_w{}^{0})^2}\right\} \Big/ \Big[\, 6\sum_{u=1}^{n_u} u(p_{1,u} v_1{}^{u-1} - p_{2,u} v_2{}^{u-1})$$

$$+ 3(v_1 - v_2) \sum_{u=1}^{n_u} u(u-1)(p_{1,u} v_1{}^{u-2} + p_{2,u} v_2{}^{u-2})$$

$$- v_1 v_2 \sum_{u=1}^{n_u} u(u-1)(u-2)(p_{1,u} v_1^{u-3} - p_{2,u} v_2^{u-3})] \qquad (4.109)$$

when X_w^0, Y_w^0, X_z^0, Y_z^0, $p_{1,u}$ and $p_{2,u}$ are given in advance, $\chi_{12}{}^0{}_{SP}$ and $\chi_{12}{}^0{}_{NE}$ are functions of the single variable v_2 (or v_1) (note $v_1 + v_2 = 1$);

$$\chi_{12}{}^0{}_{SP} = \chi_{12}{}^0{}_{SP}(v_2), \qquad (4.108')$$

$$\chi_{12}{}^0{}_{NE} = \chi_{12}{}^0{}_{NE}(v_2). \qquad (4.109')$$

SC can be directly calculated from $\chi_{12}{}^0$, estimated using eqn (4.109') in the range of $0 < v_2 < 1$. v_2 at CSP ($v_2{}^c$) can be determined as v_2 at $\chi_{12}{}^0{}_{SP} = \chi_{12}{}^0{}_{NE}$. For this purpose, we define A function by eqn (4.110).

$$A = \chi_{12}{}^0{}_{SP}(v_2) - \chi_{12}{}^0{}_{NE}(v_2) \qquad (4.110)$$

to calculate v_2 and $\chi_{12}{}^0$ at $A = 0$ (i.e., $v_2{}^c$ and $\chi_{12}{}^c$) using the interhalving method.

Computer experiments were carried out under the following conditions: The distribution of DP of the original polymers, Schulz-Zimm type or Wesslau type distributions, X_w^0 (and Y_w^0) $= 50 \sim 1\times10^6$ (and $50\sim 1\times 10^4$), X_w^0/X_n^0 (and Y_w^0/Y_n^0) $= 1\sim 5$, $p_{1,1} = -2 \sim 4$ and $p_{1,2} = p_{1,3} = \cdots = 0$, $p_{2,1} = p_{2,2} = \cdots = 0$.

X_z^0 (or Y_z^0), needed for calculation of eqns (4.45), (4.56) and (4.58), is evaluated from X_w^0 (or Y_w^0) and X_w^0/X_n^0 (or Y_w^0/Y_n^0), if the type of MWD is given in advance, using the following equations.

$$\left.\begin{aligned} X_z^0 &= 2X_w^0 - X_n^0 = X_w^0\{2 - (\frac{X_w^0}{X_n^0})^{-1}\} \\ Y_z^0 &= 2Y_w^0 - Y_n^0 = Y_w^0\{2 - (\frac{Y_w^0}{Y_n^0})^{-1}\} \end{aligned}\right] \text{for Schulz-Zimm type distribution} \qquad (4.111)$$

$$\left.\begin{aligned} X_z^0 &= \frac{(X_w^0)^2}{X_n^0} = X_w^0(\frac{X_w^0}{X_n^0}) \\ Y_z^0 &= \frac{(Y_w^0)^2}{Y_n^0} = Y_w^0(\frac{Y_w^0}{Y_n^0}) \end{aligned}\right] \text{for Wesslau type distribution} \qquad (4.112)$$

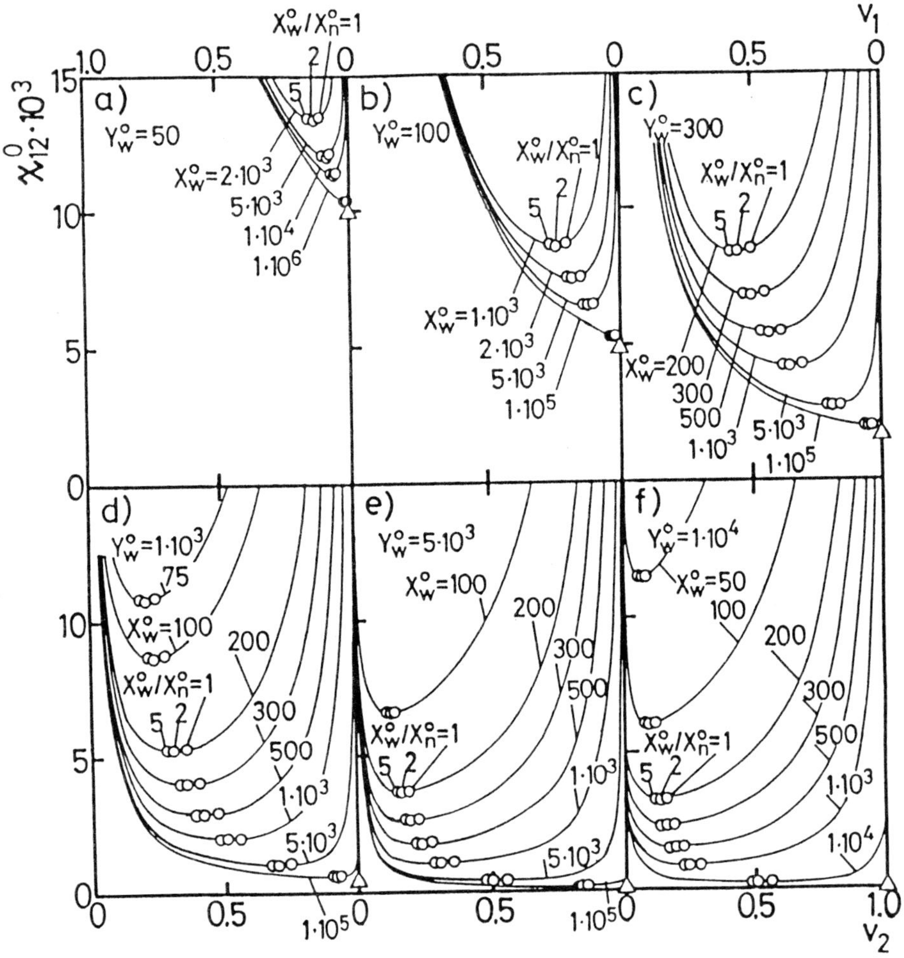

Fig. 4.2 Effects of the molecular weight distribution and the weight average degree of polymerization of the original polymer 1 on SC and CSP: Original polymers 1 and 2, Schulz-Zimm type distribution; $Y_w{}^0/Y_n{}^0=2$; a) $Y_w{}^0=50$, b) $Y_w{}^0=100$, c) $Y_w{}^0=300$, d) $Y_w{}^0=1\times10^3$, e) $Y_w{}^0=5\times10^3$ and f) $Y_w{}^0=1\times10^4$; $p_{1,u}=p_{2,u}=0$; ○, CSP and △, Flory θ point.

Figs 4.2 and 4.3 show the effect of $X_w{}^0$ and $X_w{}^0/X_n{}^0$ on the SC and the CSP for a binary mixture of a multicomponent polymer 1 with the Schulz-Zimm type (Fig. 4.2) or the Wesslau type (Fig. 4.3) molecular weight distribution and a multicomponent polymer 2 with the Schulz-Zimm type molecular weight distribution ($Y_w{}^0/Y_n{}^0=2$). $\chi_{12}{}^0$ at the ordinate axis can be converted to

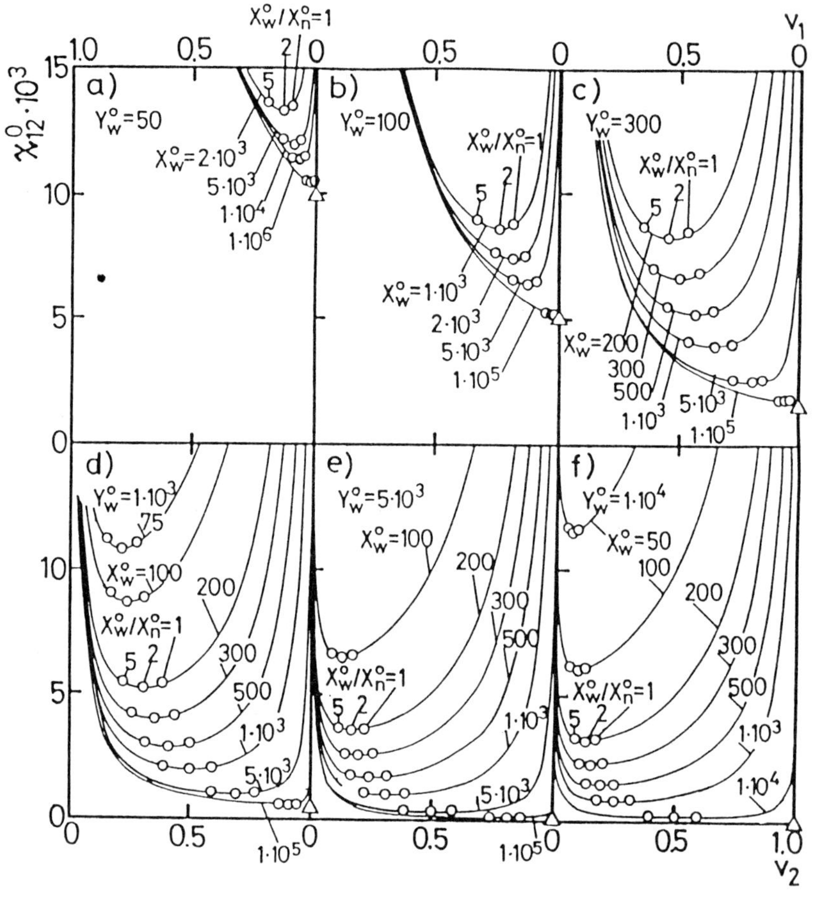

Fig. 4.3 Effects of the molecular weight distribution and the weight average degree of polymerization of the original polymer 1 on SC and CSP: Original polymer 1, Wesslau type distribution and original polymer 2, Schulz-Zimm type distribution; $Y_w{}^0/Y_n{}^0=2$; a) $Y_w{}^0=50$, b) $Y_w{}^0=100$, c) $Y_w{}^0=300$, d) $Y_w{}^0=1\times10^3$, e) $Y_w{}^0=5\times10^3$ and f) $Y_w{}^0=1\times10^4$; $p_{1,u}=p_{2,u}=0$; ○, CSP and △, Flory θ point.

temperature through the relation,

$$\chi_{12}{}^0=a_{12}{}^0+b_{12}{}^0/T \tag{4.113}$$

where $a_{12}{}^0$ and $b_{12}{}^0$ are parameters independent of temperature. In the figures, the polydispersity of polymer 2 is kept constant

($Y_w{}^0/Y_n{}^0 = 2$) and $Y_w{}^0$ is varied from 50 (Figs 4.2a and 4.3a) to 1×10^4 (Figs 4.2f and 4.3f). The solid line is SC and the unfilled circle is CSP. Note that the concentration dependence of χ_{12} is neglected in the figures ($p_{1,u} = p_{2,u} = 0$). As is evident from eqn (4.57), SC is unambiguously determined by $X_w{}^0$ and $Y_w{}^0$ alone. Then with an increase in $X_w{}^0/X_n{}^0$ from 1 to 5 for the given combination of $X_w{}^0$ and $Y_w{}^0$, SC dose not change, but CSP moves on SC to the polymer 1 side, resulting in an increase in the polymer 1 volume fraction at CSP $v_1{}^c$. The effect of $X_w{}^0/X_n{}^0$ on the degree of shift of CSP is significantly larger in Fig. 4.3 than in Fig. 4.2. Kamide et al. already observed for a quasi-ternary system consisting of polymer 1, polymer 2 and a single solvent, the similar phenomena that broadening of polydispersity of one polymer makes CSP to shift to the side of the polymer (ref. 3).

In the case of $X_w{}^0 \neq Y_w{}^0$, CSP shifts on SC to a larger extent when the original polymer with larger molecular weight changes its polydispersity than when the polymer with smaller molecular weight changes its polydispersity: For example, compare ($X_w{}^0$, $Y_w{}^0$) = (5×10^3, 100) in Fig. 4.2b with ($X_w{}^0$, $Y_w{}^0$) = (100, 5×10^3) in Fig. 4.2e. In other words, the effect of polydispersity on CSP is more significant in polymers with a larger molecular weight and this tendency is more remarkable for the Wesslau type polymer than for the Schulz-Zimm type polymer: For example, compare ($X_w{}^0$, $Y_w{}^0$) = (5×10^3, 100) in Fig. 4.3b with ($X_w{}^0$, $Y_w{}^0$) = (100, 5×10^3) in Fig. 4.3e. As $X_w{}^0$ and $Y_w{}^0$ increase the two-phase region gets wider. In the case of $Y_w{}^0 \ll X_w{}^0$ (see, Figs 4.2a, 4.2b, 4.3a and 4.3b), SC is similar to that for polymer 1 / single solvent system, where polymer 2 is approximated as a solvent. In the case of $Y_w{}^0 \gg X_w{}^0$ (see, Figs 4.2e, 4.2f, 4.3e and 4.3f), SC is similar to that for polymer 2 / single solvent. Attention should be paid to the fact that the scale of the ordinate axis in Figs 4.2 and 4.3 is very small, covering $0 \leq \chi_{12}{}^0 \leq 0.015$ and when SC of a polymer 1 / polymer 2 system is plotted on the same $\chi_{12}{}^0$ scale as a polymer / solvent system, the shape of the former is expected to be notably different from that of the latter. Fig. 4.4 is an example, where SC of polymer 1 / polymer 2 systems with $Y_w{}^0 = 1\times 10^4$ and $X_w{}^0 = 1\sim 100$ is plotted on a usual $\chi_{12}{}^0$ scale ranging from 0 to 0.6. As $X_w{}^0$ decreases, SC progressively changes its shape from flat to sharp, accompanied by a rapid increase in $\chi_{12}{}^c$. In the case of $X_w{}^0 = 1$ (and $Y_w{}^0 = 1\times 10^4$), SC coincides completely with that of polymer / solvent system as follows;

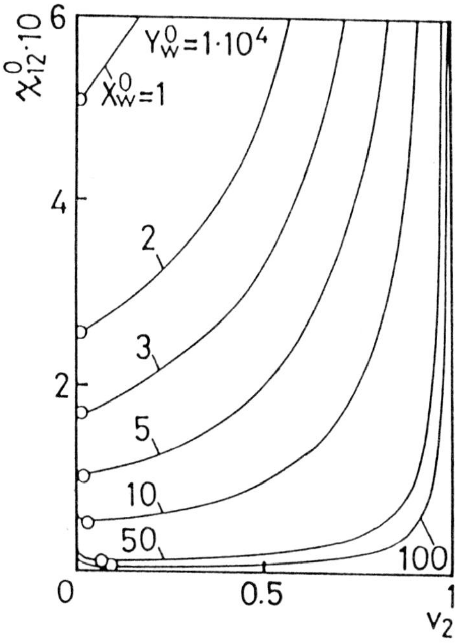

Fig. 4.4 Effect of the weight average degree of polymerization of the a polymer with a smaller average molecular weight on the SC and the CSP (unfilled circle): Original polymers 1 and 2, Schulz-Zimm type distribution; $X_w^0/X_n^0=2$, $Y_w^0/Y_n^0=2$; $X_w^0=1\sim100$, $Y_w^0=1\times10^4$; $p_{1,u}=p_{2,u}=0$.

The spinodal condition of polymer / single solvent system, where χ is independent of the concentration and the polymer molecular weight, is given by

$$\frac{1}{X_w v_p}+\frac{1}{1-v_p}-2\chi_{00}=0 \tag{4.114}$$

where X_w is the weight average degree of polymerization of the polymer, v_p, the polymer volume fraction, χ_{00} is thermodynamical interaction parameter between a polymer and a solvent.

By putting $Y_w^0=1$, $v_1=v_p$, $v_2=1-v_p$ and $\chi_{12}=\chi_{00}$ into the equation for a spinodal condition of a polymer 1 / polymer 2 system (eqn (4.37)), we obtain eqn (4.114).

In the similar manner, eqns (4.46)-(4.48) for the polymer 1 / polymer 2 system reduce to the well-known equations,

$$\chi_{00}{}^{c} = \frac{1}{2X_w}\{X_w (X_z)^{-1/2} + 1\} (X_z{}^{1/2} + 1), \qquad (4.115)$$

$$v_0{}^{c} = \frac{X_w (X_z)^{-1/2}}{1 + X_w (X_z)^{-1/2}}, \qquad (4.116)$$

$$v_p{}^{c} = \frac{1}{1 + X_w (X_z)^{-1/2}}. \qquad (4.117)$$

Then, the computer experiments for the polymer 1 / polymer 2 system, where $X_w{}^0$ is taken as 1, proved to correspond completely to those for the polymer / solvent system.

We obtain $\chi_{12}{}^{c} = 0.51$ for the case of $X_w{}^0 = 1$ and $Y_w{}^0 = 1 \times 10^4$ and $\chi_{12}{}^{c} = 0.00605$ for the $X_w{}^0 = 100$ and $Y_w{}^0 = 1 \times 10^4$ system, indicating that the phase equilibrium of the polymer 1 / polymer 2 system occur, unless polymer 1 or 2 is much smaller than the others, in an absolutely different $\chi_{12}{}^0$ region from the polymer / solvent system.

The reason why $\chi_{12}{}^{c}$ values are grossly different in these two systems can be understood by a comparison of eqn (4.46) with the equation (eqn (4.118)) for $\chi_{12}{}^{c}$ in a polymer / solvent system, derived by putting $Y_w{}^0 = 1$ and $Y_z{}^0 = 1$ into eqn (4.46),

$$\chi_{12}{}^{c} = \frac{1}{2X_w{}^0}\left(1 + \frac{X_w{}^0}{(X_z{}^0)^{1/2}}\right) (1 + (X_z{}^0)^{1/2}). \qquad (4.118)$$

That is, inspection of eqn (4.46) and eqn (4.118) shows that $\chi_{12}{}^{c}$ for a polymer 1 / polymer 2 system given by eqn (4.46), is $1/Y_w{}^0$ times smaller than $\chi_{12}{}^{c}$ for a polymer / solvent system given by eqn (4.118).

In the case of $X_w{}^0 = Y_w{}^0$, SC becomes symmetrical with respect to the v_2 (or v_1) axis and CSP is located at around $v_2 = 0.5$ (when both polymers have the same MWD (i.e., same MWD type and $X_w{}^0/X_n{}^0 = Y_w{}^0/Y_n{}^0$), $v_1{}^{c} = v_2{}^{c} = 0.5$ is rigorously realized). With a decrease in $X_w{}^0$ and $Y_w{}^0$ under the condition of $X_w{}^0 = Y_w{}^0$, the phase diagram approaches that of a low-molecular weight solvent mixture. In Fig. 4.5, this is more systematically illustrated for a system in which both original polymers have the Schulz-Zimm type distribution. Here, $X_w{}^0$ ($= Y_w{}^0$) is varied in the range of $200 \sim 5 \times 10^4$. In the case of a larger $X_w{}^0$ ($= Y_w{}^0$), SC has the shape of a deep bath tab with a wide flat bottom (i.e., $\chi_{12}{}^0$ of SC remains almost constant over a wide range of v_2) and with a decrease in $X_w{}^0$ ($= Y_w{}^0$), SC gradually becomes sharp. $\chi_{12}{}^c$ for $X_w{}^0 = Y_w{}^0$ is given by

$$\chi_{12}{}^c = 2/X_w{}^0. \tag{4.119}$$

Eqn (4.119) is derived from eqn (4.46) by assuming that both polymers have the same polydispersity. Eqn (4.119) means that when polymer 1 and polymer 2 have the same weight-average DP ($X_w{}^0 = Y_w{}^0$) and the same polydispersity ($X_w{}^0/X_n{}^0 = Y_w{}^0/Y_n{}^0$), $\chi_{12}{}^c$ for the system is reversely proportional to $X_w{}^0$ ($= Y_w{}^0$), rapidly increasing with a decrease in $X_w{}^0$.

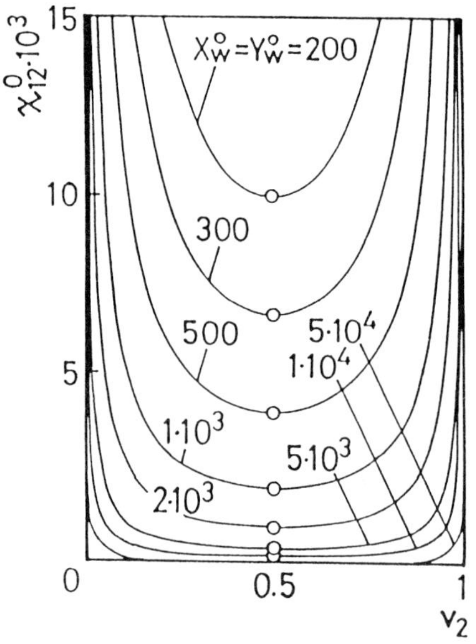

Fig. 4.5 SC and CSP of polymer 1 / polymer 2 system, where the two original polymers have the same weight average degree of polymerization (i.e., $X_w{}^0 = Y_w{}^0$), the same molecular weight distribution (Schulz-Zimm type) and the same polydispersity ($X_w{}^0/X_n{}^0 = Y_w{}^0/Y_n{}^0 = 2$); $X_w{}^0$ ($= Y_w{}^0$), shown on curve; $p_{1,u} = p_{2,u} = 0$; solid line, SC; unfilled circle, CSP.

In the case of $X_w{}^0 \neq Y_w{}^0$, $v_2{}^c$ is located at the higher concentration region of a polymer with a lower molecular weight: For example, in Fig. 4.2a, 4.2b and Fig. 4.3a, 4.3b, where $X_w{}^0 > Y_w{}^0$ always holds, $v_2{}^c$ is larger than 0.5. Note that $v_2{}^c$ depends on the weight- (and Z-) average DP ratio of polymer 1 to polymer 2,

$$v_2{}^c = \frac{1}{[(X_z{}^0/Y_z{}^0)^{1/2}(Y_w{}^0/X_w{}^0)+1]}, \tag{4.120}$$

Eqn (4.120) is simply derived from eqn (4.48). Eqn (4.120) is rewritten, using eqns (4.111) and (4.112), in the forms,

$$v_2{}^c = \frac{1}{[\{\frac{2-(Y_w{}^0/Y_n{}^0)^{-1}}{2-(X_w{}^0/X_n{}^0)^{-1}}\cdot(\frac{Y_w{}^0}{X_w{}^0})\}^{1/2}+1]}$$ for Schulz-Zimm type distribution, (4.121)

$$v_2{}^c = \frac{1}{[\{\frac{(X_w{}^0/X_n{}^0)}{(Y_w{}^0/Y_n{}^0)}\cdot(\frac{Y_w{}^0}{X_w{}^0})\}^{1/2}+1]}$$ for Wesslau type distribution. (4.122)

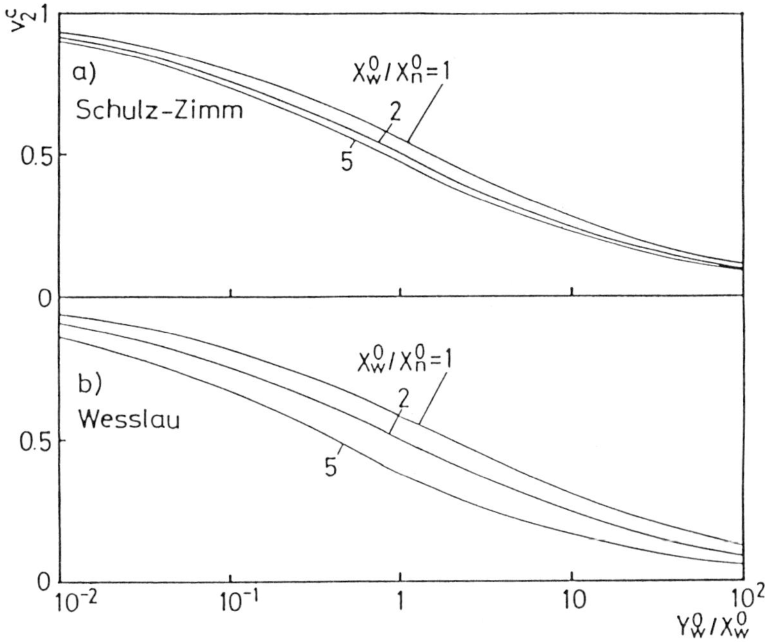

Fig. 4.6 Critical polymer 2 volume fraction $v_2{}^c$ plotted as a function of the ratio of the weight average degree of polymerization of polymer 1 and polymer 2 $Y_w{}^0/X_w{}^0$: Molecular weight distribution of polymer 1 and polymer 2, a) Schulz-Zimm, b) Wesslau; $Y_w{}^0/Y_n{}^0=2$; $X_w{}^0/X_n{}^0$, shown on the curve; $p_{1,u}=p_{2,u}=0$.

Then, it becomes clear that in the case where the polydispersity of the polymers is given in advance, $v_2{}^c$ is a single function of the ratio of the weight-average DP of the two polymers $(Y_w{}^0/X_w{}^0)$ and in addition, for the same $(Y_w{}^0/X_w{}^0)$, the same $v_2{}^c$ is obtained irrespective of $X_w{}^0$ and $Y_w{}^0$.

Fig. 4.6 shows the relationship of $Y_w{}^0/X_w{}^0$ and $v_2{}^c$. Here, the polydispersity of polymer 2, $Y_w{}^0/Y_n{}^0$ is kept constant $(=2)$ and that of polymer 1, $X_w{}^0/X_n{}^0$ is taken as 1, 2 and 5. In the case of $(X_w{}^0/X_n{}^0)=(Y_w{}^0/Y_n{}^0)=2$, $v_2{}^c=0.5$ is realized at $X_w{}^0=Y_w{}^0$ and $v_2{}^c>0.5$ in the range of $(Y_w{}^0/X_w{}^0)<1$ and $v_2{}^c<0.5$ in the range $(Y_w{}^0/X_w{}^0)>1$. In the case of $(X_w{}^0/X_n{}^0)\neq(Y_w{}^0/Y_n{}^0)$, CSP shifts to the direction of increasing the volume fraction of a polymer with a narrower distribution. The effect of polydispersity is more remarkable in the Wesslau type polymer than in the Schulz-Zimm type polymer.

In Figs 4.2 and 4.3, regardless of the type of MWD (Schulz-Zimm or Wesslau), CSP converges to a point, denoted as the unfilled triangle, when $X_w{}^0$ diverges to infinity. $\chi_{12}{}^0$ at this point corresponds clearly to Flory theta temperature of polymer / solvent system and denoted hereafter as $\chi_{12}{}^\theta$.

Fig. 4.7 shows the dependence of $X_w{}^0/X_n{}^0$ on $v_1{}^c$, $v_2{}^c$ and $\chi_{12}{}^c$ for a given combination of $X_w{}^0$ and $Y_w{}^0$. With an increase in $X_w{}^0/X_n{}^0$, $v_1{}^c$ increases and reversely $v_2{}^c\,(=1-v_1{}^c)$ decreases and $\chi_{12}{}^c$ is almost independent of $X_w{}^0/X_n{}^0$ and only dependent on the combination of $X_w{}^0$ and $Y_w{}^0$.

Fig. 4.8 shows $X_w{}^0$-dependence of $v_1{}^c$, $v_2{}^c$ and $\chi_{12}{}^c$. In the figure, the solid line is where polymers 1 and 2 have a Schulz-Zimm type MWD and the broken line, the case where polymers 1 and 2 have a Wesslau type MWD. In all cases, as $X_w{}^0$ is increased, both $v_1{}^c$ and $\chi_{12}{}^c$ decrease monotonically and $v_2{}^c$ increases. At the limit of infinite molecular weight (i.e., $X_w{}^0\to\infty$), $v_1{}^c\to 0$, $v_2{}^c\to 1$ and $\chi_{12}{}^c\to\chi_{12}{}^\theta$, where $\chi_{12}{}^\theta$ is given by eqn (4.123), one derived directly from eqn (4.37).

$$\chi_{12}{}^\theta=\frac{1}{2Y_w{}^0}. \tag{4.123}$$

For example, for $Y_w{}^0=50$, $\chi_{12}{}^\theta$ is estimated to be 0.01 and for $Y_w{}^0=1\times10^4$, $\chi_{12}{}^\theta$ is 5×10^{-5} as shown by the dotted line in the figure.

The sign of $\chi_{12}{}^0$ on SC, $\chi_{12}{}^0{}_{SP}$ is evidently determined by the sign of the denominator S in eqn (4.108),

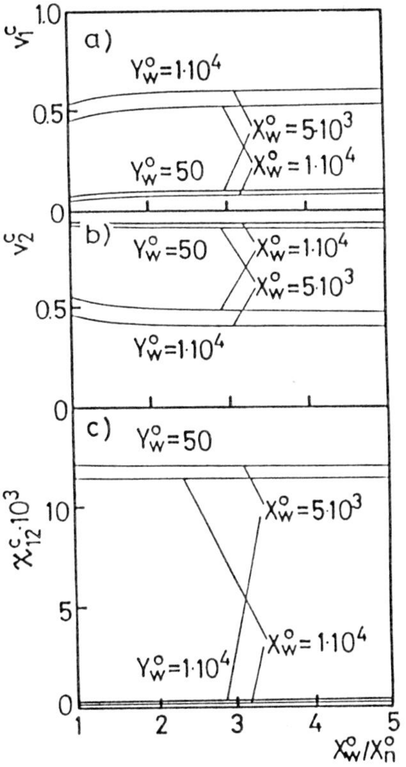

Fig. 4.7 Plots of $v_1{}^c$, $v_2{}^c$ and $\chi_{12}{}^c$ against $X_w{}^0/X_n{}^0$ of quasi-binary solutions consisting of multicomponent polymers 1 and 2: Original polymers 1 and 2, Schulz-Zimm type distribution; $Y_w{}^0/Y_n{}^0 = 2$; $Y_w{}^0 = 50$ and 1×10^4; $X_w{}^0 = 5 \times 10^3$ and 1×10^4; $p_{1,u} = p_{2,u} = 0$.

$$S = 2\{1 + \sum_{u=1}^{n_u} (p_{1,u} v_1{}^u + p_{2,u} v_2{}^u)\}$$

$$+ 2(v_1 - v_2) \sum_{u=1}^{n_u} u(p_{1,u} v_1{}^{u-1} - p_{2,u} v_2{}^{u-1})$$

$$- v_1 v_2 \sum_{u=1}^{n_u} u(u-1)(p_{1,u} v_1{}^{u-2} + p_{2,u} v_2{}^{u-2}) . \qquad (4.124)$$

In other words, $\chi_{12}{}^0{}_{SP} > 0$ at $S > 0$ and $\chi_{12}{}^0{}_{SP} < 0$ at $S < 0$. Note

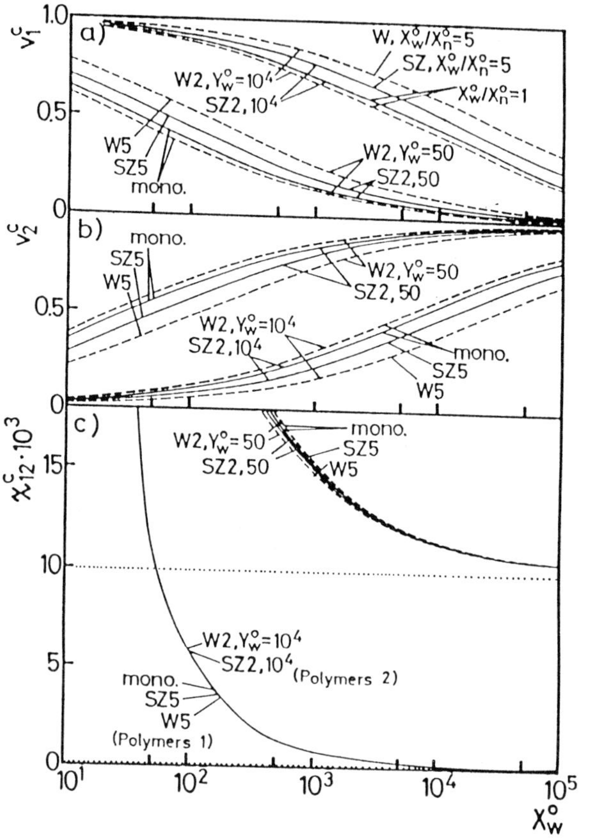

Fig. 4.8 Plots of v_1^c [a)], v_2^c [b)] and χ_{12}^c [c)] against X_w^0 of quasi-binary solutions consisting of multicomponent polymers 1 and 2: Original polymer 1, Schulz-Zimm, $X_w^0/X_n^0=5$ (SZ5), Wesslau type distribution $X_w^0/X_n^0=5$ (W5) or monodisperse polymers ($X_w^0/X_n^0=1$); Original polymer 2, Schulz-Zimm $Y_w^0/Y_n^0=2$ (SZ2) (solid line) or Wesslau $Y_w^0/Y_n^0=2$ (W2) (broken line) type distribution, $Y_w^0=50$ or 1×10^4; dotted line, $\chi_{12}^c=\chi_{12}^0$.

that $\chi_{12}^0{}_{sp}$ becomes plus and minus infinitive at singular point $S=0$.

If $p_{1,1}\neq 0$, $p_{2,1}=p_{1,u}=p_{2,u}=0$ $(u\geq 2)$, eqn (4.124) reduces to

$$S=2(1+2p_{1,1}-3p_{1,1}v_2). \qquad (4.125)$$

Then, the range of $p_{1,1}$, in which $S\neq 0$ is realized over a whole range of v_2 $(0<v_2<1)$, is calculated to be $-0.5\sim1.0$. This range is actually observed. We cannot always predict with complete confidence the existence of the singular point because eqn (4.49) is only a phenomenological equation.

Fig. 4.9 shows the effect of $p_{1,1}$ on SC and CSP for polymer 1 ($X_w^0 = 300$, $X_w^0/X_n^0 = 2$) / polymer 2 ($Y_w^0 = 300$, $Y_w^0/Y_n^0 = 2$) system. Here, $p_{1,1}$ was varied from 1.0 to -0.5 and all concentration-dependence parameters other than $p_{1,1}$ were taken as zero. The molecular weight distribution type for polymer 1 and polymer 2 were the same (i.e., Schulz-Zimm or Wesslau types). With an increase in $p_{1,1}$ the two-phase separation region gradually expands and SC shifts to the higher concentration side of polymer 2 at $p_{1,1} < 0$ and to the lower concentration side at $p_{1,1} > 0$. In other words, if $p_{1,1}$ is low, polymer 1 dissolves into polymer 2, even in the large χ_{12}^0 region (i.e., at a low temperature), to a high concentration. v_2^c is 0.5 at $p_{1,1} = 0$, irrespective of the molecular weight distribution type, but $v_2^c > 0.5$ at $p_{1,1} < 0$ and $v_2^c < 0.5$ at $p_{1,1} > 0$. Note that CSP shifts from the peak point of SC if $p_{1,1} \neq 0$.

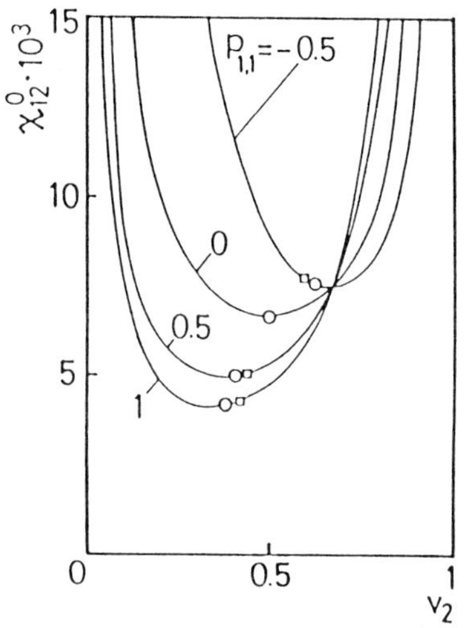

Fig. 4.9 Effect of the concentration-dependence parameter on SC and CSP: Molecular weight distribution type of original polymers 1 and 2, Schulz-Zimm or Wesslau type distribution; $X_w^0/X_n^0 = Y_w^0/Y_n^0 = 2$, $X_w^0 = Y_w^0 = 300$; $p_{1,1} = -0.5$, 0, 0.5 and 1, $p_{1,2} = p_{1,3} = \cdots = 0$, $p_{2,1} = p_{2,2} = \cdots = 0$; ○, CSP in the case when the molecular weight distribution type of original polymers 1 and 2 is Schulz-Zimm; □, CSP in the case when the molecular weight distribution type of original polymers 1 and 2 is Wesslau type distribution.

In Fig. 4.9 all SC, obtained for a variety of $p_{1,1}$, intercept at a common point. This can be interpreted by the equation of spinodal conditions: By putting all concentration-dependence parameters other than $p_{1,1}$ in eqn (4.108) to zero, $\chi_{12}{}^0{}_{SP}$ (i.e., $\chi_{12}{}^0$ on SC) can be given by

$$\chi_{12}{}^0{}_{SP} = \left[\frac{1}{(1-v_2)X_w{}^0} + \frac{1}{v_2 Y_w{}^0}\right] / 2\left[1 + p_{1,1}(2-3v_2)\right]. \tag{4.126}$$

In eqn (4.126), the denominator becomes 2 at $v_2 = 2/3$, irrespective of $p_{1,1}$. Then, SC for the systems consisting of the same $X_w{}^0$ and $Y_w{}^0$ is a function of variables v_2 and $p_{1,1}$, passing through a single point (in this case, $v_2 = 2/3$ and $\chi_{12}{}^0 = 0.0075$).

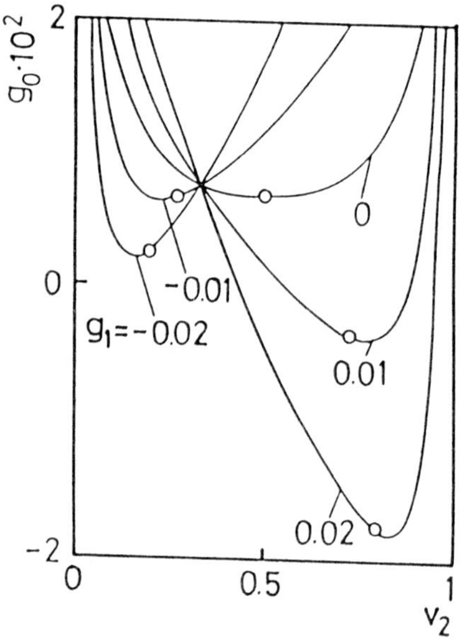

Fig. 4.10 Effect of the concentration-dependence parameter on SC and CSP obtained by Koningsveld et al.'s method: Original polymers 1 and 2, Schulz-Zimm type distribution; $X_w{}^0 = Y_w{}^0 = 300$, $X_w{}^0/X_n{}^0 = Y_w{}^0/Y_n{}^0 = 2$; $g_1 = -0.02$, -0.01, 0, 0.01 and 0.02, $g_2 = g_3 = \cdots = 0$; ○, CSP.

Fig. 4.10 shows the effect of g_1 on SC and CSP, evaluated by the Koningsveld et al.'s method (eqns (4.60) and (4.61)), for polymer 1 ($X_w{}^0 = 300$, $X_w{}^0/X_n{}^0 = 2$) / polymer 2 ($Y_w{}^0 = 300$, $Y_w{}^0/Y_n{}^0 = 2$) system. Here, $g_2 = g_3 = \cdots = g_{nu} = 0$ was assumed. An interception of all SC, obtained for various g_1, at a single point was also observed and this phenomenon can be explained as follows: Eqn (4.60) can be rewritten, by putting $g_2 = g_3 = \cdots = 0$, in the form,

$$g_0 = \frac{1}{2}\left(\frac{1}{v_1 X_w{}^0} + \frac{1}{v_2 Y_w{}^0}\right) + g_1 (1 - 3v_2) \,. \tag{4.127}$$

Eqn (4.127) indicates that all SC for a given combination of $X_w{}^0$ and $Y_w{}^0$ pass through a single point at $v_2 = 1/3$, irrespective of g_1. With an increase in g_1, $v_2{}^c$ increases progressively and g_0 at CSP (referred to as $g_0{}^c$) increases in the range of $g_1 < 0$, then decreases in the range of $g_1 > 0$ after passing through a maximum at $g_1 = 0$.

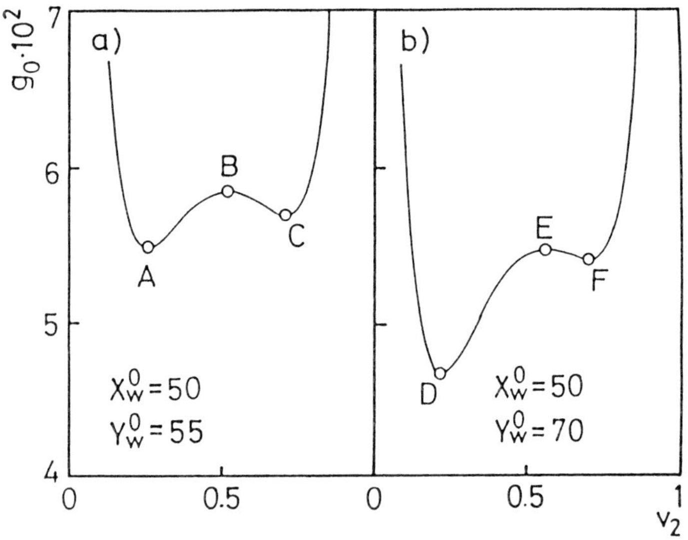

Fig. 4.11 Recalculation results of SC and CSP of polymer 1 / polymer 2 system using the Koningsveld et al.'s method: Original polymers 1 and 2, monodisperse; $X_w{}^0 = 50$; a) $Y_w{}^0 = 55$, b) $Y_w{}^0 = 70$; $g_1 = -0.04$, $g_2 = 0.04$, $g_3 = g_4 = \cdots = 0$; ○, CSP.

Fig. 4.11 shows SC and CSP, calculated from eqns (4.60) and (4.61) using the conditions employed in Koningsveld et al.'s paper (ref. 4) (polymer 1 and 2, monodisperse polymer; $X_w{}^0 = 50$, $Y_w{}^0 = 55$ a) or 70 b), $g_1 = -0.04$, $g_2 = 0.04$). The results on SC and CSP, recalculated here, are in excellent agreement with those of Fig. 4.4 of ref. 4, which were shown without detailed explanations of the calculation procedure. Both Fig. 4.11a and b show the existence of double-peaked SC and triple CSP [for $X_w{}^0 = 50$ and $Y_w{}^0 = 55$, $(v_2{}^c, g_0{}^c) = (0.262, 0.0547)$ A, (0.521, 0.0582) B and (0.719, 0.0567) C; for $X_w{}^0 = 50$ and $Y_w{}^0 = 70$, $(v_2{}^c, g_0{}^c) = (0.217, 0.0465)$ D, (0.561, 0.0546) E and (0.698, 0.0539) F].

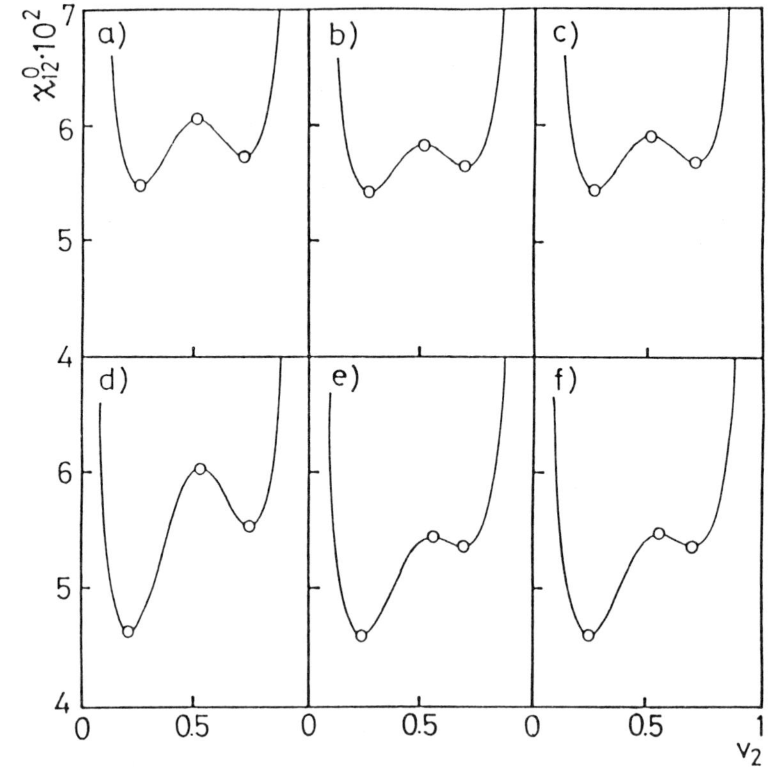

Fig. 4.12 SC and CSP calculated using concentration-dependence parameters $p_{2,1}$ and $p_{2,2}$ obtained from CSP in Fig. 4.11 under the same condition as Fig. 4.11. Fig. 4.12a-f correspond to the CSP of A-F in Fig. 4.11 respectively: Original polymers 1 and 2, monodisperse; $X_w{}^0=50$; a)-c) $Y_w{}^0=55$, d)-f) $Y_w{}^0=70$; a) $p_{2,1}=-0.7319$, $p_{2,2}=0.7319$, b) $p_{2,1}=-0.6870$, $p_{2,2}=0.6870$, c) $p_{2,1}=-0.7054$, $p_{2,2}=0.7054$, d) $p_{2,1}=-0.8608$, $p_{2,2}=0.8608$, e) $p_{2,1}=-0.7623$, $p_{2,2}=0.7623$, f) $p_{2,1}=-0.7416$, $p_{2,2}=0.7416$; $p_{1,1}=p_{1,2}=\cdots=0$, $p_{2,3}=p_{2,4}=\cdots=0$; ○, CSP.

Next, we determined $p_{1,1}$, $p_{1,2}$, $p_{2,1}$ and $p_{2,2}$, corresponding to g_1, g_2 employed in calculation of SC and CSP in Fig. 4.11 by using eqns (4.65), (4.66) and using these $p_{i,j}$ (here $p_{1,1}=p_{1,2}=0$), we calculated SC and CSP by our method (eqns (4.57) and (4.58)) for the same system as in Fig. 4 of ref. 4. Note that these SC and CSP can be calculated corresponding to three CSP in Fig. 4.11: Fig. 4.12a ($p_{2,1}=-0.7319$, $p_{2,2}=0.7319$), 4.12b ($p_{2,1}=-0.6870$, $p_{2,2}=0.6870$) and 4.12c ($p_{2,1}=-0.7054$, $p_{2,2}-0.7054$) are constructed using $g_0{}^c$ at the points A, B and C in Fig. 4.11a and Fig. 4.12d ($p_{2,1} = -0.8608$, $p_{2,2} = 0.8608$), 4.12e ($p_{2,1} = -0.7623$, $p_{2,2} = 0.7623$) and 4.12f ($p_{2,1} = -0.7416$, $p_{2,2} = 0.7416$) are constructed using $g_0{}^c$ at the points D, E and F in Fig. 4.11b. Evidently, Fig. 4.12a, b and c coincide with Fig. 4.11a and Fig. 4.12d, e and f with Fig. 4.11b. Therefore it is confirmed that in calculation of SC and CSP, Kamide et al.'s method (eqns (4.57) and (4.58)) is equivalent to the Koningsveld et al.'s method (eqns (4.60) and (4.61)), both are mutually exchangeable and give the same results.

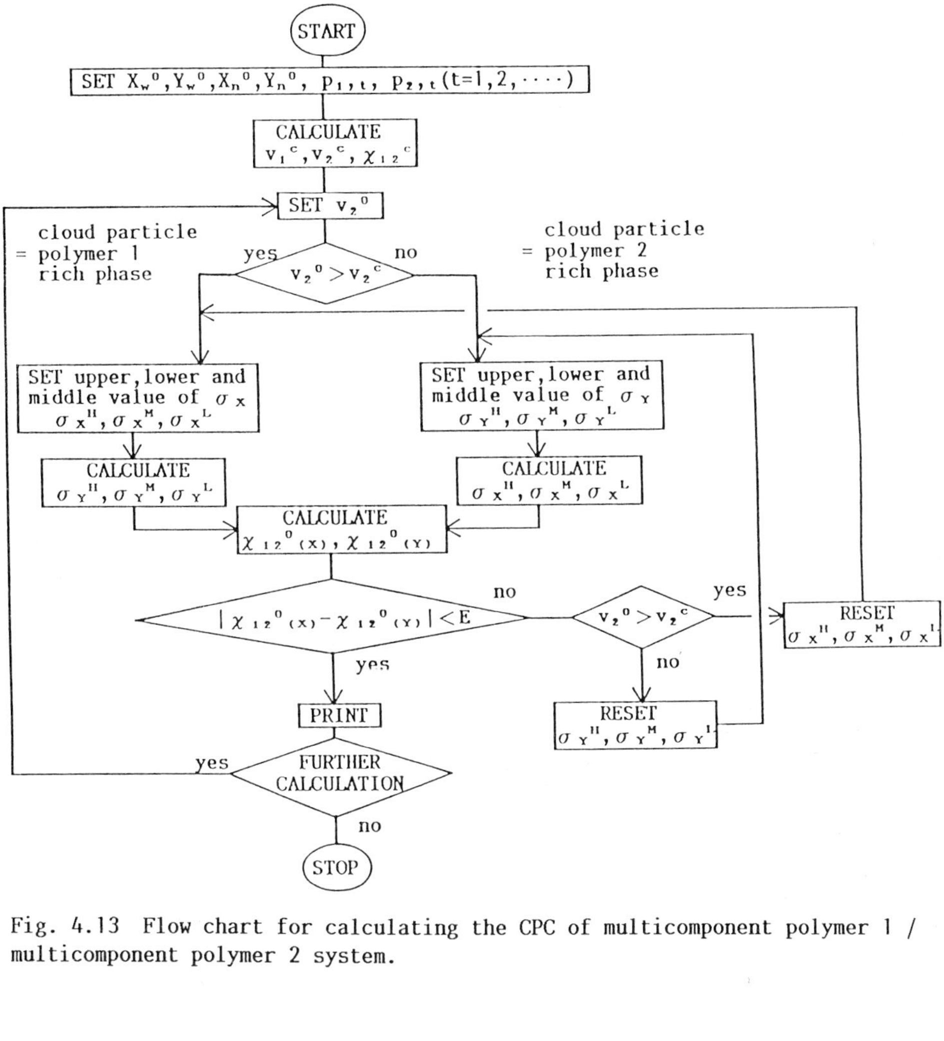

Fig. 4.13 Flow chart for calculating the CPC of multicomponent polymer 1 / multicomponent polymer 2 system.

4.4.2 Cloud point curve (ref. 7)

Fig. 4.13 shows an outline of a flow chart for a computer experiment on CPC for multicomponent polymer 1 / multicomponent polymer 2 systems. The simulation process is divided into 6 steps:

1) Set the following parameters in advance;
   (a) the normalized distribution of DP of the two original polymers, $g_0(X_i)$ and $g_0(Y_j)$ and accordingly, their number- and weight-average DP, $X_n{}^0$, $Y_n{}^0$, $X_w{}^0$ and $Y_w{}^0$,
   (b) the concentration-dependence parameters $p_{1,u}$ and $p_{2,u}$ $(u=1,\cdots,n_u)$.
2) Calculate CSP ($v_1{}^c$, $v_2{}^c$ and $\chi_{12}{}^c$) using eqns (4.57) and (4.58).
3) Set the initial volume fractions of polymer 1 and polymer 2, $v_1{}^0$ and $v_2{}^0$. In the case of $v_1{}^0 < v_1{}^c$ ($v_2{}^0 > v_2{}^c$), set three values of σ_X under the condition of $0 < v_{1(1)} < 1$ and denote, from the larger value, as $\sigma_X{}^H$, $\sigma_X{}^M$ and $\sigma_X{}^L$. Calculate σ_Y corresponding to $\sigma_X{}^H$, $\sigma_X{}^M$ and $\sigma_X{}^L$ using eqn (4.104); $\sigma_Y{}^H$, $\sigma_Y{}^M$ and $\sigma_Y{}^L$. In the case of $v_1{}^0 > v_1{}^c$ ($v_2{}^0 < v_2{}^c$), set three values of σ_Y under condition of $0 < v_{2(2)} < 1$ and denote, from the larger value, as $\sigma_Y{}^H$, $\sigma_Y{}^M$ and $\sigma_Y{}^L$. Calculate three σ_X ($\sigma_X{}^H$, $\sigma_X{}^M$ and $\sigma_X{}^L$) corresponding to three σ_Y by use of eqn (4.103).
4) Calculate, by eqn (4.106), three F (F^H, F^M, F^L) for three combinations of (σ_X, σ_Y). In the case of $v_1{}^0 < v_1{}^c$, reset ($\sigma_X{}^H$, $\sigma_X{}^L$) by ($\sigma_X{}^H$, $\sigma_X{}^M$) for $F^H \cdot F^M < 0$ and reset ($\sigma_X{}^H$, $\sigma_X{}^L$) by ($\sigma_X{}^M$, $\sigma_X{}^L$) for $F^M \cdot F^L < 0$. In the case of $v_1{}^0 > v_1{}^c$, reset ($\sigma_Y{}^H$, $\sigma_Y{}^L$) by ($\sigma_Y{}^H$, $\sigma_Y{}^M$) for $F^H \cdot F^M < 0$ and reset ($\sigma_Y{}^H$, $\sigma_Y{}^L$) by ($\sigma_Y{}^M$, $\sigma_Y{}^L$) for $F^M \cdot F^L < 0$.
5) Repeat steps 3 and 4 until the absolute magnitude of $F(|F|)$ becomes less than 1×10^{-10} ($\equiv E$), which is sufficiently small so that eqn (4.106) is regarded as solved by the interhalving method. σ^X (σ^Y) in this case is $\sigma_X{}^{CP}$ ($\sigma_Y{}^{CP}$).
6) Calculate $v_{(k)}{}^{CP}(X_i)$ and $v_{(k)}{}^{CP}(Y_j)$ ($k=1$ or 2) by putting $\sigma_X{}^{CP}$ and $\sigma_Y{}^{CP}$ estimated in step 5 into eqns (4.98a)-(4.99b) and eqns (4.100a)-(4.101b) and estimate $X_n{}^{CP}{}_{(k)}$, $Y_n{}^{CP}{}_{(k)}$, $v_1{}^{CP}{}_{(k)}$ and $v_2{}^{CP}{}_{(k)}$ ($k=1$ or 2) using eqns (4.76a)-(4.76d) from $v_{(k)}{}^{CP}(X_i)$ and $v_{(k)}{}^{CP}(Y_j)$ ($k=1$ or 2) and $\chi_{12}{}^{CP}$ using eqn (4.73) (or eqn (4.74)) from $X_n{}^{CP}{}_{(k)}$, $Y_n{}^{CP}{}_{(k)}$, $v_1{}^{CP}{}_{(k)}$ and $v_2{}^{CP}{}_{(k)}$ ($k=1$ or 2).

Print out the results. Start a new calculation (i.e., repeat steps 3-6) for another initial concentration.

The original polymers were assumed to have the following distributions of DP [$g_0(X_i)$ and $g_0(Y_j)$]:

(i) Schulz-Zimm (SZ) type distribution
(ii) Wesslau (W) type distribution

Under the condition of X_w^0 (and Y_w^0) $= 50 \sim 1 \times 10^7$, X_w^0/X_n^0 (and Y_w^0/Y_n^0) $= 1 \sim 5$, $p_{1,1} = -0.4 \sim 1.0$, computer experiments were carried out in order to calculate CPC and the shadow curve. For comparison with CPC and the shadow curve, SC and CSP were also calculated using eqns (4.57) and (4.58).

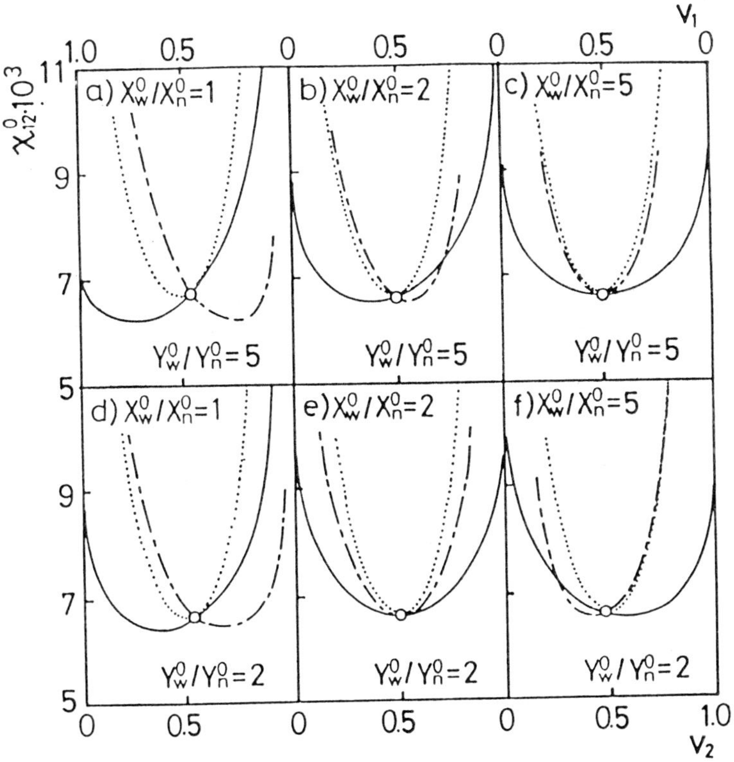

Fig. 4.14 Effect of polydispersity expressed by X_w^0/X_n^0 and Y_w^0/Y_n^0 of the polymers on the CPC and shadow curve of quasi-binary system consisting of multicomponent polymers 1 and 2: Original polymer, Schulz-Zimm type distribution; $X_w^0 = Y_w^0 = 300$, $Y_w^0/Y_n^0 = 5$ [a), b) and c)]; $Y_w^0/Y_n^0 = 2$ [d), e) and f)]; a) and d), $X_w^0/X_n^0 = 1$; b) and e), $X_w^0/X_n^0 = 2$; c) and f), $X_w^0/X_n^0 = 5$; solid line, CPC; chain line, shadow curve; dotted line, SC; open circle, CSP; $p_{1,u} = p_{2,u} = 0$.

Fig. 4.14 shows the effect of polydispersity of the original polymers, having SZ type distributions, on CPC, shadow curve, SC and CSP. Here, $X_w{}^0$ and $Y_w{}^0$ are taken as 300 and the calculations were carried out for all the combinations of $X_w{}^0/X_n{}^0 = 1$, 2 and 5 and $Y_w{}^0/Y_n{}^0 = 2$ and 5. The concentration dependence for χ_{12} was neglected. In the case where polymer 1 and polymer 2 have the same polymolecularity (i.e., $X_w{}^0/X_n{}^0 = Y_w{}^0/Y_n{}^0 = 5$ c) and 2 e)), CPC, shadow curve and SC are symmetrical with respect to v_1 (accordingly, v_2) and all these curves contact at their vertexes, which coincide with a single point (CSP), and have the same common tangential line. Comparison of Fig. 4.14c with 4.14e reveals that SC is absolutely independent of polymer polymolecularity, but CPC for $X_w{}^0/X_n{}^0 = Y_w{}^0/Y_n{}^0 = 2$ is slightly sharper than that for $X_w{}^0/X_n{}^0 = Y_w{}^0/Y_n{}^0 = 5$. In the case where the polydispersities of the two polymers, expressed by $X_w{}^0/X_n{}^0$ and $Y_w{}^0/Y_n{}^0$, are different, CPC as well as the shadow curve are unsymmetrical: For $X_w{}^0/X_n{}^0 < Y_w{}^0/Y_n{}^0$ (Fig. 4.14a, b and d), CPC shifts to the side of polymer 1, whose breadth in the molecular weight distribution (MWD) is narrower, and the shadow curve moves to the side of polymer 2, whose MWD breadth is broader than polymer 1. For $X_w{}^0/X_n{}^0 > Y_w{}^0/Y_n{}^0$ (Fig. 4.14f), the directions of the shift of CPC and shadow curves become opposite as compared with the case of $X_w{}^0/X_n{}^0 < Y_w{}^0/Y_n{}^0$. Even in these cases ($X_w{}^0/X_n{}^0$ / $Y_w{}^0/Y_n{}^0$), CPC crossed with SC at CSP.

Fig. 4.15 shows the effect of the types and breaths of MWD of the original polymers on CPC (Fig. 4.15a-d) and the shadow curve (Fig. 4.15e-h). In the figure, SC and CSP are also shown for comparison. Fig. 4.15a and e, Fig. 4.15b and f, Fig. 4.15c and g, and Fig. 4.15d and h make pairs, respectively. The combinations of MWD type of original polymers are: (polymer 1, polymer 2) = (SZ, SZ) in Fig. 4.15a and e, (W, SZ) in Fig. 4.15b and f, (SZ, W) in Fig. 4.15c and g, (W,W) in Fig. 4.15d and h. All the calculations were performed under the condition of $X_w{}^0 = Y_w{}^0 = 300$, $Y_w{}^0/Y_n{}^0 = 2$ and $X_w{}^0/X_n{}^0 = 1$, 2 and 5. Although SC, as denoted in the figure by the broken line, determined by $X_w{}^0$ and $Y_w{}^0$ alone, CPC (solid line), the shadow curve (chain line) and CSP (unfilled circle) depend remarkably on the types and breadths of MWD of the original polymers: As $X_w{}^0/X_n{}^0$ gets larger, CSP and the vertex of the shadow curve shift to a lower v_2 side and the vertex of CPC (i.e., the threshold point) to a higher v_2 side. The degree of shift is usually larger in W than in SZ. In the case where both polymer 1 and polymer 2 have W-type MWD with the same breadth [(polymer 1,

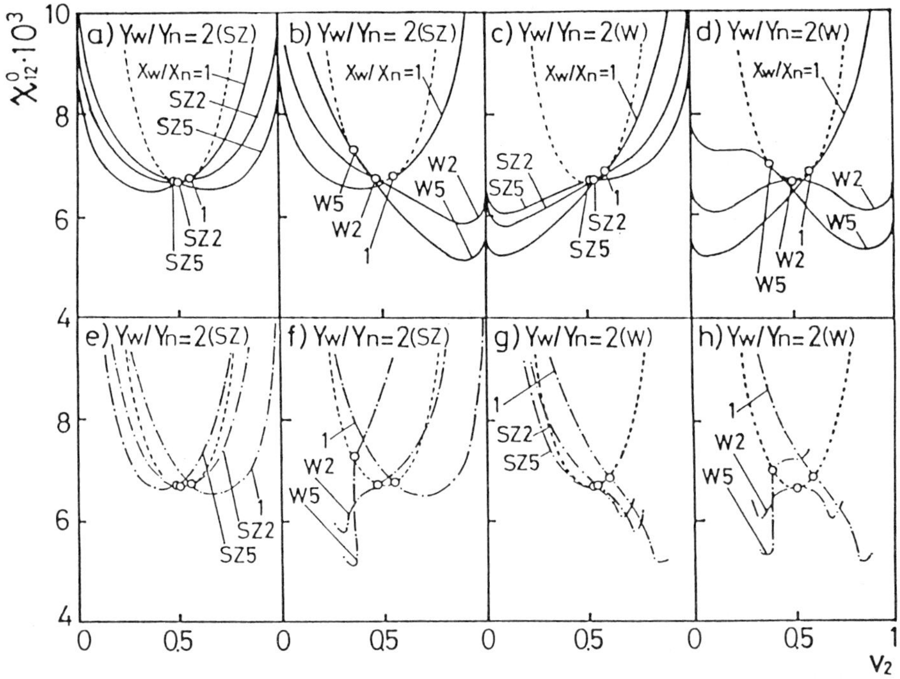

Fig. 4.15 Effect of molecular weight distribution and polydispersity expressed by $X_w{}^0/X_n{}^0$ and $Y_w{}^0/Y_n{}^0$ of the polymers on the CPC [a)-d)] and shadow curve [e)-f)] of quasi-binary system consisting of multicomponent polymers 1 and 2: Original polymer 2, Schulz-Zimm [a), e) and b), f)]; Wesslau [c), g) and d), h)]; $Y_w{}^0/Y_n{}^0=2$; $Y_w{}^0=300$. Original polymer 1, Schulz-Zimm [a), e) and c), g)]; Wesslau [b), f) and d), h)]; $X_w{}^0=300$. Solid line, CPC [a)-d)]; chain line, shadow curve [e)-h)]; broken line, SC; open circle, CSP; $p_{1,u}=p_{2,u}=0$.

polymer 2) = (W2, W2)], CPC and the shadow curve reveal double-peak curves. In the case of (polymer 1, polymer 2) = (W5, W5), CPC and the shadow curve have a larger shoulder.

Fig. 4.16 shows the effect of $X_w{}^0$ and $Y_w{}^0$ on CPC, the shadow curve, SC and CSP. Here, the polydispersities of the two original polymers are taken as the same; $X_w{}^0/X_n{}^0 = Y_w{}^0/Y_n{}^0 = 2$. CPC behaves similar to SC, broadening the two-phase coexisting region with an increase in $X_w{}^0$ and $Y_w{}^0$. The shadow curve lies between CPC and SC. CPC, the shadow curve and SC become unsymmetrical, shifting to the side of the polymer with a smaller DP if $X_w{}^0 \neq Y_w{}^0$. For example, in Fig. 4.16a, $X_w{}^0$ is always larger than $Y_w{}^0$ (= 100) and all the

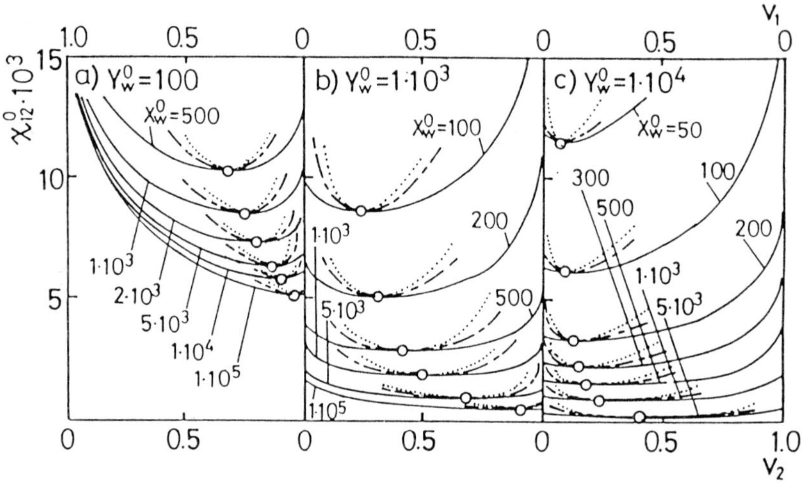

Fig. 4.16 Effect of X_w^0 (and Y_w^0) on the CPC and shadow curve of quasi-binary systems consisting of multicomponent polymers 1 and 2: Original polymer, Schulz-Zimm type distribution; $X_w^0/X_n^0 = Y_w^0/Y_n^0 = 2$; a) $Y_w^0 = 100$, b) $Y_w^0 = 1 \times 10^3$, c) $Y_w^0 = 1 \times 10^4$. Solid line, CPC; chain line, shadow curve; dotted line, SC; open circle, CSP; $p_{1,u} = p_{2,u} = 0$.

curves incline to the polymer 2 side. In other words, the phase diagrams shown in Fig. 4.16a for the polymer blend system are very similar to those for polymer / single solvent system (refs 8,9). Note that the scale of the ordinate axis in Fig. 4.16 is very small, covering only $0 < \chi_{12}^0 < 0.015$ as compared with the range of χ_{12}^0, usually observed in two-phase equilibrium for the polymer / single solvent system. And if the phase diagrams of the polymer 1 / polymer 2 system and the polymer / solvent system are plotted on the same scale of χ_{12}^0, they should have absolutely different shapes.

For a system of binary polymer mixtures having larger average molecular weight, all the curves in question conform to flat and metastable region, included by CPC and SC, becomes narrower.

Fig. 4.17 shows the effect of X_w^0 $(= Y_w^0)$ on CPC, CSP and SC for polymer 1 $(X_w^0/X_n^0 = 2)$ / polymer 2 $(Y_w^0/Y_n^0 = 2)$ systems. Here, both polymers have Schulz-Zimm type distribution and all concentration-dependence parameters are taken as zero. CPC as well as SC is symmetrical, contacting each other at CSP $(v_2^c = 0.5)$ located at the threshold point of SC and CPC. As X_w^0 $(= Y_w^0)$ gets

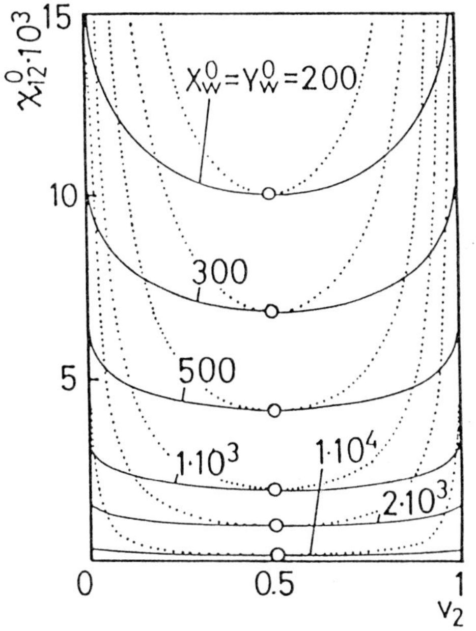

Fig. 4.17 Effect of the weight average degree of polymerization on CPC, SC and CSP in the case of $X_w{}^0=Y_w{}^0$: Original polymers 1 and 2, Schulz-Zimm type distribution; $X_w{}^0/X_n{}^0=Y_w{}^0/Y_n{}^0=2$; $X_w{}^0=Y_w{}^0=200$, 300, 500, 1×10^3, 2×10^3 and 1×10^4; $p_{1,u}=p_{2,u}=0$. Solid line, CPC; dotted line, SC; open circle, CSP.

larger, the two-phase separation region widens to a larger extent, resulting in a lowering of mutual solubility between polymer 1 and polymer 2. And as $X_w{}^0$ ($=Y_w{}^0$) increase, both CPC and SC become flat and accordingly, the metastable region, defined as the region sandwiched between CPC and SC, become narrower rapidly. This predicts that the blend of two high-molecular weight polymers has a high probability of spinodal decomposition. Similar calculations, though not in systematical manner, were made by McMaster (ref. 5) for the monodisperse polymer 1 / monodisperse polymer 2 system.

Fig. 4.18 shows the effect of the concentration-dependence parameter $p_{1,1}$ of χ_{12} on CPC, shadow curve, SC and CSP. Here, the calculations were made under the conditions of $X_w{}^0=Y_w{}^0=300$, $X_w{}^0/X_n{}^0=Y_w{}^0/Y_n{}^0=2$, $p_{1,1}=-0.4\sim1.0$, and other concentration-dependence parameters are zero. As $p_{1,1}$ increases, both $v_2{}^c$ and $\chi_{12}{}^c$ decrease. At $p_{1,1}=0$, (and, of course, $X_w{}^0=Y_w{}^0$) CPC, the

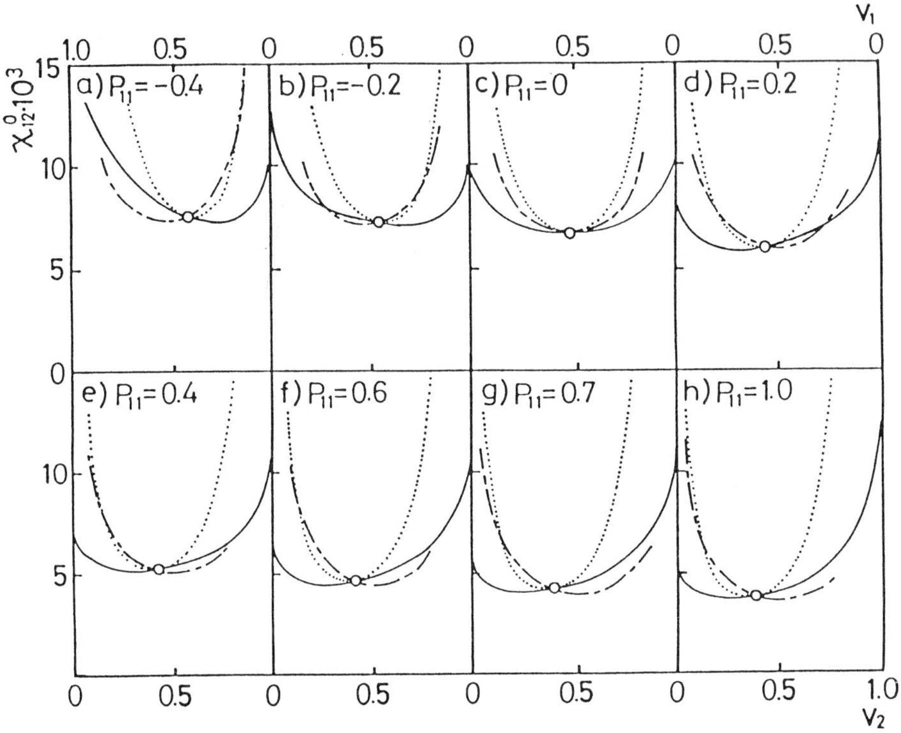

Fig. 4.18 Effect of the concentration dependence of χ_{12} parameter on CPC and shadow curve of quasi-binary systems consisting of multicomponent polymers 1 and 2: Original polymer, Schulz-Zimm type distribution ($X_w{}^0 = Y_w{}^0 = 300$ and $X_w{}^0/X_n{}^0 = Y_w{}^0/Y_n{}^0 = 2$). Solid line, CPC; chain line, shadow curve; dotted line, SC; open circle, CSP. $p_{1,2} = p_{1,3} = \cdots = p_{2,1} = p_{2,2} = \cdots = 0$.

shadow curve and SC are symmetrical. In the case of $p_{1,1} > 0$, CPC shifts to the polymer 1 side and the shadow curve to the polymer 2 side (Fig. 4.18d-h), and in the case of $p_{1,1} < 0$, CPC shifts to the polymer 2 side and the shadow curve to the polymer 1 side (Fig. 4.18a and b).

Fig. 4.19 is another representation showing the concentration-dependence parameter $p_{1,1}$ on CPC, SC and CSP, which are reconstructed from Fig. 4.18. As $p_{1,1}$ increases CPC, SC and CSP shift to smaller a $\chi_{12}{}^0$ region and two-phase separation region becomes wider. In other words, with increasing $p_{1,1}$ the mutual solubility of polymer 1 and polymer 2 decreases. In the case of $p_{1,1} < 0$, CSP shifts to the polymer 1 side of the threshold point and in the case of $p_{1,1} > 0$, CSP shifts to the polymer 2 side of the threshold point. As explained previously, all SC, obtained under the same conditions, except $p_{1,1}$, intercept at the single point.

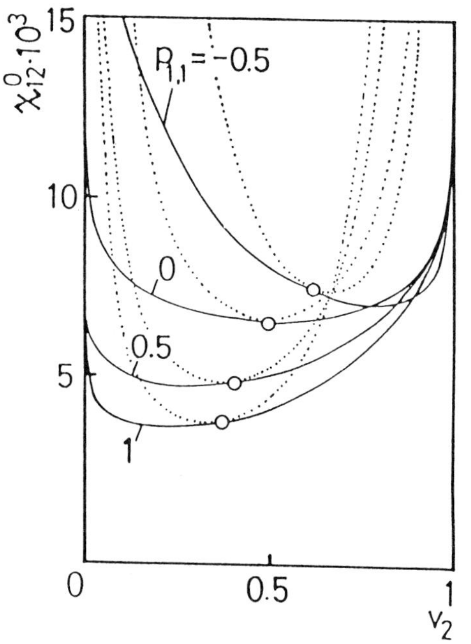

Fig. 4.19 Effect of the concentration dependence of χ_{12} parameter on CPC and SC of quasi-binary systems consisting of multicomponent polymers 1 and 2: Original polymer, Schulz-Zimm type distribution; $X_w{}^0 = Y_w{}^0 = 300$, $X_w{}^0/X_n{}^0 = Y_w{}^0/Y_n{}^0 = 2$; $p_{1,1} = -0.5$, 0, 0.5 and 1; $p_{1,2} = p_{1,3} = \cdots = 0$; $p_{2,1} = p_{2,2} = \cdots = 0$. Solid line, CPC; dotted line, SC; open circle, CSP.

4.5 EXPERIMENTAL DETERMINATION OF POLYMER-POLYMER INTERACTION PARAMETER χ_{12}

In this chapter a parameter, polymer-polymer interaction parameter χ_{12} is newly introduced. Up to now various methods have been proposed for determining the χ_{12} parameter experimentally as follows:

a) Melting point depression (ref. 10)

$$\frac{1}{T_m}-\frac{1}{T_m{}^0}=\frac{\tilde{R}V_{2u}}{4\Delta H_{2u}V_{1u}}\chi_{12}(1-v_2)^2 \tag{4.128}$$

where T_m, the melting point of polymer blend, $T_m{}^0$, the equilibrium melting point of polymer 2, V_{1u} and V_{2u}, the molar volume of the repeating unit of polymer 1 and 2, respectively, ΔH_{2u}, the enthalpy of fusion per mol of repeating unit of polymer 2. From the plot of $1/T_m - 1/T_m{}^0$ vs. $(1-v_2)^2$, χ_{12} can be evaluated, provided that other quantities such as V_{1u}, V_{2u} and H_{2u} are known in advance.

b) Vapor sorption (ref. 11)

$$\ln a_1 = \ln v_0 + (1-v_0) + (\chi_{01}v_1 + \chi_{02}v_2)(1-v_0) - \chi_{12}v_1v_2 \tag{4.129}$$

where a_1 the activity of the solvent in the solution of polymer 1 / polymer 2 / solvent, χ_{01}, the interaction parameter between solvent and polymer 1, χ_{02}, the interaction parameter between solvent and polymer 2. χ_{01} and χ_{02} can be determined by vapor sorption method with combination of eqn (4.129) with following equation.

$$a_1 = \ln v_0 + v_1 + \chi_{01}v_1{}^2 \tag{4.130}$$

or

$$a_1 = \ln v_0 + v_2 + \chi_{02}v_2{}^2 \tag{4.130'}$$

for the system of polymer 1 (or polymer 2) in solvent.

c) Inverse-phase gas chromatography (ref. 12)

Interaction parameter between solvent and polymer 1, χ_{01}, is given by

$$\chi_{01} = \ln\frac{273.2\tilde{R}\bar{v}_1}{P_0{}^0 V_g{}^0 V_0} - \left(1 - \frac{V_1}{V_0}\right) - \frac{P_0{}^0}{\tilde{R}T}(B_{11} - V_0) \tag{4.131}$$

where $\bar{v}_1$, the specific volume of polymer, $P_0{}^0$, the vapor pressure of the probe (solvent) in the column, $V_g{}^0$, the specific retention volume of the probe, V_0 and V_1, the molar volume of probe and polymer, respectively, B_{11}, second virial coefficient of probe in the vapor phase. Probe is typically a small volatile molecule and eqn (4.131) gives the interaction parameter at infinite dilution. Interaction parameter between solvent and polymer 2, χ_{02} is given in the same manner.

Interaction parameter between the probe and the polymer blend, $\chi_{0(12)}$ is given by

$$\chi_{0(12)} = \ln\frac{273.2\tilde{R}(w_1\bar{v}_1 + w_2\bar{v}_2)}{P_0{}^0 V_g{}^0 V_0} - \left(1 - \frac{V_0}{V_1}\right)v_1 - \left(1 - \frac{V_0}{V_2}\right)v_2 - \frac{P_0{}^0}{\tilde{R}T}(B_{11} - V_0) \tag{4.132}$$

where w_1 and w_2 are the weight fraction of polymer 1 and 2, respectively, V_2 is the molar volume of polymer 2. From eqns (4.131) and (4.132), the interaction parameter between polymer 1 and 2 is given by the relation:

$$\chi_{0(12)} = v_1\chi_{01} + v_2\chi_{02} - v_1 v_2 \chi_{12} V_1/V_2 . \tag{4.133}$$

d) Neutron scattering (ref. 13)

In the small-angle neutron scattering method for determining the χ_{12}, the deuterated polymer is needed and it acts as solute dispersed in solvent which is undeutrated polymer. The second virial coefficient A_2 is determined by

$$\frac{Kv_2}{R(s)} = \frac{1}{M_2 P(s)} + 2A_2 w_2 \tag{4.134}$$

where R(s), the Rayleigh ratio, P(s), the intermolecular function, $s = (4\pi\lambda\sin(\theta/2))$, θ is the scattering angle, λ, the wave-length of the neutron radiation, w_2, the weight fraction of the deuterated polymer, M_2, the molecular weight of the deuterated polymer. χ_{12} is determined from A_2.

e) Coexistence curve (ref. 14)

$$\chi_{12}=\frac{\frac{1}{V_1}\ln\frac{v_1'}{v_1}+\frac{1}{V_2}\ln\frac{v_2'}{v_2}+(\chi_{02}-\chi_{01})(v_0'-v_0)}{(v_2'-v_2)+(v_1'-v_1)} \tag{4.135}$$

where V_1 and V_2 are the molar volume of polymer 1 and 2, respectively, v_0, v_1 and v_2 are the volume fraction of solvent, polymer 1 and polymer 2 in the first phase, respectively, v_0', v_1' and v_2' are the volume fraction of the second phase. χ_{01} and χ_{02} are obtained from independent measurements on the binary systems.

Although eqn (4.135) is the equation for solution of monodisperse polymers and solvent, the same equation can also be obtained for solution of multicomponent polymers and solvent. Substitution of eqns (5.8) and (5.9) into eqns (4.136) and (4.137), respectively, gives eqns (4.138) and (4.139).

$$\Delta\mu_{Xi(1)}=\Delta\mu_{Xi(2)}, \tag{4.136}$$

$$\Delta\mu_{Yj(1)}=\Delta\mu_{Yj(2)}, \tag{4.137}$$

$$\begin{aligned}\sigma_{Xi}= &(v_{1(1)}-v_{1(2)})+(v_{2(1)}-v_{2(2)})\\ &+(\frac{v_{1(2)}}{X_{n(2)}}-\frac{v_{1(1)}}{X_{n(1)}})+(\frac{v_{2(2)}}{Y_{n(2)}}-\frac{v_{2(1)}}{Y_{n(1)}})\\ &-\chi_{12}\{(v_{2(2)}-v_{2(1)})-(v_{1(2)}v_{2(2)}-v_{1(1)}v_{2(1)})\}\\ &-\chi_{01}\{(v_{0(2)}-v_{0(1)})-(v_{0(2)}v_{1(2)}-v_{0(1)}v_{1(1)})\}\\ &+\chi_{02}(v_{0(2)}v_{2(2)}-v_{0(1)}v_{2(1)}),\end{aligned} \tag{4.138}$$

$$\begin{aligned}\sigma_{Yj}= &-(v_{1(1)}-v_{1(2)})-(v_{2(1)}-v_{2(2)})\\ &-(\frac{v_{1(2)}}{X_{n(2)}}-\frac{v_{1(1)}}{X_{n(1)}})-(\frac{v_{2(2)}}{Y_{n(2)}}-\frac{v_{2(1)}}{Y_{n(1)}})\\ &+\chi_{12}\{(v_{1(2)}-v_{1(1)})-(v_{1(2)}v_{2(2)}-v_{1(1)}v_{2(1)})\}\\ &+\chi_{02}\{(v_{0(2)}-v_{0(1)})-(v_{0(2)}v_{2(2)}-v_{0(1)}v_{2(1)})\}\end{aligned}$$

$$- \chi_{01}(v_{0(2)}v_{1(2)} - v_{0(1)}v_{1(1)}), \quad (4.139)$$

where the suffix (1) means polymer 1 lean-phase (= polymer 2 rich-phase) and the suffix (2) means polymer 1 rich-phase (= polymer 2 lean-phase), σ_{Xi} and σ_{Yj} are the partition coefficient of polymer 1 and 2, respectively, defined by following formula:

$$\sigma_{Xi} \equiv \frac{1}{X_i} \ln \frac{v_{Xi(2)}}{v_{Xi(1)}}, \quad (4.140)$$

$$\sigma_{Yj} \equiv \frac{1}{Y_j} \ln \frac{v_{Yj(1)}}{v_{Yj(2)}}. \quad (4.141)$$

Summation of eqns (4.138) and (4.139) gives eqn (4.142).

$$\chi_{12} = \frac{\sigma_{Xi} + \sigma_{Yj} + (\chi_{01} - \chi_{02})(v_{0(2)} - v_{0(1)})}{(v_{1(2)} - v_{1(1)}) - (v_{2(2)} - v_{2(1)})} \quad (4.142)$$

Eqn (4.142) reduces to eqn (4.135) when the two polymers are monodisperse.

f) Light scattering (ref. 15)

The second virial coefficient A_2, obtained from light scattering measurement on polymer 1 / solvent system, is related to the interaction parameter χ_{01} between polymer 1 and solvent as follows:

$$A_2 = \frac{\bar{v}_1}{V_0}\left(\frac{1}{2} - \chi_{01}\right) \quad (4.143)$$

where $\bar{v}_1$ is the specific volume of polymer and V_0 is the molar volume of solvent. In the same manner, the interaction parameter χ_{02} between polymer 2 and solvent can also be obtained. Light scattering measurement on polymer 1 / polymer 2 / solvent system gives A_2 of this system, and the interaction parameter χ_{12} between polymer 1 and polymer 2 can be obtained through use of the relation:

$$A_2 = \frac{\bar{v}_1\bar{v}_2}{2V_0}(1 - \chi_{01} - \chi_{02} - \chi_{12}). \quad (4.144)$$

g) Optical theta condition (ref. 16)

Combination of the theory of light scattering from multicomponent solution (refs 17,18) and the Flory-Huggins expression for the free energy of mixing two monodisperse polymers 1 and 2 with a solvent (see, for example, eqn (5.6)) gives eqns (4.146)-(4.148) under the condition given by eqn (4.145).

$$\psi_1 X_1 x_1 + \psi_2 X_2 x_2 = 0, \tag{4.145}$$

$$\frac{K^* V_0 v}{R_0} = A' + B'v + O(v^2), \tag{4.146}$$

$$A' = (\psi_1{}^2 X_1 x_1 + \psi_2{}^2 X_2 x_2)^{-1}, \tag{4.147}$$

$$B' = 2\psi_1 \psi_2 X_1 X_2 x_1 x_2 A'^2 \chi_{12} \tag{4.148}$$

with

$$K^* = \frac{4\pi^2 n^2}{N_A \lambda_0{}^4}, \tag{4.149}$$

$$v = v_1 + v_2, \tag{4.150}$$

$$x_1 = 1 - x_2 = v_1/v, \tag{4.151}$$

$$\psi_i = \partial n/\partial v_i \qquad (i=1,2) \tag{4.152}$$

where R_0, the forward scattering intensity excess over that of the solvent, n, refraction index of the solution, N_A, the Avogadro number, λ_0, the wavelength in vacuum, V_0, the solvent molar volume, X_1 and X_2, the degree of chain length of polymer 1 and 2, respectively. The condition of eqn (4.145) is called "optical theta condition". The initial slope of the v/R_0 vs. v plot is directly proportional to χ_{12} regardless of thermodynamic properties of the solvent. According to Fukuda et al. (ref. 16) who originally proposed an idea of an optical theta condition, this condition may be relatively easily met; one has only to find common solvent whose refractive index is between those of the polymers so that ψ_1 and ψ_2 are opposite in sign. With such a solvent, one can adjust either molecular weights or blending ratio x_i so as to satisfy eqn (4.145).

Table 4.1 collects the literature data on χ_{12} determined by

Table 4.1
Polymer-polymer interaction parameter χ_{12} for various binary polymer mixtures.

Polymer 1	Polymer 2	(Solvent)	Ratio (weight)	χ_{12}	T/℃	Method	ref.
PS[a]	PVME[b]		35/65	-0.75	30	Vapor Sorption	11
PS	PVME		45/55	-0.69	30	Vapor Sorption	11
PS	PVME		45/55	-0.60	50	Vapor Sorption	11
PS	PVME		25/75	-0.40	25	Small-Angle Neutron Scattering	20
PS	PVME		50/50	-0.30	25	Small-Angle Neutron Scattering	20
PS	PMMA[c]	Toluene	—	0.007 ~0.0393	23±1	Phase Diagram	21
PS	PMMA	Bromobenzene	—	0.007 ±0.004	30	Optical theta Condition	22
PS	PMMA	Bromobenzene					
$M_w=242\times10^4$	$M_w=219\times10^4$		—	0.0026	30	Optical theta Condition	16
$M_w=80.6\times10^4$	$M_w=82.5\times10^4$		—	0.0030	30		16
$M_w=28.3\times10^4$	$M_w=21.1\times10^4$		—	0.0040	30		16
$M_w=2.45\times10^4$	$M_w=2.45\times10^4$		—	0.0117	30		16
Atactic PS	Atactic PP	Toluene	—	0.62	30	Phase Diagram	23
PS	Polybutadiene		—	0.1	—	Phase Diagram	14

a, Polystyrene(atactic), b, Polyvinylmethylether, c, Polymethylmethacrylate.

Table 4.1 (continued-1)

Polymer 1	Polymer 2	(Solvent)	Ratio (weight)	χ_{12}	T/℃	Method	ref.
PS	Polybutadiene		—	0.418 ~0.79	150	Light Scattering	24
PS	Polybutene		—	0.1	—	Phase Diagram	14
PS	Polyisoprene	Cyclohexane	—	0.36 ~0.37	—	Light Scattering	25
PS	Poly(α-methyl styrene)	n-Butylchloride	—	-0.87 ~-0.24	30 ~60	Vapor Sorption	26
PS	Polyisobutyrene	Toluene	—	0.0080 ~0.0289	—	Light Scattering	27
		Benzene	—	0.0073 ~0.0405	—	Light Scattering	27
PS(M_w=3x10^6)	PDMS[d](M_w=4.2x10^6)	Styrene	—	0.039 ~0.063	19 ~35	Optical theta Condition	28
		Cyclohexane	—	0.036 ~0.051	25 ~52	Optical theta Condition	28
Polybutene	Silicone		—	0.2	—	Phase Diagram	14
PEMA[e]	PVF_2[f]		—	-0.13	160	Melting Point Depression	29

d, Polydimethylsiloxane, e, Polyethylmethacrylate, f,Polyvinylidenechloride.

Table 4.1 (continued-2)

Polymer 1	Polymer 2	(Solvent)	Ratio (weight)	χ_{12}	T/℃	Method	ref.
PEMA	PVF_2		—	-0.08	160	Melting Point Depression	30
PMMA	PVF_2		—	-0.29	160	Melting Point Depression	10
PMMA	PVF_2		50/50	-0.12	200	Melting Point Depression	31
PMMA	PVF_2		—	-0.20	170	Melting Point Depression	32
PMMA	PVF_2	Acetophenone	25/75	0.22	200	Inverse-Phase Gas Chromatography	33
PMMA	PVF_2	Acetophenone	50/50	-0.05	200	Inverse-Phase Gas Chromatography	33
PMMA	PVF_2	Acetophenone	75/25	-0.20	200	Inverse-Phase Gas Chromatography	33
PMMA	PVF_2	Acetophenone	90/10	-0.28	200	Inverse-Phase Gas Chromatography	33
PMMA	PVF_2	Cyclohexanol	25/75	0.01	200	Inverse-Phase Gas Chromatography	33
PMMA	PVF_2	Cyclohexanol	50/50	-0.01	200	Inverse-Phase Gas Chromatography	33

Table 4.1 (continued-3)

Polymer 1	Polymer 2	(Solvent)	Ratio (weight)	χ_{12}	T/℃	Method	ref.
PMMA	PVF_2	Cyclohexanol	75/25	-0.18	200	Inverse-Phase Gas Chromatography	33
PMMA	PVF_2	Cyclohexanol	90/10	-0.22	200	Inverse-Phase Gas Chromatography	33
Isotactic PMMA	PVF_2		—	-0.13	175	Melting Point Depression	34
Syndiotactic PMMA	PVF_2		—	-0.06	175	Melting Point Depression	34
Atactic PMMA	PVF_2		—	-0.10	175	Melting Point Depression	34
PMMA	PVF_2		25/75	-0.3	25	Small-Angle Neutron Scattering	20
PMMA	PVF_2		50/50	-0.1	25	Small-Angle Neutron Scattering	20
Polymethyl-acrylate	Polyepichloro-hydrin		—	1.47 ∼-0.09	76	Inverse-Phase Gas Chromatography	35
Polymethyl-acrylate	Polyepichloro-hydrin		—	0.49 ∼-0.07	125	Inverse-Phase Gas Chromatography	35
Phenoxy[g]	PBA[h]		—	-1.04	61	Melting Point Depression	36

g, Polyhydroxyether of Bisphenol A, h, Poly1,4-butyleneadipate.

Table 4.1 (continued-4)

Polymer 1	Polymer 2	(Solvent)	Ratio (weight)	χ_{12}	T/℃	Method	ref.
Phenoxy	PBA		—	-0.89	55	Vapor Sorption	36
Phenoxy	PEA[i]		—	-0.52	49	Melting Point Depression	36
Phenoxy	PEA		—	-0.21	55	Vapor Sorption	36
Phenoxy	PCL[j]		—	-0.42	56	Melting Point Depression	36
Phenoxy	PCL		—	-1.01	55	Vapor Sorption	36
Phenoxy	PCDS[k]		—	-0.76	55	Vapor Sorption	36
PEMA	CPE[l]	Chloroform	75/25	-0.32	80	Inverse-Phase Gas Chromatography	37
PEMA	CPE	Chloroform	50/50	-0.22	80	Inverse-Phase Gas Chromatography	37
PEMA	CPE	Chloroform	25/75	-0.44	80	Inverse-Phase Gas Chromatography	37
PEMA	CPE	Chloroform	25/75	-0.14	100	Inverse-Phase Gas Chromatography	37
PEMA	CPE	Chloroform	25/75	-0.22	120	Inverse-Phase Gas Chromatography	37
PVC[m]	PMPPL[n]		—	-0.29	90	Melting Point Depression	38

i, Polyethyleneadipate, j, Poly ε-caprolactone, k, Poly1,4-cyclohexanedimethanolsuccinate,
l, Clorinated polyethylene, m, Polyvinylchloride, n, Poly α-methyl-α-n-propyl-β-propiolactone.

Table 4.1 (continued-5)

Polymer 1	Polymer 2	(Solvent)	Ratio (weight)	χ_{12}	T/℃	Method	ref.
PVC	PPL°		—	-0.05	250	Melting Point Depression	38
PVC	PCL	Ethanol	50/50	-2.8	120	Inverse-Phase Gas Chromatography	39
PVC	PCL	Ethanol	30/70	-5.0	120	Inverse-Phase Gas Chromatography	39
PVC	PCL	Chloroform	50/50	-2.4	120	Inverse-Phase Gas Chromatography	39
PVC	PCL	Chloroform	30/70	-3.4	120	Inverse-Phase Gas Chromatography	39
PVC	PCL	MEK	50/50	-6.4	120	Inverse-Phase Gas Chromatography	39
PVC	PCL	MEK	30/70	-8.1	120	Inverse-Phase Gas Chromatography	39
PVC	PCL	Pyridine	50/50	-5.4	120	Inverse-Phase Gas Chromatography	39
PVC	PCL	Pyridine	30/70	-8.1	120	Inverse-Phase Gas Chromatography	39
PVC	PCL	Acetonitrile	50/50	-9.3	120	Inverse-Phase Gas Chromatography	39
PVC	PCL	Acetonitrile	30/70	-10.7	120	Inverse-Phase Gas Chromatography	39

o, Polypivalolactone.

Table 4.1 (continued-6)

Polymer 1	Polymer 2	(Solvent)	Ratio (weight)	χ_{12}	T/℃	Method	ref.
PVC	PCL	Fluorobenzene	50/50	-2.9	120	Inverse-Phase Gas Chromatography	39
PVC	PCL	Fluorobenzene	30/70	-4.5	120	Inverse-Phase Gas Chromatography	39
PVC	PCL	Carbontetra-chloride	50/50	3.0	120	Inverse-Phase Gas Chromatography	39
PVC	PCL	Carbontetra-chloride	30/70	2.2	120	Inverse-Phase Gas Chromatography	39
PVC	PCL	Hexane	50/50	2.8	120	Inverse-Phase Gas Chromatography	39
PVC	PCL	Hexane	30/70	1.8	120	Inverse-Phase Gas Chromatography	39
PC[p]	CDTI[q]		—	0.00	35	Vapor Sorption	40
PPO[r]	PS		—	-1.08	35	Vapor Sorption	41
D-PSAN-19[s]	PSAN-19[t]		—	0.4	25	Small-Angle Neutron Scattering	13
D-PSAN-19	PSAN-20.5		—	4.4	25	Small-Angle Neutron Scattering	13

p, Polycarbonate of Bisphenol A, q, Poly1,4-cyclohexanedimethanol-co-terephthalic acid-co-isophthalicacid, r, Polyphenyleneoxide, s, Deuterated polystyrene-co-acrylonitrile, t, Polystyrene-co-acrylonitrile.

the above methods for various binary polymer mixture. Note that Riedle and Prud'homme (ref. 19) constructed a comprehensive table on χ_{12} values from literature published until 1983.

The reliability of the methods for estimating χ_{12} and the dependence of χ_{12} on the polymer composition, the molecular weight and the stereoregularity will be discussed below:

For polyethylmethacrylate (PEMA) / polyvinylidenechloride (PVF_2) system around 160℃ Kwei et al. (ref. 29) estimated χ_{12} to be -0.13 and Imken et al. (ref. 30) obtained $\chi_{12} = -0.08$, both by the melting point depression method. These values are obviously not in good agreement. Note that by the melting point depression method, the concentration dependence of χ_{12}, if any, is not principally considered, because χ_{12} is evaluated from the slope of the melting point depression ΔT and the polymer volume fraction v_1.

Using the inverse-phase gas chromatography Dipaola-Baranyi et al. (ref. 33) determined χ_{12} for polymethylmethacrylate (PMMA) / PVF_2 system with a solvent as probe. Compared at the same composition (i.e., 50/50, wt/wt) χ_{12} was estimated as -0.05 for a probe of acetophenone and -0.01 when cyclohexanol was utilized. This suggests strongly that the solvent nature significantly influences the results. In this method, χ_{12} has a tendency to decrease with as increase in PMMA composition, regardless the solvent nature as probe.

Kwei et al. (ref. 11) determined by vapor sorption method χ_{12} at 30℃ for polystyrene (PS) / PVME system with various compositions: $\chi_{12} = -0.75$ at the weight fraction of PS $v_{PS} = 0.35$ and $\chi_{12} = -0.69$ at $v_{PS} = 0.45$. This fact indicates the existence of concentration dependence of χ_{12} for binary polymer mixture, which was also demonstrated by small-angle neutron scattering method: For PS / PVME system at 25℃ $\chi_{12} = -0.04$ at $v_{PS} = 0.25$ and -0.30 at $v_{PS} = 0.50$.

Esker et al. (ref. 26) demonstrated that similar χ_{12} value by the light scattering method were obtained for PS / polyisobutylene system when toluene and benzene were used as solvent for the sample preparation. In other words, in this case the effect of solvent, used for sample preparation, on χ_{12} seems less remarkable. Note here that toluene and benzene have not only similar chemical structure, but also not so different solution supermolecular structure.

In the optical theta condition method, insignificant solvent effect on χ_{12} was investigated by Kaddour et al. (ref. 28) for

PS / polydimethylsiloxane (PDMS) system prepared using styrene and cyclohexane. Fukuda et al. (ref. 16) also examined the molecular weight dependence of χ_{12}, estimated by their optical theta condition method for PS / PMMA system which was prepared using bromobenzene as solvent: χ_{12} decreased monotonically from 0.0117 to 0.026 with an increase in M_w from PS ($M_w = 2.45 \times 10^4$) / PMMA ($M_w = 2.45 \times 10^4$) to PS ($M_w = 242 \times 10^4$) / PMMA ($M_w = 219 \times 10^4$).

Two polymer / polymer systems were investigated using the different method: Harris et al. (ref. 36) reported different χ_{12} value for polyhydroxyether of bisphenol A(phenoxy) / poly(ε-caprolactone) system at almost the same temperature (55 and 56℃) using the vapor sorption and the melting point depression methods (i.e., $\chi_{12} = -1.01$ by the vapor sorption method and $\chi_{12} = -0.42$ by the melting point depression method). Wendorff (ref. 31) and DiPaola-Baranyi et al. (ref. 33) evaluated χ_{12} value for PMMA / PVF_2 system with constant composition (50/50 (wt/wt)) at 200℃ to be $\chi_{12} = -0.12$ by the melting point depression method and -0.01 by the inverse-phase gas chromatography method, respectively. Thus, χ_{12} varies depending on the method employed.

Roerdink et al. (ref. 34) examined the effect of stereoregularity of polymer on χ_{12} by applying the melting point depression method to the binary mixtures of PMMA with three different stereoregularities and PVF_2 and they found χ_{12} to be -0.13 for isotactic PMMA / PVF_2, -0.10 for atactic PMMA / PVF_2 and -0.06 for syndiotactic PMMA / PVF_2.

Summarizing, χ_{12} for binary polymer mixture system depends on the composition (i.e., concentrations v_1 and v_2), the molecular weight (X_n, X_w and Y_n, Y_w), the tacticity of polymers and often on the solvent nature used for sample preparation. The comparison of the methods has not yet been made comprehensively and even now is open for further study.

REFERENCES

1 R.L. Scott, J. Chem. Phys., 17 (1949) 268.
2 R. Koningsveld, H.A.G. Chermin and M. Gordon, Proc. Roy. Soc. Lond. A., 319 (1970) 331.
3 K. Kamide, S. Matsuda and H. Shirataki, Polym. J., 20 (1988) 949.
4 R. Koningsveld and L.A. Kleintjens, J. Polym. Sci., 61 (1977) 221.

5 L.P. McMaster, Macromolecules, 6 (1973) 760.
6 H. Shirataki, S. Matsuda and K. Kamide, Brit. Polym. J., to be published.
7 H. Shirataki, S. Matsuda and K. Kamide, Brit. Polym. J., to be published.
8 K. Kamide, S. Matsuda, T. Dobashi and M. Kaneko, Polym. J., 16 (1984) 839.
9 S. Matsuda, Polym. J., 18 (1986) 981.
10 T. Nishi and T.T. Wang, Macromolecules, 8 (1975) 909.
11 T.,K. Kwei, T. Nishi and R.F. Roberts, Macromolecules, 1 (1974) 667.
12 D.D. Deshpande, D. Patterson, H.P. Schreiber and C.S. Su, Macromolecules, 1 (1974) 530.
13 B.J. Schmitt, R.G. Kirste and J. Jelenic, Makro. Chem,. 181 (1980) 1655.
14 G. Allen G. Gee and J.P. Nicholson, Polymer, 1 (1960) 56.
15 L.O. Kaddour, M.S. Amasagasti and C. Strazielle, Makromol. Chem., 188 (1987) 2223.
16 T. Fukuda, M. Nagata and H. Inagaki, Macromolecules, 17 (1984) 548.
17 J.G. Kirkwood and R.J. Goldberg, J. Chem. Phys., 18 (1950) 54.
18 W.H. Stockmayer, J. Chem. Phys., 18 (1950) 58.
19 B. Riedl and R.E. Prud'homme, Polym. Eng. and Sci., 24 (1984) 1291.
20 G. Hadziioannou, R. Stein and J. Higgins, Am. Chem. Soc., Div. Polym. Chem., Polym Prepr., 24 (1983) 213.
21 W.W.Y. Lau, C.H. Burns and R.Y.H. Huang, Eur. Polym. J., 23 (1987) 37.
22 A.C. Su and J.R. Fried, Macromolecules, 19 (1986) 1417.
23 D. Berek, D. Lath and V. Durdovic, J. Polym. Sci., 16 (1967) 659.
24 R.J. Roe and W.C. Zim, Macromolecules, 13 (1980) 1221.
25 Z. Tong, Y. Einaga, H. Miyashita and H, Fujita, Macromolecules, 20 (1987) 1883.
26 S. Saeki, Y. Aoyanagi, M.Tsubokawa and T. Yamaguchi, Polymer, 25 (1984) 1779.
27 M.W.J. van der Esker and A. Vrij, J. Polym. Sci., 14 (1976) 1943.
28 L.O. Kaddour and C. Strazielle, Polymer, 28 (1987) 459.
29 T.K. Kwei, G.D. Patterson and T.T. Wang, Macromolecules, 9 (1976) 780.

30 R.L. Imken, D.R. Paul and J.W. Barlow, Polym. Eng. Sci., 16 (1976) 593.
31 J.H. Wendorff, J. Polym. Sci., Polym. Lett. Ed., 18 (1980) 439.
32 B.S. Morra and R.S. Stein, J. Polym. Sci., Polym. Phys. Ed., 20 (1982) 2243.
33 G. DiPaola-Baranyi, S.J. Fletche and P. Degre, Macromolecules, 15 (1982) 885.
34 E. Roerdink and G. Challa, Polymer, 19 (1978) 173.
35 Z.Y. Al-Saigh and P. Munk, Macromolecules, 17 (1984) 803.
36 J.E. Harris, D.R. Paul and J.W. Barlow, Polym. Eng. Sci., 23 (1983) 676.
37 C. Zhikuan and D.J. Walsh, Eur. Polym. J., 19 (1983) 519.
38 M. Aubin and R.E. Prud'homme, Macromolecules, 13 (1980) 365.
39 O. Olabisi, Macromolecules, 8 (1975) 316.
40 P. Masi, D.R. Paul and J.W. Barlow, J. Polym. Sci., Polym. Phys. Ed., 20 (1982) 15.
41 G. Morel and D.R. Paul, J. Membrane Sci., 10 (1982) 273.

Chapter 5

QUASI-TERNARY SOLUTION CONSISTING OF MULTICOMPONENT POLYMER 1, MULTICOMPONENT POLYMER 2 AND SINGLE SOLVENT SYSTEM

5.1 INTRODUCTION

In 1949, Scott (ref. 1) first studied theoretically the coexistence curve, the spinodal curve (SC) and the critical solution point (CSP) of the solution consisting of a single component polymer 1, a single component polymer 2 and a single low-molecular solvent (rigorous ternary system) to derive the approximate equations of CSP (eqn 17a-c of ref. 1). Note that these equations are only effective under the conditions of $|\chi_{01}-\chi_{02}|<1$ and $(X)^{1/2}<Y<X^2$. Here, χ_{01} (or χ_{02}) is the thermodynamic interaction parameter between a solvent and polymer 1 (or 2) and X and Y, the degree of polymerization of polymers 1 and 2, respectively. Soon after, Tompa (ref. 2) succeeded in deriving the rigorous equations giving SC and CSP for these ternary systems. Spinodal and plait point conditions obtained by him are (refs 2,3),

$$\Sigma x_i v_i - 2\Sigma x_i x_j (\chi_i + \chi_j) + 4x_0 x_1 x_2 (\chi_0\chi_1 + \chi_0\chi_2 + \chi_1\chi_2) v_0 v_1 v_2 = 0 \tag{5.1}$$

and

$$\Sigma \frac{x_i^{\,2} v_i}{(1-2\chi_i x_i v_i)^3} = 0 \qquad (i=0,1,2) \tag{5.2}$$

where

$$\chi_0 = \chi_{01} + \chi_{02} - \chi_{12}, \tag{5.3a}$$

$$\chi_1 = \chi_{01} + \chi_{12} - \chi_{02}, \tag{5.3b}$$

$$\chi_2 = \chi_{02} + \chi_{12} - \chi_{01} \tag{5.3c}$$

and χ_{12}, the thermodynamic interaction parameter between polymers 1 and 2; v_0, v_1 and v_2, the volume fractions of solvent, polymers 1 and 2; $x_0=1$, $x_1=X$, and $x_2=Y$, respectively. Scott (ref. 1) and Tompa (refs 2,3) calculated SC and CSP for a very specialized symmetrical case: $\chi_{01}=\chi_{02}$. Using the Tompa's equations (eqns (5.1)-(5.3c)), Zemann and Patterson (ref. 4) calculated SC and CSP for $\chi_{01}\neq\chi_{02}$, showing that a small difference in χ_{01} and χ_{02} has a marked effect on polymer compatibility. They also found that $\chi_{12}=0$ with keeping $\chi_{01}\neq\chi_{02}$ leads to a closed loop region of incomplete miscibility (ref. 4) although the binary mixture of polymer 1 and 2 shows complete miscibility (for example, see, Figs 1-4 of ref. 4). Hsu and Prausnitz (ref. 5) proposed a numerical procedure of calculating a coexistence curve (in other words, binodal curve); note that in this case, the binodal curve coincides with the cloud point curve (CPC)) using Gibbs' phase equilibria condition. They examined in detail the effect of asymmetry of χ parameters between a solvent and polymer 1 (or 2) (i.e., $\chi_{01}\neq\chi_{02}$) on the phase diagram and also examined the existence of a close loop immiscible region by putting χ_{12} lower than a critical value (ref. 5), semi-empirically determined. This critical value can be evaluated by eqn (5.51) in this Chapter. Recently, Šolc (ref. 6) proposed equations of CPC and CSP by introducing two new variables, η and ξ, related to the two partition coefficients, σ_X and σ_Y, through $\eta^2=\sigma_X\sigma_Y$ and $\xi^2=\sigma_Y/\sigma_X$. The overall polymer volume fraction v $(=v_1+v_2)$ at CSP (v_c) was expressed by use of CPC equation (eqn 4 of ref. 6) at the limit of σ_X (and σ_Y)$\to 0$, in a simple closed form as follows (ref. 6);

$$\left[\frac{v_c}{1-v_c}\right]^2 = x_0\frac{w_1x_1^{\,2}\xi^{-3}+w_2x_2^{\,2}\xi^{3}}{(w_1x_1\xi^{-1}+w_2x_2\xi)^3} \tag{5.4}$$

where w_1 and w_2 are the volume compositions of polymers 1 and 2 ($w_1=v_1/v$ and $w_2=v_2/v$). Šolc (ref. 7) also studied general conditions for multiple critical points using double roots of his CPC equation. All the above theoretical studies were limited to a monodisperse polymer 1 / monodisperse polymer 2 / single solvent system.

In 1970, a theoretical study on the phase equilibrium of a multicomponent polymers 1 / multicomponent polymers 2 / single solvent system was first carried out by Koningsveld, Chermin and Gordon (refs 8,9). Note that they cited an equation of a spinodal condition (eqn 12 of ref. 8) originally derived by Chermin (unpublished work) and they described an equation of neutral equilibrium condition (eqn 13 of ref. 8) without detailed derivation. As far as the author knows, Chermin's work still remains to be unpublished. Can we put our unconditional confidence in and employ for further calculations equations derived without any mathematical detail? Koningsveld et al. showed experimental prism phase diagrams (i.e., T-v_1-v_2 space; T, Kelvin Temperature) of polyethylene / isotactic polystyrene / diphenylether (ref. 8) and linear polyethylene / isotactic polypropylene / diphenylether systems (refs 8,9), and deduced, from the experimental diagrams on quasi-ternary systems and by analogy of quasi-binary (multicomponent polymers / single solvent) systems, the form of interaction parameters (χ_{kl}; kl= 01, 02 and 12) with the help of the spinodal equation (eqn 27 of ref. 8; rewritten form of eqn 12 of ref. 8) as (ref. 8);

$$\chi_{kl} = \chi_{kl,1} + \frac{\chi_{kl,2}}{T} + \chi_{kl,3}T \tag{5.5}$$

where $\chi_{kl,1}$, $\chi_{kl,2}$ and $\chi_{kl,3}$ are the temperature independent coefficients (kl= 01, 02 and 12). Unfortunately, they did not carry out systematic computer experiments based on their equations (eqns 12 and 13 of ref. 8) on spinodal and neutral equilibrium. Kamide, Matsuda and Shirataki (ref. 10) studied, as a further extension of Kamide and Matsuda's previous study (ref. 11), SC and CSP of a system consisting of multicomponent polymer 1, multicomponent polymer 2 and a single solvent.

In this Chapter, we show, based on Kamide et al's treatment (ref. 10), some detailed derivations of thermodynamic equations giving SC and the CSP, and disclose the effect of average molecular weight and molecular weight distribution of the two kinds of original polymers and of the three thermodynamical interaction χ parameters, χ_{01}, χ_{02} and χ_{12} on SC and CSP.

5.2 SPINODAL CURVE AND CRITICAL SOLUTION POINT

According to the Flory-Huggins lattice theory of the dilute polymer solutions, Gibbs' free energy change of mixing ΔG is given by (ref. 12)

$$\Delta G = \tilde{R}TL \left[v_0 \ln v_0 + \sum_{i=1}^{m_1} \frac{v_{Xi}}{X_i} \ln v_{Xi} + \sum_{j=1}^{m_2} \frac{v_{Yj}}{Y_j} \ln v_{Yj} \right.$$

$$\left. + \chi_{01} v_0 v_1 + \chi_{02} v_0 v_2 + \chi_{12} v_1 v_2 \right] \tag{5.6}$$

where $\tilde{R}$ is the gas constant, T, the Kelvin temperature, L, total number of lattice points ($\equiv N_0 + \sum_{i=1}^{m_1} X_i N_{Xi} + \sum_{j=1}^{m_2} Y_j N_{Yj}$: N_0, N_{xi} and N_{Yj}; the number of solvents, X_i-mer and Y_j-mer, respectively), v_0, the volume fraction of solvent, v_{Xi}, the volume fraction of X_i-mer of polymer 1, v_{Yj}, the volume fraction of Y_j-mer of polymer 2, v_1 and v_2, the total volume fraction of polymer 1 and the polymer 2 ($v_1 = \sum_{i=1}^{m_1} v_{Xi}$ and $v_2 = \sum_{j=1}^{m_2} v_{Yj}$), m_1 and m_2 are the total number of components of polymer 1 and 2, respectively.

The chemical potentials of solvent, polymer 1 (X_i-mer) and polymer 2 (Y_j-mer), $\Delta\mu_0$, $\Delta\mu_{Xi}$ and $\Delta\mu_{Yj}$, are directly derived from eqn (5.6). The results are (ref. 10):

$$\Delta\mu_0 = \tilde{R}T[\ln v_0 + (1-\frac{1}{X_n})v_1 + (1-\frac{1}{Y_n})v_2 + \chi_{01}v_1(1-v_0)$$

$$+ \chi_{02}v_2(1-v_0) + \chi_{12}v_1v_2], \quad (5.7)$$

$$\Delta\mu_{Xi} = \tilde{R}T[\ln v_{Xi} - (X_i - 1) + X_i(1-\frac{1}{X_n})v_1 + X_i(1-\frac{1}{Y_n})v_2$$

$$+ X_i\{\chi_{12}v_2(1-v_1) + \chi_{01}v_0(1-v_1) - \chi_{02}v_0v_2\}] \quad (i=1,\cdots,m_1), \quad (5.8)$$

$$\Delta\mu_{Yj} = \tilde{R}T[\ln v_{Yj} - (Y_j - 1) + Y_j(1-\frac{1}{X_n})v_1 + Y_j(1-\frac{1}{Y_n})v_2$$

$$+ Y_j\{\chi_{12}v_1(1-v_2) + \chi_{02}v_0(1-v_2) - \chi_{01}v_0v_1\}] \quad (j=1,\cdots,m_2). \quad (5.9)$$

Here, X_n and Y_n are the number-average of X_i and Y_j. We assume that (a) χ_{01}, χ_{02} and χ_{12} are independent of the concentration and molecular weight of the polymers, (b) the molar volume of solvent and segment of polymers 1 and 2 are the same, (c) solvent, polymer 1 and 2 are volumetrically additive and (d) the density of solvent is the same as that of polymer 1 and 2. Note that eqns (5.8) and (5.9) are strictly symmetrical with regard to an exchange of polymer 1 and polymer 2.

Thermodynamical requirement against the spinodal curve is that the second order differential of Gibbs' free energy ΔG should always be zero on the curve and then the conditions of spinodals are given by the $(m_1 + m_2) \times (m_1 + m_2)$ determinant in the form:

$$|\Delta G'| = \begin{vmatrix} \overline{\Delta G'}_{X_1X_1} & \overline{\Delta G'}_{X_1X_2} & \cdots & \overline{\Delta G'}_{X_1X_{m1}} & \overline{\Delta G'}_{X_1Y_1} & \overline{\Delta G'}_{X_1Y_2} & \cdots & \overline{\Delta G'}_{X_1Y_{m2}} \\ \overline{\Delta G'}_{X_2X_1} & \overline{\Delta G'}_{X_2X_2} & \cdots & \overline{\Delta G'}_{X_2X_{m1}} & \overline{\Delta G'}_{X_2Y_1} & \overline{\Delta G'}_{X_2Y_2} & \cdots & \overline{\Delta G'}_{X_2Y_{m2}} \\ \vdots & \vdots & & \vdots & \vdots & \vdots & & \vdots \\ \overline{\Delta G'}_{X_{m1}X_1} & \overline{\Delta G'}_{X_{m1}X_2} & \cdots & \overline{\Delta G'}_{X_{m1}X_{m1}} & \overline{\Delta G'}_{X_{m1}Y_1} & \overline{\Delta G'}_{X_{m1}Y_2} & \cdots & \overline{\Delta G'}_{X_{m1}Y_m} \\ \overline{\Delta G'}_{Y_1X_1} & \overline{\Delta G'}_{Y_1X_2} & \cdots & \overline{\Delta G'}_{Y_1X_{m1}} & \overline{\Delta G'}_{Y_1Y_1} & \overline{\Delta G'}_{Y_1Y_2} & \cdots & \overline{\Delta G'}_{Y_1Y_{m2}} \\ \overline{\Delta G'}_{Y_2X_1} & \overline{\Delta G'}_{Y_2X_2} & \cdots & \overline{\Delta G'}_{Y_2X_{m1}} & \overline{\Delta G'}_{Y_2Y_1} & \overline{\Delta G'}_{Y_2Y_2} & \cdots & \overline{\Delta G'}_{Y_2Y_{m2}} \\ \vdots & \vdots & & \vdots & \vdots & \vdots & & \vdots \\ \overline{\Delta G'}_{Y_{m2}X_1} & \overline{\Delta G'}_{Y_{m2}X_2} & \cdots & \overline{\Delta G'}_{Y_{m2}X_{m1}} & \overline{\Delta G'}_{Y_{m2}Y_1} & \overline{\Delta G'}_{Y_{m2}Y_2} & \cdots & \overline{\Delta G'}_{Y_{m2}Y_{m2}} \end{vmatrix}$$

$$= 0. \qquad (5.10)$$

Here, $\Delta G'$ is the Gibbs' free energy of mixing per unit volume defined by

$$\Delta G' = v_0 \left(\frac{\Delta \mu_0}{V_0}\right) + \sum_{i=1}^{m_1} v_{X_i} \left(\frac{\Delta \mu_{X_i}}{X_i V}\right) + \sum_{j=1}^{m_2} v_{Y_j} \left(\frac{\Delta \mu_{Y_j}}{Y_j V_0}\right) \qquad (5.11)$$

and $\overline{\Delta G'}_{kl}$ is defined as

$$\overline{\Delta G'}_{kl} = \left(\frac{\partial^2 \Delta G'}{\partial v_k \partial v_l}\right)_{T,P,v_n} \quad (k,l = X_1, X_2, \cdots, X_{m1}, Y_1, Y_2, \cdots, Y_{m2};\quad n \neq k,l). \qquad (5.12)$$

At CSP, in addition to eqn (5.10), the following neutral equilibrium (eqn (5.13)) condition should be satisfied concurrently.

$$|\Delta G''| = \begin{vmatrix} \frac{\partial|\Delta G'|}{\partial v_{X1}} & \frac{\partial|\Delta G'|}{\partial v_{X2}} & \cdots & \frac{\partial|\Delta G'|}{\partial v_{Xm1}} & \frac{\partial|\Delta G'|}{\partial v_{Y1}} & \frac{\partial|\Delta G'|}{\partial v_{Y2}} & \cdots & \frac{\partial|\Delta G'|}{\partial v_{Ym2}} \\ \overline{\Delta G'}_{X2X1} & \overline{\Delta G'}_{X2X2} & \cdots & \overline{\Delta G'}_{X2Xm1} & \overline{\Delta G'}_{X2Y1} & \overline{\Delta G'}_{X2Y2} & \cdots & \overline{\Delta G'}_{X2Ym2} \\ \vdots & \vdots & & \vdots & \vdots & \vdots & & \vdots \\ \overline{\Delta G'}_{Xm1X1} & \overline{\Delta G'}_{Xm1X2} & \cdots & \overline{\Delta G'}_{Xm1Xm1} & \overline{\Delta G'}_{Xm1Y1} & \overline{\Delta G'}_{Xm1Y2} & \cdots & \overline{\Delta G'}_{Xm1Ym2} \\ \overline{\Delta G'}_{Y1X1} & \overline{\Delta G'}_{Y1X2} & \cdots & \overline{\Delta G'}_{Y1Xm1} & \overline{\Delta G'}_{Y1Y1} & \overline{\Delta G'}_{Y1Y2} & \cdots & \overline{\Delta G'}_{Y1Ym2} \\ \overline{\Delta G'}_{Y2X1} & \overline{\Delta G'}_{Y2X2} & \cdots & \overline{\Delta G'}_{Y2Xm1} & \overline{\Delta G'}_{Y2Y1} & \overline{\Delta G'}_{Y2Y2} & \cdots & \overline{\Delta G'}_{Y2Ym2} \\ \vdots & \vdots & & \vdots & \vdots & \vdots & & \vdots \\ \overline{\Delta G'}_{Ym2X1} & \overline{\Delta G'}_{Ym2X2} & \cdots & \overline{\Delta G'}_{Ym2Xm1} & \overline{\Delta G'}_{Ym2Y1} & \overline{\Delta G'}_{Ym2Y2} & \cdots & \overline{\Delta G'}_{Ym2Ym2} \end{vmatrix}$$

$$= 0. \tag{5.13}$$

Gibbs' free energy change of mixing per unit volume $\Delta G'$ can be rewritten by combination of eqns (5.7)-(5.9) and (5.11) as:

$$\Delta G' = \left(\frac{\tilde{R}T}{V_0}\right)\left[v_0 \ln v_0 + \sum_{i=1}^{m_1} \frac{v_{Xi}\ln v_{Xi}}{X_i} + \sum_{j=1}^{m_2} \frac{v_{Yj}\ln v_{Yj}}{Y_j}\right.$$

$$\left. + \chi_{01} v_0 v_1 + \chi_{02} v_0 v_2 + \chi_{12} v_1 v_2\right]. \tag{5.14}$$

Substitution of eqn (5.14) into eqn (5.12) yields five types of $\overline{\Delta G'}_{kl}$:

$$\left(\frac{V_0}{\tilde{R}T}\right)\overline{\Delta G'}_{kl} = \frac{1}{v_0} - 2\chi_{01} \equiv M \qquad (\text{for } k \neq l,\ (k,l) = (X_i, X_j)), \tag{5.15a}$$

$$\left(\frac{V_0}{\tilde{R}T}\right)\overline{\Delta G'}_{kl} = \frac{1}{v_0} - 2\chi_{02} \equiv N \qquad (\text{for } k \neq l,\ (k,l) = (Y_i, Y_j)), \tag{5.15b}$$

$$\left(\frac{V_0}{\tilde{R}T}\right)\overline{\Delta G'}_{kl} = \frac{1}{v_0} + \chi_{12} - \chi_{01} - \chi_{02} \equiv K$$

$$(\text{for } k \neq l,\ (k,l) = (X_i, Y_j) \text{ or } (Y_j, X_i)), \tag{5.15c}$$

$$(\frac{V_0}{\widetilde{RT}})\,\Delta\bar{G}'_{kl} = \frac{1}{v_0} - 2\chi_{01} + \frac{1}{X_i v_{X_i}} \equiv M + M_i \qquad (\text{for } k = l = X_i), \tag{5.15d}$$

$$(\frac{V_0}{\widetilde{RT}})\,\Delta\bar{G}'_{kl} = \frac{1}{v_0} - 2\chi_{02} + \frac{1}{Y_j v_{Y_j}} \equiv N + N_j \qquad (\text{for } k = l = Y_j). \tag{5.15e}$$

Spinodal condition (eqn (5.10)) can be rewritten with the help of eqns (5.14), (5.15a)-(5.15e) as eqn (5.16);

$$|\Delta G'| = (\frac{\widetilde{RT}}{V_0})^{m_1+m_2} \left| \begin{array}{cccc|cccc} M+M_1 & M & \cdots & M & K & K & \cdots & K \\ M & M+M_2 & \cdots & M & K & K & \cdots & K \\ \vdots & \vdots & & \vdots & \vdots & \vdots & & \vdots \\ M & M & & M+M_{m1} & K & N & \cdots & N \\ \hline K & K & \cdots & K & N+N_1 & N & \cdots & N \\ K & K & \cdots & K & N & N+N_2 & \cdots & N \\ \vdots & \vdots & & \vdots & \vdots & \vdots & & \vdots \\ K & K & \cdots & K & N & N & \cdots & N+N_{m2} \end{array} \right|$$

$$= 0. \tag{5.16}$$

By use of addition rule of the determinant, $|\Delta G'|$ becomes

$$(\frac{V_0}{\widetilde{RT}})^{m_1+m_2} |\Delta G'| = \left| \begin{array}{cccc|cccc} M-K+M_1 & M-K & \cdots & M-K & K & K & \cdots & K \\ M-K & M-K+M_2 & \cdots & M-K & K & K & \cdots & K \\ \vdots & \vdots & & \vdots & \vdots & \vdots & & \vdots \\ M-K & M-K & & M-K+M_{m1} & K & K & \cdots & K \\ \hline 0 & 0 & \cdots & 0 & N+N_1 & N & \cdots & N \\ 0 & 0 & \cdots & 0 & N & N+N_2 & \cdots & N \\ \vdots & \vdots & & \vdots & \vdots & \vdots & & \vdots \\ 0 & 0 & \cdots & 0 & N & N & \cdots & N+N_{m2} \end{array} \right|$$

$$
+\sum_{i=1}^{m_1}\left|\begin{array}{cccccccc:ccc}
M+M_1 & M & \cdots & M & K & M & \cdots & M & K & \cdots & K \\
M & M+M_2 & \cdots & M & K & M & \cdots & M & K & \cdots & K \\
\vdots & \vdots & \ddots & \vdots & \vdots & \vdots & & \vdots & \vdots & & \vdots \\
M & M & \cdots & M+M_{i-1} & K & M & \cdots & M & K & \cdots & K \\
M & M & \cdots & M & K & M & \cdots & M & K & \cdots & K \\
M & M & \cdots & M & K & M+M_{i+1} & \cdots & M & K & \cdots & K \\
\vdots & \vdots & & \vdots & \vdots & \vdots & \ddots & \vdots & \vdots & & \vdots \\
M & M & \cdots & M & K & M & \cdots & M+M_{m1} & K & \cdots & K \\
\hdashline
K & K & \cdots & K & K & K & \cdots & K & M+N_1 & \cdots & N \\
\vdots & \vdots & & \vdots & \vdots & \vdots & & \vdots & \vdots & \ddots & \vdots \\
K & K & \cdots & K & K & K & \cdots & K & N & \cdots & N+N_{m2}
\end{array}\right|
$$

$$
=0. \tag{5.17}
$$

Note that each element of the i-th column of determinant of (i+1)-th term of eqn (5.17) is K. By moving (i,i) element from (i,i) to (1,1) position, (i+1)-th term becomes

$$
\begin{array}{l}\text{(i+1)-th term} \\ \text{of eqn (5.17)}\end{array} = (-1)^{2i-2}\left|\begin{array}{ccccc:ccc}
K & M & \cdots & & M & K & \cdots & K \\
K & M+M_1 & & & \vdots & \vdots & & \vdots \\
\vdots & \vdots & \ddots & & & & & \\
 & & & [i] & & & & \\
\vdots & \vdots & & \ddots & \vdots & \vdots & & \vdots \\
K & M & \cdots & & M+M_{m1} & K & \cdots & K \\
\hdashline
K & K & \cdots & & K & N+N_1 & \cdots & N \\
\vdots & \vdots & & & \vdots & \vdots & \ddots & \vdots \\
K & K & \cdots & & K & N & \cdots & N+N_{m2}
\end{array}\right|
$$

$$
=\left|\begin{array}{ccccc:ccc}
K & M-K & \cdots & & M-K & 0 & \cdots & 0 \\
K & M-K+M_1 & & & \vdots & \vdots & & \vdots \\
\vdots & \vdots & \ddots & & & & & \\
 & & & [i] & & & & \\
\vdots & \vdots & & \ddots & \vdots & \vdots & & \vdots \\
K & M-K & \cdots & & M-K+M_{m1} & 0 & \cdots & 0 \\
\hdashline
K & 0 & \cdots & & 0 & N+N_1 & \cdots & N \\
\vdots & \vdots & & & \vdots & \vdots & \ddots & \vdots \\
K & 0 & \cdots & & 0 & N & \cdots & N+N_{m2}
\end{array}\right| \tag{5.18}
$$

$$\text{(i+1)-th term of eqn (5.17)} = \Big[\, K \begin{vmatrix} M-K+M_1 & \cdots & & M-K \\ \vdots & \ddots & & \vdots \\ & & [i] & \\ \vdots & & \ddots & \vdots \\ M-K & \cdots & & M-K+M_{m1} \end{vmatrix}$$

$$+ \sum_{j\neq i}^{m_1} (-1)^{2j-1} K \begin{vmatrix} M-K & 0 & \cdots & & & & 0 \\ M-K & M_1 & \cdots & & & & 0 \\ M-K & 0 & \ddots & & & & \vdots \\ \vdots & \vdots & & [i] & & & \\ & & & & \ddots & & \\ & & & & & [j] & \vdots \\ \vdots & \vdots & & & & \ddots & 0 \\ M-K & 0 & \cdots & & & & M_{m1} \end{vmatrix} \Big]$$

$$\times \begin{vmatrix} N-K+M_1 & \cdots & N-K \\ \vdots & \ddots & \vdots \\ N-K & \cdots & N-K+M_{m2} \end{vmatrix}$$

$$= \Big[\, K \begin{vmatrix} M-K+M_1 & \cdots & & M-K \\ \vdots & \ddots & & \vdots \\ & & [i] & \\ \vdots & & \ddots & \vdots \\ M-K & \cdots & & M-K+M_{m1} \end{vmatrix}$$

$$- \sum_{j\neq i}^{m_1} K(K-M) \frac{1}{M_i M_j} \prod_{k=1}^{m_1} M_k \Big] \times \begin{vmatrix} N-K+N_1 & \cdots & N-K \\ \vdots & \ddots & \vdots \\ N-K & \cdots & N-K+N_{m2} \end{vmatrix} . \qquad (5.21)$$

By substitution of eqns (5.19) and (5.21) into eqn (5.17), we obtain

$$\left(\frac{\underset{\sim}{V_0}}{RT}\right)^{m_1+m_2} |\Delta G'| = \begin{vmatrix} M-K+M_1 & \cdots & M-K \\ \vdots & \ddots & \vdots \\ N-K & \cdots & M-K+M_{m1} \end{vmatrix} \times \begin{vmatrix} N+N_1 & \cdots & N \\ \vdots & \ddots & \vdots \\ N & \cdots & N+N_{m2} \end{vmatrix}$$

where [i] means that (i,i) element (in this case, $M-K+M_{m1}$) is removed. With the help of the Laplace expansion of the determinant, 1st and (i+1)-th term of eqn (5.17) become

$$\text{1st term of eqn (5.17)} = \begin{vmatrix} M-K+M_1 & M-K & \cdots & M-K \\ M-K & M-K+M_2 & \cdots & \vdots \\ \vdots & \vdots & \ddots & \\ M-K & M-K & \cdots & M-K+M_{m1} \end{vmatrix} \times \begin{vmatrix} N+N_1 & N & \cdots & N \\ N & N+N_2 & \cdots & \vdots \\ \vdots & \vdots & \ddots & \\ N & N & \cdots & N+N_{m2} \end{vmatrix} \quad (5.19)$$

and

$$\text{(i+1)-th term of eqn (5.17)} = \begin{vmatrix} K & M-K & \cdots & & M-K \\ K & M-K+M_1 & \cdots & & \vdots \\ \vdots & \vdots & \ddots & & \\ & & [i] & & \\ \vdots & \vdots & & \ddots & \vdots \\ K & M-K & \cdots & & M-K+M_{m1} \end{vmatrix} \times \begin{vmatrix} N-K+N_1 & \cdots & N-K \\ \vdots & \ddots & \vdots \\ N-K & \cdots & N-K+N_{m2} \end{vmatrix}$$

$$= [(-1)^2 K \begin{vmatrix} M-K+M_1 & \cdots & & M-K \\ \vdots & \ddots & & \vdots \\ & [i] & & \\ \vdots & & \ddots & \vdots \\ M-K & \cdots & & M-K+M_{m1} \end{vmatrix}$$

$$+ \sum_{j \neq i}^{m_1} (-1)^{j+2} (-1)^{j-1} K \begin{vmatrix} M-K & M-K & \cdots & & & & M-K \\ M-K & M-K+M_1 & \cdots & & & & M-K \\ \vdots & \vdots & \ddots & & & & \vdots \\ & & & [i] & & & \\ & & & & \ddots & & \\ & & & & & [j] & \\ \vdots & \vdots & & & & \ddots & \vdots \\ M-K & M-K & \cdots & & & & M-K+M_{m1} \end{vmatrix}]$$

$$\times \begin{vmatrix} N-K+N_1 & \cdots & N-K \\ \vdots & \ddots & \vdots \\ N-K & \cdots & N-K+N_{m2} \end{vmatrix} . \quad (5.20)$$

In deriving eqn (5.20), cofactor expansion of determinant is also utilized. Eqn (5.20) can be rewritten as eqn (5.21).

$$+\sum_{i=1}^{m_1}\left[\,K\begin{vmatrix} M-K+N_1 & \cdots & & M-K \\ \vdots & \ddots & & \vdots \\ & & [i] & \\ \vdots & & \ddots & \vdots \\ M-K & \cdots & & M-K+M_{m1} \end{vmatrix}\right.$$

$$\left.-\sum_{j\neq i}^{m_i} K(M-K)\frac{1}{M_iM_j}\prod_{k=1}^{m_1} M_k\,\right]\times\begin{vmatrix} N-K+N_1 & \cdots & N-K \\ \vdots & \ddots & \vdots \\ N-K & \cdots & N-K+N_{m1} \end{vmatrix}$$

$$=\left(\prod_{k=1}^{m_1} M_k\right)\left[1+(M-K)\sum_{k=1}^{m_1}\frac{1}{M_k}\right]\times\left(\prod_{k=1}^{m_2} N_k\right)\times\left[1+N\sum_{k=1}^{m_2}\frac{1}{N_k}\right]$$

$$+\sum_{i=1}^{m_1}\left[\,K\cdot\frac{1}{M_i}\left(\prod_{k=1}^{m_1} M_k\right)\left[1+(M-K)\left\{\sum_{k=1}^{m_1}\frac{1}{M_k}-\frac{1}{Mi}\right\}\right]\right.$$

$$\left.-K(M-K)\sum_{j/i}^{m_1}\frac{1}{M_iM_j}\left(\prod_{k=1}^{m_1} M_k\right)\right]\times\left(\prod_{K=1}^{m_2} N_k\right)\left[1+(N-K)\sum_{k=1}^{m_2}\frac{1}{N_k}\right]. \qquad (5.22)$$

In derivation of eqn (5.22) the following relation is used.

$$|D| = \begin{vmatrix} x_1 & a & \cdots & a \\ a & x_2 & & \vdots \\ \vdots & & \ddots & \\ a & \cdots & & x_n \end{vmatrix} = f(a) - a\frac{df(x)}{dx} \qquad (5.23)$$

where

$$f(x) = \prod_{i=1}^{m} (x_i - x) \qquad (5.24)$$

and

$$|D| = \prod_{i=1}^{m} (x_i - a)\left[1 + a\sum_{k=1}^{m_1}\frac{1}{x_k - a}\right]. \qquad (5.25)$$

Now, we can readily rewrite eqn (5.22) as eqn (5.16').

$$|\Delta G'| = \left(\frac{\tilde{R}T}{V_0}\right)^{m_1+m_2} \left(\prod_{i=1}^{m_1} M_i\right) \left(\prod_{j=1}^{m_2} N_j\right)$$

$$\times \left[\left(1+M\sum_{i=1}^{m_1}\frac{1}{M_i}\right)\left(1+N\sum_{j=1}^{m_2}\frac{1}{N_j}\right) - K^2\left(\sum_{i=1}^{m_1}\frac{1}{M_i}\right)\left(\sum_{j=1}^{m_2}\frac{1}{N_j}\right)\right] = 0. \qquad (5.16')$$

Then spinodal condition is given by:

$$\left(1+M\sum_{i=1}^{m_1}\frac{1}{M_i}\right)\left(1+N\sum_{j=1}^{m_2}\frac{1}{N_j}\right) - K^2\left(\sum_{i=1}^{m_1}\frac{1}{M_i}\right)\left(\sum_{j=1}^{m_1}\frac{1}{N_j}\right) = 0. \qquad (5.26)$$

The neutral equilibrium condition eqn (5.13) can be rewritten through use of eqns (5.15a)-(5.15e) as eqn (5.27)

$$|\Delta G''| = \left(\frac{\tilde{R}T}{V_0}\right)^{m_1+m_2-1} \begin{vmatrix} W_{X1} & W_{X2} & \cdots & W_{Xm1} & W_{Y1} & W_{Y2} & \cdots & W_{Ym2} \\ M & M+M_2 & \cdots & M & K & K & \cdots & K \\ \vdots & \vdots & \ddots & \vdots & \vdots & \vdots & & \vdots \\ M & M & \cdots & M+M_{m1} & K & K & \cdots & K \\ K & K & \cdots & K & N+N_1 & N & \cdots & N \\ K & K & \cdots & K & N & N+N_2 & \cdots & N \\ \vdots & \vdots & & \vdots & \vdots & \vdots & \ddots & \vdots \\ K & K & \cdots & K & N & N & \cdots & N+N_{m2} \end{vmatrix}. \qquad (5.27)$$

The cofactor expansion of eqn (5.27) is:

$$\left(\frac{V_0}{\tilde{R}T}\right)^{m_1+m_2-1} |\Delta G''| = (-1)^2 W_{X1} \left|\begin{array}{ccc:ccc} M+M_2 & \cdots & M & K & \cdots & \vdots \\ \vdots & \ddots & \vdots & \vdots & & \vdots \\ \hdashline M & \cdots & M+M_{m1} & K & \cdots & K \\ K & \cdots & K & N+N_1 & \cdots & N \\ \vdots & & \vdots & \vdots & \ddots & \vdots \\ K & \cdots & K & N & \cdots & N+N_{m2} \end{array}\right|$$

$$+\sum_{i=2}^{m_1} W_{Xi}(-1)^{i+1}(-1)^{i-2}\begin{vmatrix} M & M & \cdots & & M & K & \cdots & K \\ M & M+M_2 & \cdots & & M & K & \cdots & K \\ \vdots & & \ddots & & M & \vdots & & \vdots \\ & & & [i] & \vdots & & & \\ \vdots & & & \ddots & & \vdots & & \vdots \\ M & \cdots & & & M+M_{m1} & K & \cdots & K \\ \hline K & \cdots & & & K & N+N_1 & \cdots & N \\ \vdots & & & & \vdots & \vdots & \ddots & \vdots \\ K & \cdots & & & K & N & \cdots & N+N_{m2} \end{vmatrix}$$

$$+\sum_{i=1}^{m_1} W_{Yi}(-1)^{m_1+(i+1)}(-1)^{m_1+(i-2)}\begin{vmatrix} K & K & \cdots & K & N & \cdots & & N \\ M & M+M_2 & \cdots & M & K & \cdots & & K \\ \vdots & \vdots & \ddots & \vdots & \vdots & & & \vdots \\ M & M & \cdots & M+M_{m1} & K & \cdots & & K \\ \hline K & K & \cdots & K & N+N_1 & N & \cdots & N \\ \vdots & \vdots & & \vdots & N & \ddots & & \vdots \\ & & & & \vdots & & [i] & \\ \vdots & \vdots & & \vdots & & & & \ddots \\ K & K & \cdots & K & N & \cdots & & N+N_{m2} \end{vmatrix} \qquad (5.28)$$

Eqn (5.28) can be rewritten as eqn (5.29) with the help of the cofactor expansion, Laplace expansion and eqn (5.16').

$$\left(\frac{V_0}{RT}\right)^{m_1+m_2-1}|\Delta G''| = \left(\frac{1}{M}\prod_{k=1}^{m_1} M_k\right)\left(\prod_{k=1}^{m_2} N_k\right)\left[\left\{1+M\left(\sum_{k=1}^{m_1}\frac{1}{M_k}-\frac{1}{M_1}\right)\right\} - \left(1+N\sum_{k=1}^{m_2}\frac{1}{N_k}\right)\right.$$

$$-K^2\left(\sum_{k=1}^{m_1}\frac{1}{M_k}-\frac{1}{M_1}\right)\left(\sum_{k=1}^{m_2}\frac{1}{N_k}\right)\Big] \quad \sum_{i=2}^{m_1} W_{Xi}\Big[\frac{M}{M_1 M_i}\left(\prod_{k=1}^{m_1} M_k\right)\left(\prod_{k=1}^{m_2} N_k\right)$$

$$\times\left(1+N\sum_{k=1}^{m_2}\frac{1}{N_k}\right)-\sum_{j=1}^{m_2}\frac{K^2}{M_1 M_i}\left(\prod_{k=1}^{m_1} M_k\right)\frac{1}{N_j}\left(\prod_{k=1}^{m_2} N_k\right)\Big] - \sum_{i=1}^{m_2} W_{Y,j}\frac{K}{M_1}\left(\prod_{k=1}^{m_1} M_k\right)$$

$$\times\frac{1}{N_i}\left(\prod_{k=1}^{m_2} N_k\right)$$

$$= \frac{1}{M_1}\left(\prod_{k=1}^{m_1} M_k\right)\left(\prod_{k=1}^{m_2} N_k\right)\left[\left(\sum_{i=1}^{m_1} \frac{W_{Xi}}{M_i}\right)\left\{K^2 \sum_{j=1}^{m_2} \frac{1}{N_j} - M\left(1+M \sum_{j=1}^{m_2} \frac{1}{N_j}\right)\right\} - K\left(\sum_{j=1}^{m_2} \frac{W_{YJ}}{N_j}\right)\right]. \qquad (5.29)$$

Combination of eqns (5.29) and (5.16') yields eqn (5.27').

$$|\Delta G''| = \left(\frac{\tilde{R}T}{V_0}\right)^{m_1+m_2-1} \left(\prod_{i=1}^{m_1} M_i\right)\left(\prod_{j=1}^{m_2} N_j\right)\left(M_1 \sum_{i=1}^{m_1} \frac{1}{M_i}\right)^{-1}$$

$$\times \left[\left(\sum_{i=1}^{m_1} \frac{W_{Xi}}{M_i}\right)\left(1+N\sum_{j=1}^{m_2} \frac{1}{N_j}\right) - K^2 \left(\sum_{i=1}^{m_1} \frac{1}{M_i}\right)\left(\sum_{j=1}^{m_2} \frac{W_{YJ}}{N_j}\right)\right] = 0 \qquad (5.27')$$

where W_{Xi} and W_{YJ} are

$$W_{Xi} \equiv \left(\frac{\partial |\Delta G'|}{\partial v_{Xi}}\right)_{T,P,v_k} \quad (k \neq X_i,\ k = X_1, \cdots, X_{m1}, Y_1, \cdots, Y_{m2}), \qquad (5.30a)$$

$$W_{YJ} \equiv \left(\frac{\partial |\Delta G'|}{\partial v_{YJ}}\right)_{T,P,v_k} \quad (k \neq Y_J,\ k = X_1, \cdots, X_{m1}, Y_1, \cdots, Y_{m2}). \qquad (5.30b)$$

The neutral equilibrium condition is then given by:

$$\left(\sum_{i=1}^{m_1} \frac{W_{Xi}}{M_i}\right)\left(1+N\sum_{j=1}^{m_2} \frac{1}{N_j}\right) - K^2 \left(\sum_{i=1}^{m_1} \frac{1}{M_i}\right)\left(\sum_{j=1}^{m_2} \frac{W_{YJ}}{N_j}\right) = 0. \qquad (5.31)$$

Eqns (5.26) and (5.31) agree completely with the Koningsveld-Chermin-Gordon's equations (eqns 12 and 13 of ref. 8), respectively. The parameters in eqns (5.26) and (5.31) can be replaced with experimentally determined parameters such as v_0, v_1, v_2, χ_{01}, χ_{02}, χ_{12}, the weight- and z-average X_i (and Y_J), $X_w{}^0$ and $X_z{}^0$ ($Y_w{}^0$ and $Y_z{}^0$) and theoretical expressions finally obtained from eqns (5.26) and (5.31):

$$\left(\frac{1}{v_0} + \frac{1}{v_1 X_w{}^0} - 2\chi_{01}\right)\left(\frac{1}{v_0} + \frac{1}{v_2 Y_w{}^0} - 2\chi_{02}\right)$$

$$- \left(\frac{1}{v_0} + \chi_{12} - \chi_{01} - \chi_{02}\right)^2 = 0 \qquad (5.32)$$

and

$$[(\frac{1}{v_0{}^2}-\frac{X_z{}^0}{(v_1X_w{}^0)^2})(\frac{1}{v_0}+\frac{1}{v_2X_w{}^0}-2\chi_{02})+\frac{1}{v_0{}^2}(\frac{1}{v_0}+\frac{1}{v_1X_w{}^0}-2\chi_{01})$$

$$-\frac{2}{v_0{}^2}(\frac{1}{v_0}+\chi_{12}-\chi_{01}-\chi_{02})]\ (\frac{1}{v_0}+\frac{1}{v_2Y_w{}^0}-2\chi_{02})$$

$$+\ [(\frac{1}{v_0{}^2}-\frac{Y_z{}^0}{(v_2Y_w{}^0)^2})(\frac{1}{v_0}+\frac{1}{v_1X_w{}^0}-2\chi_{01})+\frac{1}{v_0{}^2}(\frac{1}{v_0}+\frac{1}{v_2Y_w{}^0}-2\chi_{02})$$

$$-\frac{2}{v_0{}^2}(\frac{1}{v_0}+\chi_{12}-\chi_{01}-\chi_{02})]\ (\frac{1}{v_0}+\chi_{12}-\chi_{01}-\chi_{02})=0 \qquad (5.33)$$

with

$$v_0+v_1+v_2=1. \qquad (5.34)$$

Eqn (5.32) shows good coincidence with eqn 27 of ref. 8 (Koningsveld-Chermin-Gordon's equation). Eqns (5.32) and (5.33) are evidently symmetrical with respect to the exchange of polymers 1 and 2, respectively, which is the thermodynamical requirement. We can also derive the symmetrical forms of eqns (5.32) and (5.33) as follows:

By introducing parameters Q_X and Q_Y defined by the following equations,

$$\frac{1}{Q_X}\equiv\sum_{i=1}^{m_1}\frac{1}{M_i}=v_1X_w{}^0 \qquad (5.35a)$$

$$\frac{1}{Q_Y}\equiv\sum_{j=1}^{m_2}\frac{1}{N_j}=v_2Y_w{}^0 \qquad (5.35b)$$

into eqn (5.26), we obtain

$$\begin{vmatrix} M+Q_X & K \\ K & M+Q_Y \end{vmatrix}=0. \qquad (5.36)$$

Eqn (5.36) can be rewritten as:

$$(M+Q_X-K)\ (N+Q_Y-K)+K(M+Q_X-K)+K(N+Q_Y-K)=0. \qquad (5.37)$$

Substitution of eqns (5.15a)-(5.15c), (5.35a) and (5.35b) into eqn

(5.37) yields eqn (5.32'), which is the symmetrical form of eqn (5.32).

$$\left(\frac{1}{X_w^0 v_1} + \chi_{02} - \chi_{01} - \chi_{02}\right)\left(\frac{1}{Y_w^0 v_2} + \chi_{01} - \chi_{02} - \chi_{12}\right)$$

$$+\left(\frac{1}{v_0} + \chi_{12} - \chi_{01} - \chi_{02}\right)\left(\frac{1}{X_w^0 v_1} + \chi_{02} - \chi_{01} - \chi_{12}\right)$$

$$+\left(\frac{1}{v_0} + \chi_{12} - \chi_{01} - \chi_{02}\right)\left(\frac{1}{Y_w^0 v_2} + \chi_{01} - \chi_{02} - \chi_{12}\right) = 0. \qquad (5.32')$$

The neutral equilibrium condition eqn (5.31) can also be represented as

$$\begin{vmatrix} Q_X R_X & Q_Y R_Y \\ K & N+Q_Y \end{vmatrix} = 0 \qquad (5.38)$$

where R_X and R_Y are defined by:

$$R_X \equiv \sum_{i=1}^{m_1} \frac{W_{Xi}}{M_i}, \qquad (5.39)$$

$$R_Y \equiv \sum_{i=1}^{m_2} \frac{W_{Yj}}{M_j}. \qquad (5.40)$$

Eqn (5.38) is rewritten as,

$$Q_X R_X\,[K + (N + Q_Y - K)] + \frac{Q_X R_X K (N + Q_Y - k)}{M + Q_X - K} - \frac{Q_X R_X K (N + Q_Y - K)}{M + Q_Y - K} - (Q_Y R_Y) K$$

$$= \frac{Q_X R_X}{M + Q_X - K}\,[K (M + Q_X - K) + (N + Q_Y - K)(M + Q_X - K) + K (N + Q_Y - K)]$$

$$- \frac{K}{M + Q_Y - K}\,[Q_X R_X (N + Q_Y - K) + Q_Y R_Y (M + Q_X - K)] = 0 \qquad (5.41)$$

and with the help of eqn (5.37), eqn (5.41) becomes:

$$\frac{Q_X R_X}{M + Q_X - K} + \frac{Q_Y R_Y}{N + Q_Y - K} = 0 \qquad (5.42)$$

here, $Q_X R_X$ and $Q_Y R_Y$ are calculated as,

$$\frac{Q_X R_X}{\text{const.}} = \frac{1}{v_0^{\ 2}}(M + Q_Y - K) + \frac{1}{v_0^{\ 2}}(N + Q_Y - K) - X_z^{\ 0} Q_X^{\ 2}\{(N + Q_Y - K) + K\},$$

(5.43a)

$$\frac{Q_Y R_Y}{\text{const.}} = \frac{1}{v_0^{\ 2}}(M + Q_X - K) + \frac{1}{v_0^{\ 2}}(N + Q_X - K) - Y_z^{\ 0} Q_Y^{\ 2}\{(N + Q_X - K) + K\}.$$

(5.43b)

Substitution of eqns (5.43a) and (5.43b) into eqn (5.42) gives

$$\frac{1}{v_0^{\ 2}}\left[\frac{1}{M + Q_X - K} + \frac{1}{N + Q_Y - K}\right]\left[(M + Q_X - K) + (N + Q_Y - K)\right] - \frac{X_z^{\ 0} Q_X^{\ 2}}{M + Q_X - K}$$

$$\times\{(N + Q_X - K) + K\} - \frac{Y_z^{\ 0} Q_Y^{\ 2}}{N + Q_Y - K}\{(M + Q_X - K) + K\} = 0. \qquad (5.44)$$

Eqn (5.44) can, using eqn (5.37), be calculated as:

$$\frac{1}{v_0^{\ 2}}(M + Q_X - K)^3 (N + Q_Y - K)^3 + X_z^{\ 0} Q_X^{\ 2} K^3 (N + Q_Y - K)^3$$

$$+ Y_z^{\ 0} Q_Y^{\ 2} K^3 (M + Q_X - K)^3 = 0 \qquad (5.45)$$

Substituting eqns (5.15a)-(5.15e), (5.35a) and (5.35b) into eqn (5.45), eqn (5.33') is finally obtained.

$$\frac{1}{v_0^{\ 2}}\left(\frac{1}{X_w^{\ 0} v_1} + \chi_{02} - \chi_{01} - \chi_{12}\right)^3 \left(\frac{1}{Y_w^{\ 0} v_2} + \chi_{01} - \chi_{02} - \chi_{12}\right)^3$$

$$+ \frac{X_z^{\ 0}}{(X_w^{\ 0} v_1)^2}\left(\frac{1}{Y_w^{\ 0} v_2} + \chi_{01} - \chi_{02} - \chi_{12}\right)^3 \left(\frac{1}{v_0} + \chi_{12} - \chi_{01} - \chi_{02}\right)^3$$

$$+ \frac{Y_z^{\ 0}}{(Y_w^{\ 0} v_2)^2}\left(\frac{1}{X_w^{\ 0} v_1} + \chi_{02} - \chi_{01} - \chi_{12}\right)^3 \left(\frac{1}{v_0} + \chi_{12} - \chi_{01} - \chi_{02}\right)^3 = 0.$$

(5.33')

Eqns (5.32') and (5.33') are obviously symmetrical with respect to the exchange of polymers 1 and 2, respectively.

Then, SC can be calculated from eqn (5.32) and v_1, v_2 and v_p at CSP (referred to as $v_1^{\ c}$, $v_2^{\ c}$ and $v_p^{\ c}$) can be evaluated by solving simultaneous equations (eqns (5.32)-(5.34)).

For a single component polymer 1 / single component polymer 2 / single solvent system, eqns (5.32) and (5.33) reduce to those whose X_w^0 (and X_z^0) and Y_w^0 (and Y_z^0) are replaced by X and Y, respectively.

$$\left(\frac{1}{v_0}+\frac{1}{v_X X}-2\chi_{01}\right)\left(\frac{1}{v_0}+\frac{1}{v_Y Y}-2\chi_{02}\right)-\left(\frac{1}{v_0}+\chi_{12}-\chi_{01}-\chi_{02}\right)^2 = 0 \qquad (5.32'')$$

and

$$\Big[\left(\frac{1}{v_0^2}-\frac{1}{v_X^2 X}\right)\left(\frac{1}{v_0}+\frac{1}{v_Y X}-2\chi_{02}\right)+\frac{1}{v_0^2}\left(\frac{1}{v_0}+\frac{1}{v_X X}-2\chi_{01}\right)$$

$$-\frac{2}{v_0^2}\left(\frac{1}{v_0}+\chi_{12}-\chi_{01}-\chi_{02}\right)\Big]\left(\frac{1}{v_0}+\frac{1}{v_Y Y}-2\chi_{02}\right)$$

$$+\Big[\left(\frac{1}{v_0^2}-\frac{1}{v_Y^2 Y}\right)\left(\frac{1}{v_0}+\frac{1}{v_X X}-2\chi_{01}\right)+\frac{1}{v_0^2}\left(\frac{1}{v_0}+\frac{1}{v_Y Y}-2\chi_{02}\right)$$

$$-\frac{2}{v_0^2}\left(\frac{1}{v_0}+\chi_{12}-\chi_{01}-\chi_{02}\right)\Big]\left(\frac{1}{v_0}+\chi_{12}-\chi_{01}-\chi_{02}\right)=0. \qquad (5.33'')$$

These equations agree exactly with those (eqns (5.1) and (5.2)) derived by Tompa (refs 2,3). Inspection of eqn (5.32) indicates that spinodal point is a function of X_w^0, Y_w^0, χ_{01}, χ_{02} and χ_{12} and is independent of the polydispersity of the two polymers.

5.3 EFFECTS OF MOLECULAR WEIGHT DISTRIBUTION OF POLYMERS AND THREE χ PARAMETERS ON THE SPINODAL CURVE AND CRITICAL SOLUTION POINT

Put the left hand side of eqns (5.32) and (5.33) with A and B. A and B (spinodal and neutral equilibrium equations) become the functions of two variables v_1 and v_2 by use of eqn (5.34) when χ_{01}, χ_{02}, χ_{12}, X_w^0, X_z^0, Y_w^0 and Y_z^0 are known as prerequisites. For given v_2, the spinodal equation can be represented by a single variable v_1 ($A=A(v_1)=0$) and v_1 at spinodal point can be evaluated using a single variable Newton method. By solving two-variable (v_1 and v_2) simultaneous eqns (5.32) and (5.33) with the aid of two-variable Newton method, CSP (namely v_1^c and v_2^c (and v_0^c)) can be evaluated.

The computer experiments of SC and CSP were performed by

Kamide et al. (ref. 10) under the following conditions: $\chi_{12} = -0.4 \sim 0.2$, χ_{01} and $\chi_{02} = 0 \sim 2.0$, $X_w{}^0$ and $Y_w{}^0 = 1 \sim 1\times10^4$, and $X_w{}^0/X_n{}^0$ and $Y_w{}^0/Y_n{}^0 = 1 \sim 5$. The original polymer was assumed to have the Schulz-Zimm type molecular weight distribution (MWD). Note that $X_z{}^0 - X_w{}^0 = X_w{}^0 - X_n{}^0$ and $Y_z{}^0 - Y_n{}^0 = Y_w{}^0 - Y_n{}^0$ hold in this case.

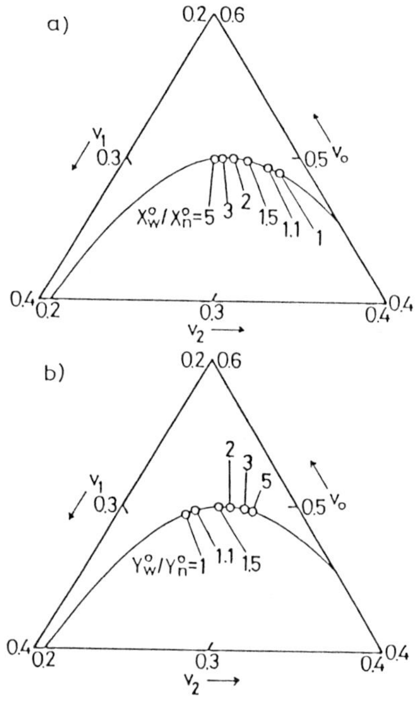

Fig. 5.1 Effects of $X_w{}^0/X_n{}^0$[a)] and $Y_w{}^0/Y_n{}^0$[b)] on spinodal curve and critical solution point of quasi-ternary systems: Original polymers 1 and 2, Schulz-Zimm type distribution ($X_w{}^0 = Y_w{}^0 = 300$); a) $Y_w{}^0/Y_n{}^0 = 2$, b) $X_w{}^0/X_n{}^0 = 2$; $\chi_{01} = 0.2$, $\chi_{02} = 0.3$, $\chi_{12} = 0.01$.

Fig. 5.1 shows the effect of the polydispersity of polymers 1 and 2 on SC and CSP of the quasi-ternary solutions. Here, the two polymers have the Schulz-Zimm distribution with $X_w{}^0$ or $Y_w{}^0 = 300$, and $\chi_{01} = 0.2$, $\chi_{02} = 0.3$ and $\chi_{12} = 0.01$ are assumed. All CSPs are exactly on the spinodals as theory predicts, moving depending on polymer polydispersity. In Fig. 5.1a, for a fixed $Y_w{}^0/Y_n{}^0$ (= 2) CSP shifts to the side of polymer 1 as $X_w{}^0/X_n{}^0$ increases from 1 to 5. Reversely, in Fig. 5.1b, for a fixed $X_w{}^0/X_n{}^0$ (= 2) CSP shifts to the side of polymer 2 as $Y_w{}^0/Y_n{}^0$ increases. Namely, CSP moves to the side of the polymer whose MWD is widened. This coincides well with the relation between CSP and the polydispersity of multicomponent polymer dissolved in a single solvent or of a multicomponent polymer in binary solvent mixture systems (ref. 11). Evidently, the CSPs for the multicomponent polymer deviate significantly from that for a single component polymer, at least in the range X_w^0/X_n^0 or $Y_w^0/Y_n^0 \gtrsim 1.05$. The CSP locates at the peak of the SC when the polymolecularities of both polymers coincide with each other (in this case $X_w^0/X_n^0 = Y_w^0/Y_n^0 = 2$).

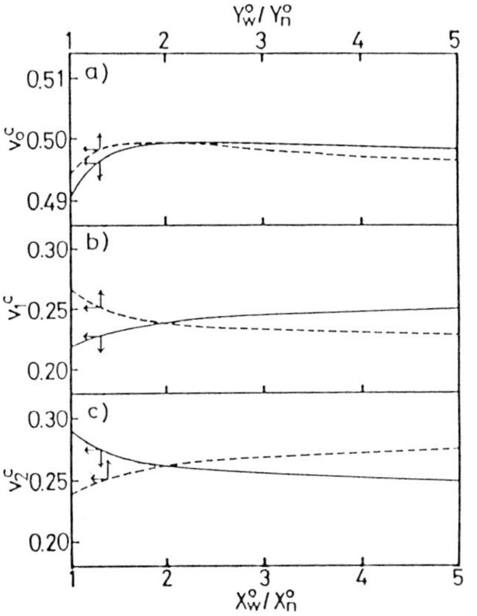

Fig. 5.2 Plots of v_0^c, v_1^c and v_2^c versus X_w^0/X_n^0 (solid line) or Y_w^0/Y_n^0 (broken line) of the original polymers 1 and 2 of quasi-ternary systems: Original polymers 1 and 2, Schulz-Zimm type distribution ($X_w^0 = Y_w^0 = 300$; solid line, $Y_w^0/Y_n^0 = 2$; broken line, $X_w^0/X_n^0 = 2$); $\chi_{01} = 0.2$, $\chi_{02} = 0.3$, $\chi_{12} = 0.01$.

Fig. 5.2 shows the effect of polymer polymolecularity on the composition at the CSP. As X_w^0/X_n^0 (solid line) or Y_w^0/Y_n^0 (broken line) increases, v_0^c increases rapidly, then decreases slowly after passing through maximum. v_1^c (or v_2^c) increases with an increase in X_w^0/X_n^0 (or Y_w^0/Y_n^0) and decreases gradually with an increase in Y_w^0/Y_n^0 (or X_w^0/X_n^0). It is of interest that the effect of polymolecularity becomes remarkable when the range X_w^0/X_n^0 (or Y_w^0/X_n^0) is less than 2.

Fig. 5.3 shows the effects of the weight-average degrees of polymerization, X_w^0 and Y_w^0, on SC (solid line) and CSP (open circle). Here, χ_{01} and χ_{12} are taken as 0.2 and 0.01, respectively, and χ_{02} is taken as 0.3 (Fig. 5.3a and b) or 1.0 (Fig. 5.3c and d). The polymers 1 and 2 are assumed to have the Schulz-Zimm type MWD ($X_w{}^0/X_n{}^0 = Y_w{}^0/Y_n{}^0 = 2$, and $Y_w{}^0 = 300$ in Fig. 5.3a and c and $X_w{}^0 = 300$ in Fig. 5.3b and d). At $\chi_{02} = 0.3$ with an increase in $X_w{}^0$ (Fig. 5.3a) the area of the single phase region decreases and CSP approaches a single point on the v_0 axis (shown as unfilled triangle) which corresponds to the Flory solvent composition of the multicomponent polymer / binary solvent mixture system (ref. 13). We denote this point as the "Flory (or θ) solution composition".

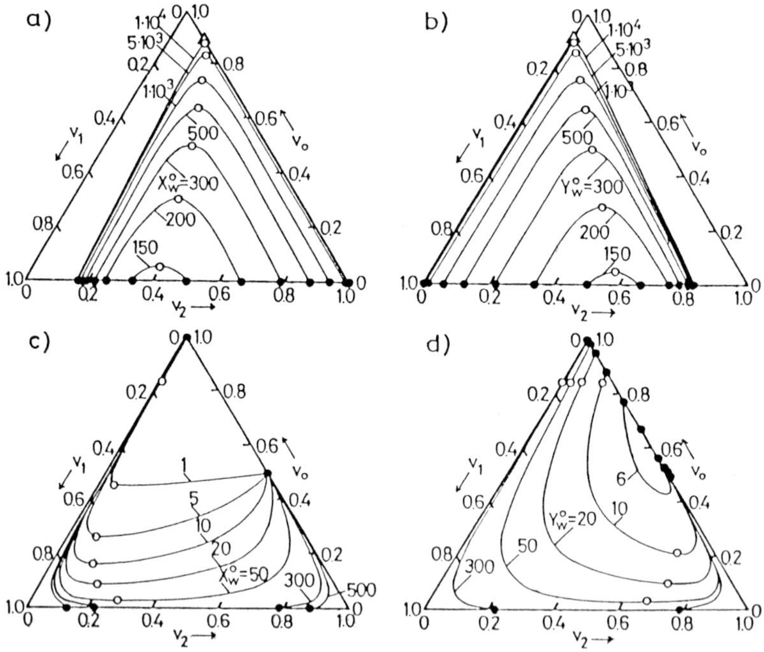

Fig. 5.3 Effect of the average molecular weight of polymer 1 [$X_w{}^0$, a) and c)] and polymer 2 [$Y_w{}^0$, b) and d)] on the spinodal curve and critical solution point of quasi-ternary systems: Original polymers 1 and 2, Schulz-Zimm type distribution [$X_w{}^0/X_n{}^0 = Y_w{}^0/Y_n{}^0 = 2$; a) and c) $Y_w{}^0 = 300$, b) and d) $X_w{}^0 = 300$]; $\chi_{01} = 0.2$, $\chi_{12} = 0.01$, a) and b) $\chi_{02} = 0.3$, c) and d) $\chi_{02} = 1.0$; ○ critical solution point, ● intersection point of the spinodal curve and composition triangle, △ Flory solution composition.

Regardless of the MWD of the original polymer, at the limit of $X_w^0 \to \infty$ (and also $X_z^0 \to \infty$), following relation holds:

$$(v_0^c, v_1^c, v_2^c) = (v_0^F, 0, v_2^F). \tag{5.46}$$

Assuming v_1^c approaches zero ($= v_1^F$) with order of $(X_w^0)^{-a}$ $(0 < a < 1)$ with increasing X_w^0; namely, $1/(v_1^c X_w^0)$ $(O((X_w^0)^{a-1}))$ $\to 0$ when $X_w^0 \to \infty$, we can obtain the equation of "Flory (or θ) solution composition" as eqn (5.47) from the equation for the spinodal condition (eqn (5.32)).

$$\left(\frac{1}{v_0} - 2\chi_{01}\right)\left(\frac{1}{v_0} + \frac{1}{v_2 Y_w^0} - 2\chi_{02}\right) - \left(\frac{1}{v_0} + \chi_{12} - \chi_{01} - \chi_{02}\right)^2 = 0 \tag{5.47}$$

and in this case $(v_0^F, v_1^F, v_2^F) = (0.8934, 0, 0.1066)$.

As the "Flory solution point" on solvent-polymer 2 axis is the CSP at the limit of $X_w^0 \to \infty$, neutral equilibrium condition (eqn (5.33)) is simultaneously satisfied regardless of v_0 and v_2 (ref. 10). Similarly, with an increase in Y_w^0 (Fig. 5.3b, $\chi_{02} = 0.3$) CSP approaches the "Flory solution composition" (unfilled triangle) which is evaluated by eqn (5.48),

$$\left(\frac{1}{v_0} + \frac{1}{v_1 X_w^0} - 2\chi_{01}\right)\left(\frac{1}{v_0} - 2\chi_{02}\right) - \left(\frac{1}{v_0} + \chi_{12} - \chi_{01} - \chi_{02}\right)^2 = 0 \tag{5.48}$$

and in this case $(v_0^F, v_1^F, v_2^F) = (0.9263, 0.0737, 0)$. Eqn (5.48) is derived in the same manner as eqn (5.47). When $\chi_{02} = 1.0$, with decrease in X_w^0 (Fig. 5.3c), the SC approaches that for a polymer / binary solvent mixture, and with a decrease in Y_w^0 (Fig. 5.3d) an interesting phenomenon, that although mutual solubility exists between polymer 1 and solvent and between polymers 1 and 2, the solution of solvent and polymer 2 does not dissolve polymer 1. This phenomenon is similar to cosolvency, already observed in the computer experiments by Matsuda and Kamide (ref. 14, section 3.4). The filled circle in Fig. 5.3a-d is the point of intersection between the composition triangle and the spinodal curve (see, eqns (5.50) - (5.52)).

Fig. 5.4 illustrates X_w^0 (or Y_w^0) dependence of v_0^c, v_1^c and v_2^c at $(\chi_{01}, \chi_{02}, \chi_{12}) = (0.2, 1.0, 0.01)$ (namely, $\chi_{01} \neq \chi_{02}$). Evidently, the systems have two CSP in the range of $X_w^0 = 1 \sim 120$

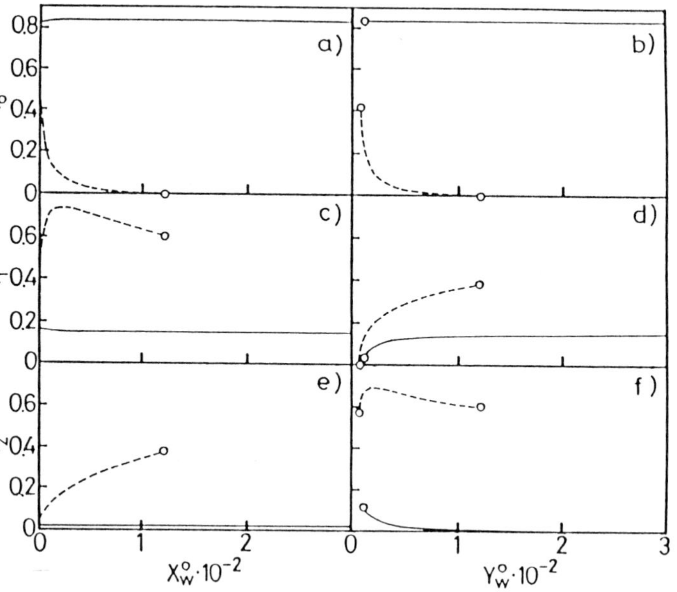

Fig. 5.4 Plots of v_0^c, v_1^c and v_2^c against X_w^0 [a), c) and d)] or Y_w^0 [b), d) and f)] of quasi-ternary solutions: Original polymers 1 and 2, Schulz-Zimm type distribution [$X_w^0/X_n^0 = Y_w^0/Y_n^0 = 2$; a), c) and e) $Y_w^0 = 300$; b), d) and f) $X_w^0 = 300$]; $\chi_{01} = 0.2$, $\chi_{02} = 1.0$, $\chi_{12} = 0.01$; solid line, critical solution point in higher v_0 region; broken line, critical solution point in lower v_0 region; ○ limiting composition of CSP within phase triangle.

for a fixed Y_w^0 (= 300) and $Y_w^0 = 10 \sim 120$ for fixed X_w^0 (= 300) (see, Fig. 5.3c and d). v_0^c, v_1^c and v_2^c found in higher v_0 region are almost constant over a wide range of X_w^0 and Y_w^0 (Fig. 5.4a-f). But in relatively small Y_w^0 region (with $X_w^0 = 300$) v_1^c decreases (and v_2^c increases) gradually with a decrease in Y_w^0 and in the case of $Y_w^0 \lesssim 8$ CSP disappears (in other words, CSP moves to the outside of the composition triangle; see, Fig. 5.3d). The open circles at the end of solid line in Fig. 5.4b, d and e show the compositions of CSP at $Y_w^0 \simeq 8$. When Y_w^0 is smaller than 5 (for fixed X_w^0 (= 300)) polymers 1 and 2 are perfectly miscible throughout the entire composition region. In the lower v_0 region, v_0^c decreases with increase in X_w^0 (or Y_w^0) [broken line in Fig. 5.4a (or Fig. 5.4b)], approaching zero at $X_w^0 \simeq 120$ (or $Y_w^0 \simeq 120$) (open circle in the figures). As X_w^0 increases at a fixed Y_w^0 (= 300), v_1^c increases first and then decreases after passing a minimum, but v_2^c increases monotonically (broken line in Fig. 5.4c and e). Y_w^0 dependence of v_1^c and v_2^c on a constant X_w^0 (= 300) is similar to their X_w^0 dependence at the constant Y_w^0, but their details are different.

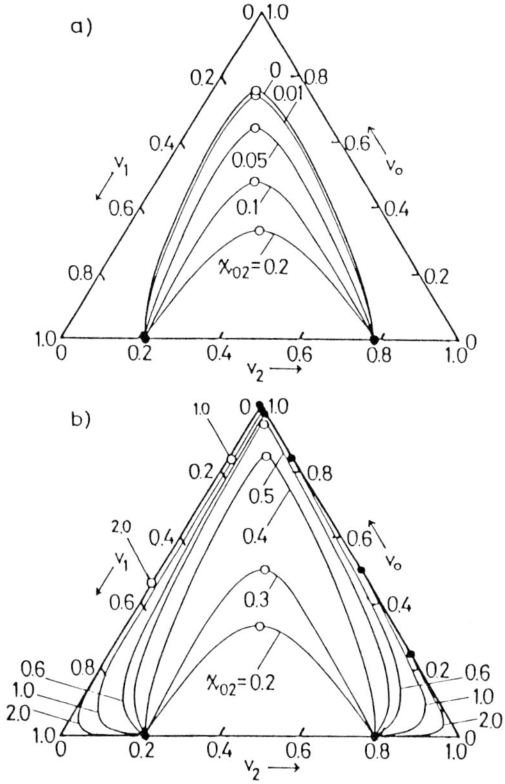

Fig. 5.5 Effect of χ_{02} on spinodal curve and CSP of quasi-ternary system consisting of multicomponent polymer 1, multicomponent polymer 2 and single solvent: Original polymers 1 and 2, Schulz-Zimm type distribution ($X_w^0 = Y_w^0 = 300$, $X_w^0/X_n^0 = Y_w^0/Y_n^0 = 2$); $\chi_{12} = 0.01$, $\chi_{01} = 0.2$; a) $\chi_{02} = 0 \sim 0.2$ and b) $\chi_{01} = 0.2 \sim 2.0$; ○ critical solution point, ● intersection point of the spinodal curve and composition triangle.

Fig. 5.5a and b show the effect of χ_{02} on the SC (solid line) and the CSP (open circle). Here, two other χ parameters are taken as constants; $\chi_{01} = 0.2$ and $\chi_{12} = 0.01$, polymers 1 and 2 are assumed to have the Schulz-Zimm type distribution with $X_w^0 = Y_w^0 = 300$ and $X_w^0/X_n^0 = Y_w^0/Y_n^0 = 2$. At $\chi_{02} = 0.2$ (= χ_{01}, in this case) the area of the single phase region becomes maximum. In other words, use of a single solvent having similar solubility as opposed to polymer 1 and polymer 2, widens effectively the single phase region. When χ_{02} ranges from 0.6 to 0.8, CSP cannot be found within the compositional triangle diagram (i.e., the critical solution point does not exist). In the range $0.8 < \chi_{02} < 2.0$, the critical solution point lies nearly on the v_1 axis.

The crossing point of SC and the three axes (v_0, v_1 and v_2 axes) of the triangle diagram (filled circle) can be theoretically calculated (eqns (5.50)-(5.52)). Multiplying both sides of eqn (5.32') by $v_0 v_1 v_2$ gives

$$v_0 \left\{\frac{1}{X_w^0} + (\chi_{02} - \chi_{01} - \chi_{12}) v_1\right\} \left\{\frac{1}{Y_w^0} + (\chi_{01} - \chi_{02} - \chi_{12}) v_2\right\}$$

$$+ v_1 \{1 + (\chi_{12} - \chi_{01} - \chi_{02}) v_0\} \left\{\frac{1}{Y_w^0} + (\chi_{01} - \chi_{02} - \chi_{12}) v_2\right\}$$

$$+ v_2 \{1 + (\chi_{12} - \chi_{01} - \chi_{02}) v_0\} \left\{\frac{1}{X_w^0} + (\chi_{02} - \chi_{01} - \chi_{12}) v_1\right\} = 0. \tag{5.49}$$

By putting $v_0 = 0$ (and, of course, utilizing $v_1 + v_2 = 1$) in eqn (5.49), we can obtain eqn (5.52). Eqns (5.50) and (5.51) are similarly derived from eqn (5.49).

$$\frac{v_0}{Y_w^0} + v_2 - 2\chi_{02} v_0 v_2 = 0, \tag{5.50}$$

$$\frac{v_0}{X_w^0} + v_1 - 2\chi_{01} v_0 v_1 = 0, \tag{5.51}$$

$$\frac{v_1}{X_w^0} + \frac{v_2}{Y_w^0} - 2\chi_{12} v_1 v_2 = 0. \tag{5.52}$$

The intersection point of SC and v_2-axis (polymer 1-2 axis) are independent of χ_{02} and are determined by the combination of X_w^0, Y_w^0 and χ_{12} (see, eqn (5.52)). In this case (v_0, v_1, v_2) = (0, 0.2113, 0.7887) and (0, 0.7887, 0.2113).

Fig. 5.6 shows the effect of χ_{02} on v_0^c, v_1^c and v_2^c. At χ_{02} $(= \chi_{01}) = 0.2$, v_0^c attains a minimum and both v_1^c and v_2^c reach a maximum. In the range of $0.55 \lesssim \chi_{02} \lesssim 0.82$, the spinodal curve intersects with the v_0-axis (solvent-polymer 2 axis) and CSP moves toward the outside of the composition triangle (see, Fig. 5.5b). The intersection point can be calculated by use of eqn

(5.50). In the region of $\chi_{02} \gtrsim 0.82$, CSP appears again near v_1-axis (solvent-polymer 2 axis) and v_1^c increases and v_0^c decreases with an increase in χ_{02} (see, Figs 5.5b and 5.6b).

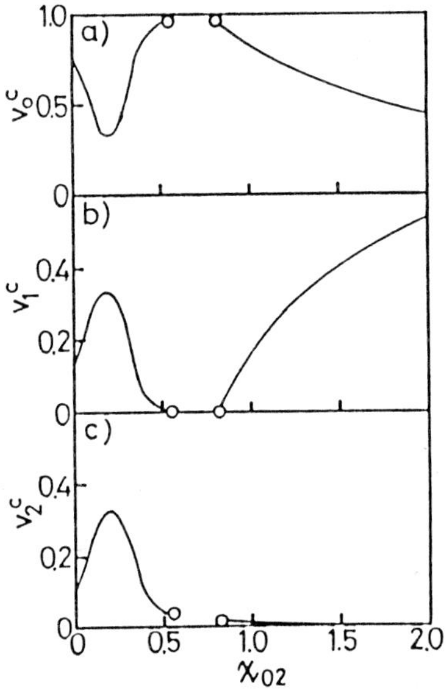

Fig. 5.6 Dependence of the composition of CSP, v_0^c, v_1^c and v_2^c on χ_{02} of quasi-ternary systems: Original polymers 1 and 2, Schulz-Zimm type distribution ($X_w^0 = 300$, $X_w^0/X_n^0 = 2$, $Y_w^0 = 300$, $Y_w^0/Y_n^0 = 2$); solid line, $\chi_{01} = 0.2$; broken line, $\chi_{01} = 1.0$.

Fig. 5.7a-d show the effect of χ_{12} on SC (solid line) and CSP (open circle). Here, χ_{01} is taken as constant (0.2) and $X_w^0 = Y_w^0 = 300$ and $X_w^0/X_n^0 = Y_w^0/Y_n^0 = 2$. In the case of $\chi_{02} = 0.3$ (Fig. 5.7a), polymers 1 and 2 are almost incompatible at $\chi_{12} = 1.0$ and accordingly there is a very narrow one-phase region whose area increases with a decrease in χ_{12}. By use of eqn (5.52), we can conclude that no two-phase region exists when $\chi_{12} < 1/150$ ($\simeq 0.0067$). When v_0 reduces zero, the system approaches the multicomponent polymers 1 / multicomponent polymers 2 system (i.e., quasi-binary polymer blend), whose thermodynamics will be discussed in detail in Chapter 4 (refs 15,16). At $\chi_{12} = 0.4$ (Fig. 5.7b) SCs are "closed loops" and two CSPs are observed when $\chi_{12} < 1/150$ ($\chi_{12} = 0$ and 0.004 in the figure), as were calculated by Zemann and Patterson (ref. 4). This phenomenon may be related to the cononsolvency (ref. 14).

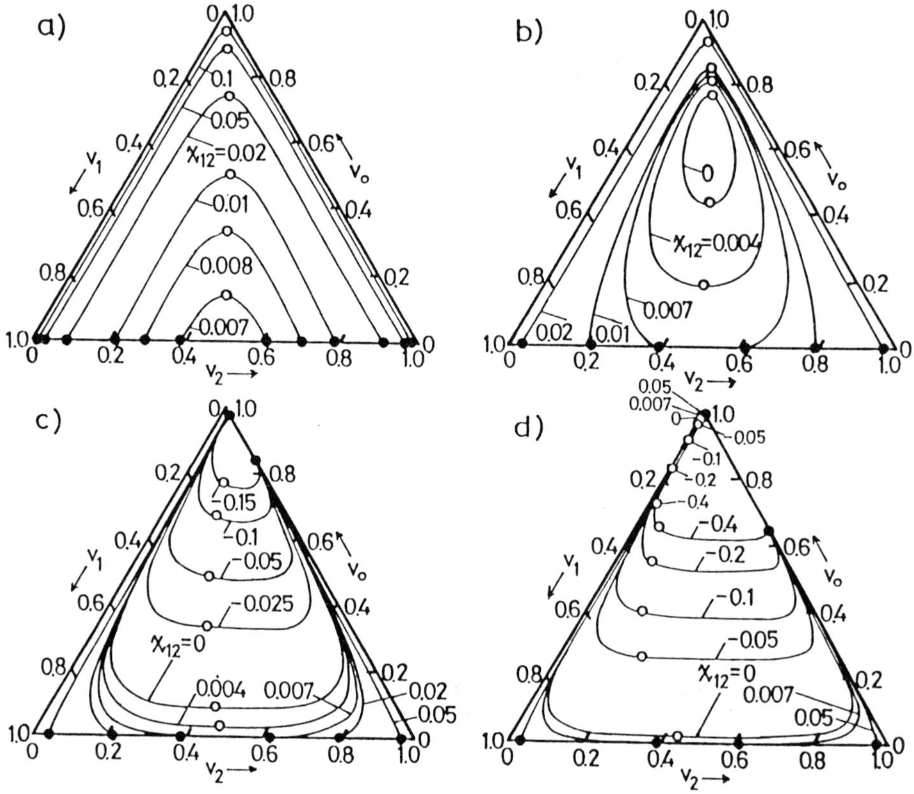

Fig. 5.7 Effects of χ_{12} on spinodal curve and CSP of quasi-ternary systems: Original polymers 1 and 2, Schulz-Zimm type distribution ($X_w{}^0 = Y_w{}^0 = 300$, $X_w{}^0/X_n{}^0 = Y_w{}^0/Y_n{}^0 = 2$); $\chi_{01} = 0.2$; a) $\chi_{02} = 0.3$, b) $\chi_{02} = 0.4$, c) $\chi_{02} = 0.6$, d) $\chi_{02} = 0.8$; ○ critical solution point, ● intersection points of the spinodal curve and axes of composition triangle graph.

At $\chi_{02} = 0.6$ and 0.8 (Fig. 5.7c and d), SCs intersect with v_0-axis (solvent-polymer 2 axis) and CSP at the higher v_0 side diminishes. Intersection points are evaluated as $(v_0, v_1, v_2) = (0.8489, 0, 0.1511)$ and $(0.9816, 0, 0.0184)$ for $\chi_{02} = 0.6$ and $(v_0, v_1, v_2) = (0.6285, 0, 0.3715)$ and $(0.9944, 0, 0.0056)$ for $\chi_{02} = 0.8$, by eqn (5.50). In the case of $\chi_{02} = 0.8$, another CSP are observed near the v_1 axis in the higher v_0 region and two CSPs approach as χ_{12} decreases.

Summarizing, the equations of spinodal and neutral equilibrium conditions (eqns (5.32) and (5.33), respectively) were derived for multicomponent polymers 1 / multicomponent polymers 2

/ single solvent systems and using these equations, SC can be calculated from χ_{01}, χ_{02}, χ_{12}, $X_w{}^0$ and $Y_w{}^0$ and CSP can be evaluated from χ_{01}, χ_{02}, χ_{12}, $X_w{}^0$, $Y_w{}^0$, $X_z{}^0$ and $Y_z{}^0$. With an increase in the breadth of the MWD of a given polymer, CSP shifted in the triangle phase diagram to the side of the polymer. The critical solvent volume fraction $v_0{}^c$ gets large with an increase in either $X_w{}^0$ and $Y_w{}^0$. In particular, at the limit of $X_w{}^0 \rightarrow \infty$ (or $Y_w{}^0 \rightarrow \infty$), CSP approaches "Flory solvent composition", which was theoretically predicted (eqns (5.47) and (5.48)). Equations giving the cross points of the polymer 1-polymer 2 axis of the triangle diagram and SC were also derived. Solving these equations, the conditions for appearance of a closed loop type two-phase region were determined.

REFERENCES

1 R. L. Scott, J. Chem. Phys., 17 (1949) 268.
2 H. Tompa, Trans. Faraday Soc., 45 (1949) 1142.
3 H. Tompa, Polymer Solutions, Butterworths Scientific Publications, London, 1956.
4 L. Zemann and D. Patterson, Macromolecules, 5 (1972) 513.
5 C.C. Hsu and J.M. Prausnitz, Macromolecules, 7 (1974) 320.
6 K. Šolc, Macromolecules, 19 (1986) 1166.
7 K. Šolc, Macromolecules, 20 (1987) 2506.
8 R. Koningsveld, H.A.G. Chermin and M. Gordon, Proc. Roy. Soc. Lond. A., 319 (1970) 331.
9 R. Koningsveld, Discuss. Faraday Soc., 49 (1970) 144.
10 K. Kamide, S. Matsuda and H. Shirataki, Polym. J., 20 (1988) 949.
11 K. Kamide and S. Matsuda, Polym. J., 18 (1986) 347.
12 P.J. Flory, Principle of Polymer Chemistry, Cornell Univ. Press, Ithaca, 1953.
13 S. Matsuda and K. Kamide, Polym. J., 19 (1987) 203.
14 S. Matsuda and K. Kamide, Polym. J., 19 (1987) 211.
15 H. Shirataki, S. Matsuda and K. Kamide, Brit. Polym. J., to be published.
16 H. Shirataki, S. Matsuda and K. Kamide, Brit. Polym. J., to be published.

Chapter 6

APPLICATION OF PHASE EQUILIBRIA: FORMATION OF POROUS POLYMERIC MEMBRANE BY PHASE SEPARATION METHOD

6.1 INTRODUCTION

Recently, polymer membranes have been and are attracting keen attention as precise media of separating materials. The separation of materials by means of polymer membranes can be readily carried out without any phase change at room temperature using a compact apparatus, which can be operated with a relatively low energy consumption as compared with other conventional methods. One of the disadvantages of the membrane separation technology seems low scale merit.

Fig. 6.1 shows the separation methods and the corresponding size of particles, which can be separated by the membranes.

In fact, some membrane systems have been successfully commercialized since 1950. However, the history of application of polymer membranes for material separation greatly predates 1950. In Table 6.1 are collected the milestone scientific discoveries and industrial achievements in the field of cellulose membrane separation: Nollet (ref. 1) is probably the first, who as early as 1748 discovered the phenomenon of osmosis by observing the pig's bladder membrane burst when it was fixed to the mouth of a bottle, in which spirit of wine was contained and then placed in water. Artificial membranes from natural and synthetic polymers appeared in the late 19th century. The first was cellulose itself and its derivative and in particular, collodion membranes (i.e., cellulose nitrate membranes), invented by Fick in 1855 (ref. 2). It is only since the 1960s, when polymer science started to progress very rapidly, that polymer membranes became the target of scientific and industrial research. Leob (ref. 3) discovered reverse osmosis in 1956 and then, developed a reverse osmosis membrane based on cellulose triacetate in the late 1950s and 1960s. Their motive force was an attempt to skim off fresh water, predicted, according to the Gibbs adsorption equation, to exist immediately above the surface of hydrophobic material contacted with saline water (ref. 4).

Among numerous methods proposed hitherto for preparing polymeric membrane such as the solvent cast method, the radiation-

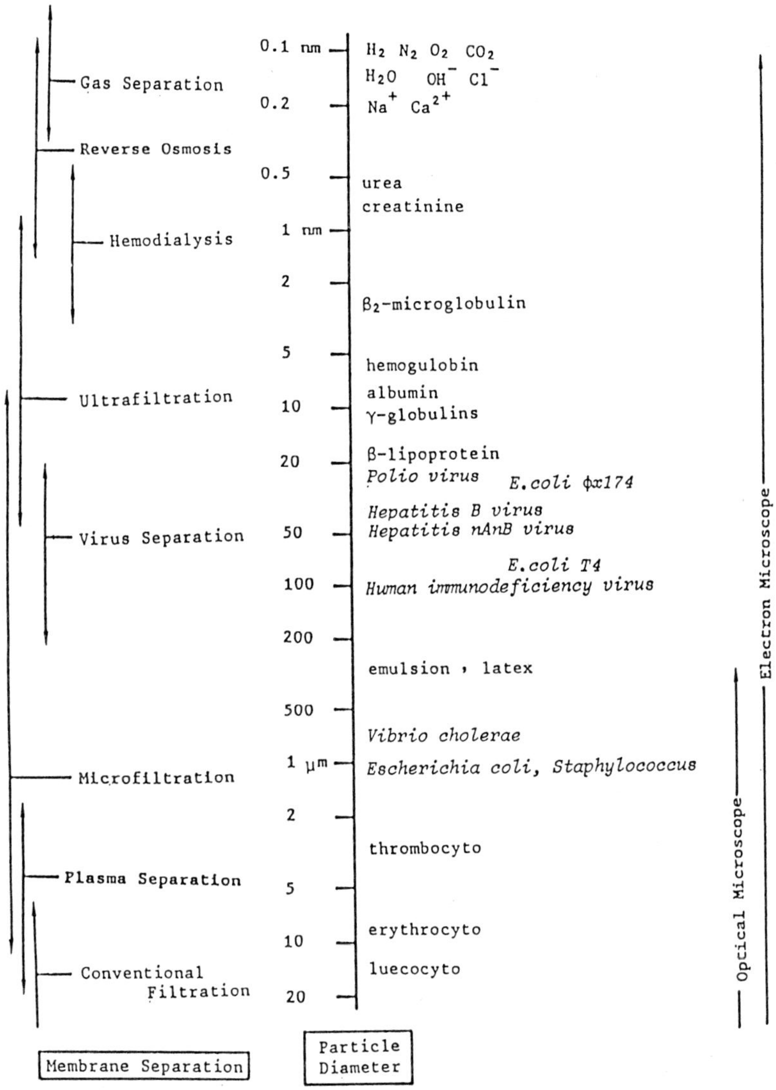

Fig. 6.1 Separation methods and size of particles to be separated.

Table 6.1
Brief history of cellulose membrane research and industry.

Year	Name	Membrane	Objects	Remarks
1748	A.J.A.Nollet	pig's bladder	aq.alcohol	osmosis
1855	A.Fick	collodion (cellulose nitrate)		surgery
1857	E.Schweitzer	Schweitzer reagent		
1867	M.Traube	copper ferrocyanide (semi-permeable membrane)		dialysis membrane
1877	W.Pfeffer	copper ferrocyanide formed on unglazed pottery	aq.sucrose	osmotic pressure
1885	J.H.van't Hoff			van't Hoff's law
1892	C.F.Cross, E.J.Bevan	viscose	fiber	
1908	E.Brandenberger	cellophane	(packing)	from viscose
1914	J.J.Abel, B.B.Turner, L.B.Rowentree	collodion	acetylsalicylic intoxication (in vitro)	albumin loss 1st hemo-dialysis
1933	W.J.Elford	collodion	virus	pore size estimation
1937	W.Thalheimer	cellophane tube	uremia (in vivo)	
1944	W.J.Kolff	cellophane	acute renal failure (lower nephrosis by crash injury)	1st clinical use
1956	S.Leob, S.Sourirajan	cellulose triacetate	saline water	reverse-osmosis
1960	W.Quinton et al.	A-V shunt	chronic renal failure	repeat-use

track-etching method, the stretching method and the sintering method, the solvent cast method is a method of vast technological importance, as it enables us to produce membrane with a wide range of mean pore sizes.

Fig. 6.2 shows correlations between casting conditions and performance of membranes prepared by the solvent casting method. It is expected that the performance of a membrane will be predominantly governed by pore characteristics and supermolecular structures, which are formed through phase separation phenomena of casting solutions, and the phase separation is unquestionably controlled by the casting conditions. Then, there will be strong sequential relationship; casting conditions $\rightleftarrows$ pore or structure formation $\rightleftarrows$ pore and supermolecular characteristics $\rightleftarrows$ performance of membrane (Fig. 6.2b). If these relations could be established quantitatively, we could design and produce effectively membranes with desired performances.

Between around 1970 and 1985 only the phenomenological

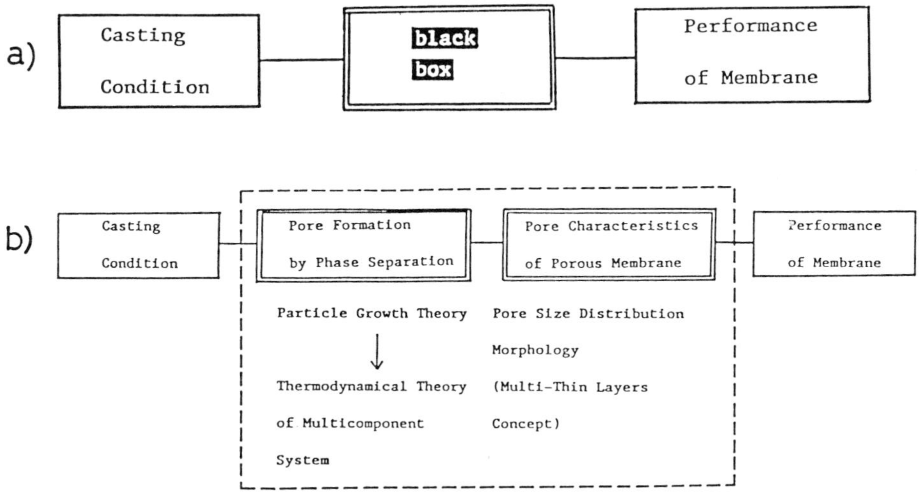

Fig. 6.2 Correlations between casting condition and performance of membrane.

correlations between preparative conditions and performance of membranes were studied not in a systematic manner (see, for example, general reviews on membrane science and technology until 1980, refs 5,6) and polymeric membranes were used commercially with only fragmental or without detailed scientific knowledge on pore characteristics such as the mean pore size, pore size distribution and porosity of the membranes. Then, there existed a large and uncoverable "black box" between the casting conditions and the performance of the membrane (Fig. 6.2a). In order to give better and quantitative understanding of the theoretical background of pore formation and to develop techniques for the pore characterization of membranes, systematic efforts were made to large extent for these ten years (refs 7-16), although even now we are far from a detailed understanding and there are many unsolved problems: For example, the physical meaning of the mean pore size, as determined by the water-flow rate methods (refs 5,7) of solvent-cast polymer membranes, consisting of multi-layers and having irregular pore shape, remains obscure.

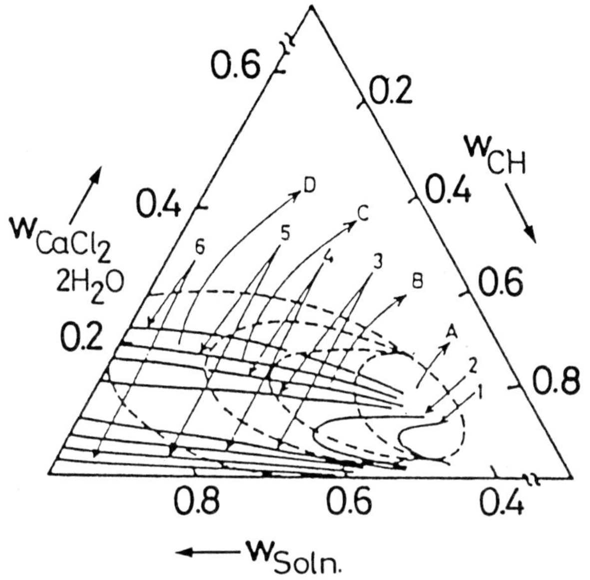

Fig. 6.3 Dependence of porosity Pr (see, eqn (6.105)) and of mean pore radius r_L, as determined by the latex method, on the composition of casting solution consisting of cellulose acetate solution (simply referred to as soln. in the figure), cyclohexanol and $CaCl_2 \cdot 2H_2O$: Soln. means the system of cellulose acetate-acetone-methanol of 20.3:100:33 by weight ratio. Curves 1, 2, 3, 4, 5 and 6 indicate the contour lines of Pr of 80, 70, 60, 50, 40 and 30%, respectively. A, B, C and D represent the regions from which the membranes of r_L larger than 0.15, 0.15~0.08, 0.08~0.02 and 0.02~0.01μm, respectively, are formed. The mean pore radius by the latex method, r_L, is determined to be equal to a radius of latex particles, at which the sharpest slope of a relation between latex particle concentration of a filtrate, and logarithm of radius of the latex particles is obtained.

In addition, studies on the pore-forming mechanism linking the solvent cast conditions and the pore characteristics (Fig. 6.2b) still remains at a very primitive stage, because the thermodynamics of the phase separation, on which the fundamentals of membrane formation by the solvent cast method are principally based, has advanced only very recently to a degree where they are applicable to the practice of membrane formation. Even now, studies on membranes are mostly concerned with preparative methods, supermolecular structure and permeability performance, and very few are devoted to the pore-forming and the permeability mechanism: Both of which are of fundamental importance (ref. 17).

Fig. 6.3 illustrates an important effect of the preparative conditions on pore characteristics such as mean pore size and porosity of membranes (refs 18,19). In this figure, cellulose diacetate membranes were prepared by (1) casting solution of cellulose diacetate (the weight-average molecular weight; $M_w = 1.05 \times 10^5$; total degree of substitution, $\langle F \rangle = 2.35$) / acetone / methanol / cyclohexanol / $CaCl_2 \cdot 2H_2O$ system on a glass plate and (2) allowing vaporization of volatile solvent at 293K for 20min, then (3) coagulating with water at 291K and (4) washing with water and dipping into methanol at 293K for 24hr, washing with methanol repeatedly and drying in vacuo. For a given composition of the system, mean pore size (full line) and porosity Pr (broken line) are determined unambiguously, if other things are equal.

Since Kamide and Manabe presented their tentative theory, no research has been done on these subjects. It is now timely desired to establish the membrane formation mechanism and to correlate it with the pore characteristics of membranes thus prepared. In this chapter, an extended summary of some pioneering work carried out by Kamide and his collaborators will be presented as an example of the application of thermodynamics of phase equilibrium. This is, we believe, the only instance in which a systematic study has been carried out on the thermodynamics of membrane formation. An exception is the works of Smolders et al. (refs 20-23), who studied the mechanism for structure formation of asymmetric membranes in cellulose acetate / acetone / water or polyurethane / dimethylformamide (DMF) / water in 1977. They explained the formation of the sponge-like structure, found in the latter system, in terms of a liquid-liquid phase separation with nucleation and growth of the polymer-lean phase. Interestingly, in the same year Kamide et al. independently proposed a theory for nuclei generation and their growth based on a liquid-liquid phase separation for membrane formation from cellulose acetate solutions (refs 18,19): They investigated in detail the compositions of two coexisting phases and two-phase volume ratio R, observing by optical and electron microscopies the existence of particles and the process of non-circular and circular pore formation.

Thereafter, Smolders et al. (ref. 21) explained the skin layer formation in the system of cellulose acetate / acetone / water in terms of rapid desolvation from the solution, showing that when solvent outflow is suppressed and nonsolvent inflow is favoured by addition of solvent to the coagulation bath, longer time is necessary for the solution to attain cloud point, resulting in obstruction of the skin layer formation to give porous structure of the membrane due to liquid-liquid phase separation. In addition, they calculated the diffusion of each component in the membrane forming process, discussing the relations between the concentration change of the components and the structure of asymmetric membranes (refs 22,23). In short, Smolders and his collaborators have confirmed their attention mainly to the formation of asymmetric membranes, whose mean pore sizes probably lie in the range of several nanometers.

Fig. 6.4 shows the phenomenological progress of phase separation of casting solution in the two typical processes. (1) Wet method: When the polymer solution is extruded through a T-shape or circular die into the coagulation bath, phase separation occurs from the surface of the extruded solution by contact with the nonsolvent. (2) Dry method: Immediately after the polymer solution is cast on the solid (glass, metal) plane surface placed in air, a volatile solvent component near the surface begins to evaporate, resulting in a significant concentration gradient, perpendicular to the solution surface. Then, the surface region of the solution becomes quickly turbid, indicating the occurrence of a phase separation by a relative decrease in good volatile solvent. In both the case of (1) and (2), the phase separation proceeds from the surface to the inner part. This strongly suggests that the detailed conditions of phase separation will change significantly along the direction of thickness of the cast solution, resulting in a less-than-negligible change in the pore characteristics of the membrane solid giving multi-layer structures, and that the pore size distribution within a hypothetical thin layer can be considered to be constant.

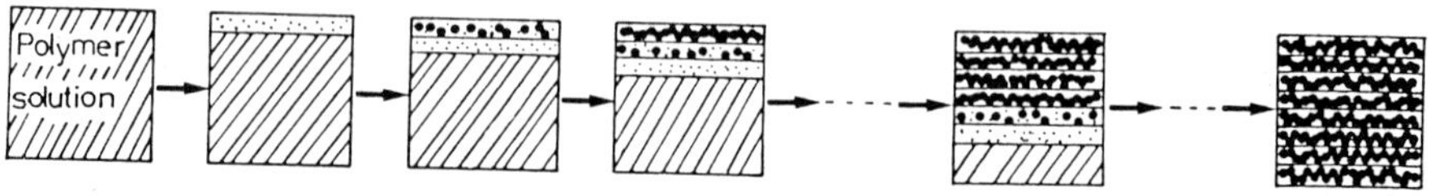

Fig. 6.4 Progress of phase separation in membrane casting process.

It has been experimentally proved that as for plane membrane, the pore characteristics change depending on the location in the thickness direction: Merin and Cheryan (ref. 24) studied the structure of plane ultrafiltration polysulfone membrane, using scanning and transmission electron microscopes and estimated the pore size distribution of a surface layer of the membrane. Based on electron microscopic observation of a cross section of plane cellulose acetate membrane, Purz et al. (ref. 25) suggested that the pore size distribution depends on the distance from the surface of a membrane. However, electron microscopic (EM) characterization of membranes is only capable of showing the surface and cross section of the membrane (refs 9-15). An accurate estimation of the pore characteristics of porous polymeric membranes requires detailed information through the thickness of the membrane. This estimation is particularly important for membranes prepared through the micro-phase separation method, since the change of pore characteristics with the distance from a membrane surface can be predicted theoretically. Applying the EM method to a series of thin sections cut off in parallel to the membrane surface (see, Fig. 6.4), Manabe et al. (ref. 15) have verified experimentally for regenerated cellulose and cellulose acetate membranes the change of pore characteristics in thickness direction, correlating changes in the phase separation condition with the thickness direction of the membranes.

First of all, the underlying mechanism of pore formation in the solvent casting method is schematically demonstrated in Fig. 6.5. Membrane formation by the solvent casting has the following striking, but not surprising, characteristics from the standpoint of phase equilibrium theory: (1) A cast solution has a relatively higher polymer concentration v_p^0 and (2) the relative weight ratio of polymer partitioned into the polymer-rich phase (i.e., gel phase), ρ_p is large (i.e., $\rho_p \geq 0.9$). (3) A system is not closed, but open and as a result, a part of solvent and nonsolvent come into and go out from the system. (4) The route to attain phase equilibrium governs significantly the membrane characteristics, which can not simply be explained by an equilibrium theory. In other words, nucleation and growth of nuclei to particles by diffusion and thereafter by amalgamation play an important role in gel membrane forming. (5) Phase separation occurs generally from the outside (or surface) of a cast solution by contacting nonsolvent vapor or liquid. (6) The detailed conditions of phase

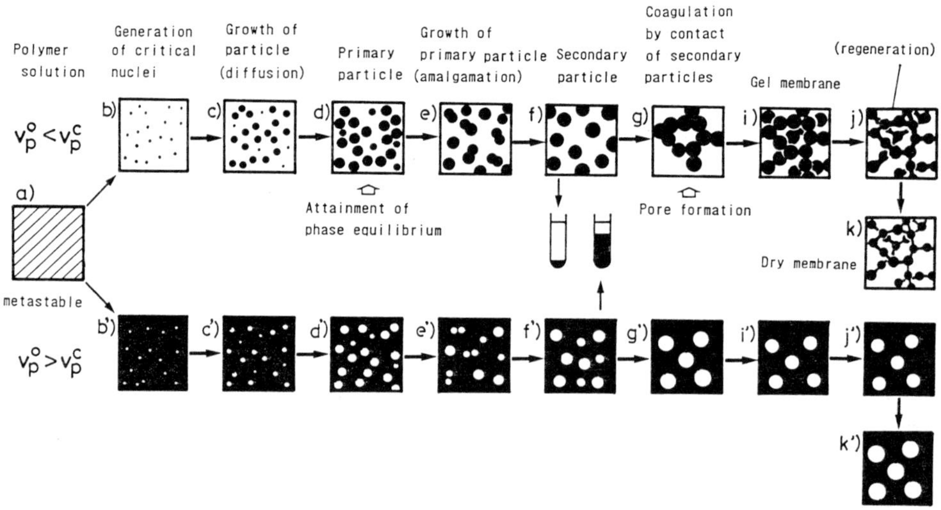

Fig. 6.5 Elementary steps in porous membrane formation by the micro-phase separation method: v_p^0, polymer volume fraction of the solution when the phase separation occurs; v_p^c, polymer volume fraction of critical solution point; steps a), g), i), j) and k) correspond to those of Fig. 6.45, respectively.

separation at inner layers of the cast solution differ remarkably from macroscopic conditions including polymer concentration and other compositions of the solution.

Depending on the initial polymer concentration [i.e., the solution concentration at the instance of phase separation, which should be strictly distinguished from the polymer concentration of the starting (in this case, casting) solution, v_p^s] v_p^0, either the polymer-rich phase or the polymer-lean phase will initially separates as the disperse phase from the solution. If $v_p^0 <$ the critical solution concentration v_p^c, the polymer-rich phase separates as small particles suspended in a medium (i.e., polymer-lean phase), and these particles grow by amalgamation. The interstitial space between particles gives pore (non-circular pore) (Fig. 6.5b-k).

When $v_p{}^0$ is larger than $v_p{}^c$, the polymer-lean phase is separated as shown in Fig. 6.5b'-k'. Since this phase has a comparatively low viscosity, the particles of the polymer-lean phase rapidly become larger by amalgamation. The aggregated polymer-lean phase particles are themselves circular, smooth pores. Even in this case, stage (g') in Fig. 6.5 should be stable in order to prepare the microporous membrane. Fig. 6.5b'-k' strongly suggests that circular pores will be found from concentrated solution. Fig. 6.6 shows typical electron micrographs of two kinds of porous membranes. The mechanism will be explained in detail and some experimental data presented to support the theory.

When a coagulated gel membrane is subjected to stretching or drawing under tension as occasionally is the case in the hollow fiber membrane manufacturing process, the morphology of the membrane, in particular, the shape of the secondary, particle changes dramatically to become microfibrils.

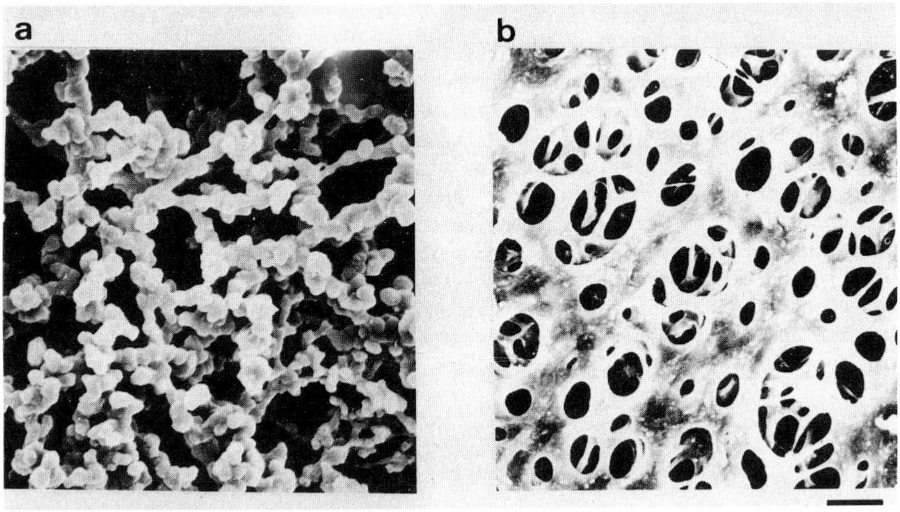

Fig. 6.6 Typical pore shapes of polymeric membranes:
a) non-circular pore; b) circular pore: Regenerated cellulose.

In short, the pore formation model in Fig. 6.5 is based on the "particle-growth concept" based on the observation of Kamide et al. in 1977 that the such small particles existed during the membrane-casting process of cellulose acetate solutions (ref. 18): They observed primary particles having a diameter of 50~ 100nm of the polymer-rich phase and their amalgamation into secondary particles having a diameter of 1μm in the casting process.

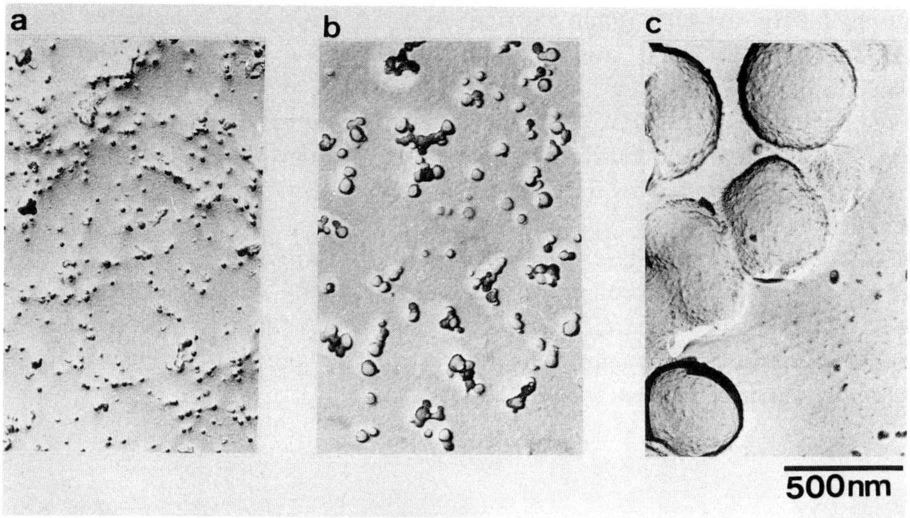

Fig. 6.7 Electron micrographs of particles in cellulose cuprammonium solution coagulated by aq. ammonia-acetone mixture (steps d, e and g in Fig. 6.5): a) primary particles (step d); b) intermediate stage of the primary particles to the secondary particles (step e); c) the secondary particles contacting with each other (step g): Starting solution (the weight fractions of cellulose, $w_{CELL}{}^{S}$; copper, $w_{CU}{}^{S}$; ammonia, $w_{NH_3}{}^{S}$ and water, $w_{H_2O}{}^{S}$ are 0.08, 0.0316, 0.1122 and 0.7762, respectively; the suffix S denote starting solution). Composition of coagulant, acetone / ammonia / water (30.00/0.56/69.44, wt/wt/wt). For electron microscopic observation, the solution cast was frozen instantaneously at 77K in liq. nitrogen and frozen solid was fractured, being carbonized and metallized concurrently at 163K under 10^{-7}mmHg; a scale bar stands for 500nm.

It is very interesting to note that as early as 1953 Kobayashi (ref. 26) speculated the optical density measurements during the addition of a nonsolvent including water for polyacrylonitrile / dimethylformamide system (and other polar organic solvents) that there was a gradual change from "molecular colloid" to "particle colloid", which is stable, often for several weeks without conversion to a precipitate. Although this particle colloid seems to correspond to the secondary particle defined in this work, he did not ascertain directly the existence of the particles. Now, porous polyacrylonitrile membranes are commercially produced on a large scale, using the solvent-casting method.

Fig. 6.7 shows the electron micrographs of cellulose cuprammonium / acetone solution cast in air (ref. 27). The existence in the solution of small particles with a diameters of ca.20nm was observed. Fig. 6.7b shows the intermediate stage of the small particles growing into the larger particles. Kamide et al. defined the smaller particles as the primary particles and the larger particles as the secondary particles. The rigorous definitions of the primary and secondary particles are given later (see, sections 6.2 and 6.3). The diameter of the particles in the intermediate stage is distributed in the range of 80nm and 200nm. Secondary particles with a diameter of 300nm to 1μm are also observed.

The particle-growth concept can be applied to any polymer membrane cast from the solution under the condition of $v_p^0 < v_p^C$. For example, Riley et al. (ref. 28) were the first to study the structure of cellulose acetate membranes used for reverse osmosis (RO) by electron microscopy (EM). They observed a dense layer with a thickness of 250nm on the RO membrane surface and concluded that there were no pores with diameters larger than 10nm, which is well detection within the capacities of EM. Later, a more detailed investigation of the surface of cellulose acetate ultra-thin membranes was made by Schultz and Asunmaa (ref. 29), who obtained EM photograph on carbon replica. They suggested that the most closely packed particles had a mean diameter of 18.8 ± 0.3nm. The radius of circles (i.e., the radius of circular pores), surrounded by these particles (see, Fig. 6.39c) was calculated as 2.13nm and the effective circle radius, obtained by subtracting the thickness of water monolayer on the particles from the apparent radius of the circle (2.13nm), was 1.85nm.

Electron micrographs of the active layer of commercially available RO membrane (Fig. 6.8) show that the active layer is composed of particles with diameters in the range 50～200nm. The porosity of the layer is low compared with the microfiltration membranes and the crevas of the particles which are in contact with each other act as pores. The cross section of the thin layer with a thickness of 2μm on the surface has particles with a mean

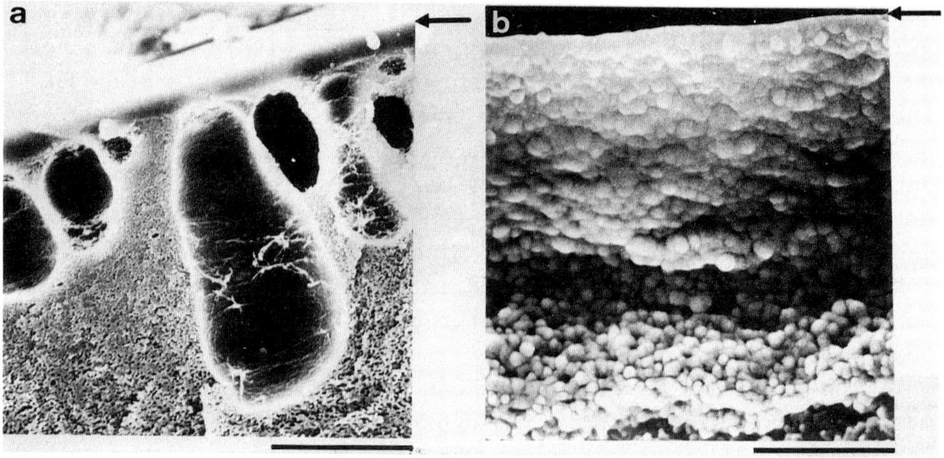

Fig. 6.8 Electron micrographs of cross-section of reverse osmosis membrane: Cellulose acetate reverse osmosis membrane assembled in a cartridge for water purification (Millipore Corp., Bedford, MA, U.S.A.) was frozen in liq. nitrogen and was cracked. A piece of the broken membrane was lyophilized at 153K for 48hr by Freeze Specimen Processing Device FD-2A (Eiko Engineer. Co., Ibaraki, Japan). Cross-section of the dried membrane was sputtered with gold in Ion Sputter JFC-1100 (JEOL Ltd., Tokyo, Japan) and was observed on a Field Emission Scanning Electron Microscope S-800 (Hitachi, Ltd., Tokyo, Japan); arrows indicate surface of the membrane.

diameter of 88.7nm, which gives an effective pore radius of 7.0 to 15.0nm (see, eqns (6.95) and (6.96)), three to seven times larger than the value estimated from the surface observation and suggesting strongly the existence of much smaller particles (with diameter of ca.20nm) on the surface.

6.2 NUCLEATION AND GROWTH OF PARTICLES

6.2.1 Nucleation

Investigation with cellulose diacetate solutions, Fig. 6.3 (refs 18,19) encouraged Kamide and Manabe to formulate a tentative theory of nucleation of critical nucleus (refs 27 and 30). Recently, a more plausible theory was advanced by Kamide et al. (ref. 31).

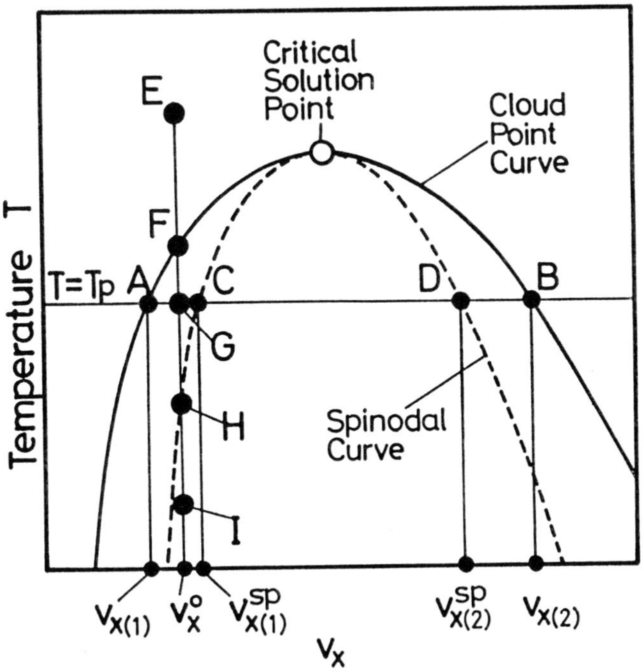

Fig. 6.9 Cloud point curve (solid line), spinodal curve (broken line) and critical solution point (open circle) of monodisperse polymer / solvent system: A and B, coexisting points (at $T=T_P$; T_P, phase separation temperature); C and D, spinodal points ($T=T_P$); E, one phase point; F, cloud point; G, metastable one phase point; H, spinodal point ($v_x=v_x^0$); I, unstable point.

Before examine the details of the theory, a general explanation of the mechanism of nucleation from the polymer solution will be presented. The generation and growth of polymer particles (nuclei) by phase separation from homogeneous solution is based principally on the same concepts used to explain the condensation of liquid droplets from super saturated vapours or the formation of ice particles from super-cooled liquids: In homogeneous polymer solutions in the metastable region of the phase diagram (point G in Fig. 6.9 or point J in Fig. 6.10) critical nuclei (see, eqn (6.11)), are formed by "concentration fluctuation" and can grow further in size spontaneously. The generation of the precipitated nuclei is always time-retarded and the nuclei thus formed grow by passing through a potential barrier (see, Fig. 6.11). On the other hand, under the appropriate conditions the precipitation occurs by passing through the critical point is instantaneous because of lack in potential barrier: This is spinodal decomposition (ref. 32).

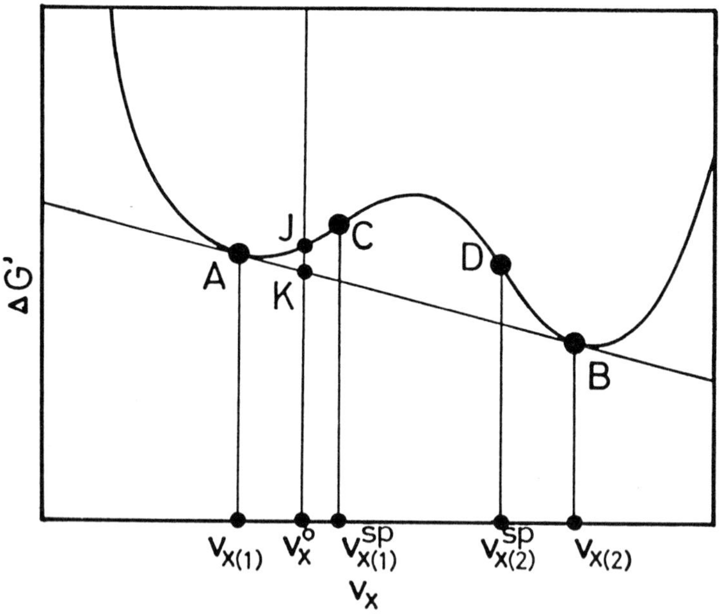

Fig. 6.10 Gibbs' free energy change ΔG' (eqn (6.3)) per unit volume at a given temperature T_P for monodisperse polymer / solvent system: A and B, coexisting points; C and D, spinodal points; J, metastable one phase point; K, point of average free energy of two coexisting phases.

Fig. 6.9 shows the cloud point curve (full line), spinodal curve (broken line), and critical solution point (open circle) for monodisperse polymer (X-mer) / single solvent system and Fig. 6.10

shows the Gibbs' free energy change $\Delta G'$ per unit volume at a given temperature T_P for X-mer / single solvent system. $v_{X(1)}$ and $v_{X(2)}$ are the polymer volume fractions of the polymer-lean and -rich phases in equilibrium and $v_{X(1)}^{SP}$ and $v_{X(2)}^{SP}$ are spinodal concentrations. In the figure, points A and B are two-phase equilibrium points and points C and D are spinodal points. If the initial polymer concentration v_X^0 lies at T_P between $v_{X(1)}$ and $v_{X(2)}$, the two-phase separation occurs. When v_X^0 lies between $v_{X(1)}$ and $v_{X(1)}^{SP}$ or between $v_{X(2)}^{SP}$ and $v_{X(2)}$, the polymer solution can exist as a metastable single phase, from which two-phase separation is initiated by formation of nuclei. In this sense, the spinodal concentration is the upper limit (or lower limit) of concentration, below (or above) which the solution can exist as a metastable single phase for $v_X^0 < v_X^C$ (v_X^C, critical polymer volume fraction) (or for $v_X^0 > v_X^C$). When the polymer solutions at point E (in Fig. 6.9) are cooled down to point F on cloud point curve, nuclei with a radius of S_{CN} (see, eqn (6.11)) are formed by thermal fluctuation, which can be regarded as concentration fluctuation only and a nucleus thus formed is considered to be in equilibrium with the region immediately surrounded by the local solution (polymer-lean phase), which can be approximated as a sphere with a radius of S_0 (see, eqn (6.14)). The extent of nucleation depends strongly on the time period spent between points F and H on the spinodal curve. As described in the previous Chapters, for a given polymer solution, cloud point curve, the spinodal curve and critical solution point can be unambiguously calculated.

Consider first the isothermal process. The activation energy of formation of a nucleus with the radius S, $\Delta\phi$ (S) is expressed in the same form as derived for nucleation from polymer melt (ref. 33):

$$\Delta\phi\ (S) = \frac{4}{3}\pi S^3 \Delta f_v + 4\pi S^2 \sigma \tag{6.1}$$

where Δf_v is the free energy change of coagulation per unit volume, and is defined as the difference between the average Gibbs' free energy of coexisting phases A and B, $\Delta\overline{G}(v_X^0)$ and the Gibbs' free energy change of mixing per unit volume $\Delta G'(v_X^0)$ (point J in Fig. 6.10; metastable phase), given by

$$\Delta f_v = \Delta\overline{G}(v_X^0) - \Delta G'(v_X^0) \tag{6.2}$$

and σ is an interfacial energy between the nucleus (i.e., polymer-rich phase for $v_x{}^0 < v_x{}^c$, $v_x{}^c$ is the critical point) and its surroundings (i.e., polymer lean phase for $v_x{}^0 < v_x{}^c$). $\Delta G'$ at point C (single phase) for monodisperse polymer / single solvent system is

$$\Delta G' = v_0 \frac{\Delta \mu_0}{V_0} + v_x \frac{\Delta \mu_x}{XV_0}$$

$$= (\frac{\tilde{R}T}{V_0}) [(1 - v_x)\ln(1 - v_x) + \frac{v_x \ln v_x}{X}$$

$$+ \chi_0 \{(v_x - v_x{}^2) + \frac{p_1}{2}(v_x - v_x{}^3) + \frac{p_2}{3}(v_x - v_x{}^4)\}] \quad (6.3)$$

where

$$\Delta \mu_0 = \tilde{R}T [\ln v_0 + (1 - \frac{1}{X}) v_x + \chi_0 (1 + p_1 v_x + p_2 v_x{}^2) v_x], \quad (6.4)$$

$$\Delta \mu_x = \tilde{R}T [\ln v_x - (X - 1)(1 - v_x)$$

$$+ X\chi_0 (1 - v_x)^2 \{1 + \frac{p_1}{2}(1 + 2v_x) + \frac{p_2}{3}(1 + 2v_x + 3v_x{}^2)\}]. \quad (6.5)$$

Here, $\Delta \mu_0$ and $\Delta \mu_x$ are the chemical potentials of solvent and polymer (i.e., X-mer), V_0, the molar volume of solvent, $\tilde{R}$, the gas constant, T, Kelvin temperature, χ_0, the polymer-solvent interaction parameter at infinite dilution, p_1 and p_2, 1st and 2nd order concentration-dependence parameters of χ. Eqns (6.3)-(6.5) are the special cases of eqns (2.112), (2.9) and (2.21), respectively. The coordinates of the points A and B [$(v_{x(1)}, \Delta G'(v_{x(1)}))$ and $(v_{x(2)}, \Delta G'(v_{x(2)}))$] are determined first by applying eqns (6.4) and (6.5) into the Gibbs' two-phase equilibrium eqns (6.6) and (6.7), respectively,

$$\Delta \mu_{0(1)} = \Delta \mu_{0(2)}, \quad (6.6)$$

$$\Delta \mu_{x(1)} = \Delta \mu_{x(2)} \quad (6.7)$$

and by substituting $v_{x(1)}$ and $v_{x(2)}$ thus calculated into eqn (6.3). For monodisperse polymer solution eqns (6.6) and (6.7) are special cases of eqns (1.21) and (1.22) for polydisperse polymer solution.

$\Delta \bar{G}(v_x{}^0)$ at point K $(v_x{}^0, \Delta \bar{G}(v_x{}^0))$ is given by the relation,

$$\Delta \bar{G}(v_x{}^0) = \frac{\Delta G'(v_{x(2)}) - \Delta G'(v_{x(1)})}{v_{x(2)} - v_{x(1)}} (v_x{}^0 - v_{x(1)}) + \Delta G'(v_{x(1)}). \quad (6.8)$$

Then, substituting eqns (6.3) and (6.8) into eqn (6.2), Δf_v can be calculated.

Of course, between the phase volume ratio R, $v_X{}^0$, $v_{X(1)}$ and $v_{X(2)}$ the relation

$$R \equiv \frac{V_{(1)}}{V_{(2)}} = \frac{v_{X(2)} - v_X{}^0}{v_X{}^0 - v_{X(1)}} \tag{6.9}$$

holds. Here, $V_{(1)}$ and $V_{(2)}$ are the volumes of the polymer-lean and -rich phases, respectively. Coordinates of the spinodal points C and D $[(v_{X(1)}{}^{SP}, \Delta G'(v_{X(1)}{}^{SP}))$ and $(v_{X(2)}{}^{SP}, \Delta G'(v_{X(2)}{}^{SP}))]$ can be calculated by solving the equation

$$\frac{1}{1-v_X} + \frac{1}{Xv_X} - \chi_0(2 + 3p_1 v_X + 4p_2 v_X) = 0 \tag{6.10}$$

for a given χ_0. Eqn (6.10) is the special case of the eqn (2.121).

The above discussion on monodisperse polymer / solvent system can easily be extended to more general systems including polydisperse polymer / solvent / nonsolvent system.

Evidently, $\Delta\phi$ is a function of S as illustrated in Fig. 6.11. When nucleation occurres from the binary polymer solution consisting of monodisperse polymer (X= 300) and a single solvent ($p_1 = 0.642$ and $p_2 = 0.19$) under the condition of $v_X{}^0 = 0.04$ and $\chi_0{}^P = 0.5285$ (accordingly, $v_{X(1)} = 0.02$ and $v_{X(2)} = 0.2948$, see, Table 6.2), $\Delta f_v = -300.988 J/m^3$. $\Delta\phi$ of Fig. 6.11 is calculated using $\Delta f_v = -300.988 J/m^3$. In calculation of $\Delta\phi$, σ was taken as $3 \times 10^{-6} J/m^2$. The radius of critical nuclei, S_{CN} is derived by applying the condition of $\partial\Delta\phi/\partial S = 0$ to eqn (6.1) to yield

$$S_{CN} = -2\sigma/\Delta f_v \tag{6.11}$$

(in this case, $S_{CN} \simeq 20$nm). When a nucleus formed at a given instant has a radius S_N larger than S_{CN}, the nucleus will continue to grow spontaneously. The activation energy of formation of critical nucleus $\Delta\phi_{CN}$ (J) is written by combining eqn (6.11) with eqn (6.1) as

$$\Delta\phi_{CN} = \frac{16}{3}\pi\sigma^3/\Delta f_v{}^2 \tag{6.12}$$

(in this case, $\Delta\phi_{CN} \simeq 5.03 \times 10^{-21}$J). It should be emphasized that in this case, although the phase equilibrium has not yet been attained over the whole solution system, R for the local equilibrium region surrounding each nucleus is considered to coincide with R for the whole system [eqn (6.9)].

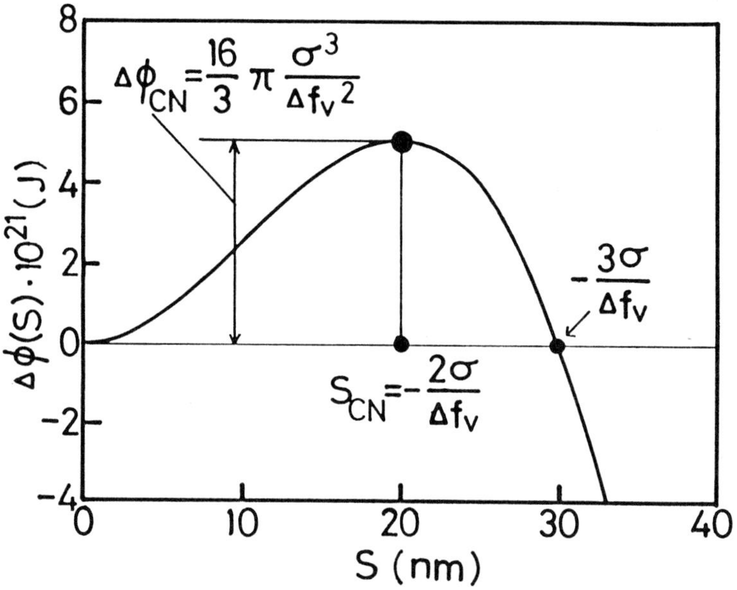

Fig. 6.11 Plot of activation energy of formation of nucleus $\Delta\phi(S)$ as a function of nucleus size S (see, eqn (6.1)). $\Delta f_v = -300.99 J/m^3$ [binary polymer solution consisting of monodisperse polymer (X=300) and a single solvent ($p_1=0.642$ and $p_2=0.19$) under the conditions of $v_X{}^0=0.04$ and $\chi_0{}^P=0.5285$ (accordingly, $v_{X(1)}=0.02$ and $v_{X(2)}=0.2948$, see, Table 6.2] and $\sigma=3\times10^{-6} J/m^3$ are assumed; in this case, $S_{CN}\simeq 20nm$ and $\Delta\phi_{CN}\simeq 5.03\times10^{-21} J$.

Fig. 6.12a and b show the cloud point curve (full line) [in a), $\chi_0{}^{CP}$ vs. $v_X{}^0$ and in b), cloud point temperature T_{CP} vs. $v_X{}^0$], spinodal curve (broken line) [in a), $\chi_0{}^{SP}$ vs. $v_X{}^0$ and in b), T_{SP} vs. $v_X{}^0$] and critical solution point (unfilled circle). Fig. 6.12c and d show $\Delta G'$ vs. $v_X{}^0$ curve and $(\Delta\bar{G}-\Delta G')$ vs. $v_X{}^0$ curve, respectively. Here, the calculations were carried out under the following conditions; $X=3\times10^3$, $p_1=0.642$, $p_2=0.190$ and $\theta=307.1K$, $\psi=0.27$ and $V_0=108.74 cm^3/mole$. These values except X and V_0 are evaluated by Kamide et al with good accuracy for atactic polystyrene / cyclohexane system (refs 34 and 35). The latter two are substituted into eqn (6.13)

$$T = \frac{\psi\,\theta}{\chi_0 + \psi - 0.5} \qquad (6.13)$$

to convert χ_0 to T. For this system, the critical solution point is $v_X{}^C = 0.0559$ and $\chi_0{}^C = 0.5048$ ($T_C = 299.76K$) (see, eqns (2.121) and (2.122)). $\Delta G'$ and $\Delta\overline{G}$ were calculated for the phase separation temperature $T_P = 297.72K$ (i.e., $\chi_0{}^P = 0.5067$). In this case, the coexisting compositions are $v_{X(1)} = 0.01$ (point A) and $v_{X(2)} = 0.1335$ (B) and the spinodal compositions are $v_{X(1)}{}^{SP} = 0.0272$ (C) and $v_{X(2)}{}^{SP} = 0.1026$ (D). $\Delta G'$ vs. $v_X{}^0$ curve is roughly linear in the range $0 < v_X{}^0 < 0.2$, but ($\Delta\overline{G} - \Delta G'$) vs. $v_X{}^0$ curve has two minimums at points A and B and two inflection points at points C and D (spinodal points).

When a solution of $v_X{}^0 = 0.0186$ ($\simeq (v_{X(1)} + v_{X(2)}{}^{SP})/2$) is cooled from a single phase state to bring about two-phase separation, the solution passes through the cloud point [$\chi_0{}^{CP}(v_X{}^0) = 0.5058$ ($T_{CP} = 298.705K$)], the phase separation point [$\chi_0{}^P(v_X{}^0) = 0.5067$ ($T_P = 297.716K$)] and the spinodal point [$\chi_0{}^{SP}(v_X{}^0) = 0.5092$ ($T_{SP} = 294.995K$)]. In this case, the depth of phase separation Δ_P is $T_{CP} - T_P = 0.989K$ and that of the spinodal point is $T_{CP} - T_{SP} = 3.71K$ and the difference between the phase separation point and the spinodal point, $T_P - T_{SP}$ is 2.721K. Tables 6.2 and 6.3 collect the value of Δf_v, $\chi_0{}^{CP}$ (and T_{CP}) and $\chi_0{}^{SP}$ (and T_{SP}), estimated for various initial polymer concentrations $v_X{}^0$ (X= 300 and 3000).

The values of σ in Table 6.2 were, for convenience, calculated from the concentration dependence of the surface free energy of cellulose acetate (the total degree of substitution, $\langle F \rangle = 0.80$) / water system at 298K (Fig. 1 of ref. 36), since similar data for polystyrene / cyclohexane are not available (ref. 36). Nevertheless, a more extensive study is necessary on (critical nucleus) interfacial free energy σ for two coexisting polymer phases, because deduction of σ from data on the surface free energy of polymer solution is not likely to be reliable then this clearly leads to a large uncertainty in the size of critical nuclei (Tables 6.2 and 6.3). However, there is hardly any doubt that it is the critical nuclei particles which are seen by electron-microscope.

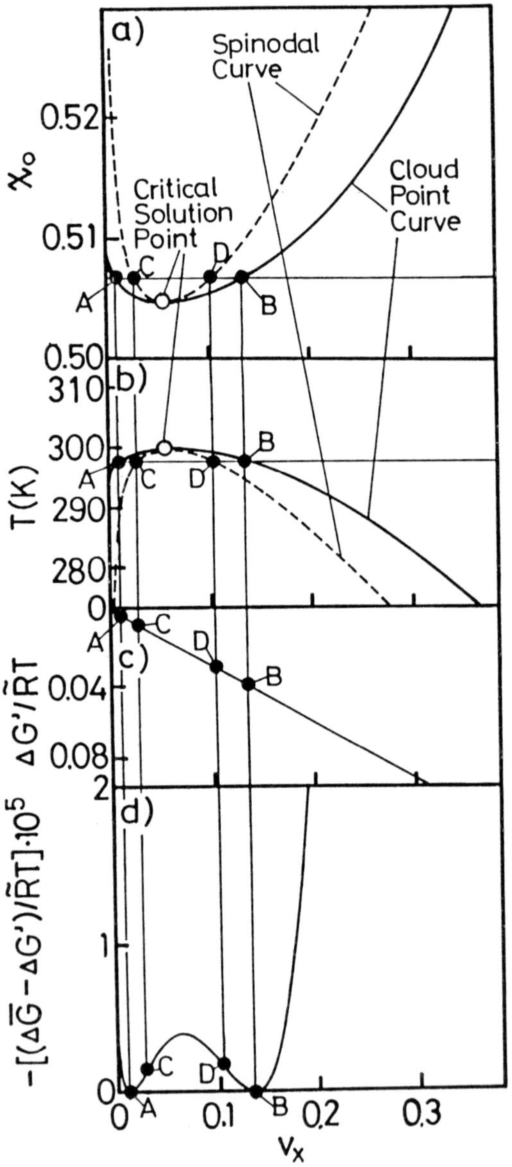

Fig. 6.12 Cloud point curve (solid line), spinodal curve (broken line) and critical solution point (open circle) of monodisperse polymer / single solvent system [a) χ_0 vs. $v_X{}^0$ plot, b) T vs. $v_X{}^0$ plot] together with $\Delta G'/\tilde{R}T$ vs. $v_X{}^0$ plot (c)) and $(\Delta\bar{G}-\Delta G')/\tilde{R}T$ vs. $v_X{}^0$ plot (d)): Original polymer; monodisperse ($X=3\times10^3$) and $p_1=0.642$, $p_2=0.190$, $\theta=307.1$K and $\psi=0.27$. $v_X{}^C=0.05588$ and $\chi_0{}^C=0.504812$ ($T_C=299.758$K) and χ_0 (and T) at phase separation point $\chi_0{}^P=0.506697$ (and $T_P=297.716$K).

TABLE 6.2

Phase separation characteristics of monodisperse low molecular weight polymer / single solvent system.*

Initial concentration	Coexisting compositions		Difference of free energy per unit volume	Interfacial free energy	Cloud point		Phase separation point		Spinodal point		Critical nucleus radii
v_X^0	$v_{X(1)}$	$v_{X(2)}$	Δf_v (J/m^3)	σ (J/m^2)	χ_0^{CP}	$T_{CP}(K)$	χ_0^P	$T_P(K)$	χ_0^{SP}	$T_{SP}(K)$	$S_{CN}(\mu m)$
					$(T_{CP}-T_P)$				(T_P-T_{SP})		
0.040	0.02	0.2947820	-300.988	1.17×10^{-3}	0.523954	280.238	0.5286859	275.798	0.5413193	264.606	7.774
	(R=12.7391)				(4.440)				(11.192)		
0.055	0.04	0.2443197	-66.343	7.86×10^{-4}	0.522022	282.092	0.5239538	300.238	0.5306858	273.964	23.69
	(R=12.621)				(1.854)				(6.274)		
0.075	0.06	0.2083209	-26.391	5.41×10^{-4}	0.520502	283.567	0.5215645	282.534	0.5238109	280.374	41.00
	(R=8.888)				(1.033)				(2.106)		
0.090	0.08	0.1791323	-4.6465	3.50×10^{-4}	0.519784	284.270	0.5202124	283.851	0.5211930	282.895	150.7
	(R=8.913)				(0.419)				(0.956)		
0.105	0.10	0.1541514	-0.32528	1.88×10^{-4}	0.519388	284.659	0.5194835	284.565	0.5197954	284.259	1157
	(R=9.830)				(0.094)				(0.306)		
0.121	0.12	0.1321907	-6.3535	4.20×10^{-5}	0.519195	284.849	0.5191965	284.848	0.5192133	284.831	1.32×10^5
	(R=11.191)				(0.001)				(0.017)		

* Polystyrene / cyclohexane system: $X=300$, $p_1=0.642$, $p_2=0.19$, $\theta=307.1K$, $\psi=0.27$.
$v_P^C=0.12604$, $\chi_0^C=0.5191813$, $T_C=284.86K$, $\sigma=4.3\times10^{-4}\ln(v_{P(2)}/v_{P(1)})$.

TABLE 6.3
Phase separation characteristics of monodisperse high molecular weight polymer / single solvent system.*

Initial concentration	Coexisting compositions		Difference of free energy per unit volume	Interfacial free energy	Cloud point		Phase separation point		Spinodal point		Critical nucleus radii
v_X^0	$v_{X(1)}$	$v_{X(2)}$	Δf_v (J/m³)	σ (J/m²)	χ_0^{CP}	T_{CP}(K) ($T_{CP}-T_P$)	χ_0^P	T_P(K)	χ_0^{SP}	T_{SP}(K) (T_P-T_{SP})	S_{CN}(μm)
0.00874	0.001	0.1942668	-70.066	0.19427	0.5069258	297.470	0.5103404	293.846	0.5190939	284.949	5.55×10^{-3}
	(R=23.970)					(3.624)				(8.897)	
0.0186	0.01	0.1335406	-11.403	0.13354	0.5057805	298.705	0.5066966	297.716	0.5092483	294.995	2.34×10^{-3}
	(R=13.365)					(0.989)				(2.721)	
0.0270	0.02	0.1067772	-2.5467	5.3387×10^{-3}	0.5053357	299.188	0.5056778	298.816	0.5067316	297.678	4.19×10^{-3}
	(R=11.397)					(0.372)				(1.138)	
0.0350	0.03	0.08838731	-0.47734	4.6927×10^{-4}	0.505063	299.484	0.5051993	299.347	0.5056194	298.880	1.97×10^{-3}
	(R=10.677)					(0.137)				(0.467)	
0.0433	0.04	0.07397497	-6.0219×10^{-2}	2.6702×10^{-4}	0.50490103	299.661	0.5049372	299.621	0.5050592	299.488	0.08868
	(R=9.295)					(0.040)				(0.133)	
0.510	0.05	0.06204242	-6.3789×10^{-4}	9.3719×10^{-5}	0.5048271	299.7412	0.5048276	299.7406	0.5048449	299.722	0.2938
	(R=11.042)					(0.0006)				(0.0186)	

* Polystyrene / methylcyclohexane system: $X=3000$, $p_1=0.642$, $p_2=0.19$, $\theta=307.1$K, $\psi=0.27$, $v_P^C=0.055883$, $\chi_0^C=0.5048121$, $T_C=299.758$K, $\sigma=4.3\times10^{-4}\ln(v_{P(2)}/v_{P(1)})$.

Fig. 6.13 shows the phase separation point (i.e., hypothetical metastable single phase, ●) as well as the cloud point curve (solid line), spinodal curve (broken line) and critical solution point (○). As the initial polymer concentration $v_x{}^0$ increases as it approaches the critical solution point, both the depth of the phase separation point from the cloud point as defined by $(T_{CP} - T_P)$ and that of the phase separation point from the spinodal point $(T_P - T_{SP})$ become significantly narrow.

The change in the state of the polymer in solution with temperature from point E to point G in Fig. 6.9 can also be investigated by the dynamic light scattering (DLS) methods, the principle of which will be given in a later section (see, section 6.3.3b). The g-average [see, for example, eqn (6.57)] hydrodynamic radius $S_{H,G}$ (see, eqn (6.61')) of phase-separated droplet (polymer-rich phase) of two well-fractionated polystyrene (PS) samples with $M_w/M_n \leq 1.05$ in cyclohexane (CH) in the temperature T range of 298～318K was investigated by the DLS method (ref. 37). The DLS measurements were carried out after PS solutions were held for at least 10hr at each temperature of measurement. Fig. 6.14a and b show a temperature dependence of S_H for PS [$M_w = 1.62 \times 10^5$, a) and 3.32×10^5, b)] in the solution (polymer concentration 3w/v%). In the figure, the Flory theta temperature θ (= 307.1K), cloud point temperature, which was determined by two methods; naked eye observation (T_{CP}(ob)) and computer simulation (T_{CP}(cal)), and the spinodal temperature, estimated by computer method T_{SP}(cal), are shown for comparison. T_{CP}(ob) is only one degree higher than T_{CP}(cal) for PS ($M_w = 1.62 \times 10^5$) / CH system. Interestingly, S_H for both systems monotonically increases with a decrease of T even above T_{CP}(ob) (or T_{CP}(cal)), indicating that at this temperature range ($\theta \simeq T_{CP}$(ob)) polymeric chains tend to aggregate together, and below T_{CP}(ob) the particle has the same magnitude in size as those of the secondary particles and is expected to be a polymer-rich phase.

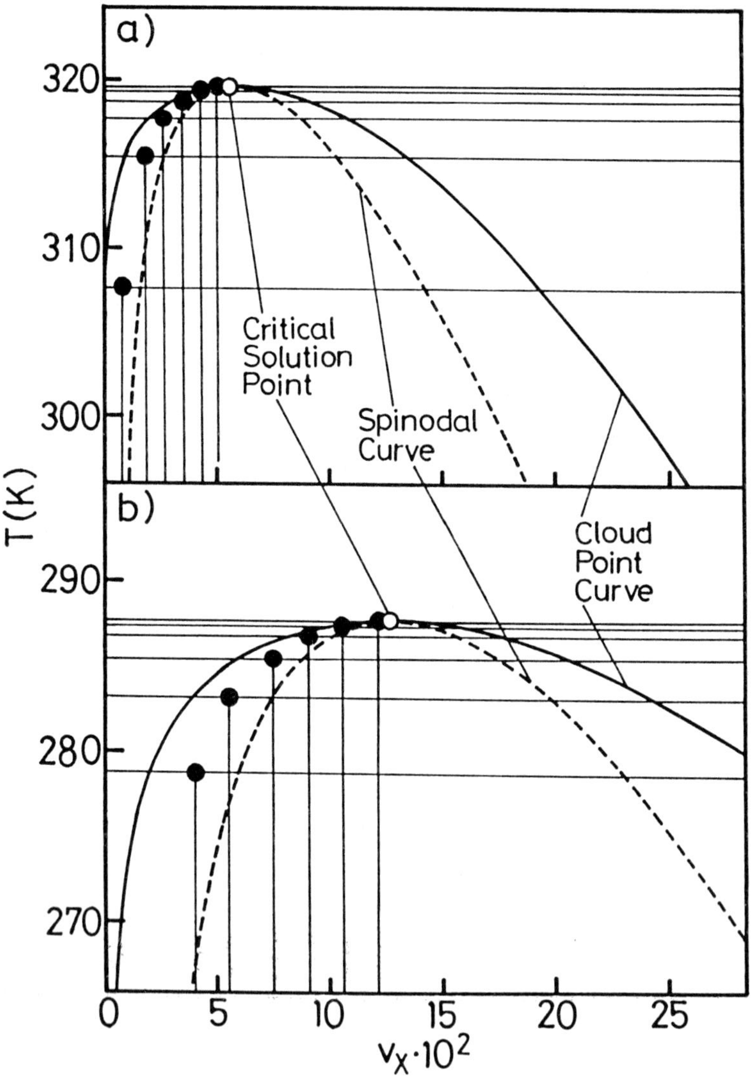

Fig. 6.13 The cloud point curve (broad solid line), spinodal curve (broken line) and critical solution point (unfilled circle) of monodisperse polymer / single solvent system: $p_1 = 0.642$, $p_2 = 0.190$, $\theta = 307.1$K and $\psi = 0.27$; ● phase separation point (hypothetical one phase). a) X = 3000 and b) X = 300.

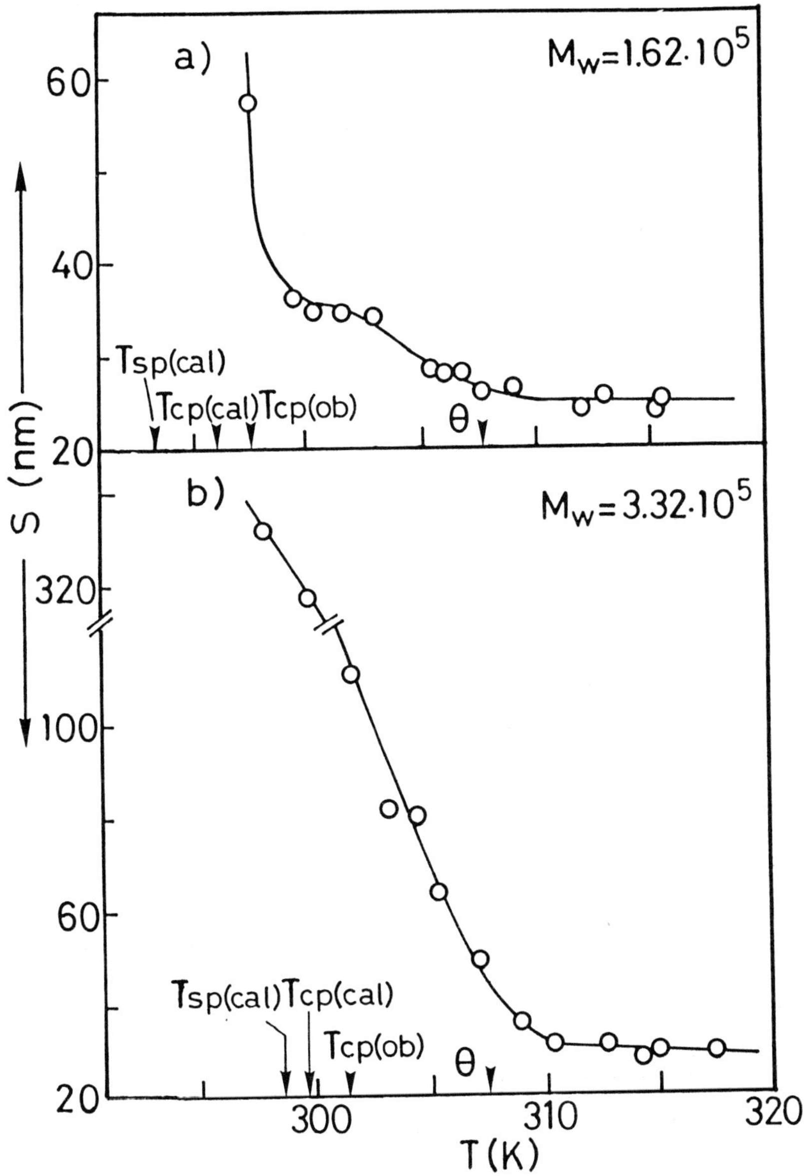

Fig. 6.14 Change in the radius of polymer particle S with temperature T for polystyrene / cyclohexane system: a) $M_w = 1.62 \times 10^5$ and b) $M_w = 3.32 \times 10^5$, initial concentration $v_p{}^0 = 0.03$. Observed cloud point temperature $T_{CP}(ob) = 297.5K$, calculated cloud point temperature $T_{CP}(cal) = 296.190K$ and spinodal temperature $T_{SP}(cal) = 292.874K$ for $M_w = 1.62 \times 10^5$ and $T_{CP}(ob) = 301.0K$, $T_{CP}(cal) = 299.556K$ and $T_{SP}(cal) = 298.550K$ for $M_w = 3.32 \times 10^5$. Theta temperature $\theta = 307.1K$.

6.2.2 Growth of nucleus (step c) to primary particle (step d)

The profile of polymer concentration around a nucleus is demonstrated in Fig. 6.15. Here, we adopt the hypothesis that thermodynamic equilibrium is attained between the nucleus ($0 < S < S_{CN}$) and its surrounding sphere ($S_{CN} < S < S_0$), but there is no equilibrium between the sphere and its outer large homogeneous phase ($S > S_0$) (Fig. 6.15a). Then, the polymer concentrations of a nucleus, its surrounding sphere and outer homogeneous phase are $v_{P(1)}$, $v_{P(2)}$ and $v_P{}^0$, respectively. It should be noted, however, that this hypothesis (i.e., "local equilibrium hypothesis") endorsed by Kamide et al. has not yet been fully verified experimentally. S_0 is related to R through the relation

$$R = \frac{S_0{}^3 - S_{CN}{}^3}{S_{CN}{}^3} \quad \left(= \frac{v_{P(2)} - v_P{}^0}{v_P{}^0 - v_{P(1)}}\right). \tag{6.14}$$

Of course, $v_{P(1)}$ and $v_{P(2)}$ can be calculated theoretically for any polymer solution (see, Chapter 2.2).

Immediately after the generation of a nucleus, the polymer molecules in the outer phase, based on the concentration difference $v_P{}^0 - v_{P(1)}$, diffuse into the sphere (Fig 6.15b). The number of the polymer molecules diffusing through the unit area of the sphere surface from the outer phase per unit time is given by solving the general equation of diffusion in the form

$$\frac{\partial v_P}{\partial t} = D\, \nabla^2 v_P \tag{6.15}$$

with

$$\nabla^2 = \frac{1}{S^2}\frac{\partial}{\partial S}\left(S^2\frac{\partial}{\partial S}\right) + \frac{1}{S^2}\left\{\frac{1}{\sin\theta}\frac{\partial}{\partial\theta}\left(\sin\theta\frac{\partial}{\partial\theta}\right) + \frac{1}{\sin^2\theta}\frac{\partial^2}{\partial\psi^2}\right\} \tag{6.16}$$

(θ and ψ are polar coordinates) under the boundary conditions of $v_P(S,0) = v_{P(1)}$ in the range of $S_{CN} < S < S_0(0)$ and $v_P(S,0) = v_P{}^0$ for $S > S_0(0)$ (Fig. 6.15a) as (ref. 38)

$$v_P(S,t) = (v_P{}^0 - v_{P(1)})\left[\frac{\phi(S_{0+}) + \phi(S_{0-})}{2} + \frac{1}{S}\left(\frac{Dt}{\pi}\right)^{1/2}\left(e^{-S_{0+}{}^2} - e^{-S_{0-}{}^2}\right)\right] + v_{P(1)} \tag{6.17}$$

where

$$\phi(x) = \frac{2}{(\pi)^{1/2}}\int_0^x e^{-u^2}du, \tag{6.18}$$

$$S_{0+} = (S_0 + S)/\{2(Dt)^{1/2}\}, \tag{6.19a}$$

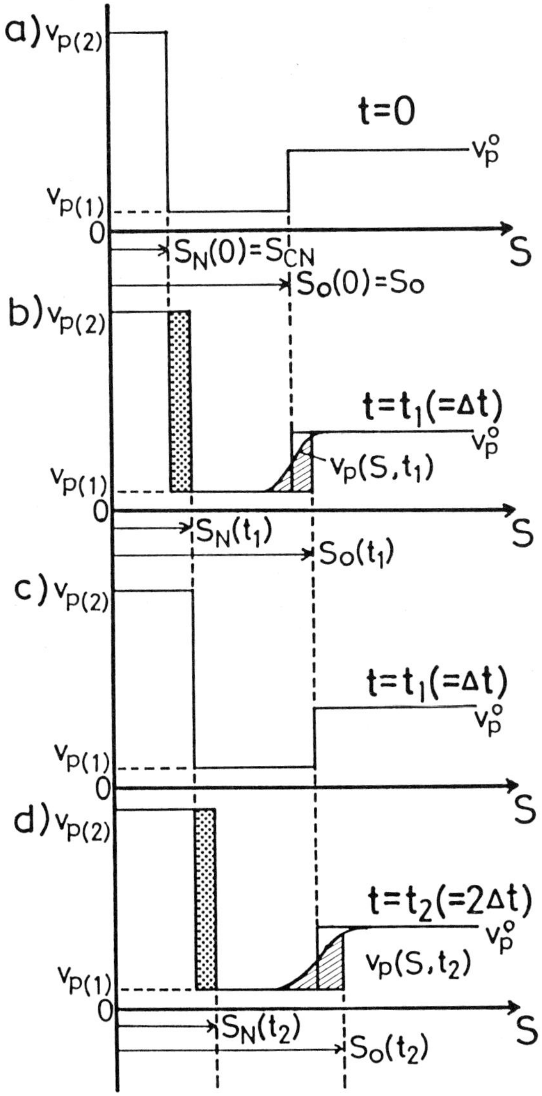

Fig. 6.15 Concentration profiles of a critical nucleus with a radius of $S_N(t)$ and its surrounding sphere with a radius of $S_0(t)$: a) the profile of polymer concentration around the critical nucleus at $t=0$ [the polymer concentrations of nucleus $(=v_{p(2)})$, its surrounding sphere $(=v_{p(1)})$ and its outer homogeneous phase$(=v_p{}^0)$]; b) increase of the radius of the nucleus (shadowed area) due to diffusion ($v_p(S,t_1)$, hatched area) from the outer phase at $t=t_1(=\Delta t)$; c) the profile of polymer concentration around the nucleus at $t=t_1$; d) increase of the radius of the nucleus (shadowed area) due to diffusion ($v_p(S,t_2)$, hatched area) from the outer phase at $t=t_2(=2\Delta t)$.

$$S_{0-} = (S_0 - S)/\{2(Dt)^{1/2}\}. \qquad (6.19b)$$

Here, ϕ (x) is the error function, D, is the diffusion coefficient and t, the growing time of nucleus by diffusion. We define $t=0$ as the instant of appearance of a critical nucleus and the concentration profile around the nucleus at $t=t_1$ ($=\Delta t$) is illustrated in Fig. 6.15b. If we can also assume that the thermodynamic equilibrium holds even at $t=t_1$ between the nucleus ($0<S<S_N(t_1)$) and the surrounding sphere ($S_N(t_1)<S<S_0(t_1)$), the diffusion of polymer molecules (hatched area in Fig. 6.15b) from the outer phase into $S_0(0)$ sphere will instantly result in an increase in the radius of the nucleus (shadowed area in Fig. 6.15b) yielding $S_N(t_1)$ [i.e., $S_{CN}(\equiv S_N(0)) \rightarrow S_N(t_1)$]. (see, Fig. 6.15b). This can be understood as nucleus growth. When the thermodynamic equilibrium between the nucleus and its surrounding sphere holds even at $t=t_2$ ($\equiv 2\Delta t$), the radius of the nucleus increases from $S_N(t_1)$ to $S_N(t_2)$ (Fig. 6.15c and d). The concentration profile of diffusion at $t=t_2$, $v_p(S,t_2)$ can be numerically evaluated in this case by solving the diffusion equation (eqn (6.15)) under the boundary conditions of $v_p(S,t_1)=v_{p(1)}$ ($S_N(t_1)<S<S_0(t_1)$) and $v_p(S,t_1)=v_p{}^0$ ($S>S_0(t_1)$).

The above treatment can be readily extended to the more general case of a nucleus with a radius of $S_N(t_i)$ and its surrounding sphere with a radius of $S_0(t_i)$, obtained at time $t=t_i$ ($\equiv i\Delta t$). $v_p(S,t_i)$ can be estimated by solving eqn (6.15) under the boundary conditions of $v_p(S,t_{i-1})=v_{p(1)}$ ($S_N(t_{i-1})<S<S_0(t_{i-1})$) and $v_p(S,t_{i-1})=v_p{}^0$ ($S>S_0(t_{i-1})$) when the following relation holds

$$R = \frac{S_0(t_i)^3 - S_N(t_i)^3}{S_N(t_i)^3} \qquad (i=0,1,2,\cdots) \qquad (6.20)$$

among $S_N(t_i)$, $S_0(t_i)$ and the phase volume ratio R determined by eqn (6.14).

An increment of polymer volume during $t=t_{i-1}$ and $t=t_i$ in the polymer-lean phase ranging from $S_N(t_{i-1})$ and $S_0(t_i)$, $\Delta V_{p(1)}$ is given by the following equation;

$$\Delta V_{p(1)}(t_i) = 4\pi \int_{S_N(t_{i-1})}^{S_0(t_i)} (v_p(S,t_i) - v_{p(1)}) S^2 dS \qquad (6.21)$$

and this brings about an instant increase in the radius of the nucleus (i.e., an increment of the polymer volume in the polymer-rich phase during $t=t_{i-1}$ and $t=t_i$, $\Delta V_{p(2)}(t_i)$). $\Delta V_{p(2)}(t_i)$ is given by:

$$\Delta V_{p(2)}(t_i) = \frac{4\pi}{3} v_{p(2)} (S_N(t_i)^3 - S_N(t_{i-1})^3). \tag{6.22}$$

Then, $|\Delta V_{p(1)}(t_i)| = |\Delta V_{p(2)}(t_i)|$ holds and

$$S_N(t_i)^3 = \frac{3}{v_{p(2)}} \int_{S_N(t_{i-1})}^{S_0(t_i)} (v_p(S,t_i) - v_{p(1)}) S^2 dS + S_N(t_{i-1})^3. \tag{6.23}$$

Solving the integral equation eqn (6.23) under the conditions of eqn (6.20), $S_N(t_i)$ and $S_0(t_i)$ are obtained. Differentiation of eqn (6.23) with respect to t gives

$$\frac{\partial S_N(t)}{\partial t} = \frac{1}{v_{p(2)} S_N(t)^2} \int_{S_{CN}}^{S_0(t)} \left(\frac{\partial v_p(S,t)}{\partial t}\right) S^2 dS. \tag{6.24}$$

Eqn (6.24) is an expression giving the growth rate of the nucleus. Kamide and Manabe (ref. 27) proposed a somewhat different approach to the estimation of the growth rate using a rather rough approximation.

The alternative method of derivation eqn (6.24) is as follows: An increment of the polymer volume in a given sphere during time interval Δt is

$$\int_{S_{CN}}^{S_0(t)} \left(\frac{\partial v_p(S,t)}{\partial t}\right) \Delta t \; 4\pi S^2 dS$$

and this yields an instantaneous increase of the radius of nucleus, given by

$$v_{p(2)} \left(\frac{\partial V_{(2)}}{\partial t}\right) \Delta t \quad \text{(here } V_{(2)} = (4\pi/3) \cdot S_N(t)^3\text{)},$$

because the nucleus and its surrounding sphere are in equilibrium. Then, we obtain

$$v_{p(2)} \cdot 4\pi S_N(t)^2 \frac{\partial S_N(t)}{\partial t} \Delta t = \int_{S_{CN}}^{S_0(t)} \left(\frac{\partial v_p(S,t)}{\partial t}\right) \Delta t \; 4\pi S_0{}^2 dS. \tag{6.25}$$

Eqn (6.25) is rewritten as eqn (6.24). $\partial v_p(S,t)/\partial t$ in eqn (6.24) can be determined by differentiation of eqn (6.17) with respect to t, in the form:

$$\frac{\partial v_p(S,t)}{\partial t} = (v_p{}^0 - v_{p(1)}) \left[\frac{1}{2(\pi)^{1/2}t}(S_{0+}e^{-S_{0+}{}^2} + S_{0-}e^{-S_{0-}{}^2}) - \frac{1}{2S}\left(\frac{D}{\pi t}\right)^{1/2}(e^{-S_{0+}{}^2} - e^{-S_{0-}{}^2}) - \frac{1}{S}\left(\frac{D}{\pi t}\right)^{1/2}(S_{0+}{}^2\, e^{-S_{0+}{}^2} - S_{0-}{}^2\, e^{-S_{0-}{}^2})\right]. \tag{6.26}$$

Fig. 6.16 demonstrates the concentration profile of a critical nucleus and its surrounding. In the figure, $v_p{}^0 = 0.04$, $v_{p(1)} = 0.02$, $v_{p(2)} = 0.3325$ (accordingly, $R = 14.625$), $\Delta f_v = -300 J/m^3$, $\sigma = 3\times 10^{-6} J/m^2$ (so $S_{CN} = 20nm$ and $S_0(0) = 50nm$), $\Delta t = 0.005\mu sec$ and $D = 2.5\times 10^{-10} m^2/sec$ were assumed. From the figure, the growth of the nucleus and the expansion of the local equilibrium region with time can be readily understood. Fig. 6.17 shows the plots of $S_N(t)$ and $S_0(t)$ against t. Under the conditions employed here, a critical nucleus grows after 0.24μsec of its birth to a particle with a radius of ca.200nm and at the same time, the local equilibrium region expands to as wide as 500nm.

As the nucleus grows in this manner, the surrounding sphere with a radius of S_0 also continues to become bigger. During the growth of the nucleus, the outer homogeneous region generates a new critical nucleus (see, Fig. 6.18). Ultimately (at a given time $t = t_p$) the original outer homogeneous phase (hatched area) will disappear completely due to consumption by expansion of the S_0 sphere and generation of critical nuclei. All the nuclei (i.e., the particles) at this instant ($t = t_p$) are conventionally defined as primary particles (Fig. 6.18).

We assume that nucleation is always absolutely sporadic and spinodal decomposition is never predominant, then the rate of production of critical nuclei per unit volume is given by (ref. 27)

$$\frac{dN_{CN}}{dt} = k_{CN} \exp(-\Delta\phi_{CN}/k_B T) \tag{6.27}$$

with

$$k_{CN} = k'_{CN}\left\{1 - \frac{N_{CN}(t)}{N_{CN}(\infty)}\right\}. \tag{6.28}$$

N_{CN} is the number of critical nuclei per unit volume (number/m³), $N_{CN}(t)$, N_{CN} at t, $N_{CN}(\infty)$, N_{CN} at $t = \infty$ and k'_{CN}, the rate constant (number/m³·sec) and $\Delta\phi_{CN}$ can be calculated using eqn (6.12). Here, nucleation is considered to be described by the same formula as that for the zeroth order reaction. However, the

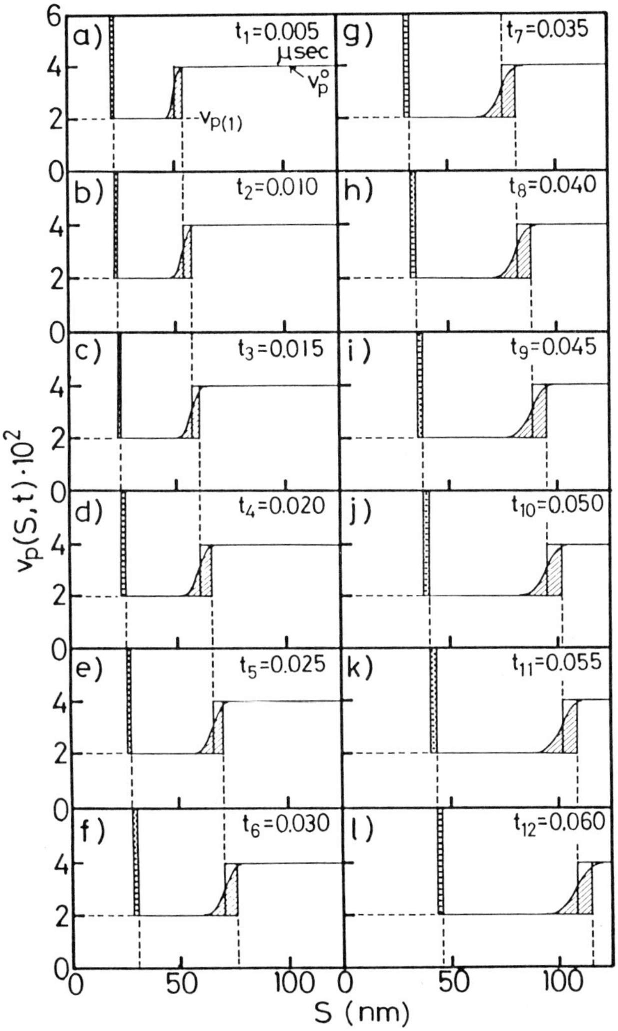

Fig. 6.16 Concentration profile of nucleus and its surroundings together with diffusion $v_p(S,t)$ (hatched area) and increase of radius of nucleus (shadowed area): Initial polymer volume fraction, $v_p^0 = 0.04$; concentration of coexisting phases, $v_{p(1)} = 0.02$ and $v_{p(2)} = 0.3325$ (accordingly, R = 14.625); difference of free energy between before and after phase separation, $\Delta f_v = -300 J/m^3$; interfacial free energy, $\sigma = 3\times10^{-6} J/m^2$ [$S_{CN} = S_N(0) = 20$nm and $S_0(0) = 50$nm]; diffusion coefficient, $D = 2.5\times10^{-10} m^2/sec$; temperature, $T_p = 300K$; a) $t_1 = 0.005$μsec; b) $t_2 = 0.010$μsec; c) $t_3 = 0.015$μsec; d) $t_4 = 0.020$μsec; e) $t_5 = 0.025$μsec; f) $t_6 = 0.030$μsec; g) $t_7 = 0.035$μsec; h) $t_8 = 0.040$μsec; i) $t_9 = 0.045$μsec; j) $t_{10} = 0.050$μsec; k) $t_{11} = 0.055$μsec; l) $t_{12} = 0.060$μsec.

nucleation does not continue without limits as was confirmed repeatedly by a large number of actual experiments if available surveys are to be believed, and it will stop at the instant when the phase equilibrium of total system is realized (i.e., $t = t_P$; see, Fig. 6.18).

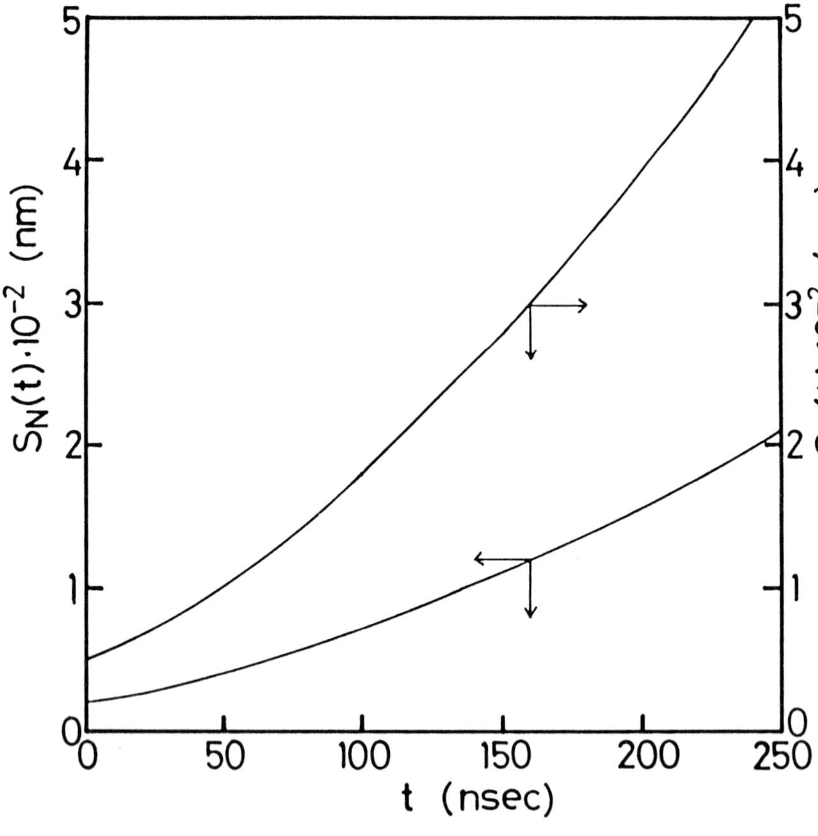

Fig. 6.17 Time dependence of radius of nucleus $S_N(t)$ and that of its surrounding sphere $S_0(t)$: Initial polymer volume fraction, $v_p{}^0 = 0.04$; concentration of coexisting phases, $v_{P(1)} = 0.02$ and $v_{P(2)} = 0.3325$ (accordingly, $R = 14.625$); difference of free energy between before and after phase separation, $\Delta f_v = -300 J/m^3$; interfacial free energy, $\sigma = 3 \times 10^{-6} J/m^2$ [$S_{CN} = S_N(0) = 20nm$ and $S_0(0) = 50nm$]; diffusion coefficient, $D = 2.5 \times 10^{-10} m^2/sec$; temperature, $T_P = 300K$.

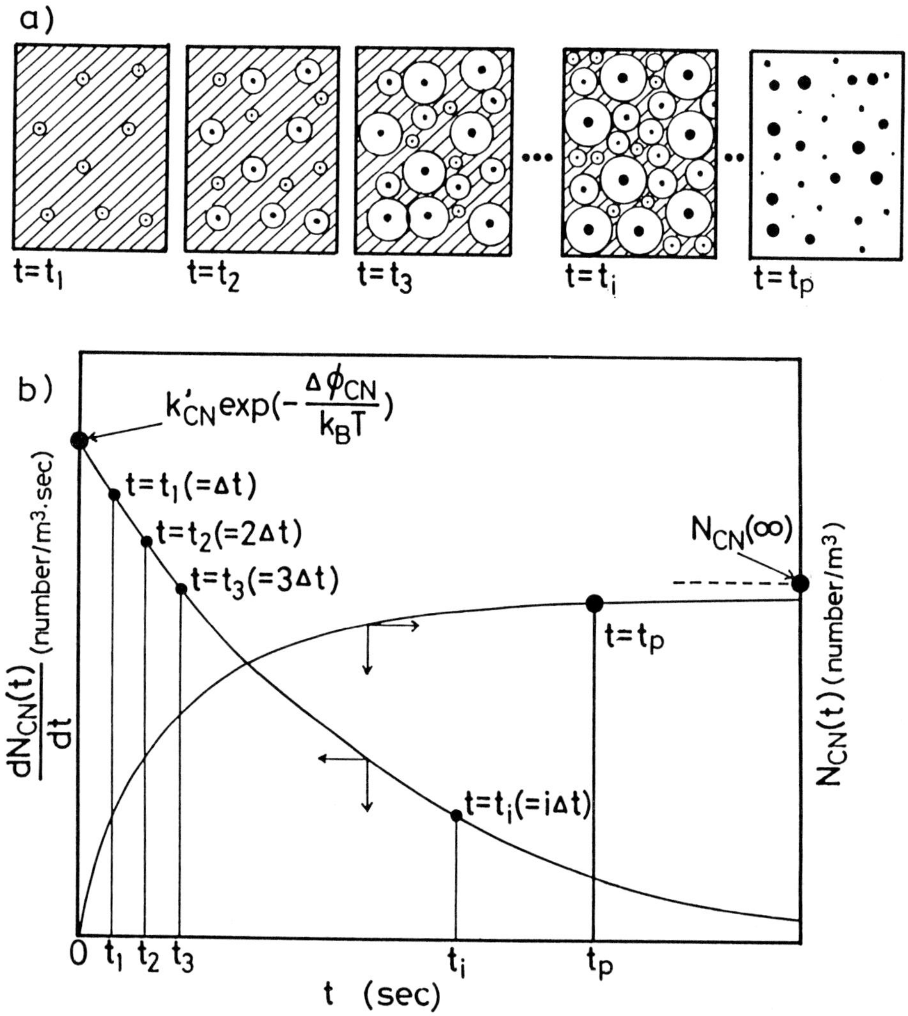

Fig. 6.18 a) Schematic representation of nucleation and growth of nuclei: ●, nucleus (polymer-rich phase); ○, surrounding sphere (polymer-lean phase); hatched area, outer homogeneous phase, b) time dependence of nucleation $N_{CN}(t)$ and rate of nucleation dN_{CN}/dt of critical nuclei.

Substitution of eqn (6.28) into eqn (6.27) yields

$$\frac{dN_{CN}(t)}{dt} = \nu \{1 - \frac{N_{CN}(t)}{N_{CN}(\infty)}\} \tag{6.29}$$

with

$$\nu \equiv k'_{CN} \exp(-\frac{\Delta\phi_{CN}}{k_B T}). \tag{6.30}$$

By solving the differential equation (eqn (6.29)), we obtain

$$N_{CN}(t) = N_{CN}(\infty)[1 - \exp(-\frac{\nu}{N_{CN}(\infty)}t)]. \tag{6.31}$$

Combination of eqns (6.29) and (6.31) gives

$$\frac{dN_{CN}(t)}{dt} = \nu \exp(-\frac{\nu}{N_{CN}(\infty)}t). \tag{6.32}$$

The total volume of S_0 spheres including growing particles at t, $V_0(t)$ is expressed as

$$V_0(t) = \int_0^t \frac{dN_{CN}(\tau)}{dt} \cdot \frac{4\pi}{3} S_0(t-\tau)^3 d\tau \tag{6.33}$$

where τ is the time of formation of nucleus and $t-\tau$ is the growing time of nucleus at t. S_0 and S_N are the functions of $t-\tau$ (i.e., $S_0 = S_0(t-\tau)$ and $S_N = S_N(t-\tau)$).

At $t = t_p$, $V_0(t) = 1$ holds. In other words, in order to estimate t_p, the following equation should be solved.

$$\int_0^{t_p} \frac{dN_{CN}(\tau)}{dt} \cdot \frac{4\pi}{3} S_0(t_p-\tau)^3 d\tau = 1. \tag{6.34}$$

Since dN_{CN}/dt can be evaluated from $N_{CN}(\infty)$ and ν [accordingly, k'_{CN}, $\Delta\phi_{CN}$ and T {see, eqn (6.30)}] by eqn (6.32) and $S_0(t_p-\tau)$ together with $S_N(t_p-\tau)$ can be calculated by solving simultaneous equation [eqns (6.20) and (6.23)], t_p can be unambiguously determined by eqn (6.34) under given phase separation conditions. Here, $N_{CN}(\infty)$ (number/m^3) should be below the number of polymer molecules (in this case, monodisperse polymer) per unit volume. For example, when weight fraction of X-mer, $w_X = 0.03$ and molecular weight of X-mer, $M = 3\times10^4$ g/mol, $N_{CN}(\infty) \leq 30\times10^3 g/(3\times10^4/N_A)g = N_A$ $(= 6.023\times10^{23})$ (number/m^3).

Fig. 6.19a and b show the effect of $N_{CN}(\infty)$ on the relations between t_p, obtained by solving an integral equation [eqn (6.34)], and k'_{CN} and on the relations between $N_{CN}(t_p)$, calculated by substituting t_p into eqn (6.31), and k'_{CN}, respectively. In these figures, $T = 300K$ and $\Delta\phi_{CN} = 5.027\times10^{-21}J$, estimated by putting $\sigma = 3\times10^{-6}J/m^2$ and $\Delta f_v = -300J/m^3$ into eqn (6.12), are used and other conditions are the same as those in Fig. 6.16. As k'_{CN} increases t_p decreases monotonically, approaching an asymptotic value which is smaller for larger

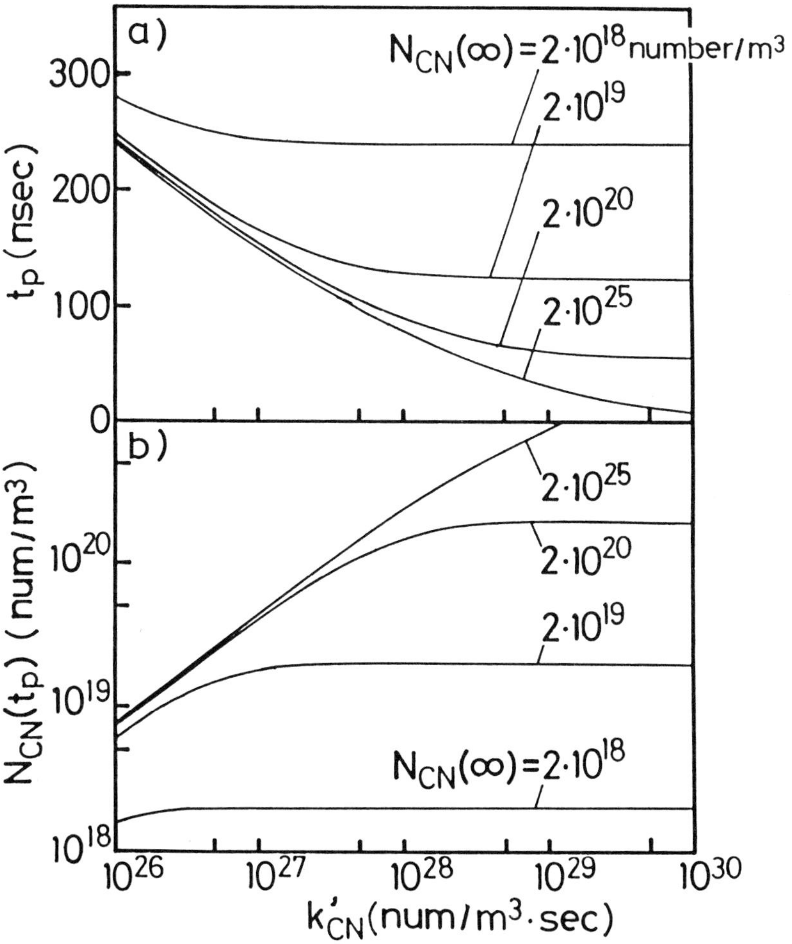

Fig. 6.19 Relations between the time needed for attainment of equilibrium t_P and rate constant of nucleation k'_{CN} (eqn (6.28)) (a) and relations between the nuclei density $N_{CN}(\infty)$ and k'_{CN}: Initial polymer volume fraction, $v_p{}^0 = 0.04$; concentration of coexisting phases, $v_{p(1)} = 0.02$ and $v_{p(2)} = 0.3325$ (accordingly, $R = 14.625$); difference of free energy between before and after phase separation, $\Delta f_v = -300J/m^3$; interfacial free energy, $\sigma = 3\times10^{-6}J/m^2$ [$S_{CN} = S_N(0) = 20nm$ and $S_0(0) = 50nm$]; diffusion coefficient, $D = 2.5\times10^{-10}m^2/sec$; temperature, $T_P = 300K$.

$N_{CN}(\infty)$. $N_{CN}(t_p)$ increases with k'_{CN}, attaining at $N_{CN}(\infty)$. Fig. 6.19 implies that in a relatively early stage of phase separation all possible nuclei generate completely and thereafter the separation proceeds by diffusional growth of nuclei.

Fig. 6.20 shows the effect of $N_{CN}(\infty)$ on the relations between the number-average or the weight-average radius of primary particles (i.e., the particles at t_p) S_n or S_w and k'_{CN}. Here, S_n and S_w are calculated through use of the relations,

$$S_n(\equiv \bar{S}_1) = \int_0^{\infty} S_1 N_{CN}(S_1)\,dS_1 \Big/ \int_0^{\infty} N_{CN}(S_1)\,dS_1 \tag{6.35a}$$

$$S_w = \int_0^{\infty} S_1^{\,2} N_{CN}(S_1)\,dS_1 \Big/ \int_0^{\infty} S_1 N_{CN}(S_1)\,dS_1 \tag{6.35b}$$

where $N_{CN}(S_1)$ (number/m^3) is the number of the primary particles with a radius of S_1. With an increase in k'_{CN} both S_n and S_w approach their asymptotic values, which strongly depend on $N_{CN}(\infty)$. This means explicitly that S_n and S_w are mainly governed by $N_{CN}(\infty)$, rather than by k'_{CN}.

Fig. 6.21a, c and e show change in $(1/\nu)dN_{CN}/dt$ (see, eqn (6.32)) with time for various combinations of $N_{CN}(\infty)$ and k'_{CN}. In the figure, t_P is indicated as an unfilled circle and beyond t_P $(1/\nu)dN_{CN}/dt$ becomes zero. $(1/\nu)dN_{CN}/dt$ decreases more rapidly with time for larger $N_{CN}(\infty)$ and larger k'_{CN}.

Fig. 6.21b, d and f show normalized size distribution of the primary particles at t_P, $N_{PP}(S_1, t_P)/N_{CN}(\infty)$. The mean pore size shifts to smaller S_1 side with an increase in $N_{CN}(\infty)$. Extremely sharp distribution of the primary particles is obtained for larger k'_{CN}, in other words, the primary particles are uniform in size. Lower k'_{CN} yields very broad size distribution of the primary particles.

Likewise, we can consider that there are at the initial stage two elementary steps: The nucleation and growth of nuclei by diffusion, which occur concurrently until $t = t_p$. We define the particles at that instant (t_p) as primary particle (PP) (see, for example, Fig. 6.18a). In this section, an isothermal process was treated, but this treatment can readily be generalized to any non-isothermal process. In that case, when the solution is cooled down to a temperature below the spinodal curve (for example, point I in Fig. 6.9), it is natural to consider that nucleation will no longer continue further.

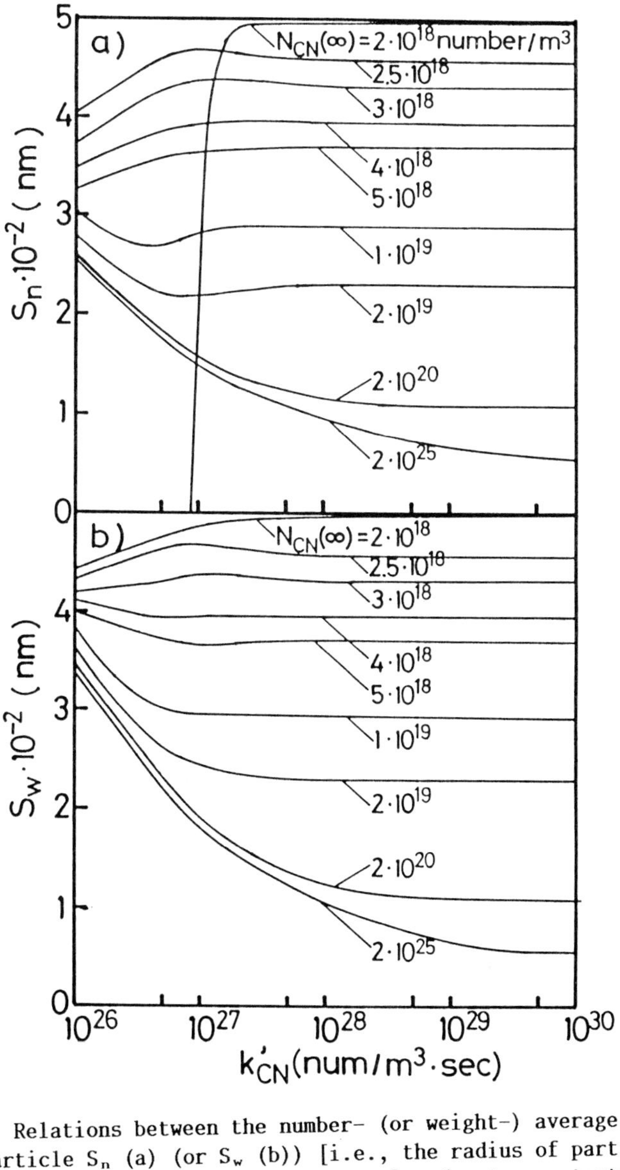

Fig. 6.20 Relations between the number- (or weight-) average radius of the primary particle S_n (a) (or S_w (b)) [i.e., the radius of particles at the time for attainment of phase equilibrium, $t=t_P$] and rate constant of nucleation k'_{CN} for various limiting nuclei density $N_{CN}(\infty)$: Initial polymer volume fraction, $v_p^0=0.04$; concentration of coexisting phases, $v_{P(1)}=0.02$ and $v_{P(2)}=0.3325$ (accordingly, $R=14.625$); difference of free energy between before and after phase separation, $\Delta f_v=-300J/m^3$; interfacial free energy, $\sigma=3\times10^{-6}J/m^2$ [$S_{CN}=S_N(0)=20nm$ and $S_0(0)=50nm$]; diffusion coefficient, $D=2.5\times10^{-10}m^2/sec$; temperature, $T_P=300K$.

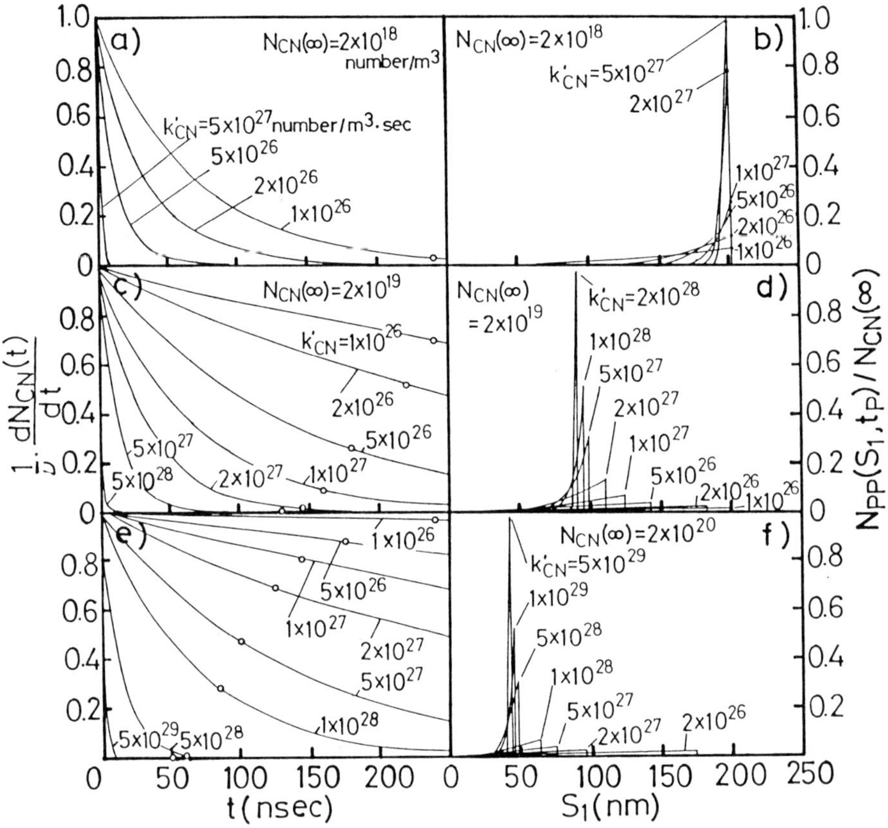

Fig. 6.21 Plots of $(1/\nu)dN_{CN}/dt$ (eqn (6.32)) as a function of time (a, c and e) and normalized size distribution of primary particle per unit volume at the time of attainment of equilibrium $t=t_P$ (b, d and f), $N_{PP}(S_1,t_P)/N_{CN}(\infty)$ for various rate constants k'_{CN}(number/m^{3}sec): a) and b) $N_{CN}(\infty)=2\times10^{18}$number/m^3; c) and d) $N_{CN}(\infty)=2\times10^{19}$number/m^3 and e) and f) $N_{CN}(\infty)=2\times10^{20}$number/m^3; $\nu=2.971\times10^{25}$number/m^{3}sec ($k'_{CN}=1\times10^{26}$), $\nu=5.942\times10^{25}$ ($k'_{CN}=2\times10^{26}$), $\nu=1.486\times10^{26}$ ($k'_{CN}=5\times10^{26}$), $\nu=2.971\times10^{26}$ ($k'_{CN}=1\times10^{27}$), $\nu=5.942\times10^{26}$ ($k'_{CN}=2\times10^{27}$), $\nu=1.486\times10^{27}$ ($k'_{CN}=5\times10^{27}$), $\nu=2.971\times10^{27}$ ($k'_{CN}=1\times10^{28}$), $\nu=1.486\times10^{29}$ ($k'_{CN}=5\times10^{29}$), other conditions are the same as those of Fig. 6.19.

6.3 GROWTH OF PRIMARY PARTICLES BY AMALGAMATION (STEP E) TO GIVE SECONDARY PARTICLES (STEP F)

6.3.1 Particle simulation approach

The primary particles (PP) collide with each other to yield larger particles which are nominated as growing particles (GP). The growth of the primary particle to the secondary particle can be theoretically investigated by using a particle Monte Carlo simulation methods to describe the following three steps (ref. 39):

1st step: Generate the primary particles at random positions in a hypothetical space (cubic body with edge of L'). This moment is defined as $t=0$. t is the time of growth of GP by amalgamation. Assume that the primary particle has the same radius S_1 of 10nm. The total number of particles N_{PP} (number/m^3) are determined from the phase volume ratio R and S_1 through use of the relation.

$$N_{PP}=\frac{1}{(4/3)\,\pi\,S_1{}^3\,(R+1)}. \qquad (6.36)$$

Here, we consider only the case where the primary particles are the polymer-rich phase [i.e., $v_P{}^0 < v_P{}^C$].

2nd step: Give a value for the velocity of displacement of the particles in advance and determine the positions of all particles randomly, using this value, after unit period of time. In other words, the particles are assumed to be engaged in a rapid random motion. Here, the unit time, for convenience, is defined as the time needed for the primary particle to move at a distance of its diameter (i.e., 20nm).

3rd step: Measure the distance between the particles. In the case when the distance between the center of gravity of the two arbitrarily chosen particles is less than the summation of the radius of each particle, these two particles are considered to have collided, yielding a single particle by amalgamation.

In the case when the movement of these particles starts after completion of generating all the primary particles, steps 2 and 3 are repeated. And in the case when the generation of the primary particles continues even after some particles already generated in the past collide with each other, all three steps are repeated until the end of the primary particle generation. The unit of time Δt is equal to that needed for calculation of a cycle of steps 1-3.

The moving velocity of the particles can be estimated by (1) energy equi-partition law or (2) the mean square displacement of brownian movement.

Case (1): The mean velocity of primary particle v_0 is given by (ref. 40)

$$\frac{1}{2}m_0 v_0^{2} = \frac{3}{2}k_B T \tag{6.37}$$

where, m_0 is the mass of the primary particle, k_B, the Boltzmann constant, T, the Kelvin temperature. Of course, v_0 is sufficiently low to avoid turbulance. We assume the density of the particle to be 1 (Kg/m^3) and then obtain $m_0 = 4/3\pi S_1^{3}$. Eqn (6.37) can be rewritten in the form,

$$v_0 = \frac{3}{2}\left(\frac{k_B T}{\pi S_1^{3}}\right)^{1/2}. \tag{6.38}$$

From eqn (6.38) the velocity of the particle is linearly proportional to $S_1^{-3/2}$. We define Δt as the time needed for a particle with radius a of S_1 to move the distance of $2S_1$, Δt is estimated from S_1 and T through use of the relation

$$\Delta t = \frac{2S_1}{v_0} = \frac{4}{3}\left(\frac{\pi S_1^{5}}{k_B T}\right)^{1/2}. \tag{6.39}$$

Putting $S_1 = 10nm$ and $T = 300K$, we obtain $\Delta t \simeq 4.5\times 10^{-10}$sec. It should be noted that, in reality, the particle moves in viscous fluid and therefore, eqns (6.37)-(6.39) hold their validity only short distances, when the viscosity effects can be neglected.

Case (2): The mean-square displacement σ^{2} of the primary particles at Δt is

$$\sigma^{2} = \frac{\Delta t k_B T}{3\pi \eta S_1} \tag{6.40}$$

where η is the viscosity of a polymer-lean phase surrounding the particles. Then Δt, defined previously, is given by

$$\Delta t = \frac{12\pi \eta S_1^{3}}{k_B T}. \tag{6.41}$$

If we put $\eta = 0.76cP$ for cyclohexane (at θ temperature (307.1K)) and $S_1 = 10nm$ and $T = 300K$, we obtain $\Delta t \simeq 7\times 10^{-6}$sec. Eqn (6.40) shows that σ is in linear proportion to $S_1^{-1/2}$, and accordingly, the particle velocity is also proportional to $S_1^{-1/2}$. Eqns (6.40) and (6.41) can not be applied in the case of short-range movement, in which the viscosity effect is neglected. Adequate relation among Δt, v and S_1 differs depending on whether the displacement of $2S_1$ (i.e., in this case, 20nm) is the distance short enough to neglect the viscosity effect or not. The right answer will be obtained by comparing the simulation with actual experiments. To ascertain the uniformity of hypothetical space, the periodic boundary condition was employed in the simulation.

Fig. 6.22 shows the plot of the ratio of number-average radius of the growing particles $\bar{S}$ to S_1, $\bar{S}/S_1$ against $t/\Delta t$, under the condition that only the primary particles move. Here, the total number of the primary particles in a given space $N_{PP}L'^3$ is 2000, all of which are assumed to have generated spontaneously at $t=0$. The size of growing particles gradually approaches the asymptotic value of $\bar{S}_2/S_1$, which is estimated to be 1.589 for $R=5$, 1.493 for $R=10$ and 1.454 for $R=20$. These are attained within the time of $t/\Delta t=10\sim 20$, i.e., $4.5\times 10^{-9}\sim 9.0\times 10^{-9}$sec in case (1) and $0.7\times 10^{-4}\sim 1.4\times 10^{-4}$sec in case (2). $\bar{S}/S_1$ are unrealistically small as compared with the experimental value ($\bar{S}/S_1>10$; see, for example, Fig. 6.7), suggesting that the conditions of calculation used in Fig. 6.22 (only the primary particles move) can not explain the real phenomenon.

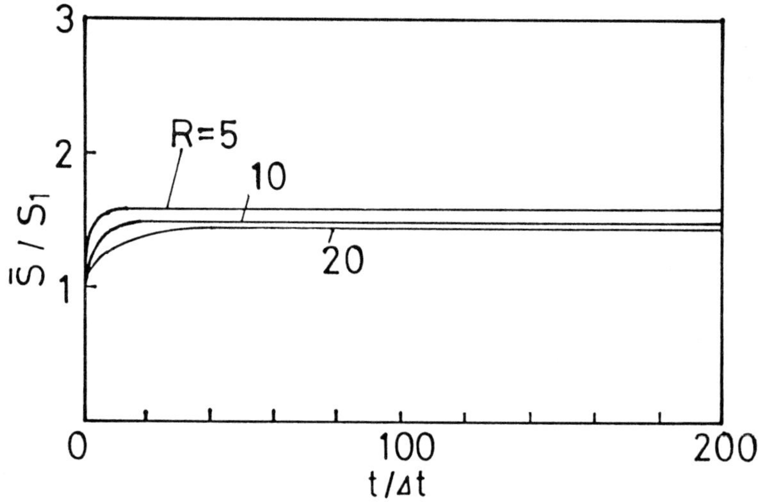

Fig. 6.22 Time-dependence of the ratio of mean radius of growing particle $\bar{S}$ to that of primary particle S_1: The case when only primary particles move. Two-phase volume ratio R is shown on the curve; when particles move according to energy equi-partition law, $\Delta t\simeq 4.5\times 10^{-10}$sec; when particles move according to brownian motion in viscous media, $\Delta t\simeq 7\times 10^{-6}$sec.

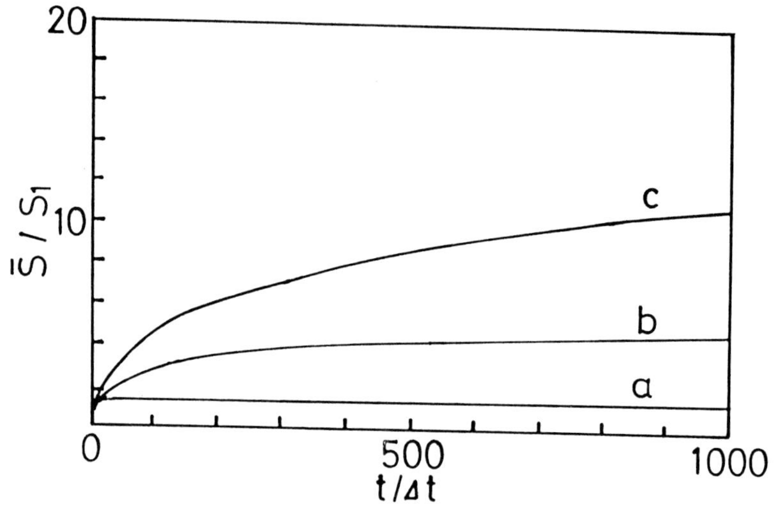

Fig. 6.23 Time-dependence of the ratio of mean radius of growing particle $\bar{S}$ to that of primary particle S_1: a, the case when only primary particles move; b, the case when particles of all sizes move according to energy equi-partition law (eqn (6.37)); c, the case when particles of all sizes move according to brownian motion in viscous media (eqn (6.40)).

Fig. 6.23 shows the plots of $\bar{S}/S_1$ against $t/\Delta t$ for the three alternative cases: a, case when only the primary particles move and cases b and c, where all the growing particles move according to the energy equi-partition law (b) (eqn (6.37)) or brownian movement (c) (eqn (6.40)). Here, $R=10$ and $N_{PP}L'^3 = 2000$ (number) were assumed. As described before, unit time Δt in case b is ca. 4.5×10^{-10}sec. The time necessary for mean particle size to attain 3.5 times as large as the primary particle is estimated from the figure to be 300 time steps ($= 300\times \Delta t$), that is 1.4×10^{-7}sec. Actually, the growth of particles finishes within several to several ten min. This means that in case b the growth rate seems too large. In case c the time needed for mean particle size of growing particles to acquire ca.6.5 times of S_1 is ca.2.1×10^{-3}sec which is estimated by assuming that the particles move in pure solvent, without containing polymer. In reality the growing particles move in viscous polymer solution and as a result the time needed for particle growth will be much longer if the viscosity of the solution η should be taken into consideration. The unit of time Δt employed for calculation is estimated from eqn (6.41), in which η is also a function of $v_{p(1)}$ and X of the polymer. The phase volume ratio R and $v_{p(1)}$ can be calculated using the coexistence curve and of course, these are not independent variables.

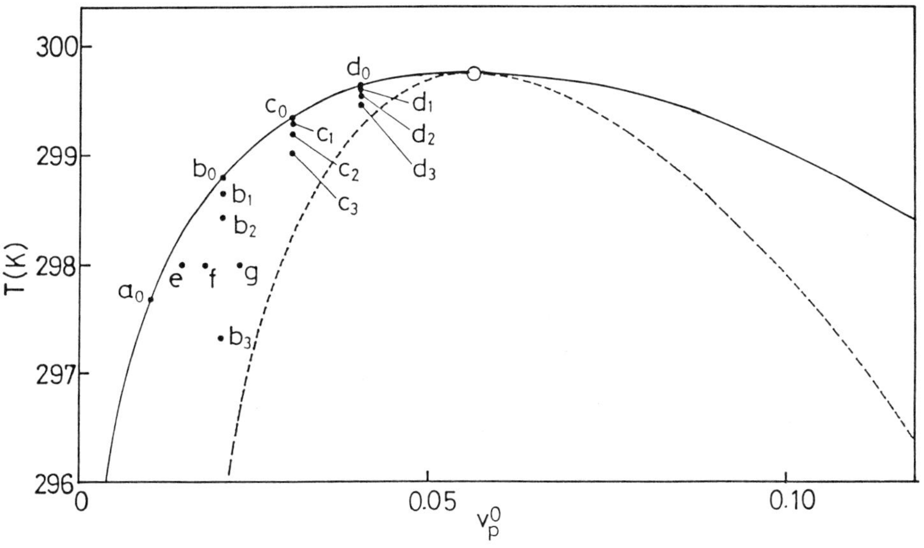

Fig. 6.24 Phase diagram of monodisperse polystyrene ($M_w = 3 \times 10^5$) / cyclohexane system: full line, coexistence curve; broken line, spinodal curve; open circle, critical solution point; filled circle, the point used for calculation of particle growth.

First, the effects of the initial polymer concentration v_p^0 and of the two-phase volume ratio R on the growth rate of particles created in phase separation process will be discussed. For this sake, we consider a system of monodisperse atactic polystyrene (the molecular weight $M = 3 \times 10^5$) / cyclohexane, whose phase diagram is shown in Fig. 6.24. In this figure, the co-existence curve (binodal curve) is represented by a full line and spinodal curve by a broken line. The growth of particles occurs in the metastable region, defined as the region between the two curves. The points in the figure are the composition of the solution, at which the solvent casting is made. The viscosity of a polymer-lean phase η is calculated using the well known empirical relation,

$$\eta = \eta_0 (1 + [\eta] v_{P(1)} + [\eta]^2 v_{P(1)}{}^2 k_H') \qquad (6.42)$$

where η_0 is the viscosity of the pure solvent, $[\eta]$, the limiting viscosity number (cm^3/g), $v_{P(1)}$, the polymer volume fraction of a polymer-lean phase. k_H', a coefficient similar to Huggins coefficient. Here, we used the values for atactic polystyrene / cyclohexane system at Flory temperature $(\theta = 307.1K)$; $\eta_0 = 0.76cP$ and $k_H' = 0.59$. $[\eta]$ is related to the weight-average molecular weight M_w through the Mark-Houwink-Sakurada equation (ref. 41)

$$[\eta] = 84.6 \times 10^{-3} M_w{}^{1/2}. \qquad (6.43)$$

Fig. 6.25a shows the ratio of the mean radius of the growing particles $\bar{S}$ to that of primary particles S_1, $\bar{S}/S_1$ as a function of time for a series of solutions, whose locations in the phase diagram are shown as points a_0, b_0, c_0 and d_0 in Fig. 6.24. Note that in this case all points are located in vicinity to the cloud point curve, corresponding to $R = 100$. Inspection of Fig. 6.25a indicates that the growth rate of the particles is larger for smaller $v_P{}^0$. That is, when R is the same, the particle growth rate is determined by $v_{P(1)}$ and is larger as $v_{P(1)}$ is smaller. This can be reasonably explained in the following manner: The mean square displacement of the particles with the same radius is inversely propotional to η (eqn (6.40)), and the mean square displacement is larger, accordingly, the frequency of collision is larger in less viscous media. In addition, the difference in the mean particle size generated at points $a_0 - d_0$ becomes less remarkable, approaching an asymptotic value. The viscosity of the solution influences the particle growth rate, especially at the initial stage and the time necessary to attain an asymptotic value, which does not depend on η.

Fig. 6.25b shows the particle growth rate of the solutions having various R and constant $v_{P(1)}$ $(= 1.20 \times 10^{-2})$ (points g, f and e in Fig. 6.24) at constant temperature. The particle grows faster for smaller R, when comparison is made at the same $v_{P(1)}$. When R is smaller, the portion of volumes occupied by the primary

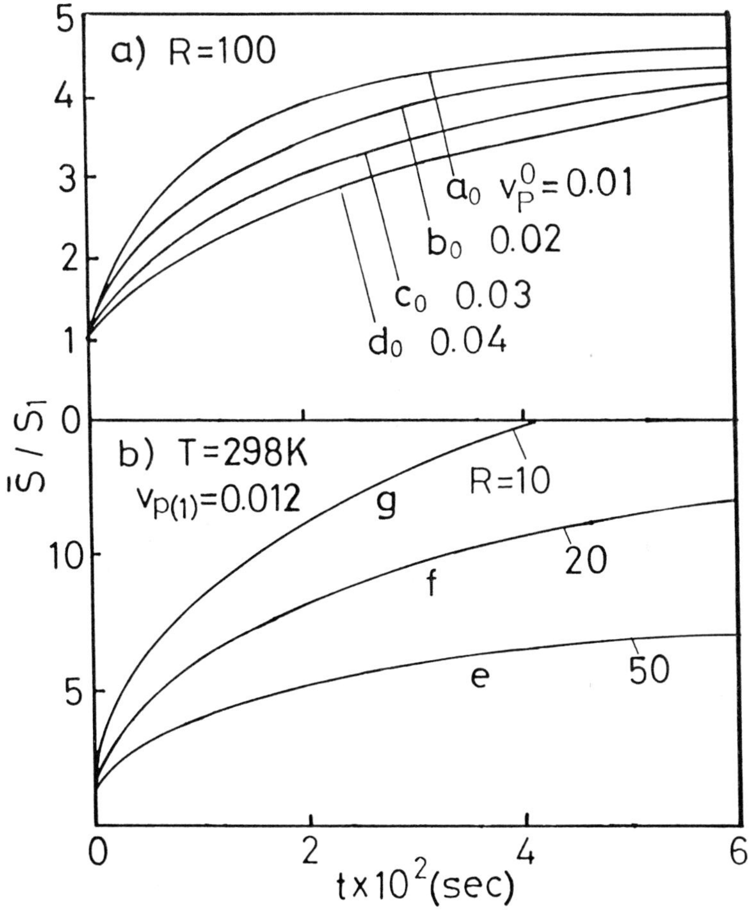

Fig. 6.25 Time-dependence of the ratio of the mean radius of growing particle $\bar{S}$ to that of primary particle S_1: a) Effect of the initial polymer volume fraction v_p^0 at constant R (=100); a_0, $v_p^0=0.01$, T=297.65K; b_0, $v_p^0=0.02$, T=298.76K; c_0, $v_p^0=0.03$, T=299.53K; d_0, $v_p^0=0.04$, T=299.65K: b) Effect of the phase volume ratio R at constant polymer volume fraction in a polymer-lean phase $v_{p(1)}$ ($=1.20\times10^{-2}$) and at constant temperature (298K); e, R=50, $v_p^0=1.45\times10^{-2}$; f, R=20, $v_p^0=1.80\times10^{-2}$; g, R=10; $v_p^0=2.31\times10^{-2}$.

particles is larger, resulting in a rapid increase of the frequency of particle-particle collision.

Fig. 6.26 shows the effect of the casting temperature [i.e., depth of phase separation, Δ_P $(= T_{CP} - T_P)$] on the mean radius of growing particles when the solutions having constant v_p^0 [0.04 for a), 0.03 for b) and 0.02 for c)] are cooled down to different casting temperatures. Note that in this case, both R and $v_{P(1)}$ vary concurrently in a very complicated manner, corresponding to a variation of the temperature, as demonstrated as points in Fig. 6.24. Interestingly, the particle growth rate is larger as the phase separation occurs at a point more close to the spinodal curve (i.e., at larger Δ_P), where the polymer concentration in a polymer-lean phase $v_{P(1)}$ is lower (and accordingly, its viscosity is lower), resulting in a larger velocity of the particles, and simultaneously, R is smaller. Comparison of Fig. 6.26a, b and c shows that even at the same R the particles grow much faster from the solutions of lower initial concentration due to lower $v_{P(1)}$ (and η). Summarizing, the growth rate of the particles is theoretically expected to be larger when the phase separation occurs under the conditions of (1) lower polymer concentration of a polymer-lean phase $v_{P(1)}$ and (2) smaller two-phase volume ratio R. Condition (1) yields the larger velocity and condition (2) makes for a larger collision frequency. These conditions will be satisfied practically, at least, when a dilute solution is quickly cooled down (or quenched) to and kept at a point in the metastable region near to the spinodal curve for a longer period.

Fig. 6.27 shows the particle size distribution at four different times when the casting condition, shown as the point d_3 in Fig. 6.24, is employed: Here, the radius of the primary particles is assumed to be absolutely uniform. At the very earliest stage (curve a in the figure), the peak of the size distribution is kept almost at S_1, but the distribution has a long tail in large particle size and as time elapses the distribution shifts in general to larger S/S_1 region. Of course, there is neither equilibrium particle size distribution nor equilibrium mean radius of the growing particles and only some approximately steady state, which depends strongly on the time scale of observation, can be realized.

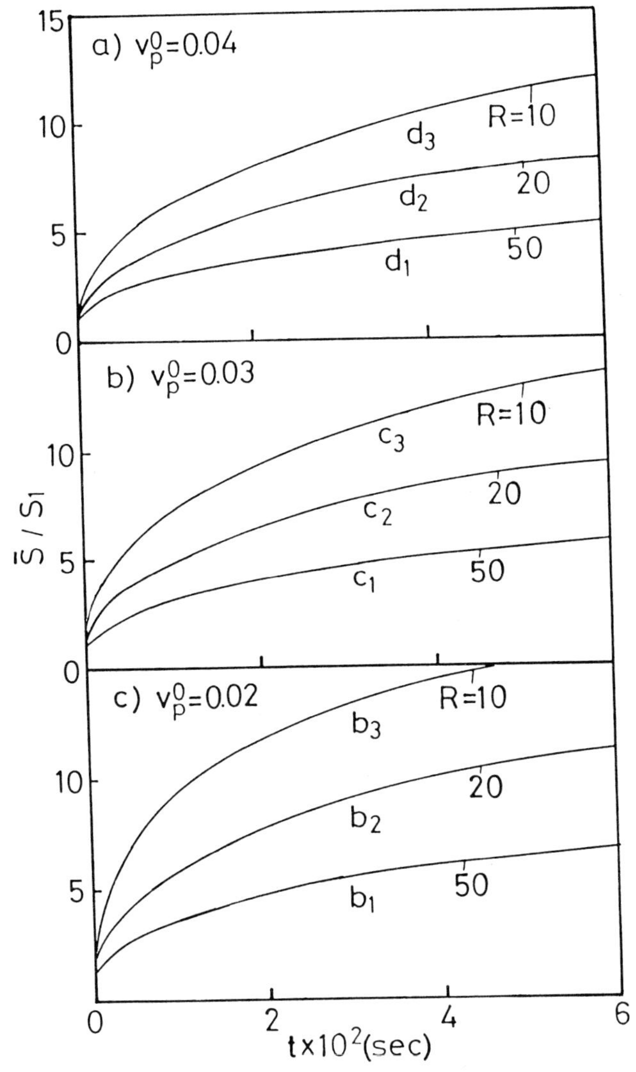

Fig. 6.26 Effect of R and v_p^0 on the relations between the mean radius of particles with time:
a) $v_p^0=0.04$; d_1, T=299.61K, R=50 and $v_{p(1)}=3.92\times10^{-2}$; d_2, T=299.58K, R=20 and $v_{p(1)}=3.79\times10^{-2}$; d_3, T=299.53K, R=10 and $v_{p(1)}=3.62\times10^{-2}$.
b) $v_p^0=0.03$; c_1, T=299.26K, R=50 and $v_{p(1)}=2.90\times10^{-2}$; c_2, T=299.15K, R=20 and $v_{p(1)}=2.62\times10^{-2}$; c_3, T=299.00K, R=10 and $v_{p(1)}=2.32\times10^{-2}$.
c) $v_p^0=0.02$; b_1, T=298.62K, R=50 and $v_{p(1)}=1.83\times10^{-2}$; b_2, T=298.41K, R=20 and $v_{p(1)}=1.56\times10^{-2}$; b_3, T=297.29K, R=10 and $v_{p(1)}=7.59\times10^{-3}$.

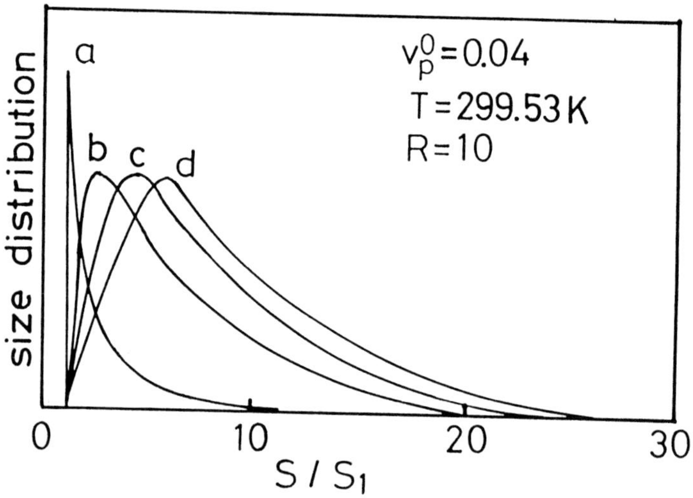

Fig. 6.27 Size distribution of growing particles: $v_p{}^0=0.04$, T=299.53K, R=10 and $v_{p(1)}=3.62\times10^{-2}$: a, 3.0×10^{-3}sec; b, 1.5×10^{-2}sec; c, 3.0×10^{-2}sec; d, 4.5×10^{-2}sec.

6.3.2 Reaction kinetics approach by the deterministic method

The growth of the primary particles can also be analyzed by the reaction kinetic approach in analogy to the polymerization reaction (in particular, propagation reaction) (ref. 42). The rate of production of the primary particles is given by eqn (6.44)

$$dN_{PP}/dt=k_CN_{CN}-\{\frac{1}{2}k_{1,1}\alpha_{1,1}N_{PP}{}^2+N_{PP}\sum_{j=2}^{\infty}k_{1,j}\alpha_{1,j}N_{GP(j)}\} \qquad (6.44)$$

with

$$N_{GP}=\sum_{j=2}^{\infty}N_{GP(j)}. \qquad (6.45)$$

Here, mean particle radius is

$$\bar{S}=\bar{S}_1(r)^{1/3}=S_1(\Sigma\, rN_{GP}/\Sigma\, N_{GP})^{1/3} \qquad (6.46)$$

where N_{PP} is the number of primary particles per unit volume, N_{CN} is taken as N_C at t_P, $N_{GP(j)}$, the number of growing particles with j primary particles (j= the number of primary particles constituting a given particle) per unit volume, k_C, the growing rate constant, $k_{1,1}$ and $k_{1,j}$, the reaction rate constants between two primary particles and between a primary particle and a growing particle with j primary particles, $\alpha_{1,1}$ and $\alpha_{1,j}$, the amalgamation probability between two primary particles and between a primary particle and a growing particle with j primary particles. $\alpha_{1,1}$ and $\alpha_{1,j}$ are usually less than unity, because not all collisions are effective in generating growing particles.

The rate of production of growing particle with j= 2 is

$$dN_{GP(2)}/dt = (1/2)\ k_{1,1}\alpha_{1,1}\ N_{PP}^2 - N_{GP(2)}\sum_{j=1}^{\infty} k_{2,j}\alpha_{2j}N_{GP(j)}. \quad (6.47)$$

If j= 2, $k_{2,j}$ becomes $(1/2)k_{2,2}$. The first term of the right-hand side in eqn (6.47) is due to formation by collisions between two primary particles and the second term corresponds to loss by collisions between a growing particle with j= 2 and another growing particle or the primary particle. If only the primary particles collide with growing particles, $k_{2,j}$ (j> 1) in eqn (6.47) becomes zero.

The production rates of growing particles with j> 3 are also given by

$$dN_{GP(3)}/dt = (1/2)\ (k_{1,2}\alpha_{1,2}N_{PP}N_{GP(2)} + k_{2,1}\alpha_{2,1}N_{GP(2)}N_{PP})$$

$$- N_{GP(3)}\sum_{j=1}^{\infty} k_{3,j}\alpha_{3,j}N_{GP(j)} \quad (6.48)$$

and

$$dN_{GP(r)}/dt = (1/2)\ \{k_{1,r-1}\alpha_{1,r-1}N_{PP}N_{GP(r-1)}$$

$$+ k_{2,r-2}\alpha_{2,r-2}N_{GP(2)}N_{GP(r-2)}$$

$$+ \cdots$$

$$+ k_{r-1,1}\alpha_{r-1,1}N_{GP(r-1)}N_{PP}\}$$

$$- N_{GP(r)}\ \{k_{r,1}\alpha_{r,1}N_{PP} + k_{r,2}\alpha_{r,2}N_{GP(2)}$$

$$+ \cdots \qquad + k_{r,\infty}\alpha_{r,\infty}N_{GP(\infty)}\}$$

$$= (1/2)\sum_{i=1}^{r-1} k_{i,r-i}\alpha_{i,r-i}N_{GP(i)}\cdot N_{GP(r-i)}$$

$$-N_{GP(r)}\sum_{j=1}^{\infty} k_{r,j}\alpha_{r,j}N_{GP(j)} \tag{6.49}$$

where, $\alpha_{i,j}$ is amalgamation probability between two growing particles constituted with i and j primary particles each. If $j=r$, $k_{r,j}$ becomes $1/2k_{r,r}$.

The reaction rate constant $k_{i,j}$ can be expressed as below

$$k_{i,j}={}_iZ_j. \tag{6.50}$$

${}_iZ_j$ is collision frequency (the number of collision per unit volume per unit time if i and j have unit concentration) and

$$k_{i,j}\simeq 4k'(S_{(i)}+S_{(j)})/(1/S_{(i)}+1/S_{(j)}) \tag{6.51}$$

with

$$k'=\tilde{R}T/(6\pi\ \eta). \tag{6.52}$$

$S_{(i)}$ and $S_{(j)}$ are the radius of the particles constituted with i and j primary particles each and η represents the viscosity of surrounding media (in this case, polymer-lean phase).

In the case, r>>1,

$$k_{r,r-1}\simeq k_{r,r}'=k'Ov_{r(r')} \tag{6.53}$$

and

$$Ov_{r(r')}\ \simeq\ 1\cdot r^{2/3}(r')^{2/3}. \tag{6.54}$$

O is the area exposed to collide, $v_{r(r')}$, the average velocity, with which particles approach the surface of the growing particle in question in steady state. Here, the radius of the primary particle is taken as unity. The primary particle, which reach the surface of a growing particle with an activation energy of amalgamation, E_{PP} or more, amalgamates with the growing particle.

When all the parameters employed here could be evaluated in advance, by solving simultaneous eqns (6.47), (6.48) and (6.49), a set of radii of $N_{GP(2)}$, $N_{GP(3)}$,···, $N_{GP(r)}$,···, $N_{GP(rmax)}$ could be determined, provided that the capacity of computer is very large and this set affords us the particle size distribution at time t. Unfortunately, since the parameters used here have not yet been fully determined by experiment, the discussion based on this theory remains qualitative and is severely limited.

The preliminary calculations show that (1) the particles continue to grow without limit yielding ultimately a single homogeneous polymer-rich phase, (2) in an intermediate stage the primary particles coexist with growing particles (this was experimentally observed (Fig. 6.29)) and (3) the apparent asymptotic value of $\overline{S}_2$ depends strongly on the assumed maximum number of primary particles in a growing particle r_{max}, which is introduced due to the limit of computer capacity.

Fig. 6.28 shows the effects of the viscosity of surrounding media η and of the two-phase volume ratio R on the growth of the particles. The calculation is made under the condition as follows: $k_C = 1$sec, $N_{PP} = 4.5 \times 10^{-3}$mol/m^3 ($= 2.7 \times 10^{21}$number/m^3) (calculated from $R = 10$ by eqn (6.36)), $k' = 1.088 \times 10^{12}$m^3/mol·sec (from $\eta = 1.2$cP, $T = 300$K), all α values $= 1$, time interval $\Delta t = 1.0 \times 10^{-6}$sec, number of calculation 100, $r_{max} = 1000$. From the figure it is clear that (1) the particles grow more slowly in viscous media at larger two-phase volume ratio, and (2) at an intermediate stage the radius of secondary particles is smaller under the above conditions (i.e., viscous media, high R).

It should be emphasized, however, that the basic principle of the reaction kinetics approach (6.3.2) is fundamentally different from the particle simulation approach (6.3.1). But both approaches give the same tendency of growth of particles.

6.3.3 Experimental observation of particle growth

The size and its distribution of the particles formed in the process of phase separation can be experimentally determined by electron micrographic observation and by light scattering measurements, which provide sound strong scientific evidence supporting the "particle growth concept".

The above two approaches ignore the possible contribution of intersurfacial properties of particles, such as interfacial free energy, ionic charge and density difference, whose role can never be underestimated in steps e) and f) in Fig. 6.5 (i.e., growing particle and secondary particle each).

(a) Microscopic observations

Fig. 6.29 shows the time dependence of the average radius of two kinds of particles as determined by electron micrograph for a cellulose cuprammonium solution coagulated in a mixture of acetone

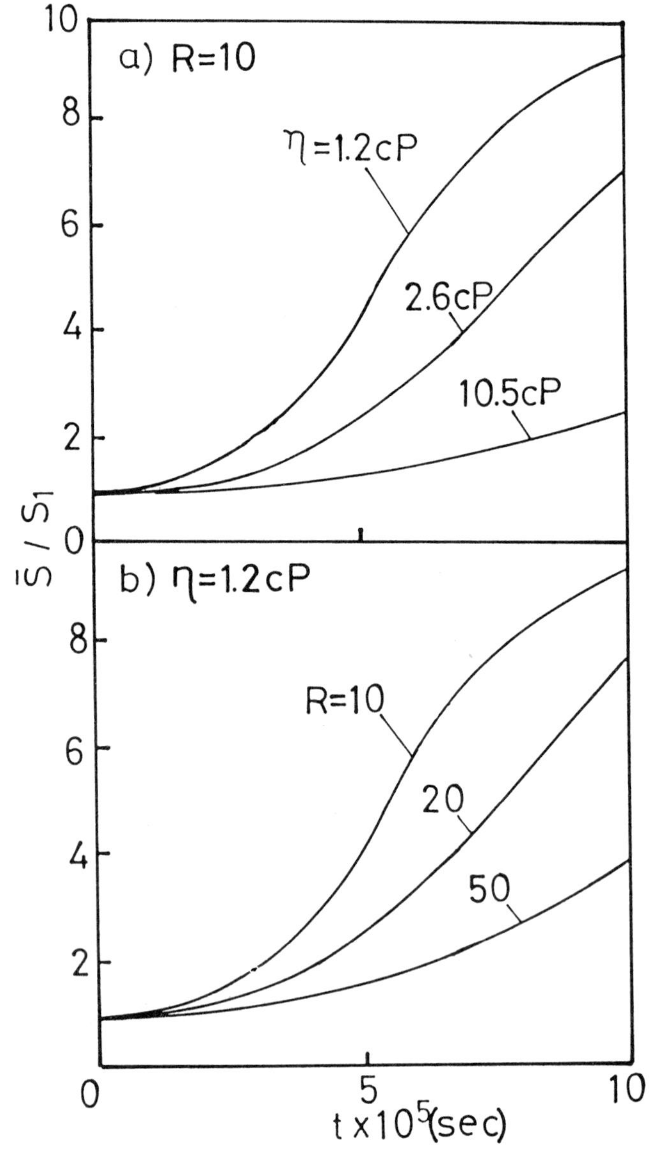

Fig. 6.28 Effect of viscosity of surrounding media η (a) and of two-phase volume ratio R (b) on the ratio of mean radius of the growing particles $\bar{S}$ to that of primary particles S_1, $\bar{S}/S_1$.

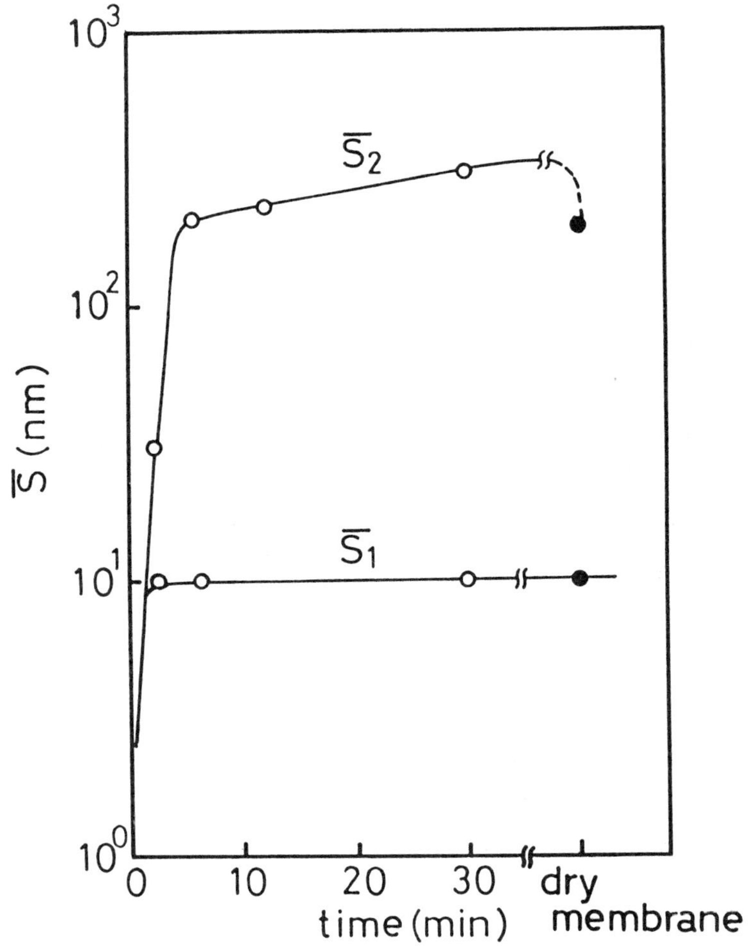

Fig. 6.29 Time-dependence of mean particle radius as determined by EM for a cellulose cuprammonium solution (the weight fractions of cellulose, $w_{CELL}{}^S$; copper, $w_{Cu}{}^S$; ammonia, $w_{NH_3}{}^S$ and water, $w_{H_2O}{}^S$ are 0.0800, 0.0316, 0.1122 and 0.7762, respectively) coagulated by mixture of acetone / ammonia / water (30.00/0.56/69.44, wt/wt/wt) at 298K (unfilled circle) and its regenerated cellulose dry membrane (filled circle): Experimental conditions are the same as those in Fig. 6.7.

/ ammonia / water (30.00/0.56/69.44, wt/wt/wt). Obviously, the primary particles grow rapidly approaching asymptotic size as the theory predicts. In this case, a secondary particle consists of ca.10^4 primary particles. The number of the primary particles is ca.2×10^{16}number/m^3 at the initial stage and at the intermediate stage where the primary particles grow to secondary particles the number of these particles is estimated to be ca.5×10^{15}number/m^3. The secondary particles with a diameter of 300nm to 1μm numbering ca.1×10^{14}number/m^3 are also observed. Note that these number densities of the particles obtained by the electron microscopic method might seriously differ from their average values of the whole system and can not be analyzed in more detail.

(b) Light scattering measurements

By dynamic light scattering (DLS), the relationships between the first order time correlation function of the electric field of the scattered light, $I_1(q,\tau)$ (where size of the scattering vector, $q=(4\pi/\lambda)\sin(\theta/2)$; λ, wave length of incident light in the solution; θ, scattering angle; τ, correlation time) and τ are obtained. The hydrodynamic diameter of the particle 2S and its distribution can be determined by the following cummulant and histogram methods, respectively.

<u>Cummulant method</u>

If the particles diffuse at random in the continuous media, I_1 is related to diffusion coefficient D_i of the particles with the same diameter $2S_i$ as follows (ref. 43):

$$I_1(q,\tau)=\sum_i g_i \exp(-q^2 D_i \tau) \tag{6.55}$$

with

$$g_i \equiv \alpha_i{}^2 N_i M_i{}^2 \tag{6.56}$$

where α_i is a polarizability per unit mass, N_i, number of the particles and M_i, mass of the each particle with diffusion coefficient D_i, respectively.

The normalized correlation function $N(q,\tau)$ is defined by

$$N(q,\tau)\equiv I_1(q,\tau)/I_1(q,0)=\sum g_i \exp(-q^2 D_i \tau)/\sum g_i$$

$$\equiv \langle \exp(-q^2 D\tau) \rangle . \tag{6.57}$$

Logarithm of $N(q,\tau)$ can be expanded by a cummulant expression,

$$\ln N(q,\tau) = 1 - k_1\tau + k_2\tau^2/2 + \cdots \qquad (6.58)$$

with

$$k_1 = q^2 \langle D \rangle_z \qquad (6.59)$$

and

$$k_2 = \langle (q^2 D)^2 \rangle_z - \langle q^2 D \rangle^2_z \ . \qquad (6.60)$$

The suffix z means the z-average. By homodyne DLS measurements we can evaluate τ dependence of I_1 and accordingly, $N(q,\tau)$ and using eqn (6.59) $\langle D \rangle_z$ is determined.

Assume that the particle as rigid sphere obeys Einstein-Stokes law,

$$2S_{H,z} = k_B T/(3\pi\ \eta_0 \langle D \rangle_z) \qquad (6.61)$$

where k_B is Boltzmann constant, η_0, solvent viscosity. Rigorously, S obtained from eqn (6.61) is not the same as z-average radius, but conveniently we denote S in eqn (6.61) as $S_{H,z}$.

Histogram method

In the histogram method, which was developed by Gulari et al. (ref. 44), I_1 is assumed to be represented in the summation form,

$$I_1(q,\tau) = \sum_{i=1}^{M'} g(\Gamma_i) \int_{\Gamma_i - \Delta\Gamma_i/2}^{\Gamma_i + \Delta\Gamma_i/2} \exp(-\Gamma_i \tau)\, d\Gamma \qquad (6.62)$$

where

$$\Gamma_i = q^2 D_i \ . \qquad (6.63)$$

M' is the number of steps in the histogram, and $\Delta\Gamma_i$ is the width of each step. The product $g(\Gamma_i)\Delta\Gamma_i$ is the intensity fraction of the light scattered from the particles with diffusion coefficient D_i and is normalized as,

$$\sum_{i=1}^{M'} g(\Gamma_i)\Delta\Gamma_i = 1. \qquad (6,64)$$

Using the method of curve fitting, such as modified Marquadt one, $g(\Gamma_i)$ is estimated from the experimentally obtained decay line of I_1. From the histogram of Γ_i, we can estimate the averaged Γ_i values (i.e., D) and then mean diameter of the particles with the aid of an Einstein-Stokes relation (eqn (6.61)). Hereafter, we denote such the averaged S obtained from the distribution function $g(\Gamma_i)$ as $S_{H,G}$:

$$S_{H,G} = k_B T / (3\pi \eta_0 \langle D \rangle_G). \tag{6.61'}$$

Here, $\langle D \rangle_G$ is the g-average self-diffusion constant.

According to Zimm's theory (ref. 45) on static light scattering of dilute polymer solution, $g(\Gamma_i)$ is written as a function of M_i, that is,

$$g(\Gamma_i) = KN_i M_i^2 P(q,S_i) \tag{6.65}$$

where K is an optical constant, $P(q,S_i)$, the scattering function. In the case of spherical particle, following relations hold.

$$M_i \propto S_i^3 \tag{6.66}$$

and

$$P(q,S_i) = (3/a^3)^2 \{\sin(a) - a \cdot \cos(a)\} \tag{6.67}$$

with parameter $a = 2qS_i$.

The weight fraction of the particles with diameter $2S_i$, $w(2S_i)$ is proportional to $N_i M_i$, therefore $w(2S_i)$ is related to $g(2S_i)$ as,

$$w(2S_i) = g(2S_i) / (2S_i)^3 P(q,S_i). \tag{6.68}$$

The number fraction $N(S_i)$ is readily derived,

$$N(S_i) = g(S_i) / (2S_i)^6 P(q,S_i). \tag{6.69}$$

Using the distribution functions $w(S)$ and $N(S)$, we can obtain the weight- and the number-average radius, S_w and S_n, respectively.

Fig. 6.30 shows a scheme of the dynamic light scattering apparatus, combined with a stopped-flow mixing apparatus designed and constructed for this purpose.

Fig. 6.31 shows the particle size distribution N(S) as a function of the time for a 0.5wt% cellulose cuprammonium solution

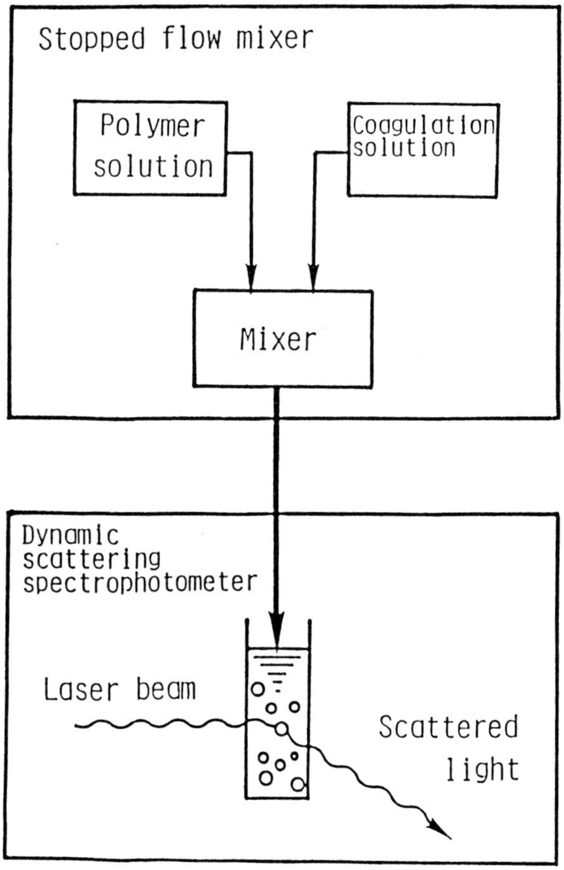

Fig. 6.30 A simplified schema of the dynamic light scattering apparatus, attached with a stopped-flow mixer: Dynamic light scattering spectrophotometer DLS-700 (Otsuka Denshi Co., Tokyo, Japan), stopped-flow mixer MC-980HD (Otsuka Denshi Co., Tokyo, Japan).

/ 25wt% aq. acetone solution system and a 0.5wt% cellulose cuprammonium solution / 3wt% aq. NaOH solution system at 293K. In this system, N(S) shifts to larger S region with time. Note that the polymer concentration (0.083wt% of the solutions used for DLS measurement) is almost 1/10 of that of the casting solutions actually employed for membrane formation.

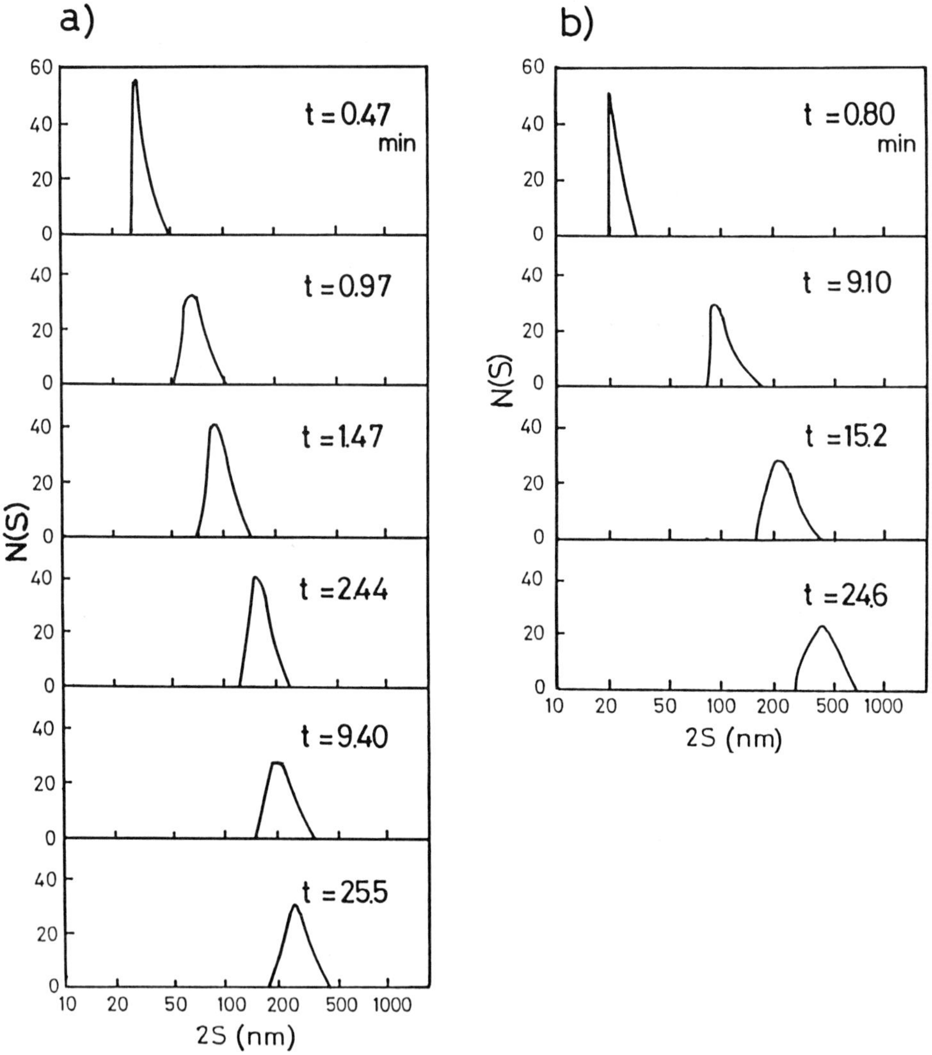

Fig. 6.31 Particle size distribution N(S) of cellulose solution in cuprammonium / acetone mixture [weight fraction of cellulose (viscosity-average degree of polymerization, $X_v = 620$); $w_{CELL} = 0.00083$; weight fraction of ammonia, $w_{NH3} = 0.04891$; weight fraction of copper, $w_{CU} = 0.00033$; weight fraction of water, $w_{H2O} = 0.74143$; weight fraction of acetone, $w_{ACETONE} = 0.2085$] at 293K [a)] and cellulose solution in cuprammonium / aq. sodium hydroxide mixture [$w_{CELL} = 0.00083$; $w_{NH3} = 0.04474$; $w_{CU} = 0.00033$; $w_{H2O} = 0.9291$; weight fraction of NaOH, $w_{NAOH} = 0.025$] at 293K [b)]: t means the past time after mixing of cellulose cuprammonium solution with aq. acetone or aq. sodium hydroxide solution.

From this figure, the mean particle size $\bar{S}(\equiv S_n)$ was calculated and plotted against time t in Fig. 6.32. $\bar{S}$ for cellulose cuprammonium / acetone solution increases rapidly, approaching to an asymptotic value (ca.110nm) within 10min, but $\bar{S}$ for cellulose cuprammonium / aq. sodium hydroxide solution continues to increase with time, exceeding 550nm or more.

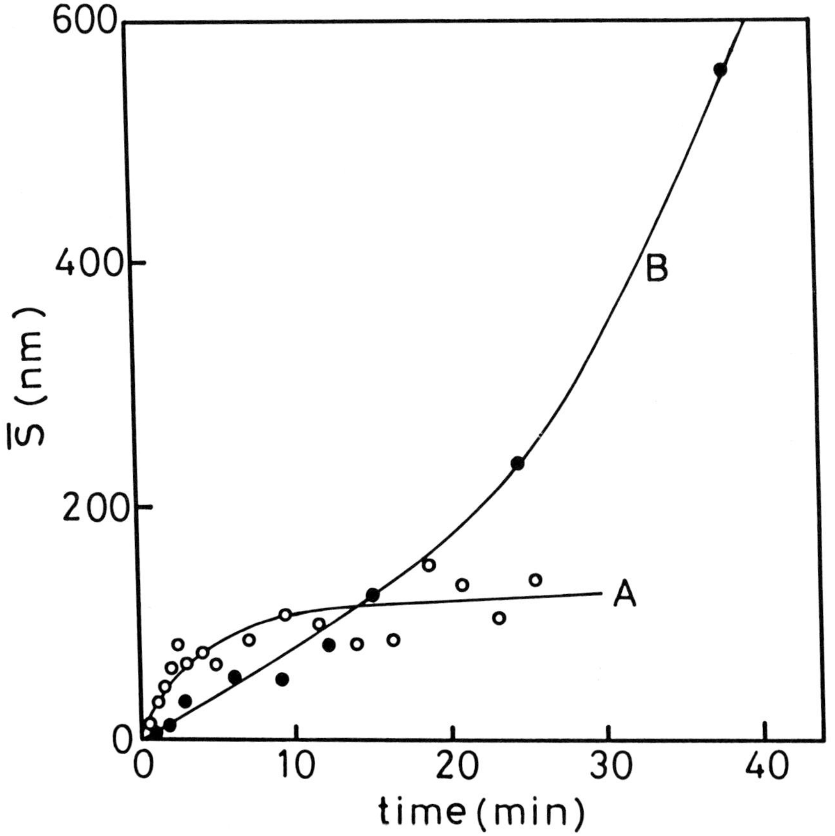

Fig. 6.32 Mean particle radius $\bar{S}$ of two cellulose solutions as functions of time after mixing with coagulants at 293K: Curve A, cellulose solution in cuprammonium / acetone mixture, whose compositions are the same as those in Fig. 6.31; curve B, cellulose solution in cuprammonium / aq. sodium hydroxide mixture, where the compositions are the same as those in Fig. 6.31.

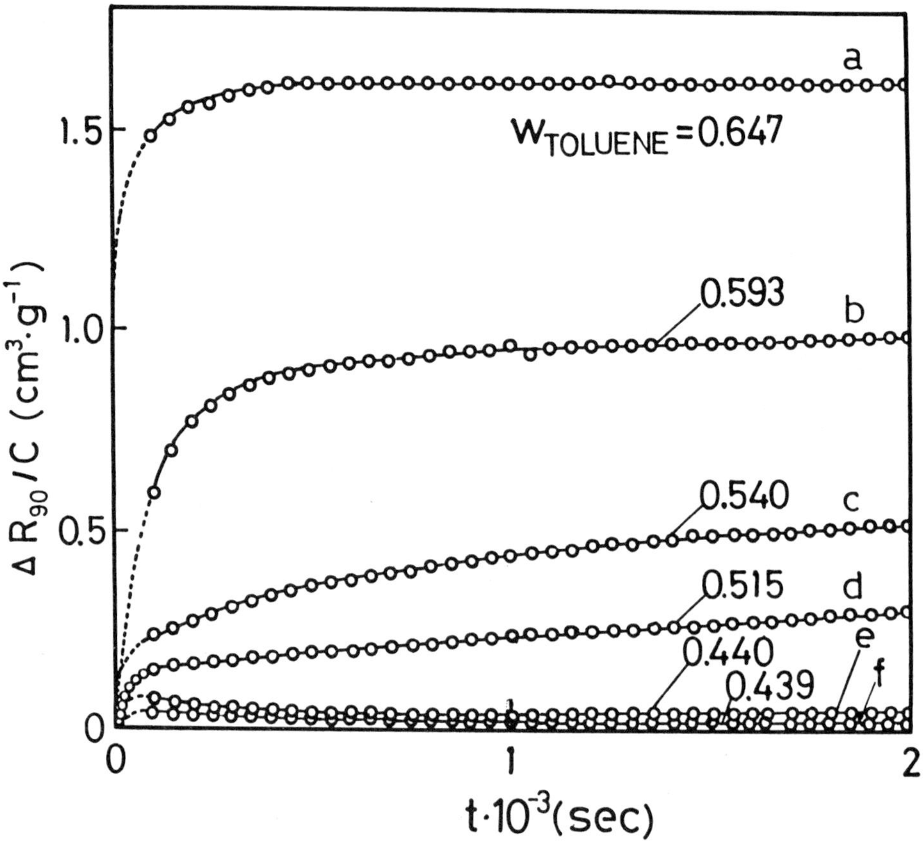

Fig. 6.33 Time dependence of the ratio of the difference of the Rayleigh ratio at 90° ΔR_{90} to polymer concentration c between polyacrylonitrile (PAN) / dimethylsulfoxide (DMSO) / toluene system and DMSO / toluene (whose composition is kept the same as in the polymer solution) at 298K: Toluene weight fraction $w_{TOLUENE}$ is shown on curve. Polymer, atactic PAN; the weight-average molecular weight $M_w = 2.0 \times 10^5$.

Using the static classical light scattering methods, the growth of the primary particles can also be predicted for synthetic polymers since these two obey the same theoretical principles. Fig. 6.33 shows the ratio $\Delta R_{90}/c$ (Here, c is the polymer concentration g/cm^3, ΔR_{90}, the difference of the Rayleigh ratio at $\theta = 90°$ at the wave length of incident light in vacuo $\lambda_0 = 633$nm between the solution and the solvent mixture) for polyacrylonitrile (atactic, the weight-average molecular weight $M_w = 2.0 \times 10^5$) / dimethylsulfoxide (DMSO) / toluene system at 298K, as a function of time (ref. 46). $\Delta R_{90}/c$ is considered to be roughly proportional to the weight of the particle (accordingly, the particle size). At $t = 0$, the solution of polymer in DMSO ($c = 9.9 \times 10^{-3} g \cdot cm^{-3}$) was mixed with toluene and ΔR_{90} was measured with DLS apparatus and results shown in Fig. 6.33. The minimum t measurable by this procedure is at present a round 1sec. The primary particles grow rapidly at an early stage, approaching an asymptotic value, which depends on the composition of solvent (DMSO) and nonsolvent (toluene). The value of $\Delta R_{90}/c$ is plotted in Fig. 6.34 against the toluene weight fraction $w_{TOLUENE}$ of the solution. The particle size increases remarkably with toluene content when $w_{TOLUENE}$ exceeds 0.51. Similar plots for PAN / DMSO / n-propanol and PAN / DMSO / water systems are also shown in the figure. The critical nonsolvent concentration, above which the particles grow without limit depends on the solvent nature, increasing in the following order: Water< n-propanol< toluene.

Therefore, it is experimentally confirmed again that in some polymer / solvent system the primary particles grow by amalgamation, approaching an asymptotic value, which is the radius of the secondary particle S_2, and this pattern of growth of particles, directly observed by DLS and electron micrographic methods (Figs 6.33 and 6.34) is theoretically reasoned. But, we do not claim that the theoretical treatment introduced here is the only reasonable explanation of the observed phenomenon.

One of the most important factors to consider when producing porous polymeric membrane is the conditions under which the secondary particles with the proper radius S_2 are obtained at pseudo-stable state and over a wide range of storage time. In other words, the extremely slow transition from stage b) to stage d) in Fig. 6.5 is an indispensable condition in the solvent-casting process of membrane technology. It is often experienced that a simple non-polar solvent system is not always adequate for this purpose. And the detailed experimental condition of obtaining metastable secondary particles are not yet fully understood. Cellulose, cellulose acetate, polyacrylonitrile and polysulfone are known to be proper for preparing porous membranes.

The secondary particles can be effectively separated without adhesion from the polymer-lean phase. Fig. 6.35 shows an electron micrograph of the dried secondary particles of regenerated cellulose by the procedure proposed by Iijima and Kamide (ref.

47). These cellulose particles are expected to find a wide commercial application such as carriers of catalyst, cell culture and cosmetics.

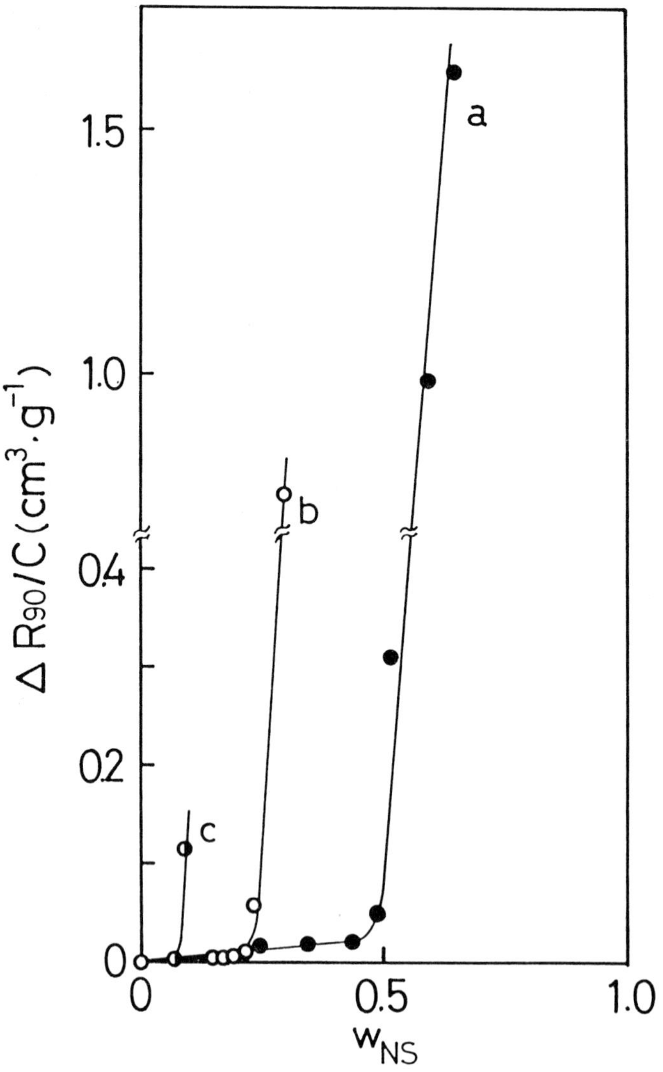

Fig. 6.34 Plot of $\Delta R_{90}/c$ obtained after 2000sec of mixing versus the nonsolvent weight fraction w_{NS} in polyacrylonitrile (PAN, $M_w = 20.0 \times 10^4$) / dimethylsulfoxide (DMSO) / nonsolvent solutions at 298K: a, toluene (●); b, n-propanol (○); c, water (◑).

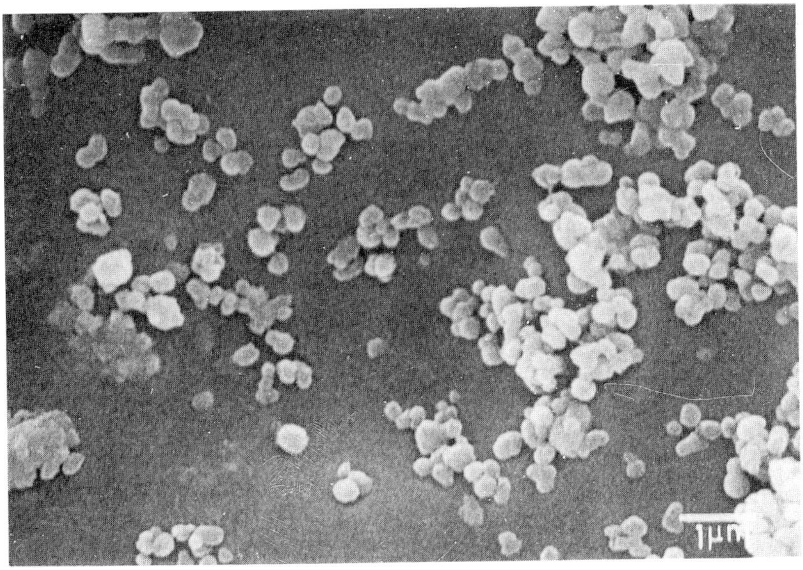

Fig. 6.35 Dry cellulose spheres separated from a mixture of the secondary particles and polymer-lean phase, which are prepared by coagulating cellulose cuprammonium solution with a 50wt% aq. acetone solution (ref. 47).

6.4 FORMATION OF PORES BY CONTACTING THE SECONDARY PARTICLES

6.4.1 Lattice theory (ref. 48)

After the growing particles approached their asymptotic size (i.e., the secondary particle) (see, Fig. 6.5), the particles contact with each other forming pores when the coagulated solution is stored without any further agitation.

Consider a hypothetical plane thickness $2S_2$ parallel to the surface of the coagulated solution and assume the solution consists of the secondary particles (i.e., polymer-rich phase) with radius of S_2 and "hypothetical particles" of polymer-lean phase, whose radius is also S_2. Let the number of the secondary particles (referred to as "polymer particle") per unit area of the plane be represented by $N_T\{1/(R+1)\}$ and that of the hypothetical polymer-lean particles (referred to as "vacant particles") by $N_T\{R/(R+1)\}$. Here, N_T is the total number of both particles on the unit area of the plane and is roughly estimated as $1/(\pi S_2{}^2)$. Obviously, a group of vacant particles contact directly with each other building a pore. The total number of pores per unit surface area (i.e., the pore density) is represented by N_P. Assume that the secondary particles and the hypothetical particles are placed randomly on a two-dimensional hexagonally close-packed lattice of the hypothetical plane (see, Fig. 6.36) to evaluate the number of distinguishable arrangements of the mixtures of the secondary particles and the hypothetical particles on a lattice. In this sense, the lattice coordination number is six.

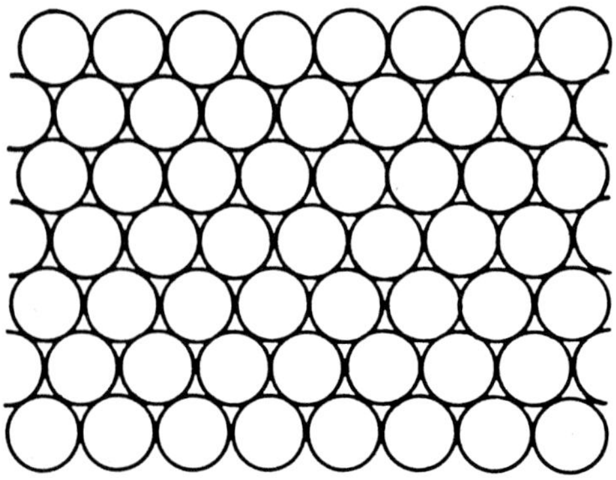

Fig. 6.36 Hexagonal closed packing lattice.

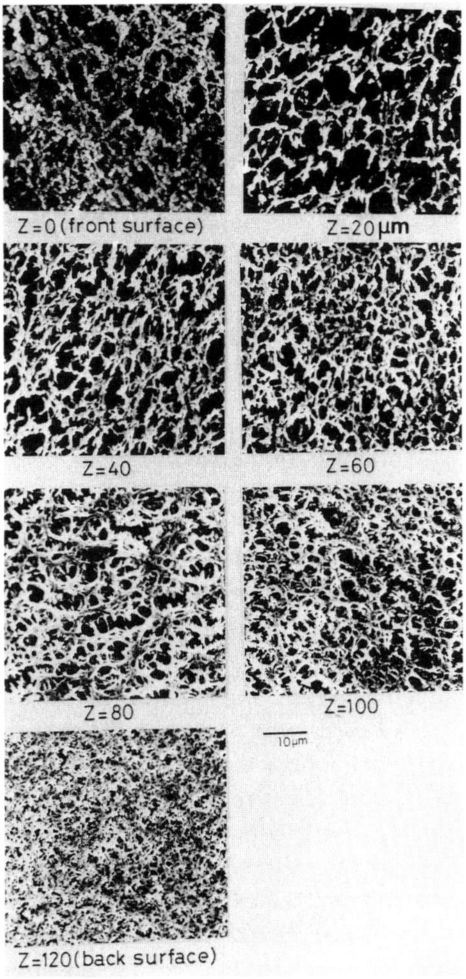

Fig. 6.37 Electron micrographs of ultra-thin sections at various distances Z(μm) from the surface of a regenerated cellulose membrane: A scale bar stands for 10μm; see, Fig. 6.38a for sample preparation.

In this theory, we assume that the pore characteristic of the hypothetical plane within the membrane, parallel to the membrane surface is kept the same. This assumption has been confirmed to be conformed to numerous experimental observations: In this case, for an electron microscopic (EM) observation, the cellulose membranes were embedded in blocks of epoxy resins and were sliced parallel to the membrane surface to give ultra-thin sections with

0.1μm thickness. These sections were prepared for each 20μm pitch of the distance Z from the front surface (see, Fig. 6.38a). Fig. 6.37 shows the electron micrographs of the ultra-thin layer section of the regenerated cellulose membrane cast from cellulose cuprammonium solution, as a function of distance Z from the top surface. On each section the small spherical particles are observed clearly, although overall supermolecular structure changes significantly depending on the distance Z. The pore size distribution N(r) was evaluated assuming that the ultra-thin section of the membrane has the straight-through cylindrical pores (Fig. 6.38a). With an increase in Z, N(r) becomes narrower and the peak value of N(r) increases. If the inner layers with the same distance Z are compared, the particular supermolecular structure is almost uniform and all the N(r) vs. r curves coincide fairly well (Fig. 6.38b). This fact definitely indicates that the inner layer at the same distance Z is formed at the same instant by phase separation under the same conditions of temperature and compositions (see, Fig. 6.4).

In the hypothetical planes of a gel membrane, a portion whose boundary is fully surrounded by the particles and which is concurrently occupied by the consecutively connected vacant particles yields a pore (Fig. 6.39a and b). Here, we neglect the crevas of the contacted polymer particles (Fig. 6.39c), which will be discussed as an inter-polymer particle pore later again.

The pore size can be approximately represented by the number, x, of vacant particles constituting a single pore. And a pore size distribution N(r) can be evaluated by translating a distribution of the number x of vacant particles constituting single pores on the hexagonal lattice sites. Consider the case when $N_T\{R/(R+1)\}$ ($=N_TL$, L is volume fraction of polymer-lean phase) vacant particles are divided into N_P cells. To ascertain the existence of N_P pores, one vacant particle is, in advance of counting, distributed to each cell. The number of ways, $W_{N_P}(N_TL-N_P)$ of partitioning the remaining (N_TL-N_P) vacant particles into N_P cells is given by

$$W_{N_P}(N_TL-N_P)=\{(N_P-1)+(N_TL-N_P)\}!/\{(N_P-1)!(N_TL-N_P)!\}. \quad (6.70)$$

After ascertaining the existence of N_P pores by distributing one vacant particle to all N_P cells, a single pore, arbitrarily chosen, is fulfilled with $(x-1)$ vacant particles further to realize the pore with x vacant particles. Next, the number of

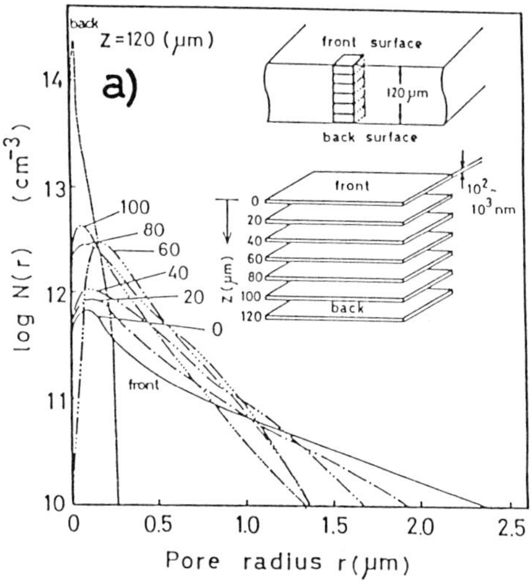

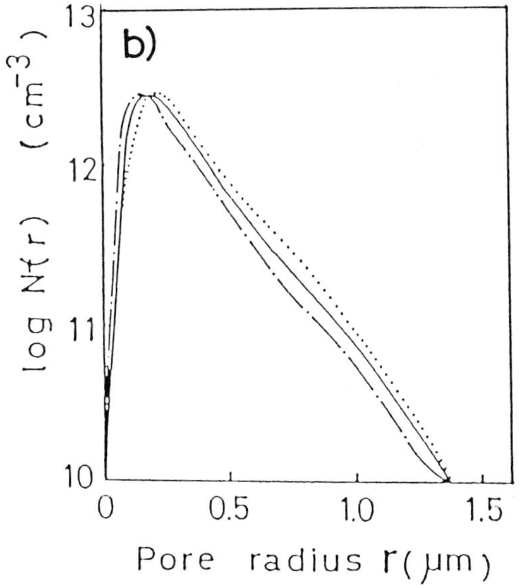

Fig. 6.38 Pore size distribution curve N(r) of ultra-thin sections for a regenerated cellulose membrane at various distances from the surface Z, evaluated from the electron micrographs as shown in Fig. 6.37: a) N(r) vs. r curves of the sections with various distances Z; in the figure, schematic representation of the preparation of the seven ultra thin sections with $10^2 \sim 10^3$nm thickness at various distance from the surface Z of 0, 20, 40, 60, 80 and 120μm is shown; b) N(r) vs. r curves of the section with Z=60μm at different locations.

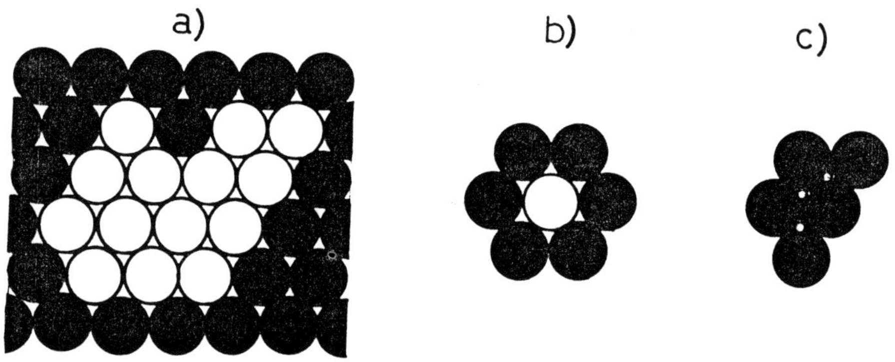

Fig. 6.39 Schematic representation of pores: a) A non-circularly growing pore ($x=14$); b) a pore having a minimum radius (i.e., $x=1$); c) three crevasses brought about by the contacted polymer particles. Filled circle, polymer particle; unfilled circle, vacant particle [a) and b)] or inter-polymer particle pore [c)].

ways, $W_{N_{P-1}}(N_TL-N_P-x+1)$, of partitioning the remaining vacant particles into (N_P-1) pores [i.e., all the pores except for the pore filled with x pore particles and note that the single pore has already x vacant particles] is given by

$$W_{N_{P-1}}(N_TL-N_P-x+1)=(N_TL-x-1)!/\{(N_P-2)!(N_TL-N_P-x+1)!\}. \tag{6.71}$$

Accordingly, when N_P pores are formed by partitioning N_TL vacant particles, the probability, $P(x)$, that x vacant particles are partitioned in a given pore, is

$$P(x)=W_{N_{P-1}}(N_TL-N_P-x+1)/W_{N_P}(N_TL-N_P)$$

$$=(N_TL-x-1)!(N_P-1)!(N_TL-N_P)!$$

$$/\{(N_P-2)!(N_TL-N_P-x+1)!(N_TL-1)!\}. \tag{6.72}$$

Eqn (6.72) can be straightforwardly simplified into (ref. 48)

$$P(x)\simeq(LN_T/N_P-1)^{-1}\cdot(1-N_P/LN_T)^x \quad \text{for } 1<LN_T/N_P,\ 1\leq x\leq LN_T-N_P+1 \tag{6.73}$$

and

$$P(x) \simeq 1 \qquad \text{for } LN_T/N_P = 1. \qquad (6.74)$$

It should be noted that $P(x)$ is normalized for the range of $x = 1$ to $(N_T L - N_P + 1)$.

LN_T vacant particles are thus partitioned on N_T sites of the hexagonal lattice in the manner so as to build up N_P pores. However, the vacant particles forming a single pore should be an assembly of vacant particles, which can take various forms under the condition that at least any particle contacts with another particle or particles directly. Consequently, this assembly can give pores with various pore shapes, which make further calculation extremely difficult. Then, for the sake of simplicity, we assume that any assembly of vacant particles is regarded as circular: On the hexagonal lattice site, the assembly of vacant particles (forming a single pore) has a strong tendency to form a hexagonal shape and the number, $f(x)$, of polymer particles needed to be surrounded fully by an assembly of x vacant particles is given by

$$f(x) = 3 + (12x - 3)^{1/2}. \qquad (6.75)$$

A polymer particle on the lattice can participate at most in the formation of three different pores (see, Fig. 6.40a and b). When a single polymer particle is directly participated to the formation of n pores, (1) the number, m, of vacant particles existing around the particle (i.e., the number of vacant nearest neighbor particles) lies between 1 and 5 (i.e., $1 \leq m \leq 5$) and (2) the probability, $P_n(m)$, that a given polymer particle is surrounded in part by m pore particles, which belong to n different pores is shown in Table 6.4. Here, we neglect the possibility that two not-directly connected vacant particles, which are the nearest neighbor of a given polymer particle, belong to a common pore. Fig. 6.40 shows some typical arrangements in two cases; $m = 4$, $n = 2$ and $m = 3$, $n = 3$. When the vacant particles gather circularly, $P_1(4) \simeq 0$ and $P_1(5) \simeq 0$ are expected.

The reciprocal of n, can be averaged over all possible arrangements of the vacant particles and polymer particles around a distinguishable polymer particle:

Table 6.4
$P_n(m)$ values for hexagonal lattice.

m \ n	0	1	2	3
0	$(1-L)^6$	/	/	/
1	/	$6L(1-L)^5$	/	/
2	/	$6L^2(1-L)^4$	$9L^2(1-L)^4$	/
3	/	$6L^3(1-L)^3$	$12L^3(1-L)^3$	$2L^3(1-L)^3$
4	/	$6L^4(1-L)^2$	$9L^4(1-L)^2$	/
5	/	$6L^5(1-L)$	/	/

/, cases in which there is no theoretical possibility of a given combination of m and n.

Fig. 6.40 Some typical arrangements of polymer particles (filled circles) and vacant particles (unfilled circles) in the nearest neighbor of a given polymer particles: a) m=4 and n=2; b) m=3 and n=3; see text for meaning of m and n.

$$\bar{n}^{-1} = \sum_{n=1}^{3} [\{\sum_{m=1}^{5} P_n(m)\} \times (1/n)] / \sum_{n=1}^{3} \{\sum_{m=1}^{5} P_n(m)\} . \qquad (6.76)$$

Then, it is clear that $\bar{n}$ is determined by the volume fraction of polymer-lean phase L (i.e., the two-phase volume ratio R).

In this manner, in order to form N_P pores, LN_T vacant particles are partitioned into N_P hexagonal assembly. The number of the assemblies, each consisting of x vacant particles, is $N_P P(x)$ and the number of polymer particles consumed in order to surround fully an assembly with x vacant particles, f(x), is given by eqn (6.75).

The total number of the polymer particles directly surrounding N_P independent pores should not exceed the total number of the polymer particles originally existing in the plane, that is

$$\sum_{x=1}^{LN_T} N_P P(x) f(x)/\bar{n} \leq N_T(1-L). \tag{6.77}$$

Here, LN_T vacant particles are consumed to build N_P pores and then the following equation holds,

$$\sum_{x=1}^{LN_T} N_P P(x) x = LN_T. \tag{6.78}$$

Eqns (6.77) and (6.78) are the boundary conditions of pore formation.

The radius r_{WET} of the pore containing x vacant particles in a wet gel membrane is given by

$$r_{WET} = x^{1/2} S_2. \tag{6.79}$$

In deriving eqn (6.79) the pore is assumed to be circular.

The volume of a polymer particle decreases to $(4\pi S_2{}^3/3) \cdot v_{P(2)} (d_{PL}/d_P')$ after drying. Here, d_{PL} and d_P' are the densities of the polymer itself and of the dried polymer particles, respectively. The radius of wet polymer particles S_2 decreases to S_2' during drying (see, Fig. 6.41):

$$S_2' = S_2 (v_{P(2)} d_{PL}/d_P')^{1/3}. \tag{6.80}$$

Accordingly, the pore radius r_{WET} corresponding to S_2 of the wet gel membrane increases to r through the relation:

$$r = r_{wet} + [1 - (v_{P(2)} d_{PL}/d_P')^{1/3}] S_2. \tag{6.81}$$

Combination of eqn (6.79) with eqn (6.81) gives

$$r = [x^{1/2} + 1 - (v_{P(2)} d_{PL}/d_P')^{1/3}] S_2. \tag{6.82}$$

The not-normalized pore size distribution N(r) is defined as

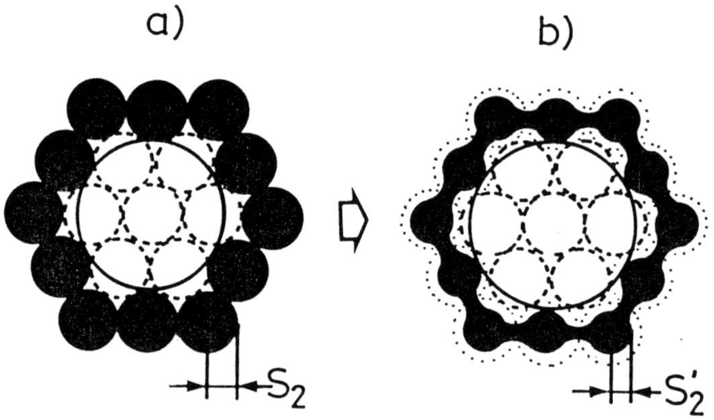

Fig. 6.41 Change in a pore size during drying treatment under constant membrane width: a) A circular pore containing seven vacant particles (i.e., $x=7$) in a wet gel membrane; filled circles, polymer particles; broken line circles, vacant particles; both kinds of particles have the same radius of S_2; b) an enlarged pore in a dry membrane; S_2', a radius of a dry polymer particle; $S_2'<S_2$ [see, eqn (6.80)].

$$N(r) = N_P P(x) \frac{dx}{dr} \tag{6.83}$$

with

$$N_P = \int_0^{\infty} N(r)\,dr \tag{6.84}$$

and

$$\frac{dx}{dr} = \frac{2}{S_2} x^{1/2}. \tag{6.85}$$

In deriving eqn (6.85) we assume that P(x) can be approximated by the continuous function and there exists one to one correspondence between x and r, in other words, all the pores in a plane have same shape (for example, circular). Eqn (6.85) is derived from eqn (6.82).

From eqns (6.73), (6.83) and (6.85), we obtain (ref. 48)

$$N(r) = (2N_P/S_2)(x-1)^{-1}(1-1/x)^{[r/S_2 - \{1-(v_{P(2)}d_{PL}/d_P')^{1/3}\}]^2} \times [r/S_2 - \{1-(v_{P(2)}d_{PL}/d_P')^{1/3}\}] \qquad (6.86)$$

with

$$\bar{x} = \frac{R}{(R+1)}\frac{N_T}{N_P}. \qquad (6.87)$$

Then, N(r) can be evaluated from N_P, R and S_2 data. Eqn (6.86) was derived by Kamide, Iijima and Matsuda (KIM) (ref. 48) and the equation can be compared with eqn (6.88), which was previously derived on the basis of a little different model by Kamide and Manabe (KM) (ref. 27), who tried first to bridge a theoretical route between the casting conditions and the pore characteristics of the membrane, although theoretical ambiguity contained in the process of derivation of eqn (6.88) is obviously much larger than that in eqn (6.86) (ref. 27).

$$N(r) = (2N_P/S_2)[(1+\bar{x})^{-1}][\bar{x}/(1+\bar{x})]^{[(r/S_2) - \{1-(v_{P(2)}d_{PL}/d_P')^{1/3}\}]^2} \times [(r/S_2) - \{1-(v_{P(2)}d_{PL}/d_P')\}^{1/3}]. \qquad (6.88)$$

In KM theory, f(x) and P(x) are given by

$$f(x) = 3 + (5+4x)^{1/2} \qquad (6.89)$$

and

$$P(x) = W_{N_P-1}(N_TL - x)/W_{N_P}(N_TL) \qquad (6.90)$$

$$= \{1/(1+LN_T/N_P)\}\{(LN_T/N_P)/(1+LN_T/N_P)\}^{x}. \qquad (6.91)$$

It should be noted that P(x) of KM theory is normalized for the range of $x=0$ to $x=N_TL$. This means that in the KM theory, cells containing no vacant particles, i.e., $x=0$, are taken into consideration as pores in the calculation of P(x). Therefore, the existence of N_P pores can not be ascertained by the KM theory.

In order to diminish the ambiguity contained in P(x) of KM theory (hereafter, referred as $P(x)_{KM}$), $P(x)_{KM}$ should be re-normalized in the range of $x=1$ to $x=N_TL$ using eqn (6.91').

$$P(x) = P(x)_{KM} / \int_1^{N_T L} P(x)_{KM}. \tag{6.91'}$$

The number-average radius of the pore r_n is calculated by substituting $N(r)$, given by eqn (6.86) or (6.88), into the relation,

$$r_n (\equiv r_1) = \int_0^\infty rN(r)\,dr / \int_0^\infty N(r)\,dr. \tag{6.92}$$

Fig. 6.42 shows the pore size distribution P(x) calculated

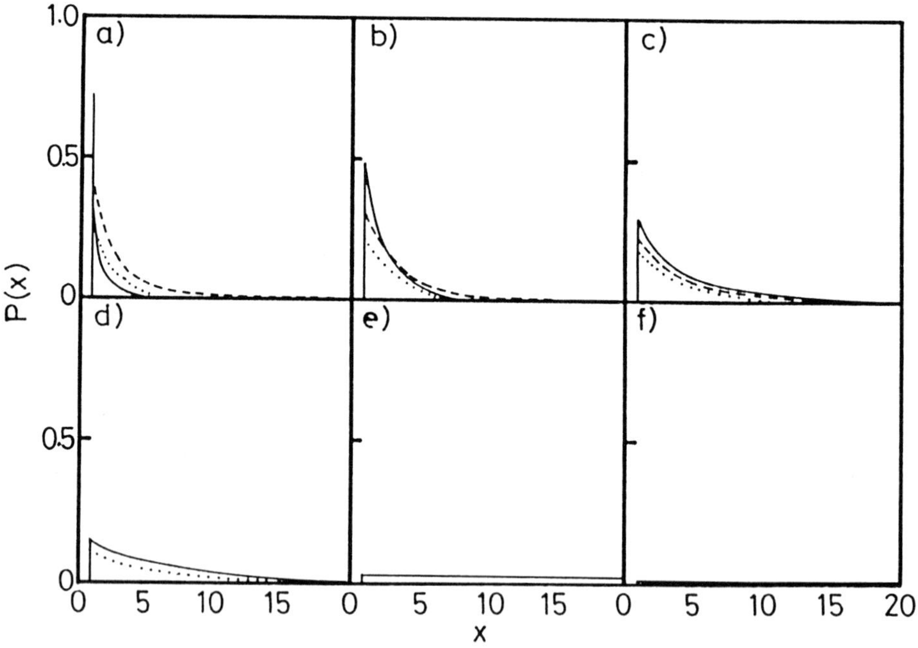

Fig. 6.42 Comparison of the pore size distribution P(x) evaluated by Kamide-Iijima-Matsuda (KIM) theory and that by Kamide-Manabe (KM) theory: The value of N_T/N_P for a given R was determined by computer experiments (40 trials in 200 lattice sites×200 lattice sites). a) R=0.11, N_T/N_P=13.85; b) R=0.25, N_T/N_P=10.35; c) R=0.43, N_T/N_P=11.34; d) R=0.67, N_T/N_P=18.55; e) R=1.0, N_T/N_P=51.35; f) R=1.5, N_T/N_P=213.9; full line, KIM theory [eqn (6.73)]; broken line, KM theory [eqn (6.91)]; dotted line, re-normalized KM theory [eqn (6.91')].

using eqns (6.73), (6.91) and (6.91') for a given combination of L (accordingly, R), N_T and N_P. Here, N_P values are chosen from computer simulation data obtained for given L and N_T. P(x) curves are in reasonably close agreement with each other and this implies that both Kamide-Iijima-Matsuda (KIM) and Kamide-Manabe (KM) theories give approximately the same P(x) vs. x curve, as anticipated (Fig. 6.42). However, a more detailed examination of Fig. 6.42 exhibits that P(x) values obtained by the KIM theory [eqn (6.73)] are larger at the small value of x ($x < 10$) than those predicted by the KM theory (eqn (6.91)), but the difference becomes insignificant at larger R.

Even if the polymer particles occupy all the sites of the hexagonal closest packing lattice (i.e., $R = 0$), there are numerous small crevasses between the polymer particles and such crevasses act as holes or pores, as in the case of reverse osmosis membranes, and are hereafter referred to as inter-polymer particle pores in order to distinguish them from the pores made from the polymer-lean phase (those pores are called vacant pores), and the crevasse in wet gel membrane becomes large during drying when the dimension of the membrane is kept constant.

Consider a polymer particle with m nearest neighbor vacant particles, belonging to n different pores and having y inter-polymer particles (see, Table 6.5). When $m = 0$ (and accordingly, $n = 0$) $y = 6$ is obtained (Table 6.5). In this manner, we can calculate y value for a given combination of m and n, as compiled in Table 6.5. The case of $m = 0$ and $n = 1$ is not theoretically realized (see, Table 6.5). In the case of $m = 3$ and $n = 3$, there is no probability of finding inter-polymer particles (i.e., $y = 0$).

Then, the average number of inter-polymer particles directly contacted with a given single polymer particle, $\bar{n}_{(i)}$ is given by

$$\bar{n}_{(i)} = \sum_{n=0}^{3} \{\sum_{m=0}^{5} P_n(m) y / (\sum_{n=0}^{3} \sum_{m=0}^{5} P_n(m))\}. \tag{6.93}$$

Total number of inter-polymer particles $N_{P(i)}$ is

$$N_{P(i)} = N_T (1 - L) \bar{n}_{(i)}. \tag{6.94}$$

An inter-polymer particle pore is formed by mutual contact of three polymer particles (Table 6.5, $m = 4$, $n = 1$) and the radius of the inter-polymer particle pore, $r_{(i)WET}$ of a wet gel membrane defined as an inscribed circle is related to S_2 through the

Table 6.5
Number (y) of inter-polymer particle pore existing around the given polymer particle and arrangements of polymer particles and vacant particles for theoretically possible combination of number, m, of vacant particles existing around the given polymer particle and number, n, of different pores existing around the given polymer particle; filled circle, polymer particle; dotted open circle, vacant particle; small open circle, inter-polymer particle pore.

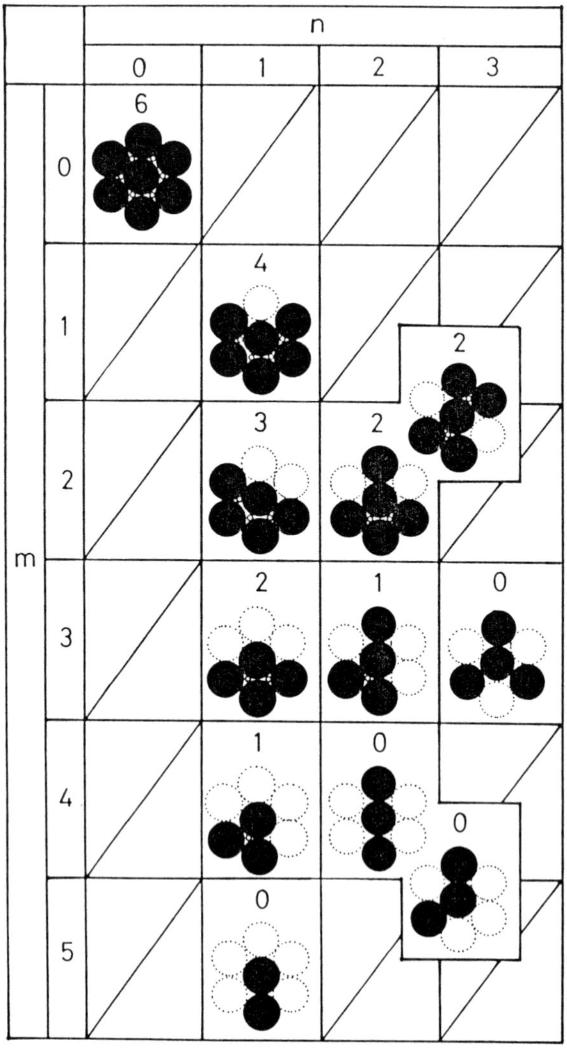

relation,

$$r_{(i)WET} = (2/3^{1/2} - 1)S_2 . \quad (6.95)$$

After drying S_2 changes to S_2' (eqn (6.80)) as shown in Fig. 6.43 and the radius of the inter-polymer particle pore of dry membrane $r_{(i)}$ is readily obtained by adding the difference between S_2 and S_2' in eqn (6.80) to $r_{(i)WET}$ in eqn (6.95) (ref. 48):

$$r_{(i)} = [2/3^{1/2} - (v_{P(2)} d_{PL}/d_P')^{1/3}] S_2 . \quad (6.96)$$

The ratio of $r_{(i)}$ to S_2, $r_{(i)}/S_2$ is roughly estimated to be 0.15, if the $v_{P(2)} \simeq d_P'/d_{PL}$ is approximated in eqn (6.96): $r_{(i)}$ is 1.5nm for $S_2 = 10$nm, 7.7nm for $S_2 = 50$nm and 31nm for $S_2 = 200$nm.

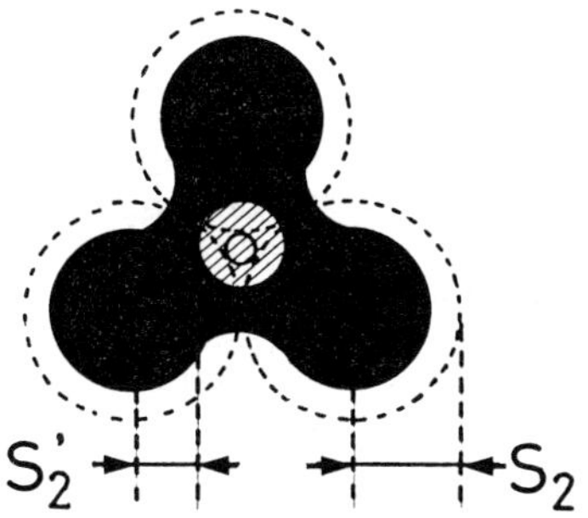

Fig. 6.43 Change in pore size of an inter-polymer particle pore after drying: Broken line circle, wet polymer particle; full line circle, wet inter-polymer particle pore; S_2, radius of a wet polymer particle; S_2', radius of a dry polymer particle; hatched area, inter-polymer particle pore after drying; blacked area, dry polymer particle.

The pore size distribution N(r) given by eqn (6.86) should be modified by taking into account $r_{(i)}$, as (ref. 48)

$$N(r) = (2N_P/S_2)(\bar{x} - 1)^{-1}(1 - 1/\bar{x})^{[r/S_2 - \{1 - (v_{P(2)} d_{PL}/d_P')^{1/3}\}]^2}$$

$$\times [r/S_2 - \{1 - (v_{P(2)} d_{PL}/d_P')^{1/3}\}]$$

$$\text{for } r \geq \{2 - (v_{P(2)} d_{PL}/d_P')\} S_2 \quad (6.97)$$

and

$$N(r) = N_T \{R/(1+R)\}\bar{n}_{(1)}/\{2-(v_{P(2)}d_{PL}/d_P')\}S_2,$$

$$\text{for } r < \{2-(v_{P(2)}d_{PL}/d_P')\}S_2. \quad (6.98)$$

Here, $\bar{x}$ is defined by eqn (6.87). Note again that eqns (6.97) and (6.98) are derived assuming that the super-particle structure (morphology) of coagulated secondary particles does not change due to collapse or contraction during the phase separation (step f of Fig. 6.5), the coagulation (step g) and drying (step k). At a latter step, its thickness changes only by a factor of $(v_{P(2)}d_{PL}/d_P')^{1/3}$. In the case of the wet method, a gel membrane is formed by dipping cast solution into a coagulating solution bath consisting of nonsolvent. In the dry method, cast solution is settled in an atmosphere of nonsolvents. In both methods, desolvation followed by partial replacement with nonsolvents and shrinkage of the gel occurs to a large extent. This is a kind of collapse of the gel structure. After being dipped, the membrane is treated with acid for generation of cellulose (if necessary) and washed and dried. Then, it is evident that the assumption employed in deriving eqns (6.97) and (6.98) contradicts greatly with the experimental facts.

Fig. 6.44 illustrates the change in the thickness of gel or dry membrane during the process of regenerated cellulose membrane formation (ref. 49). The thickness of cast solution (step a) dramatically decreases when dipped into nonsolvent bath (step b-i) or after drying (step k). For example, cellulose cuprammonium solution cast on the glass plate (Fig. 6.45 step a) contracts to give a gel membrane with a thickness of ca.1/5 of the cast solution. Therefore, several hypothetical thin planes in the cast solution with a thickness of ℓ_0 yield a new single hypothetical gel plane in the wet membrane (step i), and the new hypothetical planes contract again to give a dry membrane with a thickness of around a few tenth of ℓ_0 (step k). In other words, even if we cut off a very thin plane with thickness of $2S_2'$ from the bulk dry membrane, the thin plane can never be regarded as a hypothetical plane, in which the phase separation occurred simultaneously under the same conditions.

First, suppose that a polymer solution is cast on a plate to give a thin solution film with a thickness of ℓ_0 (Fig. 6.45a) and phase separation proceeds from the surface to the inner part of solution (Fig. 6.45g and h).

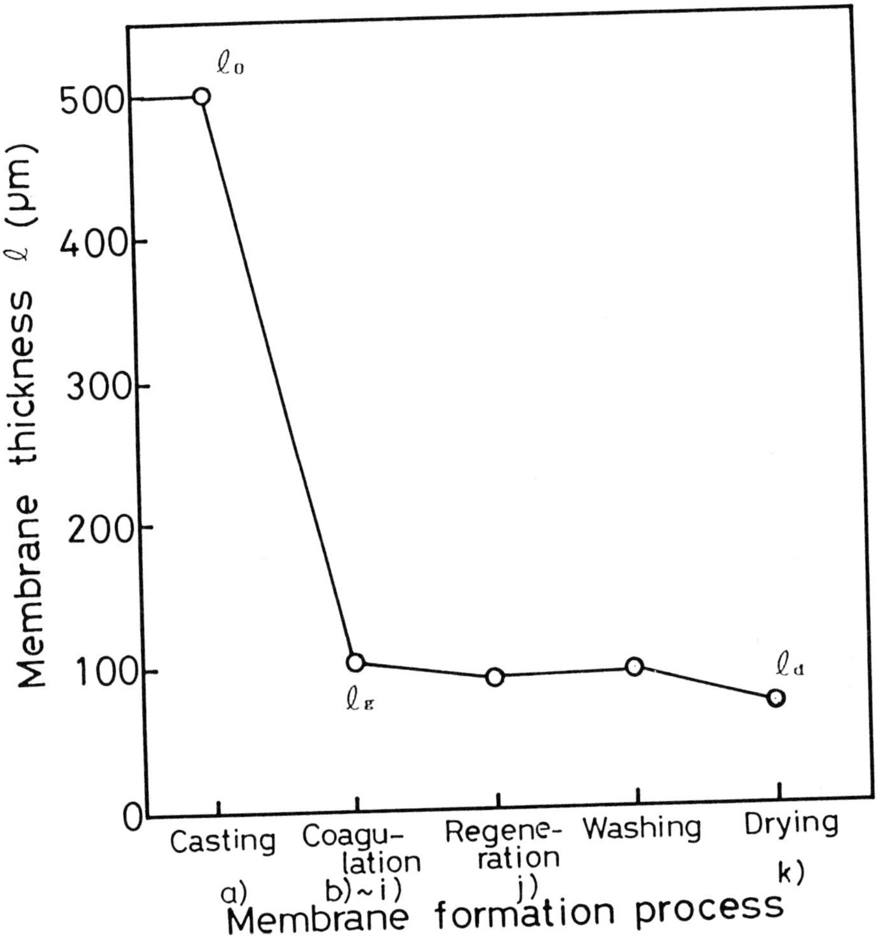

Fig. 6.44 Change in thickness ℓ of cellulose membrane during the wet process: Cast solution, cellulose cuprammonium solution [$w_{CELL}{}^S = 0.05$, $w_{CU}{}^S = 0.0198$, $w_{NH3}{}^S = 0.1752$, $w_{H2O}{}^S = 0.7550$]; coagulating solution [$w_{ACETONE} = 0.3000$, $w_{NH3} = 0.0056$, $w_{H2O} = 0.6944$]; coagulation temperature, 288K; coagulation time, 30min.

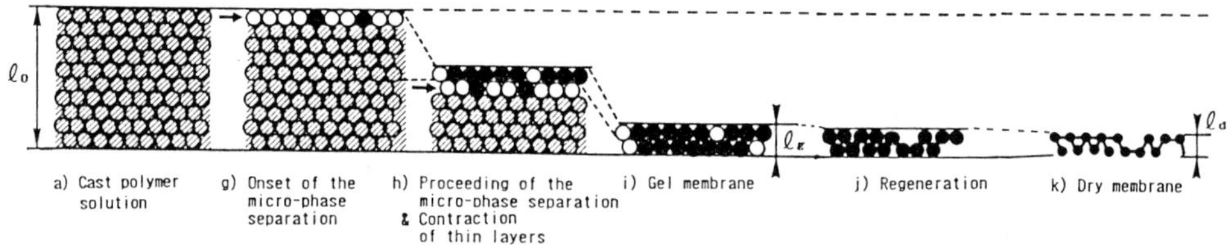

Fig. 6.45 Schematic representation of changes of membrane thickness during membrane formation process: a) Casting of polymer solution; g) starting of the micro-phase separation; h) proceeding of the micro-phase separation and contraction of thin layers; i) end of the over-all micro-phase separation; j) regeneration; k) dry membrane; ○ vacant particle; ● polymer particle; hatched area, homogeneous polymer solution not yet phase-separated; ○ in the hatched area, position of particles to be created by phase separation; ℓ_0, thickness of cast solution; ℓ_g, thickness of gel membrane; ℓ_d, thickness of dry membrane; arrows indicate a layer where the phase separation has just occurred and the layer-contraction has not occurred yet; a)-k) correspond to a)-k) steps in Fig. 6.5.

We define the volume fraction of a polymer-lean phase in a hypothetical layer, at the moment of phase separation, as $Pr_{,PS}$, which is given by the relation

$$Pr_{,PS} = R/(1+R) . \tag{6.99}$$

During the formation of a coagulated gel membrane with thickness ℓ_g (Fig. 6.45 step i), the volume fraction of vacant particles in the gel membrane, $Pr_{,GEL}$ is governed by the degree of collapse of thin layers. In the case where two hypothetical layers having the same $Pr_{,PS}$ collapse into a "single layer", it is required that a vacant particle in the upper layer is just superposed with another vacant particle in the lower layer, in order to find a vacant particle in the "single layer". The probability of occurrence of the above phenomenon (i.e., the porosity) is $\{Pr_{,PS}\}^2$. Accordingly, the porosity $Pr_{,GEL}$ of a "single layer" formed by the collapse of n hypothetical layers is given by

$$Pr_{,GEL} = \{Pr_{,PS}\}^n (= \{\frac{R}{1+R}\}^n). \tag{6.100}$$

Here, the following relation holds approximately

$$n \simeq \ell_0/\ell_g. \tag{6.101}$$

Considering an increase in volume of both vacant pores and inter-polymers particle pores (the former is estimated to be $v_{P(2)} d_{PL}/d_P'$) due to desolvation of polymer particles in drying step (Fig. 6.44), the porosity of dry membrane Pr(d) (as denoted by $Pr(d_1)$) is given by eqn (6.102)

$$Pr(d_1) = \{1-(S_2'/S_2)^3\}(1-Pr(d_1)')+Pr(d_1)'$$

$$= \{1-(v_{P(2)}(d_{PL}/d_P')\}(1-Pr(d_1)')+Pr(d_1)' \tag{6.102}$$

with

$$Pr(d_1)' = \{R/(1+R)\}^{(\ell_0/\ell_d)}. \tag{6.103}$$

S_2' is the radius of dry secondary particle (see, eqn (6.80)). In deriving eqn (6.102), it is assumed that pore density N_P does not change during drying step.

$Pr(d_1)$ should be compared with Pr(d) (denoted as $Pr(d_2)$) estimated through use of eqn (6.104)

$$Pr(d_2) = \pi N_P \{\bar{x}^{1/2}+1-(v_{P(2)}(d_{PL}/d_P')^{1/3}\}^2 S_2^2. \tag{6.104}$$

Eqn (6.104) was derived by Kamide and Manabe (ref. 27), assuming that (1) volumetric increase due to inter-polymer particle pores during drying step can be neglected, (2) no collapse occurs at coagulation step and (3) the pore characteristics of the membrane is constant, irrespective of the distance from the surface.

The porosity of dry membrane is also directly evaluated from apparent density d_A of the membrane and density of the materials constructing the membrane d_{PL}. This is denoted as $Pr(d_3)$, given by the relation,

$$Pr(d_3) = 1-(d_A/d_{PL}) \tag{6.105}$$

where $d_A = w_m/v_m$ (w_m, v_m are weight and volume of absolute dry membrane, respectively). To $Pr(d_3)$ all pores existing in the membrane contribute.

Another porosity of dry membrane $Pr(d_4)$ is determined by electron micrographs (refs 7,8),

$$Pr(d_4)\ (\equiv Pr(EM)) = \sum_i \ell_{A,i} / \sum_i \ell_i \qquad (6.106)$$

where ℓ_i is the length of i-th test lines drawn on a photograph of membrane surface and $\ell_{A,i}$ is the cut-off length, by pores, of i-th test line. $Pr(d_4)$ neglects the contribution of pores smaller than EM's resolution power limit and is suitable for the determination of the porosity of thin membrane or membrane whose whole pore characteristics can be well represented by those at the surface.

Summarizing, there are four different methods for estimating the porosity of the membrane, and it is expected that $Pr(d_1)$ corresponds to $Pr(d_3)$ and $Pr(d_2)$ to $Pr(d_4)$, respectively.

Combining eqn (6.102) with eqn (6.103) and assuming that $Pr(d_1)$ is equal to $Pr(d_3)$, we can estimate R at phase separation step from apparent density of membrane d_A and the radius of secondary particles S_2.

Table 6.6 summarizes R values thus calculated for regenerated cellulose membranes. In the table, the data obtained assuming $\ell_0/\ell_d = 1$ are also included. With an increase in cellulose concentration, R decreases gradually. When collapse of thin gel layers is neglected (i.e., $\ell_0/\ell_d = 1$) R obtained is nearly 1/10 of R estimated in considering the collapse.

Fig. 6.46 shows the pore size distribution N(r) calculated by eqns (6.86) and (6.87) using various R values for regenerated cellulose membrane, in which the following quantities are obtained; cellulose concentration, 5wt% ; $S_2 = 190$nm; $N_P = 1.96\times 10^{11}/m^2$; $Pr(d_4) = 0.301$; $\ell_0 = 500\mu m$; $\ell_d = 53\mu m$; $d_{PL} = 1.5\times 10^3 Kg/m^3$, $d_P' = 1.1\times 10^3 Kg/m^3$. In the figure, N(r), directly evaluated by the electron microscopic (EM) method (ref. 8), is also plotted by an unfilled circle. The experimental N(r) curve by EM method lies between theoretical N(r) curves with $R = 0.11\sim 1.0$, which are exceedingly small, as compared with R directly determined in the phase separation step. This great disparity can only be explained in terms of a collapse of some tens of hypothetical thin planes during coagulation and drying processes (step g-k), yeilding a hypothetical thin dry membrane (see, Fig. 6.45). In fact, the thickness of the cast solution was

Table 6.6
Degree of collapse of thin layers at coagulation step, porosity, and two-phase volume ratio R of some regenerated cellulose membranes.

	$w_{CELL} \times 10^2$						
	4	5	6	7	8	9	10
Thickness of casting soln ℓ_0 (μm)	500	500	500	500	500	500	500
Thickness of dry membrane ℓ_d (μm)	51.0	53.0	57.0	63.0	68.8	70.0	68.0
Porosity $Pr(d_3)$ *a	0.823	0.778	0.742	0.731	0.695	0.654	0.571
$Pr(d_4)$ *b	—	0.320	0.359	0.387	—	—	—
Two-phase volume ratio R							
from (eqns (6.102) and (6.103))*c	24.5	17.6	13.2	11.1	8.4	6.6	4.3
from (eqn 6.103) with $\ell_0/\ell_d=1$	2.5	1.8	1.4	1.3	1.0	0.77	0.43

Membrane preparation, some cuprammonium cellulose solutions were prepared by adding 28wt% aq. ammonia to an original solution ($w_{CELL}=0.1$, $w_{CU}=0.0395$, $w_{NH3}=0.0703$, and $w_{H2O}=0.7902$). Cast solutions were immersed into a coagulating bath ($w_{ACETONE}=0.30$, $w_{H2O}=0.70$) for 30 min at 298K. After regeneration in 2wt% aq. sulfuric acid, washing and replacement of water by acetone, membranes were dried; *a, apparent density method (eqn (6.105)); $d_{PL}=1.5\times10^3 Kg/m^3$; *b, electron micrographic method (eqn (6.106)); *c, $d_{PL}=1.5\times10^3 Kg/m^3$, $d_P'=1.1\times10^3 Kg/m^3$, $v_{P(2)}=0.4$, $Pr(d_1)=Pr(d_3)$.

observed to undergo some remarkable changes during the above steps, which can not simply be accounted for by desolvation alone (see, Fig. 6.44).

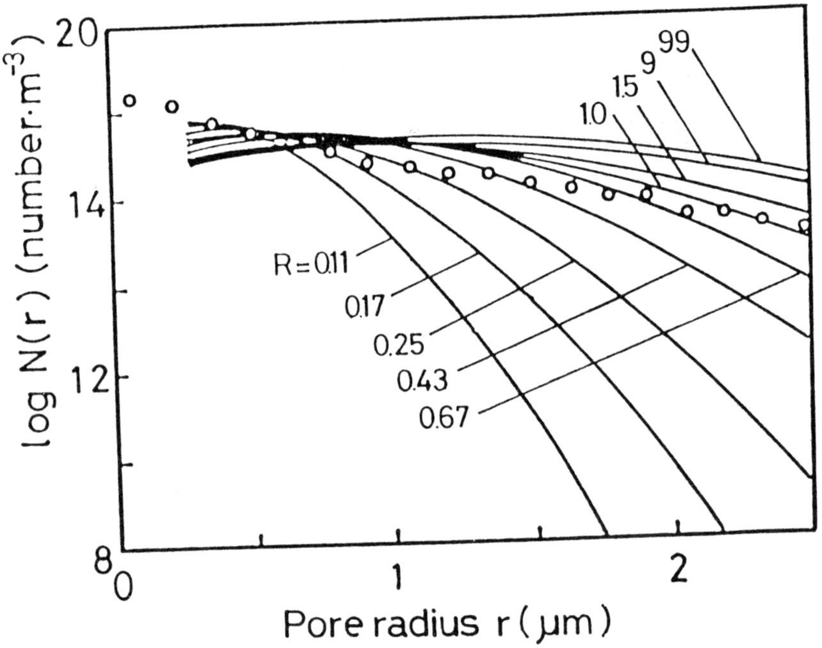

Fig. 6.46 Comparison between pore size distribution curves N(r) by Kamide-Iijima-Matsuda (KIM) theory [eqns (6.86) and (6.87)] and electron microscopic (EM) method: Numbers in the figure denote phase volume ratio R; ○ EM method [condition of preparing a regenerated cellulose membrane; cast solution, ($w_{CELL}{}^S=0.050$, $w_{CU}{}^S=0.0198$, $w_{NH3}{}^S=0.1752$, $w_{H2O}{}^S=0.755$); coagulation solution, ($w_{ACETONE}=0.300$, $w_{H2O}=0.700$); $Pr(d_3)=0.772$; $r_F=315$nm; $S_2=190$nm; $N_p=1.96\times10^{11}/m^2$, where the pore with 0.3~2.5μm are counted.]; full line, N(r) by KIM theory [eqns (6.86) and (6.87), $S_2=190$nm, $d_{PL}=1.5\times10^3$Kg/m^3, $d_P'=1.1\times10^3$Kg/m^3, $v_{P(2)}=0.4$, $N_p=1.96\times10^{11}$number/m^2].

6.4.2 Theoretical predictions of the correlation between the casting conditions and the pore characteristics of membranes

Fig. 6.47 shows the effect of the pore density N_P on normalized pore size distribution N(r) for various two-phase volume ratio R. Note that in this figure, eqns (6.86) and (6.87) were used for calculation of normalized N(r). With a decrease in N_P, N(r) becomes broader and its peak shifts to a value of larger r, irrespective of R.

Fig. 6.48 shows the effect of the two-phase volume ratio R on the pore size distribution N(r) for various radii of secondary particles S_2. With increasing R, the breadth of pore size distribution curve increases significantly with a shift of peak to the large r side. Note that the present KIM and KM theories can not determine explicitly the pore density N_P for any given set of casting conditions.

Fig. 6.49 shows the effect of the radius of the secondary particles S_2 on the pore size distribution N(r). Here, the total number of polymer particles and vacant particles per unit area N_T was varied corresponding to a variation of S_2 (i.e., $N_T = 1/(\pi S_2^2)$). The size of secondary particles does not seem to play an important role in the pore size distribution, which becomes narrower to a less large extent with an increase in S_2 under the conditions of constant N_P and constant R. It should be noted that the minimum r, r_{min}, theoretically possible to exist is determined by S_2 using eqn (6.82) with $x = 1$. r_{min} becomes larger as S_2 is larger and in the range $r < r_{min}$, $N(r) = 0$ if inter-polymer particles pores are not considered. Then, the effect of S_2 on the mean pore radius r_n will be more significant than that on N(r) (see, Fig. 6.51).

The mean pore size r_n is calculated by N(r), given by eqn (6.86) or eqn (6.88) into eqn (6.92). Fig. 6.50 shows the theoretical relations between r_n and the two-phase volume ratio R for various combinations of N_P and S_2. Fig. 6.51 shows similar relations between r_n and S_2 for given N_P and R. r_n increases with R and S_2 and in particular, the role of R is important.

Fig. 6.51 includes some experimental data on the regenerated cellulose membrane ($N_P = 0.9\times 10^{12} \sim 3.8\times 10^{12}$ number/m^2 and $S_2 = 200 \sim 300$nm) prepared from cellulose cuprammonium solution (cellulose concentration, 5 and 7wt%) coagulated in 30wt% aq. acetone solution. Here, r_n is determined by the electron

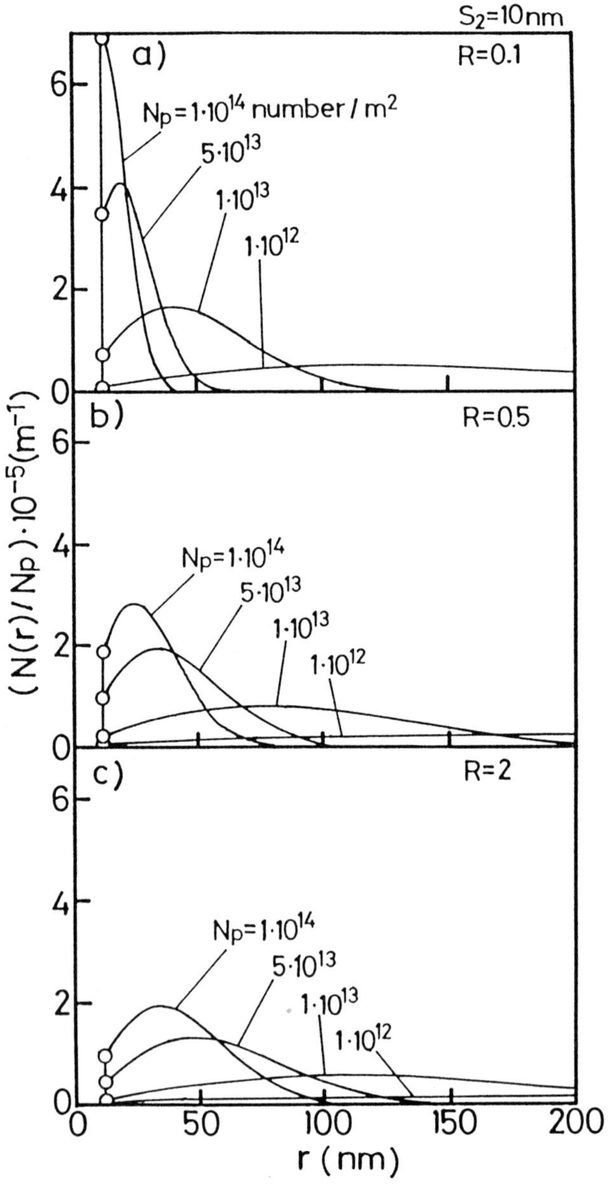

Fig. 6.47 Effect of the pore density N_p on the pore size distribution N(r) [eqns (6.86) and (6.87)]: The radius of a secondary particle $S_2 = 10$nm; unfilled circle, N(r) at the minimum pore size; a) phase volume ratio R=0.1; b) R=0.5; c) R=2.

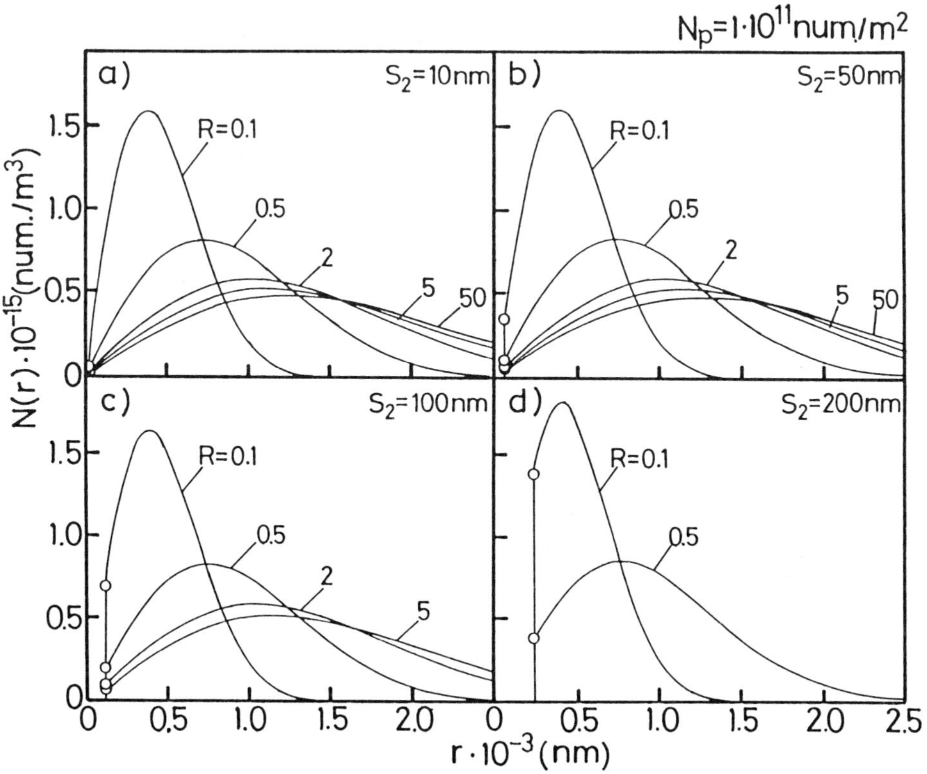

Fig. 6.48 Effect of phase volume ratio R on the pore size distribution N(r) [eqns (6.86) and (6.87)]: The pore density $N_p = 1 \times 10^{11}$ number/m²; unfilled circle, N(r) at minimum pore size; a) radius of secondary particle $S_2 = 10$nm; b) $S_2 = 50$nm; c) $S_2 = 100$nm; d) $S_2 = 200$nm.

microscopic (EM) method. The experimental data points lie on the theoretical relations if $R \lesssim 0.3$, which seems to be much smaller than that observed in phase separation experiments when the polymer concentration is smaller than the critical concentration. These unexpectedly small R values can be explained by compilation of some tens of thin gel membranes in the casting process as already confirmed before.

Obviously, the role of R is especially important in N(r). By using eqn (6.86), we can not calculate N(r), and accordingly r_n, in the range of $S_2 > 200$nm and $R < 0.2$ ($N_P = 10^{12} \sim 10^{13}/m^2$), in which LN_T/N_P becomes less than unity.

Inspection of Figs 6.50 and 6.51 leads us to the conclusion that in order to prepare porous polymeric membrane with a larger

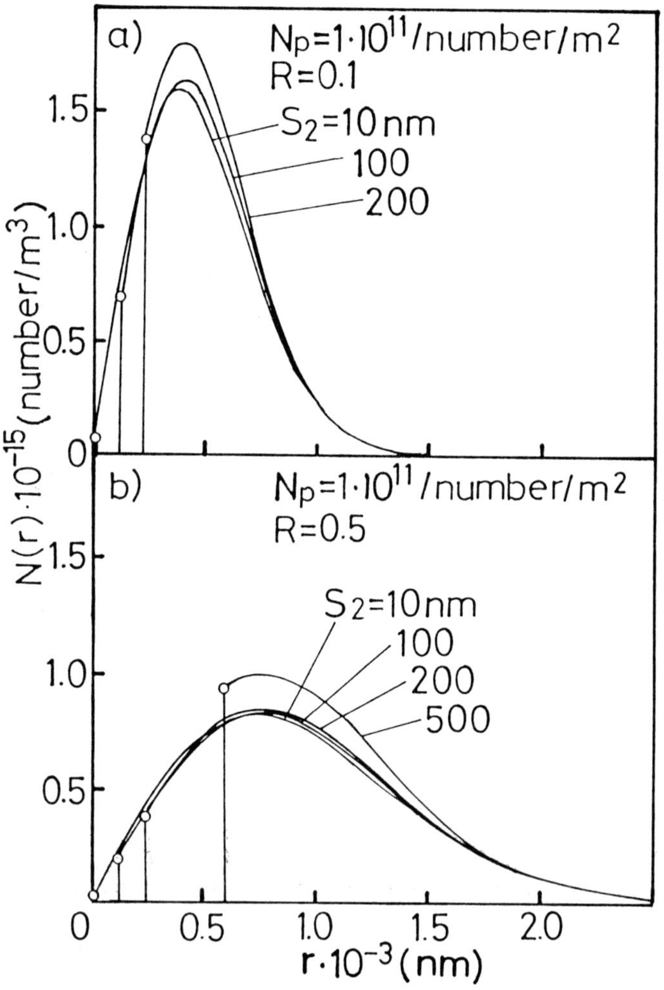

Fig. 6.49 Effect of the radius of secondary particle S_2 on the pore size distribution N(r) [eqns (6.86) and (6.87)]: The pore density $N_P = 1 \times 10^{11}$number/m²; unfilled circle, N(r) at minimum pore size; a) the phase volume ratio R=0.1; b) R=0.5.

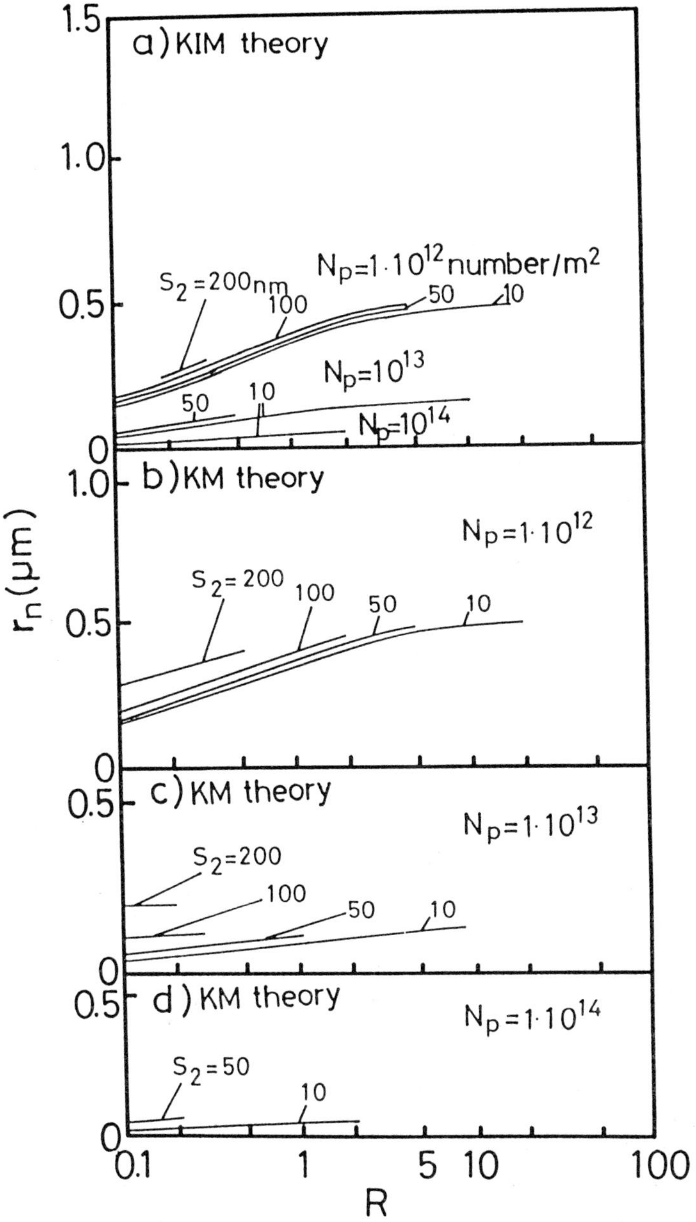

Fig. 6.50 Phase volume ratio R dependence of the mean pore radius r_n for various combinations of the radius of secondary particle S_2 and pore density N_p: a) KIM theory [eqn (6.86)]; b), c) and d) KM theory [eqn (6.88)].

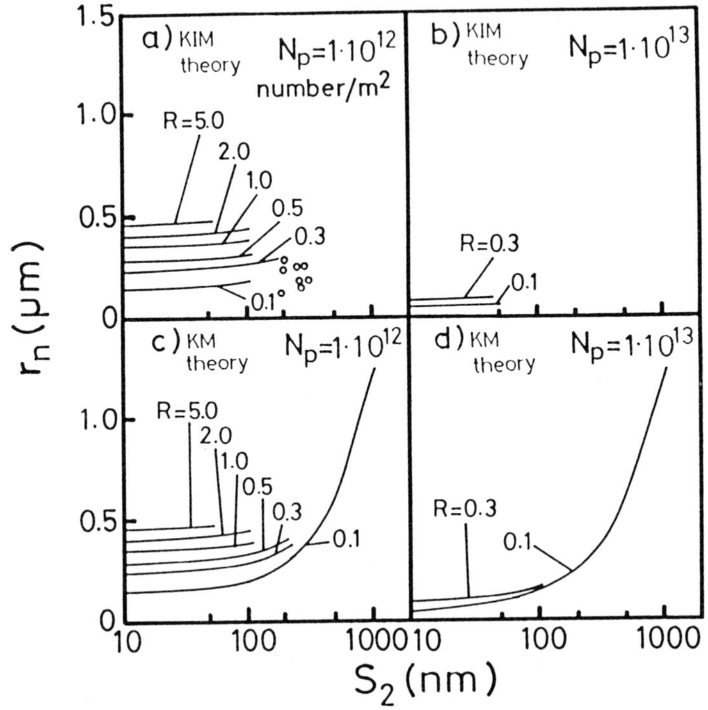

Fig. 6.51 S_2 dependence of the mean pore radius r_N: a) KIM theory [eqn (6.86)]; $N_P = 1\times10^{12}$number/m^2; ○ experimental data on regenerated cellulose membrane ($N_P = 0.9\times10^{12}\sim3.8\times10^{12}$number/m^2 and $S_2 = 200\sim300$nm) prepared by casting two cellulose cuprammonium solutions ($w_{CELL}{}^S = 0.050$, $w_{CU}{}^S = 0.0198$, $w_{NH3}{}^S = 0.1752$, $w_{H2O}{}^S = 0.7750$ and $w_{CELL}{}^S = 0.070$, $w_{CU}{}^S = 0.0277$, $w_{NH3}{}^S = 0.1332$, $w_{H2O}{}^S = 0.7691$) into coagulant ($w_{ACETONE} = 0.300$, $w_{H2O} = 0.700$);
b) KIM theory; $N_P = 1\times10^{13}$number/m^2; c) KM theory [eqn (6.88)]; $N_P = 1\times10^{12}$number/m^2; d) KM theory; $N_P = 1\times10^{13}$number/m^2.

mean pore radius, casting conditions must be selected so that a larger R and larger S_2 are realized. Theory and computer experiments on the thermodynamics of phase separation indicate that the larger phase volume ratio R will be realized when a polymer with lower weight-average degree of polymerization X_w and broader molecular distribution X_w/X_n is dissolved in a single solvent having larger p_1 or in a binary solvent mixture with smaller $\chi_{12}{}^0$, larger $\chi_{13}{}^0$, smaller $\chi_{23}{}^0$, $p_{12}\simeq0.2$ and larger

p_{13} and p_{23}, to give a dilute solution, which is phase-separated under the condition of a larger relative amount of polymer precipitated, d_p.

6.4.3 Computer simulation experiments

The process of forming porous polymer membranes can also be investigated by computer simulation using the Monte Carlo methods (ref. 50). In this approach polymer particles and vacant particles, whose number ratio is pre-determined by the two-phase volume ratio R (or the porosity Pr), are arranged randomly on a hypothetical hexagonally closest packing lattice plane and all pores formed in the plane are counted one by one to give the pore size distribution P(x), and if necessary, by assuming circular pore, P(x) is converted into N(r) (r is pore radius). Note that in the simulation the occupancy of a given site is approximated by the overall fraction of occupied lattice sites (i.e., R).

Fig. 6.52 shows typical examples of arrangements of secondary particles on a hypothetical plane. In this figure, pore particles are omitted, and not shown here. In the case of $R > 9$, all polymer particles are almost completely isolated, resulting in a single large pore together with a small number of very small pores (not shown here). Experimentally similar patterns are observed by an electron microscopic method for polymer solutions at stage g in Fig. 6.5.

Fig. 6.53 shows pore size distribution P(x) of a membrane constructed by computer experiments at various given values of R (unfilled circle). In the figure, P(x) calculated using eqn (6.73) in Kamide-Iijima-Matsuda (KIM) theory is shown as a full line for comparison. Computer experiments indicate that P(x) has a peak at $x = 1$ irrespective of R, i.e., the pore having a single vacant particle is the most popular in the solvent-cast membrane, as predicted by theory. This result conforms to the lattice theory. At $R < 0.1$, the theory is in an excellent agreement with the computer experiments over an entire range of x. However, in the range $R \simeq 0.25 \sim 0.43$ KIM theory agrees fairly well with experiments, except $x = 1$; the peak value of P(x) (in this case, P(1)) is underestimated in KIM theory. The disparity of the theory from the experiments becomes remarkably large as R becomes larger and the theory underestimates the number of small pores.

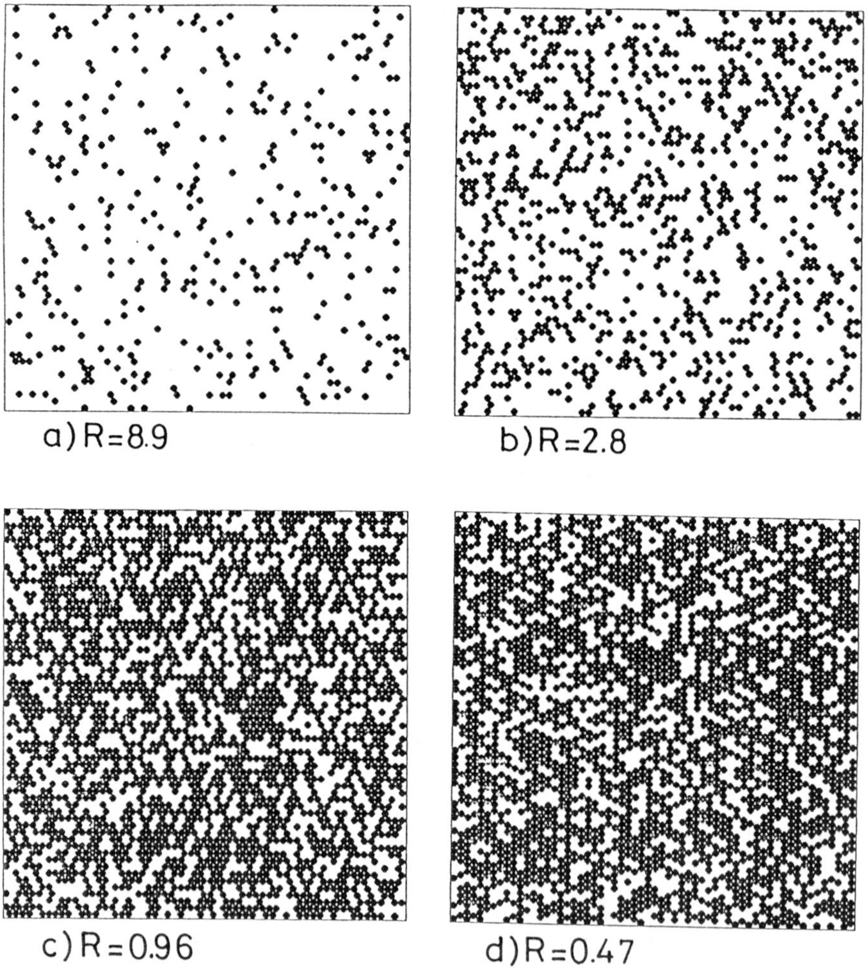

Fig. 6.52 Typical arrangements of secondary particles (black circles) on a hypothetical plane: a) R=8.9; b) R=2.8; c) R=0.96; d) R=0.47: Computer experiments.

In other words, data for pore size distribution of the membrane prepared at relatively large R can not be fitted by the rather

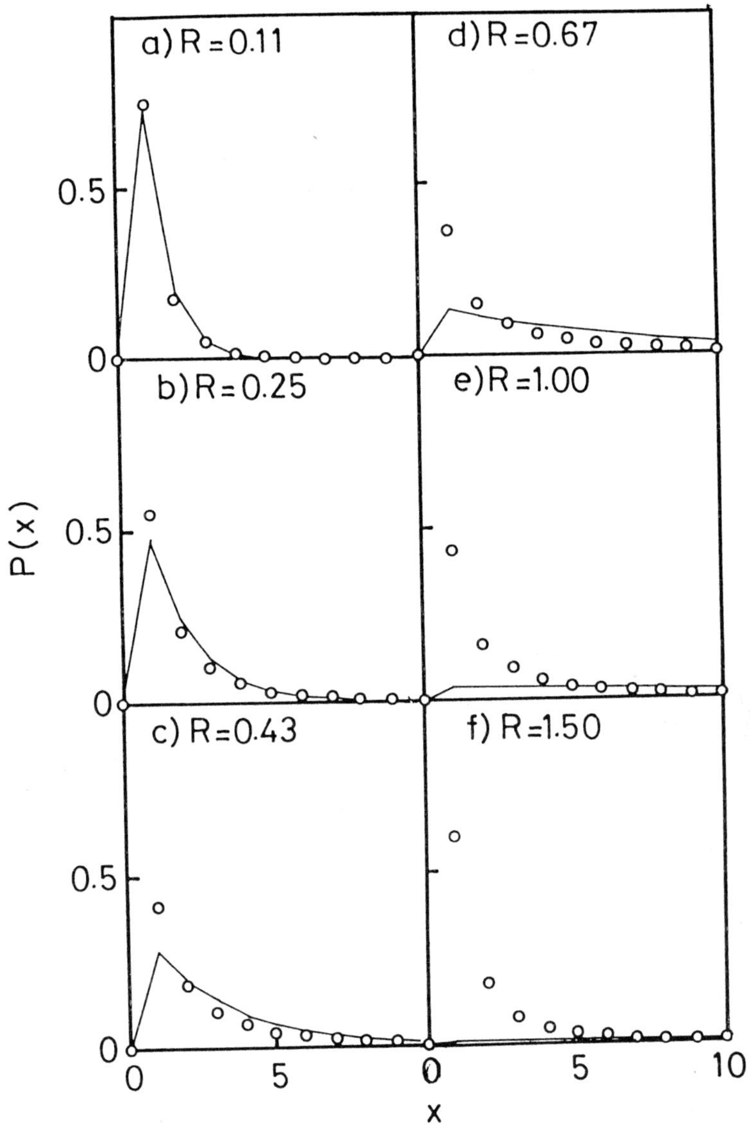

Fig. 6.53 Comparison of pore size distribution P(x) of membrane, hypothetically cast in computer experiments, (open circle) with that calculated by Kamide-Iijima-Matsuda theory using eqn (6.73) (full line): a) R=0.11; b) R=0.25; c) R=0.43; d) R=0.67; e) R=1.00; f) R=1.50; N_T/N_P determined by computer experiments for given R was employed for calculation of P(x) by eqn (6.73).

simple eqn (6.73). Although KIM theory gives only an approximate estimate of the pore size distribution at small R values (0.1~0.4), an apparent R value (R_A) estimated for actual dry membranes by $Pr(d_3)/\{1+Pr(d_3)\}$ falls in the above range and the pore characteristics of the dry membrane is proved to be well represented not by R value evaluated directly from the phase separation experiments under the same conditions as those of membrane casting, but by the apparent R value (R_A) determined for the membrane.

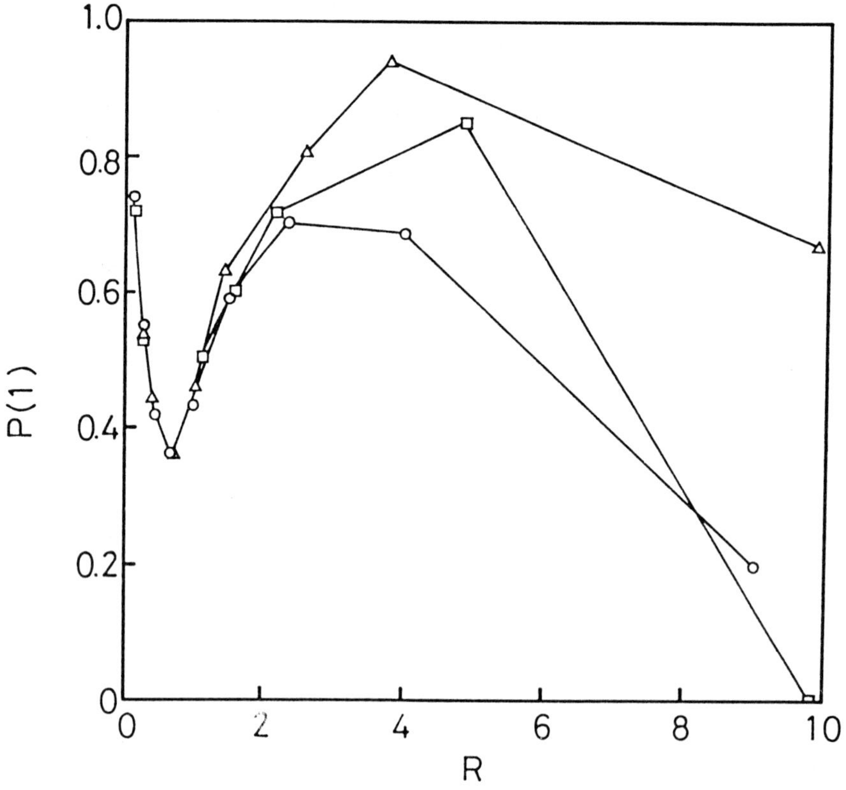

Fig. 6.54 P(1) (i.e., maximum P(x)) as a function of the two-phase volume ratio R: Computer experiments, size of hypothetical layer; 500 lattice sites ×500 lattice sites (△), 700 lattice sites ×700 lattice sites (□), 1265 lattice sites ×1265 lattice sites (○).

Fig. 6.54 shows the maximum frequency P(1) ($\equiv$ number of pores having single pore particle (i.e., x = 1) per unit area of membrane $N_{P(1)}/N_P$) as a function of R. In the figure the three cases of simulations are illustrated. In the range of R < 1, P(1) is independent of the size of the layer assumed, but in the range R > 1, P(1) is smaller as the size of the layer is larger, that is at larger R region the effect of the edge of the lattice space of the calculation can not be neglected. In the figure, the curve calculated for 1265 lattice sites $\times$ 1265 lattice sites can roughly be considered as an envelope line. P(1) changes in complicated manner with R: P(1) decreases with an increase in R first and then increases again after passing through minimum, P(x) exhibits second maximum, which is smaller than P(1), at higher R region.

As described previously, neither the KIM nor KM theories N_P can be used to unambiguously, because arrangements of polymer and vacant particles on the layer are statistical phenomena and N_P and $N_{P(x)}$ (number of pores with x vacant particles per unit area; $x \geq 1$) are determined for a given specific arrangement, in consequence, both unavoidably having some distributions. Mean values of N_P ($\bar{N}_P$) and of $N_{P(1)}$ ($\bar{N}_{P(1)}$) can be evaluated from their frequency $\tilde{P}(N_P)$ and $\tilde{P}(N_{P(1)})$ through the relations,

$$\bar{N}_P = \int N_P \tilde{P}(N_P)\, dN_P \tag{6.107a}$$

and

$$\bar{N}_{P(1)} = \int N_P \tilde{P}(N_{P(1)})\, dN_{P(1)}. \tag{6.107b}$$

Fig. 6.55 shows $\tilde{P}(N_P)$, obtained under different two-phase volume ratio R. In the figure, S_2 is taken as 50nm. As R increase $\tilde{P}(N_P)$ shifts to lower N_P side and distribution becomes narrower. Here, N_P value was determined, for a sheet of the membrane hypothetically cast on the plane of 150 lattice sites $\times$ 150 lattice sites, and the frequency distribution of N_P was evaluated on 100 sheets of the membranes.

Fig. 6.56 shows the mean pore density of total pore with various x, $\bar{N}_P$ and that of the pore with x = 1, both evaluated by the computer experiment. $\bar{N}_P$ attains maximum at $R \simeq 0.3$ and $\bar{N}_{P(1)}$ shows maximum at $R \simeq 0.1$. These pore densities diminish rapidly with an increase in R, approaching zero. The characteristic features of P(1), observed in Fig. 6.54, are well explained by R

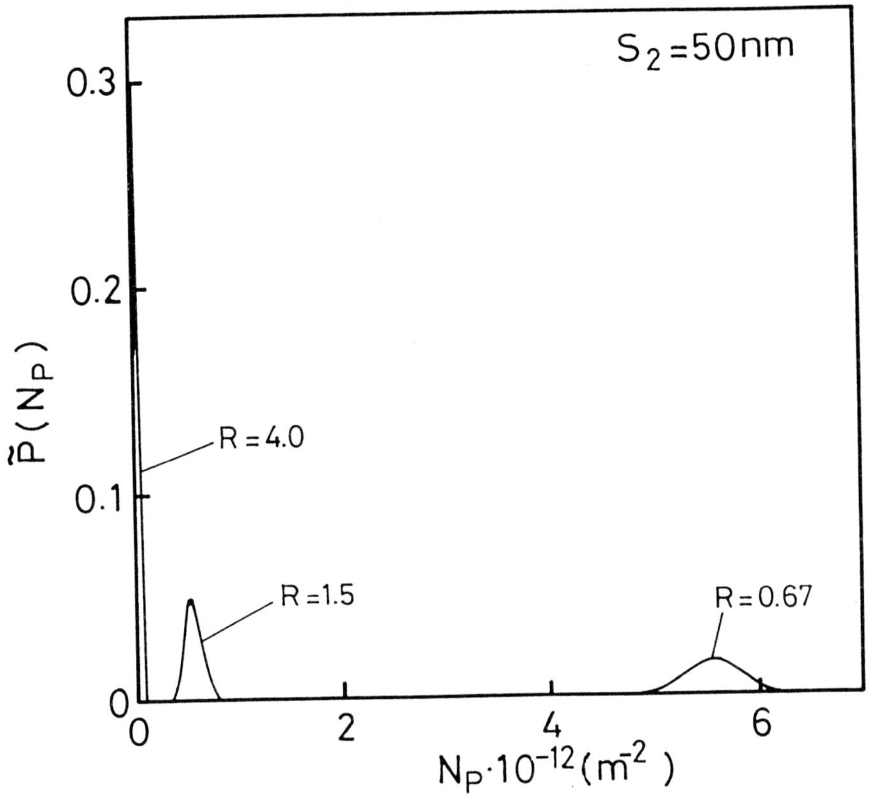

Fig. 6.55 Frequency distribution of pore density (N_P), $\tilde{P}(N_P)$ for various two-phase volume ratio R: Computer experiments; layer space, 100 lattice sites ×100 lattice sites; repetition of calculation (i.e., number of trials), 3000 times; S_2, 50nm.

dependencies of $\bar{N}_P$ and $\bar{N}_{P(1)}$.

In computer experiments N_T is unambiguously determined for a given S_2 as $1/(\pi S_2{}^2)$. Then, computer experiments are carried out under given condition of N_T and R to give N_P. The mean number of pore particles, $\bar{x}$, in a pore is defined by

$$\bar{x} = \int xP(x)\,dx. \tag{6.108}$$

Fig. 6.57 shows the plots of $\bar{x}$, thus estimated, as a function of R. In this case, the ratio N_T/N_P varies simultaneously

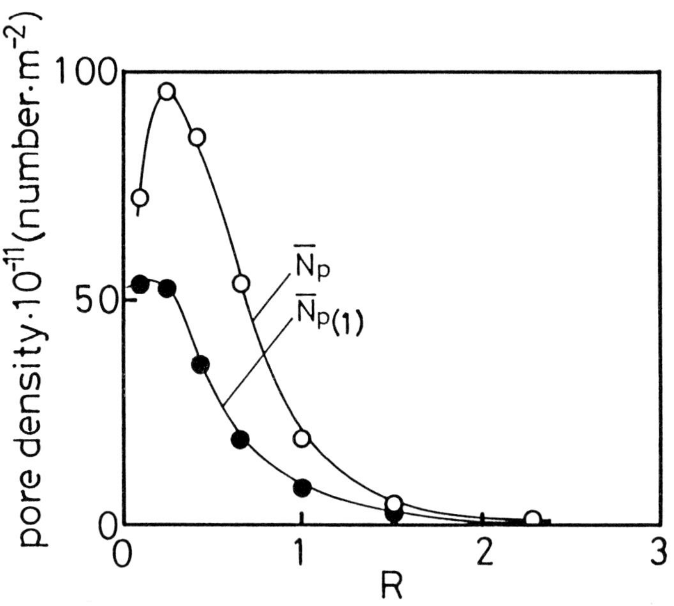

Fig. 6.56 Plots of the mean pore density of total pores, $\bar{N}_P$ and that of the pore with x = 1, $N_{P(1)}$ versus two-phase volume ratio R: Computer experiments, $S_2 = 50$nm.

corresponding to change in R. $\bar{x}$ gets larger with R, in particular the effect of R on $\bar{x}$ is remarkable in the small R region. For a given combination of N_T, N_P and R, $\bar{x}$ is also calculated by KIM theory (eqn (6.87)). The results are also included in the figure. $\bar{x}$, calculated by KIM theory is in excellent agreement with those by computer experiments, in the range $0 \lesssim R \lesssim 4$ and even in the range $4 < R < 10$, the theory agrees fairly well with experiments.

Fig. 6.58 shows the effect of radius of the secondary particles S_2 on the mean pore density (hereafter, simply expressed as N_P, if no confusion). All the relations between N_P and R, obtained for given S_2, are superposable by shifting to the direction of the vertical axis. When S_2 is increased by a factor of 10, N_P for a given R is decreased by a factor of 1/100.

Fig. 6.59 shows the relation among the mean pore radius r_n, the radius of the secondary particles S_2 and two-phase volume

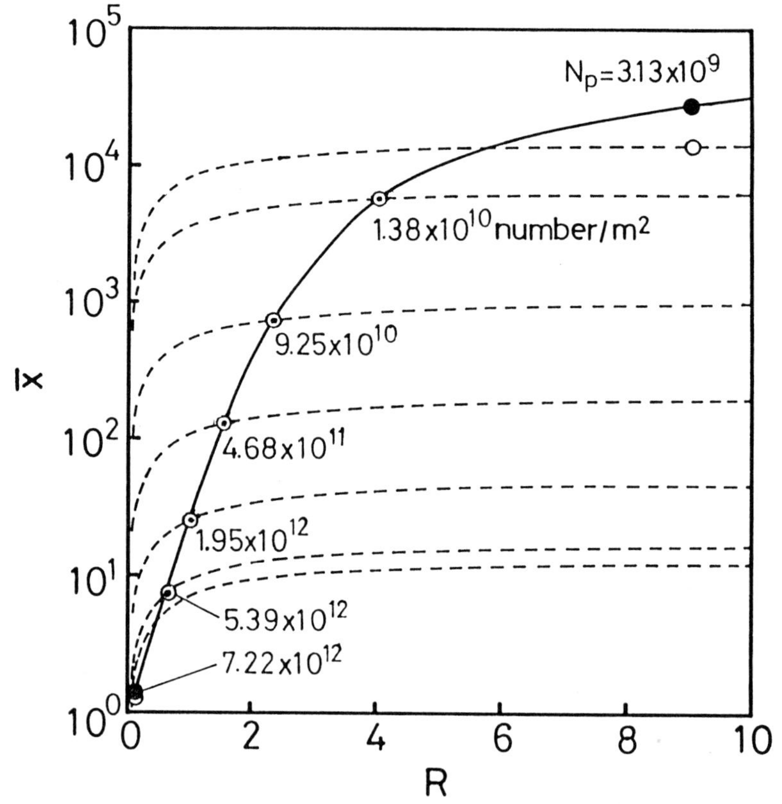

Fig. 6.57 Comparisons between the theoretical mean number of vacant particles in a pore, $\bar{x}$ (eqn (6.87)) (broken line and open mark) and $\bar{x}$, calculated by computer experiments (closed mark for given two-phase volume ratio R): In the range $R \lesssim 4$, closed mark coincides excellently with open mark and ⊙ mark is shown in the figure. Pore density N_p is shown on the curve.

ratio R. r_n increases with R. The rate of increase dr_n/dR is larger for larger S_2. Note that the r_n vs. R curve for a given S_2 is superposable.

Using the data from computer experiments we can check the validity of $P_n(m)$ values estimated in the lattice theory (see, Table 6.4). Fig. 6.60 shows the ratio of $P_n(m)$, estimated by the theory, to $P_n(m)$ of computer experiments, $P_n(m)$ (theory) / $P_n(m)$ (Comp. Exp.), as a function of R for a given combination of m and n. In the range $R < 2.3$ (i.e., $L < 0.3$) $P_0(0)$ defined in Table 6.4

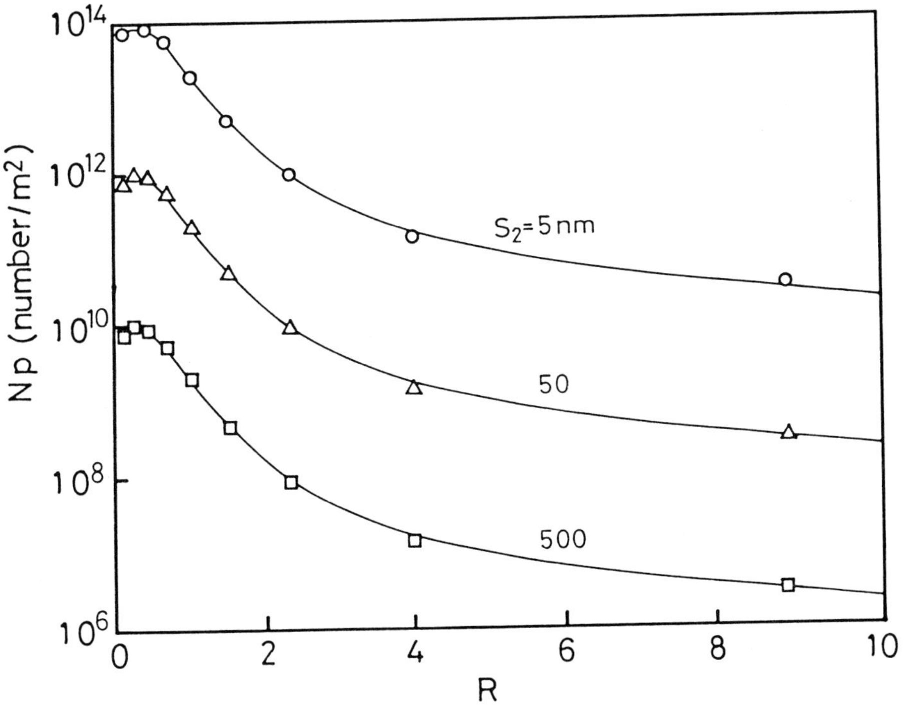

Fig. 6.58 Effect of the radius of the secondary particles S_2 on the relationship between the total number of pores per unit area of surface N_p and two-phase volume ratio R: Computer experiments.

coincides excellently with computer experiments. Deviation of data point on the ratio $P_0(0)$ at $R = 2.3$ from unity may be caused by a large scattering of data inherent to simulation experiments at higher R. The ratio of $P_1(m)$ ($m = 2$, 3 and 4) deviates from unity, increasing with an increase in R. The striking deviation of the theory from computer experiments for $m \geq 2$ is noticed at $R \geq 0.2$. In the theory, an arrangement, as illustrated in Fig. 6.40a ($m = 4$), is treated as a case of $P_2(4)$ (i.e., $m = 4$ and $n = 2$), but in computer experiments all the arrangements of polymer and vacant particles in a plane are completely considered in calculating any

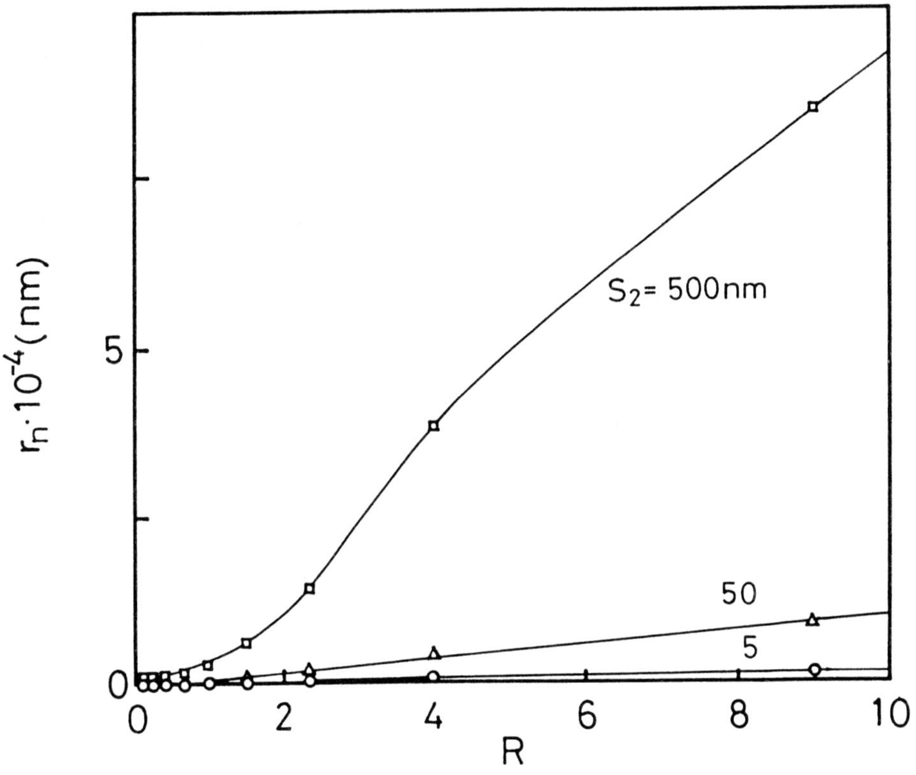

Fig. 6.59 Relationships between the mean pore radius r_n and the two-phase volume ratio R for various S_2: Computer experiments, circular pore growth was assumed.

$P_n(m)$ and then some portion of $P_2(4)$ in the theory may be rightly counted as a case of $P_1(4)$, when four vacant particles in Fig. 6.40a are connected indirectly. This indicates that the theory leads often to underestimation of $P_1(m)$ (m 2, 3 and 4), especially in the larger R region. The ratios of $P_1(m)$ (m 5 and 6) are always unity over a wide range of R. In the range $R< 0.3$, $P_2(m)$ (m= 2, 3 and 4) and $P_3(3)$, calculated by equation listed in Table 6.4 agree well with those of the experiments. The ratio of the above $P_n(m)$ deviates remarkably as R increases beyond 0.67, because $P_2(m)$ and $P_3(3)$ are significantly overestimated in the theory, in which only the nearest neighbours are taken into

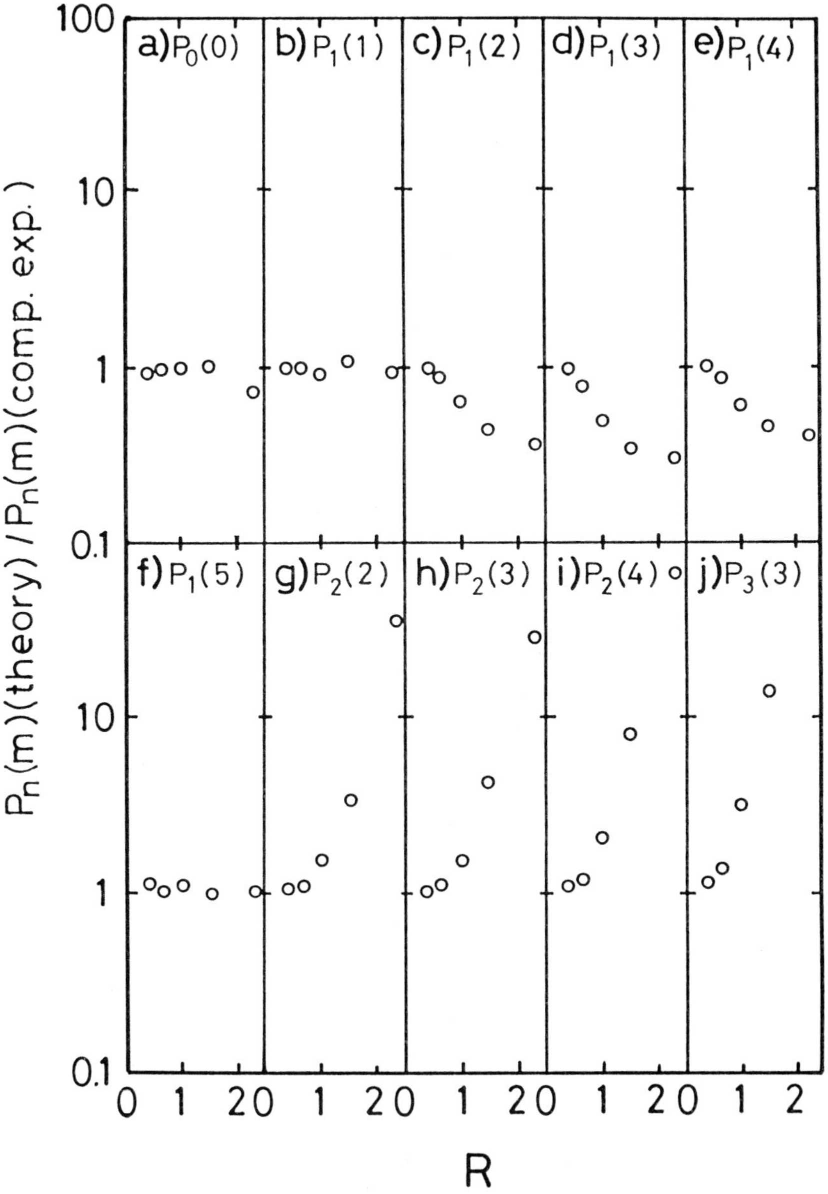

Fig. 6.60 The ratio of $P_n(m)$, estimated by the Kamide-Iijima-Matsuda theory from Table 6.4, to $P_n(m)$, determined by the computer experiments.

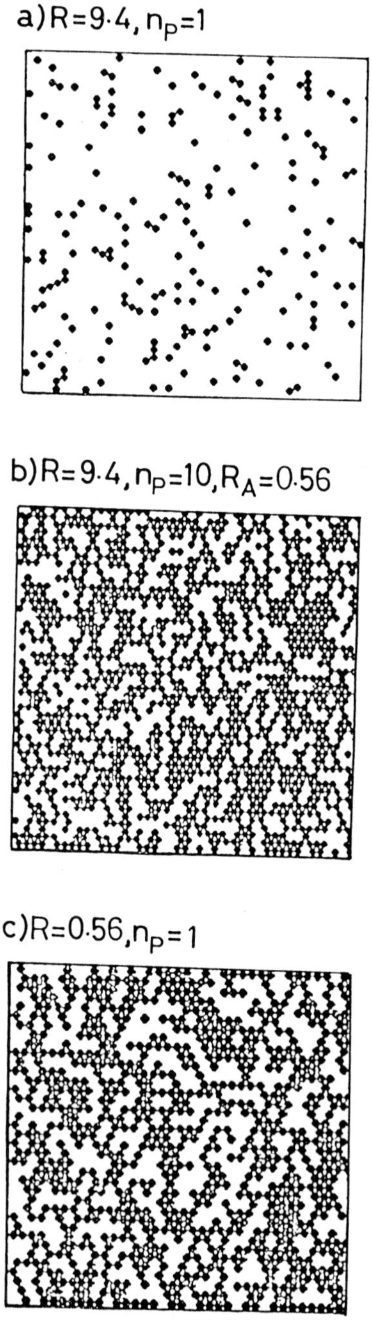

Fig. 6.61 Morphology of hypothetical thin dry planes which are made by piling up a given number (n_P) of hypothetical gel layers formed at common R (computer experiments): a) $R=9.4$, $n_P=1$; b) $R=9.4$, $n_P=10$, $R_A=0.56$; and c) $R=0.56$, $n_P=1$.

account. Therefore, we can conclude that in the larger R region, $P_n(m)$ value by the theory deviates significantly from that by the experiments. This means that the theory becomes very crude at larger R and Table 6.4 should be revised in these regions in future.

Fig. 6.61 shows some arrangements of the secondary particles when a given number (n_P) of hypothetical thin gel planes are piled up to amalgamate or to melt down to give a single dry plane, whose thickness is almost equivalent to the diameter of the dried secondary particles. In this case, apparent phase volume ratio R_A is defined by the relation (ref. 50),

$$R_A = \left\{\frac{R}{R+1}\right\}^{n_P} \left[1 - \left\{\frac{R}{R+1}\right\}^{n_P}\right]^{-1}. \qquad (6.109)$$

In deriving eqn (6.109), overlapping of secondary particles belonging to different hypothetical gel planes were assumed to be totally absent. Computer experiments show that this approximation of non-overlapping is not too crude: When 10 gel planes formed at $R = 9.4$ are piled and collapsed in computer experiments, the phase volume ratio evaluated by taking into consideration overlapping for the piled planes, R_A is 0.56, which should be compared with $R_A = 0.54$, calculated by assuming non-overlapping (eqn (6.109)). Fig. 6.61c shows morphology of a hypothetical thin plane directly formed at $R = 0.56$. This can be compared with that, formed by compilation of ten thin planes with $R = 9.4$ (Fig. 6.61b). Morphology of these two planes is absolutely indispensable.

Fig. 6.62 shows the pore size distribution P(x) of a thin dry membrane prepared by piling 10 gel planes at $R = 9$ and that of a membrane made directly at $R = 0.56$. P(x) of these membranes is practically the same.

Fig. 6.63 shows comparisons of morphology of hypothetical thin membranes made by computer experiments with that of regenerated cellulose membrane with corresponding apparent phase volume ratio R_A. In Fig. 6.63b and d only the first single layer with a thickness of $2S_2'$ was sketched and photocopied using electron micrographs of cellulose membranes. It has been ascertained that computer experiments can produce membranes with morphology very similar to actual membranes.

These findings indicate explicitly that we can accurately evaluate the morphology and pore size distribution by computer experiments using by R_A, in place of R (which is measured at the casting process). Note that in the phase separation stage

($v_p^0 < v_p^c$) R should be larger than unity, but R_A for dry membranes is usually less than unity, because some tens of hypothetical gel planes are piled up to give a single dry plane.

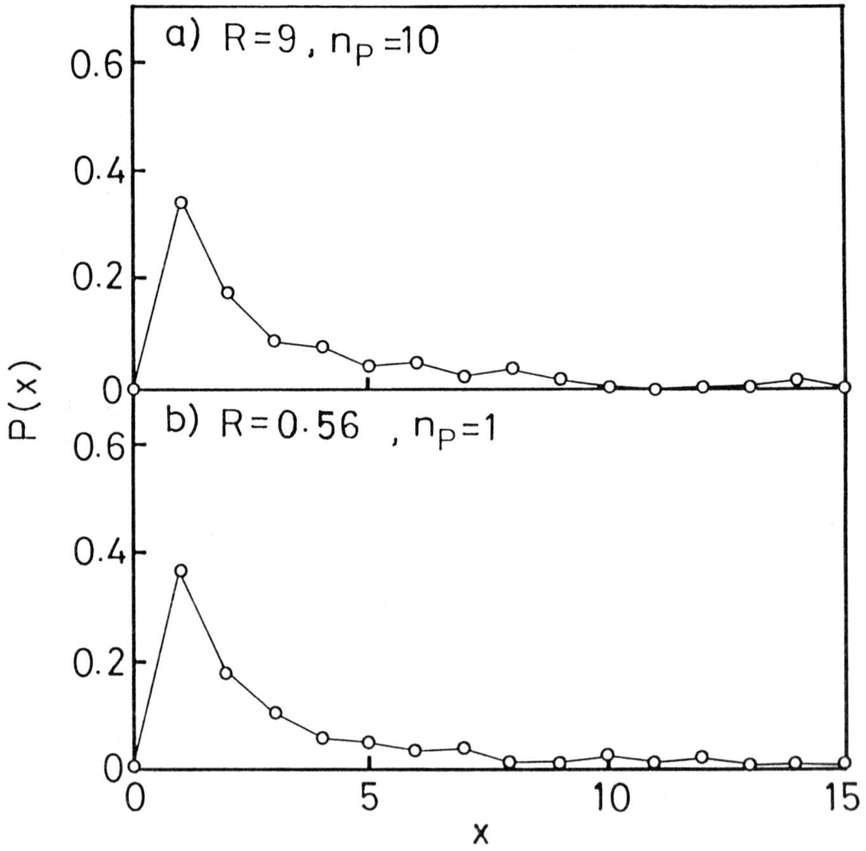

Fig. 6.62 Pore size distribution P(x) of hypothetical thin dry membrane made by piling n_P hypothetical gel planes formed at R. Computer experiments show that piling of 10 gel planes formed at R=9 gives a dry plane with R_A=0.56.

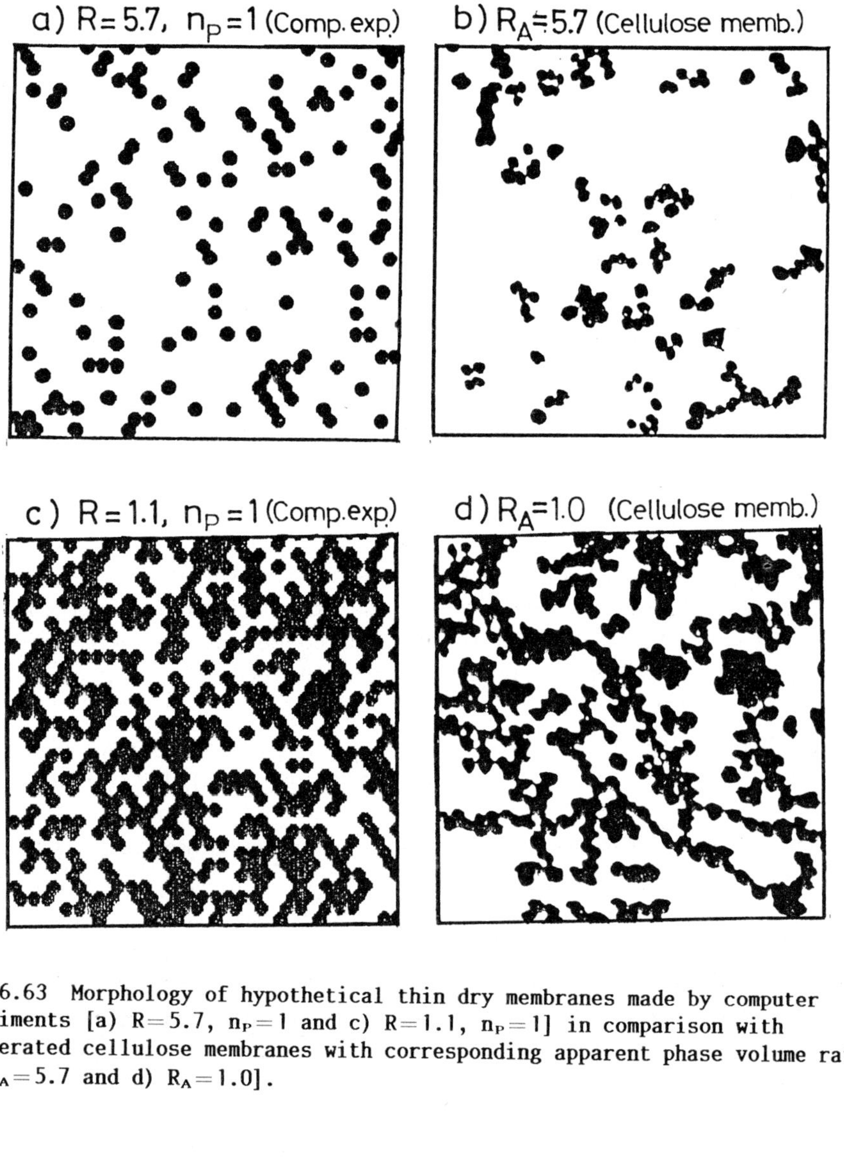

Fig. 6.63 Morphology of hypothetical thin dry membranes made by computer experiments [a) $R=5.7$, $n_P=1$ and c) $R=1.1$, $n_P=1$] in comparison with regenerated cellulose membranes with corresponding apparent phase volume ratio [b) $R_A=5.7$ and d) $R_A=1.0$].

6.4.4 Probabilities of finding through, semi-open, and isolated pores in membranes

In the preceding sections it was theoretically and experimentally shown that porous polymeric membranes prepared by the micro-phase separation method should be considered as composites, in which many hypothetical ultra-thin layers are piled up and when polymer concentration is lower than critical concentration, the ultra-thin layers are two-dimensionally composed of many small particles.

Up to now, mean pore sizes have been estimated by a water-flow-rate method (ref. 7) or a gas permeation method (ref. 13), assuming that all existing pores can be approximated with straight-through cylinders. This assumption is evidently unrealistic and unacceptable and in consequence, the physical meaning of mean pore size by a flow-rate method is unavoidably obscure. We can determine the size of two-dimensional pores and their distribution on the top and bottom surfaces and on ultra-thin layer sections of membranes by the EM method (ref. 15), but N(r) for three-dimensional pores cannot be evaluated by this method. Of course, N(r) by a bubble point method (refs 7,8) deviates significantly from the mean pore size calculated from two-dimensional N(r) by the EM method, because in the former only through pores are investigated and in the latter all possible pores such as isolated, semi-open and through pores are taken into consideration (Fig. 6.64). Then we asked a fundamental question: What is pore in a real polymer membrane?

In this section, we present an approximate theory for evaluating the existing probabilities of isolated, semi-open and through pores P_I, P_S and P_T from experimental two-dimensional porosity Pr data by the EM method and to calculate, using the theory, numerous pore characteristics including P_I, P_S and P_T as function of Pr.

Assume that (1) the membrane is a porous membrane, (2) the membrane consists of multi-layers (ref. 15), (3) the network-like structure in a given layer is constituted of secondary particles formed by amalgamating the primary particles generated in the initiation of the micro-phase separation, (4) the radius of the secondary particles is almost constant ($= S_2$), and (5) the secondary particles are packed in a given layer by hexagonal closest model. Fig. 6.64 shows a schema of a membrane model, in which the membrane is represented as a mixture of secondary

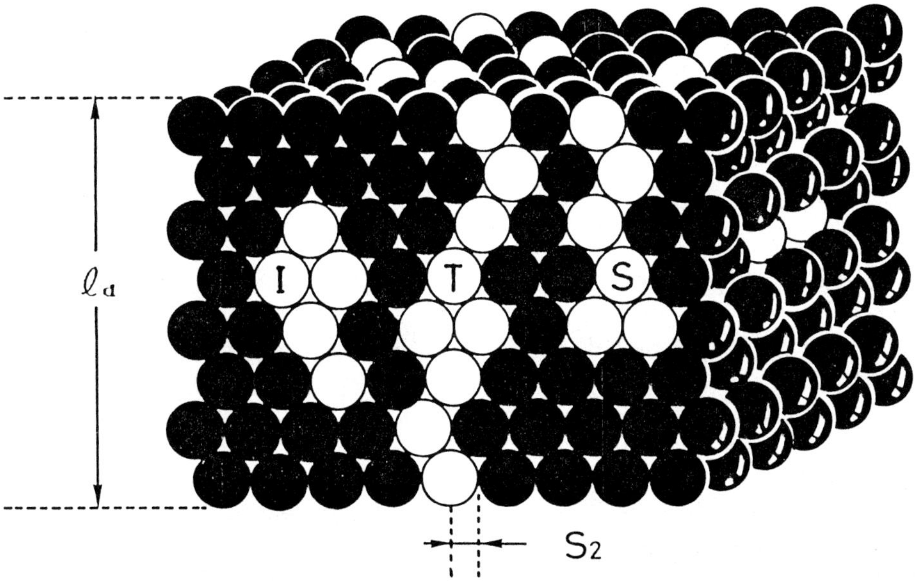

Fig. 6.64 Schematic representation of a membrane structure composed of secondary and vacant particles: Filled sphere, secondary particle; unfilled sphere, vacant particle; radius of these spheres is S_2. The symbols of I, S and T stand for isolated, semi-open and through pores, respectively. Here, we assumed $S_2 \simeq S_2'$ and no morphological change during drying.

polymer particles (filled sphere) and imaginary vacant particles (unfilled sphere), both having the same radius of S_2. Here, before washing, the polymer-lean phase is assumed to be representable by vacant particles and after washing, becomes a part of a pore. The mutual interaction energies between secondary particles, between vacant particles and between secondary and vacant particles are assumed to be equivalent to each other.

When the thickness of the membrane is ℓ_d and the radius of a particle is S_2, the total number of layers N constituting a membrane is given by

$$N = 6^{1/2}\ell_d/4S_2 . \tag{6.110}$$

Two-dimensional porosity Pr, as determined by the EM method, for a hypothetical layer in a membrane is equivalent to the summation of P_T, P_S and P_I, respectively. That is, eqn (6.111) holds

$$Pr = P_T + P_S + P_I . \tag{6.111}$$

Eqn (6.111) indicates that if two of these three probabilities and Pr are known in advance, the remaining probability can be readily evaluated.

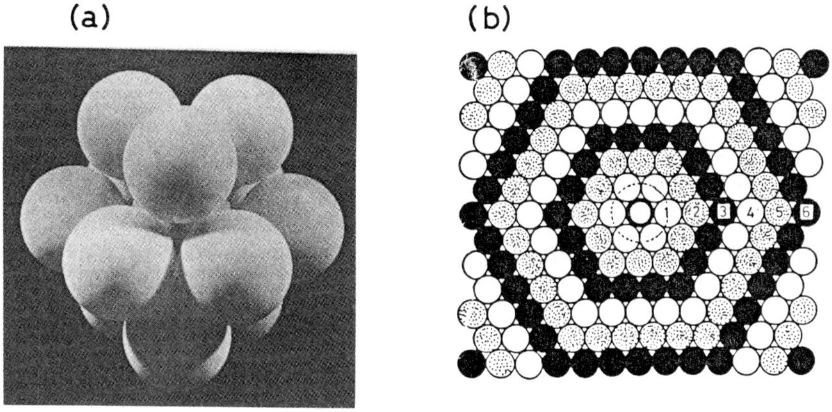

Fig. 6.65 Schematic representation of "the shell structure": a) a photograph of a model of the first shell; the particle located in the center of the shell cannot be seen; b) cross sectional view of shells; thick full line stands for a given vacant particle in the center of the shell structure; figures indicate shell order; dotted line indicates a region of the first shell having a radius of $2S_2$.

Theoretical relation between P_I and Pr

In order to derive the above relation "the shell concept" is tremendously helpful. When we pay attention to an arbitrarily chosen particle in a membrane, the total number of nearest neighbor particles packed around the particle in question is twelve (Fig. 6.65a). These twelve particles make a kind of spherical shell surrounding the given particle and here after we define this shell as the first shell. The second, third,...etc. shells can also be defined in a similar manner (Fig. 6.65b). The radius of the first shell R_1 is approximately $2S_2$ as shown as the broken curve in Fig. 6.65b, and in general, the radius of the i-th shell R_i is approximately related to S_2 through the relation,

$$R_i \simeq 2S_2 (2/3)^{1/2} (i-1) + 2S_2 . \qquad (6.112)$$

The total number of particles constructing the i-th shell, $M_{(i)}$, can be approximated as the ratio of the surface area of the shell to the cross sectional area of the secondary particle:

$$M_{(i)} = 4\pi R_i^2 / \pi S_2^2 . \qquad (6.113)$$

Combination of eqn (6.112) with eqn (6.113) gives

$$M_{(i)} = 16\{(2/3)(i-1)^2 + 2(2/3)^{1/2}(i-1) + 1\} . \qquad (6.114)$$

Among $M_{(i+1)}$ particles belonging to the (i+1)-th shell, $m_{1(i)}$ particles are located as the nearest neighbouring particles to a given particle existing in the i-th shell. We assume, as noted, hexagonal closest packing, and then $m_{1(i)}$ is given by

$$m_{1(i)} = 3M_{(i+1)}/M_{(i)} \quad \text{for } M_{(i)} \neq \infty . \qquad (6.115)$$

At the limit of $M_{(i+1)} = \infty$ (i.e., $R_{i+1} = \infty$), eqn (6.115) can be simply reduced to

$$m_{1(i)} = 3 \qquad \text{for } M_{(i)} = \infty . \qquad (6.115')$$

The maximum number of particles in the (i+1)-th shell, contacting directly with $\ell_{(i)}$ particles, which are chosen arbitrarily from total particles in the i-th shell (Here, we assume $\ell_{(i)} \ll M_{(i)}$), $m_{\ell(i)}$ is

$$m_{\ell(i)} = 3\ell_{(i)} M_{(i+1)}/M_{(i)} \qquad \text{for } \ell_{(i)} < M_{(i)}/3. \qquad (6.116)$$

Since $m_{\ell(i)}$ should be an integer, we round off the fraction in the right side of eqn (6.116) to integer.

The total number of secondary and vacant particles in a unit volume of a membrane N_0 is calculated by the approximate relation,

$$N_0 = 3/\{4(\pi S_2^3)\} . \qquad (6.117)$$

An isolated pore should be surrounded completely by secondary particles. First, we consider cases in which an isolated pore is constituted with a vacant particle, two vacant particles and three vacant particles. Considering the probability that twelve possible seats around a given particle are occupied by vacant particles or secondary particles, we derive equations of the total number of isolated pores in the unit volume of the membrane for each case. Finally, a general expression to give the total number of the isolated pores having n_0 vacant particles in the unit volume of the membrane is derived.

The total number of isolated pores, formed by only one vacant particle, in a unit volume of a membrane, N_1, is given as the products of N_0 and the probability that a particle chosen arbitrarily in a membrane is vacant (= Pr) and the probability that all the twelve neighboring particles against that vacant particle are secondary particles ($= (1-Pr)^{12}$), that is

$$N_1 = N_0 Pr \cdot q^{12} \tag{6.118}$$

with

$$q = 1 - Pr. \tag{6.119}$$

The total number of isolated pores, which are formed by two consecutive vacant particles in a unit volume of a membrane, N_2, is given by the product of (1) N_0 and (2) the combinatory, ${}_{12}C_1$, to choose one particle from twelve particles in the first shell and (3) the probability that the particle chosen is thus vacant (= Pr) and (4) the probability that all remaining eleven particles in the above-mentioned first shell are secondary particles ($= q^{11}$) and (5) the probability that the second shell particles contacting directly the first shell vacant particle are all secondary particles ($= q^{m1(1)}$). That is,

$$N_2 = N_0 \cdot {}_{12}C_1 \cdot Pr^2 q^{11} q^{m1(1)}. \tag{6.120}$$

The total number of isolated pores, each made of three consecutive vacant particles in a unit volume of a membrane, N_3, is approximately given by

$$N_3 = N_0 \{{}_{12}C_2 \cdot Pr^3 q^{10} q^{m1(1)} q^{m1(1)}$$

$$+ {}_{12}C_1 \cdot {}_{m1(1)}C_1 \cdot Pr^3 q^{11} q^{m1(1)-1} q^{m1(2)}\}. \qquad (6.121)$$

The total number of isolated pores, each made of n_0 consecutive vacant particles in a unit volume of a membrane, N_{n_0} for $n_0 < 13$ is derived in a similar manner as in deriving eqns (6.118), (6.119), (6.120) and (6.121). The result is

$$N_{n_0} = N_0 \cdot q^{(13-n_0)} (1-q)^{n_0} \sum_{j_{(1)}=0}^{12} {}_{12}C_{j(1)} \cdot A_{j(1)} \quad \text{for } n_0 < 13 \qquad (6.122)$$

where $A_{j(1)}$ is the probability that $j_{(1)}$ vacant particles among n_0 consecutively connected particles are found in the first shell and ${}_{12}C_{j(1)}$ stands for the combinatory of choice of $j_{(1)}$ from twelve. $A_{j(1)}$ is given by the following recurrent equation:

$$A_{j(1)} = \sum_{j_{(2)}(2)=1}^{m_{1(1)}} {}_{m1(1)}C_{j_{(2)}(2)} \cdot A_{j_{(1)}(1),j_{(2)}(2)} q^{m1(1)-j_{(2)}(2)}$$

$$+ q^{m1(1)}. \qquad (6.123)$$

$A_{j_{(1)}(1),j_{(2)}(2)}$ is very complicated and hereafter we employ $A_{j(1),j(2)}$.

$$A_{j(1),j(2)} = 0 \qquad \text{for } j_{(2)} = 0, \qquad (6.124)$$

$$A_{j(1),j(2)} = \sum_{j_{(3)}=1}^{m_{1(2)}} {}_{m1(2)}C_{j(3)} \cdot A_{j(1),j(2),j(3)} q^{m1(2)-j(3)} + q^{m1(2)}$$

$$\text{for } j_{(2)} \neq 0. \qquad (6.124')$$

Generalized form of $A_{j_{(1)}(1),j_{(2)}(2)}$ in eqn (6.123) should be $A_{j_{(1)}(1),\cdots j_{(i-1)}(i-1),j_{(i)}(i),\cdots}$, and its much simpler expression is $A_{j(1),\cdots j(i-1),j(i)}$.

$A_{j(1),\cdots j(i-1),j(i),\cdots}$ is the probability that $j_{(1)},\cdots,j_{(i-1)},j_{(i)},\cdots$ vacant particles, all belonging to a pore consisting of n_0 consecutively connected particles, are found in the first,···, the (i-1)-th, the i-th··· shells, respectively, as given by

$$A_{j(1),\cdots,j(i-1),j(i),j(i+1),\cdots} = 0$$

$$j_{(i+1)} = j_{(i+2)} = \cdots = j_{(m)} = 0 \qquad \text{for } j_{(i)} = 0. \qquad (6.125)$$

In general,

$$A_{j(1),j(2),\cdots,j(m-1)} = \sum_{j_{(m)}=1}^{m_{(m-1)}} {}_{m\ell(m-1)}C_{j(m)} \cdot A_{j(1),j(2),\cdots,j(m)}$$

$$\times q^{m\ell(m-1)-j(m)} + q^{m\ell(m-1)} \qquad \text{for } j_{(i)} \neq 0. \qquad (6.126)$$

The maximum order $\tilde{m}$ of the shells, in which at least one vacant particle of any isolated pore having n_0 sequential vacant particles is found, can be determined by eqn (6.127):

$$\sum_{i=1}^{\tilde{m}} j_{(i)} = n_0 - 1. \qquad (6.127)$$

Obviously, $\tilde{m}$ is less than n_0:

$$\tilde{m} \leq n_0 - 1. \qquad (6.128)$$

The total number of vacant particles contributing to formation of the isolated pores, made of n_0 consecutive vacant particles is $n_0 N_{n0}$ and the summation of $n_0 N_{n0}$ over all possible n_0 in a unit volume of membrane, coincides with P_I:

$$P_I = \sum_{n_0=1}^{N_0 Pr} n_0 N_{n0} / N_0 \qquad (6.129)$$

Using eqns (6.122)-(6.127) and eqn (6.129), we can calculate P_I as a function of Pr (accordingly, q).

Theoretical relation between P_T and Pr

In order to derive a theoretical relation of P_T as a function of Pr, the multi-thin layer model (Fig. 6.64) is more convenient than the shell concept employed in the preceding section since through pores must penetrate, even if by a roundabout route, from the front surface to the back surface of a membrane. When a particle in the first layer is a vacant particle, three vacant or secondary particles in the 2nd layer contact directly to the above mentioned vacant particle and the three particles in the 2nd layer, seven vacant or secondary particles in the 3rd layer contact directly. This situation is represented by the photograph in Fig. 6.66a.

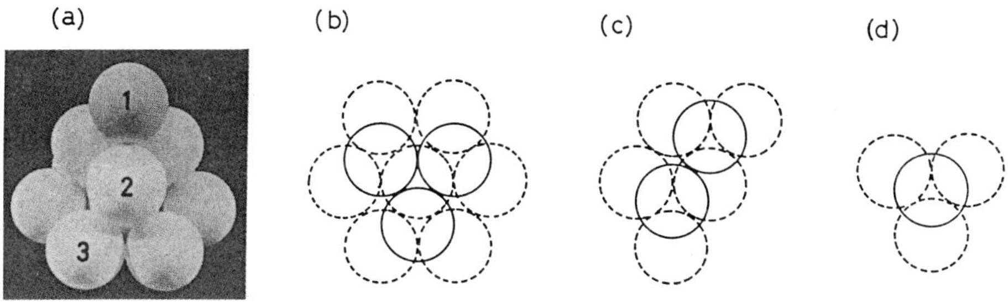

Fig. 6.66 Some configurational patterns of the arrangement of vacant and secondary particles when a site of the first layer is occupied by a vacant particle: Full line, vacant or secondary particle on the second layer; broken line, vacant or secondary particle on the third layer; a) a photograph of part of the layer structure model; figures indicate layer order; only four 3rd layer particles are seen; b) configuration of seven particles on the third layer neighboring with three particles on the second layer; c) configuration of five particles on the third layer neighboring with two particles on the second layer; d) configuration of three particles on the third layer neighboring with one particle on the second layer.

When all three particles in the 2nd layer are vacant particles, seven 3rd layer particles contact directly these vacant particles (Fig. 6.66b) and when two of these three 2nd layer particles are vacant particles, five 3rd layer particles contact directly these vacant particles (Fig. 6.66c) and when only one 2nd layer particle is a vacant particle three 3rd layer particles contact this 2nd layer particle (Fig. 6.66d).

Suppose an arbitrarily chosen pore in the k-th layer, which consists of consecutive $\ell_{[k]}$ vacant particles and these particles can directly contact at most $m_{[k][k+1]}$ particles in the (k+1)-th layer. Hereafter we employ a simpler expression $m_{[k+1]}$ instead of $m_{[k][k+1]}$. After some tedious calculation, we obtain an approximate relation between $m_{\ell[k]}$ and $\ell_{[k]}$:

$$m_{\ell[k+1]} \simeq \{1 + 2\ell_{[k]} + (12\ell_{[k]} - 3)^{1/2}\}/2 \tag{6.130}$$

Here, $m_{\ell[k]}$ is an integer and then the value given by the left hand side of eqn (6.130) should round off the fraction.

Consider a case when a membrane consists of two thin layers and all the pores on the first layer are constituted with only one vacant particle. Then, the existing probability of the through pore, penetrating through two neighbouring layers $P_{T[2]}$ is given by the summation of these three probabilities: The probability that one of three 2nd layer particles, contacting directly the vacant particle on the first layer is vacant ($= Pr \cdot {}_3C_1 \cdot Pr \cdot q^2$) and the probability that two of those three 2nd layer particles mentioned above are vacant ($= Pr \cdot {}_3C_2 \cdot Pr^2 \cdot q$) and the probability that all those three particles in the second layer are vacant ($= Pr \cdot {}_3C_3 \cdot Pr^3$). Then, $P_{T[2]}$ is given by

$$P_{T[2]} = Pr \cdot {}_3C_1 \cdot Pr \cdot q^2 + Pr \cdot {}_3C_2 \cdot Pr^2 \cdot q + Pr \cdot {}_3C_3 \cdot Pr^3 \tag{6.131a}$$

$$= Pr \sum_{i=1}^{3} {}_3C_i \cdot Pr^i \cdot q^{3-i} = Pr(1 - q^3)\,. \tag{6.131b}$$

By generalizing the above procedure established for a membrane with two thin layers to the case when a membrane consists of k thin layers, we obtain the existing probability of the through pore $P_{T[k]}$:

$$P_{T[k]} = Pr \sum_{\ell_{[2]}=1}^{m_{\ell[2]}} {}_{m\ell[2]}C_{\ell[2]} \cdot Pr^{\ell[2]} \cdot q^{m\ell[2]-\ell[2]} \sum_{\ell_{[3]}=1}^{m_{\ell[3]}} \{{}_{m\ell[3]}C_{\ell[3]}$$

$$\times Pr^{\ell[3]} \cdot q^{m\ell[3]-\ell[3]} \sum \cdots \sum_{\ell_{[k-1]}=1}^{m_{\ell[k-1]}} \{{}_{m\ell[k-1]}C_{\ell[k-1]} \cdot Pr^{\ell[k-1]}$$

$$\times q^{m\ell[k-1]-\ell[k-1]} \sum_{\ell_{[k]}=1}^{m_{\ell[k]}} \{{}_{m\ell[k]}C_{\ell[k]} \cdot Pr^{\ell[k]} \cdot q^{m\ell[k]-\ell[k]}\}\} \cdots \}\,. \tag{6.132}$$

When k equals N, then $P_{T[N]}$ corresponds to P_T.

Consider the consecutive x vacant particles in the first layer, constituting a part of a through pore. The existing probability of a through pore, which has x vacant particles on the first layer, $P_T(x)$ can be approximated by

$$P_T(x) = (\pi S_2^2) \cdot x \cdot M(x) [1 - \{1 - (P_{T[N]}/Pr)\}^x]. \tag{6.133}$$

When we consider one vacant particle in the first layer, the probability that this particle does not become part of a through pore is $\{1 - (P_{T[N]}/Pr)\}$. The probability that none of x vacant particles in the first layer becomes a part of through pores is $\{1 - (P_{T[N]}/Pr)\}^x$. The probability that any of x consecutive vacant particles in the first layer become a part of a through pore is $[1 - \{1 - (P_{T[N]}/Pr)\}^x]$. When we define $P(x)$ as the frequency distribution function of x in the first layer, $P(x)$ can be easily transformed into a pore radius distribution function $N(r)$ in the first layer, evaluated independently, through the relations:

$$P(x)\,dx = N(r)\,dr, \qquad r = x^{1/2} S_2. \tag{6.134}$$

The mean value of the existing probability of a straight-through pore, whose pore end consists of x vacant particles, $P_T(x)$, averaged over all possible x as designated by P_T, is calculated by

$$P_T = \pi S_2^2 \int x P(x) [1 - \{1 - (P_{T[N]}/Pr)\}^x]\,dx. \tag{6.135}$$

Fig. 6.67 shows dependence of $P_{t[k]}$ on layer order k constituting a membrane with a given $Pr(d_4)$. In the range $Pr(d_4) > 0.9$, $P_{t[k]}$ remains practically constant over the entire range of k.

In the range $Pr(d_4) < 0.8$, $P_{T[k]}$ decreases first, approaching a limiting value ($P_{T[\infty]}$) with increasing k. In the fourth column of Table 6.7 are $P_{T[\infty]}$ values estimated from Fig. 6.67. As expected, $P_{T[\infty]}$ is larger for membranes with larger Pr. k value giving a level-off value of $P_{T[k]}$, designated here as k_m, can be roughly estimated from Fig. 6.67. Here, we regard for convenience $P_{T[k]}$ as $P_{T[\infty]}$ if the condition $P_{T[k]} \geqq P_{T[k-1]} \times 0.9999$ holds. This condition can be rewritten in the form

$$(k-1)^{1/2} q^{k+1} \leqq 0.0001. \tag{6.136}$$

k_m values calculated thus are summarized in the second column of Table 6.7.

With an increase in Pr, k_m decreases rapidly and monotonously.

In the fifth and sixth column of Table 6.7, P_S and P_I for

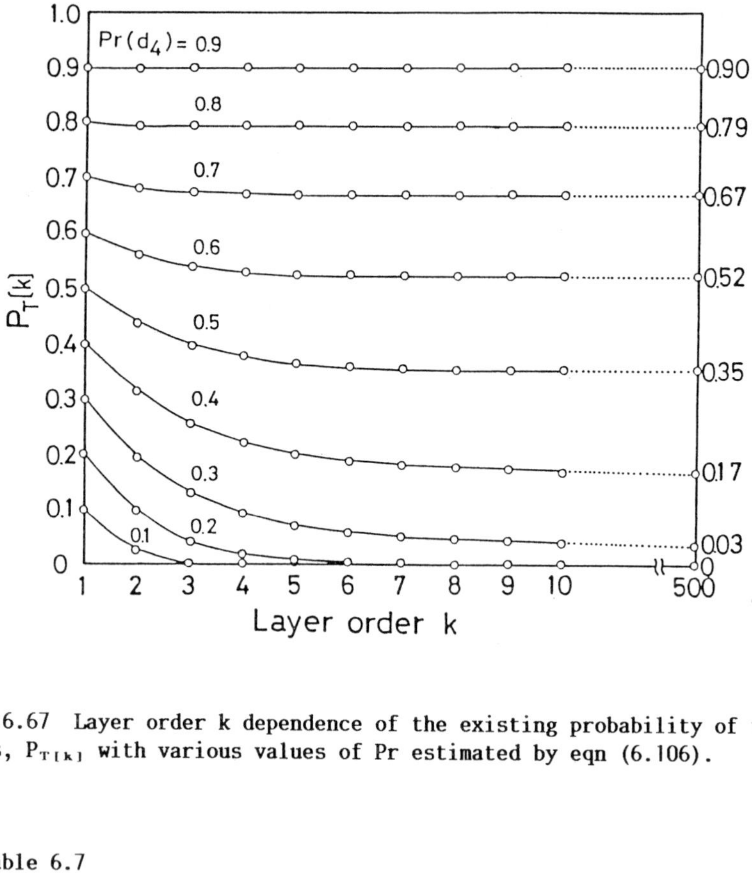

Fig. 6.67 Layer order k dependence of the existing probability of through pores, $P_{T[k]}$ with various values of Pr estimated by eqn (6.106).

Table 6.7
Pore shape characteristics determined theoretically as a function of Pr.

Pr	k_m	$\ell_{[km]}$	$P_{T[\infty]}$	P_S	P_I	P_T/Pr	y	ε
0.1	109	104.6	0.0000	0.0317	0.0683	0.0000	1.04	1.009
0.2	49	53.4	0.0001	0.1840	0.0159	0.0005	1.09	1.012
0.3	30	35.6	0.0321	0.2665	0.0014	0.107	1.13	1.014
0.4	20	26.4	0.168	0.232	0.000	0.42	1.19	1.017
0.5	15	20.6	0.352	0.148	0.000	0.704	1.24	1.019
0.6	11	16.5	0.524	0.076	0.000	0.873	1.32	1.024
0.7	8	12.4	0.669	0.031	0.000	0.956	1.43	1.029
0.8	6	10.9	0.791	0.009	0.000	0.989	1.61	1.039
0.9	4	8.5	0.899	0.001	0.000	0.999	2.04	1.060

given Pr are listed. Here, P_I was calculated using eqns (6.122)-(6.127) and (6.129) and P_S from $1-P_{T[\infty]}-P_I$. Hereafter, $P_{T[\infty]}$ is simply denoted as P_T when there is no confusion.

For a regenerated cellulose membrane, S_2 and ℓ_d were found to be 0.2μm and 200μm, respectively. From these values, the total number of thin layers constituting a membrane, N, was estimated using eqn (6.110) as 612. This membrane has a porosity of 0.469 by the EM method. Therefore, inspection of Fig. 6.67 and Table 6.7 leads to the conclusion that for this cellulose membrane N is almost 35 times larger than k_m (= 18) and P_T was calculated to be 0.294 from the relations between Pr and $P_{T[\infty]}$ (= P_T).

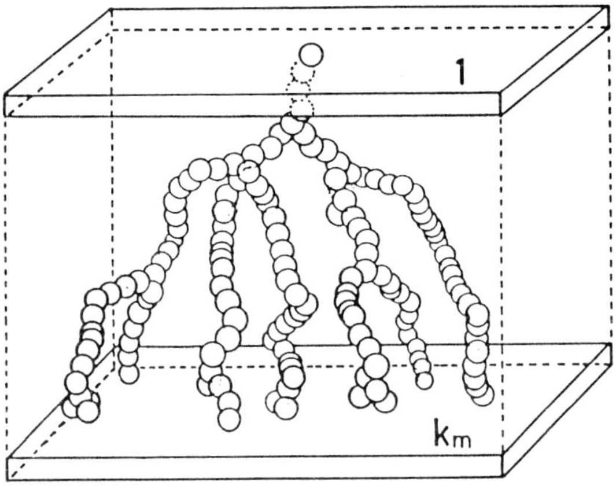

Fig. 6.68 Schematic representation of a through pore. A through pore has one vacant particle on the first layer and many branching pores, consisting of $\ell_{i[km]}$ consecutive vacant particles in the k_m-th layer; unfilled sphere, vacant particle.

As shown in Fig. 6.68, there should be many branching pores in the k_m-th layer, connected to a given vacant particle on the first layer. Considering the total number of vacant particles constituting one of the above-mentioned k_m-th layer branching pores, $\ell_{i[km]}$, we define $\ell_{[km]}$ as the summation of $\ell_{i[km]}$ over the k_m-th layer by eqn (6.137).

$$\ell_{[km]} = \sum_i \ell_{i[km]} \tag{6.137}$$

$\ell_{[km]}$ is the total number of vacant particles in the k_m-th layer, connected to a given vacant particle on the first layer.

Not all the $\ell_{[km]}$ vacant particles have to be in contact with each other in the k_m-th layer, but when the $\ell_{[km]}$ vacant particles are assumed to be consecutive in the k_m-th layer, $m_{\ell\,[km][km+1]}$ (hereafter we employ $m_{\ell\,[km+1]}$, instead) may be approximately equal to the summation of the maximum number of particles in the (k_m+1)-th layer, contacting directly consecutive $\ell_{i\,[km]}$ particles in the k_m-th layer over the (k_m+1)-th layer, $\sum_i m_{\ell i\,[km][km+1]}$ as eqn (6.138).

$$m_{\ell\,[km+1]} = \sum_i m_{\ell i\,[km][km+1]}. \qquad (6.138)$$

Accordingly, $m_{\ell\,[km+1]}$ should satisfy the condition given by eqn (6.139):

$$q^{m_{\ell\,[km+1]}} \leq 0.0001. \qquad (6.139)$$

Eqn (6.139) was derived considering that at least one of the vacant particles on the k_m-th layer connected to a given vacant particle on the first layer connects directly with vacant particle(s) on the (k_m+1)-th layer without fail if all these vacant particles build up a through pore.

Combination of eqn (6.139) with eqn (6.130) on the condition of eqn (6.138) enables us to calculate $\ell_{[km]}$ as a function of Pr. Results are shown in the third column of Table 6.7. For example, $\ell_{[km]}$ and k_m values of a membrane with Pr = 0.5 are estimated from Table 6.7 to be 20.6 and 15, respectively. This means that one vacant particle on the first layer is indirectly connected to 20.6 vacant particles on the 15-th layer by forming branches.

Now consider the case in which a vacant particle located on the k-th layer arbitrarily chosen connects directly, on average, to y vacant particles on the (k+1)-th layer. If y can be regarded as constant, an independent of k, y can be calculated from k_m and $\ell_{[km]}$ by the relation

$$y^{km-1} = \ell_{[km]} \qquad (6.140)$$

y values obtained thus for various Pr are presented in the 8th column of Table 6.7.

Fig. 6.69 shows Pr dependence of P_T, P_S and P_I of a membrane

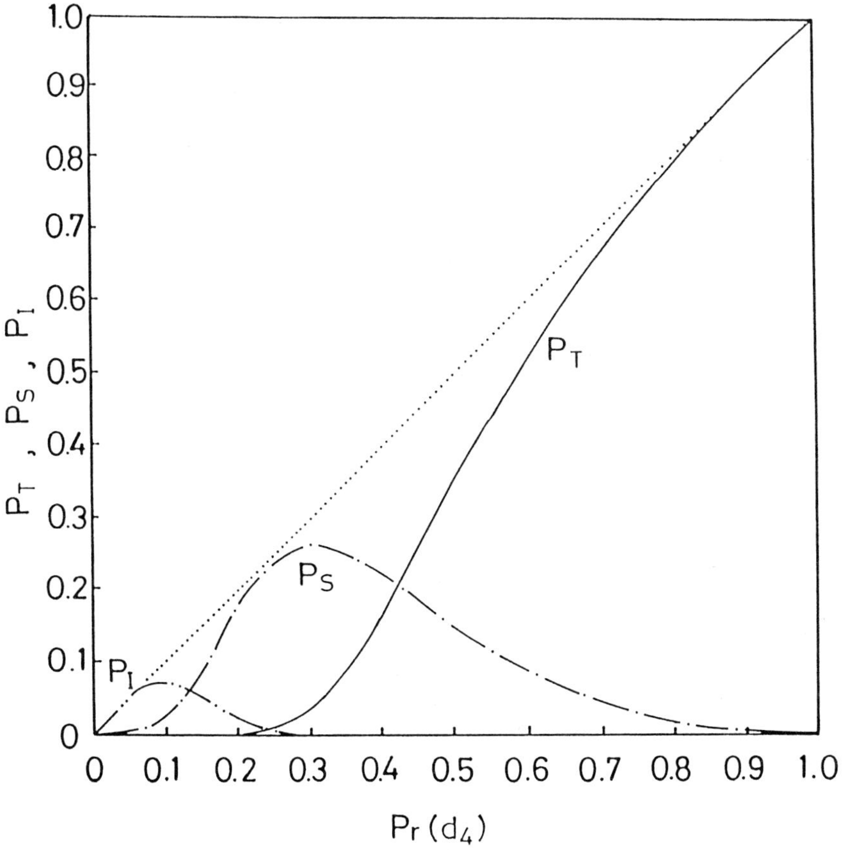

Fig. 6.69 Pr dependence of P_T, P_S and P_I at N=500.

with N= 500. For Pr> 0.4, through pores are most abundantly observed and for 0.15< Pr< 0.4, the semi-open pores are dominant and for Pr< 0.15, most of the pores are the isolated pores.

The water flux J is related only with the through pore. Then the mean pore radius r_F evaluated from J (see, eqn (6.145)) and porosity should be improved to r_{FE} given as follows

$$r_{FE} = r_F \, (Pr/P_T)^{1/2} \, \varepsilon^{1/2}. \quad (6.141)$$

Here, the tortuous coefficient ε is given by

$$\varepsilon = \{1 + (3/4)^2 \cdot 6\ell_{[km]}/Pr(1/4k_m)^2\}. \qquad (6.142)$$

The value of S_2 for a typical regenerated cellulose plane membrane was 200nm and ℓ_d was 200µm. By substitution of these S_2 and ℓ_d values into eqn (6.110), N was determined to be 612. $Pr(d_4)$ was 0.469 and $2r_F$ was 204nm. P_T and ε calculated using these experimental values are 0.294 and 1.018, respectively. The calculated value of $2r_{FE}$ given in eqn (6.141) is 257nm. The mean pore size of $2(r_3 \cdot r_4)^{1/2}$ from the EM photographs of this membrane was 250nm. The coincidence between $2r_{FE}$ and $2(r_3 \cdot r_4)^{1/2}$ is surprisingly excellent.

Validity of the approximate theory described above can be checked by computer experiments based on the Monte Carlo method (ref. 51).

The results show that the patterns of branching pores as shown in Fig. 6.68 are unacceptable and through pores make a very complicated and more closely entangled network structure.

The effect of the porosity on P_T, P_S and P_I are qualitatively unquestionable and similar curves with those in Fig. 6.69 are obtained, although both P_I and P_S become negligibly small at Pr beyond 0.4 in computer experiments.

6.5 CRITICAL POINTS OF CASTING SOLUTION / NONSOLVENT SYSTEM AND PORE SHAPE OF MEMBRANES CAST FROM A SOLUTION

6.5.1 Cellulose cuprammonium / acetone aq. solution

In the practical membrane casting process, much more complicated systems such as polymer / ternary solvent mixture and polymer / quaternary solvent mixture systems are often employed. It should be noted that for four and higher multicomponent systems the critical point is not a single point: In ternary systems consisting of a multicomponent polymer in a binary solvent mixture, the critical point can be represented not by a point but by a curve in a regular triangular prism, in which an axis of ordinate is temperature and composition of the system is defined on a point in the triangle.

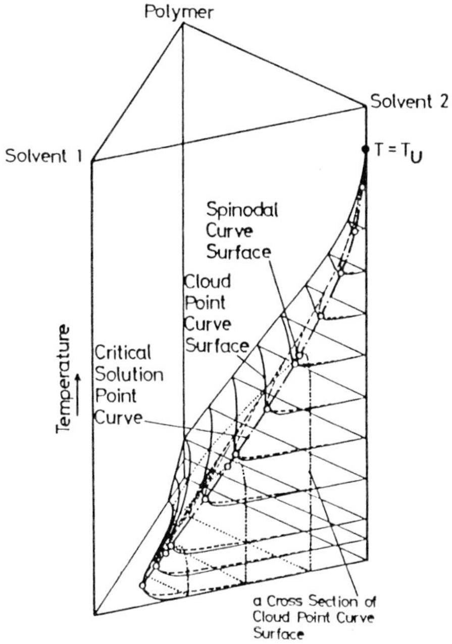

Fig. 6.70 Schematic representation of phase diagrams of quasi-ternary system consisting of multicomponent polymers dissolved in a binary solvent mixture: Unfilled circle, upper critical solution point; full line, cloud point curve at constant temperature; broken line, spinodal curve at constant temperature; dotted line, a cross section of the cloud point surface with a constant volume fraction of solvent v_1 plane; T_U, temperature above which the system is a single phase for an entire composition.

Fig. 6.70 demonstrates effects of temperature on phase diagrams, including cloud point curve, spinodal curve and upper critical solution point, for a quasi-ternary system consisting of multicomponent polymers in binary solvent mixtures (ref. 52). In the figure, the full and broken lines are the cloud point curve and spinodal curve at constant temperature, respectively. The chain line is the critical solution point curve. The cross section of cloud point curve surface with a constant volume fraction of solvent 1, v_1 plane is shown as a dotted line, whose shape is very similar to the cloud point curve of a quasi-binary system of multicomponent polymers in a single solvent (for example, see, Fig. 2.78). Above a specific temperature T_U, the quasi-ternary system is a single phase over an entire composition.

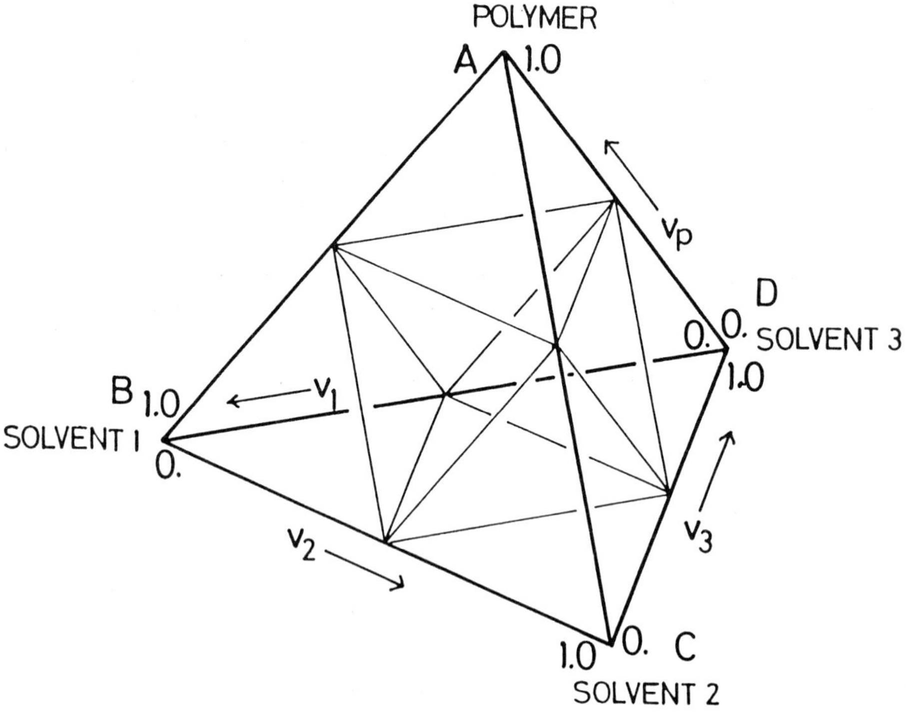

Fig. 6.71 Regular triangular pyramid graph of multicomponent polymer / solvent 1 / solvent 2 / solvent 3 system.

The phase diagram of the system of polymers in a ternary solvent mixture is usually represented by a regular triangular pyramid (Fig. 6.71) and cloud point curve, spinodal curve and the critical solution point in a binary solvent system become a cloud point curve surface, spinodal curve surface and critical solution point curve, respectively. At each point in the figure, the equation $v_1 + v_2 + v_3 + v_p = 1$ (v_1, v_2, v_3 and v_p are the volume fractions of solvent 1, solvent 2, solvent 3 and polymer, respectively) is satisfied.

The spinodal curve surface, critical solution point curve and cloud point curve surface for quasi-quaternary system, can be drawn as illustrated in Fig. 6.72. Fig. 6.72a is the case where both solvents 1 and 2 are good solvents and solvent 3 is nonsolvent. In the figure the surface enclosed by the full line is the cloud point curve surface and the area enclosed by the dotted line is the spinodal curve surface (hatched surface) and the chain line is the critical solution point curve. Surface ABD or surface ACD corresponds to phase diagrams of quasi-ternary systems and each plane has a single critical point (open circle). Here, there exists a critical solution point in a triangular cross section (AED), in which the ratio $v_2/(v_1 + v_2)$ is kept constant ($\overrightarrow{AE}$, $\overrightarrow{ED}$ and $\overrightarrow{DA}$ in surface of AED are coordinate axes of $v_1 + v_2$, v_3 and v_p, respectively). By changing the ratio $v_2/(v_1 + v_2)$ from 0 to 1, cloud point curve surface, spinodal curve surface and critical solution point curve can be drawn. Fig. 6.72b is the case where solvent 1 is a good solvent and both solvents 2 and 3 are nonsolvents. In this case two nonsolvents are assumed to be mutually miscible. Fig. 6.72c is the case where solvent 2 is a good solvent and solvents 1 and 3 are nonsolvents. Cosolvency occurs between two nonsolvents 1 and 3. In this case there are two cloud point curve surfaces, two spinodal curve surfaces, and two critical solution point curve surfaces. Fig. 6.72d is the case where all the three solvents (solvents 1, 2 and 3) are nonsolvents. In this case, cosolvency occurs between solvent 1 and solvent 2. Phase diagram within the surface ABC (which corresponds to quasi-ternary system) is just the same phase diagram as that of cosolvency observed for polymer / binary solvent mixture system having two critical points (open circles). For cuprammonium cellulose / acetone / water system, acetone, ammonia and water are considered to be nonsolvents, and this corresponds to the case of Fig. 6.72d.

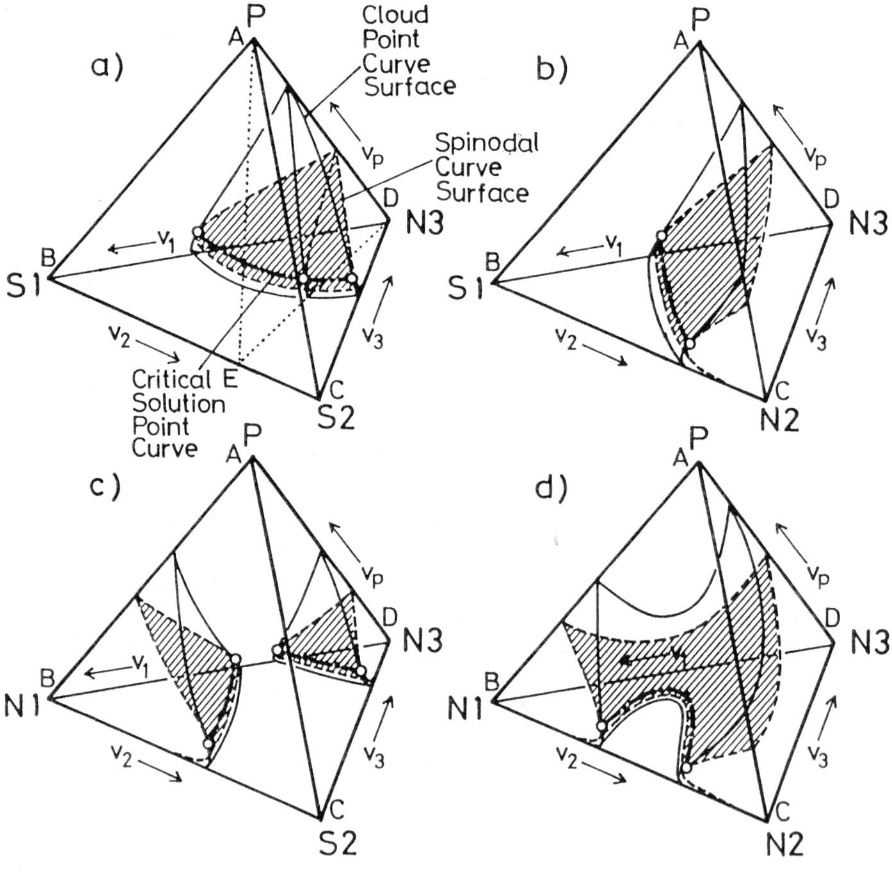

Fig. 6.72 Cloud point curve surface (enclosed by a full line), spinodal curve surface (enclosed by a broken line and hatched) and critical solution point curve (chain line) of quasi-quaternary system consisting of multicomponent polymer / solvent 1 / solvent 2 / solvent 3 system: a) solvent 1, 2=good solvent and solvent 3=nonsolvent, b) solvent 1=good solvent and solvent 2, 3=nonsolvent. c) solvent 1, 3=nonsolvent and solvent 2=good solvent, d) solvent 1, 2, 3=nonsolvent.

Whether or not experimental verification of theoretical correlation between pore shape (circular and the uncircular) and casting conditions as illustrated in Fig. 6.5 is possible depends principally on determination of the critical points of the casting solution. One of the most popular practical porous polymeric membranes materials is cellulose, regenerated from cellulose cuprammonium solution. For example, the cellulose membrane shown in Fig. 6.6 was made from the five components system consisting of cellulose, copper, ammonia, water and acetone. We denote the weight fractions of these components as w_{CELL}, w_{CU}, w_{NH3}, $w_{ACETONE}$ and w_{H2O}. Then, one obtains:

$$w_{CELL} + w_{CU} + w_{NH3} + w_{ACETONE} + w_{H2O} = 1. \tag{6.143}$$

It has been widely confirmed that in cellulose cuprammonium solution cellulose forms a complex with copper and ammonia, which dissolves in aq. ammonia (ref. 53) and the weight (or molar) ratio of cellulose to copper is nearly maintained constant ($w_{CELL}:w_{CU} \simeq 5:2$). Hence, we assume that the above dissolved state can be represented mainly by three χ parameters, such as $\chi_{CELL-CU}$, χ_{CU-NH3} and $\chi_{CELL-NH3}$ in phase separation of cuprammonium cellulose / water / acetone system.

Phase diagram of five components system (quasi-quintuple system) can be fully represented only in four dimensional space, but when the system with a constant ratio of $w_{CELL}/(w_{CELL} + w_{CU})$ is concerned, cloud point curve surface, spinodal curve surface and critical solution point curve can be drawn in regular triangular pyramid (however, shadow curve and coexistence curve can not be represented within a cross section of the constant $w_{CELL}/(w_{CELL} + w_{CU})$). Three starting cellulose (the viscosity-average degree of polymerization $X_v = 500$) cuprammonium solutions (referred hereafter to as SS1, SS2 and SS3, respectively), whose w_{CELL} are 0.05, 0.07 and 0.08, respectively, are prepared by diluting a solution having the composition of $w_{CELL} = 0.1000$, $w_{CU} = 0.0395$, $w_{NH3} = 0.0703$ and $w_{H2O} = 0.7902$ with 28wt% aq. ammonia and these solutions have always constant $w_{CELL}/(w_{CELL} + w_{CU})$ (= 0.717). In Table 6.8 are collected the composition of mixtures of the above cellulose cuprammonium solutions and aq. acetone (acetone concentration, 30~98wt%) as coagulant at the phase separation step and the phase ratio R (after phase separation) at 298± 1K.

TABLE 6.8
The composition of cuprammonium cellulose / acetone / water system at phase separation in Fig 6.72 and the two-phase volume ratio R at the points.

Acetone Conc. (wt%) in coagulant	Starting solution S_1 (Cellulose concentration 5wt%) Weight fraction					Phase volume ratio
	$W_{CELL}X10^2$	$W_{ACETONE}X10^2$	$W_{CU}X10^2$	$W_{NH3}X10^2$	$W_{H2O}X10^2$	R
(Starting soln.)	5.00	—	1.98	17.52	75.50	—
30	0.92	24.45	0.37	3.24	71.02	10.50
40	2.00	24.00	0.79	7.01	66.21	4.80
45	2.47	20.23	0.95	8.66	65.14	3.67
50	2.76	22.40	1.09	9.67	64.07	3.25
60	3.22	21.40	1.27	11.28	62.86	3.22
70	3.43	22.02	1.36	12.00	61.19	2.33
80	—	—	—	—	—	—
98	3.96	20.29	1.57	13.89	60.29	2.00

Acetone Conc. (wt%) in coagulant	Starting solution S_2 (Cellulose concentration 7wt%) Weight fraction					Phase volume ratio
	$W_{CELL}X10^2$	$W_{ACETONE}X10^2$	$W_{CU}X10^2$	$W_{NH3}X10^2$	$W_{H2O}X10^2$	R
(Starting soln.)	7.00	—	2.77	13.32	76.91	—
30	1.26	24.61	0.50	2.39	71.24	38.40
40	2.92	23.30	1.15	5.55	67.05	2.25
45	—	—	—	—	—	—
50	3.76	23.10	1.49	7.16	64.47	1.47
60	4.43	22.07	1.75	8.42	63.30	1.42
70	4.77	22.28	1.89	9.08	61.98	0.78
80	—	—	—	—	—	—
98	5.44	21.78	2.15	10.36	60.26	0.77

Acetone Conc. (wt%) in coagulant	Starting solution S_3 (Cellulose concentration 8wt%) Weight fraction					Phase volume ratio
	$W_{CELL}X10^2$	$W_{ACETONE}X10^2$	$W_{CU}X10^2$	$W_{NH3}X10^2$	$W_{H2O}X10^2$	R
(Starting soln.)	8.00	—	3.16	11.22	77.62	—
30	1.06	23.99	0.63	2.25	71.53	13.6
40	3.49	22.53	1.38	4.90	67.70	1.38
45	—	—	—	—	—	—
50	4.61	21.16	1.82	6.47	65.98	0.87
60	5.42	19.36	2.14	7.60	65.48	0.33
70	5.78	19.41	2.28	8.11	64.41	0.625
80	6.10	19.00	2.41	8.55	63.93	0.638
98	6.25	18.16	2.58	9.14	63.60	0.825

Fig. 6.73 shows the phase diagram speculated from Table 6.8. Here, $\overrightarrow{BC}$, $\overrightarrow{CD}$, $\overrightarrow{DB}$ and $\overrightarrow{DA}$ are the axes of $w_{ACETONE}$, w_{NH3}, w_{H2O} and $w_{CELL}+w_{CU}$. The distance of cloud point from the phase separation point (i.e., the depth of phase separation, Δ_P) can be assumed to be equal with each other. Here, Δ_P is defined by the relation

$$\Delta_P{}^2 \equiv [(w_{CELL}{}^{CP}+w_{CU}{}^{CP})-(w_{CELL}{}^{P}+w_{CU}{}^{P})]^2+(w_{NH3}{}^{CP}-w_{NH3}{}^{P})^2$$

$$+(w_{ACETONE}{}^{CP}-w_{ACETONE}{}^{P})^2+(w_{H2O}{}^{CP}-w_{H2O}{}^{P})^2. \qquad (6.144)$$

Compositions of cloud point and phase separation point are given by $[(w_{CELL}{}^{CP}+w_{CU}{}^{CP}),\ w_{NH3}{}^{CP},\ w_{ACETONE}{}^{CP},\ w_{H2O}{}^{CP}]$ and $[(w_{CELL}{}^{P}+w_{CU}{}^{P}),\ w_{NH3}{}^{P},\ w_{ACETONE}{}^{P},\ w_{H2O}{}^{P}]$, respectively.

The surface enclosed by the full line in Fig. 6.73 is a deduced cloud point curve surface and the surface enclosed by the broken line is a deduced spinodal curve surface. The chain line is a critical point curve. The figure shows the existence of two two-phase separation regions, resulting in cosolvency. Point E in the figure is the limit below which aq. ammonia solution can exist stably ($w_{NH3} \simeq 33$wt%) and the hereafter, phase diagram is discussed in the range of $w_{NH3}/(w_{NH3}+w_{H2O}) \lesssim 0.33$ (i.e., within the triangle prism ABCE).

Fig. 6.74a is a magnification of part of Fig. 6.73. In this figure, the data points obtained by using 40~98wt% aq. acetone as nonsolvent are shown for comparison. The phase separation point (filled mark) is near to cloud point estimated (unfilled mark) (that is, the depth of phase separation is narrow) and accordingly, Δ_P is considered to be nearly constant. The dotted line simply connects the data points, obtained by using aq. acetone with the same acetone concentration and this is considered to lie within constant Δ_P surface. By projecting a contour line of $R=1$ onto cloud point curve surface the critical solution point curve (chain line) can be determined. Fig. 6.74b shows a contour line of R in the surface of equal depth of phase separation.

Fig. 6.75 shows the line of intersection between a deduced cloud point curve surface and triangular surface AFG (see, Fig. 6.73). In the figure, the bold full line is cloud point curve and on the branch of cloud point curve, in which $w_{CELL}+w_{CU}$ is higher than the corresponding critical concentration, R is always zero and on the branch, in which $w_{CELL}+w_{CU}$ is less than the corresponding critical concentration, R is infinite. The fine full

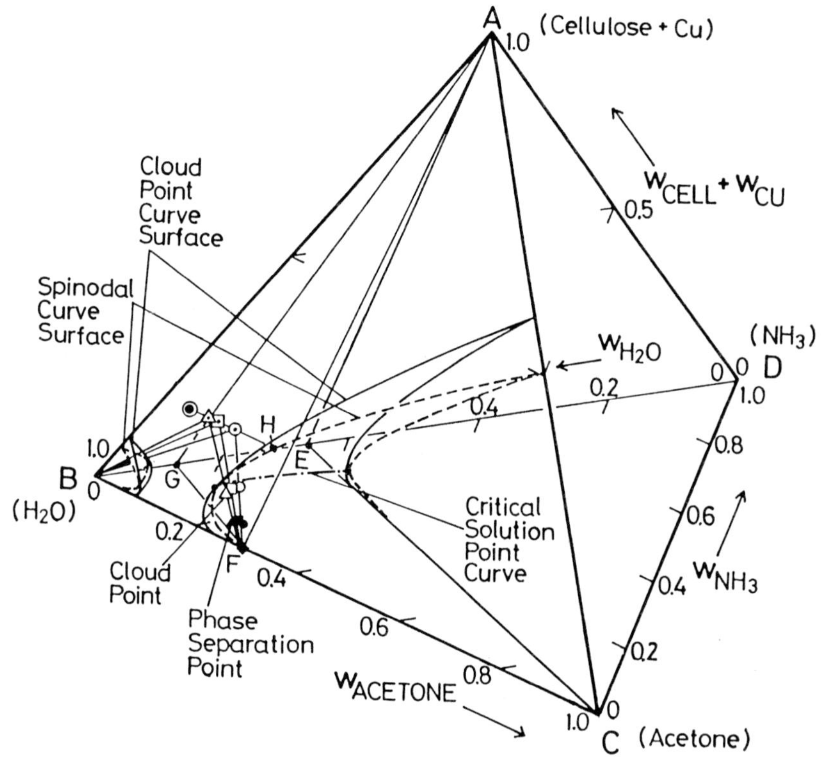

Fig. 6.73 Cloud point curve surface (enclosed by full line), spinodal curve surface (enclosed by broken line) and critical solution point curve (chain line) of cuprammonium cellulose / acetone / water system: ◉ cuprammonium cellulose ($X_v = 500$) solution, [$w_{CELL}{}^S = 0.10$]; ⊙ starting solution, S_1 [$w_{CELL}{}^S = 0.050$]; ⊡ starting solution, S_2 [$w_{CELL}{}^S = 0.070$]; ◬ starting solution, S_3 [$w_{CELL}{}^S = 0.080$]; ●, ■, ▲ phase separation point when 30wt% aq. acetone (point F) [$w_{ACETONE} = 0.30$, $w_{H_2O} = 0.70$] is used as coagulant; ○, □, △ cloud points; A, cellulose (the viscosity-average degree of polymerization $X_v = 500$) + Cu; B, water; C, acetone; D, ammonia; E, 33wt% aq. ammonia; G, 12.6wt% aq. ammonia; H, 28wt% aq. ammonia.

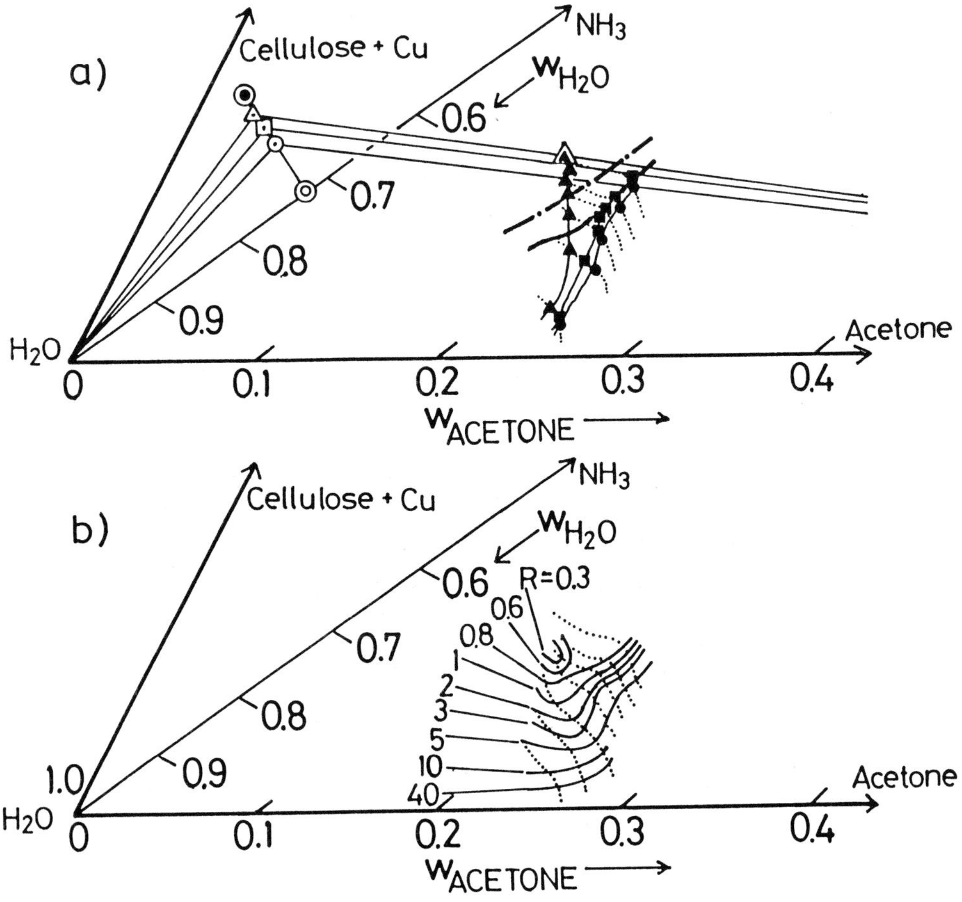

Fig. 6.74 Phase diagram of cuprammonium cellulose / acetone / water system: a) Starting solution; S_1, ⊙ [$w_{CELL}{}^S = 0.05$]; S_2, ⊡ [$w_{CELL}{}^S = 0.07$]; S_3, ◬ [$w_{CELL}{}^S = 0.08$]; ●, ■, ▲ phase separation points from the starting solutions with $w_{CELL}{}^S = 0.05$, 0.07 and 0.08, respectively; ◉ cellulose ($X_v = 500$) cuprammonium solution [$w_{CELL}{}^S = 0.10$]; ◎ aq. ammonia solution [$w_{NH_3} = 0.28$, $w_{H_2O} = 0.78$]; chain line, critical solution point curve; full line, contour line of phase volume ratio R = 1; b) contour line of R on the surface of equal depth of phase separation (Δ_P = const., full line); dotted line, linked line of the phase separation point, obtained when aq. acetone coagulant with the same acetone concentration is utilized.

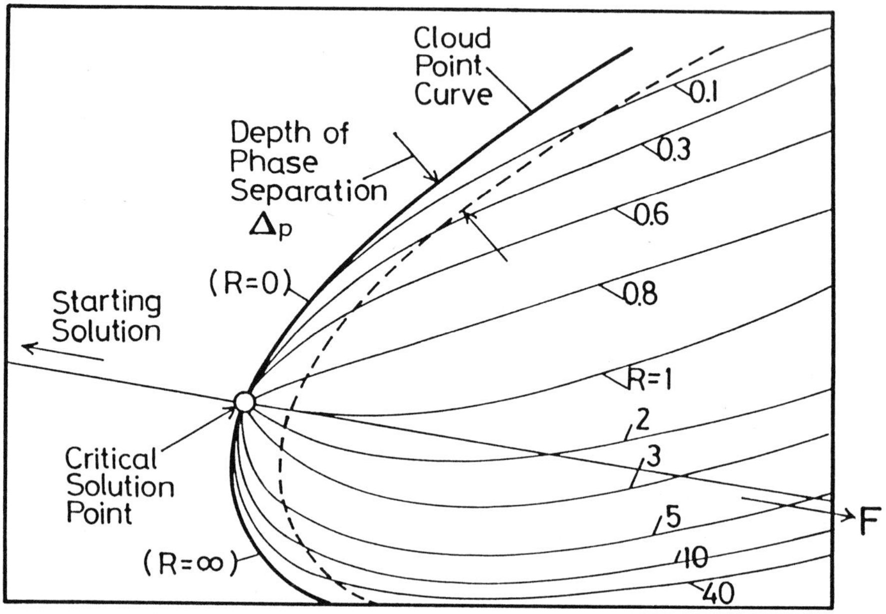

Fig. 6.75 Cross section of AFG surface [aq. acetone coagulant ($w_{ACETONE}=0.30$, $w_{H2O}=0.70$)] of cloud point curve surface and critical solution point curve in Fig. 6.73: Bold full line, cloud point curve; fine full line, contour line of phase volume ratio R ($=0.1\sim40$); broken line, equal phase separation depth Δ_P curve; unfilled circle, critical solution point; filled circle, intersecting point of contour R curve and contour Δ_P curve.

line is an estimated contour line of R and the dotted line is an estimated contour line of Δ_P. The cross points of these two contour lines are shown by a full circle. By the experiments (Table 6.8) the phase ratio at the cross point can be determined. Projection of phase separation point at $R=1$ onto the cloud point curve yields the critical solution point.

Fig. 6.76 shows the calculation of the cloud point curve (χ_0^{CP} vs. v_P^0 curve; bold full line) and the contour line of R (χ_0^P vs. v_P^0 curve at constant R; fine full line) of quasi-binary system consisting of multicomponent polymer and a single solvent. Here, the Schulz-Zimm type molecular weight distribution [see, eqns (2.59)-(2.61)] is assumed for the polymer (the weight-average molar volume ratio of polymer to solvent, $X_w^0=2117$ and

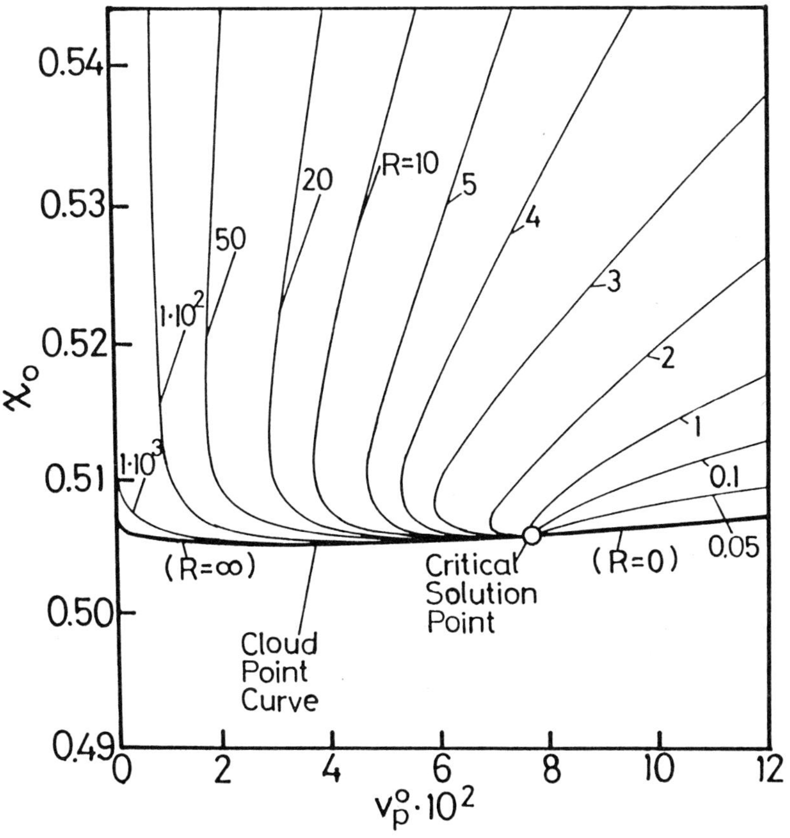

Fig. 6.76 Comparison of the polymer-solvent interaction parameter at infinite dilution χ_0 [see, eqn (2.4)] at cloud point (χ_0^{CP}) vs. v_p^0 curve (broad solid line) with χ_0 vs. v_p^0 curve at constant phase volume ratio R line (fine solid line): Original polymer, Schulz-Zimm type distribution ($X_w^0=2117$, $X_w^0/X_n^0=2.8$); $p_1=0.643$, $p_2=0.200$, $k_0=0$ (ref. 35); unfilled circle, critical solution point.

$X_w^0/X_n^0=2.8$). The calculations are carried out under the conditions of $p_1=0.643$, $p_2=0.200$ ($p_3=p_4=\cdots=0$) and $k_0=0$. The critical solution point is shown as an unfilled circle. The estimated contour line of R in Fig. 6.75 agrees fairly with the results of rigorous computer experiments shown in Fig. 6.76.

Fig. 6.77 shows the relations between the two-phase volume ratio R and acetone weight fraction $w_{ACETONE}$ when 70wt% aq.

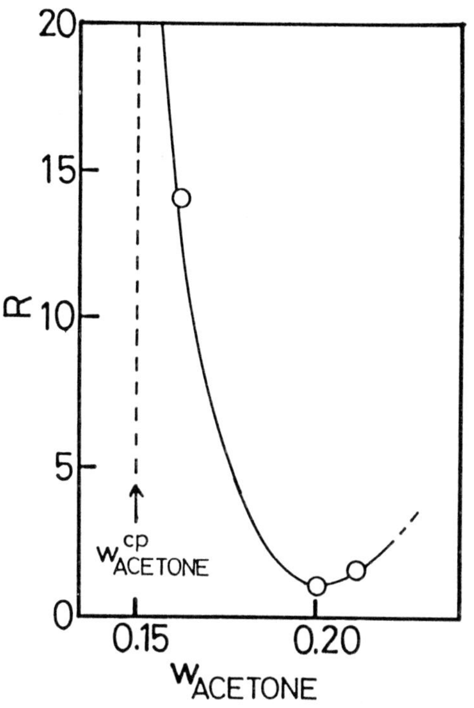

Fig. 6.77 Change in the phase volume ratio R with the weight fraction of acetone $w_{ACETONE}$ in cellulose cuprammonium solution / acetone system: Here, to the cellulose cuprammonium solution ($w_{CELL}{}^{S}=0.090$, $w_{CU}{}^{S}=0.0356$, $w_{NH3}{}^{S}=0.0913$ and $w_{H2O}{}^{S}=0.7831$), aq. acetone solution ($w_{ACETONE}=0.70$, $w_{H2O}=0.30$) was added as a coagulant at 298K; estimated compositions of cloud point (denoted by an arrow in the figure), ($w_{CELL}{}^{CP}=0.0707$, $w_{CU}{}^{CP}=0.0280$, $w_{NH3}{}^{CP}=0.0717$, $w_{H2O}{}^{CP}=0.6796$, $w_{ACETONE}{}^{CP}=0.15$).

acetone is added slowly to cellulose cuprammonium solution ($w_{CELL}{}^{S}=0.090$, $w_{Cu}{}^{S}=0.0356$, $w_{NH3}{}^{S}=0.0913$, $w_{H2O}{}^{S}=0.7831$, the superscript S means the starting solution) at 298K in order to cause the phase separation. The broken line in the figure is cloud point, whose composition is as follows; $w_{CELL}{}^{CP}=0.0707$, $w_{CU}{}^{CP}=0.0280$, $w_{NH3}{}^{CP}=0.0717$, $w_{H2O}{}^{CP}=0.6796$ and $w_{ACETONE}{}^{CP}=0.15$, the superscript CP means cloud point. R becomes infinite at cloud point. In this case, phase separation proceeds through the region below the critical point curve as shown in Fig. 6.73 or Fig. 6.74. With an increase in $w_{ACETONE}$, R decreases and then increases after passing through the minimum R_{min} at $w_{ACETONE} \simeq 0.20$. Similar behavior is also observed in the phase separation of a multicomponent polymer / single solvent systems (Fig. 2.15) and a multicomponent polymer / binary solvent mixture systems (Figs 3.16 and 3.18). Using eqn (6.144) we can calculate the depth of phase separation Δ_P: From composition of the solutions at R_{min} ($w_{CELL}{}^P = 0.0643$, $w_{CU}{}^P = 0.0254$, $w_{NH_3}{}^P = 0.0652$, $w_{H_2O}{}^P = 0.6451$ and $w_{ACETONE}{}^P = 0.2000$, the superscript P means phase separation point) Δ_P is estimated to be 0.0615. There is a possibility that using R, estimated from the porosity of dry membranes ($Pr(d_3)$; see, eqn (6.105)) and the composition of the solution at phase separation, Δ_P at the coagulation step can be evaluated from Fig. 6.75.

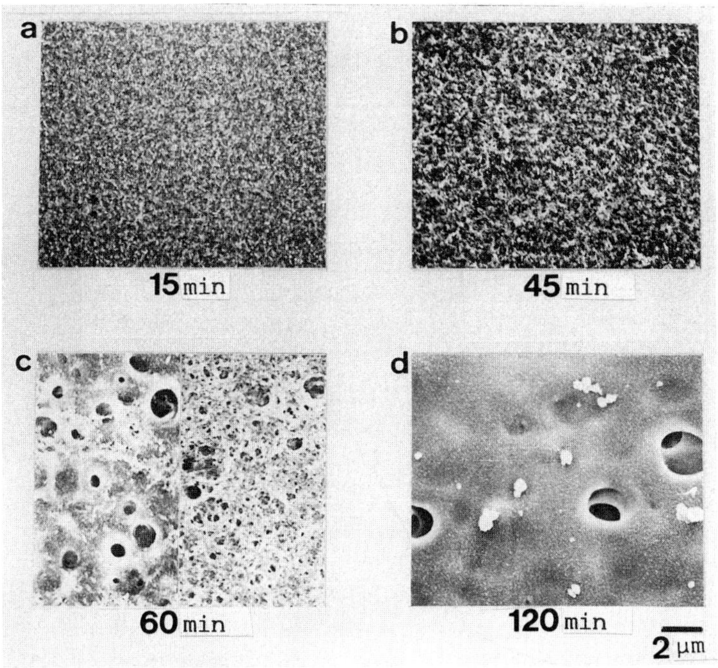

Fig. 6.78 Scanning electron micrographs of the front surface of regenerated cellulose membranes prepared from cellulose cuprammonium solution by contacting with acetone vapor phase (the dry method): Composition of atmosphere, acetone/H_2O = 98/2 (wt/wt); composition of cellulose (the viscosity-average degree of polymerization, $X_v = 620$) casting solution, ($w_{CELL} = 0.0800$, $w_{CU} = 0.0316$, $w_{NH_3} = 0.112$, $w_{H_2O} = 0.776$); a) 15min; b) 45min; c) 60min; d) 120min after casting; phase separation temperature T_P, 298 ± 2K; c) shows two different locations; a scale bar stands for 2µm.

When the cellulose cuprammonium aq. solution of concentration $w_{CELL} = 0.08$ is placed in an atmosphere of acetone vapor, phase separation occurs at a point denoted by ▲ in Fig. 6.74a, along with line and throughout the process w_{CELL} is always higher than the critical concentration (ref. 54). The gel membrane thus prepared was regenerated after various casting times and dried.

Fig. 6.78 shows electron micrographs of the membrane surfaces cast using different time periods (15, 45, 60 and 120min) and thereafter, regenerated in aq. sulfuric acid and dried. Evidently circular pores appear after 60min and in addition they gradually grow with an elapse of casting time.

Fig. 6.79 shows time-dependence of mean pore size r_n of circular pores, evaluated from photographs in Fig. 6.78. Growth of circular pores is explicitly slow as compared with that of non-circular pores.

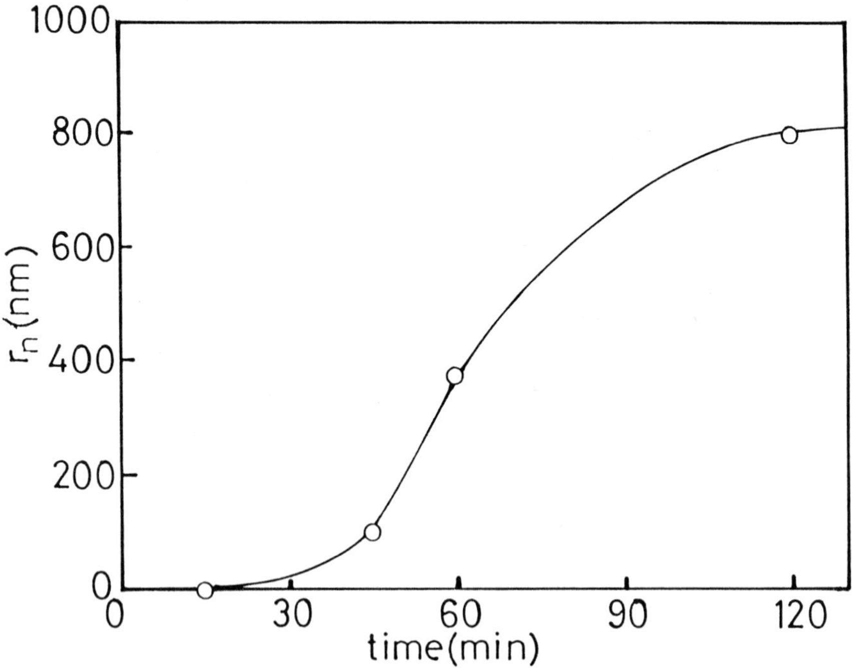

Fig. 6.79 Relationship between mean pore size r_n of the membrane surface and phase separation time (step g-i in Fig. 6.45): Regenerated cellulose membranes, prepared from starting cellulose cuprammonium solution with cellulose concentration $w_{CELL}^{S} = 0.08$ coagulated in an atmosphere of acetone vapor (dry method): In the case, phase separation proceeds at w_{CELL} higher than critical cellulose concentration. Phase separation temperature $T_P = 298 \pm 2K$.

Mechanical properties, such as tensile strength and tensile elongation of membranes, which are the most important characteristics controlling the easiness of handling and the condition of practical operations in membrane separation, are strongly correlated with pore characteristics.

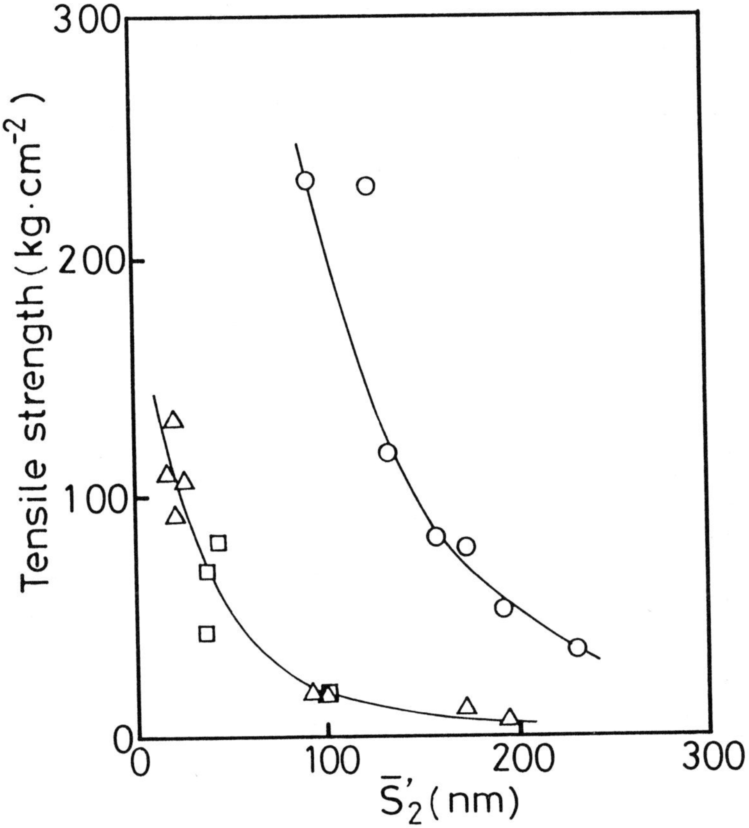

Fig. 6.80 Relations between tensile strength and mean-radius of secondary particles $\bar{S}_2'$ of cellulose and its derivative dry membranes: ○ regenerated cellulose; □ cellulose nitrate; △ cellulose mixed esters (probably, nitrate and acetate; regenerated cellulose membranes were prepared on laboratory scale by Kamide et al. and all cellulose derivative membranes were commercial products.

Fig. 6.80 shows the relations between the tensile strength and the mean radius of secondary particles $\bar{S}_2'$ on the surface of regenerated cellulose membranes prepared in our laboratory and commercially available cellulose derivative membranes, which consist of secondary particles and in consequence have, without exception, uncircular pores (ref. 55). All data points can be classified into two master curves; one for the regenerated cellulose and another for the cellulose derivatives (nitrate and acetate). Each curve indicates that the tensile strength decreases with increasing $\bar{S}_2'$ and cellulose membranes have much stronger tensile strength, compared at the same $\bar{S}_2'$, than cellulose derivative membranes and this can be reasoned mainly in terms of the degree of development of hydrogen bonding between the secondary particles: In cellulose membranes hydrogen bondings are developed well but, in cellulose derivative membranes, hydrogen

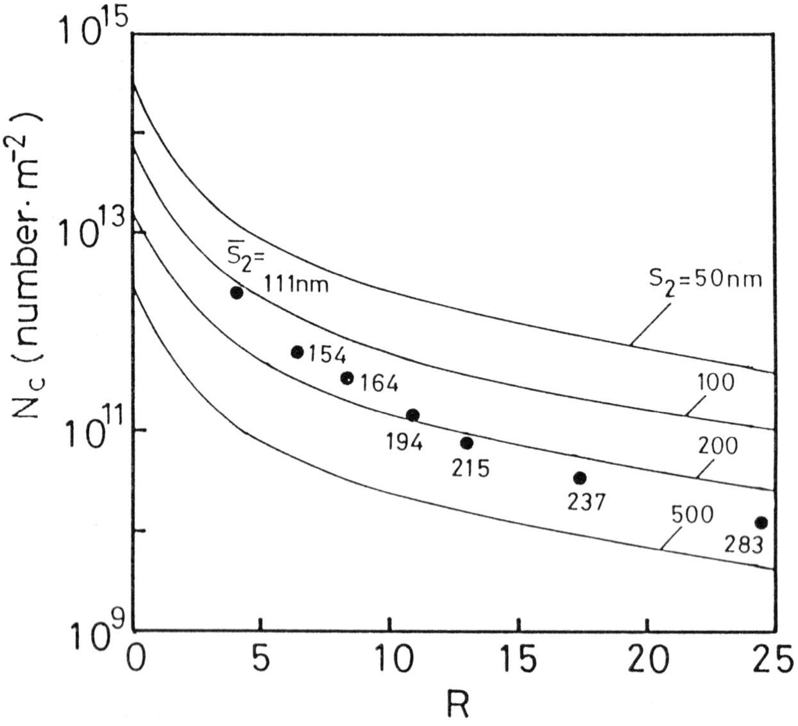

Fig. 6.81 Dependence of the total number of contacts between the nearest neighbor secondary particles per unit surface of membrane, N_C on the two-phase volume ratio R and the radius of secondary particles S_2: Full line, computer simulations; filled circles, experiments on regenerated cellulose membrane; number on circle means $\bar{S}_2'$(nm).

bondings are, more or less, less developed because some of the three hydroxyl groups in the gluco-pyranose rings are substituted in cellulose derivatives.

The dependence of tensile strength on S_2 is qualitatively considered to be strongly governed by the total numbers of contacts between nearest secondary particles per unit surface of the membrane N_C. N_C was calculated by computer experiments for a given combination of R and S_2.

Fig. 6.81 shows the semi-log plots of N_C against two-phase volume ratio R relations for various S_2. N_C decreases with an increase in S_2 and R. In the figure, the experimental data points for regenerated cellulose membranes are also included as filled circle. Here, the mean radius of secondary particles of the wet gel membrane $\bar{S}_2'$, converted from $\bar{S}_2$ for dry membrane using eqn (6.80), is indicated on curve. The experimental data are shown as

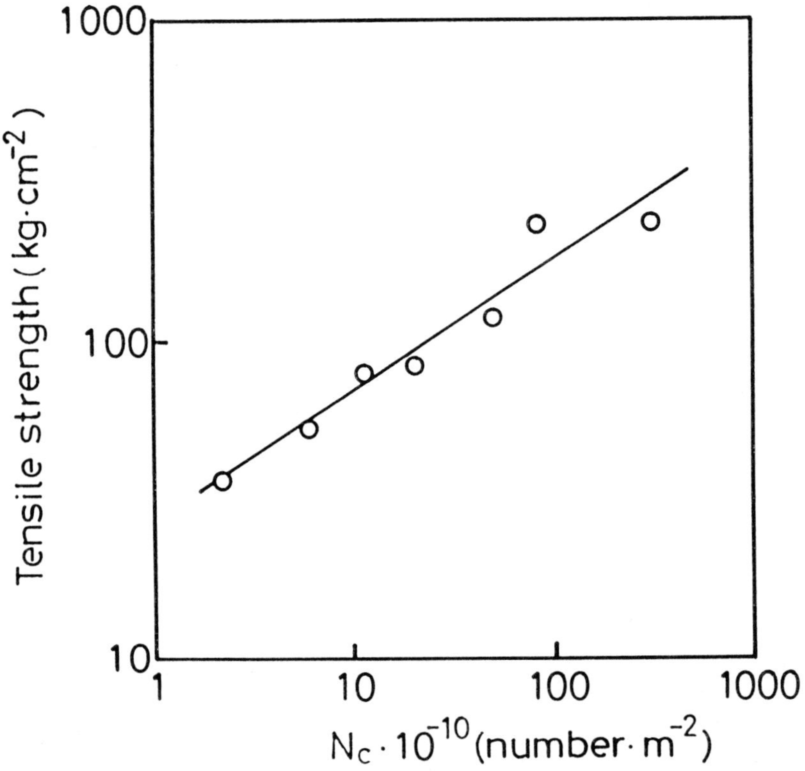

Fig. 6.82 Dependence of tensile strength of porous cellulose membranes on the total number of contacts between the nearest neighbor secondary particles N_C.

filled circles. Numbers on circles are S_2 in nm. It is clear that this series of membranes are produced under wide conditions such as low R-small S_2 to high R-large S_2.

Fig. 6.82 shows the relations between the tensile strength and the total number of contacts of the nearest neighbor secondary particles per unit area N_C of regenerated cellulose membranes. Thus, it is experimentally confirmed that the tensile strength of porous membranes can be interpreted in terms of N_C, if the membranes are composed of the same materials.

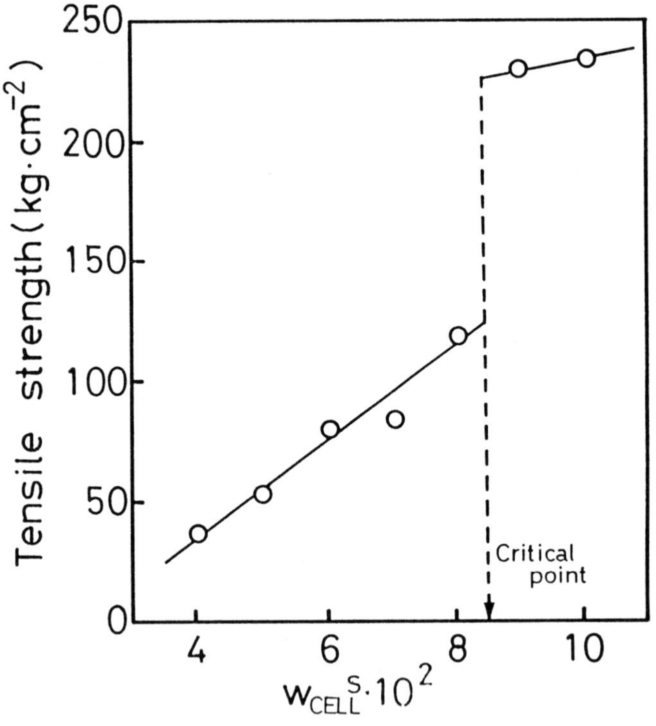

Fig. 6.83 Dependence of tensile strength of regenerated cellulose membranes on the weight fraction of cellulose w_{CELL} of casting solution (cellulose cuprammonium solution:
$w_{CELL}{}^S=0.04$, $w_{CU}{}^S=0.0158$, $w_{NH3}{}^S=0.1961$, $w_{H2O}{}^S=0.7481$;
$w_{CELL}{}^S=0.05$, $w_{CU}{}^S=0.0198$, $w_{NH3}{}^S=0.1752$, $w_{H2O}{}^S=0.7550$;
$w_{CELL}{}^S=0.06$, $w_{CU}{}^S=0.0237$, $w_{NH3}{}^S=0.1541$, $w_{H2O}{}^S=0.7622$;
$w_{CELL}{}^S=0.07$, $w_{CU}{}^S=0.0277$, $w_{NH3}{}^S=0.1332$, $w_{H2O}{}^S=0.7691$;
$w_{CELL}{}^S=0.08$, $w_{CU}{}^S=0.0316$, $w_{NH3}{}^S=0.1122$, $w_{H2O}{}^S=0.7762$;
$w_{CELL}{}^S=0.09$, $w_{CU}{}^S=0.0356$, $w_{NH3}{}^S=0.0913$, $w_{H2O}{}^S=0.7831$;
$w_{CELL}{}^S=0.10$, $w_{CU}{}^S=0.0395$, $w_{NH3}{}^S=0.0703$, $w_{H2O}{}^S=0.7902$):
Composition of coagulant, $w_{ACETONE}=0.3$, $w_{NH3}=0.0056$ and $w_{H2O}=0.6944$.

Fig. 6.83 shows the plot of the tensile strength of cellulose membrane, cast and regenerated from cuprammonium cellulose system using aq. acetone as coagulant, against the weight fraction of cellulose w_{CELL}^{S} of the cast solution (i.e., starting solution). As w_{CELL}^{S} increases, the tensile strength increases approximately linearly up to the critical solution point (in this case, $w_{CELL}^{C} \simeq 0.085$) and in the vicinity of the critical point the tensile strength increases discontinuously. Above the critical point the tensile strength increases gradually with the cellulose concentration. The dramatical change in the morphology of membranes cast at the critical solution point (see, Fig. 6.83) may be, of course, closely correlated with the dependence of tensile strength on the cellulose concentration; in the range $w_{CELL} >$ the critical point (w_{CELL}^{C}) the membrane can be approximated with bulk thin film with punched circular pores, which yields stronger mechanical properties, as compared with a composite membrane of small particles (see, Figs 6.64 and 6.83).

6.5.2 Poly(acrylonitrile / methylacrylate) copolymer / aq. nitric acid solution

Polyacrylonitrile(PAN) or acrylonitrile(AN) / methylacrylate(MA) copolymer are also frequently employed as a starting materials for polymer membranes. Since these kind of polymer dissolves in polar or ionic, but is a relatively simple solvent, it is quite easy to correlate the pore shape of membranes with a phase diagram for polymer solutions (ref. 56).

Fig. 6.84 shows a phase diagram of a quasi-ternary system consisting of AN/MA (92/8 (wt/wt), the weight-average molecular weight $M_w = 13.7 \times 10^4$) in aq. nitric acid at 273 ± 0.1K. $\overrightarrow{AB}$, $\overrightarrow{BC}$ and $\overrightarrow{CA}$ are axes of H_2O, HNO_3 and AN/MA copolymer, respectively, whose weight fractions are denoted by w_{H_2O}, w_{HNO_3} and $w_{AN/MA}$. The composition of the starting solution (point F in the figure) is ($w_{H_2O}^{S}$, $w_{HNO_3}^{S}$, $w_{AN/MA}^{S}$) = (0.277, 0.563, 0.16). Here, 67wt% aq. nitric acid (point G in the figure) is used as a binary mixture to form a starting solution. Aq. nitric acid becomes nonsolvent when w_{HNO_3} is relatively low. To the starting solution at point F aq. nitric acid ($w_{HNO_3} = 0.36$ (point D) or 0.42 (point E)) is added dropwisely to cause two phase separation.

Table 6.9 summarizes the compositions of cloud point ($w_{H_2O}^{CP}$, $w_{HNO_3}^{CP}$, $w_{AN/MA}^{CP}$) and of polymer-rich phases (points H and I for

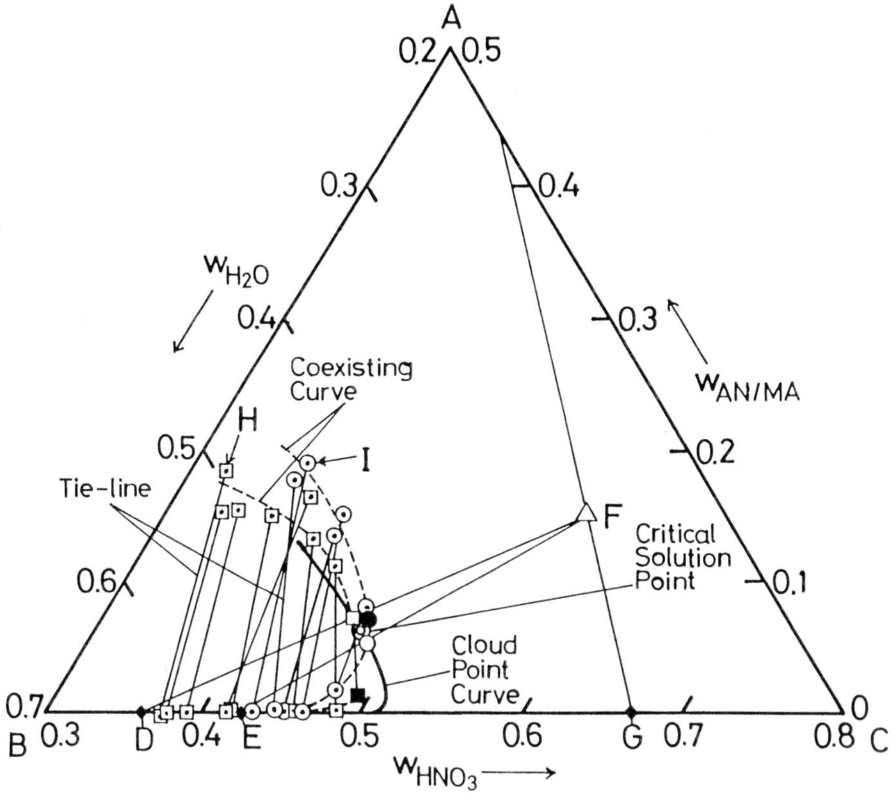

Fig. 6.84 Phase diagram of poly (acrylonitrile / methylacrylate) (AN/MA = 92/8, wt/wt) copolymer (the weight-average molecular weight, $M_w = 13.7 \times 10^4$) / water / nitric acid solutions: Phase separation temperature $T_P = 273 \pm 0.1$K; point F, starting solution ($w_{AN/MA}{}^S = 0.16$, $w_{H_2O}{}^S = 0.277$, $w_{HNO_3}{}^S = 0.563$); □ and ○, cloud points; ■ and ●, shadow points; ⊡ and ⊙, composition of coexisting phases (i.e., polymer-lean and -rich phases) after phase separation for aq. nitric acid coagulant with $w_{HNO_3} = 0.36$ and 0.42, respectively. Tie-line and coexistence curve are represented by the fine full line and the broken line.

Table 6.9
Composition of cloud point and phase-separated gel (polymer-rich phase) of acrylonitrile(AN) / methylacrylate(MA) (AN/MA=92/8, wt/wt, $M_w=13.7\times10^4$) system at 273±0.1K.

HNO_3 conc. of coagulant	Cloud point			Phase-separated gel			Phase volume
(H_2O/HNO_3 mixed nonsolvent)	$w_{H_2O}^{CP}$	$w_{HNO_3}^{CP}$	$w_{AN/MA}^{CP}$	$w_{H_2O(2)}$	$w_{HNO_3(2)}$	$w_{AN/MA(2)}$	ratio (R)
36wt%(D in Fig. 6.84)	0.463	0.457	0.080	0.50	0.32	0.18 (H)	0.67
42wt%(E in Fig. 6.84)	0.464	0.481	0.055	0.44	0.37	0.19 (I)	5.67

Starting solution: $w_{H_2O}^{S}=0.277$, $w_{HNO_3}^{S}=0.563$, $w_{AN/MA}^{S}=0.16$.
Critical point: $w_{H_2O}^{C}=0.4675$, $w_{HNO_3}^{C}=0.4675$, $w_{AN/MA}^{C}=0.065$.

aq. nitric acid coagulants with $w_{HNO_3} = 0.36$ and 0.42, respectively) ($w_{H_2O(2)}$, $w_{HNO_3(2)}$, $w_{AN/MA(2)}$) and phase volume ratio R at phase separation points. In Fig. 6.84 symbols ■ and ● are the estimated shadow points and the fine full line connecting the cloud point (□ and ○) and the shadow point (■ and ●) is a limiting tie-line (fine full line), corresponding to $w_{HNO_3} = 0.36$ and 0.42, respectively. The cloud point curve is a bold full line in the figure. Critical solution point for this system (as denoted by ◎ in the figure) is estimated to be $w_{H_2O}{}^c = 0.4675$, $w_{HNO_3}{}^c = 0.4675$ and $w_{AN/MA}{}^c = 0.065$, respectively. When a starting

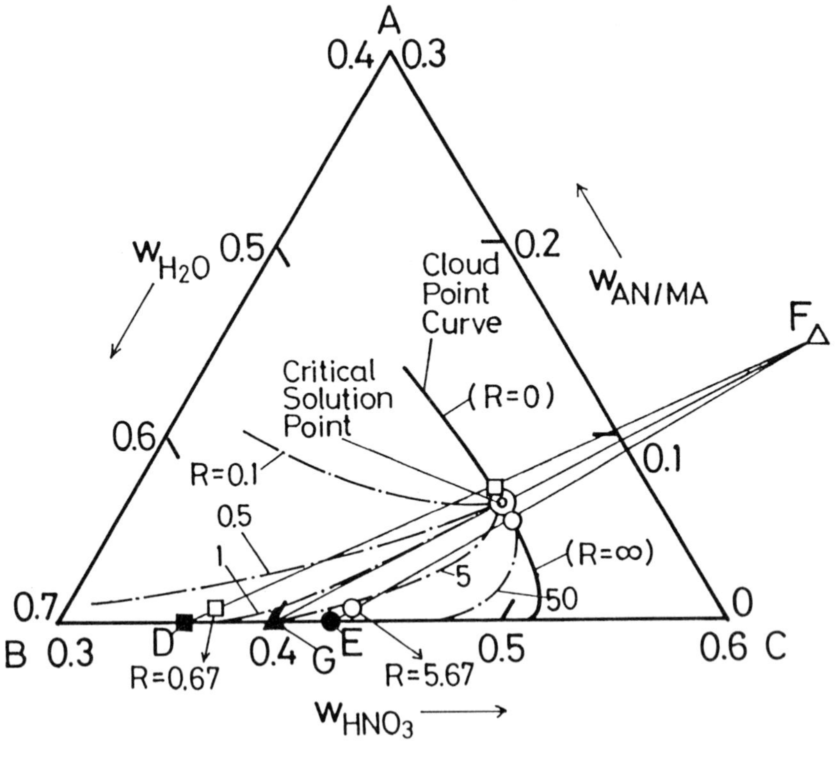

Fig. 6.85 Cloud point curve (broad full line) of poly(acrylonitrile / methylacrylate) (AN/MA = 92/8, wt/wt) copolymer (the weight-average molecular weight, $M_w = 13.7 \times 10^4$) / water / nitric acid solutions: Chain line, contour line of phase volume ratio R; ◎ critical solution point; fine full line, phase separation line connecting the compositions of the starting solution [F(△)] and aq. nitric acid coagulants [D(■), E(●) and G(▲)]. Number on curve means R value.

solution at point F is mixed with a binary mixture of H_2O / HNO_3 with $w_{HNO_3} = 0.40$ (i.e., 40wt% aq. nitric acid) phase separation occurs passing through the critical point.

Fig. 6.85 shows the contour line of R, estimated using the data in Table 6.9, together with the cloud point curve (Fig. 6.85) for AN/MA copolymer dissolved in aq. nitric acid at 273± 0.1K. The contour line of $R = 1$ passes through the critical point as the theory predicts. In the case where w_{HNO_3} of a nonsolvent mixture is less than 0.40 (i.e., $w_{HNO_3} < 0.40$), polymer weight fraction at the cloud point $w_{AN/MA}^{CP}$ is always higher than the critical polymer weight fraction $w_{AN/MA}^{C}$ (i.e., $w_{AN/MA}^{CP} > w_{AN/MA}^{C}$). So that the solution (starting from F) mixed with binary mixture of H_2O / HNO_3 coagulant with $w_{HNO_3} < 0.40$ always passes through cloud point curve branch of $w_{AN/MA}^{CP} > w_{AN/MA}^{C}$ (and $R = 0$ at cloud point curve) and enters into a two-phase equilibrium region, in which $R < 1$ is attained for small depth of phase separation (i.e., near cloud point curve). Reversely, in the case where w_{HNO_3} of nonsolvent mixture is higher than 0.40 (i.e., $w_{HNO_3} > 0.40$), $w_{AN/MA}^{CP} < w_{AN/MA}^{C}$ so that the solution (starting from F) mixed with binary coagulant mixture passes through the cloud point curve branch of $w_{AN/MA}^{CP} < w_{AN/MA}^{C}$ (and $R = \infty$ at cloud point curve) and enters into a two-phase equilibrium region of $R > 1$.

AN/MA copolymer membranes were prepared by coagulating the solution at point F $[(w_{H_2O}^{S}, w_{HNO_3}^{S}, w_{AN/MA}^{S}) = (0.277, 0.563, 0.160)]$ in Fig. 6.84 (or in Fig. 6.85) with 36, 39 and 45wt% aq. nitric acid coagulants at 273± 0.1K. When the concentration of nitric acid in the coagulant is 36 or 39wt%, the phase separation proceeds through the cloud point, on which $w_{AN/MA}^{CP} > w_{AN/MA}^{C}$ holds and when the concentration is 45wt%, the phase separation proceeds through the cloud point of $w_{AN/MA}^{CP} < w_{AN/MA}^{C}$.

Fig. 6.86 shows scanning electron micrographs of the surface of the AN/MA copolymer gel membranes. The surface of the gel membrane coagulated with a 36wt% aq. nitric acid is smooth and highly dense and that of the gel coagulated with 39wt% aq. nitric acid has numerous circular pores with an average diameter of ca.150nm. In contrast to this, the surface of gel membrane coagulated with 45wt% aq. nitric acid has noncircular pores formed by the secondary particles with average diameter $S_2 =$ ca.500nm.

Fig. 6.87 shows bright field image of transmitted electron microscopy of the cross section of coagulated AN/MA copolymer gel used in Fig. 6.86. Large capillary holes are formed in the inner part of the gel membrane when 36wt% aq. nitric acid is employed

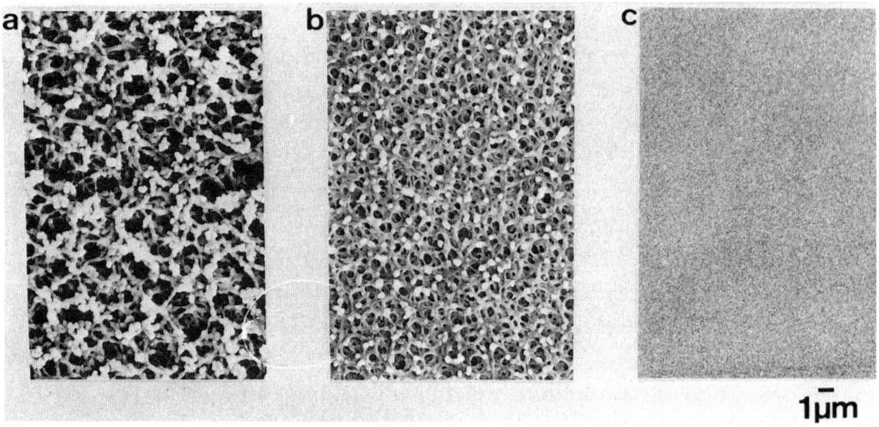

Fig. 6.86 Morphology of coagulated gel of acrylonitrile(AN) / methylacrylate(MA) copolymer (AN/MA=92/8, wt/wt; the weight-average molecular weight $M_w=13.7\times10^4$)-67wt% aq. nitric acid solution ($w_{AN/MA}{}^S=0.160$, $w_{HNO_3}{}^S=0.563$, $w_{H_2O}{}^S=0.277$) coagulated by aq. nitric acid with various concentrations: a) 45wt%; b) 39wt%; c) 36wt% aq. nitric acid; a scale bar stands for 1μm.

as coagulant. In this case the particles of a polymer-lean phase, brought about by phase separation, amalgamate to create capillary holes. Small pores, which are tracks of the particles of a polymer-lean phase independently isolated, are observed in the inner part of the membrane when 39wt% aq. nitric acid is used as coagulant. In contrast, no pores were observed in the inner part of the membrane coagulated with 45wt% aq. nitric acid. In this manner, the pore structure of the membrane can be quite reasonably interpreted by phase separation thermodynamics of polymer solutions.

Gel membranes as shown in Figs 6.86 and 6.87 are equivalent to an intermediate in the fiber spinning process. Fig. 6.88 shows scanning electron micrographs of a) the membrane prepared by coagulating a solution of AN/MA copolymer ($M_w=13.7\times10^4$), dissolved in 67wt% aq. nitric acid, with a coagulant of 42wt% nitric acid, b) the membrane, obtained by stretching as-coagulated membrane [a)] to the ratio of 3 in a water bath at 318K and c) the membrane, obtained by stretching as-coagulated membrane [a)] to the ratio of 8 in a water bath at 318K. Secondary gel particles,

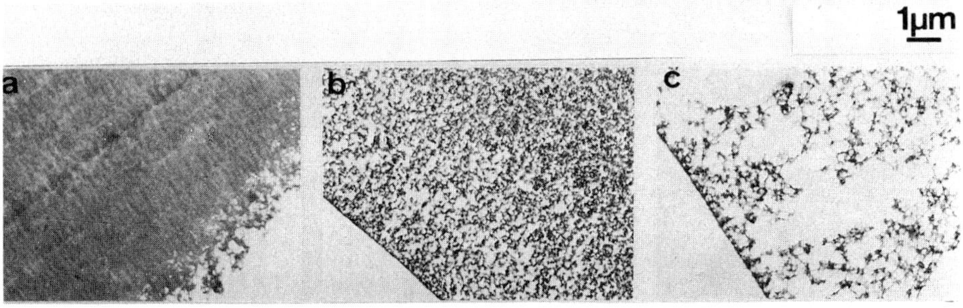

Fig. 6.87 Bright field images of transmission electron micrographs of a cross section of coagulated gel of acrylonitrile(AN) / methylacrylate(MA) copolymer (AN/MA=92/8, wt/wt, weight-average molecular weight, $M_w=13.7\times10^4$, polymer concentration, 16wt%)-67wt% aq. nitric acid solution ($w_{AN/MA}{}^{S}=0.160$, $w_{HNO_3}{}^{S}=0.563$, $w_{H_2O}{}^{S}=0.277$), coagulated by aq. nitric acid with various concentrations: a) 45wt%, b) 39wt%, c) 36wt% aq. nitric acid; a scale bar stands for 1µm.

observed in non-stretched, as-coagulated gel membrane, are drawn, finally to yield microfibrils, fully oriented parallel to the fiber axis. This is a characteristic of fiber structures.

Depending upon whether the polymer solution is phase separated above or below its critical solution point the supermolecular structure of the fiber prepared varies remarkably. With an ordinary wet spinning methods used for industrial purpose in order to achieve adequate spinnability and productivity, the fiber is spun, passing through a route above critical point (procedure A; line FD in Fig. 6.85). In this case, the polymer-lean phase is the dispersed phase and because of the strong coagulating force of the coagulating bath, a thin skin layer having a thickness of 0.1µm to several µm is instantly formed and desolvation then proceeds by dialysis through this skin layer, resulting in numerous voids inside the fiber. In addition, the skin layer inhibits dispersion of dye molecules at the dyeing step. It is also thought that this skin layer causes a reduction in the physical properties of the fiber, such as softness. Furthermore, the presence of a voids often results in defects of the physical properties, such as occurrences of a devitrification phenomenon, lack of softness and reduction of dyeability (see, for example, ref. 57). In contrast to this, when the fiber is spun, passing through a route below the critical point (procedure B; line FE in Fig. 6.85), the coagulated gel, which has no skin

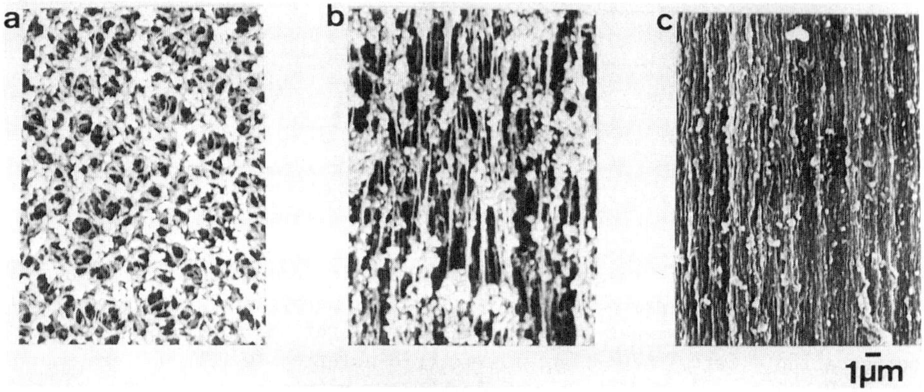

Fig. 6.88 Morphological change of gel particles (in this case, the secondary particles) in the drawing process of acrylonitrile(AN) / methylacrylate(MA) copolymer (AN/MA=92/8, wt/wt); the weight-average molecular weight, $M_w = 13.7\times10^4$: Casting solution ($w_{AN/MA}{}^S = 0.160$, $w_{HNO_3}{}^S = 0.563$, $w_{H_2O}{}^S = 0.277$): Coagulating conditions; coagulating reagents, 42wt% aq. nitric acid, 273.15K, 30min: Drawing conditions, 42wt% aq. nitric acid, 318K. a) Immediately after coagulation, b) ×3 stretched, c) ×8 stretched: All samples were washed with water and acetone after coagulation [a)] or drawn [b) and c)] and dried in vacuo; a scale bar stands for 1μm.

structure, consists of a larger number of small gel particles, like the secondary particles and the voidless, closely packed fibers are formed by drawing gel particles with quick desolvation.

The effect of phase separation conditions on the supermolecular structure and physical properties of fibers will be demonstrated briefly below (ref. 58).

A copolymer comprising 92wt% of acrylonitrile and 8wt% of methylacrylate was dissolved in an aq. 67wt% solution of nitric acid at 273K to obtain a spinning dope having a polymer concentration of 16wt%. The dope was extruded into a coagulating bath at 278K through a spinneret having 100 orifices, each having a diameter of 0.7mm. In the procedure A (line FD in Fig. 6.84 or 6.85), a coagulating bath consisted of 36wt% nitric acid and the undrawn fiber was washed with water, then subjected to draw at ratio of 8 in hot water at 369K. In the procedure B (line FE in Fig. 6.84 or 6.85), a coagulating bath contained 42wt% nitric acid and the coagulated fiber was taken out at a draw ratio of 8 at 343K. Subsequently, drawn fiber was washed with water and then drawn at a draw ratio of 1.5 in hot water at 369K, sufficiently dried in hot air at 403K. Fiber produced by procedure B has a surface consisting of particulate and/or microfibrillar structures having a width of 0.01 to 0.3μm and a length of 0.05 to 0.5μm and fibrillar structures formed by aggregation of stretched secondary particulate and/or microfibrillar structures having a width of 0.3 to 0.4μm and a length of at least 90μm (Fig. 6.89). However, fiber produced by procedure A (conventional procedure) has no such feature.

Table 6.10 illustrates some of the physical properties of the AN/MA copolymer filaments, prepared by the above two methods, and of carpets made from these fibers. Evidently, AN/MA copolymer filaments, prepared by procedure B, have superior properties over that by procedure A, in particular the former have shown a significant improvement in durability, abrasion resistance, fibrillation resistance and compression elastic recovery. In addition, the occurrence of fly wastes is reduced in the fiber produced by procedure B. The fiber properties depends strongly on what route of phase separation is employed in the wet-spinning process.

The above experimental facts obtained for AN/MA copolymer / aq. nitric acid system were observed also for many other AN/MA copolymer / solvent systems (ref. 57) and, we believe, for any polymer solutions. It is clear that the underlying fundamental principles of phase separation in the early stage of the fiber-wet spinning process are just the same as those for the membrane-casting process: That is, the thermodynamics of phase equilibrium of polymer solution apply to both equally well.

Fig. 6.39 Scanning electron micrograph of a surface of wet-spun fiber of acrylonitrile(AN) / methylacrylate(MA) copolymer (92/8, wt/wt): Dope (starting solution), ($w_{AN/MA}{}^{s}=0.160$, $w_{HNO_3}{}^{s}=0.563$, $w_{H_2O}{}^{s}=0.277$); coagulant bath, 42wt% aq. nitric acid; total draw ratio, $\times 12$.

Table 6.10
Some physical properties of acrylonitrile / methylacrylate copolymer ($AN/MA=92/8$, wt/wt, $M_w=13.7\times10^4$) fibers spun from 67wt% aq. nitric acid solution (the polymer concentration, 16wt%) using two different procedures.

Physical properties		[Measuring method]	Method A (Route FD in Fig.6.84)*a	Method B (Route FE in Fig.6.84)*b
Single filament	Denier(g/9000m length)	JIS*g L1013	10.0	10.0
	Tensile strength(g/d)	JIS L1069	3.2	4.0
	Tensile elongation(%)	JIS L1069	56.5	54.1
	Loop strength(g/d)	JIS L1070	3.6	6.8
	Loop elongation(%)	JIS L1070	32.0	44.0
	Elastic recovery at 3% elongation (%)	JIS L1073	86.0	95.0
	Young modulus (Kg/mm^2)	JIS L1073	270.0	483.0
carpet*c	Abrasive weight loss after 1000 times abrasion testing (%)*d	JIS L1021	34.5	19.4
	Decrease in thickness of pile under dynamic loading (%)*e	JIS L1021	32.9	22.9
	Abrasive weight loss at practical test (corridor, 2×10^4 persons)*f (%)		24.9	14.4

*a, Coagulation bath, 36wt% aq. nitric acid; drawing, hot water at 369K; draw ratio, ×8.

*b, Coagulation bath, 42wt% aq. nitric acid; drawing, (1st step) 42% aq. nitric acid at 343K; draw ratio, ×8; (2nd step) hot water at 369K draw ratio ×1.5.

*c, Tufted carpet with pile length of 7mm; 1/10 gauge; number of metric yarn count 1/5(Nm).

*d, Relative weight loss of pile portion of the carpet.

*e, The height reduction ratio of compression elastic recovery by applying a load of $0.2Kg/cm^2$ 1000 times.

*f, Relative weight loss of pile portion of the carpet, which was laid on the floor of corridor and on which 2×10^4 persons walked.

*g, Japanese industrial standard.

6.6 EXPERIMENTAL RELATIONS BETWEEN THE SOLVENT-CASTING CONDITIONS AND PORE CHARACTERISTICS OF MEMBRANES

Fig. 6.90 shows the effect of the weight fraction of cellulose in a starting solution, $w_{CELL}{}^{S}$ on the two-phase volume ratio R, calculated from $Pr(d_3)$ using eqns (6.102) and (6.103) and cited from Table 6.6, and on the mean radius of secondary

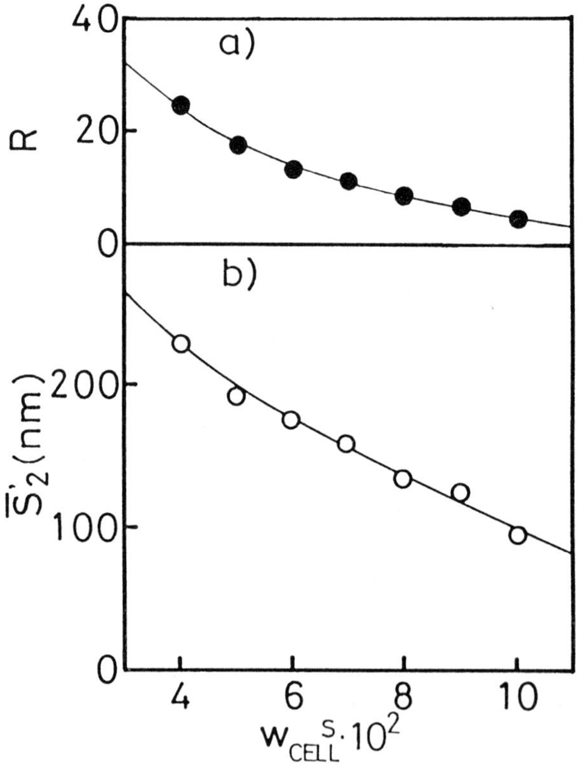

Fig. 6.90 Effect of the weight fraction of cellulose in starting solution, $w_{CELL}{}^{S}$ on two-phase volume ratio R when phase separation occurs: R was indirectly calculated from porosity of a dry membrane $Pr(d_3)$ through use of eqns (6.102) and (6.103) [a)] and effect of $w_{CELL}{}^{S}$ on the mean-radius of secondary particles S_2' of the dry membrane surface cast from the solutions [b)]: Cast solution, cellulose cuprammonium solution; coagulating solution, ($w_{ACETONE} = 0.3000$, $w_{NH_3} = 0.0056$, $w_{H_2O} = 0.6944$); temperature, 298K; R data were cited from Table 6.6.

particles $\bar{S}_2'$, observed on the surface of a dry membrane, regenerated from cellulose cuprammonium solution. With a decrease in w_{CELL}^S, both R and $\bar{S}_2'$ increase concurrently, yielding larger pores. The effect of the polymer concentration on R and $\bar{S}_2'$, observed experimentally, is in good agreement with that obtained by the particle simulation approach (see, Fig. 6.25).

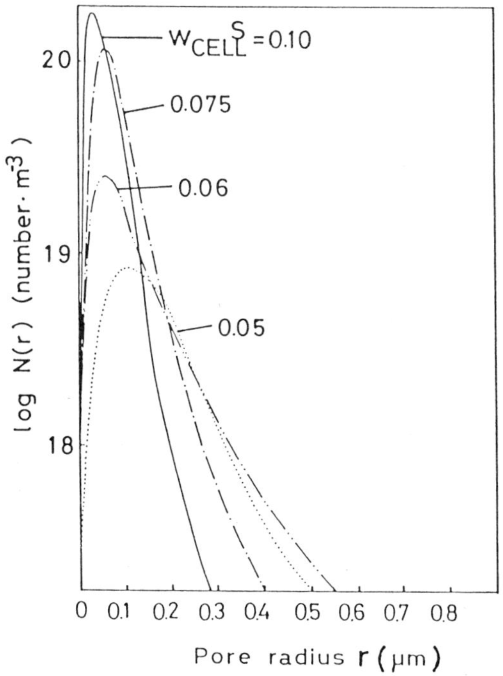

Fig. 6.91 Effect of the weight fraction of cellulose of starting solution w_{CELL}^S on N(r) of the back surface (i.e., surface opposite to front surface, from which phase separation proceeds) of regenerated cellulose membrane. The numbers in the figure indicate w_{CELL}^S values; casting solution, cellulose cuprammonium solution ($w_{CELL}^S = 0.05 \sim 0.10$); coagulating solution, ($w_{ACETONE} = 0.3000$, $w_{NH_3} = 0.0056$, $w_{H_2O} = 0.6944$); temperature, 298K.

Fig. 6.91 shows the effect of the weight fraction of cellulose in starting solution, w_{CELL}^S on pore size distribution N(r), as determined by the electron microscopic method on the surface of cellulose membranes regenerated from cellulose cuprammonium solutions. As w_{CELL}^S becomes lower pore size distribution gets broader and the peak of the distribution shifts to the larger r. The effect of polymer concentration on N(r) can be well interpreted in terms of S_2 and R, and Fig. 6.91 can be compared with Fig. 6.90.

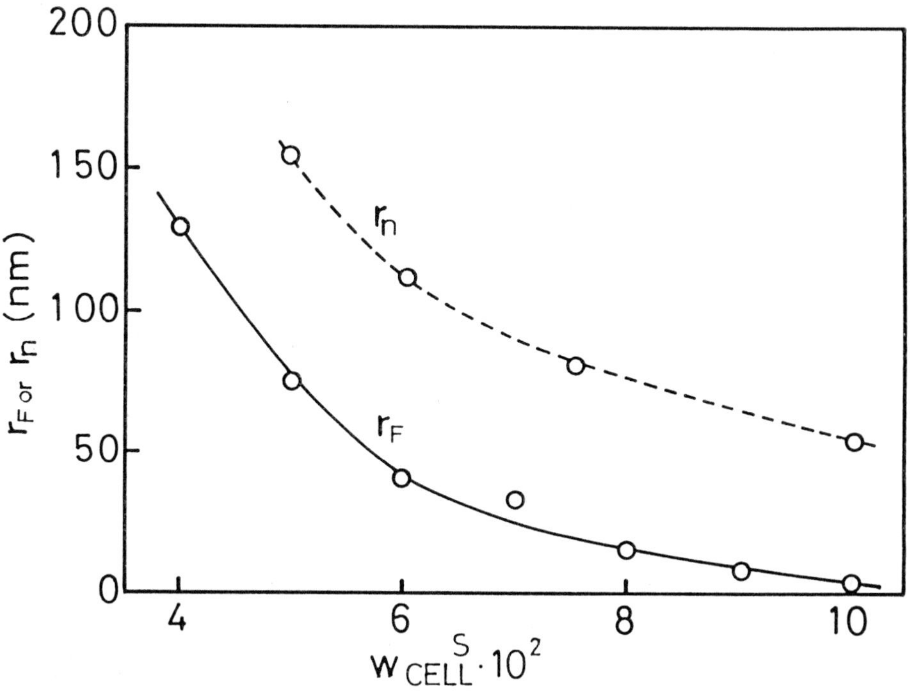

Fig. 6.92 Effect of the weight fraction of cellulose of a starting solution, w_{CELL}^S on the mean pore radius of the membrane r_F as determined by water-flow rate method (full line, eqn (6.145)) and r_n estimated from pore size distribution N(r) in Fig. 6.91 (broken line, eqn (6.92)); casting solution, cellulose cuprammonium solution ($w_{CELL}^S = 0.04 \sim 0.10$); coagulating solution, ($w_{ACETONE} = 0.3000$, $w_{NH3} = 0.0056$, $w_{H2O} = 0.6944$); temperature, 298K.

Fig. 6.92 shows mean pore size, evaluated by the water-flow rate method, r_F as a function of weight fraction of cellulose, w_{CELL}^S in cellulose cuprammonium solution. Here, r_F (nm) is calculated through the relation,

$$r_F = (v \cdot \ell_d \cdot \eta_w)^{1/2} / (\Delta P \cdot A \cdot Pr(d_3))^{1/2} = (r_3 \cdot r_4)^{1/2} \tag{6.145}$$

where v is water-flow rate ($cm^3 \cdot min^{-1}$), ℓ_d, membrane thickness (μm), η_w, viscosity of water (cP), ΔP, loaded pressure (mmHg), A, membrane surface area (m^2) and $Pr(d_3)$, porosity, r_3 and r_4 are some averages of pore radius defined by Kamide and Manabe as $r_3 = \int r^3 N(r)dr / \int r^2 N(r)dr$ and $r_4 = \int r^4 N(r)dr / \int r^3 N(r)dr$ (ref. 7). Kamide and Manabe showed that r_F is equivalent to $(r_3 \cdot r_4)^{1/2}$ (ref. 7). In the figure, r_n estimated from N(r) in Fig. 6.91 by eqn (6.92) is also included. r_n decreases with increasing weight fraction of cellulose as is the case of r_F. Note that r_n corresponds to the number of average pore size on the surface of a membrane and usually r_F changes depending on the distance from the surface and in contrast to this, r_F reflects the pore characteristics of the whole membrane.

Fig. 6.93 shows the plot of the mean pore radius r_n by the electron microscopic method as a function of the mean radius of secondary particles $\bar{S}_2'$ on the regenerated cellulose membrane surface. Neglecting the possible variation of casting conditions such as R and polymer concentration, r_n is approximately proportional to $\bar{S}_2'$.

Fig. 6.94 shows the dependence of the radius of secondary particles $\bar{S}_2'$, porosity $Pr(d_3)$, estimated by eqn (6.105), ratio of thickness of the dry membrane ℓ_d to cast solution ℓ_0 and mean pore radius by the water flow method r_F, estimated by eqn (6.145), of regenerated cellulose membranes on coagulation temperature. At higher coagulation temperature larger secondary particles are formed and interestingly, the gel membrane is relatively rigid and more resistant to collapse during the coagulation step (in other words, the degree of "melt down" is small) and as a result membranes with larger porosity and larger pores can be prepared, provided that all conditions other than the temperature are kept constant.

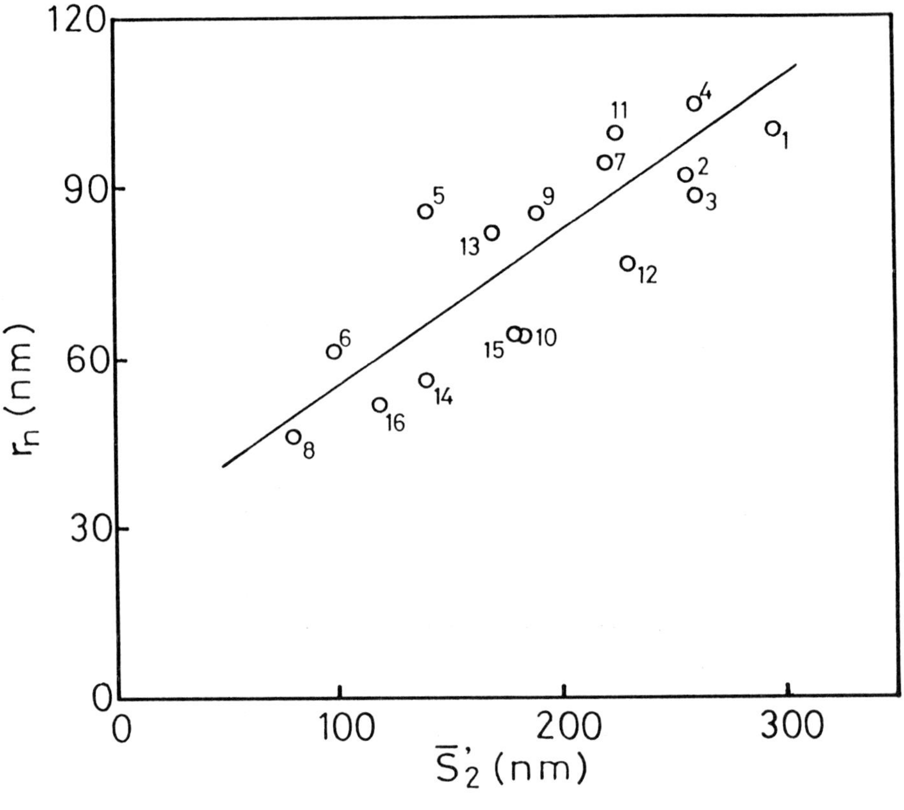

Fig. 6.93 Relationship between the mean radius of secondary particles $\bar{S}_2'$ observed on the front surface of cellulose membranes, cast and regenerated from cuprammonium cellulose / acetone mixture and mean pore radius r_n of the front surface of the membrane by the electron microscopic method: Compositions of casting solutions (starting solution), ($w_{CELL}^S = 0.07$, $w_{CU}^S = 0.0277$, $w_{NH3}^S = 0.1332$, $w_{H2O}^S = 0.7691$); ($w_{CELL}^S = 0.06$, $w_{CU}^S = 0.0231$, $w_{NH3}^S = 0.149$, $w_{H2O}^S = 0.768$); ($w_{CELL}^S = 0.05$, $w_{CU}^S = 0.0198$, $w_{NH3}^S = 0.1752$, $w_{H2O}^S = 0.7551$); preparing conditions (w_{CELL}^S of cast solution, $w_{ACETONE}$ and w_{NH3} of coagulant), 1= [0.07, 0.3000, 0.0056]; 2= [0.05, 0.3000, 0.0056]; 3= [0.07, 0.3000, 0]; 4= [0.06, 0.2500, 0.0056]; 5= [0.06, 0.3000, 0]; 6= [0.06, 0.2000, 0.0056); 7= [0.07, 0.2000, 0.0056]; 8= [0.05, 0.1500, 0.0056]; 9= [0.07, 0.2500, 0.0056]; 10= [0.05, 0.3000, 0]; 11= [0.05, 0.2500, 0.0056]; 12= [0.06, 0.2000, 0.0056]; 13= [0.07, 0.3000, 0.0056]; 14= [0.06, 0.1500, 0.0056]; 15= [0.05, 0.3000, 0]; 16= [0.06, 0.3000, 0.0056].

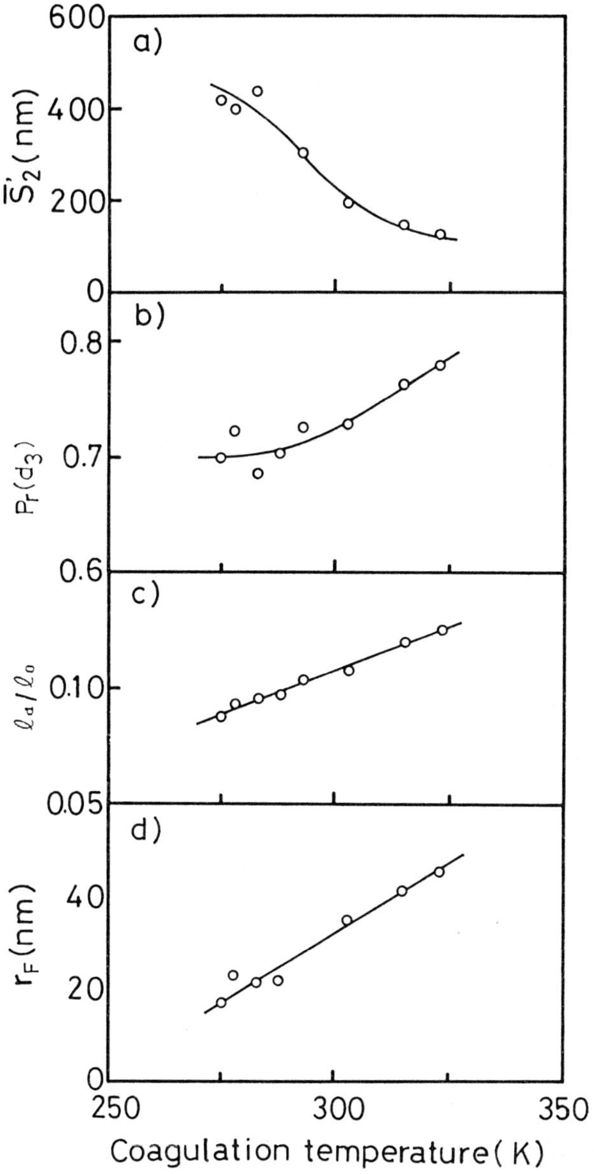

Fig. 6.94 Effect of coagulation temperature on the mean radius of secondary particles S_2', [a)]; porosity by the apparent density method $Pr(d_3)$ (eqn (6.105)), [b)]; the ratio of thickness of dry membrane ℓ_d to that of the casting solution ℓ_0, [c)]; the mean pore radius determined by the water-flow rate method r_F (eqn (6.145)), [d)]; of the cellulose membrane, cast and regenerated from a cellulose cuprammonium / acetone mixture: Compositions of casting solution ($w_{CELL}{}^S=0.0600$, $w_{CU}{}^S=0.0181$, $w_{NH3}{}^S=0.1490$, $w_{H2O}{}^S=0.7730$); composition of coagulant ($w_{ACETONE}=0.3000$, $w_{NH3}=0.0056$, $w_{H2O}=0.6944$); coagulating time=30min; thickness of casting solution, $\ell_0=500\mu m$.

6.7 FORMATION OF MULTI-LAYERED STRUCTURE OF MEMBRANES

In an earlier section (6.4.1) the multi-layered structure of a solvent-cast porous polymer membrane was demonstrated by using electron microscopy. The existence of multi-layers can also be shown by relations between parameters, such as porosity $Pr(d_4)$ and mean pore size r_f and Z/ℓ_d (Z is the distance from front surface of membrane, ℓ_d, thickness of dry membrane) of cellulose diacetate and regenerated cellulose membranes (Fig. 6.95). Here, $Pr(d_4)$, r_3 and r_4 (eqn (6.145)) were estimated from electron micrographs of ultra-thin sections at various Z of regenerated cellulose membrane (Fig. 6.37) and cellulose acetate membrane (Fig. 7 of ref. 15). In cellulose acetate membrane both $Pr(d_4)$ and $(r_3 \cdot r_4)^{1/2}$ increase with Z, but in the regenerated membrane, Z dependence is just the reverse. This fact leads to the prediction that the local polymer concentration of the solution, corresponding to a thin-layer section at a distance Z, increases with Z in the case of regenerated cellulose membrane and decreases with Z in the case of cellulose acetate membranes. This prediction was explicitly confirmed by phase-separation experiments and the mobility of molecules and ions (ref. 15).

Now, for the case of regenerated cellulose membranes, a profile can be drawn of the compositions of the casting and phase-separated solutions for various times elapsed after contact with the coagulation bath (Fig. 6.96). Fig. 6.96a shows a schematic representation of change in composition as the function of distance from the interfacial surface of the casting solution immediately after casting. In Fig. 6.96b the concentration of cellulose near $Z=0$ of the casting solution w_{CELL} $(Z=0)$ becomes higher than w_{CELL} of the original solution (i.e., $w_P{}^S = 0.06$, P is polymer) due to transportation of water molecules from casting solution to coagulation solution. Concentrations of water (w_{H_2O}), ammonia (w_{NH_3}) and acetone $(w_{ACETONE})$ near $Z=0$ of the casting solution change due to differences in their diffusional transportation velocity. When $w_{ACETONE}$ $(Z=0)$ is below ca.6wt%, two-phase separation occurs. Polymer-rich phase builds up a new boundary to the casting solution as shown tentatively in Fig. 6.96c. Polymer-rich phase appears first as primary particles, which grow to secondary particles during the process on phase separation and a network structure is formed by collision of particles (composition of polymer-rich phase is omitted in Fig. 6.96). Component molecules in the polymer-lean phase mix

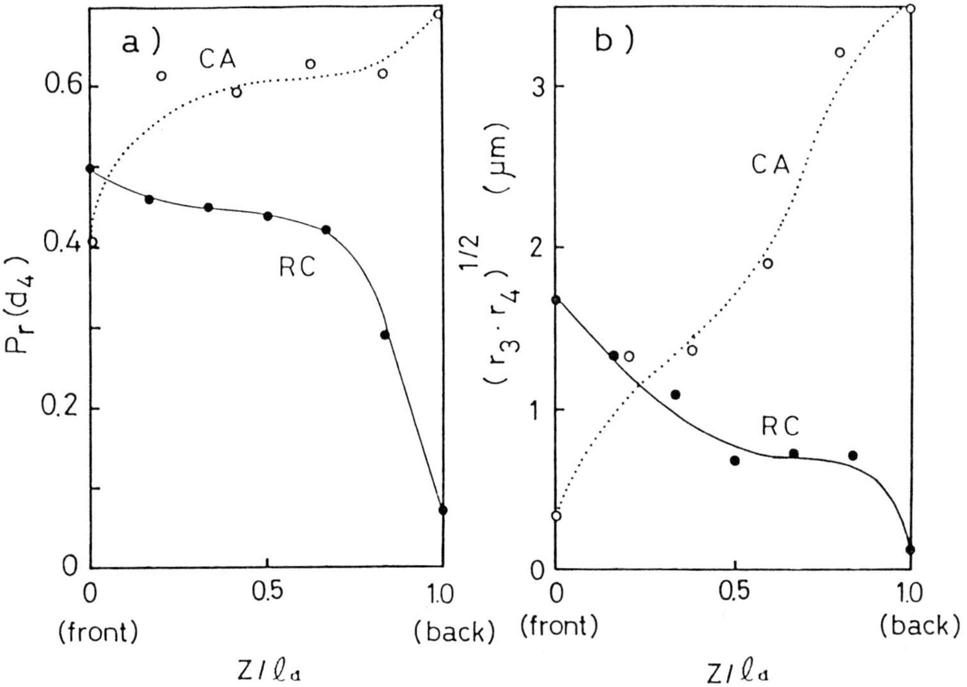

Fig. 6.95 Changes in pore characteristics with the ratio of distance from membrane surface Z to membrane thickness ℓ_d: a) Cellulose membrane (RC) regenerated from cellulose cuprammonium / acetone mixture; compositions of cast solution ($w_{CELL}{}^S=0.0600$, $w_{CU}{}^S=0.0181$, $w_{NH3}{}^S=0.1490$, $w_{H2O}{}^S=0.7730$); composition of coagulant ($w_{ACETONE}=0.3000$, $w_{NH3}=0.0056$, $w_{H2O}=0.6944$); $\ell_d=120\mu m$; b) cellulose diacetate membrane (CA) prepared from solution of cellulose acetate (the total degree of substitution ⟪F⟫ of 2.35) in acetone / methanol (MT) / cyclohexanol (CH) / $CaCl_2 \cdot 2H_2O$ (CC) system by evaporating volatile solvents; composition of cast solution, ($w_{CA}{}^S=0.081$, $w_{ACETONE}{}^S=0.510$, $w_{MT}{}^S=0.128$, $w_{CH}{}^S=0.232$, $w_{CC}{}^S=0.049$); $\ell_d=150\mu m$.

instantaneously with the coagulation solution. The composition of the coagulation solution after mixing can be demonstrated in Fig. 6.96c.

The w_{CELL} of the casting solution increases gradually with time at the same Z because the total flux of water and ammonia from the casting solution is always larger than that of acetone from the coagulation solution. The time when the phase separation

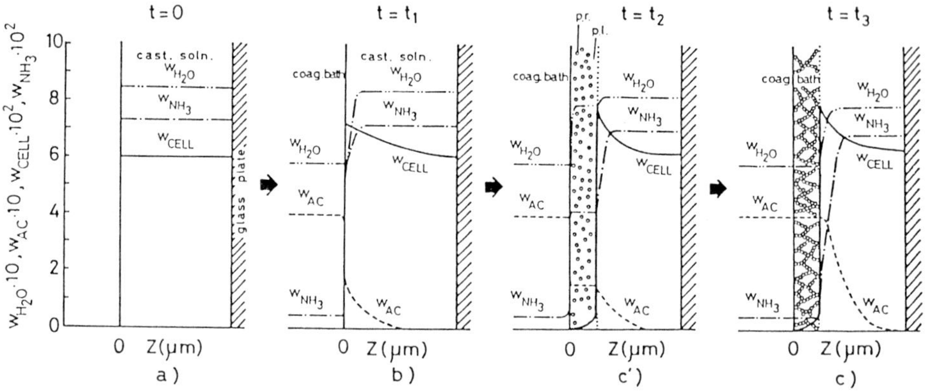

Fig. 6.96 Changes in the profile of the composition of a cast solution during the formation of a cellulose membrane regenerated from cellulose cuprammonium / acetone mixture as a function of the distance from membrane surface Z with time t after immersing into coagulating solution: a) Immediately after immersing into coagulating solution ($t=0$); b) at an arbitrary time ($t=t_1$) before the phase separation; c') immediately after the occurrence of the phase separation near the boundary between casting solution and coagulating solution ($t=t_2$); c) after mixing of the polymer-lean phase and cast solution ($t=t_3$); w_{H2O}, concentration of H_2O; w_{NH3}, concentration of NH_3; w_{CELL}, cellulose concentration; w_{AC} ($=w_{ACETONE}$), concentration of acetone; p.r., polymer-rich phase; p.l., polymer-lean phase.

occurs is controlled mainly by the concentration of ammonia w_{NH3}. Phase separation proceeds from the front surface ($Z=0\mu m$) to the back surface (in this case, $Z=120\mu m$ for a regenerated membrane) with time. With an increase in Z the initiation time of phase separation becomes long and then phase separation occurs at a higher value of w_{CELL}. From this, we can conclude definitely that when the total flux of water and ammonia from the casting solution to the coagulation solution is always larger than that of acetone from the coagulation solution, the mean pore radius and porosity decrease with increasing Z (as in the case of a regenerated membrane).

The characteristics of the phase separation in cellulose acetone membrane formation have been disclosed by Kamide et al. (ref. 18) who demonstrated that on the basis of the composition data of polymer-lean phase and vapor pressure of component solvents at 298K, the dependence of w_{CELL} on Z for cellulose acetone can be estimated.

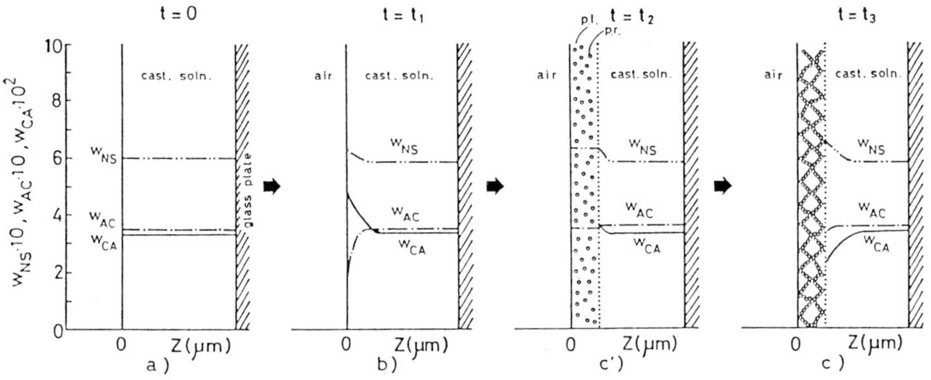

Fig. 6.97 Changes in the concentration of cellulose acetate (w_{CA}), acetone ($w_{ACETONE}$) and nonsolvent (w_{NS}) in a casting solution during the formation of a cellulose acetate membrane as a function of the distance from membrane surface Z with time t after casting: a) Immediately after casting on a glass plate ($t=0$); b) at an arbitrary time ($t=t_1$) before phase separation; c') immediately after the occurrence of phase separation ($t=t_2$) near the boundary of casting solution; c) after mixing of the polymer-lean phase and casting solution ($t=t_3$).

Fig. 6.97 shows the Z dependence of concentrations of cellulose acetate (w_{CA}), acetone ($w_{ACETONE}$) and nonsolvents (w_{NS}) in polymer-lean phase and casting solution at a given time t. Here, the nonsolvents are methanol, calcium chloride, water and cyclohexanol. Until t_1, no phase separation occurs (Fig. 6.97b). Acetone (a good solvent in this case) evaporates from the casting solution resulting in an increase in w_{CA} and w_{NS}, which is pronounced at the interfacial boundary between the casting solution and air and w_{CA} at $t=t_1$ is larger than that at $t=0$. The increase gives rise to phase separation at $t=t_2$ (Fig. 6.97c). After phase separation occurs, the polymer-rich phase generates many fine particles (primary particles (ref. 27)) and these particles finally contact each other. The solvent molecules in the polymer-lean phase mix with the casting solution at $t=t_3$ and then the profile of Fig. 6.97c is obtained. This mixing initiates the decrease in w_{CA} at the casting solution. Since the volume of polymer-lean phase increases as phase separation proceeds, w_{CA} decreases with time. This is the case for cellulose acetate membranes and w_{CA} decreases with Z. Therefore, the mean pore radius and porosity of the resultant cellulose acetate membrane increase with Z, unlike the regenerated cellulose membrane.

When hollow fiber membranes, which are more popularly applied in industry than plane membranes, are produced by contacting with coagulation solution at outer and inner surfaces of the casting solution, micro-phase separation may occur simultaneously from both surfaces resulting in a symmetrical pattern of pore characteristics with respect to $Z/\ell_d = 0.5$. In the case of regenerated cellulose hollow fiber membranes, an increase in cellulose concentration may occur at the inside of the membrane and in consequence, the cellulose concentration at the instant, when the micro-phase separation occurs, may exceed the critical concentration at the central position ($Z/\ell_d \simeq 0.5$). In order to disclose the multi-layer nature of hollow fiber membranes their sectional method, originally proposed by Manabe et al. for plane membrane (refs 8,15), needs great skill and is extremely elaborate. Recently, Kamide et al. established an alternative improved method for evaluating pore characteristics of hollow fiber as a function of the distance from the membrane surface by slicing off a thin section film (i.e., aslant sliced thin sectional (ASTS) method (ref. 9)). In fact, it is observed that in some regenerated cellulose membranes, the pores at the central position are obviously circular, although the pores at the surface parts are non-circular (Fig. 6.98). For regenerated membrane, there are two minimums in $2(r_3 \cdot r_4)^{1/2}$ and one maximum in the porosity.

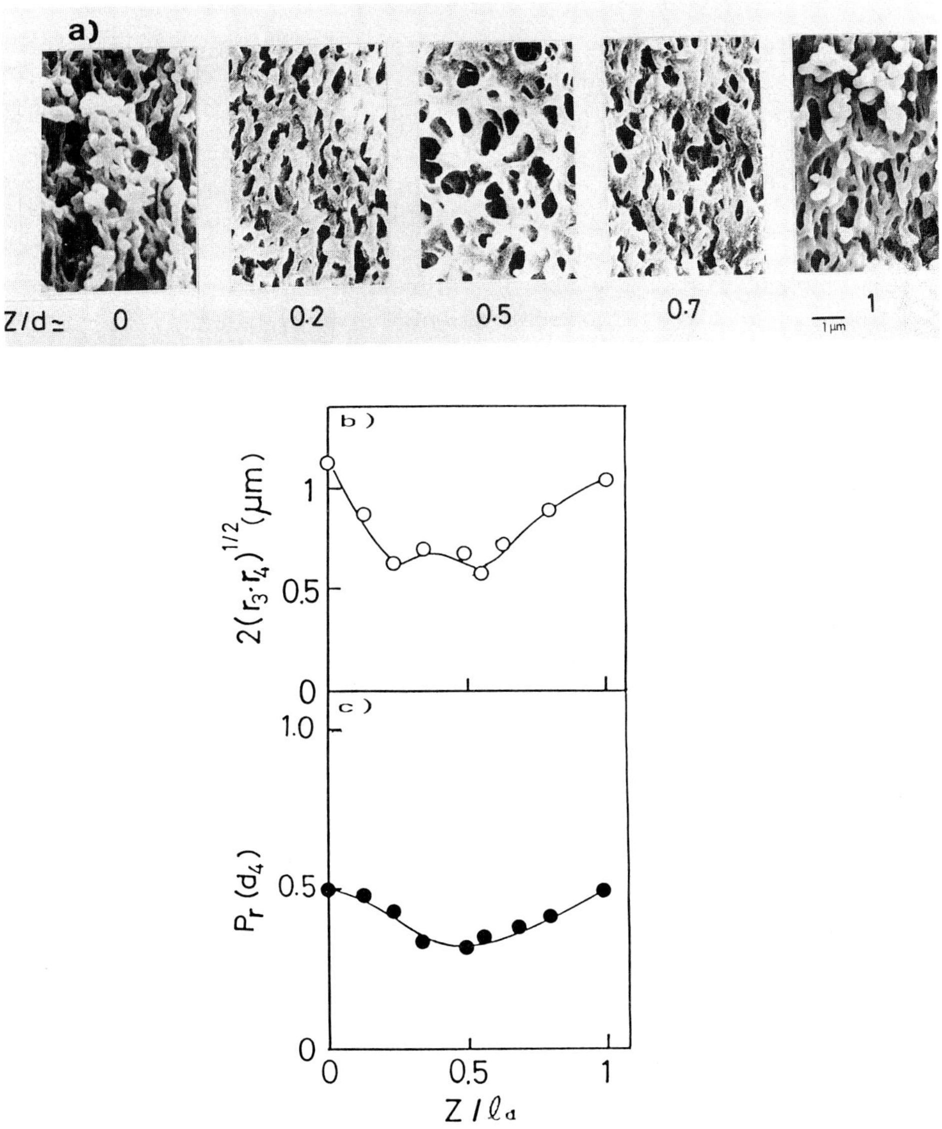

Fig. 6.98 Electron micrographs of inner parts of a regenerated cellulose hollow fiber membrane observed at various positions Z/ℓ_d (Z, distance from an inner surface of a hollow fiber membrane; ℓ_d, membrane thickness) by ASTS (aslant sliced thin sectional) method (ref. 9), [a)]; Z/ℓ_d dependence of the mean pore diameter $2(r_3 \cdot r_4)^{1/2}$ (eqn (6.145), [b)]; porosity estimated by the electron microscopic method of the membrane $Pr(d_4)$ (eqn (6.106), [c)]: Mean pore diameter of the membrane estimated by water flow method (eqn (6.145)), $r_F = 20.1$μm; a scale bar stands for 1μm.

6.8 SOME OF THE PERFORMANCES OF MULTI-LAYERED MEMBRANES

6.8.1 Complete exclusion of disease viruses in human blood by membrane filtration

Although it is not intended to describe the application of polymer membranes in this work, we will introduce briefly the some notable performances characteristics of the products, which are closely correlated to the supermolecular structure of membranes (ref. 54).

A characteristic feature of filtration through multi-layered membranes is an easy attainment of unbelievably high efficiency by a single filtration operation due to numerously repeated separations corresponding to the number of thin layers within a membrane. It should be pointed out that the pore size distribution for a given thin-layer is, as described in an earlier section, determined by the phase-separation mechanism governing the pore formation of the layer and then, it is never extremely sharp. But repetition of separation through thin-layers enables us to improve separation efficiency almost without limit, which allows complete removal of the disease virus (having diameter in the range 25～200nm), such as acquired immune deficiency syndrome (AIDS) virus (≡human immunodeficiency virus; HIV) and hepatitis B virus, contaminated in human blood occasionally to a concentration of 10^9 virus particles/cm^3. Quite recently, the above viruses have been attracting enormous public attention and the establishment of a method for the effective removal of these viruses, of which we will briefly describe certain aspects, is urgently called for, and a clinical therapy or in fractionated plasma derivatives manufacturing.

The movement of the virus in porous membrane during filtration is directly observable by electron microscopy (ref. 59). Viruses are usually considered to be monodisperse. Virus filtration through porous cellulose membrane, regenerated from cuprammonium solution, provides a powerful model in studying the underlying mechanism of material separation by porous membranes.

Fig. 6.99a shows a transmission EM photograph of Escherichia coli phage T4 captured by regenerated cellulose membranes with mean pore radius r_F, as determined by the water-flow method (eqn (6.145)), of 17.5nm. Fig. 6.99b shows the concentration distribution of T4 phage within a membrane. In the process of membrane formation, the micro-phase separation proceeds simultaneously from outer and inner surfaces of the membrane.

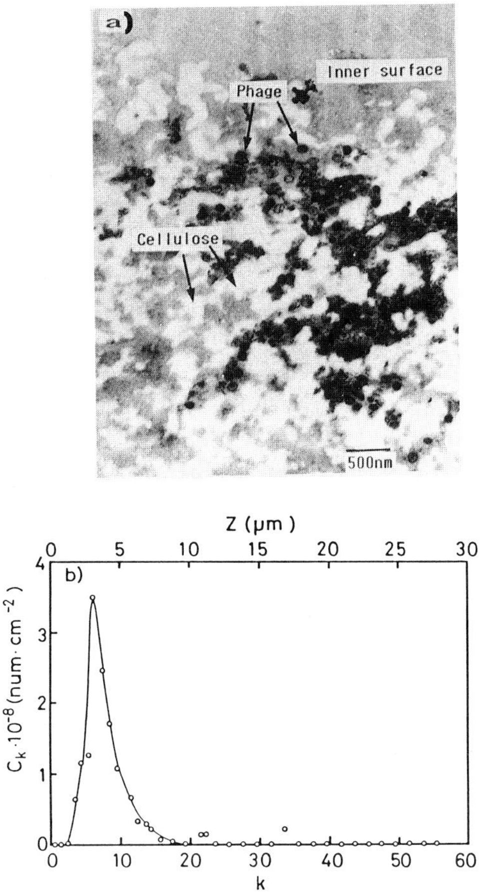

Fig. 6.99 Transmission electron micrograph of T4 phage captured by cellulose hollow fiber membrane, regenerated from cellulose cuprammonium solution-acetone mixture, [a)]; the number of viruses captured by the unit area of k-th group of thin layers with total thickness of 0.5μm during filtration of 6min, C_k, as a function of k (k is proportional to the distance from the inner surface of hollow fiber Z) in the membrane, [b)]: Mean pore size r_F (eqn (6.145)), 17.5nm; membrane thickness, $\ell_d = 28$μm; porosity, $Pr(d_3)$ (eqn (6.105)) $= 0.504$; inner radius of hollow fiber, 137.5μm; virus concentration in filtrand, $A_0 = 6.1 \times 10^9$ PFU (plaque forming unit)/cm^3; amount of filtrate, 1.0mℓ/cm^2; loaded pressure, $\Delta P = 200$mmHg; a scale bar stands for 500nm. Sample preparation for TEM observation: Immediately after filtration of phage solution, a hollow fiber was fixed with a 2wt% glutaraldehyde / paraformaldehyde mixture in a cacodylate buffer (pH=7.4) for 30min, and was postfixed with a 2wt% osmic acid for 2hr. Then, the fiber was dehydrated with acetone and embedded in Epon 812. An ultra-thin section with a thickness of ca.100nm of the embedded fiber was stained with uranylacetate and lead citrate.

Then, the porous structure is the same at the same distance from the center, changing gradually from both surfaces. Analysis of the membrane formation process and of the membrane structure shows that a thin layer has a thickness of $0.1 \sim 1.0 \mu m$.

Consider a plane membrane consisting of a number of thin layers stacked in parallel and assume that secondary polymer particles with radius S_2 and vacant particles of the same size are randomly arranged on a closely-packed hexagonal lattice of each layer. The number of thin layers constituting a membrane N with thickness ℓ_d is given by eqn (6.110).

The capture of viruses by porous membranes is described as the probability process (ref. 59): Assign numbers to thin layers from the front surface, to which the solution containing viruses is fed, to the back surface. The number of viruses supplied to a unit surface area of i-th layer for the time interval δt from a unit area of (i-1)-th layer is designated as δA_{i-1}, and similarly, the number of viruses passed through a unit area of i-th layer for δt and accordingly, supplied to the (i+1)-th layer for that time interval as δA_i and the number of viruses captured by unit area of i-th layer for δt as δC_i. Then, we obtain the relation

$$\delta A_{i-1} = \delta A_i + \delta C_i . \tag{6.146}$$

The probability that a given virus, entered into i-th layer, is captured by that layer, γ_i is defined by the relation,

$$\gamma_i = \delta C_i / \delta A_{i-1} . \tag{6.147}$$

Next, consider a group of $(\ell + 1)$ consecutive thin layers ranging from k-th to $(k+\ell)$-th. The number of viruses passed through $(k+\ell)$-th layer for time interval δt, $\delta A_{k+\ell}$ is related to the number of viruses supplied to k-th layer for the same time interval through the relation,

$$\delta A_{k+\ell} = (1-\gamma_k)(1-\gamma_{k+1}) \cdots (1-\gamma_{k+\ell}) \delta A_{k-1} . \tag{6.148}$$

Eqn (6.148) is straightforwardly derived from eqns (6.146) and (6.147).

The probability that a given virus, entered into a group consisting of $(\ell+1)$-th thin layers between k-th and $(k+\ell)$-th layers, is captured by that group, $\gamma_{k,k+\ell}$ is given by

$$\gamma_{k,k+\ell} = (1-\gamma_k)(1-\gamma_{k+1})\cdots(1-\gamma_{k+\ell}). \quad (6.149)$$

In the case of $k=1$ and $k+\ell=N$, eqn (6.148) reduces to

$$\delta A_N = (1-\gamma_1)(1-\gamma_2)\cdots(1-\gamma_N)\,\delta A_0 \quad (6.150a)$$

$$= (1-\gamma_{1,k1})(1-\gamma_{k1+1,k2})(1-\gamma_{k2+1,k3})\cdots(1-\gamma_{kn+1,N})\,\delta A_0. \quad (6.150b)$$

Eqn (6.150b) corresponds to the case when all the thin layers in a membrane were divided into n+1 groups. δA_N and δA_0 are total number of viruses in the filtrate for δt and that of the supplying solution, respectively. Both δA_N and δA_0 can be determined with good accuracy by the plaque forming method (ref. 60). δA_N is also estimated from δA_0 and $\gamma_{ki,ki+\ell}$ $(i=1,\cdots,n)$ data using eqn (6.150b). Here, $\gamma_{ki,ki+\ell}$ can be independently determined by direct counting of the number of viruses captured in a given group of thin layers. For this purpose, the electron microscopic method is useful.

The rejection coefficient of the membrane against the virus ϕ is defined by

$$\phi = \log(\delta A_0/\delta A_N). \quad (6.151)$$

Then, there are two methods for estimating ϕ for given combination of membrane and virus. The above discussion can be readily extended to hollow-fiber type membranes with a slight modification.

Using γ_i values for a regenerated cellulose membrane calculated from the virus distribution curve in Fig. 6.99b and δA_0 by the plaque forming method, ϕ was estimated for the membrane to be 5.25, which agrees fairly well with the value estimated from δA_N and δA_0, both by the plaque forming method (4.02). It is now evident that even if the probability of capturing viruses in each layer γ_i is low (0.1, in this case), sufficient high rejection coefficient can be attained if the number of the layers exceeds almost 100. This is a physical explanation why regenerated cellulose membranes prepared by phase-separation method have such a high performance.

6.8.2 Fine separation of blood components and exclusion of contaminants from blood

Human blood is a very complicated aq. mixture of ions, metabolites, proteins, blood cells and platelet and sometimes it is contaminated by toxic materials and disease viruses. Among them, some specific materials (or viruses) can be separated from the blood by membrane dialysis or filtration method as illustrated as follows (Fig. 6.100).

(1) Exclusion of ions and metabolites from blood: hemodialysis (pore diameter $2r_n = 2\sim 3$nm)

(2) Separation of serum proteins and blood cells from blood: plasma separation (pore diameter $2r_n = 200$nm)

(3) Fractionation of serum proteins such as albumin and globulin (pore diameter $2r_n = 10\sim 20$nm)

(4) Removal of disease viruses from blood (see, section 6.8.1): (pore diameter $2r_n = 30\sim 100$nm)

(a) Hemodialysis membrane

The artificial kidney is an apparatus for purifying blood of patients, whose renal function is notably lost due to chronical renal insufficiency or for intoxication. A very short overview of the history of artificial kidneys is given below (ref. 61): In 1914 Abel et al. (ref. 62) employed collodion (cellulose nitrate) membrane to remove acetyl salicylic acid contaminated in blood by dialysis (i.e., hemodialysis) as a treatment for acetylsalicylic acid intoxication. This is the first application of the hemodialysis and they named this "artificial kidney", a widely used terminology. The most fatal short point of collodion membranes is imperfect impermeability against blood proteins; about 95～99% of albumin passed during the dialysis. In 1937, Thalheimer (ref. 63) used cellophane tubes, invented by Brandenberger in 1908 and mass-production started in the 1920s as packing material, for hemodialysis to reduce azotemia. Kolff (ref. 64), at the end of World War II, first applied clinically cellophane-dialyzer, newly designed by him, as a new way of treating uremia. Cellophane used by him was, of course, industrial grade for sausage packing and control of the pore size of the membrane was impossible and unavoidably the cellophane membrane had a wide pore size distribution. It is important to notice that the first artificial membrane as a separation media is cellulose and its first application was found in medical treatment.

Repeated and prolonged clinical application of artificial

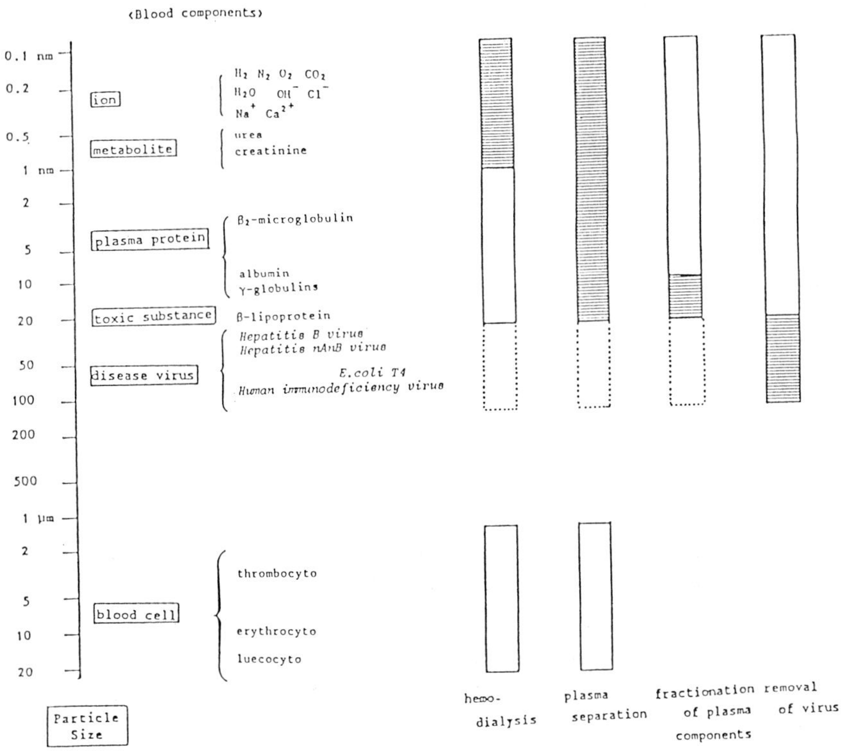

Fig. 6.100 Several membrane technologies for separation of blood components including plasma proteins and blood cells and for exclusion of contaminants such as metabolites, toxic substances and disease viruses in human blood: Hatched area, materials to be separated.

kidneys with cellophane as a therapy of chronical renal failure (chronical hemodialysis) become possible after an important invention of arterio-venous fistula shunt by Quinton et al. (ref. 65) in 1960, enabling us to carry out extracorporeal circulation. Before this invention, artificial kidneys had been limited to the treatment of acute renal insufficiency only. Then, artificial kidneys were widely popularized as a clinical and repeated treatment for chronical renal failure and for this purpose, cellulose membranes, regenerated from cuprammonium solution and saponified cellulose triacetate were commercialized in the 1960s.

Membrane of synthetic polymers such as polyacrylonitrile (PAN), polymethylmethacrylate (PMMA), ethylene-vinyl alcohol copolymer (EVA) were developed for artificial kidneys in the 1970s, but their position in the artificial kidney market remains at present, at a comparatively low level even today. In this sense, artificial kidneys were initiated and then developed along with development of regenerated cellulose membranes, by which the type of dialysis apparatus was invented: In the 1960s a coil type artificial kidney was put into practice and in the 1970s plate type and hollow-fiber type artificial kidneys were commercialized. During a short time, these three types were utilized. But, the coil type, then the plate type fall from use in the late 1970s-early 1980s. At present, more than 90% of hemodialyzers commercially available are the hollow-fiber type. Cellulose has played always a leading role in the technical development of the hollow-fiber membrane: Saponified cellulose acetate (SCA) was first used for cellulose hollow-fiber membrane (Cordis-Dow, 1967). Then, in 1973 Asahi Chemical Industry developed a cuprammonium cellulose hollow fiber membrane. Cordis-Dow commercialized cellulose acetate hollow fiber membrane in 1977. During these fifteen years, regenerated cellulose membrane has made a prominent improvement in the following points: (1) pin hole, (2) break-down during dialysis, (3) membrane thickness (100μm→7μm), (4) purity (low molecular weight non cellulosic components) and (5) sterilization method (ethylene oxide, steam). World market of artificial hemodialysis apparatus is expanding at a rate of 5%/year and world consumption estimated is 3×10^7 unit/year in 1988. Among them, cuprammonium cellulose membrane occupies 66%, cellulose acetate membrane, 15% (ref. 61).

As the number of patients receiving long-term hemodialysis treatment is rapidly increasing, new chronic hemodialysis-associated syndromes presumably caused by accumulation in the blood of impermeable metabolites through conventional dialysis membrane have been reported: For example, carpal tunnel syndrome amyloid protein, deposited in the synovium or the transverse carpal ligament, pressed perineural tissue of the medium nerve, resulting in acute pain (amyloidosis). This syndrome develops at a probability of 70% in patients receiving long-term (more than 15 years) hemodialysis treatment and is suspected to be the chronic hemodialysis-associated amyloidosis and examined energetically in recent years. Amyloid responsible for this syndrome is β_2-microglobulin (β_2-MG) (ref. 66) with the molecular weight of

11800 (a molecule is approximately 4.5nm× 2.5nm× 2.0nm in size) (ref. 67).

A cuprammonium cellulose hollow fiber membrane commercially available for hemodialysis was believed to have the mean pore diameter of 2~3nm by the water-flow rate method (200mmHg), which is too small to allow complete permeation or ultrafiltration of β_2-MG. In the framework of traditional cuprammonium cellulose process it was said the impossible to produce membranes with larger pore sizes. Recently, a new cuprammonium cellulose hollow fiber membrane with larger mean pore sizes (4~9nm) was developed and sieving coefficient S_C (a ratio of a solute concentration of a filtrate to that of a filtrand) of β_2-MG was improved up to 0.4~0.7 and by applying this high performance cuprammonium cellulose membrane for long-term hemodialysis, serum β_2-MG level in the patients decreased effectively.

When a regenerated cellulose membrane is employed as a separation media in extracorporeal circulation of blood, number of leukocyte per unit volume of blood was found to decrease temporarily down to 20~40% of the initial value in 15min after starting the circulation (i.e., transient leukopenia). This phenomenon is closely correlated with activation of complements and some one pointed out that, in order to avoid a transient leukopenia, synthetic polymer (i.e., hydrophobic) membranes seem rather preferable to a cellulose membrane. But this has proved to be not true at all: Akizawa et al. (ref. 68) disclosed that activation of complements by cellulose is caused by a direct activation of the second path of complement (ref. 69) due to contact of the free hydroxyl group on the membrane with human blood. And a new technique was successfully developed very recently to suppress the activation of complement: The free OH groups on the membrane surface were masked completely with cationic polymers (ref. 70), which effectively prevent direct contact of the OH group with blood. In the new membrane, masking polymer builds a nearly mono-molecular layer on membrane surface and is fixed by forming hydrogen bonding, ionic bonding and hydrophilic bonding to the surface of the membrane. The new membrane has a sufficient performance in this respect. This kind of study is now rapidly advancing to improve the surface properties of cellulose membrane. Although there has been considerable controversy in the past about the risks and benefits of usage of a regenerated cellulose membrane in artificial kidneys, the above described strong scientific evidence indicate the unquestionable advantage of cellulose membranes. The copper content remained in the commercially available cuprammonium cellulose membrane is kept at a very low level ($\ll$ 10ppm) and the serum copper of the patients receiving long-term hemodialysis using cuprammonium cellulose membranes never exceeds the upper limit of the normal value (ref. 71).

(b) Ultrafiltration membrane: Ultrafiltration (i.e., hemofiltration)-type artificial kidney

From a view point of renal physiological biology, living kidney evolved from dialysis-type (invertebrate animal) $\rightarrow$ filtration-type (fish) $\rightarrow$ filtration-reabsorption-type (higher vertebral animal) (ref. 72). This is due to historical necessity corresponding to a change in living circumstances and living patterns. In other words, the environments govern the pattern of mass transfer of metabolites. In majority of commercially available artificial kidneys, the mass transfer is controlled by diffusion mechanism (dialysis), indicating that an artificial kidney used at the present can only be regarded as an "artificial kidney for protozoan".

Kamide, Sato and their collaborators, since early 1970s, (refs 72-74) have carried out systematic studies on filtration-type kidneys using a cellulose diacetate (total degree of substitution $\langle\langle F \rangle\rangle = 2.46$) membrane, having the mean pore radius of 0.01～several μm. They established that (1) the adequate mean pore radius of the cellulose acetate membrane, suppressing the concentration of serum proteins in the filtrate to 1/10 of that in the fresh blood, is ca.90nm, (2) parallel flow is strongly recommended for use, (3) the components except for proteins in blood are concentrated by filtration: Total protein, $\times$ 0.05; blood urea nitrogen (BUN), $\times$ 1.80; uremic acid, $\times$ 1.30; Na^+, $\times$ 1.20; Cl^-, $\times$ 1.40; K^+, $\times$ 1.75, (4) for removal of water from the patient, this filtration-type can be applied clinically, but the balance of water and metabolites including urea, creatinine and uric acid is, as anticipated, not satisfactory.

The filtration-type artificial kidney is a kind of artificial glomerulus and a cellulose acetate membrane corresponds to basement membrane in a human kidney. Therefore, if some artificial tubule could be invented in the future, the artificial human kidney in a true sense could be realized as an ultimate resolution. A hemofiltration with pre-filtrate or post-filtrate alimentation is practised to add conventional, but never complete function of reabsorption for practical use. This type of artificial kidney has the two functions: 1) the removal of metabolite solutes, together with water by filtration (hemofiltration) and 2) the infusion of aq. electrolyte solutions as fluid supplementation. Fluid supplement must be absolutely sterile because it is infused into the body. The amount of fluid supplied is 20～40ℓ per one hemodialysis treatment, which should be balanced very precisely with that of filtrate, the apparatus is expensive. At present, only 1% of total patients suffering from chronic renal insufficiency are treated with this type of artificial kidney. Filtration/dialysis-type artificial kidneys (hemodiafiltration) were first studied in 1974 by Sato (ref. 75) together with a hemofiltration/adsorption-type artificial kidney. In filtration/dialysis-type artificial kidney metabolites are removed from the blood by filtration as well as by dialysis. This artificial kidney needs only 5～10ℓ of dialysate per each treatment and is very gradually becoming popular. As membranes for filtration-type artificial kidneys, PAN hollow fibers, PMMA hollow fibers, cellulose acetate plane membranes and polystyrene hollow fibers have been commercialized.

(c) Hemofiltration

In 1977, the possibility of plasma separation by a hollow fiber membrane was first shown by Yamazaki et al. (ref. 76). Separation of blood components such as red and white cells by membrane (i.e., plasma separation) has been for these 10 years extensively studied as an application of industrial ultrafiltration technology. A new membrane method was expected to exchange human plasma more promptly and readily as compared with conventional centrifuge method. A cellulose acetate plasma separation membrane has been manufactured and commercialized for this purpose. The cellulose acetate hollow fiber membrane with 0.1～0.4μm in pore size is used as a plasma separation membrane for the treatment of intractable ascites and utilized in hepatic support systems (plasma perfusion detoxication) or plasma exchange for acute hepatic failure. Furthermore, the cellulose acetate membrane has been experimentally evaluated and clinically used in renal liver and immuno diseases, not only to remove toxic macromolecular substances, but also to supply necessary nutrients (ref. 77). Besides cellulose derivatives, non-cellulosic plasma separation membranes (PMMA, polyvinylalcohol, polyethylene and polypropylene) have been commercialized up to now.

As a result of development of the plasma separation membrane, a double-filtration plasmapheresis, in which the selective removal of large molecular weight, disease-related materials is carried out using the membrane, is now established as a new therapy. In this treatment, a plasma separation membrane with mean pore size of 0.1～0.4μm is called as a primary membrane and a membrane with mean pore sizes of 10～80nm for fractionation of plasma components, as a secondary membrane (Fig. 6.101).

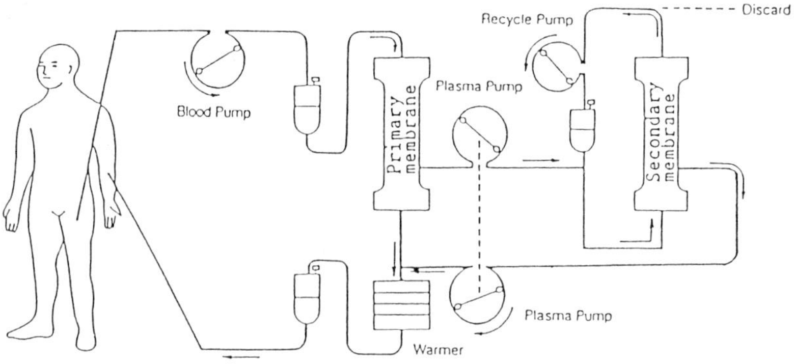

Fig. 6.101 Schematic representation of double-filtration plasmapheresis.

Here, a number of particular essential points concerning membrane formation are discussed. Phenomenologically complicated process of membrane pore formation and wet spinning of fibers can be treated as a target of quantitatively theoretical study of thermodynamics, in which various important concepts are proposed and the validity of some of them are experimentally verified, although this chapter is mere a milestone in the understanding of porous polymeric membrane.

REFERENCE

1 See, for example, A.J. Ihde, The Development of Modern Chemistry, Chapt. 15, 15.4, Harper & Row, Pub., New York, 1964.
2 A. Fick, Ann. Physik. Chem., 94 (1855) 59.

3 S. Leob, Synthetic Membranes, Desalination, Am. Chem. Soc. Sympo. Ser., American Chemical Society, 1981, No. 153, p. 1.
4 S. Sourirajan, in M. Malaiyandi, O. Kutowy, F. Talbot (Eds), Proceedings of the International Membrane Conference on the 25th Anniversary of Membrane Research in Canada, 1986, Chapter 1, p. 3.
5 R.E. Kesting, Synthetic Polymeric Membranes, 2nd ed., John Wiley & Sons, 1985.
6 H.K. Lonsdale, J. Memb. Sci., 10 (1982) 81.
7 K. Kamide and S. Manabe, Characterization Technique of Straight-Through Porous Membrane, in A.R. Cooper (Ed.), Ultrafiltration Membranes and Applications, Plenum Publishing Corp., 1980, p. 173-202.
8 S. Manabe, Y. Shigemoto and K. Kamide, Polym. J., 17 (1985) 775.
9 K. Kamide, S. Nakamura, T. Akedo and S. Manabe, olym. J., 21 (1989) 241.
10 K. Kamide, S. Manabe and T. Matsui, Kobunshi Ronbunshu, 34 (1977) 299.
11 T. Nohmi, S. Manabe, K. Kamide and T. Kawai, Kobunshi Ronbunshu, 34 (1977) 729.
12 T. Nohmi, S. Manabe, K. Kamide and T. Kawai, Kobunshi Ronbunshu, 35 (1978) 509.
13 S. Manabe, K. Kamide, T. Nohmi and K. Kawai, Kobunshi Ronbunshu, 37 (1981) 405.
14 K. Kamide and S. Manabe, Sen-i Gakkaishi, 37 (1981) 361.
15 S. Manabe, Y. Kamata, H. Iijima and K. Kamide, Polym. J., 19 (1987) 391.
16 S. Manabe, H. Iijima and K. Kamide, Polym. J., 20 (1988) 307.
17 K. Kamide, Nisshinbo International Conference, Tokyo, 1988.
18 K. Kamide, S. Manabe, T. Matsui, F. Sakamoto and S. Kajita, Kobunshi Ronbunshu, 34 (1977) 205.
19 K. Kamide, in Soc. Polym. Sci. Japan (Ed.), Polymer Alloys, 1981, Chapter 4.
20 D.M. Koenhen, M.H.V. Mulder and C.A. Smolders, J. Appl. Polym. Sci., 21 (1977) 199.
21 J.G. Wijmans, J.P.B. Baaij and C.A. Smolders, J. Memb. Sci., 14 (1983) 263.
22 A.J. Reuvers, J.W.A. van den Berg and C.A. Smolders, J. Memb. Sci., 34 (1987) 45.
23 A.J. Reuvers and C.A. Smolders, J. Memb. Sci., 34 (1987) 67.
24 U. Merin and M. Cheryan, J. Appl. Polym. Sci., 25 (1980) 2139.

25 H.A. Purz, C. Rosin, B. Phillips, K.J. Stiiller and P.V. Zgglinicki, Acta Polymerica, 32 (1981) 726.

26 H. Kobayashi, Kagaku, 23 (1953) 315.

27 K. Kamide and S. Manabe, in D.R. Lloyd (Ed.), Materials Science of Synthetic Membranes, ACS Symposium Series, 1985, No. 269, p. 197.

28 R.L. Riley, J.O. Gardner and U. Merten, Science, 143 (1964) 801.

29 R.D. Schultz and S.K. Asunmaa, Ordered Water and the Ultrastructure of the Cellular Plasma Membrane, in J.F. Danielli, A.C. Riddiford and M. Rosenberg (Eds), Recent Progress in Surface Science, Academic, New York, N.Y., 1970, Vol. 3, p. 291-332.

30 K. Kamide and S. Manabe, in M. Malaiyandi, O. Kutowy and F. Talbot (Eds), Proceedings of the International Membrane Conference on the 25th Anniversary of Membrane Research in Canada, National Research Council Canada, 1986, Chapter 2, p. 83.

31 K. Kamide, S. Matsuda and H. Iijima, unpublished results.

32 See, for example, J.W. Cahn, J. Chem. Phys., 42 (1965) 93.

33 See, for example, L. Manderkern, Crystallization of Polymers, McGrow-Hill Book Co., 1964, Chapter 8.

34 K. Kamide and S. Matsuda, Polym. J., 16 (1984) 825.

35 K. Kamide, S. Matsuda, T. Dobashi and M. Kaneko, Polym. J., 16 (1984) 839.

36 K. Kamide, T. Akedo and M. Saito, Polymer Preprint, Japan, 38 (1989) 1160.

37 K. Kamide and M. Saito, unpublished results.

38 See, for example, H. Margenau and G.M. Murphy, The Mathematics of Physics and Chemistry, D. Van Nostrand Co., 1943, Chapter 7.

39 K. Kamide, H. Iijima, H. Shirataki and H. Hanahata, unpublished results.

40 See, for example, W.J. Moore, Physical Chemistry, third ed., Prentice-Hall, Inc., 1960, Chapter 30.

41 H. Inagaki, H. Suzuki, M. Fujii and T. Matsuo, J. Phys. Chem., 70 (1966) 1718.

42 K.S. Gandhi and S.V. Babu, Macromolecules, 13 (1980) 791.

43 B.J. Berne and R. Pecora, Dynamic Light Scattering, John Wiley & Sons, New York, 1976, Chapter 8, p. 195.

44 E. Gulari, Y. Tsunashima and B. Chu, J. Chem. Phys., 70 (1979) 3965.

45 B. Zimm, J. Chem. Phys., 16 (1948) 1093.
46 K. Kamide, Y. Miyazaki and Y. Shimaya, unpublished results.
47 H. Iijima and K. Kamide, Jpn. Kokai Tokkyo (Patent) Koho JP 61, 211, 342 (1986).
48 K. Kamide, H. Iijima and S. Matsuda, unpublished results.
49 K. Kamide, M. Iwata and H. Iijima, unpublished results.
50 K. Kamide, A. Kataoka and H. Iijima, unpublished results.
51 K. Kamide, H. Shirataki and H. Iijima, unpublished results.
52 K. Kamide and S. Matsuda, Polym. J., 18 (1986) 347.
53 See, for example, Th. Lieser and H. Swiatkowski, Liebigs Ann. Chem., 538 (1939) 110.
54 K. Kamide, H. Iijima and M. Iwata, unpublished results.
55 K. Kamide, M. Iwata, A. Kataoka and H. Iijima, unpublished results.
56 K. Kamide, Y. Miyazaki and S. Matsuda, unpublished results.
57 K. Kamide, Y. Honda and S. Kajita, U.S.P., 4, 663, 232, (1987). S. Kajita and K. Kamide, Jpn. Kokai Tokkyo Koho JP, 61,119,707 [86,119,707], 61,138,709 [86,138,709], 61,138,710 [86,138,710], 61,138,711 [86,138,711], 61,138,712 [86,138,712], 61,138,713 [85,138,713].
58 K. Kamide, Y. Miyazaki and S. Kajita, unpublished result.
59 K. Kamide, S. Sogawa and H. Iijima, Polym. Preprint Jpn., 38 (1989) 323.
60 See, for example, L.S. Kukera and Q.N. Myrvik (Eds), Fundamentals of Medical Virology, 2nd ed., Philadelphia, Lea & Febiger, 1985, Chapter 2; R.Y. Stanier et al. (Eds), The Microbial World, 5th ed., Englewood Cliffs, N.J., Perntice-Hall, 1986, Chapter 9; Res. Institute of Medical Science (Ed.), The Experimental Microbiology, 5th ed., Tokyo, Maruzen, 1976, Chapter 30.
61 K. Kamide, Nisshinbo International Conference, Tokyo, 1988.
62 J.J. Abel, L.G. Rowntree and B.B. Turner, J. Pharmacol. Exp. Ther., 5 (1914) 275.
63 W. Thalheimer, Proc. Soc., Exptl. Biol. Med., 37 (1938) 641.
64 W.J. Kolff and H.T.J. Berk, Acta Med. Scand., 117 (1944) 121.
65 W. Quinton, D. Dillard and B.H. Scribner, Transact. A. S. A. I. O., 12 (1960) 207.
66 F. Gegyo, T. Yamada, S. Odani, Y. Nakagawa, M. Arakawa, T. Kunitomo, H. Kataoka, M. Suzuki, Y. Hirasawa, T. Shirahama, A.S. Cohen and K. Schmid, Biochem. Biophys. Res. Commun., 129 (1985) 701.
67 J.W. Becker and G.N. Reeke Jr., Proc. Natl. Acad. Sci. USA, 82 (1985) 4225.

68 T. Akizawa, T. Kitaoka, S. Koshikawa, T. Watanabe, K. Imamura, T. Turumi, Y. Suwa and S. Eiga, Trans. Am. Soc. Artif. Intern. Organs, 32 (1986) 76.
69 See, for example, B. Alberts, D. Bray, J. Lewis, M. Raff, K. Roberts and J.D. Watson (Eds), Molecular Biology of the Cell, Garland Publishing, Inc., New York & London, 1983, Chapter 17.
70 E. Corretge, A. Kishida, H. Konishi and Y. Ikada, Grafting of Poly(ethylene Glycol) on Cellulose Surfaces and the Subsequent Decrease of the Complement Activation, in C. Migliaresi et al. (Eds), Polymers in Medicine III, Elsevier Science Publishers B. V., Amsterdam, 1988, p. 61.
71 M. Ohnishi, J. Osaka City Med. Cent., 35 (1986) 189.
72 K. Kamide, S. Manabe, K. Sato, K. Hamada and S. Miyazaki, Jpn. J. Artif. Organs, 4 (1975) 220.
73 K. Kamide, S. Manabe, K. Sato, K. Hamada, H. Okunishi and S. Miyazaki, Jpn. J. Artif. Organs, 4 (1975) 171.
74 K. Kamide, S. Manabe, K. Hamada, H. Okunishi and S. Miyazaki, Jpn. J. Artif. Organs, 5 (1976) 139.
75 K. Sato, Jpn. J. Artif. Organs, 3 (1974) 422.
76 Z. Yamazaki, Y. Fujimori, K. Sanjo, Y. Kojima, M. Sugiura, T. Wada, N. Inoue, T. Sakai, T. Oda, N. Kominami, U. Fujisaki and K. Kataoka, Jpn. J. Artif. Organs, 2 (1978) 273.
77 Z. Yamazaki, N. Inoue, Y. Fujimori, T. Takahara, T. Oda, K. Ide, K. Kataoka and Y. Fujisaki, Biocompatibility of Plasma Separation of an Improvement Cellulose Acetate Hollow Fiber, in H.G. Sieberth (Ed.), Plasma Exchange Plasmapheresis-Plasma Separation, Sehattauer, Stuttgart-New York, 1980, p. 45-51.

Chapter 7

CONCLUDING REMARKS

Throughout this book we employed as a basic tenet the modified Flory-Huggins theory, in which the concentration- and the molecular weight-dependencies of the thermodynamic interaction parameter χ are reasonably considered. Now, we can present the following sound experimental evidences supporting the validity of the modified Flory-Huggins theory at least for non-polar polymer / non-polar solvent(s) systems.

(1) The concentration dependence parameters p_1 (and p_2) (see, eqn (2.3)) determined by systematic analysis on phase equilibria coincide well with those by critical solution points (see, Table 2.4) and moreover, p_1 values estimated by these methods are in good agreement with other methods such as osmometry and light scattering (see, Table 2.4). Of course, the former methods are unquestionably more accurate.

(2) The values of the Flory entropy parameter ψ_0 estimated by the second virial coefficient method are significantly smaller than those obtained by the critical point method by Shultz and Flory (eqn (2.227)) in which the concentration- and molecular weight-dependencies of χ parameter are neglected. But, when p_1 and p_2 are taken into account the ψ_0 values estimated from the critical solution points agree well with those by the second virial coefficient method (see, Table 2.21).

(3) The Flory enthalpy parameter at infinite dilution κ_0, evaluated from ① temperature dependence of vapor pressure and membrane osmotic pressure, ② critical solution temperature T_c and critical solution concentration $v_p{}^c$ for a series of solutions of polymers, ③ temperature dependence of the second virial coefficient in the vicinity of the Flory theta temperature θ and ④ calorimetry yields a single master curve against the weight- or number-average molecular weight M_w or M_n for polymer / solvent system and excellent agreement was confirmed between κ_0 values at θ deduced by various methods when M_w or M_n was the same (see, Fig. 2.110).

(4) A significant disparity of thermodynamic parameters including p_1 and p_2, observed for atactic polystyrene in aromatic solvents from those theoretically predicted by the modified Flory-Huggins theory (eqns (2.9) and (2.21)) and those estimated for atactic polystyrene in aliphatic solvents can be explained by the characteristic features of the dissolved state of polystyrene in aromatic solvents: The polystyrene phenyl ring is stacked parallel to the solvent phenyl ring for polystyrene / aromatic solvent systems (see, Figs 2.105d and 2.106b).

Unfortunately, the serious drawback of the computer simulation technique is that the results obtained are severely restrictive and individually specific. Thus only abundant and systematic accumulation of numerous cases can lead us to generally acceptable concepts on thermodynamics of the phase equilibrium of polymer solutions. In the course of long-term study (1963-1989) on thermodynamics of phase equilibria of polymer solutions, computer experiments have been carried out about 2×10^4 times by Kamide and his coworkers. If we try to get the same knowledge purely, as that by the computer experiments, from actual experiments, it will take more than 620 years, provided that all experiments are without exception, successful.

GLOSSARY OF SYMBOLS

The following list includes only symbols which are of general importance.

A	the left hand side of eqn (3.25) (phase equilibrium condition of solvent)
a	parameter used in eqn (6.67)
a	temperature independent parameter in eqn (2.6) of χ_0
a'	temperature independent parameter in eqn (2.7)
a_0	activity of solvent
A_2	second virial coefficient
$A_{j(i)}$	probability that $j_{(i)}$ vacant particles among n_0 consecutively connected vacant particles are found in the i th shell
$A_{j(1)}$	probability that $j_{(1)}$ vacant particles among n_0 consecutively connected vacant particles are found in the first shell
$A_{j(1),j(2),\cdots j(i-1),j(i)}$	probability that $j_{(1)},j_{(2)},\cdots J_{(i-1)},j_{(i)}$ vacant particles, all belonging to a pore consisting of n_0 consecutively connected particles, are found in the first, second, ···, (i-1)-th, i-th shell, respectively
B	the left-hand side of eqn (3.26) (phase equilibrium condition of nonsolvent)
B'	the long range intermolecular interaction parameter
b	temperature independent parameter in eqn (2.6)
b'	temperature independent parameter in eqn (2.7)
C	difference between calculated and given ρ_p defined by eqn (3.36)
c	weight concentration
C_H	C correspond to R_H
C_L	C correspond to R_L
C_M	C correspond to R_M
c'	temperature independent parameter in eqn (2.7)
D	difference between calculated and assumed $X_{n(2)}$ defined by eqn (3.37)
D	the diffusion coefficient
d_A	apparent density of membrane; the ratio of weight

	and volume of absolute dry membrane ($d_A = w_m/v_m$)
D_i	diffusion coefficient of the particles with the same radius S_i
$<D>_G$	the g-average self-diffusion constant
d_P	density of polymer
d_P'	density of the dried polymer particle
d_{PL}	density of the polymer itself
d_S	density of solvent
$<D>_Z$	z-average of D_i
E	relative error of analytical fractionation defined by eqn (2.102)
E_1	allowance error defined by $\|C(R^a)\| < E_1$
E_2	allowance error defined by $\|D(X_{n(2)}{}^a)\| < E_2$
E_3	allowance error defined by $\|V_p{}^0 - V_p{}^{0g}\| < E_3$
E'	relative error of analytical fractionation defined by eqn (2.107)
f(x)	number of polymer particles needed to be surrounded fully by an assembly of x vacant particles
$f_{X(1)}$	probability of the X-mer in a polymer-rich phase
$f_{X(2)}$	probability of the X-mer in a polymer-lean phase
g	thermodynamic interaction parameter defined by eqn (2.24)
g_0	temperature dependent term of g defined by eqn (2.25)
g_{00}	temperature independent term of g_0 in eqn (2.25)
g_{01}	temperature independent term of g_0 in eqn (2.25)
g_i	parameter defined by eqn (6.56)
g(M)	molecular weight distribution (MWD)
$g_{(1)}(X)$	MWD of the polymer in the polymer-lean phase
$g_{(2)}(X)$	MWD of the polymer in the polymer-rich phase
$g_0(X_i)$	a summation of $g_{(1)}(X_i)$ and $g_{(2)}(X_i)$
h	polydispersity parameter of Schulz-Zimm type distribution defined by eqn (2.61)
i	order of shells
I_1	the first order time correlation function of the electric field of the scattered light
$I(M_{w,j})$	cumulative weight distribution of j-th fraction
J	water flux
$j_{(i)}$	total number of vacant particles in the i-th shell
K	quantity defined by eqn (5.15c)
K	an optical constant
k	layer order

k'	molecular weight-independent parameter defined by eqn (2.5)
k'	Huggins coefficient
k_H'	parameter defined by eqn (6.42)
k_0	molecular weight-dependence parameter of k' in eqn (2.5)
k_1	parameter defined by eqn (6.59)
k_2	parameter defined by eqn (6.60)
k_B	Boltzmann constant
k_{CN}	growing rate constant
k'_{CN}	rate constant of nucleation
$k_{i,j}$	reaction rate between particle with i primary particles and particle with j primary particles
k_m	minimum k giving the level-off $P_{t[k]}$ value
L	volume fraction of polymer-lean phase
ℓ_0	thickness of hypothetical thin plane of the cast solution
$\ell_{A,i}$	the cut-off length, by pores, of i-th test line
ℓ_d	thickness of a membrane
ℓ_g	thickness of gel membrane
ℓ_i	the length of i-th test line drawn on a photograph of membrane surface
$\ell_{(i)}$	number of particles which are chosen arbitrarily from the i-th shell
$\ell_{i[km]}$	number of the consecutive vacant particles belonging to one of the branching pores at the k_m-th layer, which are connected to a given vacant particle on the first layer
$\ell_{[k]}$	number of consecutive vacant particles in the k-th layer
$\ell_{[km]}$	number of the vacant particles at the k_m-th layer, which are connected to a given vacant particle on the first layer $(= \sum_i \ell_{i[km]})$
M	quantity defined by eqn (5.15a)
M'	the number of steps in histogram
M_i	quantity defined by eqn (5.15d)
M_i	mass of the particle with diffusion coefficient D_i
M_n	the number-average molecular weight
M_s	molecular weight of solvent
M_w	the weight-average molecular weight
M(i)	total number of the particles constituting the i-th shell

$M(x)$	frequency distribution function of pore which is made of x consecutive vacant particles
m	total number of components existing in the polymer
m_1	total number of components existing in the polymer 1
m_2	total number of components existing in the polymer 2
$\tilde{m}$	the maximum order of shells
$m_{(i)}$	the maximum number of the particles in the (i+1)-th shell in contact directly with $\ell_{(i)}$ particles, which are chosen arbitrarily from all particles in the i-th shell
$m_{\ell i[km][km+1]}$	the maximum number of the particles in the (k_m+1)-th layer in contact directly with consecutive $\ell_{i[km]}$ particles of the i-th pore in the k_m-th layer
$m_{\ell[k]}$	the maximum number of the particles in the k-th layer, contacting directly consecutive $\ell_{[k-1]}$ vacant particles in the (k-1)-th layer, $m_{\ell[k-1][k]}$
$m_{\ell[km][km+1]}$	the maximum number of the particles in the (k_m+1)-th layer, contacting directly consecutive $\ell_{[km]}$ vacant particles in the k_m-th layer
$m_{1(i)}$	total number of the partices, belonging to the (i+1)-th shell and located as the nearest neighbouring particle compared to a given particle existing in the i-th shell
N	quantity defined by eqn (5.15b)
N	total number of layers constituting a membrane
N_0	molar number of solvent
N_0	total number of secondary and vacant particles in a unit volume of a membrane
n_0	number of the consecutively connected vacant particles, which constitute isolated pores
n_0	molar fraction of solvent
N_1	total number of the isolated pores, which are formed by only one vacant particle, in a unit volume of a membrane
N_1	molar number of solvent 1
N_2	molar number of solvent 2
N_2	total number of the isolated pores, which are formed by two consecutive vacant particles in a unit volume of a membrane

N_3 total number of the isolated pores, each made of three vacant particles in a unit volume of a membrane

N_{3a} total number of the isolated pores constituted by three vacant particles, two of which directly connect to each other, in a unit volume of a membrane

N_{3b} total number of the isolated pores constituted by three vacant particles, two of which are not directly connected to each other, in a unit volume of a membrane

N_{3c} total number of the isolated pores constituted by three vacant particles, one is in the first shell and another is in the second shell, in a unit volume of a membrane

N_C total number of contact between nearest secondary particles per unit surface of the membrane

N_{CN} number of critical nuclei per unit volume

N_{GP} number of growing particles per unit volume

N_i number of particles with diffusion coefficient D_i

N_J quantity defined by eqn (5.15e)

N_{n0} total number of the isolated pores, each made of n_0 consecutive vacant particles in a unit volume of a membrane

N_P number of pores in one layer

N_{PP} number of primary particles per unit volume

$N(r)$ pore radius distribution function

n_P a given number of hypothetical thin gel planes

n_q highest order of concentration-dependence parameter of χ_{13}

n_r highest order of concentration-dependence parameter of χ_{23}

n_s highest order of concentration-dependence parameter of χ_{12}

N_T total number of polymer and vacant particles

n_t total number of fraction in a run

N_{Xi} molar number of X_i-mer

n_{Xi} molar fraction of X_i-mer

N_{YJ} molar number of Y_J-mer

O the area exposed to collide

P pressure

P vapor pressure of solution

P_0 vapor pressure of solvent composition in solution

$P_0{}^0$	vapor pressure of pure solvent
$p_{12,1}$	1st order parameter of $p_{12,s}$
$p_{13,1}$	1st order parameter of $p_{13,q}$
$p_{23,1}$	1st order parameter of $p_{23,r}$
p_J, p_n	concentration-dependence parameter of χ (or χ_1)
$p_{1,u}$	concentration (v_1)-dependence parameter of χ_{12}
$p_{2,u}$	concentration (v_2)-dependence parameter of χ_{12}
$p_{12,s}$	concentration-dependence parameter of χ_{12}
$p_{13,q}$	concentration-dependence parameter of χ_{13}
$p_{23,r}$	concentration-dependence parameter of χ_{23}
P_I	existing probability of isolated pore
$\tilde{P}(N_P)$	frequency of N_P
$P_n(m)$	the probability that a given polymer particle is surrounded in part by m pore particles, which belong to n different pores
Pr	two-dimensional porosity by the EM method in a unit area of a membrane surface or hypothetical thin-layer
Pr(d)	the porosity of dry membrane
$Pr_{,GEL}$	volume fraction of vacant particles in the gel membrane
$Pr_{,PS}$	volume fraction of polymer-lean phase in a hypothetical layer at the moment of phase separation
P_S	existing probability of a semi-open pore
P_T	existing probability of a through pore
$P_{T[k]}$	existing probability of the through pore when a membrane consists of k thin layers and all the pores on the first layer are constituted by only one vacant particle
$P_T(x)$	existing probability of the though pore, which ends at a given pore and made of x consecutive vacant particles in the first layer
$P_{T[2]}$	existing probability of the through pore, penetrating through two neighbouring layers when a membrane is consisted of two thin layers and all the pores on the first layer are constituted by only one vacant particle
$P_{T[\infty]}$	$P_{t[k]}$ at the limit of $k \to \infty$
P(x)	pore size distribution
q	q-th order of $p_{13,q}$
q	the scattered vector
q	probability that a secondary particle exists, which

	is equal to $1-Pr$
R	two-phase volume ratio $(=V_{(1)}/V_{(2)})$
r	pore radius
r	r-th order of $p_{23,r}$
$\tilde{R}$	gas constant
r_3	the third mean pore radius
r_4	the fourth mean pore radius
r_F	mean pore radius determined by the water-flow-rate method
R_A	an apparent two-phase volume ratio estimated for the actual dry membrane
R^a	the assumed value of R
r_{FE}	improved mean pore radius obtained by considering ε, Pr and P_T
R_i	radius of the i-th shell
R_H	high value of R^a
R_L	low value of R^a
r_L	mean pore radius
R_M	middle value of R^a
r_n	the mean pore radius
R_p	resolving power [eqn (2.72)]
r_{WET}	the radius of pore containing x vacant particles in a wet gel membrane
S	radius of particle
s	order of $p_{12,s}$
S_0	radius of local equilibrium region
S_{0+}	quantity defined by eqn (6.19a)
S_{0-}	quantity defined by eqn (6.19b)
S_1	radius of primary particle
S_2	radius of secondary and vacant particles
S_{CN}	radius of critical nucleus
S_H	hydrodynamic radius of particle
$S_{H,G}$	the g-average hydrodynamic radius of particle
$S_{H,Z}$	the z-average hydrodynamic radius of particle
S_i	particle radius with diffusion coefficient D_i
S_N	radius of nucleus
S_n	the number-average radius of primary particle
S_w	the weight-average radius of primary particle
$S_{(i)}$	radius of particle constituted with i primary particles
T	Kelvin temperature
T_c	critical solution temperature

T_{CP}	cloud point temperature
T_P	phase separation temperature
T_{SP}	spinodal point temperature
T_U	temperature above which the system is a single phase for an entire composition
V	total volume of the solution
v	overall polymer volume fraction ($\equiv v_1 + v_2$)
v	water-flow rate
$\bar{v}$	specific volume of polymer
V_0	volume of starting solution ($\equiv V_1{}^0 + V_3{}^0$)
v_0	volume fraction of solvent
$v_0(X_i)$	volume fraction of X_i-mer in the starting solution
V_1	total volume of solvent 1 (or polymer 1)
v_1	the volume fraction of solvent 1 (or polymer 1)
V_2	total volume of solvent 2 (or polymer 2)
v_2	the volume fraction of solvent 2 (or polymer 2)
$V_{(1)}$	volume of polymer-lean phase
$V_{(2)}$	volume of polymer-rich phase
$v_{1(1)}$	volume fraction of solvent 1 in polymer-lean phase
$v_{1(2)}$	volume fraction of solvent 1 in polymer-rich phase
$v_{2(1)}$	volume fraction of solvent 2 in polymer-lean phase
$v_{2(2)}$	volume fraction of solvent 2 in polymer-rich phase
$v_1{}^{(1)}$	v_1 of spinodal point on solvent 1-polymer axis ($v_2 = 0$)
$v_2{}^{(2)}$	v_2 of spinodal point on solvent 2-polymer axis ($v_1 = 0$)
$v_{(1)}(X_i)$	volume fraction of X_i-mer in polymer-lean phase
$v_{(2)}(X_i)$	volume fraction of X_i-mer in polymer-rich phase
$v_{(1)}{}^{CP}(X_i)$	$v_{(1)}(X_i)$ at cloud point ($= v_{Xi(1)}$)
$v_{(2)}{}^{CP}(X_i)$	$v_{(2)}(X_i)$ at cloud point ($= v_{Xi(2)}$)
$v_{1(1)}{}^*$	$v_{1(1)}$ at $\rho_p = 1.0$
$v_{1(2)}{}^*$	$v_{1(2)}$ at $\rho_p = 1.0$
$v_{2(1)}{}^*$	$v_{2(1)}$ at $\rho_p = 1.0$
$v_{2(2)}{}^*$	$v_{2(2)}$ at $\rho_p = 1.0$
$V_1{}^0$	volume of solvent 1 of starting solution
$V_3{}^0$	volume of polymer of starting solution
$v_1{}^c$	v_1 at critical point
$v_2{}^c$	v_2 at critical point
$v_0{}^F$	Flory solvent (or solution) composition of v_0
$v_1{}^F$	Flory solvent (or solution) composition of v_1
$v_2{}^F$	Flory solvent (or solution) composition of v_2
v_i	volume fraction of X_i-mer ($= v_{Xi}$)

v_m	volume of absolute dry membrane
v_p	overall polymer volume fraction
$v_{p(1)}$	polymer volume fraction in polymer-lean phase
$v_{p(2)}$	polymer volume fraction in polymer-rich phase
$v_p^{(1)}$	v_p of spinodal point on solvent 1-polymer axis ($v_2 = 0$)
$v_p^{(2)}$	v_p of spinodal point on solvent 2-polymer axis ($v_1 = 0$)
v_p^0	initial (overall) polymer volume fraction
v_p^{0g}	given value of polymer volume fraction
v_p^c	polymer volume fraction of critical solution point
v_p^s	polymer volume fraction of the starting solution
v_p^{sc}	v_p^s corresponds to critical solution point
$v_{p,c}^0$	v_p^0 corresponds to the minimum value of σ and R
$v_{p(1)}*$	$v_{p(1)}$ at $\rho_p = 1.0$
$v_{p(2)}*$	$v_{p(2)}$ at $\rho_p = 1.0$
v_X	the volume fraction of single component X-mer of polymer (or polymer 1)
v_X^0	initial X-mer concentration
v_X^c	critical volume fraction of X-mer
$v_{X(1)}$	X-mer volume fraction of polymer-lean phase
$v_{X(1)}^{SP}$	X-mer spinodal concentration of polymer-lean phase
$v_{X(2)}$	X-mer volume fraction of polymer-rich phase
$v_{X(2)}^{SP}$	X-mer spinodal concentration of polymer-rich phase
v_{Xi}	the volume fraction of X_i-mer of polymer (or polymer 1)
$v_{Xi(1)}$	volume fraction of X_i-mer in polymer-lean phase ($= v_{(1)}(X_i)$)
$v_{Xi(2)}$	volume fraction of X_i-mer in polymer-rich phase ($= v_{(2)}(X_i)$)
v_Y	the volume fraction of single component Y-mer of polymer 2
v_{Yj}	the volume fraction of Y_j-mer of polymer 2
W_1	the weight fraction of component 1 in binary polymer mixture belonging to the same chemical homologue
W_2	the weight fraction of component 2 in binary polymer mixture belonging to the same chemical homologue
w_1	the composition of polymer 1 ($w_1 = v_1/v$)
w_2	the composition of polymer 2 ($w_2 = v_2/v$)
$w_{ACETONE}$	weight fraction of acetone
$w_{ACETONE}^{CP}$	weight fraction of acetone at cloud point
$w_{ACETONE}^{P}$	weight fraction of acetone at phase separation point

$W_{ACETONE}^{S}$	weight fraction of acetone in starting solution
$W_{AN/MA}$	weight fraction of acrylonitrile / methylacrylate (AN/MA) copolymer
$W_{AN/MA}^{C}$	weight fraction of AN/MA copolymer at critical solution point
$W_{AN/MA}^{CP}$	weight fraction of AN/MA copolymer at cloud point
$W_{AN/MA}^{S}$	weight fraction of AN/MA copolymer in starting solution
W_{CA}^{S}	weight fraction of cellulose diacetate in cast solution
W_{CC}^{S}	weight fraction of $CaCl_2 \cdot 2H_2O$ in cast solution
W_{CH}^{S}	weight fraction of cyclohexanol in cast solution
W_{CELL}	weight fraction of cellulose
W_{CELL}^{CP}	weight fraction of cellulose at cloud point
W_{CELL}^{P}	weight fraction of cellulose at phase separation point
W_{CELL}^{S}	weight fraction of cellulose in starting solution
W_{CU}	weight fraction of copper
W_{CU}^{CP}	weight fraction of copper at cloud point
W_{CU}^{P}	weight fraction of copper at phase separation point
W_{CU}^{S}	weight fraction of copper in starting solution
W_{H2O}	weight fraction of water
W_{H2O}^{C}	weight fraction of water at critical solution point
W_{H2O}^{CP}	weight fraction of water at cloud point
W_{H2O}^{P}	weight fraction of water at phase separation point
W_{H2O}^{S}	weight fraction of water in starting solution
W_{HNO3}	weight fraction of nitric acid
W_{HNO3}^{C}	weight fraction of nitric acid at critical solution point
W_{HNO3}^{CP}	weight fraction of nitric acid at cloud point
W_{HNO3}^{S}	weight fraction of nitric acid in starting solution
W_{m}	weight of absolute dry membrane
W_{MT}^{S}	weight fraction of methanol in cast solution
$W_{N}(M)$	the number of ways of partitioning the M vacant particles into N cells
W_{NAOH}	weight fraction of NaOH
W_{NH3}	weight fraction of ammonia
W_{NH3}^{CP}	weight fraction of ammonia at cloud point
W_{NH3}^{P}	weight fraction of ammonia at phase separation point
W_{NH3}^{S}	weight fraction of ammonia in stating solution
W_{NS}	weight fraction of nonsolvent
W_{P}	weight fraction of polymer

$w_{TOLUENE}$	weight fraction of toluene
W_{Xi}	quantity defined by eqn (5.30a)
W_{YJ}	quantity defined by eqn (5.30b)
X	the molar volume ratio of single component X-mer (polymer or polymer 1) to solvent
x	number of consecutive vacant particles in the first layer forming a given pore
X_a	X for f_X (or $f_{X(2)}$) $= 0.5$
X_i	the molar volume ratio of X_i-mer (of polymer or polymer 1) to solvent
X_{max}	maximum component of $g_0(X)$
X_{max}'	effective maximum component of $g_0(X)$
X_{min}	minimum component of $g_0(X)$
X_{min}'	effective minimum component of $g_0(X)$
X_n	number-average of X_i $(i=1,\cdots,m)$
X_n^0	X_n of original polymer (or polymer 1)
$X_{n(1)}$	X_n of polymer in polymer-lean phase
$X_{n(2)}$	X_n of polymer in polymer-rich phase
$X_{n(2)}^a$	the assumed value of $X_{n(2)}$
X_p	value of X_i at peak of molecular weight distribution
X_v	viscosity-average degree of polymerization
X_w	weight-average of X_i $(i=1,\cdots,m)$
X_w^0	X_w of the original polymer (or polymer 1)
$X_{w(1)}$	X_w of polymer in polymer-lean phase
$X_{w(2)}$	X_w of polymer in polymer-rich phase
X_z	z-average of X_i $(i=1,\cdots,m)$
X_z^0	X_z of the original polymer (or polymer 1)
Y	function of g_1 and g_2 defined by eqn (2.229)
Y	the molar volume ratio of single component Y-mer (polymer 2) to solvent
Y	yield of fractionation
y	average number of vacant particles on the (k+1)-th layer, with which a vacant particle, arbitrarily chosen, on the k-th layer is connected directly
$Y_{1.1}$	Y for $X_w/X_n \leq 1.1$
$Y_{1.05}$	Y for $X_w/X_n \leq 1.05$
Y_J	the molar volume ratio of Y_J-mer (of polymer 2) to solvent
Y_n	number-average of Y_J
Y_n^0	Y_n of original polymer 2
Y_w	weight-average of Y_J
Y_w^0	Y_w of original polymer

Y_z	z-average of Y_j
$Y_z{}^0$	Y_z of original polymer
Z	distance from the front surface of membrane
${}_iZ_j$	collision frequency
α	the linear expansion factor
α	first term of g parameter in eqn (2.31)
α_i	polarizability per unit mass
$\alpha_{i,j}$	amalgamation probability between particle with i primary particles and particle with j primary particles
β	temperature dependent factor of g defined by eqn (2.32)
β'	polydispersity parameter in eqn (2.63) $(= (2\ln(X_w{}^0/X_n{}^0)^{1/2})$
β_{00}	temperature independent parameter of eqn (2.32)
β_{01}	temperature independent parameter of eqn (2.32)
Γ	gamma function
γ	temperature and concentration independent parameter in eqn (2.31)
Γ_i	parameter defined by eqn (6.63)
γ_i	the probability that a given virus, entered into i-th layer, is captured that layer
$\gamma_{k,k+\ell}$	the probability that a given virus, entered into a group consisting of $\ell+1$ thin layers between k-th and $(k+\ell)$-th layers, is captured
δ	parameter defined by eqn (2.228)
δ	the solubility parameter
δA_i	the number of viruses passed through a unit area of i-th layer for δt
δC_i	the number of viruses captured by unit area of i-th layer
Δf_v	the free energy change of coagulation per unit volume
$\Delta\bar{G}$	the average Gibbs' free energy change of coexisting two phases
ΔG	Gibbs' free energy change of mixing
ΔG^{id}	ideal Gibbs' free energy change of mixing
ΔG^E	excess Gibbs' free energy change of mixing
$\Delta G'$	Gibbs' free energy change of mixing per unit volume
$\Delta G_{12}{}^E$	excess Gibbs' free energy change of solvent 1-2
$\Delta G_{13}{}^E$	excess Gibbs' free energy change of solvent 1-polymer

ΔG_{23}^{E}	excess Gibbs' free energy change of solvent 2-polymer
ΔH	heat change of mixing
ΔH^{0}	partial molar heat change of dilution with respect to solvent
ΔH^{id}	ideal heat change of mixing (= 0)
Δ_{P}	the depth of phase separation
ΔP	loaded pressure
ΔR_{90}	the difference of the rayleigh ratio at $\theta = 90°$
ΔS_{0}	partial molar entropy change of dilution with respect to solvent
ΔS^{id}	ideal entropy change of mixing
ΔS_{0}^{comb}	combinatory entropy term of ΔS_{0}
$\Delta V_{P(1)}$	increment of polymer volume during unit time in polymer-lean phase
$\Delta V_{P(2)}$	increment of polymer volume during unit time in polymer-rich phase
$\Delta \mu_{0}$	the chemical potential of solvent
$\Delta \mu_{1}$	the chemical potential of solvent 1
$\Delta \mu_{1}$	the chemical potential of solvent 2
$\Delta \mu_{Xi}$	the chemical potential of polymer 1 (X_i-mer)
$\Delta \mu_{YJ}$	the chemical potential of polymer 2 (Y_J-mer)
$\Delta \mu_{0(1)}$	$\Delta \mu_{0}$ in polymer-lean phase
$\Delta \mu_{0(2)}$	$\Delta \mu_{0}$ in polymer-rich phase
$\Delta \mu_{1(1)}$	$\Delta \mu_{1}$ in polymer-lean phase
$\Delta \mu_{1(2)}$	$\Delta \mu_{1}$ in polymer-rich phase
$\Delta \mu_{2(1)}$	$\Delta \mu_{2}$ in polymer-lean phase
$\Delta \mu_{2(2)}$	$\Delta \mu_{2}$ in polymer-rich phase
$\Delta \mu_{Xi}$	chemical potential of X_i-mer
$\Delta \mu_{Xi(1)}$	$\Delta \mu_{Xi}$ in polymer-lean phase
$\Delta \mu_{Xi(2)}$	$\Delta \mu_{Xi}$ in polymer-rich phase
$\Delta \mu_{0}^{E}$	excess chemical potential of solvent
$\Delta \mu_{0}^{id}$	ideal chemical potential of solvent
$\Delta \mu_{Xi}^{E}$	excess chemical potential of X_i-mer
$\Delta \mu_{Xi}^{id}$	ideal chemical potential of X_i-mer
$\Delta \phi_{CN}$	the activation energy of formation of critical nucleus
$\Delta \phi$ (S)	the activation energy of formation of nucleus with radius S
ε	fractionation efficiency defined by eqn (2.68)
ε	tortuous coefficient
ε'	fractionation efficiency defined by eqn (2.69)

η	variable related to the two partition coefficients ($\eta^2 = \sigma_X \sigma_Y$)
η	viscosity of solution
η_0	viscosity of the pure solvent
$[\eta]$	the limiting viscosity number
η_w	viscosity of water
θ	Flory theta temperature
θ	scattering angle
κ	Flory enthalpy parameter
κ_0	κ at infinite dilution
$\kappa_1, \kappa_2, \cdots$	1st, 2nd, ··· order concentration-dependence parameter of κ
λ	wave length of incident light in the solution
λ^*	selective absorption coefficient
μ_1	first moment of differential molecular weight distribution g(M)
ν	parameter defined by eqn (6.30)
ξ	variables related to the two partition coefficients ($\xi^2 = \sigma_X/\sigma_Y$)
π	osmotic pressure
ρ	fraction size
ρ_p	relative amount of polymer partitioned in polymer-rich phase
ρ_s	relative amount of polymer partitioned in polymer-lean phase
$\rho_{p,min}$	ρ_p corresponds to the minimum X_w/X_n of polymer in polymer-rich phase
ρ_p^g	ρ_p as initial condition
τ	time of formation of nucleus
τ	correlation time
σ	partition coefficient
σ	interfacial energy per unit area
σ_0	1st term of σ in eqn (2.35)
σ_{01}	molecular weight-dependent coefficient of σ in eqn (2.35)
σ^2	the mean-squre displacement of particle
σ_H	σ correspond to R_H
σ_i	partition coefficient of X_i-mer [eqn (2.35)]
σ_L	σ correspond to R_L
σ_M	σ correspond to R_M
σ_X	partition coefficient of monodisperse X-mer
σ_Y	partition coefficient of monodisperse Y-mer

σ'	standard deviation of molecular weight distribution of polymer
$\sigma_{(1)}'$	σ' of polymer in polymer-lean phase
$\sigma_{(2)}'$	σ' of polymer in polymer-rich phase
σ_i'	σ' of the original polymer
σ^*	σ at $\rho_P = 1$
χ	thermodynamic interaction parameter between polymer and solvent
χ_0	χ at infinite dilution
χ_0^C	χ_0 at critical solution point
χ_0^{CP}	χ_0 on cloud point curve
χ_0^P	χ_0 at phase separation point
χ_0^{SP}	χ_0 on spinodal curve
χ_{00}	temperature dependent factor of χ
$\chi_{00,h}$	enthalpy term of χ_{00} [eqn (2.219)]
$\chi_{00,s}$	entropy term of χ_{00} [eqn (2.218)]
χ_{01}	thermodynamic interaction parameter between solvent and polymer 1
χ_{02}	thermodynamic interaction parameter between solvent and polymer 2
χ_{12}	thermodynamic interaction parameter between polymer 1 and polymer 2
χ_{13}	thermodynamic interaction parameter between solvent 1 and polymer
χ_{23}	thermodynamic interaction parameter between solvent 2 and polymer
χ_i	thermodynamic interaction parameter between X_i-mer and solvent
χ_{kl}	temperature dependent thermodynamic interaction parameter between k and l
$\chi_{kl,1}$	1st term of χ_{kl}
$\chi_{kl,2}$	temperature dependent ($\sim 1/T$) coefficient of χ_{kl}
$\chi_{kl,3}$	temperature dependent ($\sim T$) coefficient of χ_{kl}
χ_0^c	χ_0 at critical solution point
χ_{12}^0	temperature dependent factor of χ_{12}
χ_{13}^0	temperature dependent factor of χ_{13}
χ_{23}^0	temperature dependent factor of χ_{23}
ϕ	the rejection coefficient of the membrane against the virus
ψ	Flory entropy parameter
ψ_0	ψ at infinite dilution

AUTHOR INDEX

The number directly after the initial(s) of the author is the page number on which the author (or his book) is mentioned in the text. Numbers in brackets are reference numbers. Numbers underlined give the page on which the complete reference is listed.

SUBJECT INDEX

The Many Islands of Polynesia

AMOS P. LEIB

ILLUSTRATED WITH MAPS AND PHOTOGRAPHS

A WORLD BACKGROUND BOOK

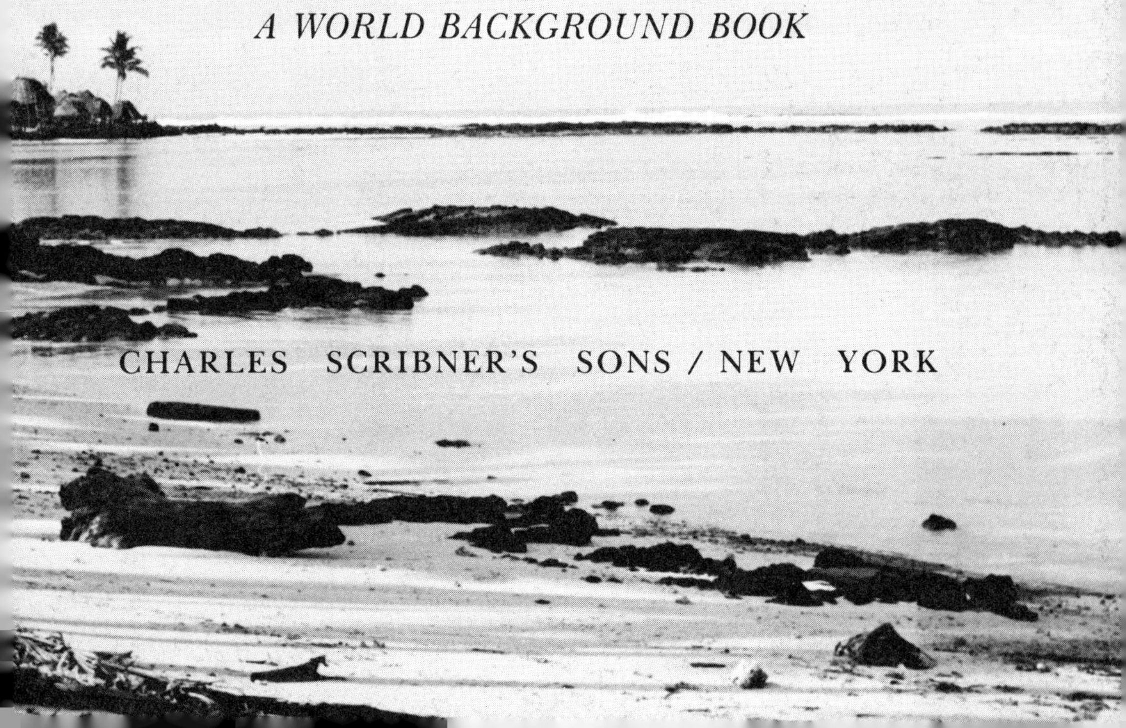

CHARLES SCRIBNER'S SONS / NEW YORK

A—8.72 [H]

PRINTED IN THE UNITED STATES OF AMERICA
Library of Congress Catalog Card Number 71–37184
SBN 684–12805–5 (trade cloth, RB)

ILLUSTRATION ACKNOWLEDGMENTS

Grateful acknowledgment is made for permission to use the following.

Photographs: page 160, courtesy of Alexander and Baldwin, Inc.; pages 25 and 101, courtesy of Bernice P. Bishop Museum; page 203, by Dr. Richard Chesher; page 153, courtesy of Dole Photo; page 66, courtesy of Government of American Samoa, Office of Tourism; pages 39, 42, 43, 52, 54, courtesy of Government Tourist Department, Tahiti; pages 154, 158, 176, 182, 189, 191, 206, courtesy of Hawaii Visitors Bureau; page 147, courtesy of John Warnecke and Associates; pages 19, 66, 86, 89, 90, 93, 94, 98, 109, 152, by Amos P. Leib; page 204, courtesy of Makai Range, Inc.; pages 36, 46, 163, courtesy of Matson Lines; pages 119, 123, 126, 131, 134, 135, 138, courtesy of New Zealand Information Service; page 73, courtesy of Pan Am; titlepage and pages 59, 77, 82, 91, 106, courtesy of Qantas; page 15, courtesy of States Steamship Company; page 72, by Captain A. G. Shearer, Fiji Airways Ltd.; page 170, courtesy of United States Coast Guard, Public Information Office, Honolulu; page 157, courtesy of United States Navy. *Maps:* pages 4–5, 35, 70, 115, 143, rendered by Instructional Resources Service Center, University of Hawaii, Honolulu, Hawaii.